Quantum Communications and Measurement

Quantum Communications and Measurement

Edited by

V. P. Belavkin
University of Nottingham
Nottingham, England

O. Hirota
Tamagawa University
Tokyo, Japan

and

R. L. Hudson
University of Nottingham
Nottingham, England

Plenum Press • New York and London

Library of Congress Cataloging-in-Publication Data

International Workshop on Quantum Communications and Measurement (1994
: Nottingham, England)
Quantum communications and measurement / edited by V.P. Belavkin,
O. Hirota, and R.L. Hudson.
p. cm.
"Proceedings of an International Workshop on Quantum
Communications and Measurement, held July 11-16, 1994, in
Nottingham, England"--T.p. verso.
Includes bibliographical references and index.
ISBN 0-306-45128-X
1. Quantum optics--Congresses. 2. Optical communications-
-Congresses. 3. Quantum electronicy--Congresses. I. Belavkin,
V. P. II. Hirota, O. (Osamu), 1948- . III. Hudson, R. L. (Robin
Lyth), 1960- .
QC446.15.I68 1995
621.382'7--dc20 95-32969
CIP

Proceedings of an International Workshop on Quantum Communications and Measurement, held July 11–16, 1994, in Nottingham, England

ISBN 0-306-45128-X

A Division of Plenum Publishing Corporation
233 Spring Street, New York, N. Y. 10013

10 9 8 7 6 5 4 3 2 1

Printed in the United States of America

ORGANIZING COMMITTEE

V P Belavkin and R L Hudson University of Nottingham, United Kingdom

O Hirota Tamagawa University of Tokyo, Japan

ADVISORY BOARD

Professor H Araki Research Institute of Mathematical Sciences, Kyoto University, Kyoto, Japan

Professor C W Gardiner Department of Physics, University of Waikato, Hamilton, New Zealand

Professor R J Glauber Physics Department, Lyman Laboratory, Harvard University, Cambridge, Massachusetts 02138

Professor H Haken Institut für Theoretische Physik, Universität Stuttgart, Pfaffenwaldring 57/4, 70550 Stuttgart, Germany

Professor K Haubold Physics Department, Universität Oldenburg, Fachbereich 8, Oldenburg D-2900, Germany

Professor J Klauder Department of Physics, University of Florida, 215 Williamson Hall, Gainesville, Florida 32611

Professor P L Knight Optics Section, The Blackett Laboratory, Imperial College, Prince Consort Road, London SW7 2BZ, United Kingdom

Professor J T Lewis School of Theoretical Physics, Dublin Institute for Advanced Studies, 10 Burlington Road, Dublin 4, Ireland

Professor S Reynaud Laboratoire de Spectroscopie Hertzienne, Université Pierre et Marie Curie, BP 74, F-75252 Paris Cedex 05, France

Professor R L Stratonovich Physics Department, Moscow State University, Moscow 117234, Russia

INVITED SPEAKERS

L Accardi University of Rome II

A Barchielli University of Milan

S L Braunstein Weizmann Institute of Science, Rehovot

P Busch University of Cologne

C Fabre University P M Curie, Paris

P L Knight Imperial College, London

K Haubold University of Oldenburg

K Kasai Ministry of Posts and Telecommunications, Kobe

G J Milburn University of Queensland

M Ohya Science University of Tokyo

M Ozawa Nagoya University

K R Parthasarathy Indian Statistical Institute (Delhi Centre)

I C Percival Queen Mary and Westfield College, London

A Rimini University of Pavia

C M Savage Australian National University, Canberra

P Staszewski N Copernicus University, Toruń

H P Yuen Northwestern University, Evanston, Illinois

PREFACE

The International Workshop on Quantum Communications and Measurement was held at the University of Nottingham from July 10–16, 1994. It followed the successful meeting on Quantum Aspects of Optical Communications in Paris in November 1990. This time the conference was devoted to mathematical, physical and engineering aspects of quantum noise, signal processing and quantum information in open systems, quantum channels, and optical communications. It brought research workers in the experimental and engineering aspects of quantum optics and communication systems into contact with theoreticians working in quantum probability and measurement theory. The workshop was attended by more than 130 participants from 22 different countries. The largest groups [after the UK (31)] were from Japan (19) and from Russia (14). The subjects discussed included the mathematical foundations of quantum communication systems, experiments and devices, the problem of collapse and continuous measurement, quantum input and output processes, causality and nondemolition observation, squeezed states, quantum jumps, state diffusion and spontaneous localization, filtering and control in quantum systems, and new quantum optical phenomena and effects, including nonclassical light.

These new mathematical and physical ideas were stimulated by recent advances in generation and detection of light with low quantum noise and the development of techniques for trapping a single atom over an extended period of time, making it possible to observe individual quantum phenomena at the macroscopic level. The behaviour of an individual atom or a quantum particle under continuous observation is different from the statistical behaviour described by an "a priori", given, master equation. Such a dynamics was first derived for the purposes of filtering quantum communications (see the translation included in this volume from Russian of Belavkin's 1980 paper) and can be nonmixing but it is irreversible, extremely chaotic, and cannot be described by a standard Schrödinger equation. The appropriate "a posteriori" stochastic wave equation was recently obtained from the standard quantum stochastic Schrödinger equation and is now taken as the primary equation in many new phenomenological quantum theories of quantum jumps, state diffusion, continuous or permanent reduction, and spontaneous localization. This conference made it clear that such theories do not contradict either the dynamical (Hamiltonian) principle of quantum mechanics or the Copenhagen (probabilistic) interpretation, but fill the gap between these two conflicting pictures. They are derived from the Hamiltonian principle and probabilistic interpretation by a standard quantum filtering method, as the result of conditioning with respect to the quantum trajectories of continuous nondemolition observation. Thus, the main quantum measurement problem of the dynamical derivation for the collapse of the reduced state vector has been finally solved as these proceedings claim in at least three papers.

With the new stochastic quantum theory, the dynamical possibilities of quantum communication systems can be studied consistently, using standard detection and estimation methods of random signal processing. With the new light sources available, the possibility of using low-temperature quantum noise light in optical communications is nearing reality. The experimental and quantum-engineering problems of the optimal processing of quantum light in optical communication were actively discussed at this meeting.

The workshop covered a wide range of topics in quantum probability and measurement, from fundamental problems of quantum mechanics to its applications in the newly developing quantum technologies.

The workshop brought together leading figures in the field in the pleasant surroundings of Hugh Stewart Hall on Nottingham University campus. We are grateful to the local organizers from the Conference Office and the Hall for their helpfulness and hospitality.

We are grateful to our sponsors: Hitachi, the International Scientific Foundation, the International Communication Foundation, KDD, Mitsubishi, the London Mathematical Society, the Engineering and Physical Sciences Research Council, and the European Union HCM Scheme. Above all, the workshop could never have taken place without the generous support of Tamagawa University. We have had full collaboration from Plenum Press in the publication of this volume. Finally, the success of the meeting owes much to the excellent, meticulous, and always patient work, both before and afterwards, of our Conference Secretary, Krystyna Glowczewska.

V P Belavkin
O Hirota
R L Hudson

CONTENTS

Part I.
Quantum States
&
Input-Output Processes

THE WORLD OF QUANTUM NOISE AND THE FUNDAMENTAL OUTPUT PROCESS

V. P. Belavkin[1], O. Hirota[2], and R. L. Hudson[1]

[1]Department of Mathematics, University of Nottingham, UK
[2]Research Center of Quantum Communications,
Tamagawa University, Tokyo, Japan

Abstract

An introduction is given to the stationary theory of quantum stochastic processes of second order, which includes KMS processes like the equilibrium finite temperature quantum noise given by the Planck spectral formula. It is shown that for each stationary noise there exists a natural output process which is identical to the noise in the infinite temperature limit, and flipping with the noise if the time is reversed at finite temperature. A canonical Hilbert space representation of the quantum noise and the fundamental output process is established and a decomposition of their spectra is found. A brief explanation of quantum stochastic integration with respect to the input-output processes is given using only correlation functions. This provides a mathematical foundation for linear stationary filtering transformations of quantum stochastic processes. It is proved that the colored quantum stationary noise and its time-reversed version can be obtained in the second order theory by a linear nonadapted filtering of the standard vacuum noise uniquely defined by the canonical creation and annihilation operators on the spectrum of the input-output pair.

Keywords :
Quantum Noise, Output Process, Quantum Correlations, Time Reversion, Spectral Analysis, Stochastic Integration, Linear Filtering.

Quantum Communications and Measurement
Edited by V. P. Belavkin *et al.*, Plenum Press, New York, 1995

1 INTRODUCTION

In this paper we develop a correlation theory of stationary quantum noise, give its spectral analysis and classification and extend the theory of quantum stochastic integration [1] to colored quantum noise [2]. The typical example of such noise is given by the Planck spectral formula. It shows that finite temperature equilibrium quantum noise is not white in contrast to classical equilibrium noise given by the Nyquist spectral formula. The weak coupling limit and rotating wave approximation [2, 3] make it possible to approximate the equilibrium quantum noise by canonical pairs of noncommutative white noises in a narrow spectral band near a resonant frequency.

Although such an approximation is sufficient in quantum optics and in many other practical cases, it is not satisfactory from a purely theoretical point of view because it does not correctly predict the simplest quantum stochastic motion of a free Brownian particle in an equilibrium environment, when the resonant frequency is zero.

In order to describe the output stochastic processes of quantum filters and other devices of quantum measurement and communications as functional transformations of the equilibrium noise, we also need a generalisation of the notion of quantum output fields [4–6] in the framework of stochastic integration with respect to the colored quantum noise.

The well developed quantum stochastic calculus and noncommutative extension of Ito integration to vacuum noise with canonical commutation relations in the time domain [6, 7] is not directly applicable to such cases, neither is the calculus and integration with respect to stochastic models of temperature quantum noise with flat spectrum [5, 8].

But we can use the methods of quantum stochastic integration in the spectral domain, in which the stationary quantum noise is δ-correlated with a certain modification. The latter is required because the colored quantum processes are not frequency-stationary and need not be adapted in the spectral representation. Although non-stationary and non-adapted theories of quantum stochastic calculus and integration have been already established [9–11], here we will use the much more simple approach of mean square integration which is sufficient for the non-adapted linear transformations. The corresponding classical theory of stochastic integration was developed by Wiener before the Ito integration theory and is based on the possibility of representing any (colored) process stationary in the wide sense as a linear filtering integration of the standard white noise.

Here we shall prove that any quantum process stationary in the wide sense can be also obtained from a standard one by filtering. In fact, we will show that it can be obtained by mean square integration with respect to a canonical pair of orthogonal quantum integrators describing the standard zero temperature (vacuum) noise in the second order. Although such a possibility is known for the Gaussian care using a Bogolubov transformation, it has not been realized in the theory of quantum noise and integration because it doubles degrees of freedom of the noise in such representation.

We will remove this unpleasant feature by deriving the fundamental output process for a given quantum noise as the canonically time-reversed noise, which commutes with the input noise and has maximal possible correlations with it. In the classical(infinite temperature) case, the fundamental output process coincides with the noise. In the quantum case it gives the best possible nondemolition filtering and time continuous indirect observation of the noise.

The fundamental output process for a given Gaussian quantum noise in the framework of quantum stochastic calculus was introduced in [12] and the possibility of its nondemolition observation was demonstrated for Markovian models of interaction with a quantum open system in the quantum theory of filtering (see the recent survey [13] and papers cited therein).

To explain the idea of the fundamental output process, let us represent a canonical Bose-noise annihilation and creation pair b, $b^\dagger$ with non-zero temperature correlations

$$\langle b^\dagger b\rangle = n, \quad \langle bb^\dagger\rangle = n+1$$

by the linear combination $b = \sqrt{n+1}a + \sqrt{n}c$ of the zero temperature pairs a, $a^\dagger$ and c, $c^\dagger$ with

$$[a, a^\dagger] = 1, \quad [c^\dagger, c] = 1, \quad [a^\dagger, c] = [a, c] = 0.$$

Then the combination $\tilde{b} = \sqrt{n}a + \sqrt{n+1}c$ defines a canonical pair $\tilde{b}$, $\tilde{b}^\dagger$ of output creation and annihilation operators with correlations

$$\langle \tilde{b}^\dagger \tilde{b}\rangle = n+1, \quad \langle \tilde{b}\tilde{b}^\dagger\rangle = n$$

which commutes with the noise, $[\tilde{b}, b] = [\tilde{b}, b^\dagger] = 0$, and provides the maximal mutual correlations $\langle \tilde{b}b^\dagger\rangle = \sqrt{n(n+1)}$, corresponding to correlation coefficient $r = 1$.

In the classical case, the latter would mean that $\tilde{b} = b$, but this is not the case: $\tilde{b}$ commutes with $b^\dagger$ but b does not. Nevertheless $\tilde{b}$ is uniquely defined for the noise b by the described properties as the time-reversed noise, which in the real arrow of time plays the role of the fundamental output process. Moreover, the zero temperature pair a, c is uniquely determined by the pair $(b, \tilde{b})$, being given by

$$a = \sqrt{n+1}b - \sqrt{n}\tilde{b}, \quad c = \sqrt{n+1}\tilde{b} - \sqrt{n}b.$$

2 QUANTUM CORRELATIONS AND REVERSED PROCESSES

Here we sketch of a second order theory of quantum noise and linear filtering of stationary signals, initiated more than 20 years ago in the pioneer paper [14]. A quantum stochastic process in second order is determined by the zero mean values $\langle x\rangle = 0$ and finite variances alone $\langle x^2\rangle < \infty$ of self-adjoint quantum variables $x = x^\dagger$ and also by their complex correlations $\langle xx'\rangle \neq \langle x'x\rangle$. Representing an invertible correlation kernel as a bracket product, we first describe the a quantum noise by real vectors $|x\rangle = |x\rangle^\sharp$ in a complex Euclidean space equipped with the involution $|z\rangle \mapsto |z\rangle^\sharp$. The latter corresponds to the Hermitian conjugation $z \mapsto z^\dagger = \sum \bar{\zeta}^j x_j$ for complex combinations $z = \sum \zeta^j x_j$ of the noise variables ${x_j}^\dagger = x_j$ and is defined by

$$\left|\sum \zeta^j x_j\right\rangle^\sharp = \sum \bar{\zeta}^j |x_j\rangle, \quad |z\rangle^{\sharp\sharp} = |z\rangle, \quad \zeta^j \in \mathbf{C}.$$

Let us consider a discrete time sequence $(x_j)_{j\in\mathbf{Z}}$ of quantum variables $(\cdots, x_{-1}, x_0, x_1, \cdots)$, with the time ordered correlations

$$K_{ij} = \langle x_i\ x_j\rangle \equiv \langle x_i|x_j\rangle\,, \quad i > j \in \mathbf{Z}$$

which we use to define a scalar product of the corresponding vectors $|x_j\rangle$. In contrast to the classical theory, these correlations need not be real valued because time reversal in quantum mechanics corresponds to the complex conjugation:

$$K_{ji} = \langle x_j\ x_i\rangle = \overline{\langle\ x_i\ x_j\rangle} = \overline{K}_{ij}.$$

Although the reversed correlations are not "observable" in the real arrow of time, they are defined by complex conjugation in such a way that the Hermitian matrix $K = [K_{ij}]$ is positive definite:

$$\zeta^{\dagger}K\zeta := \sum \overline{\zeta}^{i}\,\langle x_j\ x_i\rangle\,\zeta^{j} = \left\langle z^{\dagger}\ z\right\rangle \geq 0,$$

as is the conjugate = transpose $\overline{K} = [\overline{K}_{ij}] = [K_{ji}] = \widetilde{K}$:

$$\zeta^{\dagger}\widetilde{K}\zeta := \sum \overline{\zeta}^{i}\,\langle x_j\ x_i\rangle\,\zeta^{j} = \left\langle\ z\ z^{\dagger}\right\rangle \geq 0.$$

Here for the simplicity we shall restrict ourselves to the case when the matrices K and $\widetilde{K}$ commute as in the stationary case

$$K_{ij} = k_{i-j}, \quad k_{-j} = \overline{k}_j, \quad \forall j \in \mathbf{Z}$$

(see also [12] for a more general, noncommutative case). We claim in this case that $\overline{K}$ can be treated as the correlation matrix

$$\widetilde{K}_{ij} = \langle \widetilde{x}_i\ \widetilde{x}_j\rangle \equiv \langle \widetilde{x}_i|\widetilde{x}_j\rangle, \quad \forall i \geq j \in \mathbf{Z}$$

of another self-adjoint sequence $\widetilde{x}_j = (\cdots, \widetilde{x}_{-1}, \widetilde{x}_0, \widetilde{x}_1, \cdots)$, which is linearly related to $x = (x_j)_{j\in\mathbf{Z}}$. The sequence $\widetilde{x}$ describes the fundamental output process as reversed noise by the self-adjoint quantum variables $\widetilde{x} = \widetilde{x}^{\dagger}$ which will also be represented by real vectors $|\widetilde{x}\rangle$ with respect to another involution; $|\widetilde{x}\rangle^{\flat} = |\widetilde{x}\rangle$ uniquely defined in the case of invertible K. This suggests a natural norm

$$\|\zeta\|^2 = \left\langle z^{\dagger}\ z\right\rangle + \left\langle z\ \ z^{\dagger}\right\rangle = \zeta^{\dagger}\left(\ K + \widetilde{K}\right)\zeta$$

in the space $\mathcal{E}$ of test sequences $\zeta = \{\zeta^i\}$ for the simultaneous treatment of linear combinations $|z\rangle = \sum \zeta^j |x_j\rangle$ of the $\sharp$ - real vectors $|x_j\rangle = |x_j\rangle^{\sharp}$, and the combinations $|z\rangle = \sum \zeta^j |\widetilde{x}_j\rangle$ of $\flat$ - real vectors $|\widetilde{x}_j\rangle = |\widetilde{x}_j\rangle^{\flat}$, representing $\widetilde{x}_j = \widetilde{x}_j^{\dagger}$.

Correlation matrix

$$R = \left(K\ \overline{K}\right)^{1/2} = \left(\overline{K}\ K\right)^{1/2} = \overline{R}.$$

corresponding to the zero commutators and maximal correlations between x and $\widetilde{x}$:

$$R_{ij} = \langle x_i\ \widetilde{x}_j\rangle \equiv \langle x_i|\widetilde{x}_j\rangle\,, \quad \forall i > j \in \mathbf{Z}$$

If K is invertible, this output process is strongly correlated with the noise, corresponding to the correlation coefficient $r = 1$, and in the classical case $\widetilde{K} = K$ such a sequence $(\widetilde{x}_j)$ is always identified with (x_j). The invertible case corresponds to the thermal noise which we call standard if $R_{ij} = \delta_{ij}$, that is if the inverse noise $(\widetilde{x}_j)$ is δ-correlated with (x_j) such that $\widetilde{K} = K^{-1}$. The opposite situation, when $(\widetilde{x}_j)$ is uncorrelated with $(x_j) : R_{ij} = 0$ such that $\widetilde{K}K = 0$ is possible only in the quantum case. It corresponds to the vacuum noise which we call standard if $K + \widetilde{K} = I$.

Let us prove the existence of the time-reversed sequence $\widetilde{x} \neq x$ in the quantum case $\widetilde{K} \neq K$, and find the adjoint involution $|z\rangle \mapsto |z\rangle^\flat$,

$$\left| \sum \zeta^j \widetilde{x}_j \right\rangle^\flat = \sum \overline{\zeta}^j |\widetilde{x}_j\rangle, \quad |z\rangle^{\flat\flat} = |z\rangle, \quad \zeta^j \in \mathbf{C}$$

describing the Hermitian conjugation $z \mapsto z^\dagger = \sum \overline{\zeta}^j \widetilde{x}_j$ for the complex combinations $z = \sum \zeta^j \widetilde{x}_j$ in the case of invertible K.

The commuting matrices K and $\widetilde{K}$ have common right eigen vectors (generalized row-vectors), represented by (unnormalised) complex eigen sequences $\boldsymbol{u} = (u_j)_{j\in\mathbf{Z}}$,

$$\sum_{i\in\mathbf{Z}} u_i K_{ij} = \kappa u_j, \qquad \sum_{i\in\ bfZ} u_i \widetilde{K}_{ij} = \widetilde{\kappa} u_j, \quad j \in \mathbf{Z}$$

with eigenvalues κ and $\widetilde{\kappa} \geq 0$, corresponding to the complex conjugate eigen sequences $\overline{\boldsymbol{u}} = (\overline{u}_j)_{j\in\mathbf{Z}}$:

$$\sum_{i\in\mathbf{Z}} \overline{u}_i K_{ij} = \widetilde{\kappa} \overline{u}_j, \quad \sum_{i\in\mathbf{Z}} \overline{u}_i \widetilde{K}_{ij} = \kappa \overline{u}_j, \quad j \in \mathbf{Z}.$$

Indeed, let K and $\widetilde{K}$ have a common eigen-row-vector $\boldsymbol{u}$ with eigenvalues $\kappa \geq 0$ and $\widetilde{\kappa} \geq 0$ respectively. Because $\widetilde{K}$ is the complex conjugate matrix of K, $\overline{\boldsymbol{u}}$ is also a common eigenvector, but with the real eigenvalue $\widetilde{\kappa}$ for K and κ for $\widetilde{K}$. Two cases arise:

$\kappa = \tilde{\kappa}$ In this case we may assume without loss of generality that $u_j = \overline{u}_j$, $\forall j \in \mathbf{Z}$.

$\kappa \neq \tilde{\kappa}$ In this case the eigenvectors $\boldsymbol{u}$, $\overline{\boldsymbol{u}}$ are orthogonal and so we have $\sum u_j^2 = 0$.

This means that the index set Ω of the eigen sequences $\langle\nu| := (u_j(\nu))_{j\in\mathbf{Z}}$, $\nu \in \Omega$ is equipped with a natural flip

$$\nu \mapsto -\nu, \quad -(-\nu) = \nu, \quad \widetilde{\kappa}(\nu) = \kappa(-\nu)$$

such that $\overline{\boldsymbol{u}}(\nu) = \boldsymbol{u}(-\nu)$ and the (generalised) Plancherel measure $\mathrm{d}\nu$ on Ω is invariant under this flip. The complete (with respect to $\mathrm{d}\nu$) orthogonal set $\{\boldsymbol{u}(\nu)|\nu \in \Omega\}$ can be chosen in such a way that $u_j(\nu) = u_j(-\nu)$ iff $\kappa(\nu) = \kappa(-\nu)$ and $\sum_{j\in\mathbf{Z}} u_j^2(\nu) = 0$ iff $\kappa(\nu) \neq \kappa(-\nu)$.

We shall assume without loss of generality that the Plancherel measure $\mathrm{d}\nu$ on the null subset $\mathrm{N}_0 = \{\nu : \kappa(\nu) = 0 = \widetilde{\kappa}(\nu)\}$ is zero, such that its support Ω is identified with the union $\mathrm{N}_+^\perp \cup \mathrm{N}_-^\perp$, where

$$\mathrm{N}_+^\perp = \{\nu : \ \kappa(\nu) > 0\}, \mathrm{N}_-^\perp = \{\nu : \widetilde{\kappa}(\nu) > 0\}$$

are the complementary subsets of the null sets N_+ and N_- for κ and $\widetilde{\kappa}$.

Let us prove that the pairs $x_j = \{x_j(\nu) \mid \nu \in \Omega\}$, $\widetilde{x}_j = \{\widetilde{x}_j(\nu) \mid \nu \in \Omega\}$ of complex amplitudes

$$x_j(\nu) = \kappa(\nu)^{1/2} u_j(\nu) \equiv \langle \nu | x_j \rangle, \quad \widetilde{x}_j(\nu) = \widetilde{\kappa}(\nu)^{1/2} u_j(\nu) \equiv \langle \nu | \widetilde{x}_j \rangle$$

give the canonical spectral representation of the quantum noise vectors $|x_j\rangle$ and define the output noise vectors $|\widetilde{x}_j\rangle$ by the inversion $z^*(\nu) = \overline{z}(-\nu)$ as $\widetilde{x}_j = x_j{}^*$.

Indeed, by virtue of the completeness of $\{\boldsymbol{u}(\nu) \mid \nu \in \Omega\}$ it follows that

$$\langle x_i | x_j \rangle := \int \overline{x}_i(\nu) x_j(\nu) \mathrm{d}\nu = \int \kappa(\nu) \overline{u}_i(\nu) u_j(\nu) \mathrm{d}\nu = K_{ij}$$

$$\langle \widetilde{x}_i | \widetilde{x}_j \rangle := \int \overline{x}_i{}^*(\nu) x_j{}^*(\nu) \mathrm{d}\nu = \int \widetilde{\kappa}(\nu) \overline{u}_i(\nu) u_j(\nu) \mathrm{d}\nu = \widetilde{K}_{ij}$$

$$\langle x_i | \widetilde{x}_j \rangle := \int \overline{x}_i(\nu) x_j{}^*(\nu) \mathrm{d}\nu = \int \rho(\nu) \overline{u}_i(\nu) u_j(\nu) \mathrm{d}\nu = R_{ij}$$

where we also used the invariance of the measure on Ω with respect to the flip $\nu \mapsto -\nu$.

Let us denote by $\mathcal{D}_\sharp \subset L^2(\Theta)$ and $\mathcal{D}_\flat \subset L^2(\Theta)$ the dense subspaces

$$\mathcal{D}_\sharp = \left\{ z\colon \lambda^{1/2} z \in L^2(\Omega) \right\}, \mathcal{D}_\flat = \left\{ z : \lambda^{-1/2} z \in L^2(\Omega) \right\}$$

of square integrable complex amplitudes $z \in L^2(\Omega)$ with the support in

$$\Theta = \{\nu : \kappa(\nu) \neq 0 \neq \widetilde{\kappa}(\nu)\} = \mathrm{N}_+^\perp \cap \mathrm{N}_-^\perp,$$

where $\lambda(\nu)^{1/2} = \left[\widetilde{\kappa}(\nu)/\kappa(\nu)\right]^{1/2}$ and $\lambda(\nu)^{-1/2} = \left[\kappa(\nu)/\widetilde{\kappa}(\nu)\right]^{1/2}$ are both well defined as positive operators by the multiplications in $L^2(\Theta)$. The involution $|z\rangle^\sharp = |z^\dagger\rangle$ can be defined on the subspace $\mathcal{D}_\sharp \subseteq L^2(\Omega)$, generated by the complex amplitudes $z = \sum \zeta^j x_j$ with $z^* \in \mathcal{D}_\flat$ and the adjoint involution $|z\rangle^\flat = |z^\dagger\rangle$ can be defined on the subspace $\mathcal{D}_\flat \subseteq L^2(\Omega)$, generated by $z = \sum \zeta^j \widetilde{x}_j$ with $z^* \in \mathcal{D}_\sharp$ as

$$z^\sharp(\nu) = \lambda(\nu)^{-1/2} z^*(\nu), \quad z^\flat(\nu) = \lambda\,(\nu)^{1/2} z^*(\nu), \quad \nu \in \Theta.$$

This simply follows from the identities

$$x_j^\sharp(\nu) = \lambda(\nu)^{-1/2} x_j^*(\nu) = x_j(\nu), \quad \nu \in \mathrm{N}_-^\perp$$

$$\widetilde{x}_j^\flat(\nu) = \lambda(\nu)^{1/2} \widetilde{x}_j^*(\nu) = \widetilde{x}_j(\nu), \quad \nu \in \mathrm{N}_+^\perp$$

for the canonical generating sequences $\{x_j\}$ and $\{\widetilde{x}_j\}$.

In the case of nonzero temperature, when $\Theta = \Omega$ and $\mathcal{D}_\sharp$, $\mathcal{D}_\flat$ are dense in $L^2(\Omega)$, the spectral representations of quantum noise and the output process are connected by the spectral linear filters

$$\widetilde{x}(\nu) = \lambda(\nu)^{1/2} x(\nu), \quad x(\nu) = \lambda(\nu)^{-1/2} \widetilde{x}(\nu).$$

The canonical vector realization $x_j, \widetilde{x}_j$ of $|x_j\rangle, |\widetilde{x}_j\rangle$ in the time representation:

$$x_j{}^i = \int \overline{u}_i(\nu) x_j(\nu) \mathrm{d}\nu, \qquad \widetilde{x}_j{}^i = \int \overline{u}_i(\nu) \widetilde{x}_j(\nu) \mathrm{d}\nu$$

is given by the matrix elements of the square roots $X = K^{1/2}$ and $\widetilde{X} = \widetilde{K}^{1/2}$

$$x_j{}^i = X_{ij}, \quad \widetilde{x}_j{}^i = \widetilde{X}_{ij}, \quad i, j \in \mathbf{Z}$$

such that $\widetilde{x}^i_j = \overline{x}_j{}^i$ as $\widetilde{X} = \overline{X}$. They are thought as the $\ell^2(\mathbf{Z})$ columns

$$x_j = \left(x_j{}^i\right)^{i\in\mathbf{Z}}, \qquad \widetilde{x}_j = \left(\overline{x}_j{}^i\right)^{i\in\mathbf{Z}}$$

with the scalar products $\langle x_i|x_j\rangle = x_i^\dagger x_j, \langle\widetilde{x}_i|\widetilde{x}_j\rangle = \widetilde{x}_i^\dagger\widetilde{x}_j$:

$$x_i{}^\dagger x_j = \sum_{k\in\mathbf{Z}} \overline{X}_{ki}X_{kj} = K_{ij} = \sum_{k\in\mathbf{Z}} X_{ki}\overline{X}_{kj} = \widetilde{x}_j{}^\dagger\widetilde{x}_i, \quad x_i{}^\dagger\widetilde{x}_j = R_{ij}.$$

Hence the time reversion $x \mapsto \widetilde{x}$ is represented by the usual complex conjugation $\widetilde{x} = \overline{x}$, $\overline{x}_j = (\,\overline{x}_j{}^i)^{i\in\mathbf{Z}}$, which is indeed the transformation : $\overline{x}_j{}^i = x_i{}^j$ due to the self-adjointness of the square roots of $K = K^\dagger$ and so $\overline{K} = \overline{K}^\dagger$. Although the vectors $x_j, \widetilde{x}_j$ seem to be complex, they are self-adjoint $x_j{}^\sharp = x_j$, $\widetilde{x}_j{}^\flat = \widetilde{x}_j$ with respect to the different involutions which are given in the case $\Theta = \Omega$ by $z^\sharp = L^{-1/2}z^*$, $\widetilde{z}^\flat = L^{1/2}\widetilde{z}^*$, where

$$L^{1/2}x_j = \sum L^{1/2}{}_k x_j{}^k = \widetilde{x}_j, \qquad L^{-1/2}\widetilde{x}_j = \sum L^{-1/2}{}_k x_j{}^k = x_j$$

are the linear input-output and reversed filters in the time representation. The operator $L = \overline{K}K^{-1}$ with the matrix elements

$$L_{ij} = \int \overline{u}_i(\nu)\lambda(\nu)u_j(\nu)\mathrm{d}\nu, \qquad \lambda(\nu) = \widetilde{\kappa}(\nu)/\kappa(\nu)$$

is characterized by the properties $L > 0, \overline{L} = L^{-1}$ and is called the modular operator in this invertible case. It can be described by the adjoint involutions as $L|z\rangle = |z\rangle^{\sharp\flat}$, and its algebraic analogue is the main object of study in Tomita-Takesaki theory. Note that if K is not invertible, the canonical realizations of x and $\widetilde{x}$ are still connected by the isometric involution $z \mapsto z^*$ although it may not be described by the linear filtering because the adjoint involutions $\sharp, \flat$ may not be represented in $\mathcal{E}$. In contrast to the classical theory the constructed canonical realization $\widetilde{x}_j$ of the reversed quantum process does not coincide with x_j and can be even orthogonal to the real quantum noise, if $\mathrm{N}_+ \cup \mathrm{N}_- = \Omega$.

In the case of a stationary time sequence $x_j = \sqrt{\varepsilon}x(t_j)$, $t_j = \varepsilon j$ the spectrum Ω is in the interval $[-\varepsilon^{-1}/2, \varepsilon^{-1}/2] \subset \mathbf{R}$, the flip $\nu \mapsto -\nu$ is the usual reflection, and

$$u_j(\nu) = \varepsilon^{1/2}\exp\{-2\pi\mathrm{i}\nu t_j\}.$$

The canonical vector realizations of the stationary sequences (x_j) and $(\,\widetilde{x}_j)$ is given in the time representation by the matrix elements $x_j{}^i = \chi_{j-i}$, $\widetilde{x}_j{}^i = \widetilde{\chi}_{j-i}$, where the complex sequences

$$\begin{aligned} \chi_j &= \varepsilon^{1/2}\int_\Omega \kappa(\nu)^{1/2}\exp\{-2\pi\mathrm{i}\nu t_j\}\mathrm{d}\nu \\ \widetilde{\chi}_j &= \varepsilon^{1/2}\int_\Omega \widetilde{\kappa}(\nu)^{1/2}\exp\{-2\pi\mathrm{i}\nu t_j\}\mathrm{d}\nu \end{aligned}$$

are connected by the usual time reflection $\widetilde{x}(t_j) = x(-t_j)$ with respect to $t = 0$. If K is invertible, the time reflection is described by the stationary input-output and reversed complex filters

$$\chi_{-i} = \sum_{j\in\mathbf{Z}} l^{1/2}{}_{i-j}\chi_j, \qquad \chi_i = \sum_{j\in\mathbf{Z}} l^{-1/2}{}_{i-j}\chi_{-j},$$

where $l^{1/2}{}_{i-j} = L^{1/2}{}_{ij}$, $\quad l^{-1/2}{}_{i-j} = L^{-1/2}{}_{ij}$, are given by the integrals

$$l_\theta{}^{1/2}{}_j = \varepsilon \int_\Theta \lambda(\nu)^{1/2} \exp\{2\pi \mathrm{i}\nu t_j\}\mathrm{d}\nu, \quad l_\theta{}^{-1/2}{}_j = \varepsilon \int_\Theta \lambda(\nu)^{-1/2} \exp\{2\pi \mathrm{i}\nu t_j\}\mathrm{d}\nu$$

over $\Theta \subseteq \Omega$ in the general case.

3 SPECTRAL DECOMPOSITION AND LINEAR FILTERING

Let us consider the stationary quantum noise $x : t \mapsto x(t)$ with continuous time $t \in \mathbf{R}$. It can be treated as the limit as $\varepsilon \to 0$ of a stationary sequence $\varepsilon^{-1/2}x^\varepsilon{}_j$ with correlations $k^\varepsilon{}_j = \varepsilon k(t)$, given by a complex positive-definite function

$$k(t) = \int_{-\infty}^{\infty} \exp\{\, 2\pi \mathrm{i}\nu t\}\, \kappa(\nu)\mathrm{d}\nu = \langle x(t)|x(0)\rangle\,.$$

This may exist only as a generalized function(distribution) if the support $\mathrm{N}_+^\perp \subseteq \mathbf{R}$ of the spectral density κ is unbounded. The reversed noise $\widetilde{x} : t \mapsto \widetilde{x}(t)$, corresponding to the spectral density $\widetilde{\kappa}(\nu) = \kappa(-\nu)$, is described upto second order by the autocorrelation function

$$\widetilde{k}(t) = \int_{-\infty}^{\infty} \exp\{2\pi \mathrm{i}\nu t\}\, \widetilde{\kappa}(\nu)\mathrm{d}\nu = \langle \widetilde{x}(t)|\widetilde{x}(0)\rangle$$

and by the symmetric cross-correlation function

$$r(t) = \int_{-\infty}^{\infty} \exp\{\, 2\pi \mathrm{i}\nu t\}\, \rho(\nu)\mathrm{d}\nu = \langle x(t)|\widetilde{x}(0)\rangle\,,$$

where $\rho(\nu) = [\widetilde{\kappa}(\nu)\kappa(\nu)]^{1/2}$, so that r is the convolutional square root of

$$\left[\widetilde{k} * k\right](t) = \int_{-\infty}^{\infty} k(s-t)k(s)\mathrm{d}s.$$

If $\widetilde{k} * k = 0$, i.e. if $\widetilde{\kappa}\kappa = 0$, and so $\mathrm{N}_+^\perp \cap \mathrm{N}_-^\perp = \emptyset$ we have stationary vacuum noise. The noise is called standard if $k(t) + \widetilde{k}(t)$ is the Dirac δ-function, i.e. if $\widetilde{\kappa} + \kappa = 1$.

Such a noise is purely nonclassical because the condition $\widetilde{\kappa}\kappa = 0$ for a symmetric function $\widetilde{\kappa} = \kappa$ is only possible in the trivial case $\mathrm{N}_+^\perp = \emptyset = \mathrm{N}_-^\perp$ when $x = 0 = \widetilde{x}$.

In the general case the subset $\Theta = \mathrm{N}_+^\perp \cap \mathrm{N}_-^\perp$ of Ω is not empty, but the complement $\Theta^\perp = \Omega \setminus \Theta$ can be decomposed into the disjoint union of

$$\mathrm{N}_- = \{\nu \in \Omega | \widetilde{\kappa}(\nu) = 0\} \subseteq \mathrm{N}_+^\perp, \quad \mathrm{N}_+ = \{\nu \in \Omega | \kappa(\nu) = 0\} \subseteq \mathrm{N}_-^\perp, \quad \mathrm{N}_+ \cap \mathrm{N}_- = \emptyset.$$

This means that the quantum noise and its time-reversed version can be uniquely decomposed in the correlation theory into sums $x_o + x_\theta, \widetilde{x}_o + \widetilde{x}_\theta$ of uncorrelated vacuum $(x_o, \widetilde{x}_o)$ and thermal $(x_\theta, \widetilde{x}_\theta)$ components. In the spectral representation they are given by

$$x_\circ(t) = P_-^\perp x(t), \quad \widetilde{x}_\circ(t) = P_+^\perp \widetilde{x}(t),$$
$$x_\theta(t) = P_\theta x(t), \quad \widetilde{x}_\theta(t) = P_\theta \widetilde{x}(t), \qquad \forall t \in \mathbf{R}$$

where $P_+^\perp$ and $P_-^\perp$ are the projectors on $L^2(\mathrm{N}_+)$ and $L^2(\mathrm{N}_-)$ and $P_\theta = P_+P_-$ is the orthoprojector on the subspace $L^2(\Theta)$. The orthoprojectors $P_+ = 1 - 1_+$ and $P_- = 1 - 1_-$ define the best input-output and output-input estimates

$$P_+\widetilde{x}(t) = 1_\theta\widetilde{x}(t) = \lambda_\theta{}^{1/2}x(t),$$

$$P_-x(t) = 1_\theta x(t) = \lambda_\theta{}^{-1/2}\widetilde{x}(t),$$

where 1_θ, defined by $1_\theta(\nu) = 1$ if $\nu \in \Theta$, $1_\theta(\nu) = 0$ if $\nu \notin \Theta$, is the characteristic function of $\Theta = \mathrm{N}_+^\perp \cap \mathrm{N}_-^\perp$, 1_+ and 1_- are the characteristic functions of N_+ and N_-, and

$$\lambda_\theta(\nu)^{-1/2} = (\sigma(\nu)/\widetilde{\sigma}(\nu))\, 1_\theta(\nu)$$
$$\lambda_\theta(\nu)^{1/2} = (\widetilde{\sigma}(\nu)/\sigma(\nu))\, 1_\theta(\nu).$$

In the time representation

$$x(t) = \{\chi(t-s)|s \in \mathbf{R}\}, \quad \widetilde{x}(t) = \{\widetilde{\chi}(t-s)|s \in \mathbf{R}\}$$

$$\chi(t-s) = \int_\Omega \kappa(\nu)^{1/2} \exp\{2\pi\mathrm{i}\nu(s-t)\}\, \mathrm{d}\nu = \widetilde{\chi}(s-t)$$

$x_\theta(t)$ and $\widetilde{x}_\theta(t)$ are obtained by the replacing the interval Ω of integration by Θ, $x_\circ(t)$ by N_- and $\widetilde{x}_\circ(t)$ by N_+. The optimal input-output and output-input filters are given in the Fourier representation by the complex stationary linear nonadapted integrals

$$[P_+\widetilde{x}(t)]\,(s) = \int_{-\infty}^{\infty} l_\theta{}^{1/2}(s-r)\chi(t-r)\mathrm{d}r, \quad [P_-x(t)]\,(s) = \int_{-\infty}^{\infty} l_\theta{}^{-1/2}(s-r)\widetilde{\chi}(t-r)\mathrm{d}r.$$

Here $l_\theta{}^{1/2}(s-t) = L_\theta{}^{1/2}(s,t)$, $l_\theta{}^{-1/2}(s-t) = L_\theta{}^{-1/2}(s,t)$ are the complex positive definite generalized functions

$$l_\theta{}^{1/2}(t) = \int_\Theta \lambda_\theta(\nu)^{1/2} \exp\{2\pi\mathrm{i}\nu t\}\, \mathrm{d}\nu, \quad l_\theta{}^{-1/2}(t) = \int_\Theta \lambda_\theta(\nu)^{-1/2} \exp\{2\pi\mathrm{i}\nu t\}\, \mathrm{d}\nu$$

characterized by the modular property

$$l_\theta{}^{1/2}(-t) = \bar{l}_\theta{}^{1/2}(t) = l_\theta{}^{-1/2}(t).$$

The case $\mathrm{N}_+ = \emptyset = \mathrm{N}_-$ corresponds to the purely thermal noise which is called standard, if $\widetilde{\kappa}\kappa = 1$ in a given spectral region $\Omega \subseteq \mathbf{R}$. If $\Omega = \mathbf{R}$, it can be written as $\left[\,\widetilde{k} * k\right](t) = \delta(t)$ in terms of the Dirac correlation function $r(t) = \delta(t)$.

The thermal noise is called white if the spectrum κ and thus also $\widetilde{\kappa}$ is flat: $\kappa(\nu) = \sigma^2 = \widetilde{\kappa}(\nu)$, $\forall\nu \in \Omega$. White noise coincides with its time-reversed version and is essentially classical (at least in second order). There exists just one such standard noise $x = w = \widetilde{x}$ described in the second order theory by the correlation function $k(t) = \delta(t) = \overline{k}(t)$.

In the quantum case there are many standard thermal noises with δ-correlated reversed version $\widetilde{\chi}(t) = \chi(-t)$. They are parameterised by a modular spectral function

$$\lambda(\nu) > 0, \ \widetilde{\lambda}(\nu) = \lambda(-\nu) = \lambda(\nu)^{-1}, \ \forall\nu \in \Omega$$

giving the standard correlation functions k, $\overline{k}$ as the convolution square roots

$$k(t) = l^{-1/2}(t), \ \overline{k}(t) = l^{1/2}(t), \ \forall t \in \mathbf{R}.$$

Any stationary quantum process $y : t \to y(t)$ and its reverse $\widetilde{y} : t \to \widetilde{y}(t)$ described upto second order by the correlation function

$$\langle y(t)y(0)\rangle = \int \exp\{2\pi \mathrm{i}\nu t\}\sigma(\nu)^2 \mathrm{d}\nu = \langle \widetilde{y}(0)\widetilde{y}(t)\rangle$$

can be obtained by stationary filtering of standard quantum noises x and $\widetilde{x}$ with spectral densities

$$\kappa(\nu) = 1_-(\nu) + \lambda_\theta(\nu)^{-1/2}, \quad \widetilde{\kappa}(\nu) = 1_+(\nu) + \lambda_\theta(\nu)^{1/2}$$

and $\rho(\nu) = 1_\theta(\nu)$. In the canonical vector representation

$$y(t) = \{y(t,\nu)|\nu \in \Omega\}, \quad \widetilde{y}(t) = \{\widetilde{y}(t,\nu)|\nu \in \Omega\}$$

the filtering is given by

$$y(t,\nu) = f(\nu)x(t,\nu), \ y^*(t,\nu) = f(\nu)x^*(t,\nu)$$

where $\overline{f} = f = \widetilde{f}$ is a real symmetric transmission function $f : \Omega \to \mathbf{R}_+$ of the complex amplitudes

$$x(t,\nu) = \exp\{2\pi \mathrm{i}\nu t\}x(\nu), \ x^*(t,\nu) = \exp\{2\pi \mathrm{i}\nu t\}x^*(\nu)$$

$$x(\nu) = 1_-(\nu) + \lambda_\theta(\nu)^{-1/4}, \ x^*(\nu) = 1_+(\nu) + \lambda_\theta(\nu)^{1/4}.$$

The transmission function f is uniquely defined on the space $L^2(\Omega)$ of $(y, \widetilde{y})$ by

$$f(\nu) = [\sigma(\nu)\widetilde{\sigma}(\nu)]^{1/2} 1_\theta(\nu) + [\sigma(\nu) \vee \widetilde{\sigma}(\nu)]^{1/2} 1_\theta^\perp(\nu).$$

In the time representation $y(t) = \{\psi(t-s)\}$, $\widetilde{y}(t) = \{\widetilde{\psi}(t-s)\}$, given by the Fourier integrals

$$\psi(-s) = \int y(\nu)\exp\{2\pi \mathrm{i}\nu s\}\mathrm{d}\nu = \widetilde{\psi}(s),$$

the filters are written as real nonadapted stationary transformations

$$\psi(t-s) = \int_{-\infty}^{\infty} \phi(s-t)\chi(t-r)\mathrm{d}r = \widetilde{\psi}(s-t)$$

where

$$\phi(t) = \int f(\nu)\exp\{2\pi \mathrm{i}\nu t\}\mathrm{d}\nu.$$

A typical example of the stationary quantum thermal noise is given by the Planck spectral density

$$\kappa(\nu) = h\nu(\mathrm{e}^{\beta h\nu} - 1)^{-1}, \ \widetilde{\kappa}(\nu) = h\nu(1 - \mathrm{e}^{-\beta h\nu})^{-1}.$$

In this case $\widetilde{\kappa}(\nu) - \kappa(\nu) = h\nu > 0$ if $\nu > 0$, $\lambda(\nu) = \exp\{\beta h\nu\} \neq 0 \ \forall \nu \in \mathbf{R}$, and $\lambda(\nu) \neq 1$ for $\nu \neq 0$, and $\lambda(\nu) = 1$ as for classical white noise only if $\beta = 0$. It describes equilibrium Bose noise in which the electromagnetic (optical) field is in a temperature state of temperature $\theta = 1/\beta$. The modular function for such noise,

$$l(t) = \int \exp\{2\pi \mathrm{i}\nu t\}\exp\{\beta h\nu\}\mathrm{d}\nu = \delta\left(t + \frac{\hbar}{\mathrm{i}}\beta\right) = \exp\left\{\frac{\hbar}{\mathrm{i}}\beta\frac{\mathrm{d}}{\mathrm{d}t}\right\}\delta(t),$$

is extremely singular, as is the correlation function

$$k(t) = \frac{\hbar}{\mathrm{i}}\left(\exp\left\{\frac{\hbar}{\mathrm{i}}\beta\,\frac{\mathrm{d}}{\mathrm{d}t}\right\} - 1\right)^{-1}\frac{\mathrm{d}}{\mathrm{d}t}\,\delta(t)$$

($\hbar = h/2\pi$) with complex conjugate

$$\overline{k}(t) = \frac{\hbar}{\mathrm{i}}\left(1 - \exp\left\{-\frac{\hbar}{\mathrm{i}}\beta\frac{\mathrm{d}}{\mathrm{d}t}\right\}\right)^{-1}\frac{\mathrm{d}}{\mathrm{d}t}\,\delta(t)$$

for the time-reversed Planck noise corresponding to the density $\widetilde{\kappa}$.

Although the cross correlation function

$$r(t) = \frac{\hbar}{2}\left[\sin\left(\frac{\hbar}{2}\beta\,\frac{\mathrm{d}}{\mathrm{d}t}\right)\right]^{-1}\frac{\mathrm{d}}{\mathrm{d}t}\delta(t)$$

is less singular, due to the integrability of the spectral density

$$\rho(\nu) = h\nu[\exp\{\beta h\nu/2\} - \exp\{-\beta h\nu/2\}]^{-1},$$

quantum stochastic integrals corresponding to linear stationary filters for such noise are not defined in the usual sense. Indeed, the standard quantum pair $(x, \widetilde{x})$ corresponding to the Planck density have the singular correlation functions

$$l^{-1/2}(t) = \delta(t + \mathrm{i}\hbar\beta/2), \qquad l^{1/2}(t) = \delta(t - \mathrm{i}\hbar\beta/2)$$

and the cross correlation function $r(t) = \delta(t)$. To obtain the Planck noise y and its time-reversed version $\widetilde{y}$ one should integrate the standard quantum noise x and $\widetilde{x}$ with the non singular filtering function

$$f(t) = \int [h\nu/2\mathrm{sh}(\beta h\nu/2)]^{1/2}\exp\{2\pi\mathrm{i}\nu t\}\mathrm{d}\nu.$$

But the result of the integration remains singular, corresponding to the singularity of k and $\overline{k}$.

4 QUANTUM STOCHASTIC INTEGRATION

In quantum theory the noise x is thought of as represented by a continuous family $X : t \mapsto X(t)$ of (usually self-adjoint) operators $X(t) = X(t)^{\dagger}, t \in \mathbf{R}$ with zero mean values $\langle\phi|X(t)|\phi\rangle = 0$. Here ϕ is a quantum state, which can be always represented by a unit vector (probability amplitude) $|\,\phi\rangle \in \mathcal{H}, \langle\phi\mid\phi\rangle = 1$ in a proper Hilbert space $\mathcal{H}$. The quantum correlations $K(s,t) = k(s-t)$ are defined for a stationary state ϕ by the scalar product in $\mathcal{H}, k(t) = \langle\phi\,|X(t)X(0)|\,\phi\rangle$. If $k(t)$ is complex, the operator $X(t) = X(t)^{\dagger}$ will not commute with $X(0) = X(0)^{\dagger}$. The time-reversed correlations $\widetilde{k}(t) = k(-t)$ are "observable" in the real arrow of time only for the time-reversed noise $\widetilde{x}$ which gives a natural representation in the same Hilbert space $\mathcal{H}$ an output process $\widetilde{X} : t \mapsto \widetilde{X}(t)$, for which $\widetilde{k}(t) = \langle\phi\,|\widetilde{X}(t)\widetilde{X}(0)|\,\phi\rangle$ with the same $\phi \in \mathcal{H}$.

The vector processes $t \mapsto |\,x(t)\rangle$, $t \mapsto |\widetilde{x}\,(t)\rangle$ are obtained from the operator representations in $\mathcal{H}$ by the identifications

$$|\,x(t)\rangle = X(t)|\,\phi\rangle\,, \quad |\widetilde{x}\,(t)\rangle = \widetilde{X}(t)|\,\phi\rangle$$

in the minimal subspace generated by the set $\left\{X(t)|\,\phi\rangle\,,\widetilde{X}(t)|\,\phi\rangle\right\} \subset \mathcal{H}$. In a classical theory this gives a one to one correspondence between vectors and operators of multiplication in the total space $\mathcal{H}$. This is not the case in quantum probability theory as one can observe for vacuum correlations, for which it is possible that $|z\rangle = Z|\phi\rangle = 0$ for annihilation operators $Z \neq 0$. This is why the vector integration theory is not sufficient in the quantum case; we need an operator form of quantum stochastic integration.

According to the nondemolition (causality) principle of quantum theory [12] the input observables $X(0)$ must commute with the fundamental output observables $\widetilde{X}(t)$ for all $t \leq 0$ and they must commute also for $t \geq 0$ because the noise X can be regarded as the output process for the time-reversed noise $\widetilde{X}$. At the correlation level this is expressed in terms of the reflection symmetry

$$\langle\phi\,|X(0)\widetilde{X}(t)|\,\phi\rangle = r(t) = \langle\phi\,|\widetilde{X}(t)X(0)|\,\phi\rangle\,, \quad \forall t \in \mathbf{R}$$

for the cross correlation function $r(t) = (\widetilde{k} * k)^{1/2}(t)$.

Now we are going to develop a second order theory of quantum stochastic integration with respect to colored quantum noise and its fundamental output process. The realization of such theory was given in [10] in terms of a representation of the canonical commutation relations in the symmetric Fock space $\mathcal{H}$. This provides a one to one correspondence between second order theory and the Gaussian theory of quantum noise with respect to the vacuum state ϕ.

Let us denote by ζ the complex test functions $t \mapsto \zeta(t)$, for which the "integrals"

$$X\zeta = \int_{-\infty}^{\infty} X(t)\zeta(t)\mathrm{d}t, \quad \widetilde{X}\zeta = \int_{-\infty}^{\infty} \widetilde{X}(t)\zeta(t)\mathrm{d}t$$

are expected to be well defined as self-adjoint operators in $\mathcal{H}$ if $\overline{\zeta} = \zeta$. If ζ is given as a Fourier integral

$$\zeta(t) = \int \overline{c}(\nu)\exp\{2\pi\mathrm{i}\nu t\}\,\mathrm{d}\nu \equiv \widehat{c}(t),$$

then the operator-valued functionals $\zeta \mapsto X\zeta$, $\zeta \mapsto \widetilde{X}\zeta$ can be treated as quantum stochastic integrals

$$X\zeta = \int \overline{c}(\nu)\mathrm{A}(\mathrm{d}\nu), \quad \widetilde{X}\zeta = \int \overline{c}(\nu)\tilde{\mathrm{A}}(\mathrm{d}\nu)$$

for the involute (self-inverse) Fourier transforms $c = \widehat{\zeta}$ of $\zeta = \widehat{c}$:

$$c(\nu) = \int_{-\infty}^{\infty} \exp\{2\pi\mathrm{i}\nu t\}\,\overline{\zeta}(t)\mathrm{d}t \equiv \widehat{\zeta}(\nu).$$

The quantum stochastic integrators A, $\widetilde{\mathrm{A}}$, defined in [14] as operator-valued measures on the spectral intervals $\Delta \subset \Omega$, satisfy the following properties:
(i) σ - additivity:

$$\mathrm{A}\left(\sum \Delta_\mathrm{i}\right) = \sum \mathrm{A}(\Delta_\mathrm{i}), \quad \widetilde{\mathrm{A}}\left(\sum \Delta_\mathrm{i}\right) = \sum \widetilde{\mathrm{A}}(\Delta_\mathrm{i})$$

for any disjoint union $\sum \Delta_\mathrm{i} = \Delta$;
(ii) flip-self-adjointness:

$$\mathrm{A}(\Delta)^\dagger = \mathrm{A}(-\Delta), \quad \widetilde{\mathrm{A}}(\Delta)^\dagger = \widetilde{\mathrm{A}}(-\Delta)$$

where $-\Delta = \{-\nu | \nu \in \Delta\}$ for any Δ;
(iii) self-orthogonality if $-\Delta \perp \Delta'$:

$$\begin{aligned}
\langle \phi | \mathrm{A}(\Delta')\mathrm{A}(-\Delta) | \phi \rangle &= \int_{\Delta\cap\Delta'} \kappa(\nu)\mathrm{d}\nu \\
\langle \phi | \tilde{\mathrm{A}}(\Delta')\tilde{\mathrm{A}}(-\Delta) | \phi \rangle &= \int_{\Delta\cap\Delta'} \tilde{\kappa}(\nu)\mathrm{d}\nu \\
\langle \phi | \tilde{\mathrm{A}}(\Delta')\mathrm{A}(-\Delta) | \phi \rangle &= \int_{\Delta\cap\Delta'} \rho(\nu)\mathrm{d}\nu
\end{aligned}$$

where $\rho(\nu) = [\tilde{\kappa}(\nu)\kappa(\nu)]^{1/2}$ for a given stationary state ϕ. The σ - additivity makes it possible to define the quantum stochastic integrals as integral sums for all functions ζ with piecewise constant integrable $\widehat{\zeta}$; the second condition makes the $X\zeta$ and $\widetilde{X}\zeta$ self-adjoint, if $\widehat{\zeta}^* = \widehat{\zeta}$ and so $\overline{\zeta} = \zeta$, and the third means that the pair $\mathrm{A}(\Delta)^\dagger, \tilde{\mathrm{A}}(\Delta)^\dagger$ is uncorrelated with $\mathrm{A}(\Delta'), \tilde{\mathrm{A}}(\Delta')$ if $\Delta \cap \Delta' = \emptyset$. The latter can be symbolically written in the form of a multiplication table

$$\mathrm{A}(\mathrm{d}\nu')\mathrm{A}(\mathrm{d}\nu)^\dagger = \delta_{\nu\nu'}\kappa(\nu)\mathrm{d}\nu, \quad \tilde{\mathrm{A}}(\mathrm{d}\nu')\mathrm{A}(\mathrm{d}\nu)^\dagger = \delta_{\nu\nu'}\rho(\nu)\mathrm{d}\nu$$

$$\mathrm{A}(\mathrm{d}\nu')\tilde{\mathrm{A}}(\mathrm{d}\nu)^\dagger = \delta_{\nu\nu'}\rho(\nu)\mathrm{d}\nu, \quad \tilde{\mathrm{A}}(\mathrm{d}\nu')\tilde{\mathrm{A}}(\mathrm{d}\nu)^\dagger = \delta_{\nu\nu'}\tilde{\kappa}(\nu)\mathrm{d}\nu$$

where $\delta_{\nu\nu'} = 0$ if $\nu \neq \nu'$, $\delta_{\nu\nu'} = 1$ if $\nu = \nu'$.

Note that this table contains also the products

$$\mathrm{A}(\mathrm{d}\nu)^\dagger\mathrm{A}(\mathrm{d}\nu') = \delta_{\nu\nu'}\tilde{\kappa}(\nu)\mathrm{d}\nu, \quad \tilde{\mathrm{A}}(\mathrm{d}\nu)^\dagger\mathrm{A}(\mathrm{d}\nu') = \delta_{\nu\nu'}\rho(\nu)\mathrm{d}\nu$$

$$\mathrm{A}(\mathrm{d}\nu)^\dagger\tilde{\mathrm{A}}(\mathrm{d}\nu') = \delta_{\nu\nu'}\rho(\nu)\mathrm{d}\nu, \quad \tilde{\mathrm{A}}(\mathrm{d}\nu)^\dagger\tilde{\mathrm{A}}(\mathrm{d}\nu') = \delta_{\nu\nu'}\kappa(\nu)\mathrm{d}\nu$$

obtained by the flip $\nu' \mapsto -\nu, \nu \mapsto -\nu'$, and also at $\nu' = -\nu$

$$\mathrm{A}(\mathrm{d}\nu)\mathrm{A}(\mathrm{d}\nu) = 0 = \tilde{\mathrm{A}}(\mathrm{d}\nu)\tilde{\mathrm{A}}(\mathrm{d}\nu), \quad \forall \nu.$$

A realization of this table can be given in the Fock space $\mathcal{H}$ for any state vector $| \phi\rangle = \mathcal{H}$ [10].

Using this multiplication table we obtain the isometry property

$$\langle \phi | Z^\dagger Z | \phi \rangle = \int \left| \widehat{\xi}(\nu)\kappa(\nu)^{1/2} + \widehat{\eta}(\nu)\kappa(\nu)^{1/2} \right|^2 \mathrm{d}\nu = \langle z | z \rangle$$

for the map $| z\rangle = Z| \phi\rangle \mapsto z = \widehat{\xi}\kappa^{1/2} + \widehat{\eta}\tilde{\kappa}^{1/2}$ for the operators

$$Z = X\overline{\xi} + \widetilde{X}\overline{\eta}$$

acting on the state vector ϕ into the spectral representation $z(\nu) = \langle \nu | z\rangle$ of

$$| z\rangle = \int_{-\infty}^{\infty} \left(\overline{\xi}(t) | x(t)\rangle + \overline{\eta}(t) | \tilde{x}(t)\rangle \right) \mathrm{d}t.$$

Thus it is possible to extend the quantum stochastic integral in the mean square sense of $||Z\phi||^2 < \infty$ from the functions (ξ, η) with simple $(\hat{\xi}, \hat{\eta})$ to any pair of the functions (ξ, η), for which the complex amplitude z is square integrable: $z \in L^2(\Omega)$. But to ensure also that $||Z^\dagger\phi||^2 < \infty$,we must add the condition of square integrability $z^\star \in L^2(\Omega)$ for the amplitude $z^\star = \hat{\xi}^*\kappa^{1/2} + \hat{\eta}^*\tilde{\kappa}^{1/2}$, that is, we define the quantum stochastic integral Z together with $Z^\dagger$ only for the test functions $\xi, \eta \subset \mathcal{E}$.

In the classical case, $\widetilde{\kappa} = \kappa$ when $z^{\star} = z^{*}$, the square integrabilities of z and z^{*} are equivalent to that of $\hat{\xi}$ and $\hat{\eta}$ with respect to the symmetric measure $\widetilde{\kappa}\,\mathrm{d}\nu = \kappa\,\mathrm{d}\nu$.

In the purely quantum case of vacuum noise it is not so, and the conditions $z, z^{\star} \in L^2(\Omega)$ are equivalent to the square integrability of $\hat{\xi}$ and $\hat{\eta}$ with respect to $\kappa \vee \widetilde{\kappa}\,\mathrm{d}\nu$. In spectral representation the vacuum noise X together with the reversed process $\widetilde{X}$ is described by the operator-valued measures

$$\mathrm{A}^{+}(\Delta) = \begin{cases} \widetilde{\mathrm{A}}(\Delta), \Delta \subset \mathrm{N}_{+}^{\perp} \\ \mathrm{A}(\Delta), \Delta \subset \mathrm{N}_{-}^{\perp} \end{cases}, \quad \mathrm{A}_{-}(\Delta) = \begin{cases} \mathrm{A}(\Delta), \Delta \subset \mathrm{N}_{+}^{\perp} \\ \widetilde{\mathrm{A}}(\Delta), \Delta \subset \mathrm{N}_{-}^{\perp} \end{cases}$$

of creation $\mathrm{A}^{+}(\Delta)$ and annihilation $\mathrm{A}_{-}(\Delta)$,

$$\langle \phi \,|\mathrm{A}^{+}(\Delta) = 0, \qquad \mathrm{A}_{-}(\Delta)|\, \phi\rangle = 0$$

which uniquely define the canonical measures

$$\mathrm{A}_{0}^{+}(\Delta) = \int_{\Delta} [\kappa \vee \widetilde{\kappa}]^{-1/2}(\nu)\mathrm{A}^{+}(\mathrm{d}\nu),$$

$$\mathrm{A}_{-}^{0}(\Delta) = \int_{\Delta} [\kappa \vee \widetilde{\kappa}]^{-1/2}(\nu)\mathrm{A}_{-}(\mathrm{d}\nu),$$

on the intervals $\Delta \subset \Omega$ for the given state $|\,\phi\rangle \in \mathcal{H}$. The canonical pair $(\mathrm{A}_{0}^{+}, \mathrm{A}_{-}^{0})$ of the creation and annihilation measures is characterised by the properties of flip-adjointness

$$\mathrm{A}_{0}^{+}\,(\Delta)^{\dagger} = \mathrm{A}_{-}^{0}(-\Delta), \quad \mathrm{A}_{-}^{0}(\Delta)^{\dagger} = \mathrm{A}_{0}^{+}(-\Delta)$$

and orthogonality if $\Delta \cap \Delta' = \emptyset$:

$$\langle \phi\,|\mathrm{A}_{-}^{0}(\Delta')\mathrm{A}_{0}^{+}(-\Delta)|\,\phi\rangle = \nu(\Delta \cap \Delta'), \quad \langle \phi\,|\mathrm{A}_{0}^{+}(\Delta')\mathrm{A}_{-}^{0}(-\Delta)|\,\phi\rangle = 0,$$

$$\langle \phi\,|\mathrm{A}_{0}^{+}(\Delta')\mathrm{A}_{0}^{+}(-\Delta)|\,\phi\rangle = 0, \quad \langle \phi\,|\mathrm{A}_{-}^{0}(\Delta')\mathrm{A}_{-}^{0}(-\Delta)|\,\phi\rangle = 0.$$

This table can be written also symbolically as

$$\mathrm{A}_{-}^{0}(\mathrm{d}\nu')\mathrm{A}_{-}^{0}(\mathrm{d}\nu)^{\dagger} = \delta_{\nu\nu'}d\nu, \quad \mathrm{A}_{0}^{+}(\mathrm{d}\nu')\mathrm{A}_{0}^{+}(\mathrm{d}\nu)^{\dagger} = 0,$$

$$\mathrm{A}_{0}^{+}(\mathrm{d}\nu')\mathrm{A}_{-}^{0}(\mathrm{d}\nu)^{\dagger} = 0, \quad \mathrm{A}_{-}^{0}(\mathrm{d}\nu')\mathrm{A}_{0}^{+}(\mathrm{d}\nu)^{\dagger} = 0,$$

for all $\nu, \nu' \in \Omega$.

Let us prove that an arbitrary (not necessary vacuum) stationary (in the second order sense) quantum process Y together with its time-reversed version $\widetilde{Y}$ can be obtained by quantum stochastic integration with respect to the canonical pair $(\mathrm{A}_{0}^{+}, \mathrm{A}_{-}^{0})$ over $\Omega = \mathbf{R}$.

Indeed, in general the pair

$$Y(t) = \int \exp\{-2\pi \mathrm{i}\nu t\}\,\mathrm{B}(\mathrm{d}\nu), \quad \widetilde{Y}(t) = \int \exp\{-2\pi \mathrm{i}\nu t\}\,\widetilde{\mathrm{B}}(\mathrm{d}\nu)$$

of operator-valued distributions $\zeta \mapsto Y\zeta, \widetilde{Y}\zeta$ is given by a pair $(\mathrm{B}, \widetilde{\mathrm{B}})$ of operator-valued orthogonal flip-self-adjoint measures with multiplication table

$$\mathrm{B}(\mathrm{d}\nu)\mathrm{B}(\mathrm{d}\nu)^{\dagger} = \sigma(\nu)^2\mathrm{d}\nu, \quad \widetilde{\mathrm{B}}(\mathrm{d}\nu)\mathrm{B}(\mathrm{d}\nu)^{\dagger} = \widetilde{\sigma}(\nu)\sigma(\nu)\mathrm{d}\nu,$$

$$\mathrm{B}(\mathrm{d}\nu)\widetilde{\mathrm{B}}(\mathrm{d}\nu)^{\dagger} = \sigma(\nu)\widetilde{\sigma}(\nu)\mathrm{d}\nu, \quad \widetilde{\mathrm{B}}(\mathrm{d}\nu)\widetilde{\mathrm{B}}(\mathrm{d}\nu)^{\dagger} = \widetilde{\sigma}(\nu)^2\mathrm{d}\nu.$$

Such measures can be obtained as quantum stochastic integrals

$$\mathrm{B}(\Delta) = \int_\Delta \sigma(\nu)\mathrm{A}^0_-(\mathrm{d}\nu) + \tilde{\sigma}(\nu)\mathrm{A}^+_0(\mathrm{d}\nu)$$

$$\widetilde{\mathrm{B}}(\Delta) = \int_\Delta \tilde{\sigma}(\nu)\mathrm{A}^0_-(\mathrm{d}\nu) + \sigma(\nu)\mathrm{A}^+_0(\mathrm{d}\nu)$$

from the canonical pair $(\mathrm{A}^+_0, \mathrm{A}^0_-)$ of flip-adjoint annihilation and creation measures. Moreover, in the nonclassical case $\sigma(\nu) \neq \tilde{\sigma}(\nu), \forall \nu \in \Omega$, corresponding to quantum noise of finite temperature, the canonical pair $\mathrm{A} = (\mathrm{A}^+_0, \mathrm{A}^0_-)$ is uniquely defined over Ω by the pair $(\mathrm{B}, \widetilde{\mathrm{B}})$ from

$$\left(\tilde{\sigma}(\nu)^2 - \sigma(\nu)^2\right)\mathrm{A}^+_0(d\nu) = \tilde{\sigma}(\nu)\mathrm{B}(\mathrm{d}\nu) - \sigma(\nu)\widetilde{\mathrm{B}}(\mathrm{d}\nu),$$

$$\left(\tilde{\sigma}(\nu)^2 - \sigma(\nu)^2\right)\mathrm{A}^0_-(\mathrm{d}\nu) = \tilde{\sigma}(\nu)\widetilde{\mathrm{B}}(\mathrm{d}\nu) - \sigma(\nu)\mathrm{B}(\mathrm{d}\nu).$$

This means that the direct and reversed quantum stochastic integrals $Y\zeta, \widetilde{Y}\zeta$ can be written for the Fourier integrals of $\zeta = \widehat{c}$ as

$$Y\zeta = \int f^\nu_+ \mathrm{A}^+_0(\mathrm{d}\nu) + f^-_\nu \mathrm{A}^0_-(\mathrm{d}\nu) = \mathrm{A}(\tilde{\sigma}c^*, \sigma c^*)$$

with $f^-_\nu = \tilde{\sigma}(\nu)\bar{c}(-\nu), f^\nu_+ = \sigma(\nu)\bar{c}(-\nu)$ and

$$\widetilde{Y}\zeta = \int f^\nu_+ \mathrm{A}^+_0(\mathrm{d}\nu) + f^-_\nu \mathrm{A}^0_-(\mathrm{d}\nu) = \mathrm{A}(\sigma c^*, \tilde{\sigma}c^*)$$

with $f^-_\nu = \sigma(\nu)\bar{c}(-\nu), f^\nu_+ = \tilde{\sigma}(\nu)\bar{c}(-\nu)$.

To obtain these integrals in the time representation as linear stationary filters of the adjoint pair $X = (X^+, X_-)$ of the canonical creation and annihilation distributions

$$X^+_t = \int \exp\{2\pi \mathrm{i}\nu t\}\,\mathrm{A}^+_0(\mathrm{d}\nu), \qquad X^t_- = \int \exp\{2\pi \mathrm{i}\nu t\}\,\mathrm{A}^0_-(\mathrm{d}\nu)$$

we can use the Fourier-Parseval identity

$$\mathrm{A}(\widehat{\varphi}^-, \widehat{\varphi}_+) = X(\overline{\varphi}^-, \overline{\varphi}_+).$$

The latter is given by the standard creation and annihilation integrators

$$X^+_0(dt) = X^+_t \mathrm{d}t, \quad X^0_-(\mathrm{d}t) = X^t_- \mathrm{d}t$$

in the time representation as the standard quantum stochastic integral

$$X(\varphi^-, \varphi_+) = \int_{-\infty}^{\infty} \varphi^t_+ X^+_0(\mathrm{d}t) + \varphi^-_t X^0_-(\mathrm{d}t).$$

Applying this to the quantum stochastic integral in the spectral domain with the general integrands

$$f^-_\nu = \widehat{\xi}(\nu)\tilde{\sigma}(\nu) + \widehat{\eta}(\nu)\sigma(\nu) = \widehat{\varphi}^-_\nu$$
$$f^\nu_+ = \widehat{\xi}(\nu)\sigma(\nu) + \widehat{\eta}(\nu)\tilde{\sigma}(\nu) = \widehat{\varphi}^\nu_+$$

corresponding to $Y\overline{\xi} + \widetilde{Y}\overline{\eta} = \mathrm{A}(f^-, f_+)$, we obtain $Y\overline{\xi} + \widetilde{Y}\overline{\eta} = X(\overline{\varphi}^-, \overline{\varphi}_+)$, where the components

$$\varphi^-_t = (\widehat{\sigma} * \widetilde{\xi})(-t) + (\widehat{\sigma} * \eta)(t) = \widehat{f}^-_t$$

$$\varphi_+^t = (\widehat{\sigma} * \xi)(t) + (\widehat{\sigma} * \widetilde{\eta})(-t) = \widehat{f}_+^t$$

are given by the convolutions

$$(\widehat{\sigma} * \zeta)(t) = \int_{-\infty}^{\infty} \sigma(t-r)\,\zeta(r)\,\mathrm{d}r$$

with transition function

$$\widehat{\sigma}(t) = \int \exp\{2\pi \mathrm{i}\nu t\}\,\sigma(\nu)\mathrm{d}\nu.$$

Thus, the generalized processes Y and $\widetilde{Y}$ are obtained by the stationary filters

$$Y(t) = \int_{-\infty}^{\infty} \sigma(s-t) X_-^0(\mathrm{d}s) + \int_{-\infty}^{\infty} \sigma(t-s) X_0^+(\mathrm{d}s)$$

$$\widetilde{Y}(t) = \int_{-\infty}^{\infty} \sigma(t-s) X_-^0(\mathrm{d}s) + \int_{-\infty}^{\infty} \sigma(s-t) X_0^+(\mathrm{d}s)$$

corresponding to the quantum stochastic integral representation

$$Y\zeta = \int_{-\infty}^{\infty} (\widehat{\sigma} * \zeta)(s) X_0^-(\mathrm{d}s) + \int_{-\infty}^{\infty} (\widehat{\sigma} * \widetilde{\zeta})(-s) X_0^+(\mathrm{d}s),$$

$$Y\zeta = \int_{-\infty}^{\infty} (\widehat{\sigma} * \widetilde{\zeta})(-s) X_-^0(\mathrm{d}s) + \int_{-\infty}^{\infty} (\widehat{\sigma} * \zeta)(s) X_0^+(\mathrm{d}s)$$

for the test functions $\widetilde{\zeta}(t) = \zeta(-t)$ with the square integrable $\widehat{\zeta}\sigma$ and $\widehat{\zeta}\widetilde{\sigma}$.

CONCLUSION

Thus we proved that for each quantum stationary noise there exists a fundamental output process which replaces the noise if the time is reversed. The nondemolition observation of the quantum noise via the measurement of the fundamental output process provides the best mean square filtering of the noise and in the classical limit completely eliminates this noise.

The input and output stationary processes can be decomposed in the second order into an orthogonal pair, consisting of vacuum and thermal noises which are orthogonal to the vacuum and connected by a reversible input-output (modular) filter. The time-continuous representation of the stationary linear filters requires a second order theory of quantum stochastic integration.

The direct application of standard quantum stochastic integration with respect to the quantum white noise integrators is not possible for equilibrium (KMS) states because of the non-Markovian character of the corresponding standard quantum processes x and $\widetilde{x}$.

But it is possible to modify this approach for the spectral domain where the stationary quantum processes are δ-correlated. Using this approach in the second order, we proved that they can be canonically decomposed into the superpositions of the vacuum pair of direct and reversed noises and can be obtained by a linear filtering from the standard vacuum noises x and $\widetilde{x}$.

In general the spectral quantum stochastic integrates need not be adapted and homogeneous with respect to the frequency shifts and the treatment of quantum nonlinear filters and other transformations of quantum equilibrium noise requires non stationary and non adapted theory of quantum stochastic integration [10].

Acknowledgment

The first author (VPB) is grateful for the hospitality and support of the Tamagawa University where this paper was partially written.

References

[1] Hudson R.L. and Parthasarathy K.R., Quantum Ito's formula and stochastic evolution, Commun. Math. Phys., 93, 301-323, 1984.

[2] Gardiner C.W., Quantum Noise, Springer-Verlag, 1991.

[3] Accardi L., Frigerio A., and Lu Y.G., An invitation to the weak coupling and low density limits, In Quantum probability and related topics VI, World Scientifics, Singapore, 3-61, 1991.

[4] Belavkin V.P. Quantum Filtering of Markov Signals with White Quantum Noise, Radiotechnika and Electronika, 25, 1445-1453, 1980. The translation from Russion is published in this volume.

[5] Collett M.J. and Gardiner C.W., Input and output in damped quantum system: quantum stochastic differential equation and master equation, Phys. Rev. A 31, 3761-3774, 1985.

[6] Belavkin V.P., A quantum stochastic calculus in Fock space of input and output nondemolition processes, Springer-Verlag LNM 1442, 99-125, 1990.

[7] Parthasarathy K.R., An introduction to quantum stochastic calculus, Birkhäuser, 1992.

[8] Hudson R.L. and Lindsay J.M., *A noncommutative martingale representation for non-Fock quantum Brownian motion*, J. Funct. Anal., **61**, 202-221, 1985.

[9] Belavkin V.P., A quantum nonadapted Ito formula and stochastic analysis in Fock scale, J. Funct. Anal., 102, 414-447, 1991.

[10] Belavkin V.P., Chaotic states and stochastic integration in quantum systems, Russian Mathematical Survey, 1, 1-50, 1992.

[11] Lindsay J.M., Noncausal stochastic integration, Probability Theory and Related Topics, 1993.

[12] Belavkin V.P., Continuous non-demolition observation, quantum filtering and optimal estimation, In: Quantum Aspects of Optical Communication, Springer-Verlag, LNP378, 151-163, 1991.

[13] Belavkin V.P., Nondemolition principle of quantum measurement theory, Foundation of Physics, 1994.

[14] Belavkin V.P., *Optimal linear randomized filtration of quantum boson signals*, Problems of Control and Information Theory, **3**(1), 47–62, 1974.

GEOMETRY OF QUANTUM STATES[1]

Samuel L. Braunstein[†] and Carlton M. Caves

Center for Advanced Studies,
Department of Physics and Astronomy,
University of New Mexico, Albuquerque, NM 87131–1156
[†]Current address: Department of Chemical Physics,
Weizmann Institute of Science, 76 100 Rehovot, Israel

1 INTRODUCTION

In general relativity the dynamics of a system is related to the geometry of space-time in an intuitively beautiful form: space-time tells matter how to move, and matter tells space-time how to curve. In this paper we consider a natural geometry on the space of quantum states (density operators), with very different consequences. In particular, this geometry does not describe the dynamics, or evolution, of a quantum system; rather it places limits on our ability to distinguish one state from another through measurements.

2 GEOMETRY ON THE PROBABILITY SIMPLEX

In this section we review Wootters's [1] derivation of the distinguishability metric on the space of probability distributions, the probability simplex. Wootters asks how well we can distinguish distributions, given N samplings. Take the n-tuple $\vec{p} \equiv \{p^1, \ldots, p^n\}$ as a discrete probability distribution. After N samplings from this distribution, we have a frequency f^j for each possible outcome $j = 1, \ldots, n$. These

[1]This work was supported in part by the Office of Naval Research (Grant No. N00014-93-1-0116).

frequencies are distributed according to the multinomial distribution

$$\text{prob}(f^1, \ldots, f^n) = \frac{N!}{\prod_{j=1}^{n}(Nf^j)!} \prod_{k=1}^{n} (p^k)^{Nf^k} , \tag{1}$$

which as N approaches infinity asymptotes towards

$$e^{-N\sum_j f_j \ln(f_j/p_j)} \simeq \exp\left(-\frac{N}{2}\sum_j \frac{(f^j - p^j)^2}{p^j}\right) , \tag{2}$$

the latter form holding for $f^j \sim p^j$. If $\vec{p} + \Delta\vec{p}$ is a nearby distribution, it can be readily distinguished from $\vec{p}$ so long as $N\sum_j(\Delta p^j)^2/p^j$ is large; thus we can conclude that

$$ds^2 = \sum_j \frac{(dp^j)^2}{p^j} \tag{3}$$

is a Riemannian metric defined by the notion of asymptotic distinguishability. Notice that on the boundary of the probability simplex we should drop the singular terms in Eq. (3) in order to maintain consistency with statistical distinguishability in lower dimensions. In terms of "probability amplitudes" this metric is the natural metric on a sphere of radius 2:

$$ds^2 = 4\sum_j \left(d\sqrt{p^j}\right)^2 . \tag{4}$$

In this coördinate system it is easy to see that the shortest path between distributions $\vec{p}$ and $\vec{q}$ is a portion of a great circle; the statistical distance between $\vec{p}$ and $\vec{q}$ is $s = 2\cos^{-1}(\sum_j \sqrt{p^j q^j}\,)$, which is twice the angle separating the distributions on the sphere [1].

Just how natural is this distinguishability metric? We shall see that the Wootters distinguishability metric is also the natural metric associated with correlations between random variables. First, however, we will describe the differential geometry of the probability simplex. Introduce a set of basis vectors $\vec{e}_j$, $j = 1, \ldots, n$, so that the n-tuple for a probability distribution can be written as a vector

$$\vec{p} = \sum_j p^j \vec{e}_j \tag{5}$$

in an n-dimensional real vector space. The probability simplex is a manifold with boundary: normalization, $\sum_j p^j = 1$, defines an $(n-1)$-dimensional "plane" within the n-dimensional vector space, and non-negativity of the probabilities, $p^j \geq 0$ for $j = 1, \ldots, n$, gives the plane a boundary. A path $\vec{p}(\lambda)$ through the probability simplex has tangent vector

$$\frac{d\vec{p}(\lambda)}{d\lambda} = \sum_j \frac{dp^j}{d\lambda}\vec{e}_j , \tag{6}$$

which, as a consequence of normalization, satisfies $\sum_j dp^j/d\lambda = 0$. Thus the tangent space to the probability simplex at $\vec{p}$ is an $(n-1)$-dimensional vector space consisting of vectors $\vec{v} = \sum_j v^j \vec{e}_j$ that satisfy $\sum_j v^j = 0$. Affixing $\vec{p}$ to the tangent space at $\vec{p}$ recovers the entire original n-dimensional vector space, which consists of all vectors $\vec{m} = \sum_j m^j \vec{e}_j$, whether or not the components m^j constitute a normalized probability distribution.

Consider now the dual vector space of linear mappings that map vectors to real numbers. An element of the dual space, denoted $\tilde{A}$, is called a 1-form and acts on a vector $\vec{m}$ according to

$$\tilde{A}(\vec{m}) \equiv \left\langle \tilde{A}, \vec{m} \right\rangle = \sum_j A_j\, m^j \;, \tag{7}$$

where $A_j \equiv \langle \tilde{A}, \vec{e}_j \rangle$. Choosing a dual basis of 1-forms $\tilde{\omega}^j$, satisfying $\langle \tilde{\omega}^j, \vec{e}_k \rangle = \delta^j{}_k$, we can write an arbitrary 1-form as

$$\tilde{A} = \sum_j A_j\, \tilde{\omega}^j \;. \tag{8}$$

We see that a 1-form $\tilde{A}$ corresponds, in statistical parlance, to a random variable, A_j being the value of the random variable for alternative j and

$$\left\langle \tilde{A}, \vec{p} \right\rangle = \sum_j A_j\, p^j \equiv \langle \tilde{A} \rangle \tag{9}$$

being the mean value of $\tilde{A}$ with respect to the distribution $\vec{p}$.

There is a special 1-form $\tilde{\mathbb{1}} \equiv \sum_j \tilde{\omega}^j$. This 1-form defines the plane of the probability simplex, since $\langle \tilde{\mathbb{1}}, \vec{p} \rangle = 1$ for any probability distribution $\vec{p}$, and any tangent vector $\vec{v}$ lies in the surfaces of $\tilde{\mathbb{1}}$, since $\langle \tilde{\mathbb{1}}, \vec{v} \rangle = 0$. Any random variable can be decomposed uniquely as

$$\tilde{A} = \left\langle \tilde{A}, \vec{p} \right\rangle \tilde{\mathbb{1}} + \Delta\tilde{A} \;, \tag{10}$$

where $\Delta\tilde{A}$ has zero mean with respect to $\vec{p}$. The cotangent space to the probability simplex at $\vec{p}$ is an $(n-1)$-dimensional vector space made up of the zero-mean random variables $\Delta\tilde{A}$; by affixing $\tilde{\mathbb{1}}$ to the cotangent space, we recover the n-dimensional vector space of all 1-forms $\tilde{A}$. Because the tangent and cotangent spaces at $\vec{p}$ can be extended by affixing $\vec{p}$ and $\tilde{\mathbb{1}}$, we are somewhat sloppy about distinguishing these spaces from the spaces of all vectors and all 1-forms.

The linear structure of vectors and forms suffices to build the statistical structure of probabilities and random variables. Where then is the bilinear structure required for a metric? A metric tensor $\boldsymbol{g}_{\vec{p}}$ at point $\vec{p}$ is a symmetric bilinear "machine" that maps pairs of vectors to the reals in the same way as an inner product. Such a machine is equivalent to an (invertible) "lowering" operator $\mathcal{L}_{\vec{p}}$ that maps vectors linearly to 1-forms according to

$$\boldsymbol{g}_{\vec{p}}(\vec{m}, \vec{n}) \equiv \langle \mathcal{L}_{\vec{p}}(\vec{m}), \vec{n} \rangle = \langle \mathcal{L}_{\vec{p}}(\vec{n}), \vec{m} \rangle \equiv \boldsymbol{g}_{\vec{p}}(\vec{n}, \vec{m}) \;. \tag{11}$$

The adjective "lowering" is used because $\mathcal{L}_{\vec{p}}$ maps vectors, which have upper (contravariant) components, to 1-forms, which have lower (covariant) components. The inverse linear operator $\mathcal{R}_{\vec{p}}$, called the "raising" operator, maps 1-forms to vectors; it extends the action of the metric to pairs of 1-forms,

$$\boldsymbol{g}_{\vec{p}}(\tilde{A}, \tilde{B}) \equiv \boldsymbol{g}_{\vec{p}}\big(\mathcal{R}_{\vec{p}}(\tilde{A}), \mathcal{R}_{\vec{p}}(\tilde{B})\big) = \left\langle \tilde{A}, \mathcal{R}_{\vec{p}}(\tilde{B}) \right\rangle = \left\langle \tilde{B}, \mathcal{R}_{\vec{p}}(\tilde{A}) \right\rangle \;. \tag{12}$$

To get a bilinear metric structure, we must consider pairs of random variables. It is natural to define the metric's action on a pair of random variables to be the statistical correlation between the two variables,

$$\boldsymbol{g}_{\vec{p}}(\tilde{A}, \tilde{B}) \equiv \left\langle \tilde{A}\tilde{B} \right\rangle = \left\langle \tilde{A}\tilde{B}, \vec{p} \right\rangle = \sum_j A_j B_j\, p^j \;, \tag{13}$$

which is equivalent to defining raising and lowering operators by

$$\mathcal{R}_{\vec{p}}(\tilde{B}) = \sum_j p^j B_j \, \vec{e}_j \; , \qquad \mathcal{L}_{\vec{p}}(\vec{m}) = \sum_j \frac{m^j}{p^j} \, \tilde{\omega}^j \; . \tag{14}$$

Notice that $\mathcal{L}_{\vec{p}}(\vec{p}) = \tilde{\mathbb{1}}$ and that the tangent space is mapped to the cotangent space (i.e., a tangent vector $\vec{v}$ lowers to a zero-mean random variable).

The metric's action on vectors,

$$\boldsymbol{g}_{\vec{p}}(\vec{m}, \vec{n}) = \langle \mathcal{L}_{\vec{p}}(\vec{m}), \vec{n} \,\rangle = \sum_j \frac{m^j n^j}{p^j} \; , \tag{15}$$

leads to a line element

$$ds^2 = \boldsymbol{g}_{\vec{p}}\,(d\vec{p}, d\vec{p}) = \langle \mathcal{L}_{\vec{p}}(d\vec{p}), d\vec{p} \rangle = \sum_j \frac{(dp^j)^2}{p^j} \; , \tag{16}$$

which is identical to the line element (3) obtained from Wootters's distinguishability argument. Campbell [2] has reviewed the history of the correlation metric (15) in the statistics literature.

3 GEOMETRY ON QUANTUM STATE SPACE

The geometric structure of the probability simplex can be transferred intact to the space of quantum states (density operators). A more complete account of the transfer can be found in Ref. [3]. To begin the discussion, consider an arbitrary linear operator on an n-dimensional Hilbert space, written as a vector

$$\vec{O} = \sum_{j,k} O^{jk} \, \vec{e}_{jk} \; . \tag{17}$$

The operator vectors

$$\vec{e}_{jk} = |j\rangle\langle k| \; , \quad j, k = 1, \ldots, n \; , \tag{18}$$

which are formed from an orthonormal basis of Hilbert-space vectors $|j\rangle$, are a basis for the n^2-dimensional complex vector space of linear operators, and the matrix $O^{jk} = \langle j|\vec{O}|k\rangle$ consists of the matrix elements of the operator $\vec{O}$ in the basis $|j\rangle$.

The space of density operators

$$\vec{\rho} = \sum_{j,k} \rho^{jk} \, \vec{e}_{jk} \tag{19}$$

is a manifold with boundary: Hermiticity, $\vec{\rho} = \vec{\rho}^{\,\dagger}$ ($\rho^{jk} = (\rho^{kj})^*$), reduces the operator vector space to n^2 real dimensions; normalization, $\mathrm{tr}(\vec{\rho}) = 1$ ($\sum_j \rho^{jj} = 1$), defines an $(n^2 - 1)$-dimensional "plane" within this real vector space; and non-negativity, $\vec{\rho} \geq 0$ ($\vec{\rho}$ has non-negative eigenvalues), gives the plane a complicated boundary. By considering paths through the space of density operators, we see that the tangent space at $\vec{\rho}$ is an $(n^2 - 1)$-dimensional real vector space consisting of the traceless Hermitian operators. Affixing $\vec{\rho}$ to the tangent space at $\vec{\rho}$ recovers the n^2-dimensional vector space of all Hermitian operators $\vec{m} = \sum_{j,k} m^{jk} \, \vec{e}_{jk}$ where $m^{jk} = (m^{kj})^*$, if we restrict the vector space to be real, or of all linear operators $\vec{O}$, if we allow a complex vector space.

An arbitrary 1-form $\tilde{Q}$ acts on a vector $\vec{O}$ according to

$$\tilde{Q}(\vec{O}) \equiv \left\langle \tilde{Q}, \vec{O} \right\rangle = \sum_{j,k} Q_{kj}\, O^{jk} = \mathrm{tr}\left(\tilde{Q}\,\vec{O}\right) , \tag{20}$$

where $Q_{kj} \equiv \langle \tilde{Q}, \vec{e}_{jk} \rangle$. Thus

$$\tilde{Q} = \sum_{j,k} Q_{jk}\, \tilde{\omega}^{kj} \tag{21}$$

is an arbitrary linear operator expanded in terms of the dual basis $\tilde{\omega}^{kj} = |j\rangle\langle k|$, which satisfies $\langle \tilde{\omega}^{jk}, \vec{e}_{lm} \rangle = \mathrm{tr}(\tilde{\omega}^{jk}\vec{e}_{lm}) = \delta^j{}_l\, \delta^k{}_m$. A Hermitian 1-form $\tilde{A} = \tilde{A}^\dagger$ is an ordinary quantum observable, with

$$\left\langle \tilde{A}, \vec{\rho} \right\rangle = \sum_{j,k} A^*_{jk}\, \rho^{jk} = \mathrm{tr}\left(\tilde{A}\,\vec{\rho}\right) \equiv \langle \tilde{A} \rangle \tag{22}$$

being the (real) expectation value of $\tilde{A}$ with respect to the density operator $\vec{\rho}$.

The unit operator $\tilde{\mathbb{1}} \equiv \sum_j \tilde{\omega}^{jj}$ is a special 1-form: it defines the density-operator plane, since $\langle \tilde{\mathbb{1}}, \vec{\rho} \rangle = 1$ for any density operator $\vec{\rho}$, and any tangent vector $\vec{v}$ lies in the surfaces of $\tilde{\mathbb{1}}$, since $\langle \tilde{\mathbb{1}}, \vec{v} \rangle = 0$. Any observable can be decomposed uniquely as

$$\tilde{A} = \left\langle \tilde{A}, \vec{\rho} \right\rangle \tilde{\mathbb{1}} + \Delta \tilde{A} , \tag{23}$$

where $\Delta\tilde{A}$ has zero expectation value with respect to $\vec{\rho}$. The cotangent space at $\vec{\rho}$ is an (n^2-1)-dimensional real vector space made up of the zero-mean observables $\Delta\tilde{A}$; by affixing $\tilde{\mathbb{1}}$ to the cotangent space, we recover the n^2-dimensional vector space of all Hermitian 1-forms $\tilde{A}$ (observables), if we restrict the vector space to be real, and of all 1-forms $\tilde{Q}$, if we allow the vector space to be complex.

Balian, Alhassid, and Reinhardt [4] have stressed how quantum expectation values arise from the linear structure of observables as 1-forms acting on density operators as vectors. They proceed to define a metric on quantum state space by taking the second variation of the von Neumann entropy. In contrast, we continue the development begun in Sec. II by formulating a metric in terms of statistical correlations of quantum observables.

Letting the statistical correlation of a pair of observables be the metric's action on the pair of 1-forms yields

$$\begin{aligned} \boldsymbol{g}_{\vec{\rho}}(\tilde{A}, \tilde{B}) &\equiv \left\langle (\tilde{A}\tilde{B})_{\mathrm{sym}} \right\rangle = \mathrm{tr}\left(\tfrac{1}{2}(\tilde{A}\tilde{B} + \tilde{B}\tilde{A})\vec{\rho}\right) \\ &= \mathrm{tr}\left(\tilde{A}\,\tfrac{1}{2}(\vec{\rho}\tilde{B} + \tilde{B}\vec{\rho})\right) = \left\langle \tilde{A}, \mathcal{R}_{\vec{\rho}}(\tilde{B}) \right\rangle , \end{aligned} \tag{24}$$

which is equivalent to defining raising and lowering operators by

$$\mathcal{R}_{\vec{\rho}}(\tilde{Q}) = \tfrac{1}{2}\left(\vec{\rho}\tilde{Q} + \tilde{Q}\vec{\rho}\right) , \qquad \mathcal{L}_{\vec{\rho}}(\vec{O}) = \mathcal{R}^{-1}_{\vec{\rho}}(\vec{O}) . \tag{25}$$

Notice that $\mathcal{L}_{\vec{\rho}}(\vec{\rho}) = \tilde{\mathbb{1}}$ and that a tangent vector, i.e., a traceless Hermitian operator, lowers to a 1-form in the cotangent space, i.e., a zero-mean observable. In the orthonormal basis that diagonalizes $\vec{\rho} = \sum_j p^j |j\rangle\langle j|$, the lowering operator can be written out explicitly, yielding the line element

$$\begin{aligned} ds^2 &= \boldsymbol{g}_{\vec{\rho}}(d\vec{\rho}, d\vec{\rho}) = \langle \mathcal{L}_{\vec{\rho}}(d\vec{\rho}), d\vec{\rho} \rangle \\ &= \mathrm{tr}(d\vec{\rho}\, \mathcal{L}_{\vec{\rho}}(d\vec{\rho})) = \sum_{j,k} \frac{2}{p^j + p^k} |d\rho^{jk}|^2 . \end{aligned} \tag{26}$$

This line element has coördinate singularities on the boundary of quantum state space, but a careful analysis [3,5] shows that these singularities are no worse than—indeed, they are precisely the same as—the coördinate singularities in the Wootters metric (3).

By optimizing over all generalized quantum measurements that can be used to distinguish neighboring quantum states $\vec{\rho}$ and $\vec{\rho}+d\vec{\rho}$, Braunstein and Caves [5] obtained the quantum metric (26) as the distinguishability metric on quantum state space. This metric has also arisen in another context: it is the infinitesimal form of a distance function introduced by Bures [6], which Uhlmann [7] interpreted as a generalization of transition probabilities to mixed states and which has been investigated further by Hübner [8,9] and by Jozsa [10].

4 GEOMETRY INSIDE THE BLOCH SPHERE

We now study the geometry inside the Bloch sphere, which describes arbitrary 2×2 density operators

$$\vec{\rho} = \frac{1}{2}\left(\vec{\mathbb{1}} + \boldsymbol{x}\cdot\vec{\boldsymbol{\sigma}}\right) , \tag{27}$$

where $\boldsymbol{x} = \sum_{i=1}^{3} x^i \boldsymbol{e}_i$ is the Bloch vector and $\vec{\boldsymbol{\sigma}}$ is the vector of Pauli matrices. Here vectors in the space of linear operators are distinguished by an arrow, and 3-vectors in Bloch space are written in boldface, the 3-vector of Pauli matrices being both. The condition on states that $\mathrm{tr}(\vec{\rho}^{\,2}) \le 1$ implies that $|\boldsymbol{x}| \le 1$, with the pure states at $|\boldsymbol{x}| = 1$ completely surrounding the mixed states (this does not occur in higher dimensions).

To find the line element, we need to calculate the action of the lowering operator on a tangent vector $d\vec{\rho} = \frac{1}{2}d\boldsymbol{x}\cdot\vec{\boldsymbol{\sigma}}$. By inverting the action of the raising operator, we can show that [3]

$$\mathcal{L}_{\vec{\rho}}(d\vec{\rho}) = \frac{1}{1-|\boldsymbol{x}|^2}\left(-\left(\boldsymbol{x}\cdot d\boldsymbol{x}\right)\vec{\mathbb{1}} + \left[\left(1-|\boldsymbol{x}|^2\right)d\boldsymbol{x} + \left(\boldsymbol{x}\cdot d\boldsymbol{x}\right)\boldsymbol{x}\right]\cdot\vec{\boldsymbol{\sigma}}\right) . \tag{28}$$

Hence the line element takes the form

$$ds^2 = \frac{\left(\boldsymbol{x}\cdot d\boldsymbol{x}\right)^2}{1-|\boldsymbol{x}|^2} + d\boldsymbol{x}\cdot d\boldsymbol{x} . \tag{29}$$

The form of Eq. (29) suggests that we introduce a fourth coördinate $x^0 = \sqrt{1-|\boldsymbol{x}|^2}$. In terms of the four coördinates x^μ, $\mu = 0, 1, 2, 3$, the interior of the Bloch sphere is a three-dimensional surface defined by

$$\sum_{\mu=0}^{3}(x^\mu)^2 = 1 , \tag{30}$$

whose geometry is induced by the four-dimensional flat Euclidean line element

$$ds^2 = \sum_{\mu=0}^{3}(dx^\mu)^2 , \tag{31}$$

a result obtained by Hübner [8]. Thus the interior of the Bloch sphere has the geometry of the northern hemisphere of a unit 3-sphere. The north pole of the

3-sphere is the unit density operator $\frac{1}{2}\vec{\mathbb{1}}$, and the equator is the unit 2-sphere of pure states.

Geodesics of the metric (31) are great circles on the unit 3-sphere. Let a great circle be written as $x^\mu(s)$, where s measures arc length away from the point that has the maximum value of x^0. The unit 4-vectors $u^\mu \equiv x^\mu(s=0)$ and $v^\mu \equiv (dx^\mu/ds)|_{s=0}$ are orthogonal, with $v^0 = 0$, and the great circle is given by

$$x^\mu(s) = u^\mu \cos s + v^\mu \sin s \; . \tag{32}$$

Writing $u^0 = \cos\chi$, $u^i = n^i \sin\chi$, and $v^i = m^i$, where $\boldsymbol{n}$ and $\boldsymbol{m}$ are orthogonal unit 3-vectors, we can write the geodesic in terms of 3-vectors on Bloch space as

$$\boldsymbol{x}(s) = \boldsymbol{n} \sin\chi \cos s + \boldsymbol{m} \sin s \; . \tag{33}$$

Another useful set of coördinates is provided by the three components of the 3-vector

$$\boldsymbol{R} \equiv \frac{\boldsymbol{x}}{x^0} = \frac{\boldsymbol{x}}{\sqrt{1-|\boldsymbol{x}|^2}} \; . \tag{34}$$

Geometrically these coördinates are defined by taking a point on the unit 3-sphere and projecting it onto the 3-plane $x^0 = 1$ along a line originating at the origin. A great circle being the intersection of the unit 3-sphere with a 2-plane containing the origin, its projection onto the 3-plane $x^0 = 1$ is the intersection of the 2-plane with the 3-plane. This intersection is a straight line, as we can verify by writing the great circle (33) in the projected coördinates:

$$\boldsymbol{R}(s) = \boldsymbol{n} \tan\chi + \boldsymbol{m}\frac{\tan s}{\cos\chi} \; . \tag{35}$$

5 TRANSPORT IN QUANTUM STATE SPACE

Consider the natural vector field $\mathcal{R}_{\vec{\rho}}(\tilde{A})$ associated with a fixed observable $\tilde{A}$. It is easy to see that $\mathcal{R}_{\vec{\rho}}(\tilde{A})$ is orthogonal to the surfaces of constant expectation value $\langle \tilde{A}\rangle = \big\langle \tilde{A}, \vec{\rho}\,\big\rangle$: any tangent vector $\vec{v}$ that lies in a surface $\langle\tilde{A}\rangle = \text{const}$ satisfies $\langle \tilde{A}, \vec{v}\,\rangle = 0$, which implies

$$\boldsymbol{g}_{\vec{\rho}}\big(\mathcal{R}_{\vec{\rho}}(\tilde{A}), \vec{v}\big) = \big\langle \mathcal{L}_{\vec{\rho}}\big(\mathcal{R}_{\vec{\rho}}(\tilde{A})\big), \vec{v}\,\big\rangle = \langle \tilde{A}, \vec{v}\,\rangle = 0 \; . \tag{36}$$

Suppose we wish to construct a path that goes from the state $\vec{\rho}_0$ to another state that has a different expectation value for $\tilde{A}$. The most "efficient" way to do this is by consistently moving in the direction orthogonal to the surfaces $\langle\tilde{A}\rangle = \text{const}$; we can accomplish this by choosing

$$\frac{d\vec{\rho}}{d\lambda} = \mathcal{R}_{\vec{\rho}}(\tilde{A}) \; , \tag{37}$$

where λ is some parameter along the path so constructed, with $\vec{\rho}(\lambda = 0) = \vec{\rho}_0$. There is another compelling motivation for considering paths that satisfy Eq. (37): Braunstein and Caves [3, 5] have shown that one way to distinguish optimally between states $\vec{\rho}$ and $\vec{\rho} + d\vec{\rho}$ is to measure the observable $\mathcal{L}_{\vec{\rho}}(d\vec{\rho})$; paths that satisfy Eq. (37) are those for which this optimal observable $\mathcal{L}_{\vec{\rho}}(d\vec{\rho}) = d\lambda\,\tilde{A}$ is the same all along the path.

Due to the identity $\mathcal{R}_{\vec{\rho}}(\tilde{A}) = \mathcal{R}_{\tilde{A}}(\vec{\rho})$, we see immediately that the solution of Eq. (37) is

$$\vec{\rho}_U(\lambda) = e^{\lambda \mathcal{R}_{\tilde{A}}}(\vec{\rho}_0) = e^{\tilde{A}\lambda/2}\vec{\rho}_0\, e^{\tilde{A}\lambda/2}\,. \tag{38}$$

This solution, however, has a serious flaw: it fails to preserve normalization. Since $\mathcal{R}_{\vec{\rho}}(\tilde{A})$ does not generally lie in the tangent space at $\vec{\rho}$, it takes $\vec{\rho}$ out of the space of normalized density operators. This problem is easy to fix by projecting $\mathcal{R}_{\vec{\rho}}(\tilde{A})$ into the tangent space at $\vec{\rho}$. Since $\vec{\rho}$ is the unit vector orthogonal to the tangent space, the projected vector is

$$\begin{aligned} \mathcal{R}_{\vec{\rho}}(\tilde{A}) - \vec{\rho}\, \boldsymbol{g}_{\vec{\rho}}\big(\vec{\rho}, \mathcal{R}_{\vec{\rho}}(\tilde{A})\big) &= \mathcal{R}_{\vec{\rho}}(\tilde{A}) - \vec{\rho}\big\langle \tilde{\mathbb{1}}, \mathcal{R}_{\vec{\rho}}(\tilde{A})\big\rangle \\ &= \mathcal{R}_{\vec{\rho}}(\tilde{A}) - \vec{\rho}\big\langle \tilde{A}, \vec{\rho}\big\rangle = \mathcal{R}_{\vec{\rho}}(\Delta\tilde{A})\,. \end{aligned} \tag{39}$$

The projection is achieved by removing the expectation value of $\tilde{A}$ and dealing with the zero-mean observable $\Delta\tilde{A} = \tilde{A} - \big\langle \tilde{A}, \vec{\rho}\big\rangle\, \tilde{\mathbb{1}}$, which lies in the cotangent space at each point along the path. Thus we seek a path along the vector field $\mathcal{R}_{\vec{\rho}}(\Delta\tilde{A})$:

$$\frac{d\vec{\rho}}{d\lambda} = \mathcal{R}_{\vec{\rho}}(\Delta\tilde{A})\,. \tag{40}$$

Notice that in this formulation the parameter λ is uniquely determined. It is useful, however, to note the effect of changes in parameter. An arbitrary parameter change, i.e., $\lambda' = f(\lambda)$, leads to a location-dependent rescaling of the observable $\tilde{A}$, i.e., $\tilde{A}' = \tilde{A}/(df/d\lambda)$. An affine change in parameter, i.e., $\lambda' = a\lambda + b$, rescales the observable according to $\tilde{A}' = \tilde{A}/a$ and changes the initial point of the path.

We can easily verify that the solution of Eq. (40) is [11]

$$\begin{aligned} \vec{\rho}\,(\lambda) &= \frac{\vec{\rho}_U(\lambda)}{\big\langle \tilde{\mathbb{1}}, \vec{\rho}_U(\lambda)\big\rangle} = \frac{\vec{\rho}_U(\lambda)}{\mathrm{tr}\big(\vec{\rho}_U(\lambda)\big)} = \frac{e^{\lambda\mathcal{R}_{\tilde{A}}}(\vec{\rho}_0)}{\mathrm{tr}\big(e^{\lambda\mathcal{R}_{\tilde{A}}}(\vec{\rho}_0)\big)} \\ &= \frac{e^{\tilde{A}\lambda/2}\vec{\rho}_0 e^{\tilde{A}\lambda/2}}{\mathrm{tr}\big(e^{\tilde{A}\lambda}\vec{\rho}_0\big)} = \frac{e^{\Delta\tilde{A}\lambda/2}\vec{\rho}_0 e^{\Delta\tilde{A}\lambda/2}}{\mathrm{tr}\big(e^{\Delta\tilde{A}\lambda}\vec{\rho}_0\big)}\,. \end{aligned} \tag{41}$$

These expressions tell us how to transport a state orthogonal to surfaces of constant $\langle\tilde{A}\rangle$. It is useful to note that the inverse problem can be solved, too: given a pair of states $\vec{\rho}_0 = \vec{\rho}\,(\lambda = 0)$ and $\vec{\rho}_1 = \vec{\rho}\,(\lambda = 1)$, an observable $\tilde{A}$ which connects these states optimally in the manner of Eq. (41) is given by

$$\tilde{A} = 2\ln\Big(\vec{\rho}_0^{\,-1/2}\sqrt{\vec{\rho}_0^{\,1/2}\vec{\rho}_1\vec{\rho}_0^{\,1/2}}\,\vec{\rho}_0^{\,-1/2}\Big)\,. \tag{42}$$

Any observable differing from $\tilde{A}$ by a constant multiple of the identity operator $\tilde{\mathbb{1}}$ will also do the job. As noted above, rescaling $\tilde{A}$ according to $\tilde{A}' = \tilde{A}/a$ simply rescales the parameter, i.e., $\lambda' = a\lambda$, thus leaving $\vec{\rho}_1 = \rho(\lambda' = a)$.

We now illustrate these considerations with two examples: a quantum system with one classical degree of freedom and a two-state quantum system.

Example 1: Consider a quantum system with canonical position operator $\tilde{x}$ and canonical momentum operator $\tilde{p}$. It is interesting to note that a path orthogonal to surfaces of constant $\langle\tilde{x}\rangle$ does *not*, in general, correspond to the unitary position displacement generated by the momentum operator, i.e.,

$$\vec{\rho}\,(X) \equiv e^{-iX\tilde{p}}\vec{\rho}_0 e^{iX\tilde{p}} \neq \frac{e^{\tilde{x}\lambda/2}\vec{\rho}_0 e^{\tilde{x}\lambda/2}}{\mathrm{tr}(e^{\tilde{x}\lambda}\vec{\rho}_0)} \equiv \vec{\rho}\,(\lambda) \tag{43}$$

($\hbar = 1$). This can be seen most directly by considering the Wigner function for a density operator $\vec{\rho}$:

$$W(x,p) = \int \frac{dq}{2\pi} \langle x + \tfrac{1}{2}q | \vec{\rho} | x - \tfrac{1}{2}q \rangle e^{-ipq} . \tag{44}$$

The Wigner functions for $\vec{\rho}(X)$ and $\vec{\rho}(\lambda)$ are

$$W_X(x,p) = W_0(x - X, p) , \qquad W_\lambda(x,p) = \frac{e^{x\lambda} W_0(x,p)}{\int dx\, dp\, e^{x\lambda} W_0(x,p)} . \tag{45}$$

It is easy to show that these two Wigner functions coincide for all values of λ, with X being some function of λ, if and only if $W_0(x,p)$ factors into a Gaussian in x times a function of p.

Example 2: Consider a two-state system. The most general (nontrivial) observable, after rescaling, can be written as

$$\tilde{A} = \boldsymbol{a} \cdot \tilde{\boldsymbol{\sigma}} , \tag{46}$$

with $|\boldsymbol{a}| = 1$. We can derive the path orthogonal to surfaces of constant $\langle \tilde{A} \rangle$ from

$$\vec{\rho}(\lambda) \equiv \tfrac{1}{2}\Big(\vec{\mathbb{1}} + \boldsymbol{x}(\lambda) \cdot \vec{\boldsymbol{\sigma}}\Big) = \frac{\exp(\boldsymbol{a}\cdot\tilde{\boldsymbol{\sigma}}\,\lambda/2)\,\vec{\rho}_0\,\exp(\boldsymbol{a}\cdot\tilde{\boldsymbol{\sigma}}\,\lambda/2)}{\mathrm{tr}\Big(\exp(\boldsymbol{a}\cdot\tilde{\boldsymbol{\sigma}}\,\lambda)\,\vec{\rho}_0\Big)} . \tag{47}$$

Simplifying this expression, we find that the trajectory of the Bloch vector is given by

$$\boldsymbol{x}(\lambda) = \frac{\boldsymbol{x}_0 - \boldsymbol{a}(\boldsymbol{a}\cdot\boldsymbol{x}_0) + \boldsymbol{a}[\sinh\lambda + (\boldsymbol{a}\cdot\boldsymbol{x}_0)\cosh\lambda]}{\cosh\lambda + \boldsymbol{a}\cdot\boldsymbol{x}_0 \sinh\lambda} . \tag{48}$$

Calculating

$$x^0(\lambda) = \sqrt{1 - |\boldsymbol{x}(\lambda)|^2} = \frac{\sqrt{1 - |\boldsymbol{x}_0|^2}}{\cosh\lambda + \boldsymbol{a}\cdot\boldsymbol{x}_0 \sinh\lambda} , \tag{49}$$

we can write the Bloch-vector trajectory in terms of the projected coördinates (34) as:

$$\boldsymbol{R}(\lambda) = \frac{\boldsymbol{x}(\lambda)}{x^0(\lambda)} = \frac{\boldsymbol{x}_0 - \boldsymbol{a}(\boldsymbol{a}\cdot\boldsymbol{x}_0) + \boldsymbol{a}[\sinh\lambda + (\boldsymbol{a}\cdot\boldsymbol{x}_0)\cosh\lambda]}{\sqrt{1 - |\boldsymbol{x}_0|^2}} . \tag{50}$$

We recognize this trajectory as a geodesic (35) with

$$\boldsymbol{n}\tan\chi = \frac{\boldsymbol{x}_0 - \boldsymbol{a}(\boldsymbol{a}\cdot\boldsymbol{x}_0)}{\sqrt{1 - |\boldsymbol{x}_0|^2}} , \qquad \boldsymbol{m} = \boldsymbol{a} , \qquad \frac{\tan s}{\cos\chi} = \frac{\sinh\lambda + (\boldsymbol{a}\cdot\boldsymbol{x}_0)\cosh\lambda}{\sqrt{1 - |\boldsymbol{x}_0|^2}} . \tag{51}$$

Moreover, in these projected coördinates it is obvious that the observable of Eq. (42), which connects states $\vec{\rho}_0$ and $\vec{\rho}_1$, is specified by

$$\boldsymbol{a} \propto \boldsymbol{R}_1 - \boldsymbol{R}_0 = \frac{\boldsymbol{x}_1}{\sqrt{1 - |\boldsymbol{x}_1|^2}} - \frac{\boldsymbol{x}_0}{\sqrt{1 - |\boldsymbol{x}_0|^2}} . \tag{52}$$

CONCLUSION

The natural metric defined by correlations between observables yields a measure of statistical distinguishability between states of a system, either classical probability distributions or quantum density operators. This metric and its underlying geometry have already led to powerful results: generalized Mandelstam-Tamm uncertainty relations [5, 12] and geometrically intuitive methods to construct observables for optimal or near optimal detection in problems of quantum communication [13–15]. They promise an abundant harvest of further results as the intricate connections between geometry and quantum measurements are explored.

References

[1] W. K. Wootters, Phys. Rev. D **23**, 357 (1981). Wootters's distance is half the distance we use here.

[2] L. L. Campbell, Inform. Sciences **35**, 199 (1985).

[3] S. L. Braunstein and C. M. Caves, in *Fundamental Problems in Quantum Theory,* edited by D. Greenberger, Ann. NY Acad. Sci., to be published.

[4] R. Balian, Y. Alhassid, and H. Reinhardt, Phys. Rep. **131**, 1 (1986).

[5] S. L. Braunstein and C. M. Caves, Phys. Rev. Lett. **72**, 3439 (1994).

[6] D. J. C. Bures, Trans. Am. Math. Soc. **135**, 199 (1969).

[7] A. Uhlmann, Rep. Math. Phys. **9**, 273 (1976).

[8] M. Hübner, Phys. Lett. A **163**, 239 (1992).

[9] M. Hübner, Phys. Lett. A **179**, 226 (1993).

[10] R. Jozsa, J. Mod. Opt. special issue on quantum communication, to be published.

[11] The classical analogue [2] of this solution gives a path through the probability simplex which is called "exponential tilting" and is used for an important class of statistical approximations; see O. E. Barndorff-Nielsen and D. R. Cox, *Asymptotic Techniques for Use in Statistics* (Chapman and Hall, London, 1989) p. 105.

[12] S. L. Braunstein, C. M. Caves, and G. J. Milburn, University of New Mexico Center for Advanced Studies preprint, submitted to Phys. Rev. A.

[13] C. A. Fuchs and C. M. Caves, submitted to Phys. Rev. Lett.

[14] C. A. Fuchs and C. M. Caves, this volume.

[15] C. A. Fuchs and C. M. Caves, in *Fundamental Problems in Quantum Theory,* edited by D. Greenberger, Ann. NY Acad. Sci., to be published.

ON THE PROBLEM OF UNIVERSALITY OF QUANTUM THEORY [1]

Pavel Bóna

Department of Theoretical Physics, Comenius University
842 15 Bratislava, Slovakia
and
Department of Mathematics, Silesian University
746 01 Opava, Czech Republic

Abstract

Two aspects of the problem of "objectification" in quantum mechanics (QM) are considered:
(1) The question of description of the usually requested nonexistence of interference of macroscopically different states, and (2) the possibility of distinction of Genuine mixture (or "Gemenge") from the Elementary mixture (or "improper mixture").

Some possibilities for a necessary modification of QM are considered. Especially modifications of the set of "observables" allowing, e.g., description of quantal measurement process in framework of systems with finite number of degrees of freedom are discussed. An idealised solvable model of measurement process in quantum theory is presented. A nonlinear version of quantum mechanics (NLQM – a consistent generalisation of the Weinberg's one) describing different evolutions for Genuine and Elementary mixtures is presented. "Relative isolation" of quantal systems is defined in this connection. NLQM contains QM, classical mechanics, and some "quasiclassical" theories as its exact subtheories.

[1]Supported from Grant No. 201/93/2245 of the Czech Grant Agency.

1 INTRODUCTION

Quantum theory is formulated in terms of statistical distributions of values of observed quantities. The probability distributions of QT cannot be endowed with a "simultaneous ignorance interpretation": It cannot be reformulated in terms of Kolmogorovian model of probability theory describing statistical distributions on a *firmly given* set of elementary events (representing in applications a "background reality" – some sort of "Universe of all possible events"); Kolmogorovian model works well, however, in classical theory (CT) of physical phenomena, where an "elementary event" determines sharp values of all possible quantities. On the other hand, classical description of a "background" (resp. a "pretheory", cf. [11]) for QT seems to be a necessary prerequisite for introduction into QT such basic concepts as reference frames, basic observables, preparations, and measurements; hence, mere definition of "a microsystem" depends crucially on a use of classical concepts. No clear boundary is known in physics between phenomena described by CT, resp. QT. Moreover, interests of physicists are attracted presently by "mesoscopic phenomena" for description of which both CT and QT are useful. Hence, the requirement of *universality* of an adequate physical theory containing both quantum and classical concepts is quite actual. We expect that QT can be a basis for such a universal theory.

CT cannot be included into the traditional quantum mechanics (QM) of systems with finite number of degrees of freedom, cf. [2]: Even classical "observables" are not present in QM. Hence, QM needs some modifications (resp. extensions) to become a universal QT. We intend to consider here some questionable features of QM which could be, perhaps, modified without changing any essential and successful feature of standard QT, with the aim to reach universality of (such a minimal modification of) QT, cf. also [3].

One of those essential features of QT which should be kept in its universal modification (let us call UQT the intended universal theory) is, according to present author's opinion, the irreducibly probabilistic character of QT. Hence, also description of classical phenomena in framework of UQT should be probabilistic. This might seem to contradict to everyday experience with "individual objects" which can be observed repeatedly as "the same object" with unchanged (sharp) values of all their relevant quantities. The mentioned "relevant quantities" seem to be stable also with respect to measurements of any other quantities of the observed empirical object. (Warning: It should be distinguished between theoretically defined "physical system" specified by a given set of "observables", "states", etc., and an "empirical object" specified by some empirical procedures, some features of which – e.g. some specific observable quantities – serve as an empirical realization of the theoretical system; that "empirical object" can be used for an empirical realization of also other theoretically defined "systems" which, e.g., are described by larger sets of "observables".) It seems to be no serious problem to reformulate CT in terms of very narrow (Kolmogorovian) statistical distributions, to match just described intuition. More problems with intuition arise in trials of description of "macroscopically unstable" situations, e.g., of the process of quantal measurement. In such situations, QT leads (e.g., via a version of the "reduction postulate") to statistical distributions of values of macroscopic (classical) quantities with large dispersions, and, in the same time, QT doesn't contain any (although *intuitively desirable*) possibility of dealing with individual events (e.g., with the process of acquirement of a specific pointer position). We shall accept here such a state of affairs as granted: We believe that there are physical situations (as in the case of measurement of some quantities of a microscopic system) in which there is impossible to forecast deterministically a motion of otherwise classical objects (e.g., the pointer). The instability of situations allows us to foresee only a time evolution of statistical ensembles (in the sense of Genuine mixtures) of states of macroscopic (classical) systems, hence we cannot theoretically describe observed individual results of a measurement. This could be considered as a source of effective irreversibility (and of a loss of information contained in theoretical description of actual states) in formally time-reversible theory. (Let us mention here, however, that also in pure CT there are unstable situations, e.g. systems with unstable flows – see Anosov diffeomorphisms, etc., where the formally deterministic description is practically as stochastic, as the above mentioned quantum processes; this "practical stochasticity" is due to our restricted ability of distinction between points in a phase space corresponding to different initial conditions of mutually quickly diverging trajectories.)

Let us mention that a sort of the desired UQT can be obtained in framework of C^*-algebraic theory of infinite quantal systems (cf. [12], [13]): CT can be included there together with QT, but the unstable processes of the type of quantal measurement are lasting there for infinite time interval, hence no evolution of irreversible situations is described by those theories.

Another generalization of QM is the form of nonlinear quantum mechanics (NLQM) described in the last section of this paper: It also includes QM together with CM ($\equiv$ classical mechanics), but the problem of description of measurement is not solved yet in that framework.

2 CONTROVERSIAL "OBSERVABLES"

In standard formulations of QM, any selfadjoint operator on separable Hilbert state-space $\mathcal{H}$ of the considered system represents an "observable". No differences of observability of different observables are formulated on this general level. Hence, the set of all bounded "observables" consists of all symmetric bounded operators $\mathcal{L}(\mathcal{H})_s$. The set $\mathcal{L}(\mathcal{H})_s$ is the same for all finite-particle systems (with some modifications for systems of identical particles, or systems with absolutely conserved charges): A structureless point particle is described by the same set of abstract "observables" as a collection of 10^{25} atoms. In practice, however, some "basic observables" of a given system are picked out, e.g. generators of $2f+1$ – dimensional Weyl-Heisenberg group and a few other operators (e.g. generators of Galilean group representation) for a system with f degrees of freedom. This shows that in practice some "hierarchy of observability" of formally given operators is respected in QM.

A possible restriction can be anticipated for measurability of "observables" of multiparticle systems (let us keep in mind – for simplicity – mutually distinguishable particles only, e.g. particles placed in sites of a crystalline solid): To measure interference of two distinct permutationally invariant multiparticle states it would be necessary to measure very–high–order correlation functions, those of the order of number of particles present in the system, cf. also [3]. To illustrate this idea, let us take a large system to be described by tensor-product Hilbert space $\mathcal{H}^N := \otimes_{p=1}^N \mathcal{H}_p \ni \Phi, \Psi$; $\Phi := \otimes_{p=1}^N \varphi_p$, $\Psi := \otimes_{p=1}^N \psi_p$, with φ_p and $\psi_p \in \mathcal{H}_p$. Assume that $\|\varphi_p\| = \|\psi_p\| = 1$, and $|\langle \varphi_p | \psi_p \rangle| \leq \lambda < 1$ for all p. Let us consider the observables of the form

$$A := \sum_{p_1, p_2, \ldots p_k} X^1_{p_1} \otimes X^2_{p_2} \otimes \ldots \otimes X^k_{p_k} \otimes I_{\{N \setminus \{p_1, p_2, \ldots p_k\}\}} \tag{1}$$

with X^j_p being selfadjoint in $\mathcal{L}(\mathcal{H}_p)$ of the norm ≤ 1. Let $k(\ll N)$ be fixed. Then, for $N \to \infty$, one has

$$|\langle \Psi | A \Phi \rangle| < N^k \lambda^{N-k} \approx O(e^{-\alpha N}) \tag{2}$$

for some $\alpha > 0$. If there would not be any other observables of the considered system, then (2) would present a kind of "effective disjointness" of the states described by Ψ and Φ with respect to the chosen set of observables of the form (1). Conventional QM admits, however, observables with k in (1) of the order of N: $k \approx N$. Such "observables" (i.e. products of large number of one particle operators) seem to be practically unobservable, [4]. This might be a reason of "practical classicality" of some subsystems (determined by their observables, etc.) of large quantal systems.

The following note (although related to previous ones) could further support the requirement for extension of QT by some criteria for "observability of observables": Projection operators are used to be considered to represent "questions" in QT with "yes"–"no" answers (cf. [5], [6], [8]). Their interpretation is, however, determined by interpretation of vectors in corresponding subspaces of states in $\mathcal{H}$. But interpretation of vectors in $\mathcal{H}$ is equally problematic as the interpretation of operators: vectors correspond to "preparation procedures", whereas operators are ascribed to "registration procedures", cf. [1], [10], [11]. Hence, all systems with any finite number of degrees of freedom are described formally by "the same number of questions", what seems to be absurd from the point of view of physical intuition. (The situation is here analogous as that one occurring at "counting points" on the real line, resp. on its intervals of nonzero length: mathematically there is no difference, but endowing the reals by a metric a natural difference is created.)

Our last argument for necessity of distinction between "degrees of observability" of different formally defined "observables" is, perhaps, the most "physical" one but, simultaneously, it seems to be most difficult to be taken seriously in building of a general theory. We propose here a natural *principle of measurability* according to which the set of (really observable) "observables" is determined as the set of those selfadjoint operators which can be measured by a "realistic" apparatus. I.e., to any "observable", an apparatus can be (in principle, at least theoretically) constructed by a use of observed interactions only (electroweak, gravitational, etc.). One could pose the problem alternatively: Let us look for observables which can be measured by a given set of apparatuses (trying to construct the "realistic" ones). An (unrealistic) example of such a construction is given in the next section.

It might be superfluous to mention here, that a consistent restriction of the set of observables in QT could be a difficult task, since there are strong and naturally-looking requirements on that set, e.g. its Jordan–algebraic, or quantum logic structures, cf. [5], [16]. The NLQM of the last section represents an extension of the sets of observables, states, etc. which introduces, however, a richer structures into these sets, hence offers some possibilities to attempt to solutions to our present problem.

3 AN OBSERVABLE DETERMINED BY QUANTUM MEASUREMENTS

We shall present here a model of an infinite size apparatus (formulated rigorously in C^*-algebra ic terms, [7]) detecting a scalar particle moving on a line. This has to illustrate how a measured quantity can be defined from the interaction of the apparatus with the measured "microsystem" (this scalar nonrelativistic particle interacts with *first two* spins in the chain). The state space of a particle (modelled also as a quantum of scalar Fermi field) is simply $\mathcal{H} \equiv L^2(\mathbf{R}, dx)$. A suitably chosen element $\sigma \in L^2(\mathbf{R}, dx)$ describes also interaction with the apparatus "placed" on the line within the set $supp(\sigma) \subset \mathbf{R}$. The apparatus is modelled by a semiinfinite spin chain described by the spin creation - annihilation operators $a_n^{(*)}$ $(n \in \mathbf{Z}_+)$, and by the Hamiltonian (cf.[17])

$$H_\infty := \sum_{n \in \mathbf{Z}_+} a_n^* a_n (a_{n+1} + a_{n+1}^*) a_{n+2} a_{n+2}^*. \tag{3}$$

Let us introduce now the model which represents a nonideal measurement. The microsystem (particle) is described by the Fermi field described by annihilation operators $b(\varphi)$, with $\varphi \in L^2(\mathbf{R}, dx)$. The total Hamiltonian is now

$$H := H_\infty + d\Gamma(\hat{p}^2) + W, \tag{4}$$

where the interaction W is defined by

$$W := \gamma(a_1^* + a_1) a_2 a_2^* \otimes b^*(\sigma) b(\sigma), \tag{5}$$

with $\gamma \in \mathbf{R}$ and $\sigma \in L^2(\mathbf{R}, dx)$, $\|\sigma\| = 1$. The symbol $d\Gamma(\hat{p}^2)$ denotes the second quantized, [14], one particle Hamiltonian $\hat{p}^2$, with $\hat{p} := -i\partial_x$. Let the chain be at $t = 0$ in its "vacuum", i.e. all spins are pointing down. Let the initial state of the particle be described by an arbitrary normalized wave function $\varphi_0 \in L^2(\mathbf{R}, dx)$. (We do not use here the "incoming" asymptotics for $t \to -\infty$ of the scattering theory.) Denoting by ω_t the state (i.e. the expectation functional) of the total system in time t, and by A an arbitrary observable of that composed system, one can obtain that

$$\bar{\omega}(A) := \lim_{t \to \infty} \omega_t(A) \tag{6}$$

exists, and one arrives for suitable parameters of W at:

$$\bar{\omega} = (w\omega_1^{chain} + (1-w)\omega_0^{chain}) \otimes \omega_0^{field}, \quad w \not\equiv 0, \tag{7}$$

where ω_j^{chain} $(j = 0, 1)$ denotes the state of the chain with all spins pointing down, resp. pointing up. The field state ω_0^{field} is the Fermi vacuum. We obtained here a mixture of two disjoint states.

The probability w of switching the apparatus can be expressed explicitly in terms of dynamics of the chain, as well as of the particle-chain interaction parameters, cf. [7]. It can be written in the form

$$w = \langle \varphi_0 \mid \Lambda \varphi_0 \rangle. \tag{8}$$

Here Λ is a positive operator of the norm $\|\Lambda\| \leq 1$ acting on the Hilbert space $L^2(\mathbf{R}, dx)$. This operator represents the measured (unsharp) observable of the particle. Its explicit expression can be written in the form

$$\Lambda = \gamma^2 (F_\sigma^+)^* \hat{f} F_\sigma^+. \tag{9}$$

Here $F_\sigma^+ \in \mathcal{L}(\mathcal{H}, L^2(\mathbf{R}, dt))$, with

$$F_\sigma^+(\sigma') := \vartheta(t)\langle \beta_0 \otimes \Omega(\sigma)|\ U_t\beta_0 \otimes \Omega(\sigma')\rangle, \tag{10}$$

where $\vartheta(t) = 1$ for $t > 0$, $\vartheta(t) = 0$ for $t < 0$; $U_t := exp(-itH)$ is the total system's evolution operator, β_0 is the spin–chain "vacuum", and $\Omega(\sigma')$ is the one–particle state with the particle wave–function σ'; $\hat{f} \in \mathcal{L}(L^2(\mathbf{R}, dt))$ is the convolution operator with the function $f : t \mapsto f(t)$ determined as a diagonal matrix element of the free evolution $U_t^0 := exp(-itH_0)$ of the total system with $H_0 := H - W$:

$$f(t) := \langle \beta_1 \otimes \Omega(\sigma)|\ U_t^0 \beta_1 \otimes \Omega(\sigma)\rangle,$$

with $\beta_1 := a_1^*\beta_0$, $a_m\beta_0 := 0$ for all $m \in \mathbf{Z}_+$, $\Omega(\sigma) := b^*(\sigma)\Omega_0$, and $b(\sigma')\Omega_0 := 0$ for all $\sigma' \in \mathcal{H}$. The operator F_σ^+ can be found explicitly in terms of the free evolution and of the interaction parameters, [7], hence Λ is a known "effect" defining an "unsharp observable". Physical interpretation of this observable is "in principle" given by the measurement process just described.

4 ON DIFFERENT INTERPRETATIONS OF MIXED STATES

Let us consider a microsystem in (conventional formulation of) QM: It is characterized by irreducibility of its set $\mathcal{L}(\mathcal{H})$ of "observables". Hence, any vector $\varphi \in \mathcal{H}$ represents a pure state. Let ϱ be any density matrix on $\mathcal{H}$ different from one–dimensional projections. Then ϱ can be nontrivially decomposed in many different ways into a convex combination

$$\varrho := \sum_j \lambda_j \varrho_j \tag{11}$$

of density matrices ϱ_j. Different decompositions (11) can correspond to decompositions with respect to different, mutually incompatible (i.e. mutually noncommuting) commutative subsets of observables. Such "incompatible decompositions" cannot be both interpreted as Genuine mixtures although they are in the framework of QM *of the considered microsystem* mutually equivalent. This indicates that for a separately taken isolated microsystems the *concept of Genuine mixture (i.e. "Gemenge") is meaningless* in the framework of conventional QM.

On the other side, for classical systems one has instead of ϱ a probability measure on its phase space, and this is uniquely decomposable into its extremal (i.e. pure) components – into an integral of Dirac measures concentrated in points of the phase space.

In the case of quantal systems with (some kind of) superselection rules, e.g. in the C^*-algebra ic description of infinite systems, there are distinguished unique decompositions of states (e.g. the central decomposition, cf. [14]). In the both last mentioned situations the components of the decompositions are mutually "noninterfering", i.e. the considered systems (also in any of their empirical realizations) cannot be coupled to any other system in such a way where a quantummechanical interference would be observed between states of the composed system corresponding to different components of the decomposition.

To be more specific, let us try to express some ideas in terms of QM. Let us consider one–dimensional projections P_{φ_j} on $\mathcal{H}$ corresponding to an orthonormal set $\{\varphi_j\, , j = 1, 2, \ldots\}$ of vectors in $\mathcal{H}$. In the case if a "classical" subsystem is present in the considered system, the vectors φ_j for different values of j can correspond to different superselection sectors (e.g. to different values of some classical quantity) and, in that case, there is no "observable" $B \in \mathcal{L}(\mathcal{H})$ such that

$$\langle \varphi_j \mid B\varphi_k \rangle \neq 0, \quad \text{for } \textit{some} \;\; j \neq k. \tag{12}$$

For a microscopic system there are always $B \in \mathcal{L}(\mathcal{H})$ such that (12) is valid. Let now $\varrho_j := P_{\varphi_j}$ in (11). Then there is always possible to find (in mathematical theory of QM) a system with its Hilbert space $\mathcal{K}$ and an orthonormal basis $\{\psi_j, j = 1, 2, \ldots\} \subset \mathcal{K}$ such that ϱ is obtained as the partial trace (cf. [8]) with respect to $\mathcal{K}$ from the vector–state Ψ,

$$\Psi := \sum_j \sqrt{\lambda_j}\, \varphi_j \otimes \psi_j \in \mathcal{H} \otimes \mathcal{K}. \tag{13}$$

If, for any choice of "the system $\mathcal{K}$ ", and for any "observable R" of the "combined system $\mathcal{H} \otimes \mathcal{K}$", one has

$$\langle \Psi | \, R\Psi \rangle = \sum_j \lambda_j \, \langle \varphi_j \otimes \psi_j | \, R\varphi_j \otimes \psi_j \rangle, \tag{14}$$

we say that the states P_{φ_j} of "the system $\mathcal{H}$" are mutually noninterfering. This is the case of the mentioned belonging of different φ_j into different superselection sectors in otherwise standard QM, [18].

Hence, for macroscopic (classical) systems the concept of Genuine mixture (cf. also its rough description in [3]) makes sense: The distinguished unique decomposition of the state (let us denote it ϱ also in general situation) into mutually noninterfering components ϱ_j can be interpreted in terms of a statistical ensemble of states ϱ_j describing a collection of systems (also in their empirical meaning, if interpreted) each of which "really occurs" in one of the states ϱ_j. Let us note here that concepts like "existence", "real occurrence", or "being" of some specific "objects" seem to be of profound macroscopic (and, conceptually: of classical) character: Their empirical definition rests on repeatability of empirical measurement procedures and on reproducibility of their (individual – sharp) results with respect to all their (macroscopic) observables. If these "classical features" of quantummechanically described macrosystems would be only "effective" in the sense of considerations of preceding sections, then also such concepts as "existence" could be only effective.

On the other hand, we would like to have a concept of Genuine mixture also for microsystems, because there are situations which intuitively fit in those described above for macrosystems with classical behaviour. E.g., we can prepare a collection of microsystems by choosing during the preparation a procedure preparing the state P_{φ_j} with probability λ_j, for all j's. Let us see, however, that now we have not an ensemble of separated microsystems, but, at the best, an ensemble of couples *microsystem & preparation device*, and the "Genuine interpretation" is possible due to the presence of macroscopic devices in macroscopically distinct states correlated with the states of the microsystem in the sense of eq. (13). We can ask, however, whether there is some possibility of distinction of the last mentioned situation (in which the density matrix of the micro – ensemble was ϱ from (11)) from that one described by (13) with both subsystems being microscopic, and the partial density matrix being again the same ϱ. Since microsystems can be coupled both with microsystems, as well as with macrosystems, this question addressed to the state ϱ of a separately considered system in an instant of time has (at least in the framework of

QM) negative answer. Another possibility arises if we assume that there is present an interaction of the considered microsystem with the correlated macrosystem where interaction Hamiltonian depends on values of the macroscopic quantity distinguishing different ϱ_j in (11). Then we could ask for future time evolution of the state ϱ, and we could look for some characteristic differences between evolutions in the cases when the considered microsystem is correlated with macrosystems (with different values of the macroscopic quantity) and in the cases in which such a correlation is missing (or the microsystem's states ϱ_j are all correlated with a unique value of the macroscopic quantity determining the interaction). Since general evolution of the state (13) of the coupled system might lead to an arbitrary change of its partial trace ϱ_t, we have to restrict our interest to some class of evolutions where the posed question could obtain a reasonable meaning.

Let us choose a class of such (sub–)systems in QM (or also in UQT) which satisfy the following two conditions.
(a) Time evolution of the chosen systems (in considered global situations including also description of other relevant systems) can be described by unitary operators, i.e.

$$\varrho_t = U(t, t_o; \varrho_{t_o})\varrho_{t_o} U(t, t_o; \varrho_{t_o})^{-1}, \quad t \in \mathbf{R}, \ \varrho_{t_o} \in \mathcal{T}_{+1}(\mathcal{H}) \equiv \mathcal{S}_*. \qquad (15a)$$

The possible dependence of U on ϱ_{t_o} leads, however, to a problem: The ambiguity of the expression of state ϱ in various convex combinations could lead to ambiguity in time evolutions, if calculated as the evolution of the components of the convex decompositions. The possibility of the ϱ_{t_o} – dependence of U will be interpreted here as a possible presence of the above described correlations of microscopic states with (at least weakly interacting) macrosystem's states, cf. [21]. We shall assume further:
(b) The function $\varrho \mapsto U(t_1, t_2; \varrho)$ is defined on $\mathcal{S}_*$ for all $t_j \in \mathbf{R}$ $(j = 1, 2)$ and satisfies a 'cocycle property' (up to possible phase factors):

$$U(t_2, t_1; \varrho_{t_1})U(t_1, t_0; \varrho_{t_0}) \equiv U(t_2, t_0; \varrho_{t_0}), \ U(t, t; \varrho) \equiv I. \qquad (15b)$$

Now it is natural to define two kinds of (mixed) states of the considered (sub–)systems:
(i) Density matrices describe indecomposable quantal states evolving according to (15), and they should be dealt within the theory as a whole (their components in formal decompositions (11) have no interpretation, hence for this states eq. (11) is physically irrelevant).
(ii) States corresponding to Genuine mixtures are described by *specific decompositions* of the form (11) (here different components ϱ_j correspond to different sharp values of classical quantities of a correlated macrosystem); in these states each component of the (specific!) decomposition (11) evolves separately according to (15).

The systems satisfying the conditions (a) and (b) could be imagined intuitively as microsystems moving in an external (classical) field which can depend moreover on the microscopic state of the system. Some correspondence between actual states of microsystems and their macroscopic environment can be understood due to necessity of preparation of specified microstates by macrodevices. Since there is no evolution of quantum correlations changing spectral characteristics of density matrices of states of systems fulfilling (a) and (b), we shall call such systems to be *relatively isolated*. Genuine mixtures correspond to convex combinations (11) of microstates correlated with macroscopically different states of the "relevant macrosystems".

Systems satisfying nontrivially the above conditions are, e.g., individual quantal systems of an infinite collection of equal quantal systems interacting mutually by specific very weak, and of very long range ("mean–field", cf.[19], [20]) interactions. This describes in fact nonlinear situations in QM, and a theory of such situations is sketched in the next section. We assume that such nonlinear situation could occur in measurement – like situations, as it was mentioned also in [3].

5 ON NONLINEAR EXTENSION OF QM

Let us consider a complex Hilbert space $\mathcal{H}$ and the set $\mathcal{S}_*$ of all density matrices on it. The set $\mathcal{S}_*$ is embedded into the Banach space $\mathcal{T}(\mathcal{H})$ of all trace–class operators on $\mathcal{H}$. Let $\mathcal{U}(\mathcal{H})$ be the unitary group of $\mathcal{L}(\mathcal{H})$. $\mathcal{S}_*$ can be decomposed into orbits of $\mathcal{U}(\mathcal{H})$ with respect to the coadjoint representation:

$$Ad^*(u)\varrho \ := u\varrho\, u^*$$

for all $u \in \mathcal{U}(\mathcal{H})$, and all $\varrho \in \mathcal{S}_*$. Each orbit consists of all density matrices with the same spectral characteristics (i.e. spectrum and spectral multiplicities). Orbits $\mathcal{O}_\varrho$ corresponding to ϱ with zero eigenvalue subspaces in $\mathcal{H}$ of finite codimension form embedded submanifolds of $\mathcal{T}(\mathcal{H})$, other orbits are not submanifolds of $\mathcal{T}(\mathcal{H})$, cf. [21]. Hence, the projective Hilbert space $P(\mathcal{H})$ elements of which are one-dimensional projections in $\mathcal{H}$ (e.g., $\mathcal{H} := L^2(\mathbf{R}, dx)$) forms an orbit of the first mentioned kind. The set $\mathcal{S}_*$ will be considered here as the quantal analogy of the phase space of Hamiltonian CM. Let us consider the set $\mathcal{F}$ of real valued differentiable (for simplicity; for more general – also unbounded – cases see [21]) functions on $\mathcal{T}(\mathcal{H})$ representing *generators* of transformations of the generalized phase space $\mathcal{S}_*$. We shall define Poisson brackets on $\mathcal{S}_*$. Let for all "sufficiently nice" real-valued functions $f,\ h, \dots$ the symbols $d_\varrho f,\ d_\nu h \dots$ denote the (Frèchet) differentials at the points $\varrho,\ \nu, \dots$ of $\mathcal{S}_*$. We can consider these differentials as unambiguously defined bouned operators in $\mathcal{L}(\mathcal{H})$. Then we define the Poisson bracket of f and h by, [21],

$$\{f,h\}(\nu) := i\,Tr(\nu[d_\nu f, d_\nu h]), \tag{16}$$

where $[\cdot,\cdot]$ means the commutator of bounded operators in $\mathcal{L}(\mathcal{H})$, and Tr is the trace in the space $\mathcal{T}(\mathcal{H})$.

Time evolution given by the Hamiltonian function h is determined now in the usual way from Eq. (16). It can be shown, [21], that the evolution of density matrices can be described by the 'cocycle' of unitary operators $u_h(t,\nu)$ (with $t \in \mathbf{R},\ \nu \in \mathcal{S}_*$) satisfying Schrödinger-like (in general nonlinear) equation with the 'Hamilton operator' $d_\varrho h$:

$$i\frac{d}{dt}u_h(t,\nu) = d_{\nu(t)}h\ u_h(t,\nu). \tag{17}$$

Here $\nu(t) := u_h(t,\nu)\,\nu\,u_h^*(t,\nu)$, if $\nu(0) := \nu$; this determines time evolution of any $\nu \in \mathcal{S}_*$ under the Hamiltonian dynamics corresponding to Poisson brackets (16) (there is an ambiguity in formulation of eq. (17) for $u_h(t,\nu)$, cf.[21]). We see that the evolution in this NLQM satisfies conditions (a) and (b) of the preceding section, with $U(t_1,t_2;\varrho) \equiv u_h(t_1 - t_2, \varrho)$. If $h(\nu) \equiv Tr(\nu H)$ for some ν-independent operator H, then the Eq. (17) converts into the usual Schrödinger equation with the Hamiltonian H; the dynamics is linear in the usual sense exactly in this last mentioned case. A nonlinear Schrödinger equation of the form (17) (for vector states $\rho^2 \equiv \rho$ only) was published first in [22]; the corresponding NLQM contained, however, interpretation, work with mixtures, and other aspects different from those of the present one.

Let us now define (formal) *observables* as differentiable functions $\mathbf{f}$, $\mathbf{h}$ of two variables $(\nu; \rho)(\in \mathcal{S}_* \times \mathcal{S}_*) \mapsto \mathbf{R}$ depending on the first variable ν affinely:

$$\mathbf{f}(\lambda_1\nu_1 + \lambda_2\nu_2; \rho) = \lambda_1\mathbf{f}(\nu_1; \rho) + \lambda_2\mathbf{f}(\nu_2; \rho). \tag{18}$$

Such functions can be represented by operator–valued functions $\mathbf{F} : \rho(\in \mathcal{S}_*) \mapsto \mathbf{F}(\rho) \in \mathcal{L}(\mathcal{H})$ in the following way:

$$\mathbf{f}(\nu; \rho) = Tr(\nu\mathbf{F}(\rho)). \tag{19}$$

The proposed interpretation of these "observables" is a natural extension of the one from QM: $Tr(\rho\mathbf{F}(\rho)^n)$ equals to n–th moment of the probability distribution of the observable $\mathbf{f}$ (resp. equivalently, of $\mathbf{F}$) in the state ρ (corresponding modifications for POV – measures, [8], can be formulated easily). Classical observables are described by $\mathbf{R}$ – valued functions $f(\rho)$ on $\mathcal{S}_*$ representing the operator – valued function

$$\mathbf{F}(\rho) \equiv f(\rho)\mathbf{I} \tag{20}$$

with $\mathbf{I}$ denoting the identity operator on $\mathcal{H}$.

Elements of $\mathcal{S}_*$ can be, in accordance with the preceding section, interpreted as Elementary (hence "improper") mixtures. The Genuine mixtures (i.e. Gemenge) are proposed here to be described by probability measures on $\mathcal{S}_*$.

We claim, [21], that the presented scheme is a mathematically consistent model containing conventional QM as its subtheory: If we choose only those "generators", resp. "observables" which are of the form $f(\rho) \equiv Tr(\rho F)$, $(F \in \mathcal{L}(\mathcal{H}))$, resp. $\mathbf{f}(\nu; \rho) \equiv Tr(\nu\mathbf{F})$, $\mathbf{F} \in \mathcal{L}(\mathcal{H})$, we obtain QM (with restriction to bounded operators, cf., however, [21]). CT can be formulated in this framework with a help of the "observables" (19) with $\mathbf{F}$ from (20). The Hamilton functions h in (17) can be chosen such that the corresponding time evolution leaves some arbitrarily chosen manifolds of generalized coherent states (in the sense of Perelomov, [23]) invariant. In this way, approximative theories to QT, like the time dependent Hartree–Fock theory (cf. [24]), together with finite dimensional Hamiltonian CT are included in our NLQM as exact (i.e. obtained without any approximations) subtheories.

References

[1] P. Busch, P. J. Lahti, and P. Mittelstaedt (Eds.), *Symposium on the Foundations of Modern Physics 1993 – Quantum Measurement, Irreversibility and The Physics of Information*, Cologne, Germany 1 – 5 June 1993 (World Scientific, London, 1993).

[2] P. Mittelstaedt in [1].

[3] P. Bóna in [1].

[4] According to some estimates the size of the corresponding apparatus would be of the order (or even larger) of the size of our Universe.

[5] V. S. Varadarajan, *Geometry of Quantum Theory*, Vols. I and II, (Van Nostrand Reinhold Co., New York, 1968, 1970).

[6] G. W. Mackey, *The Mathematical Foundations of Quantum Mechanics*, (W. A. Benjamin, New York, 1963).

[7] P. Bóna, *ACTA F.R.N. Univ. Comen. PHYSICA* **XX** (1980) 65.

[8] E. B. Davies, *Quantum Theory of Open Systems*, (Academic Press, London, 1976).

[9] P. Busch, in *Classical and Quantum Systems – Foundations and Symmetries*, (Eds. H. D. Doebner, W. Scherer and F. Schroeck, Jr.), (World Scientific, Singapore, 1993).

[10] P. Busch, P. J. Lahti and P. Mittelstaedt, *The Quantum Theory of Measurement*, (Springer, Berlin, 1992).

[11] G. Ludwig, *An Axiomatic Basis for Quantum Mechanics*, Vols. I and II (Springer, Berlin, 1985 and 1987).

[12] G. Ludwig in [1].

[13] K. Hepp, *Helv. Phys. Acta* **45** (1972) 237.

[14] O. Bratteli and D. W. Robinson, *Operator Algebras and Quantum Statistical Mechanics* (Springer, New York - Berlin, 1979 and 1981), Vols. I and II.

R. Haag and D. Kastler, *J. Math. Phys.* **5** (1964) 848.

[15] P. Bóna: *On Nonlinear Quantum Mechanics*, in *Differential Geometry and Applications, Proceedings of 5 – th International Conference on Differential Geometry and Its Applications, August 24 – 28, 1992, Opava, Czechoslovakia*, Eds. O. Kowalski, and D. Krupka (Silesian University Press, Opava, 1993).

[16] P. Pták, S. Pulmannová, *Orthomodular Structures as Quantum Logics*, (Kluwer, Dordrecht, 1991).

[17] P. Bóna, *acta phys. slov.* **27** (1977) 101.

[18] The present author apologizes to readers for this way of expression where from time to time into the formalism of QM are incorporated intrinsically foreign elements as "superselection rules" in a trial to express classical features of observed empirical systems in a modified version of QT.

[19] K. Hepp and E. H. Lieb, *Helv. Phys. Acta* **46** (1973) 573.

[20] P. Bóna, *Czech. J. Phys.* **B37** (1987) 482; *J. Math. Phys.* **29** (1988) 2223; *ibid* **30** (1989) 2994.

[21] P. Bóna, *Quantum Mechanics With Mean-Field Background*, (Comenius University, Bratislava 1991), preprint Ph10-91.

[22] (a) V.P.Belavkin and V.P.Maslov, *Tereticheskaya i Matematicheskaya Fizika*, **33**, No.1 (Oct. 1977) 17; (b) S. Weinberg, *Ann. Phys.*, **194** (1989), 336.

[23] A. M. Perelomov, *Commun. Math. Phys.*, **26** (1972) 222.

[24] D. J. Rowe, A. Ryman, and G. Rosensteel, *Phys. Rev.* **A 22** (1980) 2362.

ASYMPTOTIC BEHAVIOUR OF DENSITIES FOR NELSON PROCESSES[1]

Nicola Cufaro Petroni

Dipartimento di Fisica dell' Universitá and
I.N.F.N., via Amendola 173, 70126 Bari, Italy;
E-mail CUFARO@BARI.INFN.IT

Keywords :
Nelson processes; Fokker-Planck equation.

In a recent paper[(1)] an idea of Bohm and Vigier[(2)] about the possible decay of every initial probability density function (pdf), whose evolution is ruled by the quantum Fokker-Planck equation, toward the quantum mechanical pdf has been discussed in the light of the stochastic mechanics. The causal interpretation of the quantum mechanics[(3)] is based on the idea that a non relativistic particle of mass m, whose wave function obeys the Schrödinger equation

$$i\hbar\partial_t\psi(\mathbf{r},t) = -\frac{\hbar^2}{2m}\nabla^2\psi(\mathbf{r},t) + V(\mathbf{r},t)\psi(\mathbf{r},t)\,, \qquad (1)$$

is a classical object following a continuous and causally defined trajectory with a well defined position and accompanied by a physically real wave field ψ which contributes to determine its motion. If we write down (1) in terms of the real functions $R(\mathbf{r},t)$ and $S(\mathbf{r},t)$ with

$$\psi(\mathbf{r},t) = R(\mathbf{r},t)\,\mathrm{e}^{iS(\mathbf{r},t)/\hbar} \qquad (2)$$

[1]Paper contributed to the Workshop *Quantum Communications and Measurement*; Nottingham, July 10-16, 1994.

and separate real and imaginary parts, we have

$$\partial_t R^2 + \nabla\left(R^2 \frac{\nabla S}{m}\right) = 0\,, \tag{3}$$

$$\partial_t S + \frac{(\nabla S)^2}{2m} + V - \frac{\hbar^2}{2m}\frac{\nabla^2 R}{R} = 0\,, \tag{4}$$

where $R^2(\mathbf{r},t) = |\psi(\mathbf{r},t)|^2$ is interpreted as the density of a fluid with stream velocity

$$v(\mathbf{r},t) = \frac{\nabla S}{m}\,. \tag{5}$$

It is important to remark now that, if we define

$$v_{(+)}(\mathbf{r},t) = \frac{\nabla S}{m} + \frac{\hbar}{2m}\frac{\nabla R^2}{R^2} \tag{6}$$

the continuity equation (3) takes the form

$$\partial_t R^2 = \frac{\hbar}{2m}\nabla^2 R^2 - \nabla(R^2 v_{(+)}) \tag{7}$$

so that R^2 can also be considered as a particular solution of the evolution equation of the pdf's of a Markov process (Fokker-Planck equation)

$$\partial_t f = \nu\nabla^2 f - \nabla(f v_{(+)}) \tag{8}$$

characterized by the velocity field $v_{(+)}$ and by a diffusion coefficient

$$\nu = \frac{\hbar}{2m}\,. \tag{9}$$

This points out a possible connection between the density R^2 and the pdf of a suitable Markov process describing the random motion of a classical particle. As a matter of fact this connection is not at all compulsory at this point since the causal interpretation is a deterministic theory with no randomness involved in its fundamentals so that the analogy between (7) and (8) could also be considered purely formal. That notwithstanding the causal interpretation is obliged to add some randomness to its deterministic description in order to reproduce the statistical predictions of the quantum mechanics and hence it identifies the function $R^2 = |\psi|^2$ with the pdf of an ensemble of particles. But, since this addition is made by hand, is it easy for the critics of the model to argue that "it should be possible to have an arbitrary probability distribution (a special case of which is the function $P = \delta(x - x_0)$ representing a particle in a well defined location) that is at least in principle independent of the ψ field and dependent only on our degree of information concerning the location of the particle"[(2)]. The physical idea of Bohm and Vigier was that, even if our ensemble of quantum systems is described by an arbitrary initial pdf, this will decay in time to an ensemble with pdf $|\psi|^2$, because of the random fluctuations arising from the interactions with a subquantum medium: "no matter what the initial probability distribution may have been (for example a delta function) it will eventually be given by $P = |\psi|^2$".

A more convincing connection between quantum mechanics and classical random phenomena is achieved by means of the stochastic mechanics[(4)]: here the particle position is promoted to a stochastic Markov process $\xi(t)$ defined on some probabilistic space $(\Omega, \mathcal{F}, \mathbf{P})$ and taking values (for our limited purposes) in $\mathbf{R}^3$. This process is characterized by a pdf $f(\mathbf{r}, t)$ and a transition pdf $p(\mathbf{r}, t | \mathbf{r}', t')$ and satisfies an Itô stochastic differential equation of the form

$$d\xi(t) = v_{(+)}(\xi(t), t)dt + d\eta(t) \tag{10}$$

where $v_{(+)}$ is a velocity field which plays the role of a dynamical variable not given a priori but subsequently determined on the basis of a variational principle, and $\eta(t)$ is a Brownian process independent of ξ and such that

$$\mathbf{E}(d\eta(t)) = 0\,, \qquad \mathbf{E}(d\eta(t)\, d\eta(t)) = 2\nu\,\mathbf{I}\,dt$$

where $d\eta(t) = \eta(t + dt) - \eta(t)$ (for $dt > 0$), ν is the diffusion coefficient and $\mathbf{I}$ is the 3×3 identity matrix. A suitable definition of the Lagrangian and of the stochastic action functional for the system described by the dynamical variables f and $v_{(+)}$ allows us to select, by means of the principle of stationarity of the action, the particular processes which reproduce the quantum mechanics. In this formulation the foundations to interpret R^2 as a particular solution of a Fokker-Planck equation for the pdf of Markov processes are well established and the idea proposed by Bohm and Vigier of a relaxation in time of arbitrary pdf's solutions of (8) toward the quantum mechanical pdf $|\psi|^2$ can be checked as a property of the solutions of the Fokker-Planck equations with the field $v_{(+)}$ derived according to (6) from the wave functions solutions of (1). In other words we explore the possibility that the p associated by the Nelson stochastic mechanics to a quantum state ψ can be interpreted as the origin of the Bohm and Vigier stochastic flux and we examinate if and how the solutions of (8) selected by the stochastic mechanics to reproduce the quantum predictions attract other solutions which do not satisfy the stationary stochastic action principle and hence can not be considered as describing quantum systems.

In what follows we will limit ourselves to the case of the one dimensional trajectories, so that the Markov processes $\xi(t)$ considered will always take values in $\mathbf{R}$. The set of all the probability density functions (pdf's) of the absolutely continuous real random variables defined on a probability space $(\Omega, \mathcal{F}, \mathbf{P})$ coincides with the set $\mathcal{D}$ of all the non negative functions $f(x)$ of the hypersphere of norm 1 in the Banach space $L^1(\mathbf{R})$ and hence the time dependent pdf $f(x, t)$ of the stochastic processes $\xi(t)$ will be considered as trajectories on this subset $\mathcal{D}$. For Markov processes the transition pdf's $p(x, t| y, s)$ classified by means of the initial condition $\xi(s) = y$ (with $s < t$) are particular trajectories (with non absolutely continuous initial conditions). In $\mathcal{D}$ we then introduce a metrics induced by the norm in $L^1(\mathbf{R})$:

$$\mathbf{d}(f, g) = \frac{1}{2}\int_{-\infty}^{+\infty} |f(x) - g(x)|\, dx\,.$$

Here the factor $1/2$ guarantees that we always have $0 \leq \mathbf{d}(f, g) \leq 1$: the value 1 is attained when f and g have disjoint supports, and the value 0 when they coincide (Lebesgue almost everywhere). If the stochastic processes $\xi(t)$ under examination are Markov processes (as happens in stochastic mechanics) satisfying the equation (10) with initial condition $\xi(0) = \xi_0$, their pdf will satisfy the one dimensional evolution equation

$$\partial_t f(x, t) = \nu\partial_x^2 f(x, t) - \partial_x(v_{(+)}(x, t) f(x, t))\,, \tag{11}$$

with the initial condition $f(x, 0) = f_0(x)$ if $f_0(x)$ is the pdf of ξ_0.

Definition 1: We will say that the pdf $f(x,t)$ L^1-*approximates* the pdf $g(x,t)$ (for $t \to +\infty$), and we will write

$$f(x,t) \approx g(x,t) \qquad (t \to +\infty)\,,$$

when

$$\mathbf{d}(f,g) \to 0 \qquad (t \to +\infty)\,.$$

In particular we will say that f L^1-*converges* toward g (for $t \to +\infty$) if the pdf $g(x)$ does not depend on the time t.

We will examinate next a few properties of the concept of L^1-approximation for processes satisfying the equation (10). First of all we can prove(1) the following proposition

Proposition 1: If f and g are solutions of (11), the distance $\mathbf{d}(f,g)$ is a monotonic non-increasing function of the time t.

Of course, even if this proposition states that the distance $\mathbf{d}(f,g)$ among the solutions of (11) is a non-increasing function of time, this is not enough to derive the consequence that this distance actually decreases, let alone the the fact that it is infinitesimal, when $t \to +\infty$. However this property is sufficient to prove that, since $\mathbf{d}(t)$ is a monotone and bounded function of t, the limit of $\mathbf{d}(t)$ for $t \to +\infty$ always exists and is finite. In order to examinate the conditions that are sufficient to make the distance $\mathbf{d}(f,g)$ actually tend to zero when $t \to +\infty$, let us now introduce the following definition:

Definition 2: We will say that the family of the transition pdf's $p(x,t|\,y,0)$ L^1-*approximates* the pdf $g(x,t)$ in a *locally uniform way in* y (y - l.u.) for $t \to +\infty$, and we will write

$$p(x,t|\,y,0) \approx g(x,t) \qquad y\text{ - l.u.} \qquad (t \to +\infty)\,,$$

when for every $K > 0$ and for every $\epsilon > 0$ we can find a $T > 0$ such that

$$\mathbf{d}(p,g) = \mathbf{d}(p(x,t|\,y,0), g(x,t)) < \epsilon$$

for every $t > T$ and for every y such that $|y| \leq K$.

The local uniformity in y of the L^1-approximation of the p's to a pdf g allows us now to prove(1) the following proposition:

Proposition 2: If $p(x,t|\,y,0) \approx g(x,t)$, y - l.u., $(t \to +\infty)$, where p is the transition pdf of (11) and g an arbitrary pdf, then we have that $f(x,t) \approx g(x,t)$, $(t \to +\infty)$, for every $f(x,t)$ solution of (11).

In the proof we nowhere use the hypothesis that $g(x,t)$ is a solution of (11): in fact it is enough to suppose that g is a time dependent pdf. However, even if g is not a solution of (11), the triangular inequality for the metrics $\mathbf{d}$ allows us to show(1) that all the solutions of (11) L^1-approximate one another as stated in the following corollary:

Corollary 1: If $p(x,t|\,y,0) \approx g(x,t)$, y - l.u., $(t \to +\infty)$, where p is the transition pdf of (11) and g an arbitrary pdf, then we have that $f_1(x,t) \approx f_2(x,t)$, $(t \to +\infty)$, for every $f_1(x,t)$ and $f_2(x,t)$ solutions of (11).

Hence, under the conditions of Proposition 2, all the solutions of (11) globally tend to L^1-approximate one another after a sufficiently long time. On the other hand, if we can find two solutions f_1 and f_2 of (11) such that $\mathbf{d}(f_1, f_2)$ is not infinitesimal for $t \to +\infty$, then no pdf g can be L^1-approximated y - l.u. by the family of the transition pdf's p.

In order to discuss a few examples in detail it will be useful to derive a formula to calculate the L^1-distance among the pdf's $\mathcal{N}(m, \sigma^2)$ of *normal* random variables, namely pdf's of the form

$$g_{m,\sigma}(x) = \frac{\mathrm{e}^{-(x-m)^2/2\sigma^2}}{\sigma\sqrt{2\pi}}$$

with real m and $\sigma > 0$. In the following we will indicate with the symbol

$$\Phi(x) = \frac{1}{\sqrt{2\pi}} \int_{-\infty}^{x} \mathrm{e}^{-y^2/2}\, dy$$

the usual *error function* and we will also pose

$$\mathbf{d}(a, b;\, p, q) = \mathbf{d}(g_{a,p}, g_{b,q})\,.$$

With the previous notations we can now prove[1] that, if $p > q$ we have

$$\mathbf{d}(a, b;\, p, q) = \left[\Phi\Big(\frac{x_1 - b}{q}\Big) - \Phi\Big(\frac{x_2 - b}{q}\Big)\right] - \left[\Phi\Big(\frac{x_1 - a}{p}\Big) - \Phi\Big(\frac{x_2 - a}{p}\Big)\right] \qquad (12)$$

where

$$\begin{aligned} x_1 &= \frac{aq^2 - bp^2 - qp\sqrt{(a-b)^2 + 2(q^2 - p^2)\ln(q/p)}}{q^2 - p^2} \\ x_2 &= \frac{aq^2 - bp^2 + qp\sqrt{(a-b)^2 + 2(q^2 - p^2)\ln(q/p)}}{q^2 - p^2}\,; \end{aligned}$$

if $p = q$ and $a \neq b$ we have

$$\mathbf{d}(a, b;\, p, p) = 2\Phi\Big(\frac{|b-a|}{2p}\Big) - 1\,;$$

and finally, if $p = q$ and $a = b$ we have $\mathbf{d}(a, a;\, p, p) = 0$. This will be useful since in our examples both the transition pdf's and the pdf's derived from the quantum mechanical wave functions are normal. To show this we will use the following proposition[1] which indicates a very simple way to find the fundamental solutions of a class of evolution equations (11) which contains all the situations of our future examples.

Proposition 3: If the velocity field of the evolution equation (2) has the form

$$v_{(+)}(x,t) = -b(t)x - c(t)$$

with $b(t)$ and $c(t)$ continuous functions of time, then the fundamental solutions $p(x, t|\, y, 0)$ are normal pdf's $\mathcal{N}(\mu(t), \beta(t))$ where $\mu(t)$ and $\beta(t)$ are solutions of the equations

$$\begin{aligned} \mu'(t) + b(t)\mu(t) + c(t) &= 0 \\ \beta'(t) + 2b(t)\beta(t) - 2\nu &= 0 \end{aligned}$$

with initial conditions $\beta(0) = 0$ and $\mu(0) = y$.

We will discuss now our examples for systems reduced to a single non relativistic particle with a mass m, by remembering that the connection between the quantum mechanics and the stochastic mechanics is guaranteed if the diffusion coefficient and the Planck constant satisfy the relation (9). Let us consider first of all a simple harmonic oscillator with elastic constant k and classical (circular) frequency $\omega = \sqrt{k/m}$ and two possible wave functions obeying the Schrödinger equation: the (stationary) wave function of the ground state

$$\psi_0(x,t) = \left(\frac{1}{2\pi\sigma^2}\right)^{1/4} \mathrm{e}^{-x^2/4\sigma^2}\,\mathrm{e}^{-i\omega t/2}$$

and the (non stationary) wave function of the oscillating coherent wave packet with initial displacement a

$$\psi_C(x,t) = \left(\frac{1}{2\pi\sigma^2}\right)^{1/4} \exp\left[-\frac{(x-a\cos\omega t)^2}{4\sigma^2} - i\left(\frac{4ax\sin\omega t - a^2\sin 2\omega t}{8\sigma^2} + \frac{\omega t}{2}\right)\right]$$

where we have defined

$$\sigma^2 = \frac{\nu}{\omega}.$$

From the position (2) we find for our wave functions that

$$\begin{aligned}
R_0^2(x,t) &= f_0(x,t) = \frac{\mathrm{e}^{-x^2/2\sigma^2}}{\sigma\sqrt{2\pi}} \\
R_C^2(x,t) &= f_C(x,t) = \frac{\mathrm{e}^{-(x-a\cos\omega t)^2/2\sigma^2}}{\sigma\sqrt{2\pi}} \\
S_0(x,t) &= -\frac{1}{2}\hbar\omega t \\
S_C(x,t) &= -\frac{1}{2}\hbar\omega t - \hbar\frac{4ax\sin\omega t - a^2\sin\omega t}{8\sigma^2}
\end{aligned}$$

and hence we can calculate from (6) the corresponding velocity fields

$$\begin{aligned}
v_{(+)}^0(x,t) &= -\omega x \\
v_{(+)}^C(x,t) &= -\omega x + \omega a(\cos\omega t - \sin\omega t)\,.
\end{aligned}$$

This means that f_0 and f_C are respectively of the form $\mathcal{N}(0,\sigma^2)$ and $\mathcal{N}(a\cos\omega t,\sigma^2)$, and that the fundamental solutions $p_0(x,t|\,y,0)$ and $p_C(x,t|\,y,0)$ of (11) can be calculated by means of Proposition 3 with

$$\begin{aligned}
b_0(t) &= \omega\,, & c_0(t) &= 0 \\
b_C(t) &= \omega\,, & c_C(t) &= -\omega a(\cos\omega t - \sin\omega t)
\end{aligned}$$

so that they will respectively have the form $\mathcal{N}(\mu_0(t),\beta_0(t))$ and $\mathcal{N}(\mu_C(t),\beta_C(t))$ where

$$\begin{aligned}
\beta_0(t) &= \sigma^2(1-\mathrm{e}^{-2\omega t})\,, & \mu_0(t) &= y\mathrm{e}^{-\omega t} \\
\beta_C(t) &= \sigma^2(1-\mathrm{e}^{-2\omega t})\,, & \mu_C(t) &= a\cos\omega t + (y-a)\mathrm{e}^{-\omega t}\,.
\end{aligned}$$

A second class of examples can be drawn from the wave functions of a free particle of mass m. In particular we will choose to examinate the behavior of the (non stationary) wave function of a wave packet of minimal uncertainty centered around $x = 0$ with initial dispersion $\sigma^2 > 0$:

$$\psi_F(x,t) = \left(\frac{1}{2\pi\sigma^2\chi^2(t)}\right)^{1/4} e^{-x^2/4\sigma^2\chi(t)}$$

where

$$\chi(t) = 1 + i\omega t\,, \qquad \omega = \frac{\nu}{\sigma^2}\,.$$

In this case we have from (2)

$$\begin{aligned} R_F(x,t) &= f_F(x,t) = \frac{e^{-x^2/2\sigma^2\alpha^2(t)}}{\sqrt{2\pi}\sigma\alpha(t)} \\ S_F(x,t) &= \frac{\hbar}{2}\left(\frac{\omega t x^2}{2\sigma^2\alpha^2(t)} - \arctan\omega t\right) \end{aligned}$$

where

$$\alpha(t) = |\chi(t)| = \sqrt{1+\omega^2 t^2}\,.$$

This means that f_F is normal of the form $\mathcal{N}(0, \sigma^2\alpha^2(t))$. Moreover the velocity field is

$$v_{(+)}^F(x,t) = -\frac{1-\omega t}{1+\omega^2 t^2}\omega x$$

and the fundamental solutions $p_F(x,t|\,y,0)$ of (11) can then be calculated by means of Proposition 3 with

$$b_F(t) = \frac{1-\omega t}{1+\omega^2 t^2}\omega\,, \qquad c_F(t) = 0\,,$$

so that they will have the form $\mathcal{N}(\mu_F(t), \beta_F(t))$ where

$$\begin{aligned} \mu_F(t) &= y\sqrt{1+\omega^2 t^2} e^{-\arctan\omega t} \\ \beta_F(t) &= \sigma^2(1+\omega^2 t^2)(1 - e^{-2\arctan\omega t})\,. \end{aligned}$$

We can now use (12) in order to calculate $\mathbf{d}(p_0, f_0)$, $\mathbf{d}(p_C, f_C)$ and $\mathbf{d}(p_F, f_F)$: a long but simple calculation will show that (y - l.u.)

$$p_0 \approx f_0\,, \qquad p_C \approx f_C \qquad (t \to +\infty)$$

in the examples drawn from the harmonic oscillator, but that p_F will not L^1-approximate f_F since $\mathbf{d}(p_F, f_F)$ turns out to be different from zero and still dependent on y in the limit $t \to +\infty$:

$$\begin{aligned} \mathbf{d}(p_F, f_F) \quad \to \quad & \Phi\left(e^{\pi/2}[y - \sqrt{1-e^{-\pi}}\sqrt{y^2 - \ln(1-e^{-\pi})}\,]\right) \\ & -\Phi\left(e^{\pi/2}[y + \sqrt{1-e^{-\pi}}\sqrt{y^2 - \ln(1-e^{-\pi})}\,]\right) \\ & -\Phi\left(e^{\pi/2}[y\sqrt{1-e^{-\pi}} - \sqrt{y^2 - \ln(1-e^{-\pi})}\,]\right) \\ & +\Phi\left(e^{\pi/2}[y\sqrt{1-e^{-\pi}} + \sqrt{y^2 - \ln(1-e^{-\pi})}\,]\right). \end{aligned}$$

For example, if $y = 0$ (so that both p_F and f_F will remain centered around $x = 0$ along all teir evolution) we get in the limit $t \to +\infty$:

$$\mathbf{d}(p_F, f_F) \to 2\Big[\Phi(e^{\pi/2}\sqrt{-\ln(1-e^{-\pi})}\,) - \Phi(e^{\pi/2}\sqrt{1-e^{-\pi}}\sqrt{-\ln(1-e^{-\pi})}\,)\Big] \sim 0.011\,.$$

It is also possible to show that in this case two transition pdf's with different initial conditions $y \neq y'$ will never L^1-approximate one another as $t \to +\infty$, since

$$\mathbf{d}(p(x,t|\,y,0),\, p(x,t|\,y',0)) \to 2\Phi\Big(\frac{|y-y'|}{2\sqrt{e^\pi - 1}}\Big) - 1$$

which is zero if and only if $y = y'$. Hence on the basis of the Corollary 1 we can state that every solution of the evolution equation (11) L^1-approximates the quantum mechanical pdf (for $t \to +\infty$) only in the examples of the harmonic oscillator but not in that of the free particle.

It is apparent from our examples that the Markov processes associated to the quantum mechanical wave functions by the stochastic mechanics do not always exhibit the behavior required by the Bohm and Vigier hypothesis. In fact the calculations show that, in order to recover the property of a global relaxation in time of the pdf's toward the quantum mechanical solution, we must restrict ourselves to a particular set of physical systems. The different behaviors of our examples are in fact inscribed in the form of the time dependence of the parameters of the normal pdf's involved in our calculations. It is easy to see that, in the case of the harmonic oscillator, for every real y we have (for $t \to +\infty$)

$$\begin{aligned} \mu_0(t) &\to 0\,, & \beta_0(t) &\to \sigma^2 \\ |\mu_C(t) - a\cos\omega t| &\to 0\,, & \beta_C(t) &\to \sigma^2\,. \end{aligned}$$

On the other hand $\mu_F(t)$ and $\beta_F(t)$ behave differently from the corresponding parameters of the quantum mechanical pdf f_F, since (for $t \to +\infty$)

$$|\mu_F(t) - e^{-\pi/2} y\omega t| \to 0\,, \qquad |\beta_F(t) - (1 - e^{-\pi})\sigma^2\omega^2 t^2| \to 0\,,$$

while the quantum mechanical f_F is a normal pdf which remains centered around $x = 0$ with a variance which diverges as $\sigma^2\omega^2 t^2$. In this case it is of no avail to remark that both p_F and f_F will flatten to zero when $t \to +\infty$: the relevant fact is that this flattening happens at rates different enough to make the L^1-distance remain non zero even in the limit $t \to +\infty$. It is remarkable, however, that in the formulation chosen in the original Bohm and Vigier paper no one of our three examples would have shown the correct property: our L^1-metrics plays here an important role in discriminating the well behaved systems among all the possibilities.

The fact that the Nelson transition pdf's do not always L^1-approximate one another also means that it is impossible to find a unique pdf g L^1-approximated by them independently from y, and hence that the solutions of (11) in the discussed free particle case will not globally tend to L^1-approximate one another in time. Of course nothing forbids a priori, even in this case, that particular subsets of solutions can show the tendency to mutually L^1-approximate and hence the field is open to investigations about, for instance, the possibility that some particular solution of (11) can be stable with respect to small perturbations of their initial conditions: which in some minimal sense was the essential intention of the Bohm and Vigier proposal. In any case our examples show that, at least for a significant set of systems and wave functions the Bohm and Vigier property holds in the L^1-metrics if we adopt the transition pdf suggested by the Nelson stochastic mechanics, and hence it can be surely stated that their original idea posed an interesting an physically well grounded problem. It is not possible at present to state clearly and in a general way in which cases we realize the conditions for a global (or at least local) mutual

L^1-approximation of the solutions of (11). The examples discussed show that the discriminating property is not the stationarity of the quantum mechanical wave function since also the square modulus of the non stationary, coherent, oscillating wave packet of the harmonic oscillator attracts in L^1 every other solution of (11). An indication can be perhaps found in the fact that the main difference between two systems seems to be principally in the fact that their energy spectra are very different: the harmonic oscillator has a completely discrete spectrum and the free particle a completely continuous one. Hence a first idea can be to distinguish between bound states, which exhibit the Bohm and Vigier property, and scattering states, which do not. Of course the settlement of this question will require the discussion of further examples and the investigation of more general properties. Finally it must be pointed out that we have made the very particular choice of selecting the transition pdf's of the Nelson stochastic mechanics as a good candidate to the generation of the right stochastic flux exhibiting the Bohm and Vigier property in some suitable sense. As a consequence another possible conclusion could also be that the Nelson flux is not the right candidate to represent, in the general case, the interpretative scheme of Bohm and Vigier. Hence we consider wide open the possibility that the right transition pdf's can be built in a different way. For example it is well known that in the Nelson stochastic mechanics the diffusive part of the stochastic differential equation (10) is given a priori. Hence, since the transition pdf which propagates a given time-dependent pdf $f(\mathbf{r}, t)$ is not uniquely determined (and are not, in general, observable in the stochastic mechanics), nothing forbids to find a diffusive flux, different from that of Nelson, which exhibites the Bohm and Vigier property for every possible quantum wave function. In particular a possibility lies in a generalization of the stochastic mechanics where also the diffusive part of the stochastic differential equation ruling the process is dynamically determined in a way such that the Bohm and Vigier property is always satisfied.

References

[1] N. Cufaro Petroni and F. Guerra: *Quantum mechanical states as attractors for Nelson Processes*; in press on *Found. Phys.*

[2] D. Bohm and J.-P. Vigier: *Phys. Rev.* **96** (1954) 208.

[3] D. Bohm: *Phys. Rev.* **85** (1952) 166, 180.

[4] E. Nelson: *Phys. Rev.* **150** (1966) 1079;

[5] E. Nelson: *Dynamical Theories of Brownian Motion* (Princeton U. P.; Princeton, 1967);

[6] E. Nelson: *Quantum Fluctuations* (Princeton U. P.; Princeton, 1985).

[7] F. Guerra: *Phys. Rep.* **77** (1981) 263;

[8] F. Guerra and L. Morato: *Phys. Rev.* D**27** (1983) 1774;

[9] F. Guerra and R. Marra: *Phys. Rev.* D**28** (1983) 1916;

[10] F. Guerra and R. Marra: *Phys. Rev.* D**29** (1984) 1647.

PHASE AND NOISE

Michael J. W. Hall

Department of Theoretical Physics
Australian National University
Canberra, ACT 0200, Australia

Abstract

A new derivation of the canonical description of optical phase is given, and it is shown that any covariant method of phase estimation (such as heterodyne phase detection) may be interpreted as a *noisy* measurement of canonical phase. An important example is Gaussian noise, and it is demonstrated that phase-based communication is relatively robust with respect to such noise. A second example is homogenous phase noise (including standard phase diffusion), which has interesting connections with Lorentzian lineshapes and coherent phase states.

Key-words :
phase, noise, communication, information, diffusion

1 THE CANONICAL PHASE DISTRIBUTION

Various methods have been employed to characterise/derive the canonical representation of optical phase, for example: maximum-likelihood estimation of unknown phase-shifts [1–3]; algebraic decomposition properties of the number-shift operator [4]; Cauchy sequences of finite-Hilbert-space expectation values [5]; and inclusion-ordering properties of covariant phase distributions [6] (see also eq.(17) below).

It is of interest to note here a simple alternative approach, analogous to Louisell's derivation of the position representation via momentum considerations [7]. In the latter case one postulates a set of position kets $\{|\,\mathbf{x}\rangle\}$, such that the position distribution $p(\mathbf{x}\mid\psi)$ for any state ψ satisfies

$$p(\mathbf{x}\mid\psi)=|\langle\mathbf{x}\mid\psi\rangle|^2,\qquad p(\mathbf{x}\mid\exp[-i\mathbf{P}.\mathbf{a}/\hbar]\psi)=p(\mathbf{x}-\mathbf{a}\mid\psi),\tag{1}$$

Quantum Communications and Measurement
Edited by V. P. Belavkin *et al.*, Plenum Press, New York, 1995

where **P** denotes the generator of translations (ie, the momentum operator). From (1) one has

$$\int d^n\mathbf{x} \mid \mathbf{x}\rangle\langle\mathbf{x} \mid = 1, \qquad \exp[i\mathbf{P}.\mathbf{a}/\hbar] \mid \mathbf{x}\rangle = \gamma(\mathbf{x},\mathbf{a}) \mid \mathbf{x}-\mathbf{a}\rangle, \tag{2}$$

where $\gamma(\mathbf{x},\mathbf{a})$ is an arbitrary phase factor ($\mid \gamma \mid = 1$). Expanding (2) with respect to the complete set of eigenkets $\{\mid \mathbf{p}\rangle\}$ of **P** leads to the general solution

$$\alpha(\mathbf{x}) \mid \mathbf{x}\rangle = (2\pi\hbar)^{-n/2} \int d^n\mathbf{p}\, e^{-i\mathbf{p}.\mathbf{x}/\hbar}\, \beta(\mathbf{p}) \mid \mathbf{p}\rangle \tag{3}$$

of (1), where $\alpha(\mathbf{x}), \beta(\mathbf{p})$ are arbitrary phase factors. Finally, the *canonical* position representation is obtained by the choice $\alpha \equiv \beta \equiv 1$ (or, equivalently, by absorbing α and β into $\mid \mathbf{x}\rangle$ and $\mid \mathbf{p}\rangle$ respectively).

In complete analogy with (1), it is natural to seek a representation of optical phase via a set of *phase* kets $\{\mid \phi\rangle\}$, such that the phase distribution $p(\phi \mid \psi)$ for state ψ satisfies

$$p(\phi \mid \psi) = \mid \langle \phi \mid \psi\rangle \mid^2, \qquad p(\phi \mid \exp[iN\theta]\, \psi) = p(\phi - \theta \mid \psi), \tag{4}$$

where N denotes the generator of phase-shifts (ie, the number operator). Proceeding exactly as above for position, one is directly led to the *canonical phase representation*

$$\mid \phi\rangle = (2\pi)^{-1/2} \sum_n e^{in\phi} \mid n\rangle, \tag{5}$$

where $\mid n\rangle$ denotes the n'th number state.

For the more general case where the field is described by some density operator ρ, equations (4) and (5) generalise to the form

$$p_C(\phi \mid \rho) = \langle \phi \mid \rho \mid \phi\rangle = (2\pi)^{-1} tr[\rho f_\phi], \tag{6}$$

where

$$f_\phi = 2\pi \mid \phi\rangle\langle\phi \mid = \sum_{m,n} e^{i(m-n)\phi} \mid m\rangle\langle n \mid . \tag{7}$$

2 COVARIANT PHASE DISTRIBUTIONS

A general method of phase estimation, G, will be called *covariant* [2,8] if the measurement statistics are of the form

$$p_G(\phi \mid \rho) = (2\pi)^{-1}\, tr[\rho g_\phi], \tag{8}$$

and satisfy the phase-shift property

$$p_G(\phi \mid \exp[iN\theta]\, \rho \exp[-iN\theta]) = p_G(\phi - \theta \mid \rho) \tag{9}$$

in analogy to (4). Examples of covariant phase estimation include [8]: *canonical* phase detection, with $g_\phi = f_\phi$; *random* phase detection, with $g_\phi = 1$; and *heterodyne* phase detection, with

$$\begin{aligned} g_\phi = h_\phi &= 2\int_0^\infty d|\alpha|\,|\alpha|\mid \alpha\rangle\langle\alpha \mid \\ &= \sum_{m,n} \mid m\rangle\langle n \mid e^{i(m-n)\phi}\Gamma(\frac{m+n}{2}+1)/[m!n!]^{1/2}, \end{aligned} \tag{10}$$

where $| \alpha\rangle$ denotes the Glauber coherent state of amplitude $| \alpha | e^{i\phi}$. This latter example also corresponds to the *operational* phase measurement proposed by Noh *et al* [9], in the ideal limit [10].

Covariant and canonical phase measurements are closely related. One may show from (6)–(9) that [6]

$$p_G(\phi \mid \rho) = p_C(\phi \mid T_G(\rho)), \tag{11}$$

where $T_G(\rho)$ denotes the density operator

$$T_G(\rho) = \sum_{m,n} | m\rangle\langle n | \ \langle m | \rho | n\rangle \ \langle n | g_0 | m\rangle. \tag{12}$$

The class of canonical phase distributions thus includes *all* covariant phase distributions. Further, equation (11) suggests the interpretation

$$covariant\, phase = canonical\, phase + noise. \tag{13}$$

In particular, T_G in (12) is a completely-positive trace-preserving mapping or (non-selective) *operation* [11], and hence corresponds to a linear process which outputs state $T_G(\rho)$ for input state ρ. It follows that the covariant phase estimation method G is equivalent to *first* subjecting the field to process T_G, and then making a *canonical* phase measurement. Since T_G typically smears out the canonical phase distribution it may be regarded as a source of noise — hence equation (13).

Note that for an *arbitrary* operation T, one may define a corresponding phase estimation method as above (replacing T_G by T). The associated phase distribution is then

$$p_T(\phi \mid \rho) = p_C(\phi \mid T(\rho)). \tag{14}$$

It follows from (8), (9), (11), (12), and (14) that *the class of covariant phase estimation methods is completely characterised by the set of operations* $\{T\}$ *satisfying*

$$T(e^{iN\theta}\rho e^{-iN\theta}) = e^{iN\theta}T(\rho)e^{-iN\theta}. \tag{15}$$

Simple examples include the identity operation $T_C(\rho) = \rho$ corresponding to *canonical* phase detection, and the "number-measurement" operation

$$T_R(\rho) = \sum_{n} | n\rangle\langle n | \ \langle n | \rho | n\rangle \tag{16}$$

corresponding to *random* phase detection. The inclusion-ordering property

$$T_R \leq T \leq T_C \tag{17}$$

holds for all T satisfying (15) (where $T \leq T'$ iff $T = T' \circ T''$ for some T'').

3 GAUSSIAN NOISE

Gaussian noise provides the generic model for a field subject to random linear fluctuations [12], and is represented by the operation [12,13]

$$\Gamma_v(\rho) = (\pi v)^{-1} \int d^2\alpha \, \exp(- |\alpha|^2 / v) \, D(\alpha)\rho D^\dagger(\alpha). \tag{18}$$

Here v is a variance parameter (equal to the average number of 'noisy' photons added to the state), and $D(\alpha)$ denotes the Glauber displacement operator $\exp(\alpha a^\dagger - \alpha^* a)$.

Gaussian noise satisfies (15), and thus yields an important example of covariant phase estimation.

It is of interest to note the approximate relation

$$\textit{heterodyne phase} \approx \textit{canonical phase} + \textit{Gaussian noise} \tag{19}$$

analogous to (13), holding in the high-intensity limit when $v = 1$ (*ie*, for one photon of noise). In particular, it may be shown when $v = 1$ that

$$\langle e^{ik\phi}\rangle_H = \pi^{-1}\int d^2\alpha\,\langle\alpha|\rho|\alpha\rangle\,e^{ik\phi} \tag{20}$$

$$\langle e^{ik\phi}\rangle_\Gamma = \pi^{-1}\int d^2\alpha\,\langle\alpha|\rho|\alpha\rangle\,e^{ik\phi}\,f_k(I), \tag{21}$$

where $k \in \mathcal{Z}$; $\langle\cdot\rangle_H$, $\langle\cdot\rangle_\Gamma$ denote averaging with respect to $p_H(\phi \mid \rho)$, $p_\Gamma(\phi \mid \rho)$ respectively; $\mid\alpha\rangle$ is the coherent state of amplitude $I^{1/2}e^{i\phi}$; and $f_k(I)$ has the asymptotic expansion

$$f_k(I) \sim 1 - \frac{k^2}{8I} + \frac{k^2(k^2-8)}{128I^2} + \ldots . \tag{22}$$

Thus, in the limit of high intensity $\langle I\rangle_H \gg 1$, one obtains relation (19).

Consider now a communication channel based on canonical phase detection, and subject to Gaussian noise of variance v. It will be assumed that the *coherent phase state* [3,6]

$$\mid ue^{i\phi}\rangle = (1-u^2)^{1/2}\sum_n u^n\,e^{in\phi}\mid n\rangle \tag{23}$$

is transmitted with probability $(2\pi)^{-1}$ for some fixed value of u (these states have near-optimal phase-information properties [6]). The average photon number per signal is then

$$\overline{n} = \langle u \mid N \mid u\rangle = u^2/(1-u^2)\,, \tag{24}$$

and the average information per signal is given by [6]

$$I(\overline{n}, v) = \log 2\pi + \int d\phi\, p_\Gamma(\phi \mid u)\log p_\Gamma(\phi \mid u)\,, \tag{25}$$

where from (6), (14), (15), (18) and (23)

$$p_\Gamma(\phi \mid u) = (2\pi)^{-1}\,\langle ue^{-i\phi} \mid \Gamma_v(f_0) \mid ue^{-i\phi}\rangle\,. \tag{26}$$

To calculate $I(\overline{n}, v)$ from equations (7), (18) and (23)–(26) is numerically quite difficult. However, for the case of low noise ($v \ll 1$), one may partially trade computation for algebra via the approximation

$$\langle m \mid \Gamma_v(f_0) \mid n\rangle = 1 + v\{[mn]^{1/2} + [(m+1)(n+1)]^{1/2} - m - n - 1\} + O(v^2) \tag{27}$$

following from (7) and expansion (14) of [12].

The information rate calculated via (23)–(26) and approximation (27) is plotted in Figure 1 for the case $v = 10^{-3}$. Also plotted are the corresponding maximum information rates for channels based on heterodyne detection of coherent states, homodyne detection of squeezed states, and photodetection of number states (see Ref.[12] for details). The phase-based and coherent-state channels are indistinguishable for the chosen range of parameters. It is seen that phase information is relatively *robust* with respect to Gaussian noise.

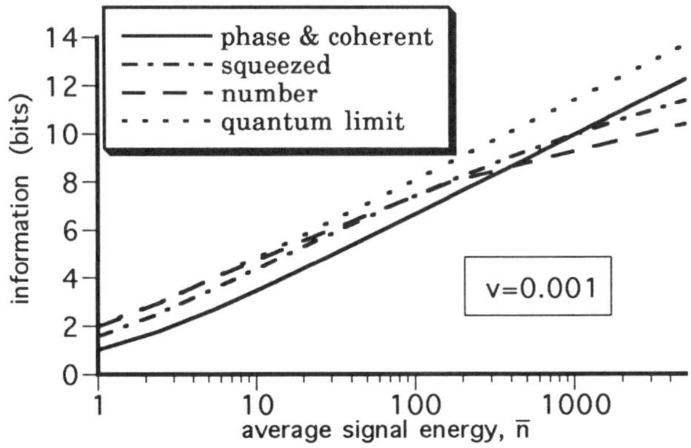

Figure 1: Information rates for quantum channels subject to Gaussian noise

4 HOMOGENOUS DISPLACEMENT NOISE

The Gaussian noise model is not always physically applicable (eg, it cannot model line-broadening processes [12]), and hence it is of interest to consider other generic noise models. Gaussian noise may in fact be viewed as a special case of *homogenous displacement noise* (or convolution semigroup noise [14]), which models random "generalised displacements" of a quantum system. In this section attention is focused on the case where the displacements are phase-shifts.

In particular, let the field be subjected to random phase shifts for a time interval t. If $p_t(\theta)$ denotes the probability of *total* phase shift θ, then the final state is described by the density operator

$$\rho_t = R_t(\rho_0) = \int d\theta\, p_t(\theta)\, \exp(iN\theta)\rho_0 \exp(-iN\theta) \tag{28}$$

(note the similarity with equation (18)). The random process will be said to represent *homogenous* phase noise if it acts uniformly in time, ie, if

$$R_s \circ R_t = R_{s+t}\,, \qquad\qquad s,t \geq 0\,. \tag{29}$$

The canonical example is *standard phase diffusion* [15] with

$$p_t(\theta) = (2\pi)^{-1}\,[1 + 2\sum_{m\geq 1} e^{-\gamma m^2 t}\, \cos m\theta]\,, \tag{30}$$

for which one finds, via (8), (9) and (28) the phase diffusion equation

$$(\partial/\partial t)\, p_G(\phi \mid \rho_t) = \gamma\, (\partial^2/\partial\phi^2)\, p_G(\phi \mid \rho_t) \tag{31}$$

for *any* covariant phase estimation method G.

The general solution to (28) and (29) is

$$p_t(\theta) = (2\pi)^{-1}\,[\,1 + 2\sum_{m\geq 1} e^{-\gamma_m t}\, \cos(m\theta - \omega_m t)\,] \tag{32}$$

where γ_m, ω_m must be chosen to ensure $p_t \geq 0$ (thus $\gamma_m \geq 0$). It is easily shown that (i) the matrix element $\langle m \mid \rho \mid n\rangle$ decoheres exponentially in time, as $\exp(-\gamma_{|m-n|}t)$;

and that (ii) the spectral density function is *Lorentzian*, with half-width γ_1. Thus Lorentzian lineshapes are not good indicators of any *particular* phase-noise model.

While the phase diffusion model (30) corresponds to the choice $\gamma_m = \gamma m^2$, $\omega_m = 0$ in (32), the "simpler" choice $\gamma_m = \gamma m$, $\omega_m = 0$ may be recognised via (6), (7) and (23) as the canonical phase distribution of the coherent phase state $| e^{-\gamma t}\rangle$. Noting property (29), the effects of the corresponding *"coherent" phase noise* model are particularly easy to calculate for coherent phase states: the canonical phase distribution of state $| u\rangle$ evolves to that of state $| ue^{-\gamma t}\rangle$. Thus, for example, the information per signal for the channel in Sec. 3 (with Gaussian noise replaced by coherent phase noise), follows immediately from (24) and equation (60) of [6] as

$$I(\overline{n}, t) = \log(1+\overline{n}) - \log(1+\overline{n}[1-e^{-2\gamma t}])\,. \tag{33}$$

Note finally that equation (29) implies the convolution relation $p_G^f = p_t * p_G^i$, relating initial and final covariant phase distribution functions via the 'noise' distribution p_t. An analogous relation $p_A^f = p_C * p_A^i$ correlates the initial and final phase distributions of an "apparatus" system with the canonical phase distribution of the field (via the evolution operator $(U^\dagger)^{N_A}$, where N_A is a number operator for the apparatus and U is the exponential phase operator) [16]. Thus phase-shift noise can formally be related to phase measurement correlations.

References

[1] C W Helstrom *Int. J. Theoret. Phys.* **11** (1974) 357; V P Belavkin *Radio Engng Electron. Phys.* **29**(1) (1976) 78.

[2] A S Holevo *Probabilistic and Statistical Aspects of Quantum Theory* (North-Holland, Amsterdam, 1982) Sec. II, IV.

[3] J H Shapiro and S R Shepard *Phys. Rev. A* **43** (1991) 3795.

[4] J M Levy-Leblond *Ann. Phys.* **101** (1976) 319.

[5] D T Pegg and S M Barnett *Phys. Rev. A* **39** (1989) 1665.

[6] M J W Hall *J. Mod. Opt.* **40** (1993) 809.

[7] W H Louisell *Quantum Statistical Properties of Radiation* (Wiley, New York, 1973) Sec. 1.10.

[8] M J W Hall and I G Fuss *Quantum Opt.* **3** (1991) 147.

[9] J W Noh, A Fougères and L Mandel *Phys. Rev. A* **45** (1992) 424.

[10] M Freyberger and W Schleich *Phys. Rev. A* **47** (1993) R30, Eq.(16).

[11] K Kraus *States, Effects and Operations* in Lecture Notes in Physics vol.190 (Springer, Berlin, 1983).

[12] M J W Hall *Phys. Rev. A* **50** (1994) 3295.

[13] M J W Hall and M J O'Rourke *Quantum Opt.* **5** (1993) 161; M J W Hall *Phys. Rev. A* **49** (1994) 42; see also A Vourdas *Phys. Rev. A* **34** (1986) 3466 and C W Helstrom *J. Math. Phys.* **20** (1979) 2063.

[14] A Kossakowski *Rep. Math. Phys.* **3** (1972) 247 Sec.4,5.

[15] See Sec. 6.7, 9.6 of Ref.[7]; also R J Glauber in *Quantum Optics and Electronics* (Gordon and Breach, New York, 1965) edited by C DeWitt *et al*, pp.63-185, Lecture XV.

[16] V P Belavkin and C Bendjaballah *Quantum Opt.* **6** (1994) 169.

ON HYPERBOLIC FLOWS AND THE PROBLEM OF CHAOS IN QUANTUM SYSTEMS[1]

Geoffrey L. Sewell[2]

Department of Physics,
Queen Mary and Westfield College
Mile End Road, London E1 4NS, England

Abstract

We briefly review the non-commutative generalisation, presented in [1], of the theory of hyperbolic dynamical systems; and then prove that hyperbolicity cannot be a paradigm for quantum chaos, except possibly in a certain asymptotic sense.

1 INTRODUCTION

In a recent work [1], the theory of hyperbolic, or Anosov, flows [2,3], which provides a paradigm for classical chaos, was generalised to non-commutative dynamical systems. The question naturally arises of whether such flows can prevail in the standard Von Neumann model [4] of (finite) quantum systems, and thus provide a paradigm for quantum chaos too. The object of this note is to demonstrate that they cannot do so, except possibly in an asymptotic sense, corresponding to the proximity of a classical limit. This is quite in line with results obtained on rather different bases about quantum chaos [5-7].

In Section 2, I shall outline the generalised theory [1] of hyperbolic flows, and in Section 3 I shall establish the above-described results concerning quantum chaos. It will be seen that the main result, i.e. Theorem 3, generalises one obtained in [1], under special conditions, for the particular case of the free dynamics of a particle on a manifold of constant negative curvature.

[1]Based on a Lecture at the International Workshop on "Quantum Communication and Measurement" held at Nottingham, 11-16 July, 1994

[2]Partially supported by European Capital and Mobility Contract No. CHRX-Ct. 92-0007

2 HYPERBOLIC FLOWS

2.1. The Classical Model. In a standard way (cf. [3]), we take our classical dynamical model, Σ_c, to be given by a triple (X, ϕ, μ), where X, the phase space, is a compact, differentiable, Riemannian manifold, $\{\phi_t | t \in \mathbf{Z}\ or\ \mathbf{R}\}$ is a one-parameter group of diffeomorphisms of X, representing the dynamics of the system, and μ is a smooth, ϕ−invariant probability measure on X, corresponding to a stationary state. We denote the tangent space at $x(\in X)$ by $T(x)$ and, for fixed t, we define $\phi_t^\star := d\phi_t : T(x) \rightarrow T(\phi_t x)$. We shall employ the following definition of hyperbolic flows, which is a restricted version of that of Anosov [2,3].

Def. 2.1. We term the dynamics of Σ_c hyperbolic if there are unit vector fields $V_1, .. ., V_m; V_{m+1}, .. ., V_n$ over X, which are linearly independent at each point of X and satisfy the following conditions.

(a) $0 < m < n = Dim(X)\ or\ Dim(X) - 1$, according to whether the time scale is discrete or continuous.

(b)

$$\phi_t^\star V_j(x) = V_j(\phi_t x) \exp(\lambda_j t) \tag{2.1}$$

where the Liapounov exponents λ are constants, with λ_j positive for $j \leq m$ and negative for $j > m$.

(c) The vector fields V_j have globally integral curves, generated by the action of one-parameter groups $\{\theta_j(s) | s \in \mathbf{R}\}$ of diffeomorphisms of X. Thus,

$$x_j(s) = \theta_j(s) x \tag{2.2}$$

is the unique global solution of the equation

$$x_j'(s) = V_j(x_j(s))\ \forall s \in \mathbf{R};\ x(0) = x \tag{2.3}$$

Comments. (1) It follows immediately from the condition (b) that, generically, orbits emanating from neighbouring points separate exponentially fast from one another. Hence the dynamics is extremely unstable, i.e. chaotic.

(2) As prototype examples of hyperbolic systems, we cite (cf. [3])

(a) the 'Arnold Cat' model, whose dynamics corresponds to iterations of an automorphism of the two-dimensional torus; and

(b) geodesic flow over a compact manifold of negative curvature.

The following key lemma, concerning hyperbolic systems, will be proved in the Appendix.

Lemma 2.2. *For any hyberbolic system,* Σ_c,

$$\phi_t \theta_j(s) \phi_t^{-1} = \theta_j(s.\exp(\lambda_j t)) \tag{2.4}$$

2.2. Algebraic Formulation of Σ_c. As a first step towards a quantum generalisation of the above model, we reformulate it as an *abelian* $W^\star$−dynamical system $(\mathcal{A}_c, \alpha_c, \omega_c)$, where $\mathcal{A}_c$, the algebra of observables, is $L^\infty(X, d\mu)$, $\{\alpha_c(t) | t \in \mathbf{Z}\ or\ \mathbf{R}\}$ is the one-parameter group of automorphisms of $\mathcal{A}_c$ induced by ϕ, i.e.

$$(\alpha_c(t) f)(x) \equiv f(\phi_t^{-1} x) \tag{2.5}$$

and ω_c is the normal state on $\mathcal{A}$ corresponding to μ, i.e.

$$\omega_c(f) \equiv \int f d\mu \tag{2.6}$$

Further, we denote by $\sigma_{c,j}(s)$ the automorphism of $\mathcal{A}$ induced by θ_j, i.e.

$$(\theta_{c,j}(s)f)(x) \equiv f(\theta_j(-s)x) \tag{2.7}$$

It follows immediately from (2.5) and (2.7) that the hyperbolicity condition (2.4) is equivalent to

$$\alpha_c(t)\sigma_{c,j}(s)\alpha(-t) = \sigma_{c,j}(s.\exp(\lambda_j t)) \tag{2.8}$$

2.3. Generalisation to Quantum Systems [1]. Now let $\Sigma = (\mathcal{A}, \alpha, \omega)$ be an arbitrary $W^\star$-dynamical system, with $\mathcal{A}$ an algebra of observables, $\{\alpha(t)|t\in\mathbf{Z}\ or\ \mathbf{R}\}$ a one-parameter group of automorphisms of $\mathcal{A}$, and ω a normal, $\alpha-$invariant state on $\mathcal{A}$. Thus, Σ provides a generalisation of the model Σ_c to quantum systems. For these, $\mathcal{A}$ is non-abelian and conforms to the canonical commutation relations.

Def. 2.3. We term the system Σ *hyperbolic* if $\mathcal{A}$ is equipped with weakly continuous one-parameter groups of automorphisms $\sigma_1(\mathbf{R}), \ldots, \sigma_m(\mathbf{R});\ \sigma_{m+1}(\mathbf{R}), \ldots, \sigma_n(\mathbf{R})$, with $0 < m < n$, such that

(a) the generators, δ_j of these groups are linearly independent, and

(b) the condition corresponding to (2.8) is fulfilled, i.e.

$$\alpha(t)\sigma_j(s)\alpha(-t) = \sigma_j(s.\exp(\lambda_j t)) \tag{2.9}$$

where the λ's are constants, with λ_j positive for $j \leq m$ and negative for $j > m$.

Comments. (1) The present definition is less restrictive than that of [1] in that it does not require the σ_j-invariance of ω.

(2) In the classical case, the derivations δ_j are those corresponding to the vector fields V_j.

(3) In general, by contrast with the classical case, Def. 2.3 does not provide a specification of n in terms of the structure of $\mathcal{A}$. We would hope that this deficiency could be remedied in the future. For present purposes, however, it suffices that $0 < m < n$.

Proposition 2.4. [1] *If Σ is hyperbolic, then the domain $D(\delta_j)$ of the derivation δ_j is stable under $\alpha(\mathbf{R})$,*

$$\alpha(t)\delta_j\alpha(-t) = \delta_j \exp(\lambda_j t) \tag{2.10}$$

and

$$||\delta_j\alpha(t)A|| = ||\delta_j A||\exp(-\lambda_j t)\ \forall A\in D(\delta_j),\ t\in\mathbf{R} \tag{2.11}$$

Comments. (1) Equation (2.10) is the differential form of (2.9), and reveals the exponential time dependence of the evolute $\alpha(t)\delta_j\alpha(-t)$ of δ_j, with Liapounov exponent λ_j.

(2) Since $\delta_j A$ represent the rate at which A moves along the orbit generated by the action of σ_j, this Proposition establishes a precise sense in which the dynamics is hyperbolic.

Examples of Hyperbolic Quantum Systems. (cf. [1].)

(1) A quantum version of the 'Arnold Cat'.

(2) A Wightman relativistic quantum field in Minkowski space, as viewed by a uniformly accelerated observer.

3 SUPPRESSION OF CHAOS BY QUANTISATION

In view of the extreme sensitivity of hyperbolic systems to perturbations of initial conditions (cf. Prop. 2.4), we shall take these as our paradigm of chaotic dynamics. The examples cited at the end of §2 thus provide us with specific models of non-commutative chaotic systems.

However, neither of these corresponds to a finite quantum system, in the standard sense [4], for the following reasons.

(1) Ex.1 is a quantised version [8] of the classical hyperbolic system given by iterations of the automorphisms of a torus, T^2. Specifically, the quantisation amounts to the replacement of T^2 by a *non-commutative* torus, in the sense of Connes [9], as defined by a Weyl algebra, $\mathcal{A}$, over the dual group, $\mathbf{Z}^2$, of T^2. The dynamics is quasi-free, and induced by the same transformation of $\mathbf{Z}^2$ as in the classical case, and the equilibrium state is the tracial one on $\mathcal{A}$. Now this model does *not* correspond a standard finite quantum system, since $\mathcal{A}$ is not a type I factor. In fact, it is either of type II_1 or a tensor product of $L^\infty(T^2)$ and the algebra of $n-by-n$ matrices, with n finite, according to whether the parameter representing the non-commutativity of the torus is irrational or rational. Thus, in the first case, $\mathcal{A}$ is isomorphic with the algebra of an infinitely extended Heisenberg ferromagnet in its infinite temperature state; and, in the second case, it is essentially classical.

(2) Ex.2 is manifestly a model of a system with an infinite number of degrees of freedom, and its algebra of observables is of type III.

In view of these observations, it is natural to ask whether chaos, as characterised by hyperbolicity, can occur in standard finite quantum systems. For example, does the hyperbolic property survive canonical quantisation in the case of the free motion of a particle on a compact manifold of constant negative curvature? The answer we shall provide to these questions is negative. In other words, we shall establish that quantum mechanics suppresses chaos. This does *not* mean that it destroys it, but simply that a weaker condition than strict hyperbolicity is needed to represent any viable definition of quantum chaos. We shall discuss this matter further in the Comment following Theorem 3.

We base our treatment of finite quantum systems on the standard model $\Sigma = (\mathcal{A}, \alpha, \omega)$, where $\mathcal{A}$ is the set, $\mathcal{L}(\mathcal{H})$, of bounded operators in a separable Hilbert space, $\mathcal{H}$, $\alpha(\mathbf{R})$ is implemented by a unitary representation of $\mathbf{R}$ in $\mathcal{H}$, generated by a Hamiltonian operator, H, with discrete spectrum, i.e.,

$$\alpha(t)A = \exp(iHt/\hbar)A\exp(-iHt/\hbar)\ \forall A \in \mathcal{A}, t \in \mathbf{R} \tag{3.1}$$

and ω is a normal, $\alpha-$invariant state on $\mathcal{A}$.

Note. The above conditions, including the discreteness of the spectrum of H, are fulfilled not only by finite systems of particles, with realistic interactions, in bounded regions of a Euclidean space or on a lattice [10, P.65], but also by the canonically quantised version of the classical hyperbolic system, given by the free dynamics of a particle on a compact manifold of constant negative curvature [11]. It was proved by [1] that the latter model is not hyperbolic, subject to the assumption that at least one of the automorphism groups σ is unitarily implementable in $\mathcal{H}$. The following theorem generalises this result to all finite systems, without any such assumption.

Theorem 3. *The above-specified model of a finite system is never hyperbolic.*

Comment. This does not rule out the possibility of a quantum chaos that is represented by a weakened form of the hyperbolicity condition of Def. 2.3. For example, it remains an open question whether the free quantum dynamics of a particle on a compact manifold of constant negative curvature fulfills the condition (2.9) in an appropriate classical limit. Since the dimensionless constant representing quantum effects in this model is $\gamma := \hbar/(mE)^{\frac{1}{2}}R$, where m and E are the mass and energy, respectively, of the particle, and R is the radius of curvature of the manifold, such a classical limit would imply that the dynamics was close to hyperbolicity when $\gamma << 1$. In this case, it could well be that the motion of the particle could still appear extremely unstable, and thus be deemed to be chaotic, over physically relevant time scales.

Proof of Theorem 3. We shall employ a *reductio ad absurdum* argument to prove that the hyperbolic property (2.9) is incompatible with the discreteness of the spectrum of H. Thus, we suppose that (2.9) is valid for one of the σ's, i.e. that

$$\alpha(t)\sigma(s)\alpha(-t) = \sigma(s.\exp(\lambda t)) \tag{3.2}$$

where λ is a real, non-zero constant, and the suffix j has been dropped. Hence, it follows from the $\alpha-$invariance of ω that

$$\int dsf(s)\langle\omega;\sigma(s)\alpha(-t)A\rangle = \int dsf(s)\langle\omega;\sigma(s.\exp(\lambda t))A\rangle \ \forall A\in\mathcal{A},\ f\in\mathbf{C}_0(\mathbf{R}),\ t\in\mathbf{R} \tag{3.3}$$

where $\mathbf{C}_0(\mathbf{R})$ denotes the continuous functions on $\mathbf{R}$ with compact support. By the dominated convergence theorem, we see that

$$R.H.S.\ of\ (3.3)\rightarrow c\omega(A),\ with\ c = \int dsf(s),\ as\ \lambda t\rightarrow -\infty \tag{3.4}$$

On the other hand,

$$L.H.S.\ of\ (3.3)\equiv\psi(\alpha(-t)A) \tag{3.5}$$

where

$$\psi = \int ds\sigma^{\star}(s)\omega, \tag{3.6}$$

and $\sigma^{\star}(s)$ is the dual of $\sigma(s)$. Thus ψ, like ω, is normal.

Now let $\{\phi_r|r\in\mathbf{N}\}$ be an orthogonal basis in $\mathcal{H}$, consisting of eigenvectors of H, and let $\{F_{mn}|m,n\in\mathbf{N}\}$ be the operators in this space defined by

$$F_{mn}\phi_r := \delta_{nr}\phi_m \tag{3.7}$$

These operators thus form a basis set for $\mathcal{A}$. Moreover, on applying (3.1) to the case where $A = F_{mn}$, it follows from (3.5) that

$$L.H.S.\ of\ (3.3) = \psi(F_{lm})\exp(-i\nu_{mn}t) \tag{3.8}$$

where ν_{mn} is the difference in the energy levels of ϕ_m and ϕ_n, in units of Planck's constant. Further, it follows from (3.1), (3.7) and the $\alpha-$invariance of ω that, if $\nu_{mn}\neq 0$, then $\omega(F_{mn})=0$. Consequently, the compatibility of (3.4) with (3.8) implies that

$$\psi(F_{mn})=0 \ when \ \nu_{mn}\neq 0$$

By (3.1) and the normality of ψ, this signifies that ψ is $\alpha-$invariant. Hence, by(3.6), together with the $\alpha-$invariance of ω and the hyperbolicity condition (3.2),

$$\int dsf(s)\langle\omega;\sigma(s)A\rangle=\int dsf(s)\langle\omega;\sigma(s)\alpha(t)A\rangle$$

$$\equiv\int dsf(s)\langle\omega;\alpha(t)\sigma(se^{\lambda t})A\rangle\equiv\int dsf(s)\langle\omega;\sigma(se^{\lambda t})A\rangle \ \forall A\in\mathcal{A}, \ f\in\mathbf{C}_0(\mathbf{R}) \ t\in\mathbf{R} \quad (3.9)$$

Now let $\{f_n\}$ be a positive sequence in $\mathbf{C}_0(\mathbf{R})$, whose integrals are all unity, and whose supports are all contained in a single compact and converge to some point u. Then, on replacing f by f_n in (3.9), and passing to the limit $n\rightarrow\infty$, we obtain the equation

$$\omega(\sigma(u)A)=\omega(\sigma(ue^{\lambda t})A) \ \forall A\in\mathcal{A}, \ t,u\in\mathbf{R}$$

Hence, replacing A by $\sigma(-u)A$ here, and defining

$$v:=u(e^{\lambda t}-1),$$

$$\omega(A)=\omega(\sigma(v)A) \ \forall A\in\mathcal{A} \ v\in\mathbf{R}, \quad (3.10)$$

which signifies that ω is $\sigma-$invariant. Consequently, since this state is also $\alpha-$invariant, it follows from [1, Th.3.3] that the spectrum of $\mathcal{H}$ covers the real line. However, this result, which stems from the supposition that Σ is hyperbolic, conflicts with the specification that the spectrum of H is discrete. We conclude therefore that Σ cannot be hyperbolic.

Appendix A

Proof of Lemma 2.2. In order to simplify the notation, we shall drop the suffixes t,j from the symbols ϕ,V,θ,λ, and define

$$\kappa=\exp(\lambda t) \quad (A.1)$$

for fixed t. Thus, by equations (2.1)-(2.3),

$$\phi^{\star}V(x)=\kappa V(\phi x); \quad (A.2)$$

and

$$x(s)=\theta(s)x \quad (A.3)$$

is the unique global solution of

$$x'(s)=V(x(s)), \ \forall s\in\mathbf{R}; \ x(0)=x \quad (A.4)$$

Hence, defining

$$x_{\phi}(s):=\phi\theta(\kappa^{-1}s)\phi^{-1}x, \quad (A.5)$$

it follows from (A.2-4) that

$$x'_{\phi}(s)=\kappa^{-1}\phi^{\star}(\phi^{-1}x)'[\kappa^{-1}s]=$$

$$\kappa^{-1}\phi^{\star}V((\phi^{-1}x)[\kappa^{-1}s]) = V(\phi(\phi^{-1}x)[\kappa^{-1}s]) = V(\phi\theta[\kappa^{-1}s]\phi^{-1}x)$$

i.e.

$$x'_{\phi}(s) = V(x_{\phi}(s)),$$

which signifies that $x_{\phi}(.)$ satisfies (A.4). Hence, by the uniqueness of the solution of that equation,

$$x_{\phi}(s) = x(s)\ \forall s{\in}\mathbf{R},$$

i.e., by (A.3) and (A.5),

$$\phi\theta(s)\phi^{-1}{\equiv}\theta(\kappa s),$$

which is the required result.

References

[1] G. G. Emch, H. Narnhofer, W. Thirring and G. L. Sewell: "Anosov Actions on Non-Commutative Algebras", J. Math. Phys. **35**, 5582 (1994)

[2] D. V. Anosov: Proc. Inst. Steklov **90**, 1 (1967)

[3] V. I. Arnold and A. Avez: "Ergodic Problems of Classical Mechanics", W. A. Benjamin, New York, 1969

[4] J. Von Neumann: "Mathematical Foundations of Quantum Mechanics", Princeton University Press, Princeton, 1955

[5] G. Lindblad: Pp. 348-360 of "Quantum Probability and Applications II", Ed. L. Accardi and W. Von Waldenfels, Lec. Notes in Mathematics 1136, Springer, 1985

[6] B. V. Chirikov, F. M. Israilev and D. L. Shepelyanski: Sov. Sci. Rev. **C2**, 209 (1981)

[7] G. Casati: Pp. 281-288 of "On Three Levels", Ed. M. Fannes, C. Maes and A. Verbeure, Nato ASI Series B 324, Plenum, New York, London, 1994

[8] F. Benatti, H. Narnhofer and G. L. Sewell: Lett. Math. Phys. **21**, 157 (1991)

[9] A. Connes: C. R. Acad. Sci A **290**, 599 (1980)

[10] D. Ruelle: "Statistical Mechanics", W. A. Benjamin, New York, 1969

[11] G. G. Emch: J. Math. Phys. **23**, 1785 (1982)

QUANTUM COHERENCES ON DIFFERENT OBSERVATION LEVELS

V.Bužek[1,2] , G.Adam[3] , and G.Drobný[1]

[1]Institute of Physics, SAS,
Dúbravská cesta 9, 842 28 Bratislava, Slovakia
[2]Department of Optics,
Comenius University, Mlynská dolina, 842 15 Bratislava, Slovakia
[3]Institute für Theoretische Physics,
Technische Universität Wien, A-1040 Vienna, Austria

Key words :
superposition states, entropy, observation level, uncertainty measure, dissipation

1 INTRODUCTION

Recently many authors have studied various nonlinear processes in which optical superpositions of coherent states can in principle be produced (for a detailed discussion see Ref.[1]). In particular, it has been shown by Yurke and Stoler [2] that in the presence of low dissipation a nonlinear Kerr-like medium may convert an initial coherent state (CS) $|\alpha\rangle$ into a quantum-mechanical superposition of two coherent states which are 180° out of phase with respect to each other. This superposition state is described by the state vector

$$|\alpha\rangle_{YS} = \frac{1}{\sqrt{2}}\left(|\alpha\rangle + e^{i\pi/2}| - \alpha\rangle\right). \tag{1.1a}$$

Another scheme for production of pure superpositions of coherent states has been proposed by Brune et al. [3]. These authors have shown that an atomic-phase detection quantum nondemolition scheme can serve for a production of superposition state of a single-mode cavity field. These superposition states are described by the following state vectors:

$$|\alpha\rangle_{even} = \frac{1}{\sqrt{2[1+\exp(-2|\alpha|^2)]}}\left(|\alpha\rangle + |-\alpha\rangle\right) \tag{1.1b}$$

and

$$|\alpha\rangle_{odd} = \frac{1}{\sqrt{2[1-\exp(-2|\alpha|^2)]}}\left(|\alpha\rangle - |-\alpha\rangle\right). \tag{1.1c}$$

The state $|\alpha\rangle_{even}$ is usually called the even coherent state, while the state $|\alpha\rangle_{odd}$ is called the odd coherent state. The even (odd) coherent state in the Fock basis is represented only by even (odd) Fock states. In other words the photon number distribution corresponding to the even (odd) coherent state exhibits significant oscillations which have their origin in quantum interference between components of the pure superposition state (1.1). Another consequence of this quantum interference is a reduction of quadrature fluctuations in the even coherent state and the Yurke-Stoler coherent state (these states exhibit the second-order squeezing as well as the fourth-order squeezing [4]). On the other hand the odd coherent state exhibit reduced fluctuations in the photon number, i.e. the odd coherent state is a sub-Poissonian state.

Nonclassical properties of the even and odd coherent states can be understood in a framework of the phase-space formalism [5,1]. Namely, coherent states represent the quantum analogue of classical points of phase space. So the even and the odd coherent states can be understood as superpositions of two "points" of the quantum-mechanical phase space. It has been shown that nonclassical effects associated with these states emerge as a direct consequence of quantum interference in phase space (see [5,1] and references therein). The superposition states (1.1) represent particular realizations of a more general superposition state

$$|\alpha\rangle_{\varphi} = \frac{1}{\sqrt{2[1+\cos\varphi\exp(-2|\alpha|^2)]}}\Big(|\alpha\rangle + e^{i\varphi}|-\alpha\rangle\Big). \tag{1.2}$$

We see that for $\varphi = 0; \pi/2$ and $\varphi = \pi$ we obtain from Eq.(1.2) expressions of the even, the Yurke-Stoler, and the odd coherent states, respectively. In general, the character of quantum interference depends on the relative phase φ (see [1,4]). In addition to pure superposition state (1.2) we can consider a statistical mixture of two coherent state described by the density operator:

$$\hat{\rho} = \frac{1}{2}|\alpha\rangle\langle\alpha| + \frac{1}{2}|-\alpha\rangle\langle-\alpha|, \tag{1.3}$$

which can be represented by two *non-interfering* "points" of the quantum-mechanical phase-space.

The question is whether by performing a particular quantum-mechanical measurement we are able to distinguish between the pure superpositions (1.1) of coherent states and corresponding mixture (1.3). Obviously, in an ideal measurement by means of which the density operator of the quantum-mechanical state is completely reconstructed this distinction is straightforward. The density operator can in principle be completely reconstructed either from the phase-space probability quasidistributions such as the Wigner function (see for instance the optical tomography method [6]) or from the complete set of moments of the annihilation and creation operators of the field mode under consideration. This kind of the measurement when all moments of the system operators are measured corresponds to the so-called "observation level $\mathcal{O}_0$" (see below). From practical point of view we are usually able to measure just a subset of the complete set of moments of the system operators. Depending on the subset of the measured moments we can define for each subset the corresponding (less comprehensive) "observation level $\mathcal{O}_k$" ($k > 0$). In this paper we address the problem whether it is possible to detect the presence of quantum coherence between component states on "observation levels $\mathcal{O}_k$" ($k > 0$). To quantify the degree of quantum coherence we will utilize entropic description related to different observation levels. In the paper we will consider also the role of dissipations.

2 MAXENT PRINCIPLE AND OBSERVATION LEVELS [7]

2.1 Uncertainty measure and entropy

The state of a quantum system can always be described by a statistical density operator $\hat{\rho}$. Depending on the system preparation, the density operator represents either a pure quantum state (complete system preparation) or a statistical mixture of pure states (incomplete preparation). The degree of deviation of a statistical mixture from the pure state can be best described by the *uncertainty measure* $\eta[\hat{\rho}]$ [7]

$$\eta[\hat{\rho}] = -k\mathrm{Tr}(\hat{\rho}\ln\hat{\rho}), \tag{2.1}$$

where k is the Boltzmann constant. If the system is prepared in a pure state it holds that $\eta[\hat{\rho}] = 0$ otherwise $\eta[\hat{\rho}] > 0$. In Eq.(2.1) the uncertainty measure has been determined through a given density operator $\hat{\rho}$. If instead of the density operator $\hat{\rho}$, expectation values G_ν of the set of operators $\hat{G}_\nu$ ($\nu = 1, \ldots, n$) are given, then the uncertainty measure associated with the measurement can be determined (see below). The set of linearly independent operators $\hat{G}_\nu$ will be referred to as an *observation level.* The operators of an observation level do not commutate necessarily. A large number of density operators can be found for any given set of expectation values $G_\nu = \langle\hat{G}_\nu\rangle$ of an observation level $\{\hat{G}\}$ such that

$$\mathrm{Tr}\ \hat{\rho}_{\{\hat{G}\}} = 1, \tag{2.2a}$$

$$\mathrm{Tr}(\hat{\rho}_{\{\hat{G}\}}\hat{G}_\nu) = G_\nu, \qquad \nu = 1, 2, ..., n. \tag{2.2b}$$

Each of these density operators $\hat{\rho}_{\{\hat{G}\}}$ can be associated with different uncertainty measures $\eta[\hat{\rho}_{\{\hat{G}\}}]$. If we wish to use only the expectation values G_ν of the chosen observation level for determining the density operator, we must select a particular density operator $\hat{\rho}_{\{\hat{G}\}} = \hat{\sigma}_{\{\hat{G}\}}$ in an unbiased manner. This density operator $\hat{\sigma}_{\{\hat{G}\}}$ must be that density operator which has the largest uncertainty measure, i.e.

$$\eta[\hat{\sigma}_{\{\hat{G}\}}] = maximum \tag{2.3}$$

with the constraints

$$\mathrm{Tr}\ \hat{\sigma}_{\{\hat{G}\}} = 1, \tag{2.4a}$$

$$\mathrm{Tr}(\hat{\sigma}_{\{\hat{G}\}}\hat{G}_\nu) = G_\nu. \tag{2.4b}$$

This is Jaynes' principle of the maximum uncertainty measure of an observation level [7]. It leads to the unbiased determination of the density operator $\hat{\sigma}_{\{\hat{G}\}}$, taking into account the existing expectation values G_ν of the observation level. As a consequence of Eq.(2.3) the following fundamental inequality holds

$$\eta[\hat{\sigma}_{\{\hat{G}\}}] = -k\mathrm{Tr}(\hat{\sigma}_{\{\hat{G}\}} \ln \hat{\sigma}_{\{\hat{G}\}}) \geq \eta[\hat{\rho}_{\{\hat{G}\}}] = -k\mathrm{Tr}(\hat{\rho}_{\{\hat{G}\}} \ln \hat{\rho}_{\{\hat{G}\}}) \tag{2.5}$$

for all possible $\hat{\rho}_{\{\hat{G}\}}$ which fulfill the Eqs.(2.2).

The variation determining the maximum of η under the conditions (2.4) leads to a generalized canonical density operator [7]

$$\hat{\sigma}_{\{\hat{G}\}} = \frac{1}{Z_{\{\hat{G}\}}} \exp\,(-\sum_\nu \lambda_\nu \hat{G}_\nu), \tag{2.6a}$$

$$Z_{\{\hat{G}\}}(\lambda_1, \ldots, \lambda_n) = \mathrm{Tr}[\exp(-\sum_\nu \lambda_\nu \hat{G}_\nu)], \tag{2.6b}$$

where λ_n are the Lagrange multipliers and $Z_{\{\hat{G}\}}(\lambda_1, \ldots \lambda_n)$ is the generalized partition function. By using the derivatives of the partition function we obtain the expectation values

$$G_\nu = \mathrm{Tr}(\hat{\sigma}_{\{\hat{G}\}} \hat{G}_\nu) = -\frac{\partial}{\partial \lambda_\nu} \ln Z_{\{\hat{G}\}}(\lambda_1, ..., \lambda_n)\,. \tag{2.7}$$

By using Eq.(2.7) the Lagrange multipliers can be expressed as functions of the expectation values

$$\lambda_\nu = \lambda_\nu(G_1, ..., G_n)\,. \tag{2.8}$$

The maximal uncertainty measure regarding an observation level $\{\hat{G}\}$ will be referred to as the entropy $S_{\{\hat{G}\}}$

$$S_{\{\hat{G}\}} \equiv \eta_{\max} = -k\mathrm{Tr}(\hat{\sigma}_{\{\hat{G}\}} \ln \hat{\sigma}_{\{\hat{G}\}}). \tag{2.9}$$

This means that for different observation levels different entropies are defined. By inserting $\sigma_{\{\hat{G}\}}$ [cf. Eq.(2.6)] into Eq.(2.9), we obtain the following expression for the entropy

$$S_{\{\hat{G}\}} = k \ln Z_{\{\hat{G}\}} + k \sum_\nu \lambda_\nu G_\nu. \tag{2.10}$$

If the dynamical evolution of the system is governed by the evolution operator $U(t, t_0)$, then time–dependent expectation values of the observation level read

$$G_\nu(t) = \mathrm{Tr}[\hat{G}_\nu \hat{U}(t, t_0) \hat{\rho}(t_0)]. \tag{2.11}$$

By using these time–dependent expectation values as constrains for maximizing the uncertainty measure $\eta[\hat{\rho}_{\{\hat{G}\}}(t)]$, we obtain the generalized canonical density operator

$$\hat{\sigma}_{\{\hat{G}\}}(t) = \frac{\exp\left(-\sum_\nu \lambda_\nu(t) \hat{G}_\nu\right)}{\mathrm{Tr}\left[\exp\left(-\sum_\nu \lambda_\nu(t) \hat{G}_\nu\right)\right]} \tag{2.12}$$

and the time–dependent entropy of the corresponding observation level

$$S_{\{\hat{G}\}}(t) = -k\mathrm{Tr}[\hat{\sigma}_{\{\hat{G}\}}(t) \ln \hat{\sigma}_{\{\hat{G}\}}(t)] = k \ln Z_{\{\hat{G}\}}(t) + k \sum_\nu \lambda_\nu(t) G_\nu(t). \tag{2.13}$$

This generalized canonical density operator does not satisfy the von Neumann equation but it satisfies an integro–differential equation [8]. The time–dependent entropy is valid for any system being arbitrarily far from equilibrium. In the case of an isolated system the entropy can increase or decrease during the time evolution.

2.2 Observation levels for the single–mode radiation field

In what follows we introduce different observation levels for a single-mode radiation field described by the annihilation and creation operators $\hat{a}$ and $\hat{a}^{\dagger}$, respectively. These operators obey the Weyl-Heisenberg commutation relation $[\hat{a}, \hat{a}^{\dagger}] = 1$.

Observation level $\mathcal{O}_0$

We denote the Fock state of the single mode radiation field by $|n\rangle$. The set of operators $|n\rangle\langle m|$ (for all n and m) is referred to as the observation level $\mathcal{O}_0$. At given time the expectation values of this observation level are the matrix elements of the density operator

$$\langle m| \hat{\rho} |n\rangle = \mathrm{Tr}[\hat{\rho} |n\rangle \langle m|], \ \forall n, m, \tag{2.14}$$

and therefore the generalized canonical density operator $\hat{\sigma}_0$ is identical to the statistical density operator $\hat{\rho}$. In this case the entropy S_0 is determined by $\hat{\rho}$, that is:

$$S_0 = -k\mathrm{Tr}(\hat{\sigma}_0 \ln \hat{\sigma}_0) = -k\mathrm{Tr}[\hat{\rho} \ln \hat{\rho}]\,. \tag{2.15}$$

The uncertainty measure corresponding to $\hat{\rho}$ is identical to the entropy of the complete observation level (i.e. an observation level containing all independent operators). The entropy given by Eq.(2.15) is called the von Neumann entropy [9]. Here we note that as a consequence of the relation

$$|n\rangle \langle m| = \lim_{\varepsilon \to 0} \sum_{h=0}^{\infty} \frac{(-\varepsilon)^h}{h!\sqrt{n!\, m!}} (\hat{a}^{\dagger})^{h+n} \hat{a}^{h+m}, \tag{2.16}$$

the observation level $\mathcal{O}_0$ can also be associated with the set of operators $\{(\hat{a}^{\dagger})^k \hat{a}^l;\ \forall k, l\}$.

Observation level $\mathcal{O}_1$

We associate the observation level $\mathcal{O}_1$ with projectors $\hat{P}_n = |n\rangle \langle n|\,,\ \forall n$. The corresponding expectation values are the probabilities P_n

$$P_n = \langle n| \hat{\rho} |n\rangle = \mathrm{Tr}[\hat{\rho} |n\rangle \langle n|], \ \forall n. \tag{2.17}$$

We note that the observation level $\mathcal{O}_1$ can also be given by the set of operators $\{(a^{\dagger})^k a^k;\ \forall k\}$. For this observation level the entropy reads

$$S_1 = -k \sum_{n=0}^{\infty} P_n \ln P_n. \tag{2.18}$$

This entropy is the so-called Shannon entropy [10].

Observation level $\mathcal{O}_2$

The observation level $\mathcal{O}_2$ will be defined by the set of operators $\{\hat{a}^{\dagger}, \hat{a}, \hat{a}^{\dagger}\hat{a}, (\hat{a}^{\dagger})^2, \hat{a}^2\}$. The corresponding expectation values are $\langle \hat{a}^{\dagger}\rangle, \langle \hat{a}\rangle, \langle \hat{a}^{\dagger}\hat{a}\rangle, \langle (\hat{a}^{\dagger})^2\rangle$ and $\langle \hat{a}^2\rangle$. In order to simplify the calculations, we introduce new operators

$$\hat{c}^{\dagger} = \hat{a}^{\dagger} - \langle \hat{a}^{\dagger}\rangle, \qquad \hat{c} = \hat{a} - \langle \hat{a}\rangle, \qquad \hat{A} = \tfrac{1}{2}(\hat{c}^{\dagger}\hat{c} + \hat{c}\hat{c}^{\dagger}) = \hat{c}^{\dagger}\hat{c} + \tfrac{1}{2} \tag{2.19}$$

and the corresponding expectation values

$$\langle \hat{c}^{\dagger}\rangle = \langle \hat{c}\rangle = 0, \tag{2.20a}$$

$$\langle (\hat{c}^{\dagger})^2\rangle = \langle (\hat{a}^{\dagger})^2\rangle - \langle \hat{a}^{\dagger}\rangle^2, \qquad \langle \hat{c}^2\rangle = \langle (\hat{a})^2\rangle - \langle \hat{a}\rangle^2, \tag{2.20b}$$

$$\langle\hat{A}\rangle = \langle\hat{a}^\dagger\hat{a}\rangle - \langle\hat{a}^\dagger\rangle\langle\hat{a}\rangle + \tfrac{1}{2}. \tag{2.20c}$$

For this observation level the entropy can be written in the following form [11]

$$S_2 = k[(\kappa+1)\ln(\kappa+1) - \kappa\ln\kappa], \tag{2.21}$$

where κ is defined by

$$\kappa = \sqrt{\langle\hat{A}\rangle^2 - \langle(\hat{c}^\dagger)^2\rangle\langle\hat{c}^2\rangle} - \tfrac{1}{2}. \tag{2.22}$$

This observation level represents a reduction of the observation level $\mathcal{O}_0$. The difference between the two observation levels is that the moments of the observation level $\mathcal{O}_0$ are independent expectation values, whereas those of the observation level $\mathcal{O}_2$

$$\langle(\hat{a}^\dagger)^k a^l\rangle = f_{k,l}[\ \langle\hat{a}^\dagger\rangle, \langle\hat{a}\rangle, \langle\hat{a}^\dagger\hat{a}\rangle, \langle(\hat{a}^\dagger)^2\rangle, \langle\hat{a}^2\rangle\ ] \tag{2.23}$$

are functions of lower moments (through which the observation level O_2 has been defined) [12].

Observation level $\mathcal{O}_3$

As the observation level $\mathcal{O}_3$ we define the set of operators $\{\hat{a}^\dagger, \hat{a}, \hat{a}^\dagger\hat{a}\}$. With the help of mean values of the operators $\hat{c}^\dagger$ and $\hat{c}$ (cf. Eq.(2.19)):

$$\langle\hat{c}^\dagger\rangle = \langle\hat{c}\rangle = 0, \tag{2.24a}$$

$$\langle\hat{c}^\dagger\hat{c}\rangle = \langle\hat{a}^\dagger\hat{a}\rangle - \langle\hat{a}^\dagger\rangle\langle\hat{a}\rangle. \tag{2.24b}$$

we obtain the entropy corresponding to the observation level $\mathcal{O}_3$ as

$$S_3 = k\left[\left(\langle\hat{c}^\dagger\hat{c}\rangle + 1\right)\ln\left(\langle\hat{c}^\dagger\hat{c}\rangle + 1\right) - \langle\hat{c}^\dagger\hat{c}\rangle\ln\langle\hat{c}^\dagger\hat{c}\rangle\right]. \tag{2.25}$$

Observation level $\mathcal{O}_4$

The observation level $\mathcal{O}_4$ corresponds to a measurement of the average photon number $\langle\hat{a}^\dagger\hat{a}\rangle = \langle\hat{n}\rangle \equiv \bar{n}$ The generalized canonical density operator of this observation level is the well-known density operator of the harmonic oscillator in the case of the thermal equilibrium. The entropy for this observation level reads

$$S_4 = k[(\bar{n}+1)\ln(\bar{n}+1) - \bar{n}\ln\bar{n}]. \tag{2.26}$$

The sets of operators corresponding to different observation levels

$$\mathcal{O}_0 = \{(\hat{a}^\dagger)^k a^l, \forall k, l\}, \tag{2.27a}$$

$$\mathcal{O}_1 = \{(\hat{a}^\dagger)^k a^k, \forall k\}, \tag{2.27b}$$

$$\mathcal{O}_2 = \{\hat{a}^\dagger, \hat{a}, \hat{a}^\dagger\hat{a}, (\hat{a}^\dagger)^2, \hat{a}^2\}, \tag{2.27c}$$

$$\mathcal{O}_3 = \{\hat{a}^\dagger, \hat{a}, \hat{a}^\dagger\hat{a}\}, \tag{2.27d}$$

$$\mathcal{O}_4 = \{\hat{a}^\dagger\hat{a}\}, \tag{2.27e}$$

are related as

$$\mathcal{O}_0 \supset \mathcal{O}_1 \supset \mathcal{O}_4, \tag{2.28a}$$

$$\mathcal{O}_0 \supset \mathcal{O}_2 \supset \mathcal{O}_3 \supset \mathcal{O}_4, \tag{2.28b}$$

which leads to relations between entropies belonging to different observation levels

$$S_0 \leq S_1 \leq S_4, \tag{2.29a}$$

$$S_0 \leq S_2 \leq S_3 \leq S_4. \tag{2.29b}$$

3 DECAY OF SUPERPOSITION STATES INTO ZERO-TEMPERATURE HEAT BATH

Usually quantum-mechanical systems are not isolated but they are influenced by their environment [13]. Therefore we will consider a single-mode light field interacting with a heat bath. For simplicity we will consider a temperature of the heat bath equal to zero. As shown in Ref.[14] this simple model reveals the most important aspects of the decay of superposition states of light. In this section we will utilize the concept of observation levels for a description of a deterioration of quantum coherences under the influence of the environment.

To be specific, we shall assume that the density operator $\hat{\rho}$ for the field mode obeys a zero temperature master equation in the Born-Markov approximation. This equation in the interaction picture can be written as:

$$\frac{\partial \hat{\rho}}{\partial t} = \frac{\gamma}{2}\left(2\hat{a}\hat{\rho}\hat{a}^\dagger - \hat{a}^\dagger\hat{a}\hat{\rho} - \hat{\rho}\hat{a}^\dagger\hat{a}\right), \tag{3.1}$$

where γ is the decay constant. Following Barnett and Knight [15] we can write the solution of the master equation (3.1) in the form

$$\hat{\rho}(t) = \exp(\hat{L}t)\exp\left[\frac{\hat{J}}{\gamma}\left(1 - e^{-\gamma t}\right)\right]\hat{\rho}(0); \tag{3.2}$$

where two operators $\hat{L}$ and $\hat{J}$ are defined by their action on the density operator, i.e.

$$\hat{J}\hat{\rho} = \gamma\hat{a}\hat{\rho}\hat{a}^\dagger; \tag{3.3a}$$

and

$$\hat{L}\hat{\rho} = -\frac{\gamma}{2}\left(\hat{a}^\dagger\hat{a}\hat{\rho} + \hat{\rho}\hat{a}^\dagger\hat{a}\right). \tag{3.3b}$$

Using the formal solution (3.2) for the $\hat{\rho}$ matrix, one can find the time-dependent expression for the density matrix of the superposition of two coherent states

$$|\Psi\rangle = A^{1/2}[|\alpha_1\rangle + |\alpha_2\rangle], \qquad \alpha_1 = \alpha \quad \alpha_2 = -\alpha; \tag{3.4}$$

at $t > 0$ (for details see Ref.[1]):

$$\hat{\rho}(t) = A\sum_{i,j=1}^{2}\langle\alpha_i|\alpha_j\rangle^{1-\mu}|\mu^{1/2}\alpha_j\rangle\langle\mu^{1/2}\alpha_i|. \tag{3.5}$$

where $\mu = \exp(-\gamma t)$.

Using the explicit expression for the density operator of the decaying superposition state we can evaluate the entropies corresponding to different observation levels.

In Fig.1 we plot the time evolution of the von Neumann entropy S_0 which corresponds to the observation level $\mathcal{O}_0$. We consider the initial state of the single-mode field to be the even CS (1.1b), the odd CS (1.1c) and the statistical mixture state (1.3). Here we note that the von Neumann entropy of the field mode which is initially prepared in the CS is equal to zero for any $t \geq 0$. From the Fig.1 it is seen that with the decay of the energy of the system the entropy of the initial statistical mixture state is monotonically decreasing. This is due to the fact that the amplitudes of the component states $|\alpha_i(t)\rangle$ exponentially decay. In the stationary limit

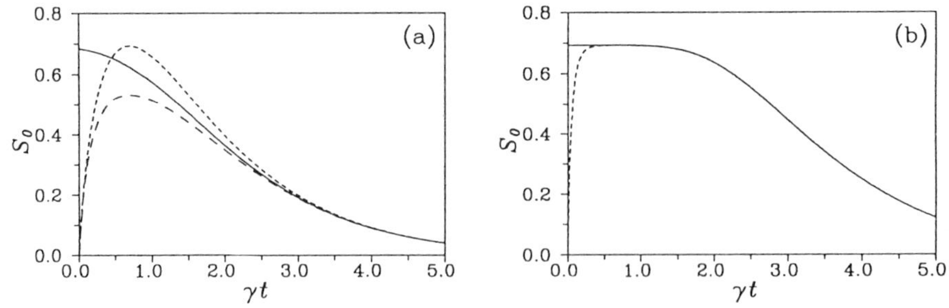

Figure 1: The time evolution of the von Neumann entropy S_0 of the field mode which decays into the zero-temperature heat bath. The initial state of the field mode is chosen to be the even CS (long-dashed line), the odd CS (short-dashed line) and the statistical mixture state (solid line). In Fig.1a the initial intensity of the field mode is chosen to unity $[\bar{n}(t=0)=1]$ in all three cases. The von Neumann entropy of the corresponding coherent state is equal to zero for any $t \geq 0$. In Fig.1b the initial intensity of the field mode is $\bar{n}(t=0)=4$. We see that with the increase of the intensity of the initial field time evolution of the von Neumann entropy of the even and odd CS is almost identical. Rapid deterioration of quantum coherences is clearly seen. The higher the intensity of the initial even (odd) CS, more rapidly quantum coherence between two component states is deteriorated and the pure state is transformed into the corresponding statistical mixture.

for $t \to \infty$, when $\alpha_i(t) \to 0$, the field mode is in the vacuum state $|0\rangle$ which is the pure state and is characterized by the zero value of the von Neumann entropy.

The initial values of the von Neumann entropy of the even and odd CS are equal to zero, because these two states are pure states, i.e. they are uniquely defined at the observation level $\mathcal{O}_0$. On the other hand the initial von Neumann entropy of the statistical mixture (1.3) is non-zero which reflects a lack of information about the state of the system. The maximum value of the entropy of the mixture of two coherent states (1.3) equals to $k \ln 2$ in the limit $\alpha \to \infty$, i.e. this mixture corresponds to a "two-state" statistical system.

The time evolution of the von Neumann entropy of the pure superposition states is dramatically different compared with corresponding mixtures. First of all, the rapid destruction of the quantum coherences due to the influence of the reservoir noise is accompanied by a rapid increase of the von Neumann entropy. As seen from Fig.1a the increase of the von Neumann entropy depends on the character of the quantum interference between two component states. In particular, in the case of the initial even CS the von Neumann entropy S_0 is *always* smaller than the entropy of the statistical mixture state. On the contrary, the interference between two component states in the case of the odd CS leads to an increase of the von Neumann entropy above the value of the corresponding mixture. After some time, when the quantum coherences are completely deteriorated the von Neumann entropy of the initial even CS, the odd CS and the statistical mixture are practically identical (in our particular case with $\bar{n}=1$ this time is approximately equal to $\gamma t \approx 3.0$).

Here we remind that the effective decay rate of the quantum coherences of the superposition of two coherent states $|\alpha\rangle \pm |-\alpha\rangle$ is proportional to $\gamma|\alpha|^2$, i.e. it depends on the "separation" between component states in the phase space [14,4]. On the contrary energy of the state is decaying with the rate γ, i.e. the initial pure state is transformed into a statistical mixture state on the time scale which is much shorter than the time scale on which the mean energy (the mean photon number) is significantly changed. This is nicely seen if we compare Fig.1a and Fig.1b which

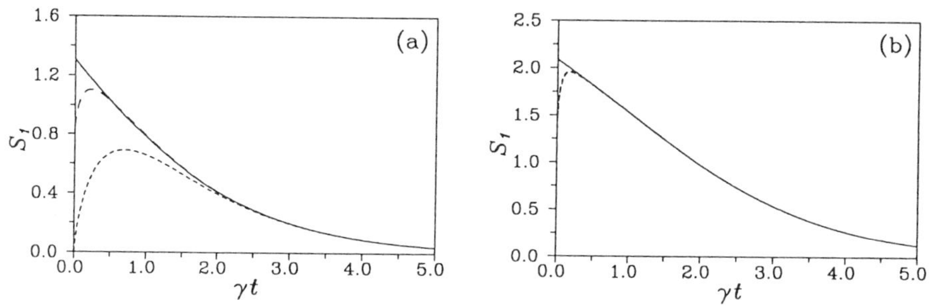

Figure 2: The time evolution of the Shannon entropy S_1 of the field mode which decays into the zero-temperature heat bath. The initial state of the field mode is chosen to be the even CS (long-dashed line), the odd CS (short-dashed line) and the statistical mixture state (solid line). In Fig.2a the initial intensity of the field mode is chosen to unity $[\bar{n}(t=0)=1]$ in all three cases. The Shannon entropy of the coherent state with the Poissonian photon number distribution is *equal* to the entropy of the mixture state under consideration which has the same photon number distribution. In Fig.2b we take $\bar{n}(t=0)=4$. We see that the Shannon entropy of the pure superposition of two coherent states is under the influence of a zero-temperature heat bath rapidly "transformed" into the Shannon entropy of the corresponding mixture. This behaviour again reflects rapid deterioration of quantum coherences between two component states.

describe the time evolution of the von Neumann entropy of the same states for two different values of the initial mean photon numbers: $\bar{n}=1$ (Fig.1a) and $\bar{n}=4$ (Fig.1b). It is interesting to note that for larger values of $\bar{n}$ the quantum coherences are deteriorated so rapidly that the von Neumann entropy of the odd CS does not exceed the entropy of the statistical mixture state. The von Neumann entropy of both the even and odd CS effectively behave identically and for very large values of $\bar{n}$ they almost instantly "converge" to the entropy of the statistical mixture.

In this paper we show that nonclassical effects at a given information level can be observed providing the entropy of the state is different from the entropy of the corresponding statistical mixture. In particular, in the case of the von Neumann entropy, i.e. on the observation level $\mathcal{O}_0$, when the "complete" information about the quantum system is reconstructed, the nonclassical effects can be observed on a very short time scale (i.e. $\gamma|\alpha|^2 t \ll 1$). As soon as the decay rate of the cavity is large enough, it is *impossible* to distinguish between the statistical mixture and the superposition state. Obviously, if the decay rate γ is equal to zero, then in principle the "complete" measurement of the state can be performed and the Wigner function (i.e. the density operator) can be reconstructed, so quantum coherences are detected completely. This means that all nonclassical effects can in principle be observed. On the other hand if some of the moments of the system operators are not considered (measured) then even though the system itself is in a pure state the available information is not sufficient to characterize this state uniquely. Therefore, the corresponding entropy is larger than zero. At this point one can ask the question whether this incomplete information is still "rich" enough to reflect the presence of quantum coherences (i.e., whether pure states can be distinct from corresponding mixtures). To answer this question we analyze in detail the time evolution of entropies S_k $(k>0)$.

In Fig.2 we plot the time evolution of the Shannon entropy S_1 (observation level $\mathcal{O}_1$) of the field mode which decays into the zero-temperature heat bath. The observation level $\mathcal{O}_1$ is sufficient for a complete description of Fock states (the

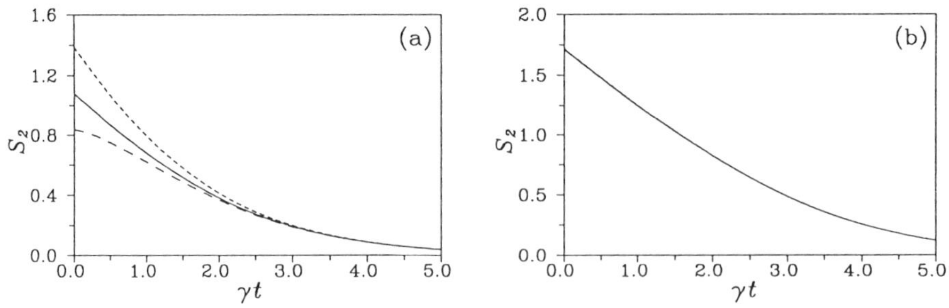

Figure 3: The time evolution of the S_2 entropy corresponding to the observation level $\mathcal{O}_2$ of the field mode which decays into the zero-temperature heat bath. The initial state of the field mode is chosen to be the even CS (long-dashed line), the odd CS (short-dashed line) and the statistical mixture state (solid line). In Fig.3a the initial intensity of the field mode is chosen to unity ($\bar{n}(t=0)=1$) in all three cases. We see that the S_2 entropy of the even CS is for any $t \geq 0$ smaller than the entropy of the corresponding mixture, i.e. in this case quantum coherence between two component states leads to a reduction of the entropy. On the contrary, for the odd CS the S_2 entropy is always larger than the entropy of the corresponding mixture state. In Fig.3b we take $\bar{n}(t=0)=4$. We see that with the increase of the intensity of the initial states under consideration the S_2 becomes insensitive with respect to the presence of a quantum coherence between two component coherent states.

Shannon entropy of any Fock state is equal to zero). As seen from Fig.2a ($\bar{n}=1$) at time $t=0$ the Shannon entropy of the odd CS is close to zero. This is due to the fact, that the odd CS with the mean photon number $\bar{n}=1$ is approximately equal to the Fock state $|1\rangle$. On the other hand the even CS with the intensity $\bar{n}=1$ can be approximated with a higher number of Fock states (at least by the states $|0\rangle$ and $|2\rangle$). Consequently, the Shannon entropy of the even CS at $t=0$ is larger than zero. The Shannon entropy of the corresponding statistical mixture which has Poissonian photon number distribution is larger than those of the pure superposition states under consideration. As soon as we take into account dissipation we find that the Shannon entropy of the statistical mixture is decaying monotonically while the Shannon entropies of the even and odd CS rapidly increase during the first instances of the time evolution (this behaviour is related to the deterioration of the quantum coherences). With the increase of the intensity of the field mode (see Fig.2b) the initial values of the Shannon entropies of the odd and even CS become almost identical. This reflect the fact, that the photon number distributions of these states are essentially identical (except that they are mutually shifted) so the same number of Fock state contribute to these pure states, and therefore the Shannon entropies of these two states are essentially equal and for large intensities $\bar{n}$ they differ by $k\ln 2$ from the Shannon entropy of the corresponding mixture. Under the influence of the zero-temperature heat bath the pure states under consideration are rapidly transformed into a corresponding statistical mixture which is reflected by the time evolution of the Shannon entropy.

Now we turn our attention to the observation level $\mathcal{O}_2$. In this case the corresponding entropy S_2 at time $t=0$ is sensitive with respect to the presence of quantum interference only for small intensities of superposition states (see Fig.3). It is interesting to note that at $t=0$ the entropy of the odd CS is larger than the entropy of the corresponding mixture state. This means that by performing a particular quantum-mechanical measurement we can determine a pure odd CS in a

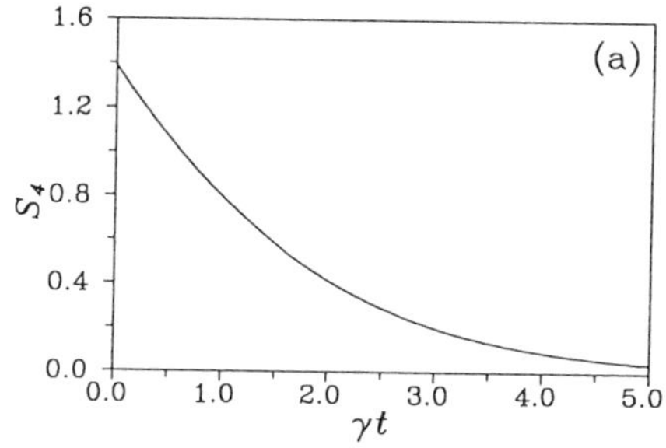

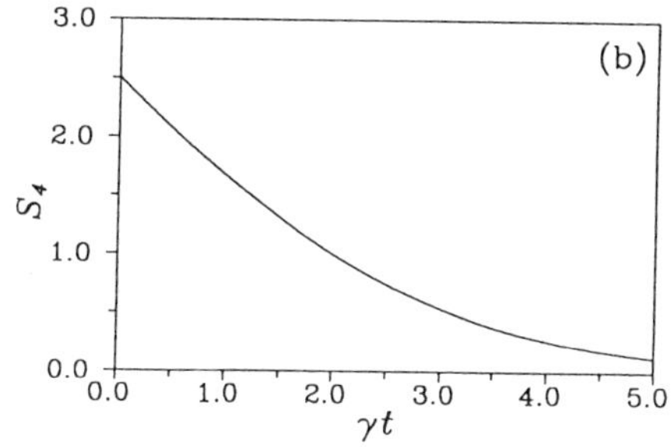

Figure 4: The time evolution of the S_4 (the thermal) entropy of the initial even CS, the odd CS and the corresponding statistical mixture state. In Fig.4a the initial mean photon number is equal to unity [i.e. $\bar{n}(t=0)=1$]. We see, that on the observation level O_4, quantum coherences between two component coherent states cannot be detected. In Fig. 4b we take $\bar{n}(t=0)=4$. We see that with the increase of the initial mean photon number the character of the time evolution of the S_4 is not changed except its maximum value is increased.

more biased way than the corresponding statistical mixture. From Fig.3a it is seen that under the influence of a zero-temperature heat bath the S_2 entropies of the three states under consideration become closer. This behaviour is more pronounced with the increase of the intensity of the field mode (compare Figs.3a and 3b).

For the given initial states the information available on the observation levels $\mathcal{O}_3$ and $\mathcal{O}_4$ is equal (this is due to the fact that $\langle \hat{a} \rangle = 0$ for any $t \geq 0$). Simultaneously we have to note that the observation level $\mathcal{O}_4$ is absolutely insensitive with respect to the presence of quantum coherences. At this observation level the entropy S_4 [see Eq.(2.26)] depends only on the mean photon number which decays as $\bar{n}(t) = \bar{n}\exp(-\gamma t)$, so the corresponding entropy is monotonically decaying during the dissipation process.

CONCLUSIONS

In the present paper we have applied the concept of observation levels and corresponding entropies for an analysis of possible measurements of quantum coherences which are established due to quantum interference between coherent components of superposition states. We have shown that entropies S_k can clearly indicate a difference between pure superposition states and corresponding statistical mixtures providing the observation level is sufficiently comprehensive and the decay rate of the single mode to the zero-temperature heat bath is small enough (such that $\gamma|\alpha|^2 t \ll 1$).

Acknowledgments
Two of us (V.B. and G.D.) would like to thank Myung Shik Kim for fruitful discussion. One of us (G.A.) would like to thank Professor J. Seke for useful discussions and comments. This work was supported in part by the Grant Agency of the Slovak Academy of Sciences Grant No. GA SAV 2/999143/93 and in part by the Jubiläumsfonds der Österreichischen Nationalbank, project no. 4814.

References

[1] V.Bužek and P.L.Knight, "Quantum interference, superposition states of light and nonclassical effects," in *Progress in optics*, ed. E. Wolf (North Holland, Amsterdam), to appear.

[2] B.Yurke and D.Stoler, *Phys. Rev. Lett.* **57**, 13 (1986).

[3] M.Brune, S.Haroche, J.M.Raimond, L.Davidovich,,N.Zagury, *Phys.Rev.A* **45**, 5193 (1992).

[4] W.P.Schleich, M.Pernigo, and Fam Le Kien, *Phys. Rev. A* **44**, 2172 (1991); K.Vogel and W.Schleich, in *Fundamental Systems in Quantum Optics*, edited by J.Dalibard, J.M.Raimond and J.Zinn-Justin (Elsevier, Amsterdam, 1990), p.713; V.Bužek, A.Vidiella-Barranco, and P.L.Knight, *Phys.Rev. A* **45**, 6570 (1992).

[5] W.Schleich and J.A.Wheeler, in *Proceeding of the First International Conference of Physics in Phase Space*, W.W.Zachary, ed. (Springer, New York, 1987), p. 200; W.Schleich, D.F.Walls, and J.A.Wheeler, *Phys. Rev. A* **38**, 1177 (1988).

[6] K.Vogel,H.Risken, *Phys. Rev. A* **40**, 2847 (1989); U.Leonhardt,H.Paul, *Phys. Rev. A* **48**, 4598 (1993); D.T.Smithey, M.Beck, M.G.Raymer, and A.Faridani, *Phys. Rev. Lett.* **70**, 1244 (1993).

[7] E. Fick and G. Sauermann, "The Quantum Statistics of Dynamic Processes", (Springer Verlag, Berlin, 1990); E.T. Jaynes, *Phys. Rev.*, **106**, 620; **108**, 171 (1957); Am. J. Phys., **31**, 66 (1963).

[8] B. Robertson, *Phys. Rev.*, **144**, 151 (1966); **166**, 175 (1967); **166**, 206 (1968); J. Seke, *Phys. Rev. A*, **21**, 2156 (1980); G. Adam, J. Seke *Phys. Rev. A*,**23**, 3118 (1981).

[9] J.von Neumann, *Mathematische Grundlagen der Quantentheorie* (Springer Verlag, Berlin, 1931).

[10] C.E.Shannon, *Bell. Sys. Techn. J.* **27**, 379 (1948); ibid. **27**, 623 (1948).

[11] G.S. Agarwal, *Phys. Rev. A*, **3**, 828 (1971).

[12] G. Adam, *Phys. Lett. A*, **171**, 66 (1992).

[13] W.H.Zurek, *Phys. Rev. D* **26**, 1862 (1982); see also in *Quantum Optics, Experimental Gravitation, and Measurement Theory*, eds. P.Meystre and M.O.Scully (Plenum Press, New York) p.87; and *Physics Today* (October), 36 (1991).

[14] D.F.Walls and G.J.Milburn, *Phys. Rev. A* **31**, 2403 (1985).

[15] S.M.Barnett amd P.L.Knight, *Phys. Rev. A* **33**, 2444 (1986).

DEGREE OF NONCLASSICAL BEHAVIOUR

N. Lütkenhaus and Stephen M. Barnett

Department of Physics and Applied Physics
University of Strathclyde
107 Rottenrow
Glasgow G4 0NG, UK.
Telephone No. (UK) 41 552 4400
Fax No. (UK) 41 552 2891
E-Mail norbert@physics.strathclyde.ac.uk

In classical physics the description of a system by a phase-space probability distribution is a useful tool. Glauber and Cahill [1,2] introduced a similar description for quantum mechanics generalizing earlier work by Wigner, Husimi and others. This description is useful for all systems which can be described in terms of bosonic or harmonic oscillator creation and destruction operators. There are two principal differences between classical and quantum phase-space. The first is, that the quantum analogue of the classical probability distribution is no longer positive definite. Hence we will refer to it as quasiprobability distribution, or quasidistribution for short. Secondly, the appropriate quasidistributions to use for calculating the expectation values of an operator depends on the ordering of the creation and annihilation operators. There are quasidistributions for all orderings continuously characterized by the ordering parameter s.

We are interested in the class of states associated with non-classical effects. It seems to be natural to seek a criterion for that by looking at their quasidistributions as these contain the complete information about the quantum state. In this paper we concentrate on the aspect of semi-definiteness of the quasidistributions. We expect, that the interval of the parameter s, in which the quasidistributions are positive semi-definite, is a measure of the degree of non-classical behaviour. We show, that there is a simple criterion for this interval to be minimized and hence for the state to be strongly non-classical. This criterion applied to the class of pure states leads to the statement, that all pure states not represented by Gaussian wavefunctions are strongly non-classical. Gaussian wave functions represent all coherent and ideal squeezed states.

Quantum Communications and Measurement
Edited by V. P. Belavkin *et al.*, Plenum Press, New York, 1995

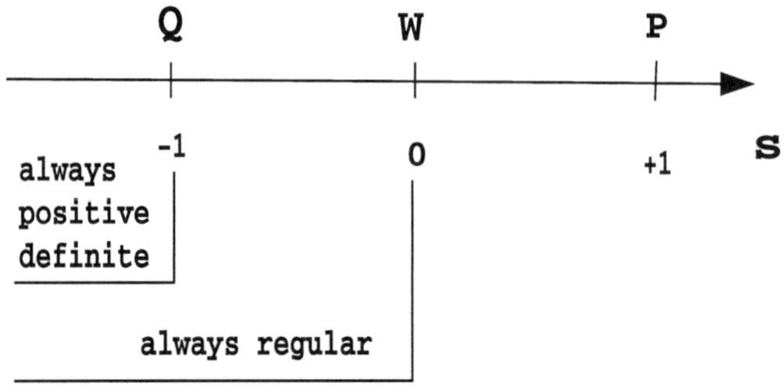

Figure 1: Main properties of the quasidistributions.

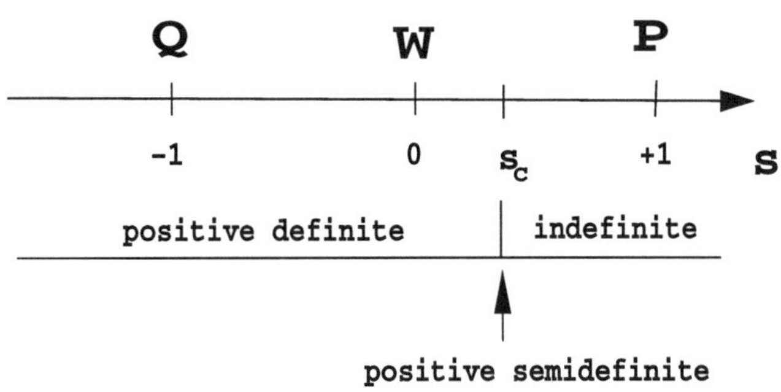

Figure 2: Regions of definiteness.

S-QUASIDISTRIBUTIONS

We restrict the results in this paper to systems described by one mode, that is one pair of creation and annihilation operators. The extension to multi-mode systems is relatively straight forward. The to unity normalised quasidistributions $W(\alpha, s)$ are defined by [1–3]

$$C(\eta, s) = \operatorname{Tr}\left\{\rho \exp\left(\eta \hat{a}^{\dagger} - \eta^{*} \hat{a}\right)\right\} \exp \frac{s}{2}|\eta|^{2} , \tag{1}$$

$$W(\alpha, s) = \frac{1}{\pi^{2}} \int_{\mathbb{R}^{2}} \exp\left(\alpha\eta^{*} - \alpha^{*}\eta\right) C_{s}(\eta)\, d^{2}\eta . \tag{2}$$

These quasidistributions are a generalization of the known P (Glauber-Sudarshan), W (Wigner) and Q quasidistributions. In figure 1 are illustrated the main properties of these s-dependent quasidistributions. They refer to normally $(s = 1)$, symmetrically $(s = 0)$, and antinormally ordered $(s = -1)$ operators respectively in the sense that

$$\left\langle a^{\dagger^{n}} a^{m} \right\rangle_{s} = \int_{\mathbb{R}^{2}} W(\alpha, s)\, \alpha^{*n} \alpha^{m}\, d^{2}\alpha , \tag{3}$$

where $'\langle\,\rangle_{s}'$ refers to the kind of ordering [1, 3] . From the definition (1, 2) follows immediately that for two different values of s, say $\tilde{s}, \overline{s}$, the characteristic functions of a state belonging to $\tilde{s}, \overline{s}$ are related by

$$C(\eta, \tilde{s}) = C(\eta, \overline{s}) \exp\left(-\frac{(\overline{s} - \tilde{s})}{2}|\eta|^{2}\right) . \tag{4}$$

If $\overline{s} > \tilde{s}$ it follows by performing a Fourier transformation (FT) in the sense of (2), that

$$W(\alpha, \tilde{s}) = W(\alpha, \overline{s}) \star \left(\frac{2}{\pi(\overline{s} - \tilde{s})} \exp\left\{-\frac{2|\alpha|^{2}}{\overline{s} - \tilde{s}}\right\}\right) , \tag{5}$$

where '$\star$' denotes the convolution defined by

$$f(\alpha) \star g(\alpha) = \int_{\mathbb{R}^{2}} f(\beta) g(\beta - \alpha)\, d^{2}\beta .$$

The convolution means, that $W(\alpha, \tilde{s})|_{\alpha=\alpha_{0}}$ is calculated from the function $W(\alpha, \overline{s})$ by averaging with a Gaussian centered at $\alpha = \alpha_{0}$. The relation (5) can be understood easily by observing that $W(\alpha, s)$ satisfies a sourceless diffusion equation with s taking the role of reverse time. [2] The diffusion process smoothes $W(\alpha, s)$ in direction of decreasing s. This allows us to prove the following theorem:

If $W(\alpha, \tilde{s}) \geq 0$ for all α and $W(\alpha_{0}, \tilde{s}) = 0$ for at least one α_{0}, then there is no $\overline{s}$ with $\overline{s} > \tilde{s}$ and $W(\alpha, \overline{s}) \geq 0$ for all α. In words this expresses, that to the 'right' of a non-negative quasidistribution with at least one zero there is no other non-negative quasidistribution. The proof is done by contradiction: Assume we have $W(\alpha, \tilde{s}) \geq 0$, $W(\alpha_{0}, \tilde{s}) = 0$ for at least one $\alpha_{0} \in \mathcal{C}$, and $W(\alpha, \overline{s}) \geq 0$, $\overline{s} > \tilde{s}$. Both quasidistributions obey (5). The Gaussian is strictly positive and by that $W(\alpha, \tilde{s})$ is strictly positive, which is a contradiction to the assumed zero $W(\alpha_{0}, \tilde{s}) = 0$ □

A second consequence of (5) is, that the parameter space of s is divided into two intervals belonging to well-behaved and not well-behaved quasidistributions respectively. The division between these parts is illustrated in figure 2. It follows from the fact that $W(\alpha, \overline{s}) \geq 0$ implies $W(\alpha, \tilde{s}) > 0$ for all $\tilde{s} < \overline{s}$, as seen in the proof above. We call the value s_{c} separating these intervals the *critical parameter* and propose to regard this value s_c as a degree of nonclassical behaviour. We have $s_c \geq -1$ since the Q-function is always non-negative. We propose to associate larger values of s_c with more nearly classical states. This idea is supported by the example of sub-Poissonian light. The property of sub-poissonian statistics turns up when the P-function, $W(\alpha, 1)$, is negative in substantial regions. (For more precision see [4].) By use of (5) with $\bar{s} = 1$ and $\tilde{s} = s_c$ we see, that $\sqrt{(1 - s_c)/2}$ is in this case the minimal width of the averaging Gaussian to smooth out the negative parts. The more substantial these parts are, the smaller is s_c.

The criterion

We specialize the first result to $\tilde{s} = -1$. Then $W(\alpha_0, -1) = Q(\alpha_0) = 0$ for at least one α_0 implies, that $W(\alpha, s)$ has negative regions for all $s > -1$. However, $Q(\alpha)$ is non-negative. This is a consequence of the identity $Q(\alpha) = \langle \alpha | \rho | \alpha \rangle / \pi$. *It follows that states represented by Q-functions with zeros have the minimal possible value of* $s_c = -1$. They are the most non-classical states in the sense stated above.

Pure states

We now specialize to the critical parameters for pure states. For a pure state $|\Psi\rangle$ the Q-function is represented by

$$Q_\Psi(\alpha) = \frac{1}{\pi} \left| \langle \Psi | \alpha \rangle \right|^2 \tag{6}$$

which is clearly non-negative. It has zeros iff $\langle \Psi | \alpha_0 \rangle = 0$ for at least one α_0 .

In preparation of a more general result we will investigate the case that the state $|\Psi\rangle$ may be represented as a finite superposition of number states $|n\rangle$,

$$|\Psi\rangle = \sum_{n=0}^{N} c_n \, |n\rangle \ . \tag{7}$$

We recall that the coherent states $|\alpha\rangle$ are represented by

$$|\alpha\rangle = \sum_{n=0}^{\infty} \frac{\alpha^n}{\sqrt{n!}} \exp\left(-\frac{|\alpha|^2}{2}\right) \, |n\rangle \ , \tag{8}$$

so that $\langle \Psi | \alpha \rangle$ simplifies to

$$\langle \Psi | \alpha \rangle = \sum_{n=0}^{N} c^*{}_n \frac{\alpha^n}{\sqrt{n!}} \exp\left(-\frac{|\alpha|^2}{2}\right). \tag{9}$$

Since $\exp\left(-\frac{|\alpha|^2}{2}\right) > 0$, $Q_\Psi(\alpha)$ has zeros iff the analytic function

$$f_N(\alpha) := \sum_{n=0}^{N} c_n^* \frac{\alpha^n}{\sqrt{n!}} \ \left(= e^{\frac{|\alpha|^2}{2}} \langle \Psi \, |\alpha\rangle \ \right) \tag{10}$$

has zeros. The fundamental theorem of algebra [5] states, that each finite polynomial in the complex variable α has at least one zero.[1] Thus $Q_\Psi(\alpha)$ has at least one zero and therefore the Q-function is the s-quasidistribution with the greatest value of the s-parameter for all finite superpositions of number states.

[1] More precisely it has N zeros, but these may be degenerate.

This result can be generalized to all pure states except coherent or squeezed states following a proof by Hudson [6]. We start at the same point as in the preparation, but now we use a infinite superposition of number states,

$$|\phi\rangle = \sum_{n=0}^{\infty} c_n \; |n\rangle \; . \tag{11}$$

That all pure states can be represented in this form follows from the completeness of the number states. Following the same arguments as before, we find as condition for zeros of the Q-function, that the function

$$f_\infty(\alpha) := \sum_{n=0}^{\infty} c_n^* \frac{\alpha^n}{\sqrt{n!}} \Big(= e^{\frac{|\alpha|^2}{2}} \langle\phi\, |\alpha\rangle \; \Big) \tag{12}$$

has zeros. The function $f_\infty(\alpha)$ is analytic in the whole complex plane. We are looking for states $|\phi\rangle$ represented by the c_n such $f_\infty(\alpha)$ has no zeros. The sum is limited by

$$\begin{aligned} \left|\sum_{n=0}^{\infty} c_n^* \frac{\alpha^n}{\sqrt{n!}}\right|^2 &\leq \sum_{n=0}^{\infty} |c_n^*|^2 \sum_{n=0}^{\infty} \frac{|\alpha|^{2n}}{n!} \\ &= 1 \; \exp\left(|\alpha|^2\right) . \end{aligned} \tag{13}$$

This characterizes the sum to be of order less than or equal two [7]. It follows from Hadamard's theorem [7], that any function, which is analytic on the whole complex plane, has no zeros and is restricted in growth to be of order two or less, is of the form

$$\exp\Big(A\alpha^2 + B\alpha + C\Big) \tag{14}$$

where A, B and C are complex constants. It is this point, which needs care in extending the result to multi-mode systems. The Hadamard theorem is proven in its general form for the single-mode case only. However, Soto and Claverie [8] have proven the restricted version needed to extend this work to multi-mode systems.

Thus we have for all values of α

$$\begin{aligned} \exp\Big(A\alpha^2 + B\alpha + C\Big) &= \sum_{n=0}^{\infty} c_n^* \frac{\alpha^n}{\sqrt{n!}} \\ &= \exp\frac{|\alpha|^2}{2} \langle\phi|\; \alpha\rangle \; . \end{aligned} \tag{15}$$

Next we show that only coherent squeezed states can satisfy this equality for all α. In the first step we observe that the function $f_\infty(\alpha)$ determines the state ϕ up to a general phase. For the proof we express the Q-function as

$$Q(\alpha) = \frac{1}{\pi} \left|\langle\phi\, |\alpha\rangle\right|^2 = \frac{1}{\pi} \exp\Big(|\alpha|^2\Big) \left|f(\alpha)\right|^2 \; . \tag{16}$$

The Q-function determines a state up to a general phase, and hence so does $f(\alpha)$.

In the second step we show that all function $f(\alpha) = \exp\left(A\alpha^2 + B\alpha + C\right)$ are generated by coherent squeezed states. The first step then implies that only coherent squeezed states generate $f(\alpha)$ of this form.

A coherent squeezed state ϕ can be represented by (see for example [9])

$$|\phi\rangle = D(\beta) S(r e^{\mathrm{i}\varphi})\, |0\rangle \qquad \beta \in \mathcal{C}, \quad r, \varphi \in \mathbb{R}\, , \tag{17}$$

with the displacement and the squeezing operator

$$D(\beta) = \exp\left(\beta a^{\dagger} - \beta^{*} a\right) \qquad S\left(re^{i\varphi}\right) = \exp\left\{\frac{1}{2}\left(re^{i\varphi}a^{\dagger^{2}} - re^{-i\varphi}a^{2}\right)\right\} . \quad (18)$$

The action of the squeezing operator on the vacuum can be simplified by using the disentangled form of this operator,

$$\begin{aligned} S\left(re^{i\varphi}\right)|0\rangle &= \exp\left\{\tanh(r)e^{i\varphi}\frac{a^{\dagger^{2}}}{2}\right\}\exp\left\{-\ln\left(\cosh(r)\right)\frac{1}{2}\left(a^{\dagger}a + aa^{\dagger}\right)\right\} \\ &\quad * \exp\left\{-\tanh(r)e^{-i\varphi}\frac{a^{2}}{2}\right\}|0\rangle \\ &= \frac{1}{\sqrt{\cosh r}}\exp\left\{\tanh(r)e^{i\varphi}\frac{a^{\dagger^{2}}}{2}\right\}|0\rangle . \end{aligned} \quad (19)$$

We use this simplification with the definition (17) in (12) which results in

$$\begin{aligned} e^{\frac{|\alpha|^{2}}{2}}\langle\phi|\alpha\rangle &= e^{\frac{|\alpha|^{2}}{2}}\langle 0|\frac{1}{\sqrt{\cosh r}}\exp\left\{\tanh(r)e^{-i\varphi}\frac{a^{2}}{2}\right\}D^{\dagger}(\beta)|\alpha\rangle \\ &= \frac{1}{\sqrt{\cosh r}}e^{\frac{|\alpha|^{2}}{2}}\exp\left\{\tanh(r)e^{-i\varphi}\frac{(\alpha-\beta)^{2}}{2}\right\} \\ &\quad * \exp\left\{\frac{1}{2}\left(\alpha\beta^{*} - \alpha^{*}\beta\right)\right\}\langle 0|\alpha-\beta\rangle \\ &= \exp\left\{\alpha^{2}\left(\frac{1}{2}\tanh(r)e^{-i\varphi}\right)\right. \\ &\quad + \alpha\left(-\beta\tanh(r)e^{-i\varphi} + \beta^{*}\right) \\ &\quad \left. - \ln\left(\sqrt{\cosh r}\right) + \frac{\beta^{2}}{2}\tanh(r)e^{-i\varphi} - \frac{|\beta|^{2}}{2}\right\} . \end{aligned} \quad (20)$$

We compare this result with the expression $\exp\{A\alpha^{2} + B\alpha + C\}$ and find the terms for A, B, C . By a variation of r, ϕ, β we find that all values of A, B can be chosen. The value of C is fixed by normalization and a given choice of phase. This can be understood by the normalization of the Q-function which can be expressed in terms of $f(\alpha)$ and a Gaussian.

As a result the critical value s_c of all pure states except the coherent and ideal squeezed states has the minimal possible value -1.This leads us to conclude, that the Gaussian states are the most nearly classical states [10].

NL is funded via the German Academic Exchange Service (DAAD) by the HSPII/AUFE programme. SMB thanks the Royal Society of Edinburgh and the Scottish Office Education Department for the award of a Research Fellowship.

References

[1] K. E. Cahill and R. J. Glauber. Ordered expansions in boson amplitude operators. *Phys. Rev.*, 177:1857–1881, 1969.

[2] K. E. Cahill and R. J. Glauber. Density operators and quasiprobability distributions. *Phys. Rev.*, 177:1882, 1969.

[3] J. Perina. *Quantum statistics of linear and nonlinear optical phenomena.* Kluwer Acad. Publ., Dordrecht, Boston, London, 1991.

[4] B. J. Dalton. Effect of internal atomic relaxation on quantum fields. *Phys. Scripta*, T12:43–50, 1986.

[5] H. A. Priestley. *Introduction to Complex Analysis.* Clarendon Press, Oxford, 1990.

[6] R. L. Hudson. When is the Wigner quasi-probability density non negative? *Rep. Math. Phys.*, 6(2):249–253, 1974.

[7] S. Saks and A. Zygmund. *Analytic functions.* Elsevier Publ. Comp., Amsterdam, London, New York, 1971.

[8] F. Soto and P. Claverie. When is the Wigner function of multi-dimensional systems nonnegative? *J. Math. Phys.*, 24(1):97–100, January 1983.

[9] R. Loudon and P. L. Knight. Squeezed light. *J. Mod. Opt.*, 34:709, 1987.

[10] I. Bialynicki-Birula. Beyond the coherent states: Gaussons of the electromagnetic field. In L.M. Narducci F. Haake and D. F. Walls, editors, *Coherence, Cooperation and Fluctuations.* Cambridge Univ. Press, Cambridge, 1986.

GENERALIZED DISPERSION RELATIONS FOR DIELECTRICS IN SQUEEZED VACUA

S.S. Hassan[1], R.K. Bullough[2], R. Saunders[3] and H.A. Batarfi[2]

[1]Ain Shams University, Mathematics Department, Cairo, Egypt.
[2]Department of Mathematics, UMIST, P.O. Box 88, Manchester M60 1QD, UK.
[3]Department of Mathematics and Physics, Manchester Metropolitan University, Manchester M1 5GD, UK.

1 INTRODUCTION

Substantially *squeezed* radiation fields are now available in a number of laboratories, and this offers an opportunity for studying the response of various physical systems to such nonclassical states of the electromagnetic field[1]. In this paper our main interest is to examine the effect of a broad-band squeezed vacuum field on the refractive index of a dielectric made up of 2-level atoms. In principle we consider two aspects of the problem: one is the effect of the squeezed vacuum on a field transmitted through the dielectric, and the refractive index, which we call $m(\omega)$ in this paper, determines the wave number of the transmitting field with respect to its wave number $k_0 = \omega c^{-1}$ in free space; the other is the wave number developing in the modes of the squeezed vacuum field in the presence of the dielectric, namely the effect of the dielectric on the squeezed vacuum. For simplicity these two calculations are performed separately as first preliminary investigations of these two problems and pressure on space means that we can do little more than state the *existence* of the calculations on the second problem in this paper. In due course we hope to combine both aspects in a connected theory. The geometry of the dielectric is important and we choose this to be a slab of finite width $L \gg k_0^{-1}$ which then

Quantum Communications and Measurement
Edited by V. P. Belavkin *et al.*, Plenum Press, New York, 1995

constitutes a Fabry-Perot cavity.

2 QUANTUM THEORY LINEAR AND NONLINEAR

The Hamiltonian for an extended system of N_a 2-level atoms coupled to the many-mode quantised electromagnetic radiation field in dipole approximation can be taken in the form[2,3]

$$H = \sum_{\mathbf{k},\lambda} \hbar\omega_k a^\dagger_{\mathbf{k},\lambda}(t) a_{\mathbf{k},\lambda}(t) + \frac{1}{2}\hbar\omega_0 \int_V \sigma_z(\mathbf{x},t)\, d\mathbf{x}$$

$$- \int_V \mathbf{p}(x,t)\cdot \mathbf{e}(x,t) d\mathbf{x} \tag{1}$$

where

$$\sigma_z(\mathbf{x},t) = \sum_{i=1}^{Na} \sigma^i_z(t)\delta(\mathbf{x}-\mathbf{x}_i) \tag{2}$$

$$\mathbf{p}(\mathbf{x},t) = \mathbf{p}\sigma_x(\mathbf{x},t) = \mathbf{p}\sum_{i=1}^{Na} \sigma^i_x(t)\delta(\mathbf{x}-\mathbf{x}_i) \ . \tag{3}$$

The $a^\dagger_{\mathbf{k},\lambda}(t), a_{\mathbf{k},\lambda}(t)$ satisfy the equal time boson commutation relations $\left[a_{\mathbf{k},\lambda}, a^\dagger_{\mathbf{k}',\lambda'}\right] = \delta_{\mathbf{k}\mathbf{k}'}\delta_{\lambda\lambda'}$ and Heisenberg representation is adopted. The σ^i_z and σ^i_x combined with the third operator σ^i_y, are Pauli spin operators satisfying at equal times the usual $su(2)$ Lie algebra for each atom i at its site $\mathbf{x}_i$ so that the collective operator densities satisfy

$$[\sigma_x(\mathbf{x},t), \sigma_y(\mathbf{x}',t)] = i\sigma_z(\mathbf{x},t)\delta(\mathbf{x}-\mathbf{x}'), \tag{4}$$

etc.

The field operator $\mathbf{e}(\mathbf{x},t)$ is transverse (div $\mathbf{e}(\mathbf{x},t) = 0$) and is given by

$$\mathbf{e}(\mathbf{x},t) = i\sum_{\mathbf{k},\lambda}\sqrt{\frac{2\pi\hbar\omega_k}{V_o}}\hat{\epsilon}_{\mathbf{k},\lambda}\left(a_{\mathbf{k},\lambda}(t)e^{i\mathbf{k}\cdot\mathbf{x}} - a^\dagger_{\mathbf{k},\lambda}(t)e^{-i\mathbf{k}\cdot\mathbf{x}}\right) \ , \tag{5}$$

where $\hat{\epsilon}_{\mathbf{k},\lambda}(\lambda = 1,2)$ are unit polarisation vectors mutually orthogonal to themselves and to $\mathbf{k}$. The volume V_o is the quantisation volume, so $V_o \to \infty$, and it contains V the macroscopic region occupied by the atoms.

The initial state of the field is first of all taken to be a single mode coherent state field of radial frequency ω together with a broad-band squeezed vacuum field of carrier frequency ω_p. We have analysed such *correlated* broad-band squeezed vacuum fields as they may be produced by an optical parametric oscillator in Refs. 4,5. Without going into any further detail here, Heisenberg's equations of motion for the atomic density operators derived from H, after taking their expectation values with certain *decorrelations* and this followed by an ensemble average over atomic sites $\mathbf{x}_i$, yield a set of c-number Bloch-types equations which in a frame rotating at frequency ω[2,3] (see also the references in Refs. 2,3) are

$$\dot{r}_+(\mathbf{x},t) + \left\{\frac{1}{2}\gamma(1+2N) + i(\omega-\omega_0)\right\} r_+(\mathbf{x},t)$$

$$+\gamma M(\mathbf{x})e^{2i(\omega_p-\omega)t} r_-(\mathbf{x},t)$$

$$= 2i\left\{E^+(\mathbf{x},t) - \Omega e^{-ik_0 z}\right\} r_3(\mathbf{x},t),\ \dot{r}_-(\mathbf{x},t) = (\dot{r}_+(\mathbf{x},t))^* \ ,$$

$$\dot{r}_3(\mathbf{x},t) + \gamma + \gamma(1+2N) r_3(\mathbf{x},t) = -i\left(E^+(\mathbf{x},t) - \Omega e^{-ik_0 z}\right) r_-(\mathbf{x},t) + c.c. \ . \qquad (6)$$

The vector $\mathbf{x}$ is any point inside V, and the quantities (densities) $r_\pm(\mathbf{x},t), r_3(\mathbf{x},t)$ are usual Bloch type densities. The ensemble average $\langle..\rangle_{\mathrm{Avr.}}$ introduces the mean number density $n \equiv \langle \sum_{i=1}^{N_a} \delta(\mathbf{x}-\mathbf{x}_i)\rangle_{\mathrm{Avr.}}$ of atoms inside V which is independent of $\mathbf{x}$ inside V ($n = 0$ outside V). The Bloch vector densities are scaled to single atoms by $r_{\pm,3}(\mathbf{x},t) = n^{-1} \ll |\sigma_{\pm,z}(\mathbf{x},t)| \gg_{\mathrm{Avr.}}$ and $\sigma_\pm$ derive as usual from $\sigma_\pm = \frac{1}{2}(\sigma_x \pm i\sigma_y)$. The c-number γ is the A-coefficient, and Ω is the amplitude of the c-number coherent state incident field of wave number $k_0 = \omega c^{-1}$: Ω is conveniently expressed as a Rabi frequency. The scalar field $E^+(\mathbf{x},t)$ is the interatomic field which has been expressed in terms of the expectation values of the sources $\sigma_\pm(\mathbf{x},t)$ and then ensemble averaged so that it is given by[2,3,6–10],

$$E^+(\mathbf{x},t) = np^2\hbar^{-1}\int_{V-v} r_+(\mathbf{x}',t)\hat{\mathbf{u}}\hat{\mathbf{u}} : \mathbf{F}(\mathbf{x},\mathbf{x}';\omega) d\mathbf{x}' \qquad (7)$$

where $\mathbf{p} = p\hat{\mathbf{u}}$ and p is the dipole matrix element for the 2-level atoms: the unit vector $\hat{\mathbf{u}}$ is taken in a fixed direction for simplicity but this should more generally (but without significant change to the present theory) be averaged isotropically over all directions as a part of the ensemble average. The field propagator $\mathbf{F}(\mathbf{x},\mathbf{x}';\omega)$ is given by $\mathbf{F}(\mathbf{x},\mathbf{x}';\omega) = (\nabla\nabla + k_0^2 U)e^{ik_0 r}r^{-1}$ and is a second rank tensor: U is the unit tensor. The length $r \equiv |\mathbf{x}-\mathbf{x}'|$ and $\mathbf{x} \neq \mathbf{x}'$. The ensemble average introduces a many-body problem of considerable complexity (c.f. e.g. Refs. 8-10). However, for present purposes we believe effects of this are small except that simply to *define* the integral in eqn. (7) we extract the small sphere v about the field point $\mathbf{x}$ and this defines the integral as a conditionally convergent integral as $v \to 0$. Reference to the Refs. 8-10 shows that this sphere is actually the source of the leading, and potentially substantial, term due to inter-atomic interaction and in many circumstances other many-body corrections are then small and can be neglected. We neglect them here, and we shall then find that v simply provides the Lorentz cavity field in this dielectric theory: it is linear in the density n and itself becomes negligible if n is small enough, as it is e.g. in vapours.

The parameters $N, M(\mathbf{x})$ are characteristic of the broad-band squeezed vacuum. The number N is proportional to the average number $\sum_{\lambda=1,2}\langle a^\dagger_{\mathbf{k},\lambda}a_{\mathbf{k},\lambda}\rangle$ of photons present in the squeezed vacuum in the modes with wave number k at resonance with the atoms: $k = \omega_0 c^{-1}$ if vacuum radiative shifts are ignored. For each mode $\mathbf{k},\lambda$ of the squeezed vacuum field $M_{\mathbf{k},\lambda} = \langle a_{\mathbf{k},\lambda}a_{\mathbf{k},\lambda}\rangle$ is non-zero and $|M_{\mathbf{k},\lambda}|^2 \leq N_{\mathbf{k}\lambda}(N_{\mathbf{k},\lambda}+1)$ for each of the modes $\mathbf{k},\lambda$. The parameter $M(\mathbf{x})$ then derives linearly from the $M_{\mathbf{k},\lambda}$ [4,5]and the actual $\mathbf{x}$-dependence is determined by the way that the squeezed vacuum is imposed on the dielectric slab. We consider two possible cases in this paper.

3 FIRST CASE OF THE SQUEEZED VACUUM FIELD: EFFECT ON THE DRIVEN DIELECTRIC

We suppose that the squeezed vacuum field has axial symmetry about the x-axis. This axis is chosen to lie in the plane of the plane parallel sided slab and so is orthogonal to the axis of the slab, the z-axis. We consider the pair of modes of the vacuum field along $\pm x$ each with the frequency of the carrier wave of the squeezed vacuum ω_p. We take these two modes as *typifying* the squeezed vacuum field with axial symmetry about the x-axis so that $M(\mathbf{x}) = 2M\cos(2\omega_p x c^{-1})$. In a previous report we considered in principle the two cases $M(\mathbf{x}) = 2M\cos(2k_0 z)$ with axial symmetry about z and $M(\mathbf{x}) = 2M\cos(2k_0 x)$ with axial symmetry about x. The choice $\omega = \omega_p$ for symmetry about z was motivated in part by a need to phase-match the incident light and an incident squeezed vacuum field. In practice if we choose $\omega = \omega_p$ we remove the $e^{2i(\omega_p - \omega)t}$ term in the equation for $\dot{r}_+(\mathbf{x}, t)$ in eqns. (6) and this makes that system solvable within the terms to be explained below. Thus finally we suppose $M(\mathbf{x}) = 2M\cos(2k_p x)$ where $k_p \equiv \omega_p c^{-1}$ and we set $\omega = \omega_p$ so that $k_p = k_0$.

If the squeezed vacuum is generated by an optical parametric oscillator there is a problem concerning correlation between pairs of modes in the squeezed vacuum and we are obliged to consider this in the work of §4. However this problem of correlation between modes need not be considered within the approximation in which we shall work in this §3.

If $M(\mathbf{x}) = M\cos(2\omega_p x c^{-1})$ and $\omega_p = \omega$ the theory is still difficult to follow through in all detail. If the incident light (along z) is a pencil of very small cross section with radius $< \frac{1}{2}c\omega_p^{-1}$, $M(\mathbf{x})$ is in effect constant across this light field. Of course a serious diffraction problem for the incident pencil now arises. We will not consider this diffraction problem here but for a first simple treatment finally assume that $M(\mathbf{x}) = M =$ constant so that the whole system, slab and field, becomes translationally invariant in directions orthogonal to z. We hope that this assumption contains enough gut physics so that for a realistic incident pencil of light the behaviours we predict for $M(\mathbf{x}) =$ constant may still be realised qualitatively.

With the assumption $M(\mathbf{x}) = M$ we can now solve the eqns. (6) by assuming the ansatz eqn. (8) below. This is written in terms of the two plane waves travelling along $\pm z$ with wave numbers $\pm m k_0$ where $m \neq 1$, and depends on $\omega \equiv \omega_p$. The quantities $r_\pm(\mathbf{x}, t)$ are envelope functions modulating carrier waves with time developments $e^{\pm i\omega t}$ (eqns. (6) are in the frame rotating at $\omega = \omega_p$). In the steady state for which $\dot{r}_{\pm,3} = 0$, we can set $r_\pm(\mathbf{x}, t) = r_\pm(\mathbf{x}, \omega)$ independent of t. Then the ansatz eqn. (8) is

$$\begin{aligned} r_+(\mathbf{x}, \omega) &= f(z)e^{-imk_0 z} + b(z)e^{imk_0 z} \\ &= r_-^*(\mathbf{x}, \omega) \quad . \end{aligned} \tag{8}$$

In this ansatz we can interpret $m \neq 1$ as the complex refractive index of the medium even though, because of the non-local field $E^+(\mathbf{x}, t)$, eqn. (7), eqns. (6) are nonlinear (and non-local). The functions $f(z)$, $b(z)$ in eqn. (8) are slowly varying functions in z (slow on the scale k_0^{-1}). We shall see that the ansatz (8) solves the nonlinear problem up to neglect of certain standing waves inside V.

Up to first order in the z-derivatives $f'(z), b'(z)$ of $f(z)$ and $b(z)$ we next find that $E^+(\mathbf{x}, t)$ becomes, in the steady state, $E^+(\mathbf{x}, \omega)$, where[3],

$$\begin{aligned} 2iE^+(\mathbf{x}, \omega) r_3(z, \omega) = \frac{i8\pi n p^2 \hbar^{-1}}{3(m^2 - 1)} \Bigg\{ &\left[\left(\frac{m^2 + 2}{3} \right) f - \frac{2imf'}{k_0(m^2 - 1)} \right] e^{-imk_0 z} \\ + &\left[\left(\frac{m^2 + 2}{3} \right) b + \frac{2imb'}{k_0(m^2 - 1)} \right] e^{imk_0 z} \Bigg\} r_3(z, \omega) \\ &+ \text{terms in } e^{\pm ik_0 z} \quad . \end{aligned} \tag{9}$$

The frequency $\omega = \omega_p$ as explained and $k_0 = k_p$. Eqn. (9) is to be substituted into the first equation of eqns. (6). We then appeal to the 'optical extinction theorem', due originally to Ewald and Oseen (see Refs. 8-11) but now in this present nonlinear context[11,12]. The extinction theorem states, in essence, that the scattering at wave number k_0 from dipole sources covering the external surfaces of V exactly cancels the harmonic field of wave number k_0 travelling along z inside V, i.e. it cancels the incident field and replaces this by dipole waves $e^{\pm imk_0z}$ ($m \neq 1$) travelling along z. In the present context these dipole waves $e^{\pm imk_0z}$ form the ansatz eqn. (8) for $r_+(\mathbf{x}, \omega)$. When $m \neq 1$ terms in $e^{\pm imk_0z}$ and those in $e^{\pm ik_0z}$ must be equated separately and from the coefficients of $e^{\pm imk_0z}$ we find in particular that

$$\eta(m)f(z) = -\gamma M b^*(z)e^{-i(m^*-m)k_0z}$$

$$\eta(m)b(z) = -\gamma M f^*(z)e^{+i(m^*-m)k_0z} \tag{10}$$

where

$$\eta(m) = \frac{1}{2}\gamma(1+2N) + i(\omega - \omega_0) - i\alpha\frac{1}{2}\gamma\frac{m^2+2}{3(m^2-1)}R_3 \ . \tag{11}$$

The number $\alpha = 8\pi np^2\hbar^{-1}\gamma^{-1}$, and the parameter R_3 is chosen self-consistently in such a way that the parameter $m(\omega)$ introduces the normal action of the slab as a Fabry-Perot cavity.

From eqns. (10) we find [3] the nonlinear dispersion relation

$$\left|\frac{1}{2}(1+2N) + i\Delta - \frac{1}{2}i\alpha\frac{m^2+2}{3(m^2-1)}R_3\right|^2 = |M|^2 \tag{12}$$

in which Δ is the scaled detuning: $\Delta = (\omega - \omega_0)/\gamma$. We have numerical results for the self-consistent nonlinear theory eqns. (10)–(12) which in principle are providing results for 'optical bistability' in a squeezed vacuum. Comparable work for the ordinary vacuum is described in Refs. 12,13. Here we linearise the problem and consider only the usual linear refractive index $m(\omega)$ but now in the squeezed vacuum.

Linearisation is achieved for a (vanishingly) weak external harmonic field imposed along z. In this case the inversion density $r_3(z,\omega)$ becomes independent of z so that $r_3 \approx -(1+2N)^{-1} \equiv R_3$. Thus eqn. (12) reduces to the generalized linear dispersion relation

$$\left[\frac{1}{2}(1+2N) + i\Delta + \frac{1}{2}i\alpha\frac{m^2+2}{3(m^2-1)}\frac{1}{(1+2N)}\right] = |M|e^{i\phi} \ . \tag{13}$$

The phase ϕ can be traced back to the phase of the squeezed vacuum modes taken in relation to the phase of the incident driving field. Thus it is a physically accessible *control parameter* and can be changed by changing the phase of the squeezed vacuum field modes.

The behaviours of the dispersion curves, $Re(m(\omega))$, as functions of the scaled detuning Δ, as well as the absorption curves $Im(m(\omega))$ as functions of Δ are shown in the Figs. 1 and 2 respectively for different values of ϕ. With $N = 0.5$, and $|M| = \sqrt{N(N+1)}$ the absorption curves of Fig. 2 narrow substantially at $\phi = \pi$ compared with $\phi = 0$ when the value of $\alpha = 10^{-3}$ (so that $n \sim 10^{11}$ atoms cm^{-3}). When $\phi = \pm\frac{1}{2}\pi$ there is on the contrary broadening with pronounced asymmetries. By increasing α to $\alpha = 5$ the Lorentz field introduces significant frequency shifts with which there are associated pronounced asymmetries.

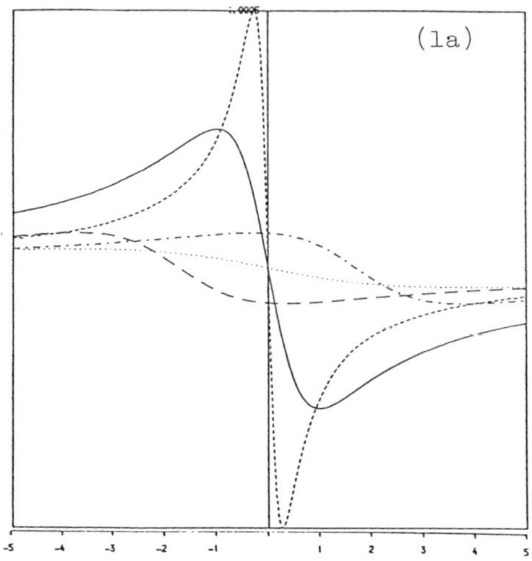

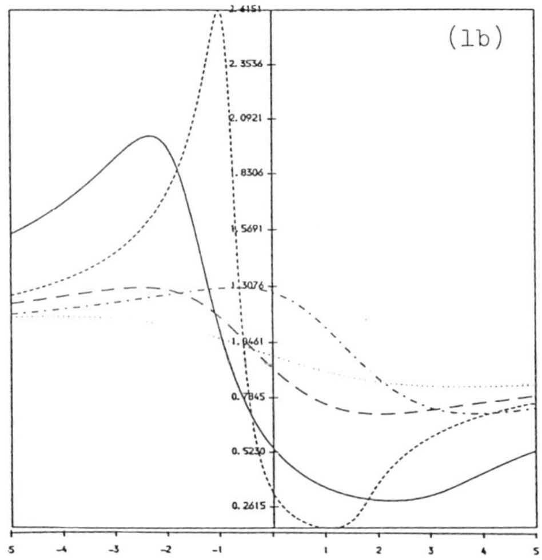

Fig. (1a): plot of $Re(m)$ against the scaled detuning Δ as abscissa for N=0, normal vacuum (full line), and $N = 0.5 : |M| = \sqrt{N(N+1)}$ with $\phi = 0, \pi, \frac{\pi}{2}, -\frac{\pi}{2}$ (the lines 'dash', 'dot', 'dash-dot' and 'large-dash' respectively); $\alpha = 10^{-3}$. **Fig. (1b)** : same as Fig. (1a) but $\alpha = 5$.

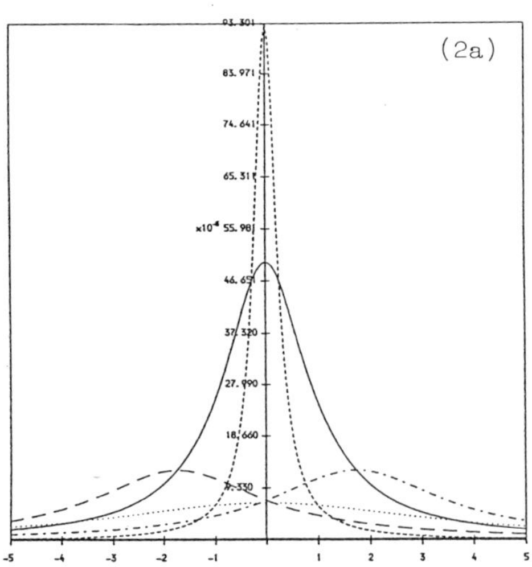

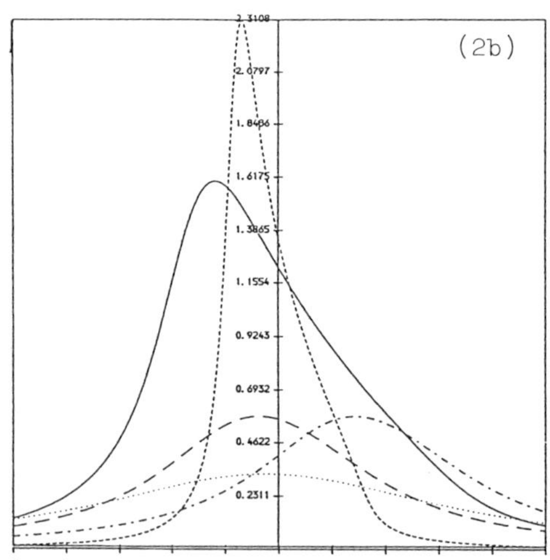

Fig. (2a): plot of $Im(m)$ against Δ. Data as in Fig. (1a). **Fig. (2b)**: same as Fig. (2a) but $\alpha = 5$.

4 SECOND CASE OF THE SQUEEZED VACUUM FIELD: EFFECT OF THE DIELECTRIC ON THE VACUUM FIELD

We now consider the case when the squeezed vacuum is a beam with axis *along z*. For simplicity here, and because we want to look at the change in mode structure of the squeezed beam, we assume there is no longer an additional coherent field along z and set $\Omega \equiv 0$ in eqns. (6). In this case, for a single uncorrelated pair of modes directed along both $\pm z$, $M(\mathbf{x}) = 2M \cos k_p z$ instead of the $2M \cos 2k_0 x$ used in §3. There is now only the frequency ω_p in the problem and eqns. (6) are taken in a frame rotating at ω_p so that the term $e^{2i(\omega-\omega_p)t}$ in eqns. (6) is unity. Since $M(\mathbf{x}) = 2M \cos 2k_p z$ this suggests that the ansatz eqn. (8) is now modified to

$$r_+(\mathbf{x}, \omega_p) = f_1(z)e^{-im(2k_p z)} + b_1(z)e^{im(2k_p z)} + f_2(z)e^{-i(m-1)(2k_p z)} + b_2(z)e^{i(m-1)(2k_p z)} \quad . \tag{14}$$

The calculations have been carried through on this basis but results must be reported elsewhere.

References

[1] See issues of *J. Opt. Soc. Am. B* **4**, No. 10 (1987); *J. Mod. Optics* **34**, Nos. 6,7 (1987). Also see articles in: "Squeezed and Non-classical Light" P. Tombesi and E.R. Pike eds. Plenum Press, New York (1989)

[2] S.S. Hassan and R.K. Bullough in: "Optical Bistability" C.M. Bowden, M. Ciftan and H.R. Robl eds. Plenum Press, New York (1981) pp. 367-404

[3] S.S. Hassan, R.K. Bullough and H.A. Batarfi in: "Studies in Classical and Quantum Nonlinear Optics" Ole Keller ed. Nova Science Publ. Inc., Commack, New York (1995)

[4] S.S. Hassan, M.R. Wahiddin, R. Saunders and R.K. Bullough "Master equation for the Dicke model in a broad-band squeezed vacuum" submitted to J. Mod. Optics (1992)

[5] R.K. Bullough, S.S. Hassan, R. Saunders and M.R. Wahiddin. To be published

[6] R. Saunders, Ph.D. thesis, University of Manchester, (1973)

[7] S.S. Hassan, Ph.D. thesis, University of Manchester, (1976)

[8] F. Hynne and R.K. Bullough, *Phil. Trans. Roy. Soc. London A* **312**, 251 (1984) and references

[9] F. Hynne and R.K. Bullough, *Phil. Trans. Roy. Soc. London A* **321**, 305 (1987)

[10] F. Hynne and R.K. Bullough, *Phil. Trans. Roy. Soc. London A* **330**, 253 (1990)

[11] R.K. Bullough and F. Hynne in: "P.P. Ewald and his dynamical theory of X-ray diffraction" D.W.J. Cruickshank, H.J. Juretschke and N. Kato eds. Oxford University Press, Oxford (1992), pp. 98-110

[12] R.K. Bullough and S.S. Hassan in: Proc. SPIE Vol. **369**, p. 257 (1982)

[13] M.N.R. Ibrahim, M.Sc. Thesis, University of Manchester, (1989)

BOSON INVERSE OPERATORS AND NEW NONCLASSICAL STATES

Anil K. Roy

Centre for Theoretical Studies
Indian Institute of Science,
Bangalore 560 012, India.

Abstract

We introduce the inverse of the hermitian operator $(\hat{a}\hat{a}^\dagger)$ and express the Boson inverse operators $\hat{a}^{-1}$ and $\hat{a}^{\dagger -1}$ in terms of the operators $\hat{a}$, $\hat{a}^\dagger$ and $(\hat{a}\hat{a}^\dagger)^{-1}$. We show that these Boson inverse operators may be realized by Susskind–Glogower's phase operators. We find a family of three two-photon annihilation operators $(\hat{a}\hat{a}^\dagger)^{-1}\hat{a}^2$, $\hat{a}(\hat{a}\hat{a}^\dagger)^{-1}\hat{a}$ and $\hat{a}^2(\hat{a}\hat{a}^\dagger)^{-1}$ and show that the eigenstates of these operators have interesting nonclassical properties including squeezing and antibunching of photons.

Keywords :
Boson operators, Phase operators, Two-photon annihilation operators, Squeezed vacuum, Antibunching of photons, Generalized coherent states.

1 INTRODUCTION OF BOSON INVERSE OPERATORS

The single mode harmonic oscillator states have been studied with the help of Boson field operators $\hat{a}$ and $\hat{a}^\dagger$. Recently their inverse operators $\hat{a}^{-1}$ and $\hat{a}^{\dagger -1}$ have also been introduced [1]. It is shown that since $\hat{a}$ and $\hat{a}^\dagger$ are singular operators, they do not possess inverses in a strict sense, but one can consider their inverses in a generalized sense. It may be observed that $\hat{a}^{-1}$ ($\hat{a}^{\dagger -1}$) is the right (left) inverse of $\hat{a}$ ($\hat{a}^\dagger$). They obey the commutation relation

$$[\hat{a}, \hat{a}^{-1}] = [\hat{a}^{\dagger -1}, \hat{a}^\dagger] = |0\rangle\langle 0|, \tag{1}$$

where $|0\rangle\langle 0|$ is the projection operator on the vacuum.

On the other hand, the operator $(\hat{a}\hat{a}^\dagger)$, being a non-singular operator, has a well defined inverse. In fact,

$$(\hat{a}\hat{a}^\dagger)^{-1} = \hat{a}^{\dagger -1}\hat{a}^{-1}, \tag{2}$$

which commutes with $(\hat{a}\hat{a}^\dagger)$ with the relation

$$(\hat{a}\hat{a}^\dagger)(\hat{a}\hat{a}^\dagger)^{-1} = (\hat{a}\hat{a}^\dagger)^{-1}(\hat{a}\hat{a}^\dagger) = \hat{I}, \tag{3}$$

where $\hat{I}$ is a unit operator. The action of $(\hat{a}\hat{a}^\dagger)^{-1}$ on a number state $|n\rangle$ is

$$(\hat{a}\hat{a}^\dagger)^{-1}|n\rangle = \frac{1}{n+1}|n\rangle, \tag{4}$$

for all non–negative integers n. Hence, one can readily express the operators $\hat{a}^{-1}$ and $\hat{a}^{\dagger -1}$ as :

$$\hat{a}^{-1} = \hat{a}^\dagger(\hat{a}\hat{a}^\dagger)^{-1}, \tag{5}$$

$$\hat{a}^{\dagger -1} = (\hat{a}\hat{a}^\dagger)^{-1}\hat{a}. \tag{6}$$

such that their actions on a number state $|n\rangle$ are given by the following relations :

$$\hat{a}^{-1}|n\rangle = \frac{1}{\sqrt{n+1}}\,|n+1\rangle, \tag{7}$$

$$\hat{a}^{\dagger -1}|n\rangle = (1-\delta_{n,o})\frac{1}{\sqrt{n}}\,|n-1\rangle. \tag{8}$$

One may note from Eqs.(7-8) that $\hat{a}^{-1}$ ($\hat{a}^{\dagger -1}$) behaves as a creation (an annihilation) operator.

The operators $\hat{a}$, $\hat{a}^\dagger$, $\hat{a}^{-1}$ and $\hat{a}^{\dagger -1}$ act irreducibly on the Hilbert space. It can, readily, be observed that contrary to $\hat{a}$ and $\hat{a}^\dagger$, $\hat{a}^{\dagger -1}$ and $\hat{a}^{-1}$ are bounded operators. Also one may find that these Boson inverse operators are related to the phase operators proposed by Susskind–Glogower. In the usual infinite dimensional Hilbert space, the phase operator proposed by Susskind and Glogower [2], $exp(i\hat{\phi})_{SG}$, is defined as:

$$exp(i\hat{\phi})_{SG}\,|n\rangle = |n-1\rangle, \qquad n > 0, \tag{9}$$

$$exp(i\hat{\phi})_{SG}\,|0\rangle = 0. \tag{10}$$

Hence, one can express $\hat{a}^{\dagger -1}$ as :

$$\hat{a}^{\dagger -1} = (\hat{a}\hat{a}^\dagger)^{-1/2}\,exp(i\hat{\phi})_{SG}. \tag{11}$$

In fact, Susskind–Glogower's phase operators give us the realization of the operators $\hat{a}^{-1}$ and $\hat{a}^{\dagger -1}$. However, these inverse operators are not related to Pegg and Barnett's phase operators [3], since Pegg and Barnett's phase operators are unitary and are defined in a finite dimensional space.

Of these four operators $\hat{a}$, $\hat{a}^\dagger$, $\hat{a}^{-1}$ and $\hat{a}^{\dagger -1}$, only $\hat{a}$ has normalizable right eigenstate (called as Coherent State) with non–zero eigenvalue. This state has extensively been studied in the literature. So, we concentrate on the combinations of these operators which exhibit two photon annihilation processes. The operator $\hat{a}^2$ is the trivial example of such two–photon annihilation operators (TAOs). One can construct a family of TAOs with the help of $\hat{a}^2$ and $(\hat{a}\hat{a}^\dagger)^{-1}$. We find that three such combinations are possible, namely, $(\hat{a}\hat{a}^\dagger)^{-1}\hat{a}^2$, $\hat{a}(\hat{a}\hat{a}^\dagger)^{-1}\hat{a}$ and $\hat{a}^2(\hat{a}\hat{a}^\dagger)^{-1}$. Their actions on a number state $|n\rangle$ are given by :

$$(\hat{a}\hat{a}^\dagger)^{-1}\hat{a}^2\ |n\rangle = \sqrt{\frac{n}{n-1}}\ |n-2\rangle, \tag{12}$$

$$\hat{a}(\hat{a}\hat{a}^\dagger)^{-1}\hat{a}\ |n\rangle = \sqrt{\frac{n-1}{n}}\ |n-2\rangle, \tag{13}$$

$$\hat{a}^2(\hat{a}\hat{a}^\dagger)^{-1}\ |n\rangle = \frac{\sqrt{n(n-1)}}{(n+1)}\ |n-2\rangle, \tag{14}$$

for all $n \geq 2$. It is readily observed that the states $|n=0\rangle$ and $|n=1\rangle$ are the eigenstates of these TAOs with zero eigenvalue. Here we wish to study the eigenvalue problem of these three operators and see if they possess normalizable right eigenstates with non–zero eigenvalue. In Sec.2, we solve their eigenvalue problem and in Sec.3, we study their nonclassical properties. We show that some of these states are new nonclassical states.

2 EIGENSTATES OF THE TAOS

We start with the first TAO $(\hat{a}\hat{a}^\dagger)^{-1}\hat{a}^2$. We write an eigenvalue equation for this operator as follows :

$$(\hat{a}\hat{a}^\dagger)^{-1}\hat{a}^2|\lambda, 1\rangle = \lambda|\lambda, 1\rangle, \tag{15}$$

where $|\lambda, 1\rangle$ is the eigenstate of the first TAO with eigenvalue λ. We have to find out the solution of this equation for a suitable complex parameter λ. We expand $|\lambda, 1\rangle$ in number state representation as :

$$|\lambda, 1\rangle = \sum_{n=o}^{\infty} C_n|n\rangle. \tag{16}$$

Substituting expression (16) in Eq.(15) and using Eq.(12), we obtain a recurrence relation for the coefficients C_n :

$$C_n = \lambda\sqrt{\frac{n-1}{n}}\, C_{n-2}. \tag{17}$$

From this, one observes that the states $|\lambda, 1\rangle$ separate into two sets, one involving only the even number states and the other involving the odd number states only. After some calculations, the expressions for the normalized eigenstates are obtained as follows :

$$|\lambda, +1\rangle = (1-|\lambda|^2)^{1/4} \sum_{n=o}^{\infty} \frac{\sqrt{(2n)!}}{2^n\ n!}\ \lambda^n\ |2n\rangle, \tag{18}$$

$$|\lambda, -1\rangle = \left[\frac{|\lambda|\sqrt{(1-|\lambda|^2)}}{\sin^{-1}|\lambda|}\right]^{1/2} \sum_{n=o}^{\infty} \frac{2^n\ n!}{\sqrt{(2n+1)!}}\ \lambda^n\ |2n+1\rangle. \tag{19}$$

Analogously, one can find the eigenstates of the remaining two TAOs. We obtain that the eigenstates of the operator $\hat{a}(\hat{a}\hat{a}^\dagger)^{-1}\hat{a}$ can be represented in the number state basis as:

$$|\lambda,+2\rangle = \left[\frac{(1-|\lambda|^2)^{3/2}}{(1-|\lambda|^2)^{1/2}+|\lambda|\sin^{-1}|\lambda|}\right]^{1/2} \sum_{n=o}^{\infty} \frac{2^n\, n!}{\sqrt{(2n)!}} \lambda^n|2n\rangle. \quad (20)$$

$$|\lambda,-2\rangle = (1-|\lambda|^2)^{3/4} \sum_{n=o}^{\infty} \frac{\sqrt{(2n+1)!}}{2^n\, n!} \lambda^n|2n+1\rangle, \quad (21)$$

and the eigenstates of the operator $\hat{a}^2(\hat{a}\hat{a}^\dagger)^{-1}$ are given by the expressions:

$$|\lambda,+3\rangle = \left[\frac{(1-|\lambda|^2)^{5/2}}{1+2|\lambda|^2}\right]^{1/2} \sum_{n=o}^{\infty} \frac{(2n+1)!}{2^n\, n!\, \sqrt{(2n)!}} \lambda^n\, |2n\rangle \quad (22)$$

$$|\lambda,-3\rangle = 2(1-|\lambda|^2)\left[3+(1+2|\lambda|^2)\frac{\sin^{-1}|\lambda|}{|\lambda|\sqrt{1-|\lambda|^2}}\right]^{-1/2} \times$$
$$\times \sum_{n=o}^{\infty} \frac{2^n\,(n+1)!}{\sqrt{(2n+1)!}} \lambda^n\, |2n+1\rangle. \quad (23)$$

One can verify that all these three sets of the eigenstates are normalizable for $|\lambda| < 1$.

It is interesting to note that $|\lambda,\pm1\rangle$, $|\lambda,\pm2\rangle$ and $|\lambda,\pm3\rangle$ states are, in fact, related. $|\lambda,\pm1\rangle$ states are the eigenstates of $(\hat{a}\hat{a}^\dagger)^{-1}\hat{a}^2$ with an eigenvalue λ. We multiply the eigenvalue equation (15) of $|\lambda,\pm1\rangle$ by $\hat{a}$ from left on both sides. We get

$$\hat{a}(\hat{a}\hat{a}^\dagger)^{-1}\hat{a}\,[\hat{a}|\lambda,\pm1\rangle] = \lambda\,[\hat{a}|\lambda,\pm1\rangle]. \quad (24)$$

Hence, we find that $\hat{a}|\lambda,\pm1\rangle$ must be proportional to $|\lambda,\pm2\rangle$ states. In a similar way, we multiply the eigenvalue equation of $|\lambda,\pm2\rangle$ states by $\hat{a}$ from left on both sides and observe that $\hat{a}|\lambda,\pm2\rangle$ must be proportional to $|\lambda,\pm3\rangle$ states. In particular, we may write some of these relations explicitly as :

$$\begin{aligned}
\hat{a}|\lambda,+1\rangle &= \lambda(\mathcal{N}_+/\mathcal{M}_-)\,|\lambda,-2\rangle,\\
\hat{a}|\lambda,-1\rangle &= (\mathcal{N}_-/\mathcal{M}_+)\,|\lambda,+2\rangle,\\
\hat{a}^{\dagger-1}|\lambda,-2\rangle &= (\mathcal{M}_-/\mathcal{N}_+)\,|\lambda,+1\rangle,\\
\hat{a}^{-1}|\lambda,+2\rangle &= (\mathcal{M}_+/\mathcal{N}_-)\,|\lambda,-1\rangle,\\
\hat{a}|\lambda,-2\rangle &= (\mathcal{M}_-/\mathcal{O}_+)|\lambda,+3\rangle.
\end{aligned} \quad (25)$$

Here $\mathcal{N}_\pm$, $\mathcal{M}_\pm$ and $\mathcal{O}_\pm$ are the normalization constants of the states $|\lambda,\pm1\rangle$, $|\lambda,\pm2\rangle$ and $|\lambda,\pm3\rangle$ respectively.

3 NONCLASSICAL PROPERTIES

Usually the nonclassical properties of any state are characterized by the manifestation of squeezing in one of the two field quadratures and antibunching of photons [4]. One can find, in literature, that a state is said to be squeezed if the uncertainty in any of the two quadratures $\hat{q}$ or $\hat{p}$ is less than that of a coherent state without violating Heisenberg's uncertainty relation. In other words,

$$\langle(\Delta\hat{q})^2\rangle \; or \; \langle(\Delta\hat{p})^2\rangle < 1/2, \quad (26)$$

with

$$\langle(\Delta\hat{q})^2\rangle \, \langle(\Delta\hat{p})^2\rangle \geq 1/4, \tag{27}$$

(assuming $\hbar = 1$) where

$$\hat{q} = \frac{\hat{a}+\hat{a}^\dagger}{\sqrt{2}}, \tag{28}$$

$$\hat{p} = \frac{\hat{a}-\hat{a}^\dagger}{\sqrt{2}\, i}, \tag{29}$$

$$\langle(\Delta\hat{q})^2\rangle = \frac{1}{2} + \langle\hat{a}^\dagger\hat{a}\rangle + Re\langle\hat{a}^2\rangle - 2(Re\langle\hat{a}\rangle)^2, \tag{30}$$

$$\langle(\Delta\hat{p})^2\rangle = \frac{1}{2} + \langle\hat{a}^\dagger\hat{a}\rangle - Re\langle\hat{a}^2\rangle - 2(Re\langle\hat{a}\rangle)^2, \tag{31}$$

and $\langle ... \rangle$ denotes the state dependent quantum mechanical average. On the other hand, antibunching of photons is associated with fluctuation in number of photons. If fluctuation in case of a particular state is less than that of a coherent state then that state is said to exhibit antibunching of photons. This is described by the negativity of the Q–parameter [5]:

$$Q = \frac{\langle(\Delta\hat{n})^2\rangle - \langle\hat{n}\rangle}{\langle\hat{n}\rangle} = \frac{\langle\hat{a}^{\dagger 2}\hat{a}^2\rangle - \langle\hat{a}^\dagger\hat{a}\rangle^2}{\langle\hat{a}^\dagger\hat{a}\rangle}. \tag{32}$$

In this section, we study these two nonclassical properties, *i.e.*, squeezing and antibunching of photons of the states $|\lambda, \pm 1\rangle$, $|\lambda, \pm 2\rangle$ and $|\lambda, \pm 3\rangle$. For this, we first calculate $\langle\hat{a}\rangle$, $\langle\hat{a}^\dagger\hat{a}\rangle$, $\langle\hat{a}^2\rangle$ and $\langle\hat{a}^{\dagger 2}\hat{a}^2\rangle$ for these states respectively. Since all these states are the eigenstates of TAO, one can readily observe that for each of them

$$\langle\hat{a}\rangle = 0. \tag{33}$$

Here we directly give the results for these states. The method followed can be seen in the ref.[1].

1. The states $|\lambda, +1\rangle$:

For these states

$$\langle\hat{a}^\dagger\hat{a}\rangle = \frac{|\lambda|^2}{(1-|\lambda|^2)}, \tag{34}$$

$$\langle(\Delta\hat{p})^2\rangle = \frac{1}{2} + \langle\hat{a}^\dagger\hat{a}\rangle - |\lambda|\cos\theta[\langle\hat{a}^\dagger\hat{a}\rangle + 1]. \tag{35}$$

where θ is the phase of the complex parameter λ. From these relations we find that squeezing in quadrature $\hat{p}$ occurs whenever

$$\cos\theta > |\lambda|. \tag{36}$$

On the other hand,

$$\langle\hat{a}^{\dagger 2}\hat{a}^2\rangle = \langle\hat{a}^\dagger\hat{a}\rangle(3\langle\hat{a}^\dagger\hat{a}\rangle + 1). \tag{37}$$

Therefore, the antibunching parameter Q is given by

$$Q = 2\langle\hat{a}^\dagger\hat{a}\rangle + 1. \tag{38}$$

We conclude that this state may show squeezing but never exhibits antibunching of photons.

2. The states $|\lambda, -1\rangle$:

For these states

$$\langle \hat{a}^\dagger \hat{a} \rangle = \frac{|\lambda|}{(1-|\lambda|^2)} \left[|\lambda| + \frac{(1-|\lambda|^2)^{1/2}}{\sin^{-1}|\lambda|} \right], \tag{39}$$

$$\langle (\Delta \hat{p})^2 \rangle = \frac{1}{2} + \langle \hat{a}^\dagger \hat{a} \rangle - |\lambda| \cos\theta [\langle \hat{a}^\dagger \hat{a} \rangle + 1] \tag{40}$$

and

$$\langle \hat{a}^{\dagger 2} \hat{a}^2 \rangle = (3\langle \hat{a}^\dagger \hat{a} \rangle + 1) \frac{|\lambda|^2}{(1-|\lambda|^2)}. \tag{41}$$

From Eqs.(39) and (40), we find that squeezing in quadrature $\hat{p}$ occurs whenever

$$\cos\theta > \frac{|\lambda| \sin^{-1}|\lambda| + (1-|\lambda|^2)^{1/2}}{\sin^{-1}|\lambda| + |\lambda|(1-|\lambda|^2)^{1/2}}. \tag{42}$$

The antibunching parameter Q is given by

$$Q = \frac{(3\langle \hat{a}^\dagger \hat{a} \rangle + 1)|\lambda|^2}{\langle \hat{a}^\dagger \hat{a} \rangle (1-|\lambda|^2)} - \langle \hat{a}^\dagger \hat{a} \rangle. \tag{43}$$

We conclude that this state may show squeezing as well as antibunching of photons.

3. The states $|\lambda, +2\rangle$:

For these states

$$\langle \hat{a}^\dagger \hat{a} \rangle = \frac{|\lambda|^2}{(1-|\lambda|^2)} \left[3 + \frac{|\mathcal{M}_+|^2}{|\mathcal{N}_-|^2} \right], \tag{44}$$

where $\mathcal{M}_+$ and $\mathcal{N}_-$ are the normalization constants of states $|\lambda, +2\rangle$ and $|\lambda, -1\rangle$ respectively.

$$\langle (\Delta \hat{p})^2 \rangle = \frac{1}{2} + \langle \hat{a}^\dagger \hat{a} \rangle - |\lambda| \cos\theta [\langle \hat{a}^\dagger \hat{a} \rangle + 2] \tag{45}$$

and

$$\langle \hat{a}^{\dagger 2} \hat{a}^2 \rangle = \frac{(5\langle \hat{a}^\dagger \hat{a} \rangle + 4)\, |\lambda|^2}{(1-|\lambda|^2)}. \tag{46}$$

From these relations we conclude that this state may show squeezing but never exhibits antibunching of photons.

4. The states $|\lambda, -2\rangle$:

For these states

$$\langle \hat{a}^\dagger \hat{a} \rangle = \frac{1+2|\lambda|^2}{(1-|\lambda|^2)}, \tag{47}$$

$$\langle (\Delta \hat{p})^2 \rangle = \frac{1}{2} + \langle \hat{a}^\dagger \hat{a} \rangle - |\lambda| \cos\theta [\langle \hat{a}^\dagger \hat{a} \rangle + 2] \tag{48}$$

and

$$\langle \hat{a}^{\dagger 2} \hat{a}^2 \rangle = \frac{(5\langle \hat{a}^\dagger \hat{a} \rangle + 4)\, |\lambda|^2}{(1-|\lambda|^2)}. \tag{49}$$

From these relations we find that squeezing in quadrature $\hat{p}$ occurs whenever

$$\cos\theta > \frac{1+2|\lambda|^2}{3|\lambda|}. \tag{50}$$

We obtain the following expression for the antibunching parameter Q:

$$Q = \frac{2|\lambda|^4 + 5|\lambda|^2 - 1}{1 + |\lambda|^2 - 2|\lambda|^4}. \tag{51}$$

Hence, we conclude that this state may show squeezing, bunching as well as antibunching of photons depending upon the eigenvalue λ.

5. The states $|\lambda, +3\rangle$:

For these states we find

$$\begin{aligned} \langle \hat{a}^\dagger \hat{a} \rangle &= |\mathcal{O}_+|^2 \sum_{n=o}^{\infty} \frac{(2n+1)!^2\, 2n}{(2n)!\, 2^{2n}\, n!^2} |\lambda|^{2n} \\ &= \frac{3|\lambda|^2(3+2|\lambda|^2)}{(1-|\lambda|^2)(1+2|\lambda|^2)}. \end{aligned} \tag{52}$$

$$\langle (\Delta \hat{p})^2 \rangle = \frac{1}{2} + \langle \hat{a}^\dagger \hat{a} \rangle - |\lambda| \cos\theta [\langle \hat{a}^\dagger \hat{a} \rangle + 3]. \tag{53}$$

Hence, we get the condition that whenever

$$\cos\theta > \frac{|\lambda|(3+2|\lambda|^2)}{1+4|\lambda|^2}, \tag{54}$$

one can find squeezing in quadrature $\hat{p}$.

Now, for the antibunching parameter Q, we get

$$\begin{aligned} \langle \hat{a}^{\dagger 2} \hat{a}^2 \rangle &= |\mathcal{O}_+|^2\, |\lambda|^2 \sum_{n=0}^{\infty} \frac{(2n+1)!^2\, (2n+3)^2}{(2n)!\, 2^{2n}\, n!^2} |\lambda|^{2n} \\ &= |\lambda|^2 \left[9 + 6\langle \hat{a}^\dagger \hat{a} \rangle + \frac{6|\lambda|^4 + 23|\lambda|^2 + 6}{(1-|\lambda|^2)\,(3+2|\lambda|^2)} \langle \hat{a}^\dagger \hat{a} \rangle \right]. \end{aligned} \tag{55}$$

We find that $\langle \hat{a}^{\dagger 2} \hat{a}^2 \rangle$ is always greater than $\langle \hat{a}^\dagger \hat{a} \rangle^2$, and hence the Q-parameter remains always positive. Thus $|\lambda, +3\rangle$ states never show antibunching of photons.

6. The states $|\lambda, -3\rangle$:

For these states

$$\langle \hat{a}^\dagger \hat{a} \rangle = \frac{|\lambda|\sqrt{1-|\lambda|^2}\,(11|\lambda|^2+4) + 3|\lambda|^2(3+2|\lambda|^2)\sin^{-1}|\lambda|}{(1-|\lambda|^2)\,\left[3|\lambda|\sqrt{(1-|\lambda|^2)} + (1+2|\lambda|^2)\sin^{-1}|\lambda|\right]}, \tag{56}$$

$$\langle (\Delta \hat{p})^2 \rangle = \frac{1}{2} + \langle \hat{a}^\dagger \hat{a} \rangle - |\lambda| \cos\theta [\langle \hat{a}^\dagger \hat{a} \rangle + 3], \tag{57}$$

Hence, one finds that the $\hat{p}$ quadrature will be squeezed whenever

$$\cos\theta > \frac{\langle \hat{a}^\dagger \hat{a} \rangle}{|\lambda|\,[3 + \langle \hat{a}^\dagger \hat{a} \rangle]}. \tag{58}$$

For the antibunching parameter, we find

$$\begin{aligned} \langle \hat{a}^{\dagger 2} \hat{a}^2 \rangle &= 8|\lambda|^2(\langle \hat{a}^\dagger \hat{a} \rangle + 1) + \frac{1}{(1-|\lambda|^2)^2} \\ \times &\left[\frac{(32|\lambda|^6 + 72|\lambda|^4 + 1)\sin^{-1}|\lambda| \;+\; (64|\lambda|^4 + 42|\lambda|^2 - 1)|\lambda|\sqrt{(1-|\lambda|^2)}}{(1+2|\lambda|^2)\sin^{-1}|\lambda| \;+\; 3|\lambda|\sqrt{(1-|\lambda|^2)}} \right]. \end{aligned} \tag{59}$$

From the numerical solution, we find that $|\lambda, -3\rangle$ states show antibunching as well as bunching of photons depending upon the value of $|\lambda|$.

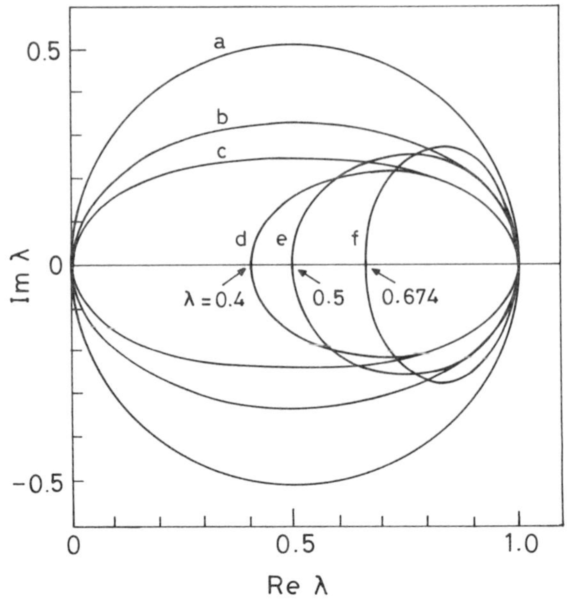

Figure 1: Regions showing squeezing in $\hat{p}$ quadrature for the states (a) $|\lambda, +1\rangle$, (b) $|\lambda, +2\rangle$, (c) $|\lambda, +3\rangle$, (d) $|\lambda, -3\rangle$, (e) $|\lambda, -2\rangle$ and (f) $|\lambda, -1\rangle$. Note that in case of odd parity states, squeezing starts from $\lambda = 0.4$ (for $|\lambda, -3\rangle$ states), 0.5 (for $|\lambda, -2\rangle$ states) and 0.674 (for $|\lambda, -1\rangle$ states). Regions for squeezing in $\hat{q}$ quadrature are just the mirror image of that for $\hat{p}$ along the $Im\lambda$ axis.

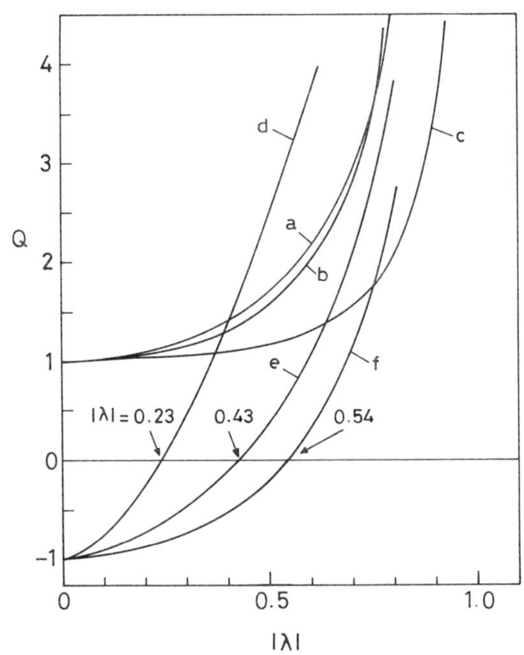

Figure 2: Behaviour of Q -parameter with respect to $|\lambda|$. Note that all the odd parity states show antibunching of photons ($Q < 0$) for some range of the eigenvalue λ.

In all these discussions, we have given various expressions related to the quadrature $\hat{p}$ only. The analogous expressions for quadrature $\hat{q}$ can also be obtained easily. We have plotted regions of squeezing in quadrature $\hat{p}$ for all these states. One may note that the state $|\lambda, -1\rangle$ starts showing squeezing from $\lambda = 0.674$, the state $|\lambda, +2\rangle$ starts showing squeezing from $\lambda = 0.5$ and the state $|\lambda, -3\rangle$ starts showing squeezing from $\lambda = 0.4$. The regions of squeezing in quadrature $\hat{q}$ are just the mirror image along the $Im\lambda$ axis of the corresponding $\hat{p}$ regions.

We have also plotted the variation of Q parameter with respect to $|\lambda|$. It is clear from the plot that all the even parity states always show bunching of photons only. Whereas, among the odd parity states the state $|\lambda, -1\rangle$ shows antibunching of photons upto the value $|\lambda| < 0.54$, the state $|\lambda, -2\rangle$ shows antibunching upto the value $|\lambda| < 0.43$ and the state $|\lambda, -3\rangle$ shows antibunching of photons upto the value $|\lambda| < 0.23$.

CONCLUSION

We conclude that Boson inverse operators resemble to Susskind-Glogower's phase operators. However, there is no any simple connection between these inverse operators and the Pegg and Barnett's phase operators. Using the well defined inverse of the operator $(\hat{a}\hat{a}^\dagger)$, we find a family of TAOs, *viz.*, $(\hat{a}\hat{a}^\dagger)^{-1}\hat{a}^2$, $\hat{a}(\hat{a}\hat{a}^\dagger)^{-1}\hat{a}$ and $\hat{a}^2(\hat{a}\hat{a}^\dagger)^{-1}$. We show that the states $|\lambda, \pm 1\rangle$, $|\lambda, \pm 2\rangle$ and $|\lambda, \pm 3\rangle$ are new nonclassical states. We have reported elsewhere [6] that of these states, $|\lambda, +1\rangle$ state is the usual squeezed vacuum and $|\lambda, -2\rangle$ state is the squeezed first excited state. These states, *i.e.,* $|\lambda, +1\rangle$ and $|\lambda, -2\rangle$ states, are also Mp(2) generalized coherent states [7] in Perelomov sense [8].

References

[1] C.L. Mehta, Anil K. Roy and G.M. Saxena, Phys. Rev. A, **46**, 1565 (1992).

[2] L. Susskind and J. Glogower, Physics, **1**, 49 (1964).

[3] D.T. Pegg and S.M. Barnett, Europhys. Lett., **6**, 483 (1988); D.T. Pegg and S.M. Barnett, Phys. Rev. A, **39**, 1665 (1989);

[4] There are various situations when the field is nonclassical even if it does not exhibit squeezing as well as antibunching of photons. For example, see, G.S. Agarwal and K. Tara, Phys. Rev. A, **46**, 485 (1992).

[5] L. Mandel, Opt. Lett., **4**, 205 (1979).

[6] C.L. Mehta, A.K. Roy and G.M. Saxena, Pramana–J. of Phys., **42**, 311 (1994).

[7] Arvind, B. Dutta, C.L. Mehta and N. Mukunda, Phys. Rev. A (1994, to appear).

[8] A.M. Perelomov, *Generalized Coherent States and their Applications*, (Springer-Verlag, 1986).

Part II.
Quantum Measurement
Problem of Collapse

MATHEMATICAL CHARACTERIZATIONS OF MEASUREMENT STATISTICS

Masanao Ozawa

Mathematical Sciences Group,
School of Informatics and Sciences
Nagoya University, Nagoya 464-01, Japan

1 INTRODUCTION

A measuring process is described by an interaction between an object to be measured and a measuring apparatus and by a subsequent reading of the outcome. By a measurement statistics we shall mean the following two elements determined by a measuring process: the probability distribution $dP(x|\rho)$ of the outcome x and the state reduction from prior states ρ to the posterior states ρ_x conditioned by the outcome x. The problem of mathematical characterizations of possible measurement statistics has a long history initiated by von Neumann's pioneering work [1] and up to now it is known as a general solution that every normalized operation valued measures $d\mathbf{X}(x)$, called an operational measure in brief, represents a realizable measurement statistics by the relation $d\mathbf{X}(x)\rho = \rho_x dP(x|\rho)$ [2], where an operation is a contraction on the space of trace class operators such that its dual is completely positive.

In this paper, we shall investigate further characterizations of a class of measurement statistics which are important from both foundational and technological points of view. A measuring process is called pure if it reduces every pure prior state to a pure posterior states. It is known that such measuring processes are characterized by the property that the state reduction $\rho \mapsto \rho_x$ decreases the entropy in average [3]. Typical operational measures representing such measuring processes are classified into two types called the von Neumann-Davies type and the Gordon-Louisell type. An operational measure of the von Neumann-Davies type is characterized by a linear isometry W from $\mathcal{H}$, the Hilbert space of the object, to $L^2(\mu, \mathcal{H})$, where μ is

Quantum Communications and Measurement
Edited by V. P. Belavkin *et al.*, Plenum Press, New York, 1995

a measure on the space of outcomes, such that the state reduction of a pure state $\rho = |\psi\rangle\langle\psi| \mapsto \rho_x = |\psi_x\rangle\langle\psi_x|$ is represented by $(W\psi)(x)d\mu(x) = \psi_x dP(x|\rho)$. An operational measure of the Gordon-Louisell type is characterized by a fixed family $\{\phi_x\}$ of state vectors such that for every prior pure state ρ the posterior states are always given by $\rho_x = |\phi_x\rangle\langle\phi_x|$. We shall show that the statistics of every pure measuring process is represented by an operational measure which is a direct sum of two (unnormalized) operational measures one of which is of the von Neumann-Davies type and the other of which is of the Gordon-Louisell type.

2 STATISTICS OF MEASUREMENTS

Throughout this paper, any quantum system is a system with finite degrees of freedom without any superselection rules and every Hilbert space is supposed to be separable so that the states of the system are described by density operators on a Hilbert space and that the observables by self-adjoint operators on the same Hilbert space.

By a *measurement* of a system $\mathbf{S}$ described by a Hilbert space $\mathcal{H}$ we shall mean an experiment described as follows. Let $\mathbf{A}$ be a quantum system, called an *apparatus*, described by a Hilbert space $\mathcal{K}$. The system $\mathbf{A}$ is coupled to the system $\mathbf{S}$ during a finite time interval from time t to $t+\tilde{\tau}$. Denote by U the unitary operator on $\mathcal{H}\otimes\mathcal{K}$ corresponding to the time evolution of the composite system $\mathbf{S}+\mathbf{A}$ from time t to $t+\tilde{\tau}$. The time t is called the *time of measurement* and the time $t+\tilde{\tau}$ is called the *time just after measurement*. Let ρ be the state of $\mathbf{S}$ at the time of measurement. In order to assure the reproducibility of this experiment, the apparatus $\mathbf{A}$ is always prepared in a fixed state σ at the time of measurement, and hence the composite system $\mathbf{S}+\mathbf{A}$ is in state $\rho\otimes\sigma$ at the time of measurement and in state $U(\rho\otimes\sigma)U^\dagger$ at the time just after measurement. At the time just after measurement, the systems $\mathbf{S}$ and $\mathbf{A}$ are separated to be noninteracting, and in order to obtain the outcome of this experiment a finite sequence $\langle \hat{B}_1, \dots, \hat{B}_n\rangle$ of compatible observables, called the *probe observables*, of the system $\mathbf{A}$ is measured by the observer presumably with an ideal macroscopic amplification of the observables $\hat{B}_1, \dots, \hat{B}_n$ to macroscopic variables $\mathbf{b}_1, \dots, \mathbf{b}_n$, called the *meter variables*, so that their joint probability distribution is the one postulated by the Born statistical formula, i.e.,

$$\begin{aligned}
&\Pr[\mathbf{b}_1 \in \Delta_1, \dots, \mathbf{b}_n \in \Delta_n |\, U(\rho\otimes\sigma)U^\dagger] \\
&\quad = \operatorname{Tr}[\{B_1(\Delta_1)\cdots B_n(\Delta_n)\otimes 1\}U(\rho\otimes\sigma)U^\dagger] \qquad (2.1)
\end{aligned}$$

for all $\Delta_1, \dots \Delta_n \in \mathcal{B}(\mathbf{R})$, where $B_1, \dots, B_n$ are the spectral measures of $\hat{B}_1, \dots, \hat{B}_n$ and $\mathcal{B}(\mathbf{R})$ stands for the Borel σ-field of the real line $\mathbf{R}$. After reading the meter observable, the observer obtains the outcome of this experiment by the *data processing* represented by a Borel function f from $\mathbf{R}^n$ to a standard Borel space $(\Lambda, \mathcal{B}(\Lambda))$, called the *outcome space*, so that the outcome variable $\mathbf{x}$ is obtained by the relation $\mathbf{x} = f(\mathbf{b}_1, \dots, \mathbf{b}_n)$.

The process of a measurement is, therefore, characterized mathematically by a 5-tuple $\mathcal{X} = [\mathcal{K}, \sigma, U, \langle \hat{B}_1, \dots, \hat{B}_n\rangle, f]$, called a *measuring process* or briefly a *measurement*, consisting of a Hilbert space $\mathcal{K}$, a density operator σ on $\mathcal{K}$, a unitary operator U on $\mathcal{H}\otimes\mathcal{K}$, a finite sequence $\langle \hat{B}_1, \dots, \hat{B}_n\rangle$ of self-adjoin operators on $\mathcal{K}$, and a Borel function f from $\mathbf{R}^n$ to a standard Borel space Λ. By the *measurement statistics* of this measurement we mean the *outcome distribution* $dP(x|\rho)$ and the *state reduction* $\rho \mapsto \rho_x$. In what follows, suppose that the system is in a state ρ at the time of measurement. The outcome distribution $dP(x|\rho)$ is the probability distribution of the outcome variable $\mathbf{x}$, and the state reduction determines the state ρ_x of the system which leads to the outcome $\mathbf{x} = x$ at the time just after measurement. The family $\{\rho_x | \ x \in \Lambda\}$ is naturally required to be a Borel family. If the measurement statistics are determined, the state ρ_Δ, at the time just after measurement, of the subensemble of samples which lead to the outcomes in $\Delta \in \mathcal{B}(\Lambda)$ is given by

$$\rho_\Delta = \frac{1}{P(\Delta|\rho)} \int_\Delta \rho_x \, dP(x|\rho), \tag{2.2}$$

provided $P(\Delta| \ \rho) \neq 0$, where the integral is a Bochner integral. The state change $\rho \mapsto \rho_\Delta$ is called the *integral state change.*

The measurement statistics of measurement $\mathcal{X}$ is determined as follows. Let $B : \mathcal{B}(\mathbf{R}^n) \to \mathcal{L}(\mathcal{K})$ be the joint spectral measure of $\hat{B}_1, \ldots, \hat{B}_n$, i.e., $B(\Delta_1 \times \cdots \times \Delta_n) = B_1(\Delta_1) \cdots B_n(\Delta_n)$ for all $\Delta_1, \ldots, \Delta_n \in \mathcal{B}(\mathbf{R})$, and $E : \mathcal{B}(\Lambda) \to \mathcal{L}(\mathcal{K})$ the spectral measure such that $E(\Lambda) = B(f^{-1}(\Delta))$ for all $\Delta \in \mathcal{B}(\Lambda)$. If the measuring process $\mathcal{X}$ is fixed, the probability distribution of the outcome variable $\mathbf{x}$ depends only on the state ρ of the system at the time of measurement and is given by

$$P(\Delta|\rho) = \mathrm{Tr}[(1 \otimes E(\Delta))U(\rho \otimes \sigma)U^\dagger] \tag{2.3}$$

for all $\Delta \in \mathcal{B}(\Lambda)$. By the statistical analysis of measurement [2,4], the integral state change of the measurement $\mathcal{X}$ is determined by the relation

$$\rho_\Delta = \frac{1}{P(\Delta|\rho)} \mathrm{Tr}_\mathcal{K}[(1 \otimes E(\Delta))U(\rho \otimes \sigma)U^\dagger], \tag{2.4}$$

where $\mathrm{Tr}_\mathcal{K}$ stands for the partial trace over $\mathcal{K}$.

In order to integrate these results, for any $\Delta \in \mathcal{B}(\Lambda)$ and $\rho \in \tau c(\mathcal{H})$, let $\mathbf{X}(\Delta)\rho$ be the trace class operator on $\mathcal{H}$ such that

$$\mathbf{X}(\Delta)\rho = \mathrm{Tr}_\mathcal{K}[(1 \otimes E(\Delta))U(\rho \otimes \sigma)U^\dagger]. \tag{2.5}$$

Then the measurement statistics is represented by

$$P(\Delta|\rho) = \mathrm{Tr}[\mathbf{X}(\Delta)\rho], \qquad \rho_\Delta = \frac{1}{\mathrm{Tr}[\mathbf{X}(\Delta)\rho]}\mathbf{X}(\Delta)\rho. \tag{2.6}$$

Equation (2.5) determines the positive linear map $\mathbf{X}(\Delta) : \rho \to \mathbf{X}(\Delta)\rho$ on $\tau c(\mathcal{H})$ and its dual $\mathbf{X}(\Delta)^*$, defined by $\mathrm{Tr}[\rho \mathbf{X}(\Delta)^* a] = \mathrm{Tr}[a\mathbf{X}(\Delta)\rho]$ for all $\rho \in \tau c(\mathcal{H})$ and $a \in \mathcal{L}(\mathcal{H})$, is a normal completely positive (CP) map on $\mathcal{L}(\mathcal{H})$. Moreover, the map $\mathbf{X} : \Delta \to \mathbf{X}(\Delta)$ from $\mathcal{B}(\Lambda)$ to the space $\mathcal{L}(\tau c(\mathcal{H}))$ of bounded linear operators on $\tau c(\mathcal{H})$ satisfies the following conditions:

1. For any $\Delta \in \mathcal{B}(\Lambda)$, $\mathbf{X}(\Delta)^*$ is a CP map on $\mathcal{L}(\mathcal{H})$.
2. For any disjoint sequence $\Delta_1, \Delta_2, \ldots$ in $\mathcal{B}(\mathbf{R})$,

$$\mathbf{X}(\bigcup_{i=1}^\infty \Delta_i) = \sum_{i=1}^\infty \mathbf{X}(\Delta_i), \tag{2.7}$$

   where the sum is convergent in the strong operator topology of $\mathcal{L}(\tau c(\mathcal{H}))$.
3. For any $\rho \in \tau c(\mathcal{H})$,

$$\mathrm{Tr}[\mathbf{X}(\mathbf{R})\rho] = \mathrm{Tr}[\rho]. \tag{2.8}$$

Mathematically, we call any map $\mathbf{X} : \mathcal{B}(\Lambda) \to \mathcal{L}(\tau c(\mathcal{H}))$ satisfying the above conditions as an *operational measure* for $(\Lambda, \mathcal{H})$, while it is called *CP map valued measure* if only conditions (1) and (2) are satisfied [5,6,2]. The operational measure $\mathbf{X}$ defined by Eq. (2.5) is called the operational measure of measurement $\mathcal{X}$. For a given operational measure $\mathbf{X}$ the relation

$$X(\Delta) = \mathbf{X}(\Delta)^* 1 \tag{2.9}$$

for all $\Delta \in \mathcal{B}(\Lambda)$ defines a probability operator valued measure (POM) for $(\Lambda, \mathcal{H})$, i.e., a positive operator valued measure $X : \mathcal{B}(\Lambda) \to \mathcal{L}(\mathcal{H})$ such that $X(\Lambda) = 1$. The POM defined by Eq. (2.9) is called the POM of $\mathbf{X}$. Note that if X is the POM of the operational measure $\mathbf{X}$ of $\mathcal{X}$, the probability distribution of the outcome variable $\mathbf{x}$ is given, as a generalization of the Born statistical formula, by

$$P(\Delta|\rho) = \mathrm{Tr}[\rho X(\Delta)] \tag{2.10}$$

for all $\Delta \in \mathcal{B}(\Lambda)$.

For a given operational measure $\mathbf{X}$ and a density operator ρ, there exists a Borel family $\{\rho_x | \ x \in \mathbf{R}\}$ of states, called the *family of posterior states* for $(\mathbf{X}, \rho)$, satisfying [7]

$$\mathbf{X}(\Delta)\rho = \int_\Delta \rho_x \,\mathrm{Tr}[d\mathbf{X}(x)\rho]. \tag{2.11}$$

Eq. (2.11) determines the family of posterior states uniquely up to almost everywhere. We call any Borel family of states satisfying Eq. (2.11) as a *version* of the family of posterior states. Thus the state reduction $\rho \to \rho_x$ of measurement $\mathcal{X}$ is characterized as the family $\{\rho_x | \ x \in \rho\}$ of posterior states for $(\mathbf{X}, \rho)$, where $\mathbf{X}$ is the operational measure of $\mathcal{X}$.

For a given operational measure $\mathbf{X}$, a measuring process $\mathcal{X}$ is called a *realization* of $\mathcal{X}$ if $\mathbf{X}$ is an operational measure of $\mathcal{X}$. An operational measure is said to be *realizable* if it has a realization. A POM is said to be *realizable* if it is the POM of a realizable operational measure. The question as to determining the all possible measurement statistics is answered by the following theorem [2].

Theorem 2.1 (General Realization Theorem) *Every operational measure is realizable and so is every POM.*

3 STATISTICS OF PURE MEASURING PROCESSES

In most of known examples arising in real physical experiments, the state reduction reduces a pure prior state ρ to a pure posterior state ρ_x for all possible outcomes x. Thus the characterization of this kind of statistics has a particular importance in applications. For this purpose, we say that an operational measure $\mathbf{X}$ is *pure* if for any pure state ρ the family $\{\rho_x | \ x \in \Lambda\}$ of posterior states for $(\mathbf{X}, \rho)$ is such that ρ_x is a pure state for almost all $x \in \Lambda$ with respect to $\mathrm{Tr}[d\mathbf{X}(x)\rho]$. A measurement is said to be *pure* if its operational measure is pure. For a pure measurement, the measurement statistics is sufficiently represented for state vectors ψ as follows:

$$\begin{aligned} \text{outcome distribution:} \quad & dP(x|\psi), \\ \text{state reduction:} \quad & \psi \mapsto \psi_x, \end{aligned}$$

where $dP(x|\psi) = dP(x| \ |\psi\rangle\langle\psi|)$ and ψ_x is such that $|\psi_x\rangle\langle\psi_x| = |\psi\rangle\langle\psi|_x$.

The pure measurements are know to have the following information theoretical characterization. Let ρ be the state of the system at the time of measurement. Then the entropy of ρ is

$$S(\rho) = -\mathrm{Tr}[\rho \log \rho]. \tag{3.1}$$

The measuring interaction changes the state of the system to the state $\mathbf{X}(\Lambda)\rho$ and this process is an irreversible dynamics which increases the entropy by $S(\mathbf{X}(\Lambda)\rho) - S(\rho) \geq 0$. The observer is, however, informed of the outcome $\mathbf{x} = x$ of the measurement and this information changes the state from $\mathbf{X}(\Lambda)\rho$ to ρ_x. This process gains the information on the system, or equivalently decrease the entropy of the system, in average by the amount given by

$$S(\mathbf{X}(\Lambda)\rho) - \int_\Lambda S(\rho_x)\,\mathrm{Tr}[d\mathbf{X}(x)\rho] \geq 0. \tag{3.2}$$

If the outcome gives enough information about the system, we can expect that this information gain compensates the dynamical entropy increase so that the total information gain is nonnegative, i.e.,

$$I(\mathbf{X}|\ \rho) = S(\rho) - \int_\Lambda S(\rho_x)\,\mathrm{Tr}[d\mathbf{X}(x)\rho] \geq 0. \tag{3.3}$$

Relation (3.3) is a quantum mechanical generalization of Shannon's fundamental inequality [8]. For a von Neumann-Lüders measurement [9], the operational measure of which is given by

$$\mathbf{X}(\Delta)\rho = \sum_{x \in \Delta} E_x \rho E_x, \tag{3.4}$$

where $\Delta \in \mathcal{B}(\mathbf{R})$ and E_x are mutually commutable projections, this fact was first conjectured by Groenewold [10] and proved by Lindblad [11]. The following theorem characterizes generally measurements which satisfy this inequality [3].

Theorem 3.1 (Generalized Groenewold-Lindblad Inequality) *An operational measure $\mathbf{X}$ is pure if and only if it satisfies $I(\mathbf{X}|\ \rho) \geq 0$ for every density operator ρ with $S(\rho) < \infty$.*

The above theorem clarifies the significance of pure measurements, but suggests little about the structure of pure measurements. In order to start the structure theory, we shall consider typical constructions of pure measurements.

Let W be a linear isometry from $\mathcal{H}$ to the Hilbert space $L^2(\Lambda, \mu, \mathcal{H})$, the L^2-space of $\mathcal{H}$-valued Borel functions on Λ with respect to a σ-finite measure μ on $\mathcal{B}(\Lambda)$. Then the relation

$$\mathbf{X}(\Delta)|\xi\rangle\langle\xi| = \int_\Delta |(W\xi)(x)\rangle\langle(W\xi)(x)|\,d\mu(x) \tag{3.5}$$

for all $\Delta \in \mathcal{B}(\Lambda)$ and $\xi \in \mathcal{H}$ defines a unique pure operational measure $\mathbf{X}$. For the measurement with this operational measure $\mathbf{X}$, the measurement statistics is given by

$$\text{outcome distribution:} \quad dP(x|\psi) = \|(W\psi)(x)\|^2\,d\mu(x),$$
$$\text{state reduction:} \quad \psi \mapsto \psi_x = \|(W\psi)(x)\|^{-1}(W\psi)(x).$$

We shall call this type of pure operational measures as the *von Neumann-Davies type*. Apart from von Neumann-Lüders measurements, a typical example arises from von Neumann's model of approximate position measurement for the system with the single degree of freedom [1, 12, 6, 3]. This model is a measuring process $\mathcal{X} = [\mathcal{K}, \sigma, U, \hat{B}, f]$ for $\mathcal{H}$ such that $\mathcal{H} = L^2(\mathbf{R})$, $\mathcal{K} = L^2(\mathbf{R})$, $\sigma = |\varphi\rangle\langle\varphi|$, $U = \exp(-i\hat{Q} \otimes \hat{P}/\hbar)$, where $\hat{Q}$ is the position of the system and $\hat{P}$ is the momentum of the apparatus, $\hat{B} = \hat{Q}$, the position of the apparatus, and $f(b) = b$. Then the operational measure of this measurement is

$$\mathbf{X}(\Delta)|\psi\rangle\langle\psi| = \int_{\Delta} |\varphi(\hat{Q} - x1)\psi\rangle\langle\varphi(\hat{Q} - x1)\psi|\, dx, \tag{3.6}$$

and $W\psi \in L^2(\mathbf{R}, dx, \mathcal{H}) = L^2(\mathbf{R}^2)$ is well-defined by

$$(W\psi)(q, x) = [\varphi(\hat{Q} - x1)\psi](q) = \varphi(q - x)\psi(q) \in L^2(\mathbf{R}^2), \tag{3.7}$$

for $\phi, \psi \in L^2(\mathbf{R})$.

Another type of pure measurements is given as follows. Let $\{\varphi_x |\ x \in \Lambda\}$ be a fixed Borel family of state vectors in $\mathcal{H}$ and X be a POM for $(\Lambda, \mathcal{H})$. Then the relation

$$\mathbf{X}(\Delta)\rho = \int_{\Delta} |\varphi_x\rangle\langle\varphi_x| \mathrm{Tr}[\rho\, dX(x)] \tag{3.8}$$

for all $\Delta \in \mathcal{B}(\Lambda)$ and $\rho \in \tau c(\mathcal{H})$ defines a unique pure operational measure $\mathbf{X}$ [13], the measurement statistics of which is given by

$$\begin{aligned} \text{outcome distribution:} \quad & dP(x|\psi) = \langle\psi|dX(x)|\psi\rangle, \\ \text{state reduction:} \quad & \psi \mapsto \psi_x = \varphi_x. \end{aligned}$$

We shall call this type of pure operational measures as the *Gordon-Louisell type* [14, 15]. A pure operational measure $\mathbf{X}$ of the von Neumann-Davies type is also of the Gordon-Louisell type if the linear isometry W is given by a family $\{A_x |\ x \in \Lambda\}$ of rank one operators on $\mathcal{H}$ such that $(W\xi)(x) = A_x\xi$ for almost all $x \in \Lambda$. A pure operational measure of the Gordon-Louisell type is said to be of the *proper Gordon-Louisell* type if it is not of the von Neumann-Davies type. A non-trivial example of a measurement of the proper Gordon-Louisell type is obtained by the unitary operator

$$U = \exp\left[-i\frac{\pi}{3\hbar\sqrt{3}}\{2(\hat{Q} \otimes \hat{P} - \hat{P} \otimes \hat{Q}) + (\hat{Q}\hat{P} \otimes 1 - 1 \otimes \hat{Q}\hat{P})\}\right] \tag{3.9}$$

together with the same $\mathcal{H}$, $\mathcal{K}$, σ, $\hat{B}$, and f as von Neumann's model. In this case, the operational measure is given by

$$\mathbf{X}(\Delta)\rho = \int_{\Delta} |\varphi_x\rangle\langle\varphi_x| \mathrm{Tr}[\rho\, dQ(x)], \tag{3.10}$$

where $\varphi_x(q) = \varphi(x - q)$ is the family of wave functions of the system and $dQ(x)$ is the spectral measure of the position of the system. For the later discussion, we say that a CP map valued measure $\mathbf{X}$ is of the von Neumann-Davies type if it is of the form of Eq. (3.5) with a bounded linear transformation $W : \mathcal{H} \to L^2(\Lambda, \mu, \mathcal{H})$, and is of the Gordon-Louisell type if it is of the form of Eq. (3.8) with a positive operator valued measure $X : \mathcal{B}(\Lambda) \to \mathcal{L}(\mathcal{H})$.

Let μ be a σ-finite measure on $\mathcal{B}(\Lambda)$. An operational measure $\mathbf{X}$ for $(\Lambda, \mathcal{H})$ is called *μ-continuous*, in symbols $\mathbf{X} \ll \mu$, if for any $\rho \in \tau c(\mathcal{H})$ the measure $\mathrm{Tr}[\mathbf{X}(\cdot)\rho]$ is absolutely continuous with respect to μ. For any operational measure $\mathbf{X}$, there exists a trace class operator $\rho_0 \in \tau c(\mathcal{H})$ such that $\mathbf{X} \ll \mathrm{Tr}[\mathbf{X}(\cdot)\rho_0]$, where ρ_0 is given, for example, by $\rho_0 = \sum_{n=1}^{\infty} 2^{-n}|\phi_n\rangle\langle\phi_n|$ with a basis $\{\phi_i|\ i = 1, 2, \ldots\}$ of $\mathcal{H}$. Let $\mathcal{M}$ and $\mathcal{N}$ be von Neumann algebras. For any normal CP map $\Phi : \mathcal{M} \to \mathcal{N}$, denote by Φ_* the predual map $\Phi : \mathcal{N}_* \to \mathcal{M}_*$ defined by $\langle a, \Phi_*\sigma\rangle = \langle \Phi a, \sigma\rangle$ for all $a \in \mathcal{M}$ and $\sigma \in \mathcal{N}_*$. Denote by $L^\infty(\Lambda, \mu, \mathcal{L}(\mathcal{H}))$ the von Neumann algebra of essentially bounded μ-measurable $\mathcal{L}(\mathcal{H})$-valued functions on Λ. The predual of $L^\infty(\Lambda, \mu, \mathcal{L}(\mathcal{H}))$ is the space $L^1(\Lambda, \mu, \tau c(\mathcal{H}))$ of Bochner μ-integrable $\tau c(\mathcal{H})$-valued functions with duality pairing

$$\langle A, F\rangle = \int_\Lambda \mathrm{Tr}[A(x)F(x)]\, d\mu(x)$$

for all $A \in L^\infty(\Lambda, \mu, \mathcal{L}(\mathcal{H}))$ and $F \in L^1(\Lambda, \mu, \tau c(\mathcal{H}))$. The μ-continuous operational measures are characterized as follows.

Theorem 4.1 *Let μ be a σ-finite measure on $\mathcal{B}(\Lambda)$. The Bochner integral formula*

$$\mathbf{X}(\Delta)\rho = \int_\Delta (\Phi_*\rho)(x)\, d\mu(x), \tag{4.1}$$

where $\rho \in \tau c(\mathcal{H})$ and $\Delta \in \mathcal{B}(\Lambda)$, sets up a one-to-one correspondence between μ-continuous operational measures $\mathbf{X}$ and unit-preserving normal completely positive maps Φ from $L^\infty(\Lambda, \mu, \mathcal{L}(\mathcal{H}))$ to $\mathcal{L}(\mathcal{H})$.

The unit-preserving normal CP map Φ in Eq. (4.1) is called the *operational distribution* of $\mathbf{X}$. Let $\mathbf{X}_1$ and $\mathbf{X}_2$ be CP map valued measures for $(\Lambda_1, \mathcal{H})$ and $(\Lambda_2, \mathcal{H})$, respectively. Suppose $\Lambda_1 \cap \Lambda_2 = \emptyset$. The *direct sum* of $\mathbf{X}_1$ and $\mathbf{X}_2$ is a CP map valued measure $\mathbf{X}$ for $(\Lambda_1 \cup \Lambda_2, \mathcal{H})$ such that $\mathbf{X}(\Delta) = \mathbf{X}_1(\Delta \cap \Lambda_1) + \mathbf{X}_2(\Delta \cap \Lambda_2)$ for all $\Delta \in \mathcal{B}(\Lambda_1 \cup \Lambda_2)$.

Theorem 4.2 *Every pure operational measure is decomposed into a direct sum of two CP map valued measures one of which is of the von Neumann-Davies type and the other of which is of the proper Gordon-Louisell type, and the decomposition is unique up to μ-almost everywhere.*

Proof. (Sketch) Let $\mathbf{X}$ be a μ-continuous pure operational measure for $(\Lambda, \mathcal{H})$, where μ is a σ-finite measure on $\mathcal{B}(\Lambda)$. From Theorem 4.1 and the Stinespring theorem, there is a separable Hilbert space $\mathcal{W}$, a nondegenerate normal *-representation $\pi : L^\infty(\Lambda, \mu, \mathcal{L}(\mathcal{H})) \to \mathcal{L}(\mathcal{W})$, and a linear isometry $W : \mathcal{H} \to \mathcal{W}$ such that $\mathbf{X}(\Delta)^* a = W^*\pi(a \otimes \chi_\Delta)W$ for all $a \in \mathcal{L}(\mathcal{H})$ and $\Delta \in \mathcal{B}(\Lambda)$. We may assume the minimality condition that $\pi[L^\infty(\Lambda, \mu, \mathcal{L}(\mathcal{H}))]W\mathcal{H} = \mathcal{W}$. By the type I representation theory, $\mathcal{W}$ admits the following direct integral decompositions:

$$\mathcal{W} = \int_\Lambda^\oplus \mathcal{H} \otimes \mathcal{K}(x)\, d\mu(x), \qquad \pi(a \otimes \Delta) = \int_\Delta^\oplus a \otimes 1(x)\, d\mu(x) \tag{4.2}$$

for all $a \in \mathcal{L}(\mathcal{H})$ and $\Delta \in \mathcal{B}(\Lambda)$ with the countable Borel partition $\{B_n|\ n = 1, \ldots, \aleph_0\}$ of Λ such that $\dim(\mathcal{K}(x)) = n$ for all $x \in B_n$. Let ξ be a vector in $\mathcal{H}$

with unit length. Put $\rho = |\xi\rangle\langle\xi|$ and $|\xi_x\rangle\langle\xi_x| = \rho_x$ Then, for any $\Delta \in \mathcal{B}(\Lambda)$ and $a \in \mathcal{L}(\mathcal{H})$, we have

$$\int_\Delta \langle\xi_x|a|\xi_x\rangle\langle\xi|d\mathbf{X}(x)^*1|\xi\rangle = \int_\Delta \langle(W\xi)(x)|a\otimes 1(x)|(W\xi)(x)\rangle\, d\mu(x). \tag{4.3}$$

Since $\mathbf{X}$ is pure, the linear functional $a \mapsto \langle(W\xi)(x)|a\otimes 1(x)|(W\xi)(x)\rangle$ is a pure positive linear functional or the zero functional for μ-almost all $x \in \Lambda$. Note that by the correspondence $\xi\otimes\eta \mapsto |\xi\rangle\langle\eta^*|$, $\mathcal{W}$ is isomorphic to the space $\sigma c(\mathcal{K}(x)^*, \mathcal{H})$ of Hilbert-Schmidt operators from $\mathcal{K}(x)^*$ to $\mathcal{H}$, so that we may identify $\mathcal{H}\otimes\mathcal{K}(x)$ with $\sigma c(\mathcal{K}(x)^*, \mathcal{H})$. Let $\mathcal{B} = \{\phi_i|\ i = 1, 2, \ldots\}$ be a basis of $\mathcal{H}$. Let $\mathcal{H}_0$ be the $\mathbf{Q}+i\mathbf{Q}$-linear span of $\mathcal{B}$. Then $\mathcal{H}_0$ is a countable dense subset of $\mathcal{H}_0$ and we can construct a $\mathbf{Q}+i\mathbf{Q}$-linear map $W_0(x)$ from $\mathcal{H}_0$ to $\sigma c(\mathcal{K}(x)^*, \mathcal{H})$ such that $W_0(x)\xi = (W\xi)(x)$ for μ-almost every $x \in \Lambda$. Then by necessary modification of values on a μ-null set, we may assume that the linear functional $a \mapsto \langle W_0(x)\xi|a\otimes 1(x)|W_0(x)\xi\rangle = \mathrm{Tr}[a(W_0(x)\xi)(W_0(x)\xi)^\dagger]$ is a pure positive linear functional on $\mathcal{L}(\mathcal{H}))$ or the zero functional for all $x \in \Lambda$ and $\xi \in \mathcal{H}_0$. Thus we have $\mathrm{rank}[(W_0(x)\xi)(W_0(x)\xi)^\dagger] \le 1$, so that $\mathrm{rank}[W_0(x)\xi] \le 1$. Let $x \in \Lambda$. For any $\xi \in \mathcal{H}_0$, there are vectors $\xi' \in \mathcal{H}$ and $\xi'' \in \mathcal{K}(x)^*$ such that $W_0(x)\xi = |\xi'\rangle\langle\xi''|$. Let $\xi \in \mathcal{H}_0$. By the additivity of $W_0(x)$, for any $\eta \in \mathcal{H}_0$, we have

$$|(\xi+\eta)'\rangle\langle(\xi+\eta)''| = |\xi'\rangle\langle\xi''| + |\eta'\rangle\langle\eta''|, \tag{4.4}$$

and hence either η' is linearly dependent on ξ' or η'' is linearly dependent on ξ''. We can further show by the additivity of $W_0(x)$ that when η varies over $\mathcal{H}_0$ either every η' is linearly dependent on ξ' or every η'' is linearly dependent on ξ''. We say that $x \in \Lambda$ is of type I if every η'' is linearly dependent on ξ'' and of type II otherwise. Let Λ_I be the set of all $x \in \Lambda$ of type I and $\Lambda_{II} = \Lambda\setminus\Lambda_I$. If $x \in B_1$ then $\dim(K(x)) = 1$ and hence x must be of type I. Let $C \in \mathcal{B}(\Lambda)$ be such that $C \subset \Lambda_I$. For any $x \in C$, there is a vector $\eta(x) \in \mathcal{K}(x)$ and a map $T(x) : \mathcal{H}_0 \to \mathcal{H}$ such that

$$W_0(x)\xi = T(x)\xi\otimes\eta(x)$$

for all $\xi \in \mathcal{H}_0$. Consider the linear subspace $\mathcal{W}_C = \pi(1\otimes\chi_C)\mathcal{W}$. By the minimality condition and the density of $\mathcal{H}_0$ in $\mathcal{H}$, the vectors of the form $\pi(a\otimes\chi_\Delta)W\xi$ with $a \in \mathcal{L}(\mathcal{H})$, $\Delta \in \mathcal{B}(\Lambda)$, $\Delta \subset C$, $\xi \in \mathcal{H}_0$ linearly generate a dense subspace of $\mathcal{W}_C$. By the relations

$$\pi(a\otimes\chi_\Delta)W\xi = \int_\Delta^\oplus aT(x)\xi\otimes\eta(x)\, d\mu(x), \tag{4.5}$$

we have $\dim(\mathcal{K}(x)) = 1$ for almost all $x \in C$. It follows that $\Lambda_I = B_1$ up to μ-null sets. For $x \in \Lambda_I$, $\mathcal{H}\otimes\mathcal{K}(x)$ can be naturally identified with $\mathcal{H}$ and we may assume

$$\int_{\Lambda_I}^\oplus \mathcal{H}\otimes\mathcal{K}(x) = L^2(\Lambda_I, \mu', \mathcal{H})$$

where μ' is the restriction of μ on $\mathcal{B}(\Lambda_I)$. Thus, for all $\Delta \in \mathcal{B}(\Lambda)$ and $\xi \in \mathcal{H}$, we have

$$\mathbf{X}(\Delta\cap\Lambda_1)|\xi\rangle\langle\xi| = \int_\Delta |(W\xi)(x)\rangle\langle(W\xi)(x)|\, d\mu(x\cap\Lambda_I).$$

For any $x \in \Lambda_{II}$, there is a vector $\eta(x) \in \mathcal{H}$ with $\|\eta(x)\| = 1$ and a map $T(x)$: $\mathcal{H}_0 \to \mathcal{K}(x)$ such that

$$W_0(x)\xi = \eta(x)\otimes T(x)\xi$$

for all $\xi \in \mathcal{H}_0$. Then we can show that

$$\mathbf{X}(\Delta \cap \Lambda_{II})|\xi\rangle\langle\xi| = \int_{\Delta} |\eta(x)\rangle\langle\eta(x)|\, \langle\xi|d\mathbf{X}(x \cap \Lambda_{II})^{*}1|\xi\rangle \tag{4.6}$$

for all $\Delta \in \mathcal{B}(\Lambda)$ and $\xi \in \mathcal{H}$. We have, therefore, proved that $d\mathbf{X}(x \cap \Lambda_I)$ is of the von Neumann-Devies type and that $d\mathbf{X}(x \cap \Lambda_{II})$ is of the Gordon-Louisell type. It is easy to verify the uniqueness of decomposition. □

References

[1] J. von Neumann. *Mathematical Foundations of Quantum Mechanics.* Princeton UP, Princeton, NJ, 1955.

[2] M. Ozawa. *J. Math. Phys.*, **25**, 79, (1984).

[3] M. Ozawa. *J. Math. Phys.*, **27**, 759, (1986).

[4] M. Ozawa. *J. Math. Phys.*, **34**, 5596, (1993).

[5] E. B. Davies and J. T. Lewis. *Commun. Math. Phys.*, **17**, 239, (1970).

[6] E. B. Davies. *Quantum Theory of Open Systems.* Academic Press, London, 1976.

[7] M. Ozawa. *Publ. RIMS, Kyoto Univ.*, **21**, 279, (1985).

[8] C. E. Shannon and W. Weaver. *The Mathematical Theory of Communication.* Univ. of Illinois Press, 1949.

[9] G. Lüders. *Ann. Physik (6)*, **8**, 322, (1951).

[10] H. J. Groenewold. *Int. J. Theor. Phys.*, **4**, 327, (1971).

[11] G. Lindblad. *Commun. Math. Phys.*, **28**, 245, (1972).

[12] E. B. Davies. *J. Funct. Anal.*, **6**, 318, (1970).

[13] M. Ozawa. *J. Math. Phys.*, **26**, 1948, (1985).

[14] J. P. Gordon and W. H. Louisell. In P. L. Kelly, B. Lax, and P. E. Tannenwald, editors, *Physics of Quantum Electronics*, pages 833–840, McGraw-Hill, New York, 1966.

[15] M. Ozawa. In P. Tombesi and E. R. Pike, editors, *Squeezed and Nonclassical Light*, pages 263–286, Plenum, New York, 1989.

STOCHASTIC DYNAMICS OF CONTINUOUSLY OBSERVED QUANTUM SYSTEMS [1]

P. Staszewski

Institute of Physics,
N. Copernicus University, Toruń, Poland

Abstract

The irreversible and stochastic behaviour of the continuously (in time) observed quantum system expressed by the reduction of the wave function cannot be described within the standard formulation of quantum mechanics. The state of a quantum system under continuous nondemolition observation evolves according to weakly-nonlinear filtering equation announced by Belavkin in 1988. This quantum stochastic differential equation of Ito type can be obtained from the unitary evolution of the compound system — "system plus measuring device" by conditioning with respect to the measuring trajectories. The equation has been applied to some significant physical problems. These include: stochastic resolution of quantum Zeno paradox for a free particle, watchdog effects for quantum particle for various cases of the diffusion nondemolition measurement and relaxation without mixing of an atom under a counting observation.

1 INTRODUCTION AND SUMMARY

The process of measurement entailing an interaction of a physical system with the measuring apparatus introduces the distortion of its free motion. While for classical systems this effect is negligible, for quantum systems the effect of measurements on their dynamics cannot be ignored. Based on two distinct laws of the temporal evolution, quantum mechanics meets serious difficulties in describing continuous (in time) measurements.

[1]Work supported in part by The State Committee for Scientific Research, under the project no. 2 0227 91 01.

To assure the repeatability of the results of the measurements of observables with discrete spectra, von Neumann proposed a law of temporal evolution valid just for instants of measurements — called *standard projection postulate.* By this postulate [1, 2], the measurement of some nondegenerate observable O with purely discrete spectrum giving the result x_k transforms the premeasurement mixed state ρ into ρ'_k, $\rho'_k = P_k \rho P_k$, where $P_k = |\psi_k >< \psi_k|$ stands for the projection onto the eigenspace of O corresponding to x_k. The state ρ'_k is normalized to the probability $p_k = \mathrm{Tr}[\rho P_k]$ of obtaining x_k. In other words, the transition $\rho \rightarrow \rho'_k$ takes place with probability p_k. Therefore, the dynamical evolution of quantum system at the instant of measurement has a stochastic character. It is rather obvious that if one were to perform measurements with high frequency (in the limit — continuously in time) the system would show stochastic irreversible behaviour for the whole period of observation.

For the system $\mathcal{S}$ undisturbed by an experiment the change of a mixed state is given by von Neumann equation which provides completely deterministic time development of $\mathcal{S}$. To each initial density operator $\rho(0)$ on the Hilbert space $\mathcal{H}$ of $\mathcal{S}$ there corresponds a unique state $\rho(t)$, $t \geq 0$, given by $\rho(t) = U(t)\rho(0)U^\dagger(t) = e^{-iHt/\hbar}\rho(0)e^{iHt/\hbar}$, where H stands for the hamiltonian of $\mathcal{S}$.

Combining these two types of evolution to describe a continuous observation within the time-interval $[0, t]$ by assuming the reduction postulate taking place at instants of ideal measurements kt/n, and the unitary evolution of $\mathcal{S}$ between the succeeding instants of measurements, one obtains in the limit $(n \rightarrow \infty)$ the so called quantum Zeno [2] paradox — suppressing the free (internal) dynamics of $\mathcal{S}$. Consecutive ideal measurements transform quantum states into the one corresponding to a single eigenvalue of the measured observable. The state change resulting from the internal (free) evolution of $\mathcal{S}$ during the time-interval of length t/n becomes negligible as $n \rightarrow \infty$, [3, 2].

A paradoxical character of the suppressing free dynamics can be visualized by examples: " ... if a [free] particle is 'continually observed' to determine whether it stays in $\mathcal{E}$ [$\mathcal{E} \subset \boldsymbol{R}^3$], then the probability of an affirmative answer is the same as the probability that the particle was in $\mathcal{E}$ initially" [4] [3] , or "An unstable particle observed continuously whether it has decayed or not will never be found to decay" [5]. As there are no isolated eigenvalues nor the corresponding atomic spectral projections in the continuum, for measurements of observables with continuous spectra the repeatability hypothesis is usually not assumed [1, 6 - 8]. The conventional projection postulate can be generalized to this case [9] by using non-normal conditional expectations. This excludes the explanation of the state change of $\mathcal{S}$ based on Hamiltonian formalism in a compound system consisting of $\mathcal{S}$ and a measuring apparatus.

Davies and Lewis [6, 7] have formulated the generalized collapse postulate in terms of the so-called operation-valued measures — *instruments* (instead of projection-valued measures) without fixing a particular rule of association between observables and operation-valued measures. The state changes given by instruments are compatible with the Hamiltonian formalism in the description of

[2] Zeno of Elea, 5th cent. B.C. Greek philosopher, was an outstanding member of a school of philosophers founded by Parmenides. Eleatic school is marked by belief in the unity of being and the unreality of motion or change.

[3] For unbounded H the study of the limit of continuous observation is no longer simple. The results of Friedman [4] allowed, in particular, to prove Zeno paradox for $H = -\frac{\hbar^2}{2m}\Delta$ acting in $\mathcal{H} = \mathcal{L}^2(\boldsymbol{R}^3)$ and a compact region $\mathcal{E} \subset \boldsymbol{R}^3$ with smooth boundary.

the measuring process in the composed system consisting of the observed system and the measuring apparatus [8]. Therefore, nonideal measurements, in particular measurements of nondiscrete observables, are possible in the instrumental approach [8, 10, 11]. In this approach, Barchielli, Lanz and Prosperi [12, 13] have resolved Zeno paradox by proving the existence of a continual limit ($\Delta t = t/n \underset{n\to\infty}{\longrightarrow} 0$) of consecutive *nonideal* measurements with decreasing accuracy (proportional to the number of measurements n).

The notion of instrument has enabled Ozawa [10, 11] to introduce the notion of quantum *posterior state* — the state of a quantum system conditioned by the result of a measurement x in a measurable set dx. This, in particular, makes it possible to describe state changes due to measurements continuous in time, conditioned by a trajectory of the observed data (stochastic processes). The *posterior dynamics* of a quantum system, i.e. the time-development of the system undergoing continuous (in time) observation is the main subject of the present paper.

Two types of processes — *counting* and *diffusion* — serve for monitoring of quantum systems. The process of photon counts provides a natural method of quantum optics for an indirect observation of radiating systems. The counting process, as a classical point stochastic process in quantum open (Markovian) systems, was introduced by Davies [14] and applications were worked out by Davies and others [14 - 20, 7]. Davies introduced an instrument on counting trajectories, describing the dynamical maps of a quantum system under the counting observation and discovered the master equation for the system density matrix. His master equation does not contain explicitly the attributes of the considered measurement, and might correspond to different measurement process (not only for that of counting-type). Davies gave a (global) solution to this equation, as an average of stochastic (posterior) operations with respect to the probability measure of photon counts, though a local description of the posterior dynamics remained unknown. This master equation was later derived by Lindblad [21] and Gorini et al. [22] in the general form as a quantum analogue of Fokker-Planck equation.

An important step towards the theory of continuous quantum measurements was the development of the characteristic-operational method [23, 12, 24 - 27]. In this approach, the differential master equation for the characteristic map of the measurement instrument for *continuous* as well as for *point processes* in quantum systems was obtained as well as the statistics of measurement data by the corresponding characteristic functional.

Recently, powerful methods of quantum stochastic calculus have been introduced into the quantum theory of open systems by Hudson and Parthasarathy [28] and Gardiner and Collett [29]. In the framework of quantum stochastic calculus all essential elements of the theory of continuous quantum measurements can be consistently described. These are: the stochastic interaction of the system in question with the measuring apparatus modelled by Bose fields, output field modes which serve for an indirect observation in quantum systems, and the corresponding quantum stochastic filtering. The method of indirect continuous observation by means of the output Bose processes first introduced by Belavkin [30, 31] has recently been developed by Barchielli and Lupieri [24, 27, 32] and Barchielli [33 - 35] on the grounds of quantum stochastic calculus, as proposed by Gardiner and Collett [29].

A local description of the dynamics of an observed quantum system, i.e. the time-development of its *state conditioned by a trajectory of the results of continuous measurement* (the posterior state) was derived by Belavkin. Belavkin's quantum filtering equation was obtained in the framework of quantum stochastic calculus of conditional expectations with respect to nondemolition measurements [36-41]. A nontrivial nondemolition observation [30], cf. also [36, 38, 41, 42] is provided by indirect measurements taken in $\mathcal{H}\otimes\mathcal{H}_R$ — the Hilbert space of a system interacting with the measuring apparatus.

The observed process $\{\,\widehat{\mathcal{Q}}(t), t \geq 0\,\}$ in $\mathcal{H}\otimes\mathcal{H}_R$ is *nondemolition* in the following sense:

for all Heisenberg systematic operators $\widehat{Z}(t)$ represented in $\mathcal{H}\otimes\mathcal{H}_R$ and for all $r \leq t$ one has

$$[\,\widehat{\mathcal{Q}}(r), \widehat{Z}(t)\,] = 0\,. \tag{1.1}$$

This means that the measurement of $\widehat{\mathcal{Q}}$ disturbs neither the present nor the future state of the system.

The observed process $\{\,\widehat{\mathcal{Q}}(t), t \geq 0\,\}$ is also assumed to be *self-nondemolition*

$$[\,\widehat{\mathcal{Q}}(r), \widehat{\mathcal{Q}}(t)\,] = 0 \quad \text{for all } r,\, t \geq 0\,. \tag{1.2}$$

Due to this property the observed process $\{\,\widehat{\mathcal{Q}}(t), t \geq 0\,\}$ can be treated as a classical one, and the measurement of $\widehat{\mathcal{Q}}$ does not meet any restrictions.

Belavkin quantum filtering equation plays the rôle analogous to that of the Schrödinger equation for unobserved quantum systems. It describes the reduction of a wave function not as a real dynamical process, but rather as the statistical evaluation of the results of the future measurements conditioned by the past observation [43, 79]. For both cases of observation — diffusion and counting — Belavkin equation can be derived within the model of unitary stochastic evolution of a compound system consisting of a quantum system in question plus Bose field in the vacuum state [38, 39, 42, 44-46]. Therefore, there is no need to postulate nonunitary and nonlinear stochastic differential equations describing the dynamical collapse (reduction) of wave function (Pearle [47-50], Gisin [51-53], Ghirardi and co-workers [54, 55], and Diósi [56-59]).

In section 2 the measurement scheme is discussed and a short presentation of the formal rules of quantum stochastic calculus is given.

In section 3 we present the derivation of the Belavkin filtering equation for a diffusion observation.

Section 4 is devoted to the application of the theory developed in previous chapters to describe the motion of a continuously observed quantum particle.

However throughout this paper we mostly deal with a diffusion observation, we would like to mention interesting applications of Belavkin's equation describing relaxation without mixing of a harmonic oscillator [44, 45] and a system with nonequidistant energy levels [73, 46, cf. also 77] under a continuous counting observation.

2 STOCHASTIC TIME-EVOLUTION OF 'THE SYSTEM PLUS APPARATUS'

2.1 Measuring apparatus

Let us assume that the measuring apparatus $\mathcal{A}$ is modelled by M-dimensional singular Bose reservoir in the vacuum state. $\mathcal{A}$ is described in terms of vector-operators

$$\boldsymbol{b}(\omega) \;=\; [b_j(\omega)]_{j=1}^{M}\,, \qquad \boldsymbol{b}^{\dagger}(\omega) \;=\; [b_j^{\dagger}(\omega)]_{j=1}^{M} \tag{2.1}$$

acting in $\Gamma(\boldsymbol{C}^M \otimes \mathcal{L}^2(\boldsymbol{R}))$, the symmetric Fock space over $\boldsymbol{C}^M \otimes \mathcal{L}^2(\boldsymbol{R})$ [60, 61]. For any ω, $\omega' \in \boldsymbol{R}$ and any pair of indices j, $l = 1, \ldots, M$ the (singular) annihilation and creation operators $b_j(\omega)$, $b_l^\dagger(\omega')$ satisfy the canonical commutation relations (CCR's)

$$\begin{aligned} [b_j(\omega), b_l^\dagger(\omega')] &= \delta_{jl}\delta(\omega - \omega'), \\ [b_j(\omega), b_l(\omega')] &= 0 = [b_j^\dagger(\omega), b_l^\dagger(\omega')]. \end{aligned} \tag{2.2}$$

The free Heisenberg evolution of the fields is generated by the hamiltonian H_A of $\mathcal{A}$

$$H_A = \hbar \sum_{j=1}^{M} \int_{-\infty}^{+\infty} \omega\, b_j^\dagger(\omega) b_j(\omega)\, d\omega. \tag{2.3}$$

Singular reservoirs model external macroscopic devices, driving open quantum systems to asymptotic states far from equilibrium [62 - 64].

2.2 Quantum system interacting with the measuring apparatus

Let us assume that a quantum system $\mathcal{S}$ with the hamiltonian H, living in the Hilbert space $\mathcal{H}$ is coupled at instant $t = 0$ to the measuring device $\mathcal{A}$. Under the assumptions: coupling is linear in the field operators, the rotating-wave approximation is made, and "coupling constants" do not depend on ω, the hamiltonian of the compound system $\mathcal{S} + \mathcal{A}$ takes the form

$$H_{S+A} = H + H_A + H_I, \tag{2.4}$$

where H_A is given by (2.3) and

$$H_I = \frac{i\hbar}{(2\pi)^{1/2}} \sum_{j=1}^{M} \int_{-\infty}^{+\infty} [L_j b_j^\dagger(\omega) - L_j^\dagger b_j(\omega)]\, d\omega. \tag{2.5}$$

In the latter equation L_j are coupling systematic operators. We assume the unitary evolution of the compound system $\mathcal{S} + \mathcal{A}$, generated by the hamiltonian H_{S+A}. By going to the interaction picture with respect to free dynamics of $\mathcal{A}$, generated by H_A (2.3), one gets

$$U(t) = \overleftarrow{T} \exp\left\{ -\frac{i}{\hbar} \int_0^t (H + H_I(s))\, ds \right\}, \tag{2.6}$$

where $\overleftarrow{T}$ stands for the chronological ordering operator. In the last formula

$$H_I = i\hbar \sum_{j=1}^{M} [L_j b_j^\dagger(t) - L_j^\dagger b_j(t)] \tag{2.7}$$

and

$$b_j(t) = \frac{1}{(2\pi)^{1/2}} \int_{-\infty}^{+\infty} e^{-i\omega t} b_j(\omega)\, d\omega. \tag{2.8}$$

Eq. (2.8) combined with CCR's (2.2) yields the following CCR's for the time-dependent field operators:

$$\begin{aligned} [b_j(t), b_l^\dagger(s)] &= \delta_{jl}\delta(t - s), \\ [b_j(t), b_l(s)] &= 0 = [b_j^\dagger(t), b_l^\dagger(s)] \qquad (j,\ l = 1, \ldots, M). \end{aligned} \tag{2.9}$$

The reservoir is initially prepared in the vacuum state, therefore

$$<b_j(t)>_v \;=\; <b_j^\dagger(t)>_v \;=\; <b_j^\dagger(t)b_l(s)>_v \;=\; 0\,, \tag{2.10}$$

$$<b_j(t)b_l^\dagger(s)>_v \;=\; \delta_{jl}\delta(t-s)\,. \tag{2.11}$$

The Hermitian and anti-Hermitian parts of $\boldsymbol{b}(t)$: $\mathrm{Re}\,\boldsymbol{b}(t) = (1/2)[\boldsymbol{b}(t) + \boldsymbol{b}^\dagger(t)]$, $\mathrm{Im}\,\boldsymbol{b}(t) = (1/2i)[\boldsymbol{b}(t) - \boldsymbol{b}^\dagger(t)]$ do not commute, but each of them has the statistical properties of (classical) standard M-dimensional white noise. Similarly as in the classical case [65], time-evolution of the system interacting with the reservoir can be described in a mathematically rigorous (and very convenient) way, in terms of a stochastic differential equation [28, 29]. One can define annihilation and creation processes

$$B_j(t) \;\stackrel{\mathrm{def}}{=}\; \int_0^t b_j(t')\,dt'\,, \quad B_j^\dagger(t) \;\stackrel{\mathrm{def}}{=}\; \int_0^t b_j^\dagger(t')\,dt'\,, \tag{2.12}$$

thus, $B_j(t)$, $B_j^\dagger(t)$ are annihilation and creation operators for the time-interval $[0,t]$. According to (2.9) the processes (2.12) satisfy the following commutation relations:

$$[B_j(t)\,,\,B_l(s)] \;=\; 0\,, \qquad [B_j(t)\,,\,B_l^\dagger(s)] \;=\; \delta_{jl}\min(t,s) \qquad (j,l=1,\ldots,M)\,. \tag{2.13}$$

Moreover, we define number processes

$$\mathcal{N}_j(t) \;=\; \int_0^t b_j^\dagger(t')b_j(t')\,dt'\,. \tag{2.14}$$

The families of operators

$$\{\boldsymbol{B}(t) \;=\; [B_j(t)]_{j=1}^M\,, \qquad \boldsymbol{B}^\dagger(t) \;=\; [B_j^\dagger(t)]_{j=1}^M\,, \qquad t\in\boldsymbol{R}_+\}\,, \tag{2.15}$$

and

$$\{\boldsymbol{\mathcal{N}}(t) \;=\; [\mathcal{N}_j(t)]_{j=1}^M\,, \qquad t\in\boldsymbol{R}_+\} \tag{2.16}$$

are quantum analogues of the standard M-dimensional Wiener diffusion process and the M-dimensional counting (Poissonian) process.

The family $\{F(t)\}$ of operators on $\Gamma(\boldsymbol{C}^M\otimes\mathcal{L}^2(\boldsymbol{R}_+))$ is called a quantum adapted process if $F(t) = \widetilde{F}(t)\otimes I_{[t}$, where $\widetilde{F}(t)$ is an operator in $\Gamma(\boldsymbol{C}^M\otimes\mathcal{L}^2(\boldsymbol{R}_+^t))$ and $I_{[t}$ is the identity operator on $\Gamma(\boldsymbol{C}^M\otimes\mathcal{L}^2(\boldsymbol{R}_{+t}))$. For adapted integrands Hudson and Parthasarathy [28] have developed a quantum stochastic integration of Ito type. Adapted intergands commute with the basic differentials $dB_j(t)$, $dB_j^\dagger(t)$, $d\mathcal{N}_j(t)$, and dt which 'point to the future'. These authors have therefore given a rigorous meaning to equations of the type

$$dM(t) \;=\; \sum_j^M \Big(F_j(t)\,d\mathcal{N}_j(t) + E_j(t)\,dB_j(t) + D_j(t)\,dB_j^\dagger(t)\Big) + C(t)\,dt\,, \tag{2.17}$$

where F, E, D, and C are adapted processes on $\mathcal{H}\otimes\Gamma$ and the fundamental integrators are of the form $I\otimes dN_j(t)$, etc.

If $M'(t)$ satisfies an equation of type (2.17), then the differential of the product $M(t)\,M'(t)$ satisfies the quantum Ito formula [28]

$$d(M(t)\cdot M'(t)) \;=\; dM(t)\cdot M'(t) + M(t)\cdot dM'(t) + dM(t)\cdot dM'(t)\,. \tag{2.18}$$

The stochastic differentials of the basic processes satisfy Hudson-Parthasarathy multiplication table

$$\begin{array}{c|cccc} & d\mathcal{N}_l(t) & dB_l(t) & dB_l^+(t) & dt \\ \hline d\mathcal{N}_j(t) & \delta_{lj} d\mathcal{N}_l(t) & 0 & \delta_{lj} dB_l^+(t) & 0 \\ dB_j(t) & \delta_{jl} dB_l(t) & 0 & \delta_{lj} dt & 0 \\ dB_j^+(t) & 0 & 0 & 0 & 0 \\ dt & 0 & 0 & 0 & 0 \end{array} \tag{2.19}$$

The dynamical evolution of the compound system (the system plus measuring apparatus) is assumed to be unitary. The unitary evolution operator, $U(t)$, satisfies the Ito quantum stochastic differential equation (QSDE), cf. [28, 29],

$$dU(t) = -KU(t)\,dt + \sum_{j=1}^{M} \left(L_j dB_j^\dagger(t) - L_j^\dagger dB_j(t)\right) U(t)\,, \qquad U(0) = I\,, \tag{2.20}$$

where

$$K = \frac{i}{\hbar} H + \frac{1}{2} \sum_{j=1}^{M} L_j^\dagger L_j \tag{2.21}$$

(with the second term on the right hand side of (2.21) being the Ito correction term). $dB_j^\dagger$-s and dB_j-s appearing in the "stochastic interaction hamiltonian" in eq. (2.20) are stochastic differentials of creation and annihilation processes (2.12).

The equation of motion for an arbitrary systematic operator $\widehat{Z}(t)$,

$$\widehat{Z}(t) = U^\dagger(t) Z U(t)\,, \qquad t \geq 0\,, \tag{2.22}$$

takes the form

$$d\widehat{Z} = \left(-\widehat{K}^\dagger \widehat{Z} - \widehat{Z}\widehat{K} + \sum_{j=1}^{M} \widehat{L}_j^\dagger \widehat{Z} \widehat{L}_j \right) dt + \sum_{j=1}^{M} \left([\widehat{L}_j^\dagger, \widehat{Z}]\, dB_j + [\widehat{Z}, \widehat{L}_j]\, dB_j^\dagger \right). \tag{2.23}$$

Eq. (2.23) follows from (2.22) by the quantum Ito formula (2.18) and Hudson-Parthasarathy multiplication table (2.19).

2.3 Output fields

QSDE (2.20) or QSDE-s (2.23) describe the distorted dynamics (in $\mathcal{H} \otimes \Gamma$) of the initially closed quantum system $\mathcal{S}$. The fields not only disturb the system, they also give a possibility of a continuous observation of $\mathcal{S}$. Let us consider the output Wiener diffusion processes

$$\{\widehat{\boldsymbol{B}}(t) = [\widehat{B}_j(t)]_{j=1}^M,\ \widehat{\boldsymbol{B}}^\dagger(t) = [\widehat{B}_j^\dagger(t)]_{j=1}^M,\ t \in \boldsymbol{R}_+\}\,. \tag{2.24}$$

Its components,

$$\widehat{B}_j(t) = U^\dagger(t) B_j(t) U(t)\,, \tag{2.25}$$

are Heisenberg operators corresponding to the input processes; in other words, they describe fields *after* interaction with $\mathcal{S}$. The annihilation processes $\widehat{B}_j(t)$ remain unchanged [35, 33] for all times $s \geq t$, i.e.

$$\widehat{B}_j(t) = U^\dagger(s) B_j(t) U(s)\,; \tag{2.26}$$

the same holds for the creation process $\widehat{\boldsymbol{B}}^{\dagger}(t)$. The property (2.26) is a consequence of two facts — QSDE (2.20) is written in the interaction picture with respect to the free dynamics of the fields and the coupling between $\mathcal{S}$ and the fields is singular. The output vector Wiener process satisfies Belavkin's nondemolition conditions (1.1), (1.2):

$$
\begin{aligned}
[\widehat{\boldsymbol{B}}(s)\,,\,\widehat{Z}(t)] &= U^{\dagger}(t)\,[\boldsymbol{B}(s)\,,\,Z]\,U(t) \;=\; 0 \qquad \forall s \leq t\,, \\
[\widehat{\boldsymbol{B}}^{\dagger}(s)\,,\,\widehat{Z}(t)] &= U^{\dagger}(t)\,[\boldsymbol{B}^{\dagger}(s)\,,\,Z]\,U(t) \;=\; 0 \qquad \forall s \leq t\,.
\end{aligned} \tag{2.27}
$$

By (2.9) the output vector "field coordinate" process

$$
\widehat{\boldsymbol{\mathcal{Q}}}(t) \;=\; \widehat{\boldsymbol{B}}(t) + \widehat{\boldsymbol{B}}^{\dagger}(t) \;=\; U^{\dagger}(t)\,\boldsymbol{\mathcal{Q}}(t)U(t) \tag{2.28}
$$

is selfnondemolition, namely

$$
[\widehat{\boldsymbol{\mathcal{Q}}}(t)\,,\,\widehat{\boldsymbol{\mathcal{Q}}}(t')] \;=\; 0 \qquad \forall t, t' \geq 0\,. \tag{2.29}
$$

Due to its Hermicity and selfcommuativity the output process $\widehat{\boldsymbol{\mathcal{Q}}}$ can be treated (in the representation in which it is diagonal) as a classical one. By (2.27) and (2.28) the measurement of $\widehat{\boldsymbol{\mathcal{Q}}}$ is nondemolition; for any systematic operator Z

$$
[\widehat{\boldsymbol{\mathcal{Q}}}(s)\,,\,\widehat{Z}(t)] \;=\; 0 \qquad \forall s \leq t\,. \tag{2.30}
$$

The measurement of $\widehat{\boldsymbol{\mathcal{Q}}}$ disturbs neither the present nor the future state of the system $\mathcal{S}$. Together with QSDE (2.23) for the unobserved process $\widehat{Z}(t)$ one can consider an analogous QSDE for $\widehat{\boldsymbol{\mathcal{Q}}}(t)$. Due to that $\widehat{Z}(t)$ becomes partially observed (as $\widehat{\boldsymbol{\mathcal{Q}}}(t)$ depends on the degrees of freedom of $\mathcal{S}$).

In the same way, the output counting process can be used for continuous monitoring of $\mathcal{S}$.

3 DERIVATION OF QUANTUM FILTERING EQUATION IN THE UNITARY MODEL OF AN INDIRECT DIFFUSION OBSERVATION

In this section we shall present the derivation of a QSDE giving the time-development of the posterior state of $\mathcal{S}$ conditioned by the continuous diffusion observation. To this end one can apply the approach via the generating map of the corresponding instrument. For one-dimensional reservoir this approach was applied by Belavkin [41], for multidimensional case by Belavkin and the author [42], and for reservoirs with possibly continuous energy spectrum by Belavkin [39]. The other possibility of deriving the quantum filtering equation is to perform the limit of quantum filtering equation for properly centered and scaled counting observation [66]. For the counting observation the corresponding evolution equation can be obtained in a similar manner [39, 45, 46].

We shall assume that initially closed quantum system $\mathcal{S}$ is coupled at $t = 0$ to M-dimensional Bose reservoir in the vacuum state $\delta_{\emptyset}$. We shall assume that the unitary evolution operator of the compound system, $U(t)$, satisfies QSDE (2.20). Information about the system $\mathcal{S}$ is contained in the output Wiener diffusion processes (2.24). The measured output vector "field coordinate" process (2.28) satisfies self-nondemolition (2.29) and nondemolition conditions (2.30).

One can observe that the measurement of $\widehat{\boldsymbol{Q}}(t)$ provides an imperfect measurement of the quantum observable $2\mathrm{Re}\,\widehat{\boldsymbol{L}}(t)$, where $\widehat{\boldsymbol{L}}(t)$ is the Heisenberg operator corresponding to the coupling systematic vector-operator $\boldsymbol{L}$:

$$\widehat{\boldsymbol{L}}(t) \; = \; U^{\dagger}(t)\widehat{\boldsymbol{L}}U(t)\,. \tag{3.1}$$

This can be seen from the QSDE for $\widehat{\boldsymbol{Q}}(t)$ (obtained in the same way as eq. (2.23))

$$d\widehat{\boldsymbol{Q}}(t) \; = \; [\,\widehat{\boldsymbol{L}}(t) \; + \; \widehat{\boldsymbol{L}}^{\dagger}(t)\,]\,dt \; + \; d\boldsymbol{Q}(t)\,, \tag{3.2}$$

in which $d\boldsymbol{Q}(t)$, the stochastic differential of the input process, plays the rôle of a random error.

Let us consider the instrument $\mathcal{J}^t$, from the algebra of operators Z of $\mathcal{S}$ into itself, with the value space Ω^t — the space of the trajectories of the observed process $\widehat{\boldsymbol{Q}}(t)$ that are stopped at t: $\boldsymbol{q}^t = \{\boldsymbol{q}(r)\,|\,r \leq t\}$. For the observed event, $d\boldsymbol{q}$, $\mathcal{J}^t$ gives the time-evolution of an initial state functional ρ to the state $\rho^t(d\boldsymbol{q}) = \rho\circ\mathcal{J}^t(d\boldsymbol{q})$ normalized to the probability $\mu^t(d\boldsymbol{q}) = \rho[\mathcal{J}^t(d\boldsymbol{q})[I]]$. The generating map Γ of the instrument $\mathcal{J}^t$ can be defined [41, cf. also 42, 39, 77] in the following way

$$<\psi|\,\Gamma(\boldsymbol{l},t)[Z]\psi> = <\psi\otimes\delta_{\emptyset}|\,\widehat{Y}(\boldsymbol{l},t)\widehat{Z}|\,\psi\otimes\delta_{\emptyset}>\,, \tag{3.3}$$

where $\boldsymbol{l}(t) = [l_j(t)]_{j=1}^{M}$ are the test functions with integrable complex components and the generating nondemolition process $\widehat{Y}(\boldsymbol{l},t)$ is defined as follows

$$\widehat{Y}(\boldsymbol{l},t) \; = \; \exp\left\{\sum_{j=1}^{M}\int_0^t l_j(r)\,d\widehat{Q}_j(r)\right\}\,. \tag{3.4}$$

Note, that the mean value of the operator $\widehat{Y}(\boldsymbol{l},t)$ is the generating functional of the Wiener diffusion process $\widehat{\boldsymbol{Q}}(t)$. Therefore, by (3.3) $<\psi|\,\Gamma(\boldsymbol{l},t)[I]\psi>$ is the generating functional for the instrument $\mathcal{J}^t$. Our aim is to find an explicit expression for the generating map Γ — this will be achieved by solving the differential equation for Γ. This equation can be found by taking the mean value with respect to $\eta = \psi\otimes\delta_{\emptyset}$ of both sides of the QSDE for the product $\widehat{Y}(t)\,\widehat{Z}(t) = U^{\dagger}(t)Y(t)ZU(t)$, where $Y(t)$ is the input process corresponding to (3.4)

$$Y(\boldsymbol{l},t) \; = \; \exp\Big\{\sum_{j=1}^{M}\int_0^t l_j(r)\,dQ_j(r)\Big\}\,. \tag{3.5}$$

By the classical Ito formula [65]

$$dY(\boldsymbol{l},t) \; = \; \sum_{j=1}^{M}\,[\,l_j(r)\,dQ_j(t) \; + \; \frac{1}{2}\,l_j^2(t)\,dt\,]\,Y(\boldsymbol{l},t)\,. \tag{3.6}$$

Then from Ito's formula (2.18) applied to the product $U^{\dagger}YZU$, QSDE (2.20), Hudson-Parthasarathy multiplication rules (2.19), formula (3.6), and by averaging of both sides of the obtained QSDE with respect to $\eta = \psi\otimes\delta_{\emptyset}$ one gets

$$<d\widehat{Y}d\widehat{Z}> = \Big\langle\widehat{\eta}(t)\Big|\sum_j\,[\frac{1}{2}l_j^2YZ + l_j(L_j^{\dagger}YZ + YZL_j) + L_j^{\dagger}YZL_j] - (K^{\dagger}YZ + YZK)\Big|\widehat{\eta}(t)\Big\rangle dt, \tag{3.7}$$

(The terms against dB_j-s and $dB_j^\dagger$-s have vanished, because for each j, $dB_j(t)\,U(t)\eta = U(t)\,dB_j(t)\,\eta = 0$). Eqs. (3.3) and (3.7) yield the forward differential equation for the generating map Γ

$$\frac{d}{dt}\Gamma[Z] = \Gamma\Big[\sum_j \big[\frac{1}{2}l_j^2 Z + l_j(L_j^\dagger Z + Z L_j) + L_j^\dagger Z\, L_j\big] - K^\dagger Z - ZK\Big]\,, \quad \Gamma(\boldsymbol{l},0)[Z] = Z\,. \tag{3.8}$$

The solution to (3.8) has the form [41, cf. also 42, 39, 77]

$$\Gamma(\boldsymbol{l},t)[Z] \;=\; \int_{\Omega^t} Y(\boldsymbol{l},\boldsymbol{q}^t)V^\dagger(\boldsymbol{q}^t)\,Z\,V(\boldsymbol{q}^t)\,d\nu(\boldsymbol{q}^t)\,, \tag{3.9}$$

where $\nu = \otimes_{j=1}^M \nu_j$ the standard product Wiener probability measure, $V(\boldsymbol{q}^t) \equiv V(t)\boldsymbol{q}$, and the stochastic propagator $V(t)$ satisfies the following QSDE

$$dV(t) \;=\; -KV(t)\,dt \;+\; \sum_j L_j V(t)\,dQ_j(t)\,, \qquad V(0) = I\,. \tag{3.10}$$

Hence, the wave function $\widehat{\varphi}(t) = V(t)\psi$ of the system $\mathcal{S}$ under the continuous nondemolition M-dimensional diffusion observation $\boldsymbol{\mathcal{Q}}$ satisfies the following QSDE

$$d\widehat{\varphi}(t) \;=\; -\Big(\frac{i}{\hbar}H + \frac{1}{2}\sum_j L_j^\dagger L_j\Big)\widehat{\varphi}(t)\,dt + \sum_j L_j\widehat{\varphi}(t)\,dQ_j(t)\,, \quad \widehat{\varphi}(0) = \psi\,. \tag{3.11}$$

Equation (3.11) plays the analogous rôle to the Schrödinger equation for the unobserved quantum system. (In (3.11) $d\boldsymbol{\mathcal{Q}}(t)$ can be replaced with $d\widehat{\boldsymbol{\mathcal{Q}}}(t)$ because in the Schrödinger picture $\boldsymbol{\mathcal{Q}}(t)$ and $\widehat{\boldsymbol{\mathcal{Q}}}(t)$ coincide.) As it was mentioned earlier, the process $\boldsymbol{\mathcal{Q}}(t)$ can be diagonalized. The $\boldsymbol{\mathcal{Q}}(t)$ appearing in (3.11) is considered as a classical M-dimensional diffusion process. The posterior wave function $\widehat{\varphi}$ is normalized to the probability density

$$p(\boldsymbol{q}^t) \;= <V(\boldsymbol{q}(t)\psi\,|\,V(\boldsymbol{q}^t)\psi> \equiv\; \widehat{p}(t)(\boldsymbol{q}) \tag{3.12}$$

of the observed process $\widehat{\boldsymbol{\mathcal{Q}}}(t)$ with respect to the standard product Wiener measure of the input process $\boldsymbol{\mathcal{Q}}(t)$. It follows from the integral representation (3.9) of (3.3) that

$$<\widehat{Y}(\boldsymbol{l},t)\,\widehat{Z}(t)> \;=\; \int_{\Omega^t} Y(\boldsymbol{l},\boldsymbol{q}^t) <V(\boldsymbol{q}^t)\psi\,|\,Z\,V(\boldsymbol{q}^t)\,\psi> d\nu(\boldsymbol{q}^t)\,. \tag{3.13}$$

The formula (3.13) gives for $Z = I$ the mean value of the output process (3.3) as the generating function of the output probability measure

$$d\mu(\boldsymbol{q}^t) \;=\; p(\boldsymbol{q}^t)\,d\nu(\boldsymbol{q}^t)\,. \tag{3.14}$$

Eq. (3.13) defines the *posterior mean value* $<Z>(\boldsymbol{q}^t)$ as

$$<Z>_t(\boldsymbol{q}^t) \;=\; <\widehat{\psi}(t)\,|\,Z\,\widehat{\psi}(t)>(\boldsymbol{q}^t) \;=\; <\psi(\boldsymbol{q}^t)\,|\,Z\,\psi(\boldsymbol{q}^t)> \;\equiv\; \widehat{z}(t)(\boldsymbol{q}) \tag{3.15}$$

in terms of the normalized posterior wave function

$$\widehat{\psi}(t)\,(\boldsymbol{q}) \;=\; \psi(\boldsymbol{q}^t)\,, \qquad \psi(\boldsymbol{q}^t) \;=\; \phi(\boldsymbol{q}^t)/p(\boldsymbol{q}^t)^{1/2}\,.$$

The normalized posterior wave function $\widehat{\psi}(t)$ satisfies the nonlinear Belavkin filtering equation

$$d\widehat{\psi}(t) \;=\; -\Big(\frac{i}{\hbar}\widetilde{H}(t) + \frac{1}{2}\sum_j \widetilde{L}_j^\dagger(t)\widetilde{L}_j(t)\Big)\widehat{\psi}(t)\,dt + \sum_j \widetilde{L}_j(t)\,d\widetilde{Q}_j(t)\,\widehat{\psi}(t)\,, \tag{3.16}$$

where $<L_j>_t = <\widehat{\psi}(t)|L_j\,\widehat{\psi}(t)>$ is the posterior mean value of L_j,

$$\widetilde{L}_j(t) \;=\; L_j \;-\; \mathrm{Re}\,<L_j>_t\,, \tag{3.17}$$

$$\widetilde{H}(t) \;=\; H \;-\; \hbar\sum_j \mathrm{Re}\,<L_j>_t\;\mathrm{Im}L_j\,, \tag{3.18}$$

and

$$d\widetilde{Q}_j(t) \;=\; dQ_j \;-\; 2\mathrm{Re}\,<L_j>_t\;dt \tag{3.19}$$

is the Ito differential of the observed commutative Wiener innovating process.

If the quantum system S is initially a Hamiltonian one, but the initial state $\sigma(0) = \rho$ is a density matrix, then the linear posterior dynamics (which does not preserve the normalization of the mixed state $\widehat{\sigma}(t)$) is given by the QSDE of the form

$$\begin{aligned} d\widehat{\sigma}(t) \;&=\; \Big(-\frac{i}{\hbar}[H\,,\,\widehat{\sigma}(t)] \;+\; \frac{1}{2}\sum_j \{[L_j\widehat{\sigma}(t)\,,\,L_j^\dagger] + [L_j\,,\,\widehat{\sigma}(t)L_j^\dagger]\}\Big)dt \\ &\quad +\; \sum_j (L_j\widehat{\sigma}(t) + \widehat{\sigma}(t)L_j^\dagger)\,dQ_j(t)\,. \end{aligned} \tag{3.20}$$

Eq. (3.20) follows easily from QSDE (3.11). The corresponding equation for the normalized posterior density matrix $\widehat{\rho}(t)$ reads

$$\begin{aligned} d\widehat{\rho}(t) \;&=\; \Big(-\frac{i}{\hbar}[H\,,\,\widehat{\rho}(t)] \;+\; \frac{1}{2}\sum_j \{[L_j\widehat{\rho}(t)\,,\,L_j^\dagger] + [L_j\,,\,\widehat{\rho}(t)L_j^\dagger]\}\Big)dt \\ &\quad +\; \sum_j (\widetilde{L}_j(t)\widehat{\rho}(t) + \widehat{\rho}(t)\widetilde{L}_j^\dagger(t))\,d\widetilde{Q}_j(t)\,. \end{aligned} \tag{3.21}$$

In the above formula the quantities marked with tildes, $\widetilde{L}_j(t)\,,\ \widetilde{Q}_j(t)$, are given by formulas (3.17) and (3.19) with $<L_j>_t = \mathrm{Tr}(\widehat{\rho}(t)L_j(t))$.

4 A QUANTUM PARTICLE UNDERGOING NONDEMOLITION DIFFUSION OBSERVATION

Bubble chamber experiments, in which the paths of ionizing particles are made by strings of vapor bubbles, are nice examples of indirect and imperfect continuous measurements. We shall study the dynamics of a quantum particle for various cases of diffusion nondemolition measurement by applying the Belavkin filtering equation. Particular attention will be paid to the case of a free observed particle.

Bubble chamber experiments, showing well-localized paths of observed quantum particles, present a great difficulty of the standard quantum mechanics. Combining projection postulate and the Schrödinger dynamics leads to quantum Zeno paradox (see Introduction). The Schrödinger equation alone describes, in the case of a free particle, the effect of spreading out of the wave packet. For a semigroup evolution equation the spreading out of the Gaussian packet is even faster (it spreads out as t^3) [42]. The presented theory, which includes the Belavkin filtering equation, offers a stochastic resolution of these difficulties of the standard quantum mechanics. The results presented in this section were published in [74] and [42] by Belavkin and the author, in [75], and in [76] by Chruściński and the author. In particular, the stochastic resolution of quantum Zeno paradox for a free particle with observed position was given in [74] and [42]. In [76] some further watchdog effects for a quantum particle (with observed momentum and position and momentum) were obtained. It was also proved that the linear Belavkin equation for a particle in a quadratic potential is uniquely relaxing.

4.1 A particle with observed position

We consider a quantum particle moving in the field of a linear force

$$\boldsymbol{F}(\boldsymbol{x}) \;=\; -\nabla\Phi(\boldsymbol{x}) \;=\; \hbar\gamma\boldsymbol{x} + m\boldsymbol{g}\,, \tag{4.1}$$

coupled to the three-dimensional Bose field in the vacuum state (measuring device), the position of which is continuously observed. We consider an indirect measurement of the particle position $\boldsymbol{X} = [X_1, X_2, X_3]$, therefore we choose the coupling vector-operator $\boldsymbol{L}$ to be proportional to $\boldsymbol{X}$,

$$\boldsymbol{L} = \left(\frac{\lambda}{2}\right)^{1/2} \boldsymbol{X}\,. \tag{4.2}$$

(We use the notation $\boldsymbol{X}\psi_x = \boldsymbol{x}\psi_x$.) The perturbed prior dynamics, cf. (2.23),

$$\begin{aligned} d\widehat{\boldsymbol{X}}(t) &= \frac{1}{m}\widehat{\boldsymbol{P}}(t)\,dt\,, \\ d\widehat{\boldsymbol{P}}(t) &= \boldsymbol{F}(\boldsymbol{x})\,dt + (2\lambda)^{1/2}\hbar\, d(\mathrm{Im}\boldsymbol{B}^\dagger(t))\,. \end{aligned} \tag{4.3}$$

describes the motion of the particle upon the sum of the deterministic force (4.1) and the stochastic (Langevin) force $\boldsymbol{f}(t) = (2\lambda)^{1/2}\hbar\,\mathrm{Im}\boldsymbol{b}^\dagger(t)$ from the Bose reservoir.

The observed nondemolition field coordinate process $\widehat{\boldsymbol{Q}}(t)$ (2.28) satisfies, due to (3.2) and (4.2), the QSDE in the form

$$d\widehat{\boldsymbol{Q}}(t) = (2\lambda)^{1/2}\,\widehat{\boldsymbol{X}}(t)\,dt + d\boldsymbol{Q}(t)\,. \tag{4.4}$$

Eq. (4.4) gives the indirect (and imperfect) measurement of the particle position. Rewriting eq. (4.4) in terms of the generalized derivatives of the processes $\widehat{\boldsymbol{Q}}(t)$ and $\boldsymbol{Q}(t)$ one has

$$\dot{\widehat{\boldsymbol{Q}}}(t) = (2\lambda)^{1/2}\,\widehat{\boldsymbol{X}}(t) + 2\mathrm{Re}\dot{\boldsymbol{B}}(t) = (2\lambda)^{1/2}\,\widehat{\boldsymbol{X}}(t) + 2\mathrm{Re}\boldsymbol{b}(t)\,, \tag{4.5}$$

in other words, the measurement of the (generalized) commutative stochastic process $\dot{\widehat{\boldsymbol{Q}}}(t)$ corresponds to the measurement of $\widehat{\boldsymbol{X}}(t)$ together with a random error represented by the standard vector white noise $2\mathrm{Re}\boldsymbol{b}(t)$. Note, that λ can be interpreted as the measurement accuracy coefficient.

Let $\widehat{\boldsymbol{q}}(t) = [\widehat{q}_j(t)]_{j=1}^3$ and $\widehat{\boldsymbol{p}}(t) = [\widehat{p}_j(t)]_{j=1}^3$ stand for posterior mean values of position and momentum of the observed particle,

$$\begin{aligned} \widehat{\boldsymbol{q}}(t) &= \int \overline{\widehat{\psi}}(t,\boldsymbol{x})\,\boldsymbol{x}\,\widehat{\psi}(t,\boldsymbol{x})\,d\boldsymbol{x}\,, \\ \widehat{\boldsymbol{p}}(t) &= \int \overline{\widehat{\psi}}(t,\boldsymbol{x})\,\frac{\hbar}{i}\,\nabla\,\widehat{\psi}(t,\boldsymbol{x})\,d\boldsymbol{x}\,. \end{aligned} \tag{4.6}$$

The Belavkin filtering equation (3.16) takes for the considered case the form

$$d\widehat{\psi}+\Big[-\frac{i\hbar}{2m}\triangle\widehat{\psi}+\Big(\frac{i}{\hbar}\Phi(\boldsymbol{x})+\frac{\lambda}{4}(\boldsymbol{x}-\widehat{q})^2\Big)\widehat{\psi}\Big]dt = \Big(\frac{\lambda}{2}\Big)^{1/2}(\boldsymbol{x}-\widehat{\boldsymbol{q}})\widehat{\psi}d\widetilde{\boldsymbol{Q}}\,, \quad \widehat{\psi}(0,\boldsymbol{x}) = \psi(\boldsymbol{x})\,, \tag{4.7}$$

where $d\widetilde{\boldsymbol{Q}}(t) = d\boldsymbol{Q}(t) - (2\lambda)^{1/2}\widehat{\boldsymbol{q}}(t)\,dt$.

The solution to eq. (4.7) corresponding to the initial state $\psi(\boldsymbol{x})$ of the form of Gaussian wave packet,

$$\psi(\boldsymbol{x}) \;=\; (2\pi\sigma_x^2)^{-3/4}\,\exp\Big\{-\frac{1}{4\sigma_x^2}(\boldsymbol{x}-\boldsymbol{q})^2 + \frac{i}{\hbar}\boldsymbol{p}\boldsymbol{x}\Big\}\,, \tag{4.8}$$

has the same form

$$\widehat{\psi}(t,\boldsymbol{x}) \;=\; \widehat{c}\,\exp\Big\{\frac{m}{2\hbar}\omega(t)(\boldsymbol{x}-\widehat{\boldsymbol{q}}(t))^2 \;+\; \frac{i}{\hbar}\,\widehat{\boldsymbol{p}}(t)\boldsymbol{x}\Big\}\,. \tag{4.9}$$

In (4.9) $\widehat{\boldsymbol{q}}(t)$, $\widehat{\boldsymbol{p}}(t)$ are posterior mean values (cf. (4.6)), $\widehat{c} \;=\; (2\tau_q^2\pi)^{-3/4}$ up to the unessential stochastic phase factor, and $\tau_q^2 \;=\; \widehat{\boldsymbol{q}^2} \;-\; \widehat{\boldsymbol{q}}^2$ is the posterior position dispersion.

Moreover, the posterior mean values $\widehat{\boldsymbol{q}}(t)$, $\widehat{\boldsymbol{p}}(t)$ fulfil linear filtration equations

$$\begin{aligned}
\widehat{\boldsymbol{p}}(t)\,dt \;-\; m\,d\widehat{\boldsymbol{q}}(t) \;&=\; \hbar\Big(\frac{\lambda}{2}\Big)^{1/2}\frac{d\widetilde{\boldsymbol{Q}}(t)}{\mathrm{Re}\,\omega(t)}\,, \qquad \widehat{\boldsymbol{q}}(0) \;=\; \boldsymbol{q}\,,\\
d\widehat{\boldsymbol{p}}(t) \;-\; m\boldsymbol{g}\,dt \;&=\; \hbar\gamma\widehat{\boldsymbol{q}}(t)\,dt - \hbar\Big(\frac{\lambda}{2}\Big)^{1/2}\frac{\mathrm{Im}\,\omega(t)}{\mathrm{Re}\,\omega(t)}\,d\widetilde{\boldsymbol{Q}}(t)\,, \qquad \widehat{\boldsymbol{p}}(0) \;=\; \boldsymbol{p}\,.
\end{aligned} \tag{4.10}$$

and $\omega(t)$ satisfies the Riccati differential equation

$$\frac{d}{dt}\,\omega(t) \;=\; i\Big[\frac{\hbar\gamma}{m} + \omega^2(t)\Big] \;-\; \frac{\hbar\lambda}{m}\,, \qquad \omega(0) \;=\; -\,\frac{\hbar}{2m\sigma_x^2}\,. \tag{4.11}$$

One can check easily that for the posterior wave function (4.9) the posterior position and momentum dispersions are given by the formulas

$$\begin{aligned}
\tau_x^2(t) \;&=\; -\,\frac{\hbar}{2m\,\mathrm{Re}\,\omega(t)}\,,\\
\tau_p^2(t) \;&=\; -\,\frac{\hbar m\,|\,\omega(t)|^2}{2\,\mathrm{Re}\,\omega(t)}\,,
\end{aligned} \tag{4.12}$$

where $\omega(t)$ is the solution to eq. (4.11). For each t the product of these dispersions satisfies the Heisenberg inequality $\tau_x^2(t)\tau_p^2(t) \geq \hbar^2/4$.

The general solution to eq. (4.11) has the form

$$\omega(t) \;=\; i\alpha\,\frac{\omega(0) \,+\, i\alpha\,\tanh{(\alpha t)}}{i\alpha \,+\, \omega(0)\,\tanh{(\alpha t)}}\,, \qquad \alpha \;=\; \Big(\frac{\hbar}{m}\Big)^{1/2}(\gamma + i\lambda)^{1/2}\,. \tag{4.13}$$

Obviously, $\lim_{t\to\infty}\omega(t) \;=\; i\alpha$, i.e. $i\alpha$ is the asymptotic stationary solution to eq. (4.11). Consequently, posterior dispersions of position and momentum tend to *finite limits independent of their initial values.* Moreover, (4.12) and (4.13) imply

$$\tau_x^2(\infty) \;=\; \frac{\hbar}{2m\,\mathrm{Im}\alpha} \;=\; \Big(\frac{\hbar}{2m}\Big)^{1/2}[(\gamma^2+\lambda^2)^{1/2} \,-\, \gamma]^{-1/2}\,, \tag{4.14}$$

$$\tau_p^2(\infty) \;=\; \frac{\hbar m\,|\alpha|^2}{2\,\mathrm{Im}\alpha} \;=\; \Big(\frac{\hbar^3 m}{2}\Big)^{1/2}\Big(\frac{\gamma^2 \,+\, \lambda^2}{(\gamma^2+\lambda^2)^{1/2} \,-\, \gamma}\Big)^{1/2}\,. \tag{4.15}$$

These are the particular watchdog effects which can be obtained from eqs. (4.14), (4.15):

(a) For $\gamma < 0$, i.e. for a harmonic oscillator, the width of the stationary Gaussian packet is smaller than that for the unobserved oscillator. If the accuracy $\lambda \to \infty$ then $\tau_x^2(\infty) \to 0$ but $\tau_p^2(\infty) \to \infty$.

(b) For $\gamma > 0$, i.e. for a linear active system (harmonic accelerator) we obtain a similar watchdog effect as in the case of a free particle, see (c).

(c) For $\gamma = 0$, i.e. for a particle in a homogeneous (gravitation or electric) field, in particular, for a free observed particle ($\boldsymbol{g} = \boldsymbol{0}$), the Gaussian packet does not spread out in time. The asymptotic posterior dispersions take the form

$$\tau_x^2(\infty) = \Big(\frac{\hbar}{2\lambda m}\Big)^{1/2}, \tag{4.16}$$

$$\tau_p^2(\infty) = \Big(\frac{\hbar^3 \lambda m}{2}\Big)^{1/2}. \tag{4.17}$$

As it follows from (4.16), the asymptotic localization of the particle in the coordinate representation is inversely proportional to its mass and the measurement accuracy λ. This means that the particle of mass zero cannot be localized by any measurement, and heavy particles ($m \to \infty$) can be localized at a point. According to the dimension of λ, $[\lambda] = (m^2 sec)^{-1}$, the measurement accuracy coefficient can be interpreted as inversely proportional to the scattering cross section and characteristic time of the transition process in a bubble chamber.

The asymptotic localization of the free quantum particle (4.16) was also obtained by other authors. Caves and Milburn [72] achieved this result in a continuous limit of the succeeding instantaneous nonideal measurements of the particle position under the assumption of the Gaussian character of the initial wave function and the Gaussian character of the instrument [13]. Diósi [58, 59] obtained (4.16) in the context of his phenomenological stochastic equation similar to eq. (3.16).

4.2 A particle with continuously observed momentum

Through this paragraph we shall work with the linear version of Belavkin's filtering equation. In order to describe the posterior dynamics of the quantum particle with continuously observed momentum one has to specify the coupling vector-operator $\boldsymbol{L}$ to be proportional to the momentum operator $\boldsymbol{P}$. Let us choose

$$\boldsymbol{L} = i\Big(\frac{\delta}{2}\Big)^{1/2} \boldsymbol{P} \tag{4.18}$$

with $\delta > 0$ being the accuracy coefficient of the nondemolition measurement of the particle momentum. Assuming that the corresponding measurement will be taken with the help of the process

$$\widehat{\mathcal{P}}(t) = i(\widehat{\boldsymbol{B}}^\dagger(t) - \widehat{\boldsymbol{B}}(t))$$

one obtains the QSDE for the observation process

$$\widehat{\mathcal{P}}(t) = d\mathcal{P}(t) + (2\delta)^{1/2}\,\widehat{\boldsymbol{P}}\,dt\,. \tag{4.19}$$

It can be verified rather easily that the underlying linear Belavkin equation takes the form

$$d\widehat{\varphi}(t) = -K\widehat{\varphi}(t)\,dt - \boldsymbol{L}\widehat{\varphi}(t)\,d\widehat{\mathcal{P}}(t)\,.$$

For a particle moving in a quadratic potential coupled to the Bose field via $\boldsymbol{L}$ given by (4.18) this equation takes the form

$$d\widehat{\varphi}(t) + \Big[- \Big(\frac{i\hbar}{2m} + \frac{\delta\hbar^2}{4}\Big)\Delta\widehat{\varphi}(t) + \frac{i}{\hbar}\Phi(\boldsymbol{x})\,\widehat{\varphi}(t)\Big]\, dt = -i\hbar\Big(\frac{\delta}{2}\Big)^{1/2}\nabla\widehat{\varphi}(t)\, d\widehat{\boldsymbol{\mathcal{P}}}(t)\,. \quad (4.20)$$

Assuming that the initial state is a Gaussian (unnormalized) wave packet, one can prove like in the case described above that the (unnormalized) solution has the form

$$\widehat{\varphi}(t,\boldsymbol{x}) \;=\; \exp\Big\{\frac{m}{2\hbar}\omega(t)(\boldsymbol{x}-\widehat{\boldsymbol{q}}(t))^2 + \frac{i}{\hbar}\widehat{\boldsymbol{p}}(t)\boldsymbol{x}\Big\}\,. \quad (4.21)$$

The posterior mean values (calculated with respect to $\widehat{\psi}(t,\boldsymbol{x}) = \widehat{\varphi}(t,\boldsymbol{x})/\|\widehat{\varphi}(t,\boldsymbol{x})\|$) satisfy the linear filtration (Hamilton-Langevin) equations

$$d\widehat{\boldsymbol{q}}(t) - \frac{1}{m}\Big(1 + \hbar\delta m\,\frac{\mathrm{Im}\omega(t)}{\mathrm{Re}\omega(t)}\Big)\widehat{\boldsymbol{p}}(t)\,dt \;-\; -\hbar\,\frac{\mathrm{Im}\omega(t)}{\mathrm{Re}\omega(t)}\Big(\frac{\delta}{2}\Big)^{1/2} d\widehat{\boldsymbol{\mathcal{P}}}(t)\,,$$
$$\widehat{\boldsymbol{q}}(0) \;=\; \boldsymbol{q}\,,$$
$$d\widehat{\boldsymbol{p}}(t) - \Big(\hbar\gamma\widehat{\boldsymbol{q}}(t) + \hbar\delta m\,\frac{|\,\omega(t)|^2}{\mathrm{Re}\omega(t)}\,\widehat{\boldsymbol{p}}(t) + m\boldsymbol{g}\Big)dt = -\hbar m\Big(\frac{\delta}{2}\Big)^{1/2}\frac{|\,\omega(t)|^2}{\mathrm{Re}\omega(t)}\,d\widehat{\boldsymbol{\mathcal{P}}}(t)\,,$$
$$\widehat{\boldsymbol{p}}(0) \;=\; \boldsymbol{p}\,. \quad (4.22)$$

The function $\omega(t)$ appearing in (4.20) satisfies the Riccati differential equation

$$\frac{d\omega(t)}{dt} - \frac{i\hbar\gamma}{m} - (\hbar\delta m + i)\,\omega(t)^2 \;=\; 0\,, \qquad \omega(0) \;=\; -\frac{\hbar}{2m\sigma_x^2}\,. \quad (4.23)$$

The general solution of eq. (4.23) reads

$$\omega(t) \;=\; \Big(\frac{\beta_1}{\beta_2}\Big)^{1/2}\frac{\omega(0) + \big(\frac{\beta_1}{\beta_2}\big)^{1/2}\tanh\big((\beta_1\beta_2)^{1/2}\,t\big)}{\big(\frac{\beta_1}{\beta_2}\big)^{1/2} + \omega(0)\tanh\big((\beta_1\beta_2)^{1/2}\,t\big)}\,, \quad (4.24)$$

where $\beta_1 = \frac{i\hbar\gamma}{m}$, $\beta_2 = -(\hbar\delta m + i)$. The posterior dispersions are given by formulas (4.12), they tend (as $t \to \infty$) to finite limits independent of their initial values

$$\tau_x^2(\infty) \;=\; \Big(\frac{3\hbar}{2m\gamma}\Big)^{1/2}\Big(\frac{1 + (\hbar\delta m)^2}{(1+(\hbar\delta m)^2)^{1/2} - 1}\Big)^{1/2}\,, \quad (4.25)$$

$$\tau_p^2(\infty) \;=\; \Big(\frac{\hbar^3 m\gamma}{6}\Big)^{1/2}\big((1+(\hbar\delta m)^2)^{1/2} - 1\big)^{-1/2}\,. \quad (4.26)$$

From these equalities one can see that if the accuracy $\delta \to \infty$ then $\tau_x^2(\infty) \to \infty$ but $\tau_p^2(\infty) \to 0$), so one obtains asymptotic localization of the particle momentum. In particular, for a free particle ($\gamma = 0$) one has $\tau_x^2(\infty) = \infty$ and $\tau_p^2(\infty) = 0$.

4.3 A joint observation of position and momentum

The presented approach to the continuous observation allows to perform a joint measurement of noncommuting observables. The directly measured quantities are components of self-adjoint processes $\widehat{\boldsymbol{\mathcal{Q}}}$ or $\widehat{\boldsymbol{\mathcal{P}}}$ (a joint measurement of them is possible provided they refer to independent modes of the field) and there is no restriction imposed on the coupling vector-operator $\boldsymbol{L}$. The components of $\boldsymbol{L}$ should be proportional to the measured operators, and the number of independent Bose fields should be equal to their number.

Our aim is to perform a joint measurement of $\boldsymbol{X}$ and $\boldsymbol{P}$ of a three-dimensional quantum particle moving in the field of a linear force (4.1). The complete measurement requires a six-dimensional Bose field $\boldsymbol{B}(t) = (\boldsymbol{B}_1(t), \boldsymbol{B}_2(t))$. Let us choose

$$\boldsymbol{L} = \left(\left(\frac{\lambda}{2}\right)^{1/2} \boldsymbol{X}, i\left(\frac{\delta}{2}\right)^{1/2} \boldsymbol{P}\right), \tag{4.27}$$

the modes marked with 1 are therefore coupled to the particle via $\boldsymbol{X}$, while those marked with 2 are coupled to the particle via $\boldsymbol{P}$. The observed process is $(\widehat{\boldsymbol{\mathcal{Q}}}(t), \widehat{\boldsymbol{\mathcal{P}}}(t))$, where

$$\widehat{\boldsymbol{\mathcal{Q}}}(t) = \widehat{\boldsymbol{B}}_1^\dagger(t) + \widehat{\boldsymbol{B}}_1(t), \qquad \widehat{\boldsymbol{\mathcal{P}}}(t) = i(\widehat{\boldsymbol{B}}_2^\dagger(t) - \widehat{\boldsymbol{B}}_2(t)). \tag{4.28}$$

One can verify rather easily that th Belavkin filtering equation takes the form

$$d\widehat{\varphi}(t) + K\widehat{\varphi}(t)\, dt = \boldsymbol{L}\widehat{\varphi}(t)\, d\widehat{\boldsymbol{\mathcal{R}}}(t), \tag{4.29}$$

where

$$\widehat{\boldsymbol{\mathcal{R}}}(t) = (\widehat{\boldsymbol{\mathcal{Q}}}(t), -i\widehat{\boldsymbol{\mathcal{P}}}(t)). \tag{4.30}$$

From (4.4) and (4.19) one obtains

$$d\widehat{\boldsymbol{\mathcal{R}}}(t) = d\boldsymbol{\mathcal{R}}(t) + ((2\lambda)^{1/2}\widehat{\boldsymbol{X}}(t), -i(2\delta)^{1/2}\widehat{\boldsymbol{P}}(t)). \tag{4.31}$$

For a particle moving in a quadratic potential, eq. (4.29) becomes

$$\begin{aligned} d\widehat{\varphi}(t) + \Big[-\Big(\frac{i\hbar}{2m} + \frac{\delta\hbar^2}{4}\Big)\Delta\widehat{\varphi}(t) + \Big(\frac{\lambda}{4}\boldsymbol{x}^2 + \frac{i}{\hbar}\Phi(\boldsymbol{x})\Big)\widehat{\varphi}(t)\Big] dt \\ = \left(\frac{\lambda}{2}\right)^{1/2} \boldsymbol{x}\widehat{\varphi}(t)\, d\widehat{\boldsymbol{\mathcal{Q}}}(t) - i\hbar\left(\frac{\delta}{2}\right)^{1/2}\nabla\widehat{\varphi}(t)\, d\widehat{\boldsymbol{\mathcal{P}}}(t). \end{aligned} \tag{4.32}$$

This equation can be solved with the method applied in previous cases. Under the assumption that the initial state is a Gaussian wave packet one obtains the unnormalized solution in the form (4.20). The posterior mean values of position and momentum fulfil now the following linear filtration equations

$$\begin{aligned} d\widehat{\boldsymbol{q}}(t) &- \frac{1}{m}\left(\frac{\lambda\hbar}{\mathrm{Re}\omega(t)}\widehat{\boldsymbol{q}} + 3\Big(1 + \hbar\delta m\frac{\mathrm{Im}\omega(t)}{\mathrm{Re}\omega(t)}\Big)\widehat{\boldsymbol{p}}(t)\right) dt \\ &= \left(\frac{\lambda}{2}\right)^{1/2}\frac{\hbar}{\mathrm{Re}\omega(t)}\, d\widehat{\boldsymbol{\mathcal{Q}}}(t) - \hbar\left(\frac{\delta}{2}\right)^{1/2}\frac{\mathrm{Im}\omega(t)}{\mathrm{Re}\omega(t)}\, d\widehat{\boldsymbol{\mathcal{P}}}(t), \quad \widehat{\boldsymbol{q}}(0) = \boldsymbol{q}, \\ d\widehat{\boldsymbol{p}}(t) &- \left(\hbar\Big(\gamma + \lambda\frac{\mathrm{Im}\omega(t)}{\mathrm{Re}\omega(t)}\Big)\widehat{\boldsymbol{q}}(t) + 3\hbar\delta m\frac{|\omega(t)|^2}{\mathrm{Re}\omega(t)}\widehat{\boldsymbol{p}}(t) + m\boldsymbol{g}\right) dt \\ &= -\hbar\left(\frac{\lambda}{2}\right)^{1/2}\frac{\mathrm{Im}\omega(t)}{\mathrm{Re}\omega(t)}\, d\widehat{\boldsymbol{\mathcal{Q}}}(t) - 3\hbar m\left(\frac{\delta}{2}\right)^{1/2}\frac{|\omega(t)|^2}{\mathrm{Re}\omega(t)}\, d\widehat{\boldsymbol{\mathcal{P}}}(t), \quad \widehat{\boldsymbol{p}}(0) = \boldsymbol{p}, \end{aligned} \tag{4.33}$$

and $\omega(t)$ satisfies the Riccati differential equation

$$\frac{d\omega(t)}{dt} + \frac{\hbar}{m}(\lambda - i\gamma) - 3(\hbar\delta m + i)\,\omega(t)^2 = 0. \tag{4.34}$$

The general solution of (4.34) can be written in the form (4.24) with $\beta_1 = \hbar(i\gamma - \lambda)$ and $\beta_2 = -3(\hbar\delta m + i)$. With the help of (4.12) one can obtain posterior dispersions for position and momentum. These dispersions tend to finite limits independent of their initial values

$$\tau_x^2(\infty) = \Big(\frac{3\hbar}{2m}(u^2+1)\Big)^{1/2} (\lambda u - \gamma + [(\gamma^2+\lambda^2)(u^2+1)]^{1/2})^{-1/2}, \quad (4.35)$$

$$\tau_x^2(\infty) = \Big(\frac{\hbar^3 m}{6(u^2+1)}\Big)^{1/2} \Big(\frac{(\gamma^2+\lambda^2)(u^2+1)}{u\lambda - \gamma + [(\gamma^2+\lambda^2)(u^2+1)]^{1/2}}\Big)^{1/2}, \quad (4.36)$$

where

$$u = \hbar\delta m. \quad (4.37)$$

In the cases $\delta = 0$ and $\lambda = 0$ eqs. (4.36) reduce to eqs. (4.14, 4.15) and eqs. (4.25), respectively. If the accuracy $\lambda \to \infty$ then eq. (4.38) yields: $\tau_x^2(\infty) \to 0$ but $\tau_p^2(\infty) \to \infty$. On the other hand, if $\delta \to \infty$ then $\tau_p^2(\infty) \to 0$ but $\tau_x^2(\infty) \to \infty$.

For a free particle ($\gamma = 0$) eqs. (4.36) reduce to

$$\tau_x^2(\infty) = \Big(\frac{3\hbar}{2m\lambda}(u^2+1)\Big)^{1/2} (u + u\,(u^2+1)^{1/2})^{-1/2} \quad (4.38)$$

$$\tau_p^2(\infty) = \Big(\frac{\hbar^3 m\lambda}{6(u^2+1)}\Big)^{1/2} \Big(\frac{u^2+1}{u + (u^2+1)^{1/2}}\Big)^{1/2}. \quad (4.39)$$

4.4 A comparison of position-momentum uncertainties for various cases of observation

In this paragraph we shall compare the product

$$\Delta = \tau_x^2(\infty)\,\tau_p^2(\infty)$$

of asymptotic dispersions in the three situations considered.

A free particle

1. A particle undergoes the continuous observation of position. By (4.16) and (4.17)

$$\Delta_1(\lambda) = \frac{\hbar^2}{2}, \quad (4.40)$$

and Δ_1 does not depend on λ.

2. A particle undergoes the continuous observation of momentum. By (4.25) one gets

$$\Delta_2(\delta) = \frac{((\hbar\delta m)^2+1)^{1/2}}{((\hbar\delta m)^2+1)^{1/2} - 1}\,\frac{\hbar^2}{2}. \quad (4.41)$$

3. A particle undergoes the continuous observation of position and momentum. By (4.36)

$$\Delta_3(\lambda,\delta) = \frac{((\hbar\delta m)^2+1)^{1/2}}{\hbar\delta m + ((\hbar\delta m)^2+1)^{1/2}}\,\frac{\hbar^2}{2}, \quad (4.42)$$

and Δ_3 is independent of λ.

From (4.40) to (4.42) one obtains

$$\frac{\hbar^2}{4} \le \Delta_3 \le \Delta_1 \le \Delta_2. \quad (4.43)$$

The position-momentum asymptotic uncertainty for a free particle with observed momentum monotonically decreases to $\hbar^2/2$ — the position-momentum uncertainty (as the momentum measurement accuracy coefficient $\delta \to \infty$). Similarly, the asymptotic position-momentum uncertainty for a particle with observed position and momentum decreases monotonically to $\hbar^2/4$ — the minimal value allowed by Heisenberg uncertainty principle.

A particle in a quadratic potential

For the case of a particle in a quadratic potential the corresponding products of dispersions take the form

$$\triangle_1(\gamma, \lambda) = \frac{(\gamma^2 + \lambda^2)^{1/2}}{(\gamma^2 + \lambda^2)^{1/2} - \gamma} \frac{\hbar^2}{2}, \tag{4.44}$$

$$\triangle_2(\delta) = \frac{((\hbar\delta m)^2 + 1)^{1/2}}{((\hbar\delta m)^2 + 1)^{1/2} - 1} \frac{\hbar^2}{2}, \tag{4.45}$$

$$\triangle_3(\gamma, \lambda, \delta) = \frac{[(\gamma^2 + \lambda^2)((\hbar\delta m)^2 + 1)]^{1/2}}{\hbar\delta\lambda m - \gamma + [(\gamma^2 + \lambda^2)((\hbar\delta m)^2 + 1)]^{1/2}} \frac{\hbar^2}{2}. \tag{4.46}$$

For each γ, λ, δ $\triangle_3 \leq \triangle_k$ ($k = 1, 2$), i.e. the asymptotic position-momentum uncertainty for the particle with continuously observed position and momentum is smaller than those resulting from the continuous observation of position and momentum alone.

The results concerning continuously observed quantum particle moving in a quadratic potential can be strengthened by relaxing the assumption of the initial state in the form of the Gaussian packet. As it was proved in [76], the linear Belavkin equation is uniquely relaxing for all three cases of observation — any initial state of the particle given by a square-intergable function relaxes to the unique (up to a stochastic phase factor) Gaussian wave packet with dispersions, (4.14), (4.25) or (4.38), respectively. Large-time behaviour of solutions of nonlinear Belavkin equation was discussed in [78].

References

[1] J. von Neumann, *Mathematical Foundations of Quantum Mechanics*, trans. by R. T. Beyer, Princeton Unive. Press, Princeton, 1955.

[2] A. Sudbery, *Quantum Mechanics and the Particles of Nature*, Cambridge Unive. Press, Cambridge, 1986.

[3] A. S. Holevo, *Quantum Probability and Quantum Statistics*, in *Progress in Science and Technology: Contemporary Problems in Mathematics*, Vol. 83, VINITI, Moscow, 1991 (in Russian).

[4] Ch. N. Friedman, Indiana Univ. Math. J. **21**, 1001 (1972).

[5] B. Misra and E. C. G. Sudarshan, J. Math. Phys. **18**, 756 (1977).

[6] E. B. Davies and J. T. Lewis, Commun. Math. Phys. **17**, 239 (1970).

[7] E. B. Davies, *Quantum Theory of Open Systems*, Academic Press, London, 1976.

[8] M. Ozawa, J. Math. Phys. **25**, 79 (1984).

[9] M. D. Srinivas, Commun. Math. Phys. **71**, 131 (1980).

[10] M. Ozawa, in *Probability Theory and Quantum Statistics*, LNM 1021, Springer, Berlin, 1983, p. 518.

[11] M. Ozawa, Publ. RIMS, Kyoto Univ. **21**, 279 (1985).

[12] A. Barchielli, L. Lanz, and G. M. Prosperi, Nuovo Cimento **72B**, 79 (1982).

[13] A. Barchielli, L. Lanz, and G. M. Prosperi, Found. Phys. **13**, 779 (1983).

[14] E. B. Davies, Commun. Math. Phys. **15**, 277 (1969).

[15] E. B. Davies, Commun. Math. Phys. **19**, 83 (1970).

[16] E. B. Davies, Commun. Math. Phys. **22**, 51 (1971).

[17] E. B. Davies, IEEE Trans. Inf. Theory **23**, 530 (1977).

[18] M. D. Srinivas and E. B. Davies, Optica Acta **28**, 981 (1980), **29**, 235 (1982).

[19] A. S. Holevo, Izv. Vuz. Mat. **26**, 3 (1982).

[20] P. Zoller, M. Marte, and D. F. Walls, Phys. Rev. A **35**, 198 (1987).

[21] G. Lindblad, Commun. Math. Phys. **48**, 119 (1976).

[22] V. Gorini, A. Kossakowski, and E. C. G. Sudarshan, J. Math. Phys. **17**, 821 (1976).

[23] V. P. Belavkin, in Proc. V Symp. Inf. Theory, Vol. 1, Moscow-Tbilisi, 1979, p. 37.

[24] G. Lupieri, J. Math. Phys. **24**, 2329 (1983).

[25] A. Barchielli, Nuovo Cimento **74B**, 113 (1983).

[26] A. Barchielli, Phys. Rev. D **32**, 347 (1985).

[27] A. Barchielli and G. Lupieri, J. Math. Phys. **26**, 2222 (1985).

[28] R. L. Hudson and K. R. Parthasarathy, Commun. Math. Phys. **93**, 301 (1984).

[29] C. W. Gardiner and M. J. Collett, Phys. Rev. A **31**, 3761 (1985).

[30] V. P. Belavkin, Radiotechnika i Elektronika **35**, 1445 (1980).

[31] V. P. Belavkin, in *Information Complexity and Control in Quantum Physics*, edited by A. Blaquiere, S. Diner, and G. Lochak, Springer, Berlin, 1987, p. 310.

[32] A. Barchielli and G. Lupieri, in *Quantum Probability and Applications II*, LNM 1136, Springer, Berlin, 1985, p. 57.

[33] A. Barchielli, Phys. Rev. A **34**, 1642 (1986).

[34] A. Barchielli, J. Phys. A: Math. Gen. **20**, 6341 (1987).

[35] A. Barchielli, in *Quantum Probability and Applications III*, LNM 1303, Springer, Berlin, 1988, p. 37.

[36] V.P. Belavkin, in *Modeling and Control of Systems in Engineering, Quantum Mechanics, Economics and Biosciences*, edited by A. Blaquiere, Springer, Berlin, 1988, p. 245.

[37] V.P. Belavkin, in *Stochastic Methods in Mathematics and Physics*, edited by R. Gielerak and W. Karwowski, World Scientific, Singapore 1989, p. 310.

[38] V.P. Belavkin, Phys. Lett. A **140**, 355 (1989).

[39] V.P. Belavkin, J. Phys. A: Math. Gen. **22**, L1109 (1989).

[40] V.P. Belavkin, Lett. Math. Phys. **20**, 85 (1990).

[41] V.P. Belavkin, J. Math. Phys. **31**, 2930 (1990).

[42] V.P. Belavkin and P. Staszewski, Phys. Rev. A **45**, 1347 (1992).

[43] V.P. Belavkin, Commun. Math. Phys. **146**, 611 (1992).

[44] P. Staszewski, A Limit Theorem for Belavkin's Filtering Equation, Proc. 5th Intern. Vilnius Conf. Prob. Theory and Math. Statistics, Vilnius 1989, p. 163.

[45] V.P. Belavkin and P. Staszewski, Rep. Math. Phys. **29**, 213 (1991).

[46] P. Staszewski and G. Staszewska, Open Sys. Information Dyn. **1**, 103 (1993).

[47] Ph. Pearle, Phys. Rev. D **13**, 857 (1976).

[48] Ph. Pearle, Phys. Rev. Lett. **53**, 1775 (1984).

[49] Ph. Pearle, Phys. Rev. D **33**, 2240 (1986).

[50] Ph. Pearle, Phys. Rev. A **39**, 2277 (1989).

[51] N. Gisin, Phys. Rev. Lett. **52**, 1657 (1984).

[52] N. Gisin, Phys. Rev. Lett. **53**, 1776 (1984).

[53] N. Gisin, Helv. Phys. Acta **62**, 363 (1989).

[54] G.C. Ghirardi, A. Rimini, and T. Weber, Phys. Rev. D **34**, 470 (1986).

[55] G.C. Ghirardi, Ph. Pearle, and A. Rimini, Phys. Rev. A **42**, 78 (1990).

[56] L. Diósi, J. Phys. A **21**, 2885 (1988).

[57] L. Diósi, Phys. Lett. A **129**, 419 (1988).

[58] L. Diósi, Phys. Lett. A **132**, 233 (1988).

[59] L. Diósi, Phys. Rev. A **40**, 1165 (1989).

[60] A. Guichardet, *Symmetric Hilbert Spaces and Related Topics*, LNM 261, Springer, Berlin, 1972.

[61] M. Reed and B. Simon, *Methods of Modern Mathematical Physics*, Vol. 1: *Functional Analysis*, Academic Press, New York, 1972; Vol. 2: *Fourier Analysis, Self-Adjointness*, Academic Press, New York, 1975.

[62] R. Alicki and K. Lendi, *Quantum Dynamical Semigroups and Applications*, LNP 286, Springer, Berlin, 1987.

[63] V. Gorini, A. Frigerio, M. Verri, A. Kossakowski, and E. C. G. Sudarshan, Rep. Math. Phys. **13**, 149 (1978).

[64] K. Hepp and E. Lieb, Helv. Phys. Acta **46**, 573 (1973).

[65] C. W. Gardiner, *Handbook of Stochastic Methods*, Springer, Berlin, 1983.

[66] A. Barchielli, V. P. Belavkin, J. Phys. A: Math. Gen. **24**, 1495 (1991).

[67] K. Kraus, Ann. Phys. **64**, 311 (1971).

[68] W. F. Stinespring, Proc. Am. Math. Soc. **6**, 211 (1955).

[69] K. Yosida, *Functional Analysis*, 2nd ed. Springer, Berlin, 1968.

[70] A. S. Holevo, *Probabilistic and Statistical Aspects of Quantum Theory*, North-Holland, Amsterdam, 1982.

[71] H. Maassen, in *Quantum Probability and Applications II*, LNM 1136, Springer, Berlin, 1985, p. 534.

[72] C. M. Caves and G. J. Milburn, Phys. Rev. A **36**, 5543 (1987).

[73] P. Staszewski and G. Staszewska, Europhys. Lett. **20**, 191 (1992).

[74] V. P. Belavkin and P. Staszewski, Phys. Lett. A **140**, 359 (1989).

[75] P. Staszewski, Stochastic Dynamics of Continuously Observed Quantum Particle, Proc. 23rd Conf. EGAS, Toruń 1991, p. 217.

[76] D. Chruściński and P. Staszewski, Physica Scripta **45**, 193 (1992).

[77] P. Staszewski, *Quantum Mechanics of Continuously Observed Systems*, N. Copernicus University Press, Toruń, 1993.

[78] V. N. Kolokol'tsov, Preprint No. 228, Institut für Mathematik, Ruhr-Universität Bochum, 1994.

[79] V. P. Belavkin, Found. Phys. **24**, 685 (1994).

ON THE DYNAMICAL INTERPRETATION FOR THE COLLAPSE OF STATE DURING QUANTUM MEASUREMENT

Rouslan L. Stratonovich [1]

Mathematics Department,
University of Nottingham, UK

Abstract

The non-selective collapse of the state of the quantum system S is interpreted as a dynamical process of interacting between S and an apparatus A. The selective collapse is interpreted dynamically as a consequence of two factors: (a) interaction between S and A and (b) the non-quantum collapse. The latter is nothing else, but essentially the well-known transition to the conditional probability distribution. In our interpretation this transition takes place when the macroscopic inexact measurement of some apparatus physical variable is made. For equivalence of the non-quantum and the quantum measurements the apparatus should be in a strongly quasi-classical state. Of course, certain mathematical difficulties should be overcome for actual constructing the dynamical interpretation. The main difficulty is the necessity of averaging of the joint probability density with respect to some apparatus or interaction variables.

1 INTRODUCTION

As is well known, the specific transformation of the state of the quantum object takes place during the quantum measurement. This transformation, called the collapse or the reduction, is introduced into the quantum theory without any relation with the quantum dynamics, where time variations of the state are gradual. Let the operator B having the discrete non-degenerate eigenvalues $\{b_j\}$ be measured.

[1] On leave of absence from Physics Department, Moscow State University, 119899 Moscow, Russia

Quantum Communications and Measurement
Edited by V. P. Belavkin *et al.*, Plenum Press, New York, 1995

When an actual eigenvalue b_l of the measured operator B becomes known to the observer, collapse $|\psi\rangle \mapsto |b_l\rangle$ or $\rho \mapsto w_l^{-1} P_l \rho P_l$ takes place. Here $w_l = |\langle b_l|\psi\rangle|^2$ or $w_l = \mathrm{Tr}(P_l \rho)$, P_l being the projector on the corresponding eigenspace. We call this collapse the *collapse proper* or the *selective* collapse.

If the actual result of the measurement is unknown or forgotten the state transformation has the form $\rho \mapsto \sum_j w_j |b_j\rangle\langle b_j|$ or $\rho \mapsto \sum_j P_j \rho P_j$. We can call it the *non-selective* collapse or the *dislapse*.

Many efforts have been made by various authors to corroborate the quantum collapse postulate with the help of the dynamic process. The general receipt for this is well known from the time of von Neumann [1]: the original quantum object S is made to interact with some auxiliary system — the apparatus A. If before the initial instant $t = 0$ systems S and A have not interacted, we can set

$$\rho(0) = \rho_{\mathrm{S}}(0) \otimes \rho_{\mathrm{A}}(0), \tag{1}$$

where $\rho_{\mathrm{S}}(0)$, $\rho_{\mathrm{A}}(0)$ are the initial density matrices of S and A correspondingly. Then after the interaction we have at $t = \tau$

$$\rho(\tau) = U_\tau \rho_{\mathrm{S}}(0) \otimes \rho_{\mathrm{A}}(0) U_\tau^\dagger. \tag{2}$$

Here U_τ is an operator of time evolution. We obtain the simplest possible dynamic interpretation of the non-selective collapse if we prove that the state

$$\rho_{\mathrm{S}}(\tau) = \mathrm{Tr}_{\mathrm{A}}\{U_\tau \rho_{\mathrm{S}}(0) \otimes \rho_{\mathrm{A}}(0) U_\tau^\dagger\} \tag{3}$$

of the quantum object at the instant τ approximately coincides with the dislapsed state $\sum_j P_j \rho_{\mathrm{S}}(0) P_j$. Note that in this interpretation no measurement of any apparatus physical quantity is made. The partial trace Tr_{A} , i. e. the trace with respect to apparatus variables appears in (3), when we are only interested in the state of the S-system state.

Of course, the relation

$$\mathrm{Tr}_{\mathrm{A}}\{U_\tau \rho_{\mathrm{S}}(0) \otimes \rho_{\mathrm{A}}(0) U_\tau^\dagger\} \approx \sum_j P_j \rho_1(0) P_j \tag{4}$$

is only valid, when certain conditions are met. It can be proved that (4) holds, if the apparatus is a simple system being in a state that is called here strongly quasi-classical. This concept is original and will be described below. According to the formula (4) expressing the simple dynamic interpretation of dislapse, the influence of the measurement of the S-system variable B of the S-system state is accounted for by the influence of the apparatus during its interaction with S.

Unlike the non-selective collapse, the selective one cannot be only interpreted by the dynamical process. The selective collapse takes place as a result of influx of information about the actual value b_l of the measured physical quantity B. It is only natural that for its dynamical corroboration the influx of information about some measured apparatus physical quantity G is necessary. In our interpretation of the selective collapse we suppose that the inexact measurement of the apparatus momentum p (or the coordinate q) is made. When the apparatus is in the strongly quasi-classical state, this measurement may be regarded both as a quantum one and as a classical one. For the case of the inexact measurement of p the measured apparatus physical quantity is

$$G = \sum_j g_j P_j^{(A)}. \tag{5}$$

Here

$$P_j^{(A)} = \theta_j(p) \tag{6}$$

are the apparatus projection operators. The functions $\theta_j(p)$ defined by

$$\theta_j(q) = \begin{cases} 1 & \text{if } q \in L_j, \\ 0 & \text{elsewhere} \end{cases} \tag{7}$$

are the indicators of intervals, into which the real axis is divided. The selective quantum collapse of the apparatus state

$$\rho_A \mapsto w_n'^{-1} P_n^{(A)} \rho_A P_n^{(A)} \tag{8}$$

corresponds to the measurement of (5).

However it is not quite consistent to corroborate the quantum collapse of the S-system state by another quantum collapse, namely collapse of the apparatus state. That is why instead of (8) it is better to treat the measurement of G as a classical one. The both treatments are possible, if the apparatus is in a strongly quasi-classical state, when the classical inexact measurement and the quantum one are approximately equivalent. Non-quantum collapse accompanies the classical measurement.

The non-quantum collapse is simply a transition from the *a priori* probability density to the *a posteriori* (or conditional) one. It is conditioned by new information acquired by the observer. Such transition is well known in the ordinary probability theory. Let, for example, a physical quantity Q be measured inexactly in the non-quantum theory and the direct observation shows that Q belongs to the set L_l. Then after the observation we should describe the state of the system by the conditional probability density

$$w(z, Q|L_l) = Cw(z, Q)\theta_l(Q), \tag{9}$$

provided that before the measurement the state was described by $w(z, Q)$. Here z is some additional variable (or variables),

$$C = \{\int w(z, Q)\theta_l(Q) dz dQ\}^{-1} \tag{10}$$

is the normalization constant (the integral in braces is supposed to be non-zero), and $\theta_l(Q)$ is defined by (7). The transition

$$w(z, Q) \overset{\text{cl}}{\mapsto} Cw(z, Q)\theta_l(Q) \tag{11}$$

is nothing else but the classical collapse.

Dynamical interpretation of the non-selective collapse immediately follows from the interpretation of the selective one. It is sufficient for this to average the selectively collapsed state with respect to observed values of the measured operator $G = \sum g_j \theta_j(Q)$.

Let $\bar{q}$, $\bar{p}$ be the non-operator coordinate and momentum. The non-quantum inexact macroscopic observation of the momentum leads to the transition

$$w(\bar{q}, \bar{p}) \overset{\text{cl}}{\mapsto} C_l w(\bar{q}, \bar{p})\theta_l(\bar{p}) \tag{12}$$

of the type (11) in A.

If the apparatus state is described by the density matrix $\rho_{\rm A}$ the above non-quantum collapse can be written

$$\mathcal{W}_{\rm A_1}\{\rho_{\rm A}\} \overset{\rm cl}{\mapsto} C_l \mathcal{W}_{\rm A_1}\{\rho_{\rm A}\}\theta_l(\bar{p}) \tag{13}$$

(in some cases $\rm A_1$ is a part of A, sometimes $\rm A_1 = A$). Here $\mathcal{W}_{\rm A_1}$ denotes the transformation to Wigner's distribution (we call it Wigner's transformation)

$$\mathcal{W}_{\rm A_1}\{\rho_{\rm A}\} = \frac{1}{2\pi\hbar}\int \exp(-i\hbar^{-1}\bar{p}v)\rho_{\rm A}(\bar{q}+\frac{v}{2},\bar{q}-\frac{v}{2})dv, \tag{14}$$

$\rho_{\rm A}(\bar{q}',\bar{q}'') = \langle\bar{q}'|\rho_{\rm A}|\bar{q}''\rangle$ is the density matrix $\rho_{\rm A}$ in the coordinate representation. Equation (14) is written for the case of one pair $(\bar{q},\bar{p})$ of conjugate apparatus dynamic variables. We will also use the reverse Wigner transformation

$$\rho_{\rm A} = \mathcal{W}_{\rm A_1}^{-1}\{w(\bar{q},\bar{p})\}. \tag{15}$$

Applying transformation $\mathcal{W}_{A_1}^{-1}$ to both sides of (13) we obtain one more form

$$\rho_{\rm A} \overset{\rm cl}{\mapsto} C_l \rho_{\rm A} * \theta_l(p). \tag{16}$$

This is the translation of the classical collapse into the quantum language. Here the following operation is denoted by $*$

$$L * M = 2\pi\hbar\mathcal{W}_{\rm A_1}^{-1}\{\mathcal{W}_{\rm A_1}[L]\mathcal{W}_{\rm A_1}[M]\}, \tag{17}$$

because $\theta_l(\bar{p}) = 2\pi\hbar\mathcal{W}_{\rm A_1}[\theta_l(p)]$. The operation $*$ and the concept of the non-quantum collapse of the type (16) will be used below (Sect. 5).

2 DIFFICULTIES OF CONSTRUCTING A DYNAMIC INTERPRETATION OF THE COLLAPSE

According to (2) the time evolution of the combined system S+A during the time period τ transforms the initial state $\rho(0) = \rho_{\rm S}\otimes\rho_{\rm A}$ of the combined system into the final state

$$\rho(\tau) = U_\tau(\rho_{\rm S}\otimes\rho_{\rm A})U_\tau^\dagger. \tag{18}$$

Now we will describe the "ideally good" dynamic interpretation of the collapse. First (2.1) should coincide with the matrix density

$$\tilde{\rho} = \sum_j |c_j|^2|b_j\rangle\langle b_j|\otimes r_j. \tag{19}$$

Here r_j is such state of the apparatus that some its observable $G_{\rm A}$ has in it the definite value g_j (i. e. $G_{\rm A}r_j = g_j r_j$) belonging to the spectrum of the non-coinciding eigenvalues. Second the equation

$$U_\tau \sum_{jk} c_j c_k^*(|b_j\rangle\langle b_k|\otimes\rho_{\rm A})U_\tau^\dagger = \sum_j |c_j|^2|b_j\rangle\langle b_j|\otimes r_j \tag{20}$$

($\mathrm{Tr}_{\mathrm{A}} r_j = 1$) must hold for every set $\{c_j\}$ with $|b_j\rangle$, ρ_{A}, U_τ, r_j independent of $\{c_j\}$. The equation $G_{\mathrm{A}} r_j = g_j r_j$ is equivalent to $P_j^{(\mathrm{A})} r_j = r_j$ or

$$(I_A - P_j^{(\mathrm{A})}) r_j = 0, \tag{21}$$

where $P_j^{(\mathrm{A})}$ is the projection operator on eigenspace corresponding to g_j (i. e. $G_{\mathrm{A}} P_j^{(\mathrm{A})} = g_j P_j^{(\mathrm{A})}$).

If eqs. (20), (21) expressing the ideal dynamic interpretation are valid we can measure the physical quantity G_{A} of the apparatus and get to know the actual value g_l of this quantity. After that we may leave only one term of the sums in (20), i.e. take the a posteriori values $\tilde{c}_j = \delta_{jl}$ instead of c_j. Then we will have from (20)

$$U_\tau(|b_l\rangle\langle b_l| \otimes \rho_A) U_\tau^\dagger = |b_l\rangle\langle b_l| \otimes r_l, \tag{22}$$

which means that we can be sure that the system S is in the state $|b_n\rangle\langle b_n|$ at $t = \tau$ and it was in this state at $t = 0$, i.e. both before and after the interaction. It is essential that the terms with $j \neq k$ are absent from the right-hand side of (20) although they are present in the left-hand side. In this respect (20) is analogous to the right-hand side of (4).

It can be shown that the dynamic interpretation (20) is impossible. Instead of (20) the equation

$$U_\tau \sum_{jk} c_j c_k^* (|b_j\rangle\langle b_k| \otimes \rho_{\mathrm{A}}) U_\tau^\dagger = \sum_j |c_j|^2 |b_j\rangle\langle b_j| \otimes r_j + \Delta\rho \tag{23}$$

should hold, where the additional operator $\Delta\rho$ cannot be very "small", namely the norm $||\Delta\rho||_1$ of $\Delta\rho$ cannot be much less than

$$||\rho(\tau)||_1 = 1. \tag{24}$$

The problem arises how one can circumvent the difficulty of $\Delta\rho$ being not small and nevertheless construct the dynamic interpretation of the collapse although not ideally good. We can point out two ways in this direction.

The first way. We can assume that U_τ and perhaps ρ_{A} depend on some non-quantum random variable λ described by the probability density $w(\lambda)$. After averaging with respect to this variable we have instead of (23)

$$\int d\lambda w(\lambda) U_\tau(\lambda) \sum_{jk} c_j c_k^* (|b_j\rangle\langle b_k| \otimes \rho_{\mathrm{A}}(\lambda)) U_\tau^\dagger(\lambda) = \sum_j |c_j|^2 |b_j\rangle\langle b_j| \otimes r_j + \Delta\rho'. \tag{25}$$

In this case $\Delta\rho'$ can be small in the sense of the norm $||...||_1$. Moreover the limit procedure $||\Delta\rho'||_1 \to 0$ is possible as some apparatus or interaction parameter σ tends to its limiting value σ_0 (for example $\sigma_0 = \infty$ or $\sigma_0 = 0$).

The method of averaging is applied by Machida and Namiki [2]. They introduce averaging with respect to a random thickness of the body that reflects the quantum particle. Actually it is important for them that the longitudinal (i.e. directed along the particle velocity) coordinate of the body be random. In one place of [3] averaging with respect to a random phase is of crucial importance.

The second way. We do not use averaging with respect to a non-quantum variable and formula (25), but demand that operators (18) and (19) be only close to each other with respect to operators belonging to some subalgebra C_0^* of the operator algebra C^* of the combined system S+A (C_0^* should include all operators of

S-system, i. e. operators of the type $D_{\rm S}\otimes I_{\rm A}$). In other words it is sufficient to demand that

$$\lim_{\sigma\to\sigma_0}\{\mathrm{Tr}[DU_\tau\sum_{jk}c_jc_k^*(|b_j\rangle\langle b_k|\otimes\rho_{\rm A})U_\tau^\dagger]-\mathrm{Tr}[D\sum_j|c_j|^2|b_j\rangle\langle b_j|\otimes r_j]\}=0 \tag{26}$$

for all $D\in C_0^*$ and all $\{c_j\}$. The limit procedure is not concretized here.

In [4] Araki includes in C_0^* operators of A that are representable in the form of the direct integral

$$D=\int D(\lambda)\mu(d\lambda) \tag{27}$$

with respect to a continuous measure μ (for instance the Lebegue measure).

Let us consider another variant of constructing the operator subalgebra. We assume that the Hilbert space $\mathcal{H}_{\rm A}$ corresponding to the apparatus is of the form $\mathcal{H}_{\rm A'}\otimes\mathcal{H}_{\rm A''}$. This means that the apparatus consists of two subsystems: $\rm A=A'+A''$. Then the total Hilbert space of the joint system is $\mathcal{H}=\mathcal{H}_{\rm S}\otimes\mathcal{H}_{\rm A'}\otimes\mathcal{H}_{\rm A''}$. We include in C_0^* all bounded operators of the type $D=D_{\rm SA'}\otimes I_{\rm A''}$, where $I_{\rm A''}$ is the identity operator on $\mathcal{H}_{\rm A''}$, $D_{\rm SA'}$ is an operator on $\mathcal{H}_{\rm S}\otimes\mathcal{H}_{\rm A'}$. Then

$$\mathrm{Tr}[D\rho(\tau)]=\mathrm{Tr}_{\rm SA'}\{D_{\rm SA'}\mathrm{Tr}_{\rm A''}\rho(\tau)\} \tag{28}$$

for all $D\in C_0^*$. Here $\mathrm{Tr}_{\rm A''}$ is the trace with respect to physical variables of A$''$. Owing to (28) equation (26) assumes the form

$$\lim_{\sigma\to\sigma_0}\{\mathrm{Tr}_{\rm SA'}[D_{\rm SA'}(\mathrm{Tr}_{\rm A''}\rho(\tau)-\mathrm{Tr}_{\rm A''}\tilde\rho)]\}=0. \tag{29}$$

The norm $\|\ldots\|_1$ entering (24) is defined by $\|A\|_1=\mathrm{Tr}(AA^\dagger)^{1/2}$. It is the norm for the linear space of operators of the density-matrix type. This norm is dual with respect to the ordinary operator norm $\|\ldots\|$ in the sense of inequality

$$|\mathrm{Tr}(D\rho_1)-\mathrm{Tr}(D\rho_2)|\le\|D\|\|\rho_1-\rho_2\|_1. \tag{30}$$

where D is arbitrary. For the case (28), (29) we apply (30) to the operators on $\mathcal{H}_{\rm S}\otimes\mathcal{H}_{\rm A'}$ denoting

$$\rho_1=\mathrm{Tr}_{\rm A''}\rho(\tau),\quad \rho_2=\mathrm{Tr}_{\rm A''}\tilde\rho. \tag{31}$$

Due to (30) for proving (29) it is sufficient to prove that

$$\lim_{\sigma\to\sigma_0}\|\rho_1-\rho_2\|_{1,\rm SA'}=0. \tag{32}$$

Passing to the matrix densities (31) and operators of the type $D=D_{\rm SA'}\otimes I_{\rm A''}$ means that an averaging is applied implicitly. This time it is averaging with respect to the quantum physical variables of subapparatus A$''$. For example, the variables of A$'$ may be these of the apparatus mass center, then other apparatus variables refer to A$''$.

If we set $\rm A''=A$, we obtain the minimal possible subalgebra C_0^* that only contains operators of the type $D=D_{\rm S}\otimes I_{\rm A}$.

The formula (26) is weaker than (20). Besides, instead of (21) it is reasonable to take weaker condition

$$\sum_{\substack{k\\k\neq j}}[\|P_k^{(\rm A)}r_j\|_1+\|P_j^{(\rm A)}r_k\|_1]\ll 1 \tag{33}$$

or

$$\sum_{\substack{k \\ k\neq j}}[\|P_k^{(A)}r_j\|_1 + \|P_j^{(A)}r_k\|_1] \rightarrow 0 \tag{34}$$

as some apparatus or interaction parameter tends to its limiting value. We remind that $I - P_j^{(A)} = \sum_{k\neq j} P_k^{(A)}$. The above weakening of the condition (21) simplifies the problem of constructing the dynamic interpretation of the quantum collapse in practice.

Several particular cases of the interaction between S and A were considered by us. The investigation showed that difference between density matrices (31) were small if the subapparatus A_1 were in a strongly quasi-classical state. In one case A_1 coincided with A′, in other cases (when A′ was absent) A_1 was a part of A″ = A.

3 THE OPERATOR-ORDERING RULES AND THE CORRESPONDING PHASE-SPACE DISTRIBUTION FUNCTIONS

The ordering rule $\mathcal{OR}$ transforms a numerical function $f(x,y)$ into an operator

$$F = \mathcal{OR}f(q,p) = \hat{L}f. \tag{35}$$

Here $\hat{L}$ is the superoperator mapping the function space on the operator space. To be the operator-ordering rule the mapping should meet an important requirement: the equation

$$\mathcal{OR}f(A,B) = f(A,B) \tag{36}$$

should hold if the commuting operators A, B are taken instead of coordinate (coordinates) q and momentum (momenta) p.

The mean value $\langle F\rangle = \mathrm{Tr}(\rho\hat{L}f)$ of the operator (35) may be represented as

$$\langle F\rangle = \int f(x,y)[\hat{L}^T\rho]_{xy}dxdy, \tag{37}$$

where $\hat{L}^T$ is the "transposed" superoperator defined by the equation

$$\mathrm{Tr}[\rho\hat{L}f] = \int f(x,y)(\hat{L}^T\rho)dxdy \tag{38}$$

valid for various ρ and f.

It is easily seen from (37) that

$$w(x,y) = [\hat{L}^T\rho]_{xy} \tag{39}$$

plays the part of the distribution function for non-operator dynamical variables $x = \bar{q}$, $y = \bar{p}$. Multiplying both sides of (39) by $\exp(iux + ivy)$ and using (38) , (35), we obtain

$$\int \exp(iux + ivy)w(x,y)dxdy = \mathrm{Tr}\{\rho[\mathcal{OR}\exp(iuq + ivp)]\}. \tag{40}$$

Besides, the formal equation

$$w(x,y) = \langle\mathcal{OR}[\delta(q-x)\delta(p-y)]\rangle \tag{41}$$

derived in the same way is valid. Function $w(x,y)$ only differs from probability density by the fact that it is not necessarily non-negative and even real.

The ordering rules may be defined with the help of the Feynman subscript λ indicating ordering: the greater the value of λ the more to the right the operator marked with λ is placed. Changing subscript λ for const $-\ \lambda$ transforms $\mathcal{OR}$ into the reverse or adjoint ordering $\overline{\mathcal{OR}}$.

In the special cases $\mathcal{S}$, $\mathcal{B}$, $\mathcal{N}$ of ordering rules considered below the formula is valid

$$\mathrm{Tr}\{[\mathcal{OR}f(q,p)][\overline{\mathcal{OR}}g(q,p)]\} = \frac{1}{(2\pi\hbar)^n}\int f(x,y)g(x,y)dxdy. \tag{42}$$

Here n is the number of pairs of conjugate coordinates and momenta. Using (42) we obtain the formula

$$w(x,y) = (2\pi\hbar)^{-n}(\hat{\hat{L}})^{-1}\rho \tag{43}$$

in addition to (39) since (42) is equivalent to $\hat{L}^T\hat{L} = (2\pi\hbar)^{-n}\hat{I}$.

We pass to some examples. The symmetrization ordering $\mathcal{OR} = \mathcal{S}$ is defined by

$$\mathcal{S}e^{bq+cp} = e^{bq+cp} \tag{44}$$

(with arbitrary b, c) or in Feynman's form

$$\mathcal{S}e^{bq+cp} = \exp\Big\{\int_0^1 d\lambda(bq+cp)_\lambda\Big\}. \tag{45}$$

Ordering $\mathcal{S}$ is self-adjoint ($\overline{\mathcal{S}} = \mathcal{S}$). In this case the distribution (39), (43) is the well known Wigner distribution. We will denote transformation $\rho \to w_{\mathcal{S}}$ in the following way

$$w_{\mathcal{S}}(x,y) = \mathcal{W}[\rho]_{xy} \tag{46}$$

(see (14)). From this and (43) we get

$$\rho \equiv \mathcal{W}^{-1}w_{\mathcal{S}} = (2\pi\hbar)^n\mathcal{S}w_{\mathcal{S}}(q,p). \tag{47}$$

Another ordering rule $\mathcal{B}$ is defined by

$$\mathcal{B}e^{cq+dp} = e^{cq}e^{dp} \tag{48}$$

or with Feynman's subscript

$$\mathcal{B}f(q,p) = f(q_2,p_1). \tag{49}$$

The rule $\overline{\mathcal{B}}$ defined by $\overline{\mathcal{B}}f(q,p) = f(q_1,p_2)$ is adjoint to $\mathcal{B}$. Corresponding distribution density is

$$w_{\mathcal{B}}(q,p) = \frac{1}{(2\pi\hbar)^n}\int e^{-iup/\hbar}\rho(q+u,q)du \tag{50}$$

with $\rho(q',q'') = \langle q'|\rho|q''\rangle$ (see [5]).

Finally, the normal ordering rule $\mathcal{N}$ is defined by

$$\begin{aligned}\mathcal{N}e^{bq+cp} = {}& \exp\{\tfrac{1}{2}[(2-\beta\gamma^*)bq + \beta\gamma^*cp + i|\beta|^2cq - i|\gamma|^2bp]\}\\ &\times\exp\{\tfrac{1}{2}[\beta\gamma^*bq + (2-\beta\gamma^*)cp - i|\beta|^2cq + i|\gamma|^2bp]\}.\end{aligned} \tag{51}$$

This formula follows from simpler equation

$$\mathcal{N}\exp(\bar{z}a + za^{\dagger}) = \exp(za^{\dagger})\exp(\bar{z}a) \tag{52}$$

with $a = (2\hbar)^{-1/2}(\beta q + i\gamma p)$, $a^{\dagger} = (2\hbar)^{-1/2}(\beta^* q - i\gamma^* p)$, $z = (\hbar/2)^{1/2}(\gamma c + i\beta d)$, $\bar{z} = (\hbar/2)^{1/2}(\gamma^* c - i\beta^* d)$. Here β, γ are arbitrary complex numbers of definite dimension restricted by condition $\mathrm{Re}\beta\gamma^* = 1$, a, $a^{\dagger}$ are the well-known operators with property

$$aa^{\dagger} - a^{\dagger}a = 1. \tag{53}$$

The adjoint (antinormal) ordering $\overline{\mathcal{N}}$ is obviously defined by

$$\overline{\mathcal{N}}e^{\bar{z}a + za^{\dagger}} = e^{\bar{z}a}e^{za^{\dagger}}. \tag{54}$$

It can be shown that

$$w_{\overline{\mathcal{N}}}(x, y) = (2\pi\hbar)^{-1}\langle\alpha|\rho|\alpha\rangle \tag{55}$$

with $\alpha = (2\hbar)^{-1/2}(\beta x + i\gamma y)$. Therefore it is the well known Q-function, $w_{\mathcal{N}}(x, y)$ being the Glauber—Sudarshan distribution. Obviously (55) is non-negative. The function $w_{\overline{\mathcal{N}}}(x, y)$ is connected with Wigner's function by

$$w_{\overline{\mathcal{N}}}(x, y) = \exp\Big[\frac{\hbar}{4}\big(|\gamma|^2\frac{\partial^2}{\partial x^2} - 2\mathrm{Im}\beta\gamma^*\frac{\partial^2}{\partial x\partial y} + |\beta|^2\frac{\partial^2}{\partial y^2}\big)\Big]w_S(x, y). \tag{56}$$

We will specify β, γ by setting $|\beta|/|\gamma| = \Delta p/\Delta q$, $\mathrm{Im}\beta^*\gamma = r$ with the coefficient r defined in Sect. 4.

Distributions corresponding to other ordering rules may be introduced as well. For example, a family of functions $w_n(x, y)$ is defined by

$$\int e^{iux+ivy}w_n(x, y)dxdy = 2^{-n}\mathrm{Tr}\{[e^{iuq/n}e^{ivp/n} + e^{ivp/n}e^{iuq/n}]^n\rho\}. \tag{57}$$

Here $n \geq 1$ is integer.

4 STRONGLY QUASI-CLASSICAL STATES OF THE SUBAPPARATUS

It is important for our dynamic interpretation of the strong collapse that the inexact measurement of the apparatus momentum be a macroscopic one. Only the macroscopic apparatus measurement can lead to the productive interpretation of the quantum collapse in S, since we cannot interpret a quantum measurement by another quantum measurement. The macroscopic (i.e. non-quantum) apparatus measurement is only possible when the apparatus is macroscopic, i.e. it consists of a great number of atoms or molecules. However we are not obliged to be interested in the details of apparatus state including joint states of all the atoms or molecules. We are only interested in the state of a simplified system A_1 whose variables can be directly observed. The coordinates and momenta of the mass center of a body, the angular coordinate and the conjugate momentum of a pointer on a fixed axis, the charge and the current in an electrical circuit can serve as examples of directly observable macroscopical variables.

In our opinion the concept of a strongly quasi-classical state of A_1 must be specified in the following way: this state can be described both in quantum and in classical terms with equal success. Of course the last statement needs some comments.

The state of a classical dynamic system is described by a probability density $w(\bar{q},\bar{p})$ in the phase space, where $\bar{q}$, $\bar{p}$ are non-operator coordinate (or coordinates) and momentum (momenta). In the non-quantum theory the mean value of any phase function $f(\bar{q},\bar{p})$ is defined in the well known way

$$\langle f(\bar{q},\bar{p})\rangle = \int f(\bar{q},\bar{p})w(\bar{q},\bar{p})d\bar{q}d\bar{p}. \tag{58}$$

In quantum theory the state of A_1 is described by a density operator ρ_{A_1} acting on $\mathcal{H}_{A_1}$. To write down the quantum generalization of (58) for the same function f, we must overcome the difficulty consisting in the fact that a function $f(\bar{q},\bar{p})$ does not determine uniquely the operator $F = f(q,p)$ because q and p do not commute. Some operator ordering rule $\mathcal{OR}$ described in the previous section must be specified to define the operator $F = \mathcal{OR}f(q,p)$. If the operator ordering rule has been chosen, the quantum generalization of (58) is obvious:

$$\langle F\rangle = \mathrm{Tr}_{A_1}\rho_{A_1}\mathcal{OR}f(q,p). \tag{59}$$

One of the necessary conditions of a strongly quasi-classical state is the approximate equality

$$\mathrm{Tr}_{A_1}\rho_{A_1}\mathcal{OR}f(q,p) \approx \int f(\bar{q},\bar{p})w(\bar{q},\bar{p})d\bar{q}d\bar{p} \tag{60}$$

of a quantum and a classical means for various f and $\mathcal{OR}$ from some sufficiently large sets $\mathcal{T}_1$, $\mathcal{T}_2$ of functions and ordering rules. Since the exact equation

$$\mathrm{Tr}_{A_1}\rho_{A_1}\mathcal{OR}f(q,p) = \int f(\bar{q},\bar{p})w_{\mathcal{OR}}(\bar{q},\bar{p})d\bar{q}d\bar{p} \tag{61}$$

holds with

$$w_{\mathcal{OR}}(\bar{q},\bar{p}) = \mathrm{Tr}_{A_1}\rho_{A_1}\mathcal{OR}[\delta(q-\bar{q})\delta(p-\bar{p})], \tag{62}$$

the condition (60) is satisfied if the difference between distributions corresponding to various operator orderings is small:

$$\mathrm{Tr}_{A_1}\rho_{A_1}\mathcal{OR}[\delta(q-\bar{q})\delta(p-\bar{p})] \approx \mathrm{Tr}_{A_1}\rho_{A_1}\mathcal{OR}'[\delta(q-\bar{q})\delta(p-\bar{p}) \tag{63}$$

for $\mathcal{OR} \in \mathcal{T}_2$, $\mathcal{OR}' \in \mathcal{T}_2$. In particular in the case of the strongly quasi-classical state we must have

$$w_S(\bar{q},\bar{p}) \approx w_B(\bar{q},\bar{p}) \approx w_{\bar{N}}(\bar{q},\bar{p}). \tag{64}$$

The density (62)can play the part of a classic probability density only when it is non-negative or is at least only slightly different from some non-negative function. This condition is met owing to equation (63) taken for $\mathcal{OR}' = \bar{N}$ because $w_{\bar{N}}(\bar{q},\bar{p})$ is non-negative (see Sect. 3).

Thus the necessary conditions (63), (64), of a strongly quasi-classical state are consequences of the following requirement: (i) the classical and the quantum definitions of the mean value must not contradict each other.

Other requirements can also be formulated. The requirement (ii) of non-contradicting entropy definitions means that the quantum and quasi-classical entropies should be approximately equal:

$$-k\mathrm{Tr}_{\mathrm{A}_1}\rho_{\mathrm{A}_1}\ln\rho_{\mathrm{A}_1} \approx -k\int w(\bar{q},\bar{p})\ln[2\pi\hbar w(\bar{q},\bar{p})]d\bar{q}d\bar{p}. \tag{65}$$

Here k is Boltzmann's constant.

The requirement (iii) of the agreement of the time evolutions means that the quantum and the classical dynamic evolutions must lead to non-contradicting results provided that the initial conditions in both theories do not contradict each other. The quantum and the classical evolutions are determined by equations

$$i\hbar\frac{\partial\rho_{\mathrm{A}_1}}{\partial t} = [H,\rho_{\mathrm{A}_1}], \qquad \frac{\partial w}{\partial t} = \frac{\partial H}{\partial q}\frac{\partial w}{\partial p} - \frac{\partial H}{\partial \bar{p}}\frac{\partial w}{\partial \bar{q}} \tag{66}$$

with $H = H_{\mathrm{A}_1}$. At $t = 0$ we set

$$w(\bar{q},\bar{p},0) = \mathrm{Tr}_{\mathrm{A}_1}\rho_{\mathrm{A}_1}(0)\mathcal{OR}[\delta(q-\bar{q})\delta(p-\bar{p})]. \tag{67}$$

Then the formula

$$w(q,p,t) \approx \mathrm{Tr}_{\mathrm{A}_1}\rho_{\mathrm{A}_1}(t)\mathcal{OR}[\delta(q-\bar{q})\delta(p-\bar{p})], \quad \mathcal{OR}\in\mathcal{T}_2 \tag{68}$$

must hold with $w(\bar{q},\bar{p},t)$ and $\rho_{\mathrm{A}_1}(t)$ varying according to their own equations.

Further we formulate the requirement (iy) of the agreement of the quantum collapse and of the classical collapse during the quantum and classical measurements: in the strongly quasi-classical state the non-trivial inexact measurement of the coordinate or momentum should be possible which can be described with equal success by both quantum and by classical theories of measurement. The quantum and the classical collapses must not contradict each other. Such measurement can be called *a macroscopic* one.

Let, for example, the apparatus momentum p be measured inexactly, i.e. one observes to which interval of the intervals $L_j = [c_j, c_{j+1})$ (or $L_j = (c_{j+1}, c_j]$ if $c_{j+1} < c_j$) the momentum belongs. This measurement is described classically by the transition from the *a priori* probability density to the *a posteriori* one

$$\tilde{w}(\bar{q},\bar{p}|L_l) = W_l^{-1}\theta_l(\bar{p})w(\bar{q},\bar{p}) \tag{69}$$

with

$$W_l = \int \theta_l(\bar{p})w(\bar{q},\bar{p})d\bar{q}d\bar{p}. \tag{70}$$

The transition

$$w(\bar{q},\bar{p}) \overset{\mathrm{cl}}{\mapsto} W_l^{-1}\theta_l(\bar{p})w(\bar{q},\bar{p}), \tag{71}$$

which can be called a *classical collapse*, takes place when the momentum is found inside the particular interval L_l. The words "the measurement can be described both by a quantum and a classical collapse" in the above statement mean that formula

$$W_l^{-1}\theta_l(\bar{p})w(\bar{q},\bar{p}) \approx \mathrm{Tr}_{\mathrm{A}_1}\{{w'}_l^{-1}\theta_l(p)\rho_{\mathrm{A}_1}\theta_l(p)\mathcal{OR}[\delta(q-\bar{q})(\delta(p-\bar{p})]\} \tag{72}$$

expressing the agreement between the quantum and classical description of the state after the observation, is valid provided that

$$w(\bar{q},\bar{p}) = \mathrm{Tr}_{\mathrm{A}_1}\{\rho_{\mathrm{A}_1}\mathcal{OR}[\delta(q-\bar{q})\delta(p-\bar{p})]\}, \tag{73}$$

i.e. provided that the agreement takes place before the measurement. Here the symmetrization operator ordering $\mathcal{S}$ can be taken as $\mathcal{OR}$. Then the last two formulas can be written as

$$W_l^{-1}\theta_l(\bar{p})w(\bar{q},\bar{p}) \approx {w'}_l^{-1}\mathcal{W}_{A_1}[\theta_l(p)\rho_{A_1}\theta_l(p)], \tag{74}$$

$$w(\bar{q},\bar{p}) = \mathcal{W}_{A_1}[\rho_{A_1}]. \tag{75}$$

It is easy to see that the right-hand side of (70) for the function (75) coincides with $\mathrm{Tr}_{A_1}[\theta_l(p)\rho_{A_1}]$. Besides $w_l' = \mathrm{Tr}_{A_1}[\theta_l(p)\rho_{A_1}]$. Therefore

$$W_l = {w'}_l. \tag{76}$$

Using (75), (76), (17) it is easily seen that (74) is equivalent with

$$\rho_{A_1} * \theta_l(p) \approx \theta_l(p)\rho_{A_1}\theta_l(p) \tag{77}$$

because both sides of (74) are Wigner's transforms of the corresponding sides of (77).

An investigation shows that all the requirements mentioned above are met if the following two necessary and sufficient conditions are satisfied (besides some other insignificant assumptions):

(a)

$$\Delta q \Delta p \sqrt{1-r^2} \gg \hbar, \tag{78}$$

(b) Wigner's distribution $w_{\mathcal{S}}(\bar{q},\bar{p})$ exists and

$$\int \left|\frac{\partial^{m+n} w_*(\xi,\eta)}{\partial \xi^m \partial \eta^n}\right| d\xi d\eta \sim 1 \quad \text{for} \quad m+n \sim 1, \tag{79}$$

where the dimensionless function $w_*(\xi,\eta)$ of dimensionless arguments is defined by equation

$$w_{\mathcal{S}}(\bar{q},\bar{p}) = \frac{(1-r^2)^{-\frac{1}{2}}}{\Delta q \Delta p} w_*\Big(2^{-\frac{1}{2}}(1-r)^{-\frac{1}{2}}\Big[\frac{\bar{q}}{\Delta q}+\frac{\bar{p}}{\Delta p}\Big], 2^{-\frac{1}{2}}(1+r)^{-\frac{1}{2}}\Big[\frac{\bar{p}}{\Delta p}-\frac{\bar{q}}{\Delta q}\Big]\Big). \tag{80}$$

Here $\Delta q = (\langle q^2\rangle - \langle q\rangle^2)^{1/2}$, $\Delta p = (\langle p^2\rangle - \langle p\rangle^2)^{1/2}$, $r = (\frac{1}{2}\langle qp+pq\rangle - \langle q\rangle\langle p\rangle)/(\Delta q\Delta p)$.

Some authors define the quasi-classical state or the wave function describing it only by the condition (a). For example, Landau and Lifschits [6] use the condition that often is equivalent to (a) for defining a quasi-classical case. Condition (a) is used in the last chapter of Bohm's book [7]. A state defined by (a) we may call *weakly* quasi-classical to distinguish it from a strongly quasi-classical state defined by both conditions (a) and (b). The condition of a weakly quasi-classical state is insufficient for obtaining our results concerning a new dynamic interpretation of quantum measurement.

A strongly quasi-classical state cannot be pure, i.e. a strongly quasi-classical wave function does not exist. Such a state can only be described by a density matrix. In fact the quantum entropy $S_{qu} = -k\mathrm{Tr}_{A_1}\rho_{A_1}\ln\rho_{A_1}$ is equal to 0 for pure state, but the quasi-classical entropy S_{cl} , which is equal to the expression on the right-hand side of (65), can be estimated (by order of magnitude) as $\ln[\hbar/(\Delta q\Delta p)]$ and hence $S_{cl} \gg 1$. This contradiction shows that the second condition (b) is violated in the case of a pure state if $\Delta q\Delta p \gg \hbar$.

Consider, for example, a quantum linear oscillator with Hamiltonian $H = H(q,p) = (p^2 + m^2\omega^2 q^2)/(2m)$. The eigenstates of H corresponds to the values $E_n = \hbar\omega(n+1/2)$. For them $\Delta q \Delta p = \hbar(n+1/2)$. Hence $\Delta q \Delta p \gg \hbar$ if $n \gg 1$. Analysis gives that Wigner's distribution of these states is of the form

$$w_S(\bar{q},\bar{p}) = \frac{(-1)^n}{\pi\hbar} L_n\Big(\frac{4}{\hbar\omega} H(\bar{q},\bar{p})\Big) \exp\Big[-\frac{2}{\hbar\omega} H(\bar{q},\bar{p})\Big], \tag{81}$$

where $L_n(x)$ are the Laguerre polynomials with specification $L_n(0) = 1$. For this distribution, according to (80), we have

$$w_*(\xi,\eta) = \frac{(-1)^n}{\pi}\Big(n+\frac{1}{2}\Big) L_n\big((2n+1)(\xi^2+\eta^2)\big) \exp\Big[-\Big(n+\frac{1}{2}\Big)(\xi^2+\eta^2)\Big]. \tag{82}$$

In this case the condition (a) is met at $n \gg 1$, but (b) is violated since the large parameter n enters the function (82) forcing her to have many oscillations. In fact, calculations show that

$$\int \Big|\frac{\partial w_*(\xi,\eta)}{\partial \xi}\Big| d\xi d\eta \gg 1 \tag{83}$$

for $n \gg 1$. Therefore the corresponding states are weakly quasi-classical and not strongly quasi-classical.

A thermodynamic equilibrium state of a linear oscillator with the density matrix $\rho = \text{const}\exp(-H/kT)$ is an example of a strongly quasi-classical state if $kT \gg \hbar\omega$. For this state

$$w_*(\xi,\eta) = (2\pi)^{-1} \exp[-(\xi^2+\eta^2)/2] \tag{84}$$

and both conditions (a), (b) are met.

5 APPLICATION OF THE NON-QUANTUM COLLAPSE TO THE INTERPRETATION OF THE SELECTIVE QUANTUM COLLAPSE

Applying transformation $\mathcal{W}_{A_1}^{-1}$ to both sides of (71) we get the formula that with the help of the operation $*$ defined by (17) may be written

$$\rho_A \overset{\text{cl}}{\mapsto} W_l^{-1} \rho_A * \theta_l(p). \tag{85}$$

Analogous transformation

$$\rho(\tau) \overset{\text{cl}}{\mapsto} W_l^{-1} \rho(\tau) * \theta_l(p) \tag{86}$$

of the joint system state takes place as a result of the macroscopic inexact measurement of the apparatus momentum. Operators on both sides of (23) are equal to $\rho(\tau)$. Therefore (86) may be written as

$$U_\tau \sum_{jk} c_j c_k^* (|l_b\rangle\langle b_k| \otimes \rho_A) U_\tau^\dagger \overset{\text{cl}}{\mapsto} W_l^{-1} \Big[\sum_j |c_j|^2 |b_j\rangle\langle b_j| \otimes [r_j * \theta_l(p)] + \Delta\rho * \theta_l(p)\Big], \tag{87}$$

this formula being valid for the arbitrary set $\{c_j\}$. The right-hand side of (87) may be written

$$W_l^{-1} \Big[|c_l|^2 |b_l\rangle\langle b_l| \otimes r_l + \delta + \Delta\rho * \theta_l(p)|\Big]. \tag{88}$$

Owing to (33) the sum

$$\delta = \sum_{\substack{j \\ j\neq l}} |c_j|^2 |b_j\rangle\langle b_j| \otimes [r_j * \theta_l(p) - r_l * \theta_j(p)] \tag{89}$$

is very small. If we take into account (34) and the equation

$$\lim_{\sigma\to\sigma_0} \mathrm{Tr}[D\Delta\rho] = 0, \qquad D \in C_0^*, \tag{90}$$

which is equivalent to (26), we obtain for all $\{c_j\}$

$$\mathrm{Tr}[DU_\tau \sum_{jk} c_j c_k^* (|b_j\rangle\langle b_k| \otimes \rho_A) U_\tau^\dagger] \overset{\mathrm{cl}}{\mapsto} \frac{|c_l|^2}{W_l} \mathrm{Tr}[D|b_l\rangle\langle b_l| \otimes r_l] + \mathrm{Tr}(D\Delta), \tag{91}$$

where $\lim \mathrm{Tr}(D\Delta) = 0$ for $D \in C_0^*$, Δ being equal to $[\delta + \Delta\rho * \theta_l(p)]/W_l$. Here lim corresponds to parameter (or parameters) of the apparatus or the interaction tending to its (their) limiting values. Of course, $|c_l|^2 = W_l$ (in (91)) due to normalization condition of the density matrix $|c_l|^2 W_l^{-1} |b_l\rangle\langle b_l| \otimes r_l$. The resulting formula (91) expresses the dynamical interpretation of the collapse proper $\sum_{jk} c_j c_k^* |b_j\rangle\langle b_j| \overset{\mathrm{cl}}{\mapsto} |b_l\rangle\langle b_l|$ with arbitrary $\{c_j\}$ because C_0^* contains all operators of the type $D_\mathrm{S} \otimes I_\mathrm{A}$, i.e. the S-system operators.

Acknowledgment

This work was supported by the Royal Society Kapitza fellowship. I am deeply grateful to Professor V. P. Belavkin for stimulating discussions.

References

[1] J. von Neumann, *Mathematische Grundlagen der Wellenmechanik* (Verlag von Springer, Berlin, 1932).

[2] S. Machida and M. Namici,
Progr. Theor. Phys., 63:1457 (1980).

[3] S. Machida and M. Namici, *Progr. Theor. Phys.*, 63:1833 (1980).

[4] H. Araki, *Progr. Theor Phys.*, 64:719, (1980).

[5] D. I. Blokhintsev, *J.Phys. (USSR)*, **2** 71, (1940).

[6] L. D. Landau and E. M. Lifshitz, Quantum Mechanics. Non-Relativistic Theory, 3d ed., Pergamon Press, Oxford 1977.

[7] D. Bohm, Quantum Theory, Prentice Hall, N. Y., 1952.

WEAKLY DISTURBING PHASE SPACE MEASUREMENTS IN QUANTUM MECHANICS

Harald Scherer and Paul Busch

Institute for Theoretical Physics,
University of Cologne,
D-50937 Cologne, Germany

Abstract

The existence of weakly disturbing quantum measurements is demonstrated by means of a phase space measurement model.

1 INTRODUCTION

Standard notions of quantum measurement theory need to be reformulated into weaker versions if general – sharp or unsharp – observables are taken into consideration [1]. In particular, the notion of an ideal measurement refers to the idea that a quantity which has a definite value should be measurable without disturbing the system in any way. Any ideal measurement of a sharp observable is repeatable and hence also value reproducible. In short, an instrument (associated with a given measurement) is ideal if it leaves unchanged (up to normalisation) any eigenstate of the measured observable. The instrument is repeatable if after its first application the probability for obtaining the same outcome upon an immediate repetition is equal to unity. Finally, an instrument is value reproducible if its application on an eigenstate yields with certainty the same result and leaves the system in a corresponding eigenstate.

As is well known, repeatable measurements do not exist for continuous observables or for unsharp observables. Accordingly there can be no ideal measurements of such quantities. Still one may conceive of procedures that just do not disturb the system much provided the property to be measured is *nearly* actualised, i.e., has probability close to unity. Such measurements are called weakly disturbing

Quantum Communications and Measurement
Edited by V. P. Belavkin *et al.*, Plenum Press, New York, 1995

(or almost ideal). They are of use for problems of establishing quantum communication channels or devising quantum nondemolition methods for detecting weak signals. From a general conceptual point of view weak disturbance is to be stipulated as a feature of measurements in a quasi-classical limit. We shall demonstrate the existence of weakly disturbing measurements in the case of certain phase space observables.

The present contribution is a report on a series of investigations into a specific model which have led to several interesting results concerning conditions for measuring jointly complementary observables such as position and momentum [2–5].

2 PHASE SPACE MEASUREMENT MODEL

2.1 Measurement scheme

We consider a phase space measurement for a particle with one degree of freedom modeled by means of a unitary map establishing a coupling between the object $\mathcal{S}$ represented by means of the Hilbert space $\mathcal{H}$ and two apparatus systems $\mathcal{A}_1$ and $\mathcal{A}_2$, also called pointers, with Hilbert spaces $\mathcal{H}_1$ and $\mathcal{H}_2$.

$$U := \exp\left\{-\frac{i}{\hbar}\lambda\widehat{Q}\otimes\widehat{P}_1\otimes\mathbf{1}_2+\frac{i}{\hbar}\mu\widehat{P}\otimes\mathbf{1}_1\otimes\widehat{Q}_2\right\}.$$

The position $\widehat{Q}$ and momentum $\widehat{P}$ of the object are coupled to the momentum $\widehat{P}_1$ of pointer 1 and the position $\widehat{Q}_2$ of pointer 2, respectively (cf. Figure 1). λ and μ

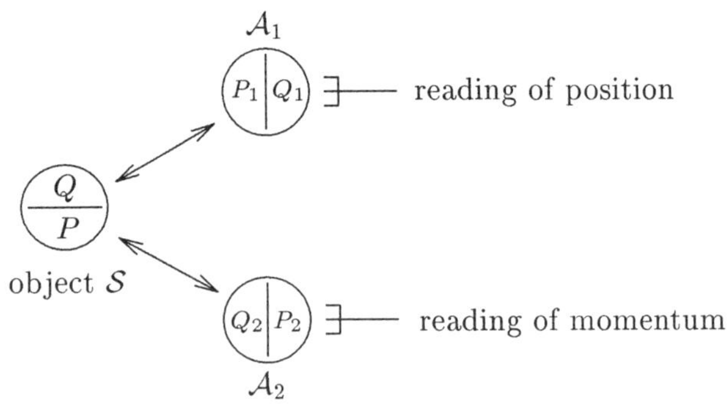

Figure 1: Measurement scheme.

are positive real numbers determining the strength of the interaction. The pointer observables shall be chosen as projection valued measures on the real Borel sets $\mathcal{B}(\mathbf{R})$,

$$\mathcal{B}(\mathbf{R})\ni\Delta_1 \quad\mapsto\quad \widehat{Q}_1(\lambda\Delta_1):=\int_{\mathbf{R}}dQ_1\,\chi_{\lambda\Delta_1}(Q_1)\;|\,Q_1\,\rangle\langle\,Q_1\,|$$

$$\mathcal{B}(\mathbf{R})\ni\Delta_2 \quad\mapsto\quad \widehat{P}_2(\mu\Delta_2):=\int_{\mathbf{R}}dP_2\,\chi_{\mu\Delta_2}(P_2)\;|\,P_2\,\rangle\langle\,P_2\,|\,.$$

where the above λ and μ appear as scaling parameters.

2.2 State transformation via measurement

Let

$$|\,\varphi,\Phi_1,\Phi_2\,\rangle := |\,\varphi\,\rangle\otimes|\,\Phi_1\,\rangle\otimes|\,\Phi_2\,\rangle\in\mathcal{H}\otimes\mathcal{H}_1\otimes\mathcal{H}_2$$

denote the state of the compound system before the measurement. The coupling U establishes correlations between the object and pointer systems which determine, in particular, the state transformation suffered by the object as well as the measured observable. The induced instrument $\mathcal{I}$ is given by the equation

$$\mathrm{tr}_{\mathcal{H}\otimes\mathcal{H}_1\otimes\mathcal{H}_2}\Big\{U \mid \varphi, \Phi_1, \Phi_2 \rangle\langle \varphi, \Phi_1, \Phi_2 \mid U^\dagger \, \widehat{A} \otimes \widehat{Q}_1(\lambda\Delta_1) \otimes \widehat{P}_2(\mu\Delta_2)\Big\}$$

$$= \mathrm{tr}_{\mathcal{H}} \Big\{\mathcal{I}(\Delta_1 \times \Delta_2) \mid \varphi\rangle\langle\varphi \mid \widehat{A}\Big\}$$

which must hold for all $|\varphi\rangle \in \mathcal{H}$, $\Delta_1 \times \Delta_2 \in \mathcal{B}(\mathbf{R}^2)$, and all linear selfadjoint operators $\widehat{A}$ on $\mathcal{H}$. The instrument reads explicitly as follows

$$\begin{aligned}
\mathcal{I}(\Delta_1 \times \Delta_2)\,\rho &= \int_{\Delta_1\times\Delta_2} dQ_1\, dP_2\, A_{Q_1P_2}\, \rho\, A^\dagger_{Q_1P_2}, \\
A_{Q_1P_2} &:= \sqrt{\tfrac{1}{2\pi\hbar}} \int_{\mathbf{R}} dq\, e^{\frac{i}{\hbar} qP_2} \int_{\mathbf{R}} d\xi\, \delta_{Q_1P_2}(\xi, q)\ | q \rangle\langle \xi |
\end{aligned}$$

for all state operators ρ. Here we have used the abbreviations

$$\begin{aligned}
\delta_{Q_1P_2}(\xi, q) &:= e^{-\frac{i}{\hbar}\xi P_2}\, \Phi_1^{(\lambda)}(Q_1 - \tfrac{1}{2}(\xi+q))\, \Phi_2^{(\frac{1}{\mu})}(\xi - q) \\
\Phi_1^{(\lambda)}(Q_1) &:= \sqrt{\lambda}\, \Phi_1(\lambda Q_1) \\
\Phi_2^{(\frac{1}{\mu})}(Q_2) &:= \sqrt{\tfrac{1}{\mu}}\, \Phi_2(\tfrac{1}{\mu} Q_2).
\end{aligned}$$

If ρ is the initial state of the object, then its state $\rho_{\mathcal{I}}(\Delta_1 \times \Delta_2)$ after the measurement with the outcome $\Delta_1 \times \Delta_2 \in \mathcal{B}(\mathbf{R}^2)$ is given through

$$\begin{aligned}
\rho_{\mathcal{I}}(\Delta_1 \times \Delta_2) &= \frac{1}{p_\rho(\Delta_1 \times \Delta_2)}\, \mathcal{I}(\Delta_1 \times \Delta_2)\,\rho \\
p_\rho(\Delta_1 \times \Delta_2) &= \mathrm{tr}_{\mathcal{H}}[\mathcal{I}(\Delta_1 \times \Delta_2)\,\rho],
\end{aligned}$$

$p_\rho(\Delta_1 \times \Delta_2)$ is the probability for obtaining the result $\Delta_1 \times \Delta_2$.

The observable a associated with the instrument $\mathcal{I}$ is the POV measure determined by the relation

$$p_\rho(\Delta_1 \times \Delta_2) = \mathrm{tr}_{\mathcal{H}}[\rho\, a(\Delta_1 \times \Delta_2)].$$

Its explicit form is

$$a: \quad \Delta_1 \times \Delta_2 \;\mapsto\; a(\Delta_1 \times \Delta_2) = \int_{\Delta_1\times\Delta_2} \frac{S_{Q_1P_2}}{2\pi\hbar}$$

where

$$\frac{1}{2\pi\hbar} S_{Q_1P_2} = A^\dagger_{Q_1P_2} A_{Q_1P_2}$$

The operators $S_{Q_1P_2}$ are state operators arising from S_{00} by means of phase space translations. It follows that the marginals of the POV measure a are unsharp position and momentum observables and that our model provides a phase space measurement with inaccuracies satisfying the Heisenberg indeterminacy relation.

2.3 Scaling properties

We consider a scale transformation for the pointer states,

$$\begin{aligned} \Phi_1^{(\lambda)}(Q_1) &\mapsto \sqrt{\alpha}\,\Phi_1^{(\lambda)}(\alpha\, Q_1) &=& \Phi_1^{(\alpha\lambda)}(Q_1) \\ \Phi_2^{(\frac{1}{\mu})}(Q_2) &\mapsto \sqrt{\tfrac{1}{\beta}}\,\Phi_2^{(\frac{1}{\mu})}(\tfrac{1}{\beta}Q_2) &=& \Phi_2^{(\frac{1}{\beta\mu})}(Q_2). \end{aligned} \tag{a}$$

For $\alpha, \beta < 1$ this leads to an increasing variance of the pointer state $\Phi_1^{(\lambda)}(Q_1)$ and a decreasing variance of the pointer state $\Phi_2^{(\frac{1}{\mu})}(Q_2)$, the latter corresponding to an increasing variance of the Fourier transform

$$\widetilde{\Phi}_2^{(\frac{1}{\mu})}(P_2) := \frac{1}{\sqrt{2\pi\hbar}} \int_{\mathbf{R}} dQ_2 \exp\left(\frac{i}{\hbar} Q_2 P_2\right) \Phi_2^{(\frac{1}{\mu})}(Q_2).$$

One can easily show that the simultaneous transformations (a) of the pointer preparations lead to the same instrument as the following transformations of the coupling constants λ and μ:

$$\lambda \mapsto \alpha\lambda, \qquad \mu \mapsto \beta\mu. \tag{b}$$

This means that the following operations are equivalent:

(a) to coarsen the pointer states

(b) to decrease the coupling constants and to compress the reading scales.

One may apply these scalings in order to introduce 'macroscopically unsharp' (as compared to $\hbar$) phase space observables which may be expected to allow weakly disturbing measurements.

2.4 Weakly disturbing measurements

The transformations (a), (b) for $\alpha, \beta < 1$ amount to choosing smaller values of the parameters λ and μ. It can be shown that the ensuing instruments are indeed more and more weakly disturbing in the following sense. Given a class of states ρ localised in phase space in a certain sense, then application of the instrument $\mathcal{I}$ yields states $\rho_{\mathcal{I}}(\Delta_1 \times \Delta_2)$ which are as close as one wishes to ρ (in trace norm metric) provided the values of λ, μ are small enough [5].

This result shall be illustrated graphically by means of a phase space representation of the states before and after measurement. To every state ρ one can assign a probability distribution $\mathcal{R}(\rho)$ on phase space, its $Q-$function

$$\mathcal{R}(\rho): \quad (Q,P) \mapsto \mathcal{R}(\rho)_{QP} := \mathrm{tr}[\rho\, S_{QP}^{(0)}]$$

where $S_{QP}^{(0)}$ are the projections on Gaussian coherent states.

We assume the object to be prepared in a Gaussian state

$$\varphi(q) = \sqrt{\frac{1}{\sqrt{\pi\hbar}\sigma}}\, e^{-\frac{1}{2\hbar}\frac{1}{\sigma^2}(q-q_0)^2}$$

with variance $\mathrm{Var}(\widehat{Q},\varphi) = (\hbar/2)\,\sigma^2$. Furthermore, we specify Gaussian states for the pointers

$$\Phi_1^{(\lambda)}(Q_1) = \sqrt{\frac{\lambda}{\sqrt{\pi\hbar}}}\, e^{-\frac{1}{2\hbar}\lambda^2 Q_1^2}, \qquad \widetilde{\Phi}_2^{(\frac{1}{\mu})}(P_2) = \sqrt{\frac{\mu}{\sqrt{\pi\hbar}}}\, e^{-\frac{1}{2\hbar}\mu^2 P_2^2}$$

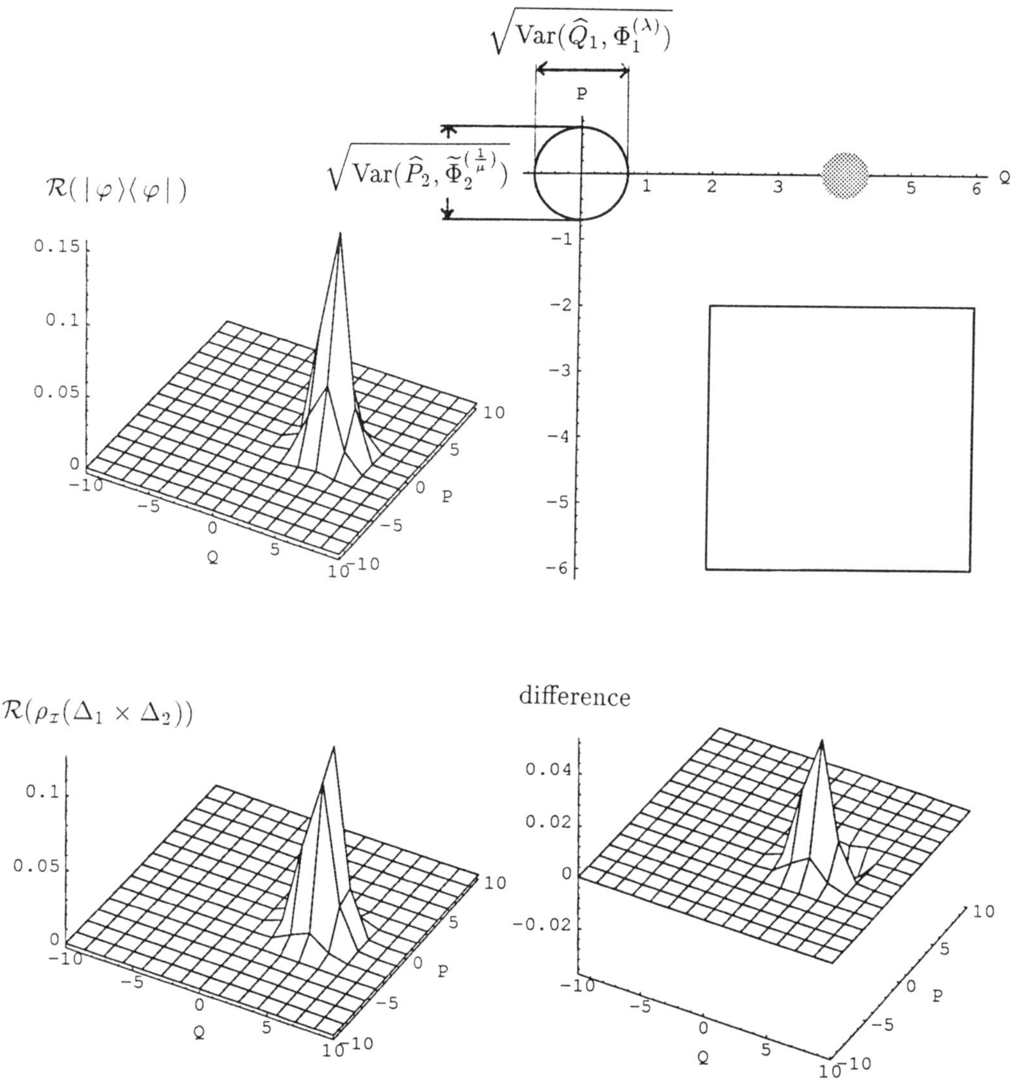

Figure 2: Phase space measurement: reference case ($\lambda = \mu = 0.5$).

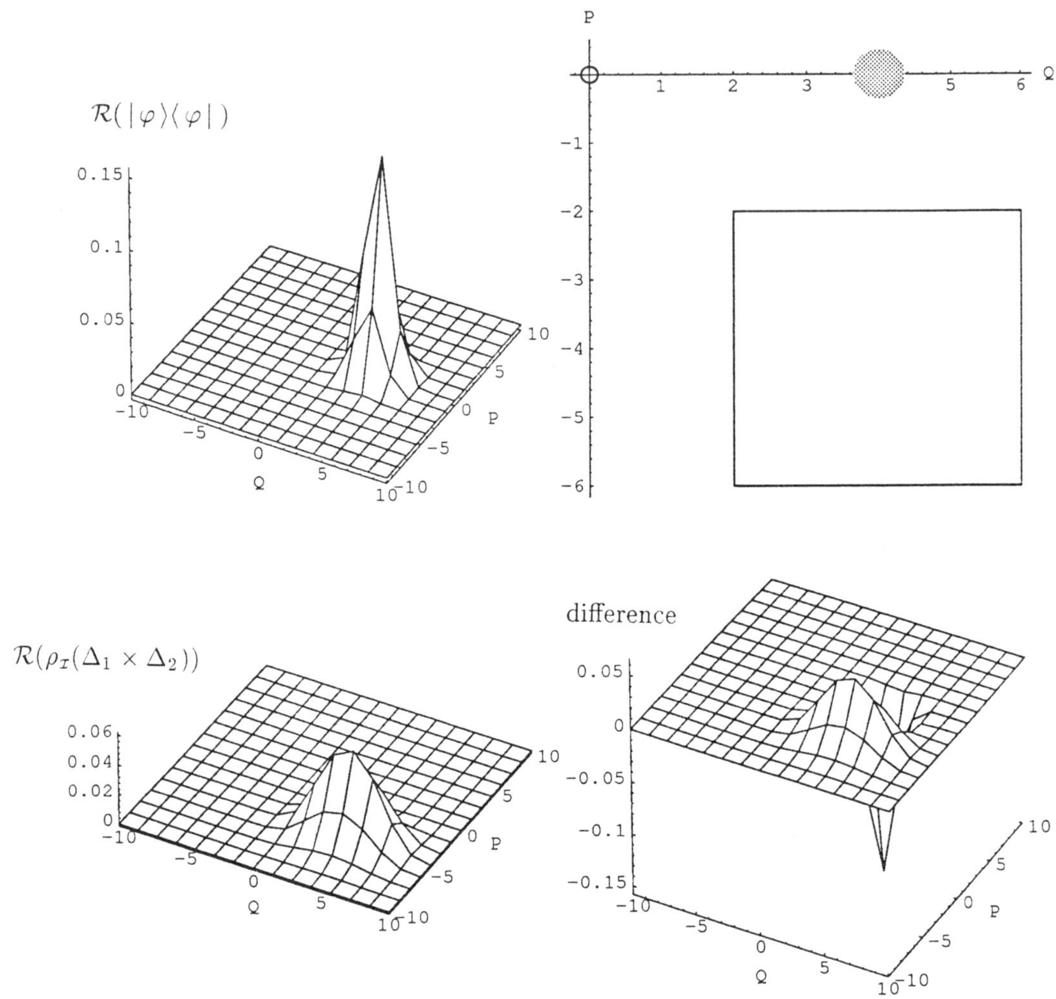

Figure 3: Phase space measurement: strongly disturbing case ($\lambda = \mu = 3$).

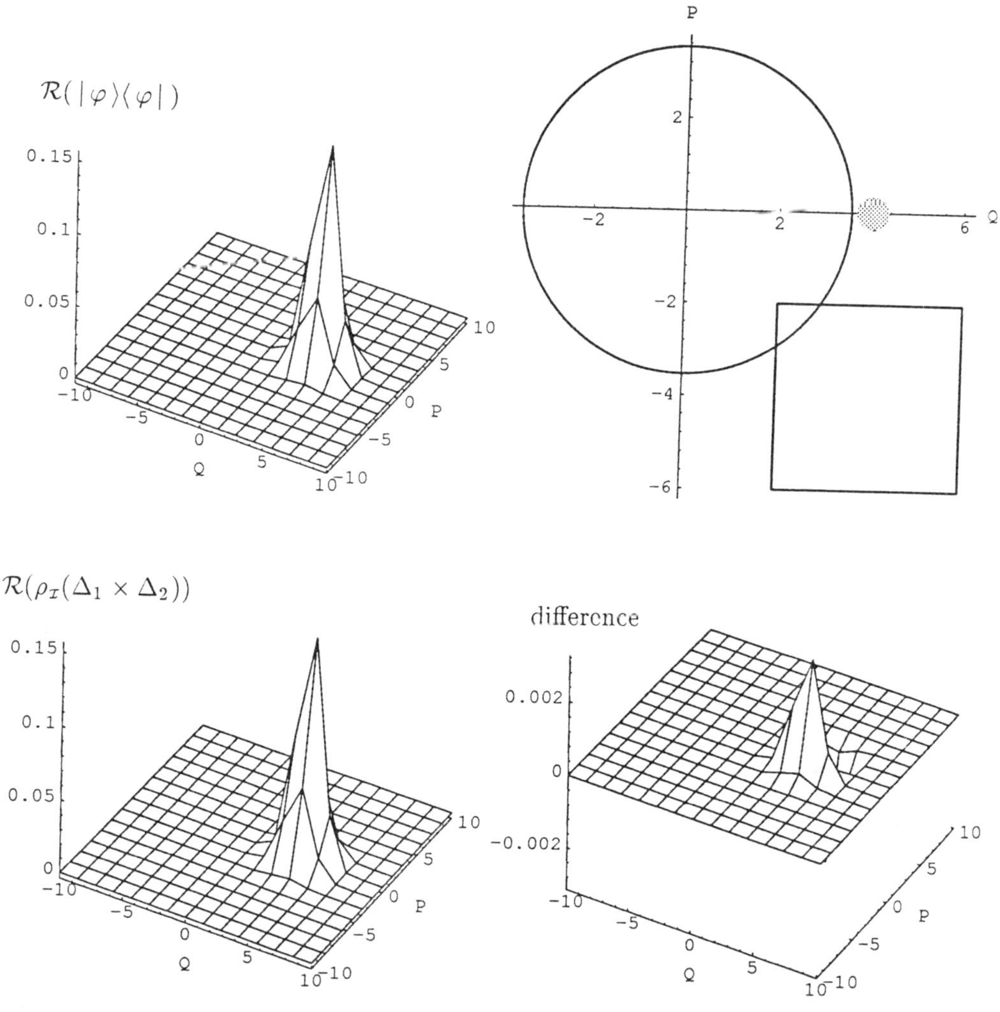

Figure 4: Phase space measurement: weakly disturbing case ($\lambda = \mu = 0.1$).

where the variances are scaled in λ and μ as follows

$$\mathrm{Var}(\widehat{Q}_1, \Phi_1^{(\lambda)}) = \frac{\hbar}{2} \frac{1}{\lambda^2}, \qquad \mathrm{Var}(\widehat{P}_2, \widetilde{\Phi}_2^{(\frac{1}{\mu})}) = \frac{\hbar}{2} \frac{1}{\mu^2}.$$

The product of these measures of inaccuracy becomes large if λ, μ are small.

The plots displayed in Figures 2–4 are based on the specifications

$$\hbar = 1, \qquad \sigma = 1, \qquad q_0 = 4.$$

The pointer unsharpnesses are indicated by means of 'error ellipses' centered around the coordinate origin. The object state is represented as a grey disc with center point $(4, 0)$. The rectangular box corresponds to the reading intervals

$$\Delta_1 = [2, 6], \qquad \Delta_2 = [-6, -2].$$

The plots provide a comparison of the initial and final object states for different values of the coupling constants λ, μ. Figure 2 shows a reference case, $\lambda = \mu = 0.5$, where the error ellipse, the object disc and the reading box are of comparable size. The initial probability distribution is smeared and shifted towards the box. In Figure 3 one has $\lambda = \mu = 3$, which corresponds to a strongly disturbing measurement: the distribution is strongly smeared and its peak is shifted close to the center of the box. Finally, choosing $\lambda = \mu = 0.1$, as indicated in Figure 4, one obtains a weakly disturbing measurement. The shift and the smearing of the distribution are quite small in this case so that the state is indeed left nearly unchanged.

3 CONCLUDING REMARKS

Our consideration of weakly disturbing measurements is based on a fairly idealised model of a joint position-momentum measurement. It is therefore interesting to note that in quantum optics realisations of phase space measurements have been proposed recently. The unitary coupling operator U can be approximately realised by means of an interaction Hamiltonian between a single-mode field and some measuring system [6]. An alternative scheme leading to a different dynamics makes use of beam splitters and homodyne detectors [7]. Weakly disturbing measurements may be applied as quantum nondemolition measurements performed on harmonic oscillators which are employed for the detection of weak signals. In such situations one may also choose the oscillator phase as an observable to be measured for monitoring the oscillator's motion. In the present model a phase observable can be derived along with an unsharp energy observable as a pair of marginals with respect to a polar coordinate representation of the phase space. Thus our scheme provides a joint measurement of phase and energy. Details can be found in [8].

References

[1] P. Busch, P. J. Lahti, P. Mittelstaedt: *The Quantum Theory of Measurement*, Springer, Berlin 1991.

[2] P. Busch: Int. J. Theor. Phys. **24** (1985) 63–92.

[3] R. Quadt: Doctoral Thesis, Köln 1992, Verlag Shaker, Aachen 1992.

[4] R. Quadt, P. Busch: Open Systems & Information Dynamics (University of Toruń) **2** (1993), in press.

[5] H. Scherer: Doctoral Thesis, Köln 1994, Verlag Shaker, Aachen, 1994.

[6] S. Stenholm: Ann. Phys. (N.Y.) **218** (1992) 233.

[7] U. Leonhardt, H. Paul: Phys. Rev. A **47** (1993) R2460–R2463.

[8] P. Busch, M. Grabowski, P. Lahti: Ann. Phys. (N.Y.) **237** (1995) 1.

QUANTUM PHASE MEASUREMENT

Zdeněk Hradil[1]

Department of Theoretical Physics
Ecole Polytechnique Fédérale de Lausanne
CH–1015, Switzerland
Email:hradil@eldp.epfl.ch

Abstract

Theory of feasible quantum phase mesurement is formulated within the framework of quantum estimation theory. Phase detection is linked to the measurement of a pair of commuting Hermitian operators. The ideal phase concepts are distinguished, among all the possible realizations as the representations of the Euclidean algebra on Hermitian operators. Possible representations of the rotational subgroup correspond to measurements with indefinite phase. Traditional "quantum phase difficulties", ascribed sometimes to the lack of uniqueness of the Hermitian phase operator, are related to the classical attempt to distinguish between quantum effect and measurement itself. In realistic quantum phase measurement, the phase statistics should be estimated using counted discrete data. Possible interferometric measurements of phase–so called operationally defined quantum phase models–are included as special realizations of presented theory.

Keywords :
Quantum phase measurement, E(2) symmetry.

Phase detection represents one of the most accurate schemes of measurement currently available and has played a major role through the history of physics. The measurements done with the Michelson interferometer triggered the developement that lead to the special theory of relativity. On the other hand, the interference plays

[1]On leave of absence from: Department of Optics, Palacký University,17. listopadu 50, 771 46 Olomouc, Czech Republic

an important role in quantum mechanics and laser interferometry of gravitation waves is expected to change our view on the Universe in the next future. A proper characterization of the quantum phase is therefore essential. This problem is as old as the quantum mechanics itself [1] and possible concepts differ substantially. These treatments may be grouped into two broad classes: those related on the definition of the phase operator (e.g. [2–5]) and theories based on the experimentally feasible schemes (e.g. [6–11]) (an up–to–date overview with more representative references may be found in the special issue of Physica Scripta [12] and review [13]). In this contribution the quantum phase detection is linked to the measurement of two commuting Hermitian operators. Different ideal quantum phase concepts will be determined by the representations of an Euclidean algebra on Hermitian operators.

Let us describe the operator picture of the phase detection. Suppose that a measurement of a pair of observables represented by commuting Hermitian operators $\hat{Y}_1, \hat{Y}_2$

$$[\hat{Y}_1, \hat{Y}_2] = 0 \tag{1}$$

is available. This measurement will be called a *feasible measurement* and will be represented in the framework of ordinary quantum mechanics by a probability operator measure (POM) [14, 15]. Operators are acting on the full Hilbert space

$$\mathcal{H} = \mathcal{H}_s \otimes \mathcal{H}_a \tag{2}$$

created by the direct product of the signal ($\mathcal{H}_s$) and apparatus ($\mathcal{H}_a$) Hilbert spaces. For the sake of brevity, any quantum state is assumed to be defined on the full Hilbert space $\mathcal{H}$. Operators acting on $\mathcal{H}_{s,a}$ spaces are interpreted as acting on $\mathcal{H}$ via tensor product with the identity operator (e.g. $\hat{a}_s \equiv \hat{a}_s \otimes \hat{1}_a$). Assuming that the operator $\hat{Y} \equiv \hat{Y}_1 + i\hat{Y}_2$ has continuous spectrum, the relations of orthogonality and completness in $\mathcal{H}$ read

$$\langle y|y'\rangle = \delta^2(y - y'), \qquad \int d^2y\, |y\rangle\, \langle y| = \hat{1}. \tag{3}$$

Here, for the sake of brevity, the notation does not show explicitly the possible degenerations of the eigenvectors $|y\rangle$ and in the following degeneration will be assumed only for the value $y = 0$. The probability distribution function (PDF) of finding the complex amplitude y by performing the measurement on a general quantum state $|\psi\rangle$ is given by

$$p(y) = |\langle\psi|y\rangle|^2. \tag{4}$$

The purpose of an arbitrary phase detection is to determine the non–random c–number displacement parameter $\theta \in (-\pi, \pi]$ entering the displacement transformation [14] of the complex amplitude $\hat{Y}$

$$\hat{Y}(\theta) = e^{-i\theta\hat{N}}\, \hat{Y}\, e^{i\theta\hat{N}}, \tag{5}$$

$\hat{N}$ being a Hermitian displacement operator and θ the true value of the phase shift.

Let us specify the physically required properties of the Hermitian operators $\hat{Y}_1, \hat{Y}_2$ and $\hat{N}$. The expectation value of $\hat{Y}$ in the limit of classical field should provide information about θ. For concreteness only models for which classically $\theta = \arg[Y]$ will be addressed. In any case the known classical dependence of $\arg[Y]$ on θ makes it possible to obtain statistical information about the phase shift via the measurement of the phase–like variable $\phi = \arg[Y]$. To assess the accuracy of the available phase data according to mathematical statistics [6, 16], the dispersion

$$D^2 = 1 - |\langle e^{i\phi}\rangle|^2 \tag{6}$$

may be used. The phase information is then included in the normalized exponential phase operator

$$\hat{R}(\theta) = \begin{cases} \hat{Y}(\theta)\,(\hat{Y}^\dagger(\theta)\hat{Y}(\theta))^{-1/2} & \text{for } \mathcal{H}_0^\perp, \\ 0 & \text{for all states in } \mathcal{H}_0. \end{cases} \tag{7}$$

Here, $\mathcal{H}_0$ is the subspace spanned by the eigenstates with zero complex amplitude — $\hat{Y}(\theta)|y=0,k\rangle = 0$, k being the quantum number specifying the degeneration — and $\mathcal{H}_0^\perp$ is the orthogonal complement of $\mathcal{H}_0$

$$\mathcal{H} = \mathcal{H}_0 \oplus \mathcal{H}_0^\perp. \tag{8}$$

The prescription (7) represents the quantum counterpart of the polar transformation of Cartesian coordinates. There is no specific phase associated with the origin, or alternatively, all values are possible. The detection of zero complex amplitude corresponds to the registration of uniformly distributed noise for the phase variable. Even if the detection of the value $y=0$ does not bring any information about the phase shift, it tells something about the quantum state, on which the measurement is performed. The overlooking of this observation, which has no classical analogy, leads effectively to the renormalization of the quantum state on which the phase measurement was performed [17].

The phase detection on $\mathcal{H}$ should be described separately on the orthogonal subspaces in (8). The respective POM's are given by marginal phase distributions of the identity operator (3). The "phase measurement" on $\mathcal{H}_0$ is described by the POM

$$\hat{\Pi}_0(\phi) = \frac{1}{2\pi}\hat{1}_0, \tag{9}$$

where $\hat{1}_0 = \sum_k |y=0,k\rangle\langle k,y=0|$ denotes the identity operator on $\mathcal{H}_0$. Obviously such measurement does not yield any information about phase shift. The measurement on $\mathcal{H}_0^\perp$ is given by the POM

$$\hat{\Pi}_\perp(\phi) = \int d^2y\,\delta(\phi - \arg y)|y\rangle\,\langle y| = \int_{0+}^{\infty} d|y|\;|y|\,||y|e^{i\phi}\rangle\,\langle |y|e^{i\phi}|, \tag{10}$$

which defines an orthogonal projector

$$\hat{\Pi}_\perp(\phi)\,\hat{\Pi}_\perp(\phi') = \delta(\phi-\phi')\hat{\Pi}_\perp(\phi). \tag{11}$$

Let us emphasize that relation analogous to (11) is not true on the full Hilbert space $\mathcal{H}$ for

$$\hat{\Pi}(\phi) = \hat{\Pi}_0(\phi) + \hat{\Pi}_\perp(\phi).$$

Significantly, the phase measurement may be described using the spectral decomposition on the Hilbert space of states with meaningful phase $\mathcal{H}_0^\perp$. Here, the Hermitian phase operator is defined by

$$\hat{\phi} = \int_{-\pi}^{\pi} d\phi\,\phi\,\hat{\Pi}_\perp(\phi), \tag{12}$$

and an arbitrary function F of the phase operator fulfills the functional identity

$$F(\hat{\phi}) \equiv \hat{F}, \tag{13}$$

where

$$\hat{F} = \int_{-\pi}^{\pi} d\phi\,F(\phi)\,\hat{\Pi}_\perp(\phi).$$

To understand the "traditional phase difficulties" we need to formulate the adequate theory on the *signal Hilbert space* $\mathcal{H}_s$ *only.* Once the quantum state of the apparatus is specified, all the operators may be reduced by performing the partial trace over the apparatus state $\hat{\rho}_a$. Consequently, the reduced POM acting only on the signal space $\mathcal{H}_s$ is

$$\hat{\mathcal{P}}(\phi) = \mathrm{Tr}_a[\hat{1}_s \otimes \hat{\rho}_a \hat{\Pi}(\phi)]. \tag{14}$$

Obviously, this reduced POM is no longer an orthogonal projector, since

$$\hat{\mathcal{P}}(\phi)\,\hat{\mathcal{P}}(\phi') \neq \delta(\phi - \phi')\hat{\mathcal{P}}(\phi).$$

The Hermitian phase operator on the reduced Hilbert space may be formally introduced as

$$\hat{\phi}_s = \int_{-\pi}^{\pi} d\phi\, \phi\, \hat{\mathcal{P}}(\phi), \tag{15}$$

but the analogue to the identity (13) is not true. This operator cannot predict the "measurable functions" of the phase operator

$$\hat{\mathcal{F}}(\phi) = \int_{-\pi}^{\pi} d\phi\, F(\phi)\, \hat{\mathcal{P}}(\phi). \tag{16}$$

Consequently, the notion of Hermitian phase operator is useful only on the *Hilbert space of states with meaningful phases* $\mathcal{H}_0^{\perp}$. Here the phase operator describes well the phase experiments since the respective POM is the orthogonal projector (11). On the *full Hilbert space* $\mathcal{H}$ or on the *signal Hilbert space* $\mathcal{H}_s$ the measurement should be represented by the respective POM, whereas the notion of "Hermitian phase operator" is redundant.

The commutator $[\hat{Y}, \hat{N}]$ determines how the observable $\hat{Y}$ is sensitive to the phase shift. In general, as follows from eqs. (6) and (7), the dispersion depends on the detected phase shift θ. This dependence may be eliminated by imposing the condition

$$[\hat{Y}, \hat{N}] = \hat{Y}, \tag{17}$$

in which case obviously $\hat{Y}(\theta) = e^{i\theta}\hat{Y}$ and the dispersion (6) is independent on the true value of phase shift. The Hermitian operators $\hat{Y}_1, \hat{Y}_2$ and $\hat{N}$ fulfilling the commutation relations (1) and (17) are generators of the Euclidean group E(2) in two dimensional space [18,19]. Measurements satisfying these relations will be called *ideal feasible measurements* and may be classified according to unitary irreducible representations of the group of rotations and translations in two dimensions. Every group element $g(\mathbf{a}, \alpha)$ is determined by the displacement vector $\mathbf{a} = (r\cos\phi, r\sin\phi)$ and by the angle of rotation α. The representation of this group may be given on the Hilbert space $\mathcal{D}$ of infinitely differentiable functions $f(\mathbf{x})$ defined on the unit circle $\mathbf{x} = (\cos\varphi, \sin\varphi)$. Introducing the notation $f(\mathbf{x}) \equiv f(\varphi)$, a representation of group E(2) is given by

$$T_R(g)f(\varphi) = e^{iRr\cos(\varphi-\phi)}f(\varphi - \alpha), \tag{18}$$

R being an arbitrary real number $R \geq 0$. The infinitesimal generators are simply the operators

$$\hat{A}_1 = R\cos\varphi, \quad \hat{A}_2 = R\sin\varphi, \quad \hat{A}_3 = -\frac{d}{d\varphi}, \tag{19}$$

fulfilling the commutation relations

$$[\hat{A}_1, \hat{A}_2] = 0, \;\; [\hat{A}_1, \hat{A}_3] = \hat{A}_2, \;\; [\hat{A}_2, \hat{A}_3] = -\hat{A}_1.$$

The relation (18) gives for $R \neq 0$ the unitary irreducible representation of group E(2). If $R = 0$, the representation (18) has the form of a reducible representation of the rotational group, which is a subgroup of E(2),

$$T_0(g)f(\varphi) = f(\varphi - \alpha) \tag{20}$$

and may be further decomposed as the direct sum of one–dimensional unitary irreducible representations

$$T_0 = \sum_{k=-\infty}^{\infty} \oplus T_{0k}, \quad T_{0k} = e^{ik\alpha}, \tag{21}$$

where k is an integer $-\infty < k < \infty$. The representations T_R and T_{0k} exhaust all the unitary irreducible representations of E(2).

Realistic schemes for interferometric measurement of phase shift are intimately connected with the $SU(1,1)$ and $SU(2)$ symmetries [9, 19, 20]. The fact that the group E(2) may be obtained by contractions of these Lie groups [18, 19] establishes the connection between the experimentally measured and the ideal phase concepts. In particular, the multimode homodyne phase shift detection proposed and performed by Noh, Fougères and Mandel [11] approximates well an ideal phase measurement even in ordinary experimental conditions. The limiting phase concept may be characterized by the complex amplitude

$$\hat{Y}_{SW} = \hat{a}_s + \hat{a}_a^\dagger \tag{22}$$

with the displacement operator $\hat{N} = \hat{a}_s^\dagger \hat{a}_s - \hat{a}_a^\dagger \hat{a}_a$, $\hat{a}_{a,s}$ being the annihilation operators of the signal and auxiliary modes. This realization of ideal phase measurement is known as Shapiro and Wagner phase concept [8]. Another realization of ideal phase measurement–representation of E(2) algebra, was recently introduced by Ban [21] as phase operator in relative–number–state basis.

Realistic description of quantum phase detection should take into account discrete nature of detected data. Phase shift should be estimated on the basis of performed measurement of complex amplitude yielding discrete spectrum. Measurable statistics $p_k(\theta) = |\langle y_k e^{-i\theta\hat{N}} \psi\rangle|^2$ encompasses information about induced phase shift. Registration of value y_k may be interpreted as inferred phase distribution

$$P_k(\phi) = \frac{1}{C_k} p_k(\phi), \quad C_k = \int_{-\pi}^{\pi} p_k(\theta) d\theta$$

and standard methods of mathematical statistics may be used for evaluation of phase data.

It is well–known that the fundamental groups of symmetry are reflected in quantum mechanics. Particularly, quadrature operators of generalized coordinate and impuls $(\hat{X}_1, \hat{X}_2)$ fulfil the Heisenberg operator algebra $[\hat{X}_1, \hat{X}_2] = i$. The phase shift $e^{i\eta\hat{X}_1}$ (i.e. the x_2 coordinate) can than be detected using $\hat{X}_2$ observable. Similarly, the ideal phase measurement may be associated with the representation of Euclidean algebra $(\hat{Y} = \hat{Y}_1 + i\hat{Y}_2, \hat{N})$ on Hermitian operators $[\hat{Y}, \hat{Y}^\dagger] = 0, \quad [\hat{Y}, \hat{N}] = \hat{Y}$. Phase shift $e^{i\theta\hat{N}}$ can be detected using complex observable $\hat{Y}$. Ideal phase detection may be approximated by realistic phase concept based on the counting of discrete data providing that corresponding quantum estimation theory is used. Presented formulation encompasses all the ordinary ideal phase concepts discussed for example in review [13] linking them to the feasible measurement schemes. Phase shift detection may be equivalently characterized as measurement of rotation angle. Formulation on signal Hilbert space only may be obtained by projection. Traditional phase difficulties, such as lack of orthogonality of operator measure or loss of "canonical" commutational relation, appear. This indicates that any realistic quantum phase concept should incorporate description of measurement process, too. In accordance with pragmatic interpretation of quantum mechanics, it does not seem to be reasonable to distinguish between measured effect and measurement itsef.

This work was supported by the Swiss National Optics Program.

References

[1] P. A. M. Dirac, Proc. R. Soc. London Ser. A **114**, 243 (1927).

[2] L. Susskind and J. Glogower, Physics **1**, 49 (1964).

[3] P. Carruthers and M. M. Nieto, Rev. Mod. Phys. **40**, 411 (1968).

[4] D. T. Pegg and S. M. Barnett, Phys. Rev. A **39**, 1665 (1989).

[5] M. J. Hall, Quantum Opt. **3**, 7 (1991).

[6] A. Bandilla and H. Paul, Ann. Phys., (Lpz)**23**, 323 (1969).

[7] C. M. Caves, Phys. Rev. D **23**, 1693 (1981).

[8] J. H. Shapiro and S. M. Wagner, IEEE J. Quantum Electron. QE-**20**, 803 (1984).

[9] B. Yurke, Samuel L. McCall and J. R. Klauder, Phys. Rev. A **33**, 4033 (1986).

[10] S. L. Braunstein and C. M. Caves, Phys. Rev. **A 42**, 4115 (1990).

[11] J. W. Noh, A. Fougères and L. Mandel, Phys. Rev. A **45**, 424 (1992).

[12] Physica Scripta, Vol. **T48**, (1993).

[13] A. Lukš, V. Peřinová, Quantum Opt. **6**, 125 (1994).

[14] C. W. Helstrom, *Quantum Detection and Estimation Theory* (Academic Press 1974, New York) chap. 8.

[15] H. P. Yuen, Phys. Lett. **91A**, 101 (1982).

[16] R. C. Rao, *Linear Statistical Inference and Its Applications* (New York, Wiley, 1973) p.175.

[17] Z. Hradil, Phys. Rev. A **47**, 4532 (1993).

[18] N. J. Vilenkin, *Special Functions and Theory of Group Representation* (Moscow, Nauka 1965); in Russian.

[19] Y. S. Kim and M. E. Noz, *Phase Space Picture of Quantum Mechanics: Group Theoretical Approach* (Singapore, World Scientific 1991).

[20] A. Vourdas, Phys. Rev. A **41**, 1653 (1990).

[21] M. Ban, J. Math. Phys. **32**, 3077 (1991).

QUANTUM PHENOMENOLOGY WITH THE PATH INTEGRAL APPROACH

T. Calarco,[1,2] R. Onofrio,[1,3] C. Presilla,[4] and L. Viola[1,3]

[1]Dipartimento di Fisica "G. Galilei",
Università di Padova, Via Marzolo 8,
Padova, Italy 35131
[2]INFN, Laboratori Nazionali di Legnaro,
Legnaro, Italy 35020
[3]INFN, Sezione di Padova, Padova, Italy 35131
[4]Dipartimento di Fisica and INFN,
Università di Roma "La Sapienza",
Piazzale A. Moro 2, Roma, Italy 00185

Abstract

A quantum measurement model based upon restricted path-integrals allows us to study measurements of generalized position in various one-dimensional systems of phenomenological interest. After a general overview of the method we discuss the cases of a harmonic oscillator, a bistable potential and two coupled systems, briefly illustrating their applications.

PACS: 03.65.Bz, 06.30.-k, 74.50.+r
Keywords: Quantum measurements, Bell inequalities

Recent efforts in high precision experiments have shown that the fundamental limitations to a measurement due to quantum mechanics play a crucial role to develop more and more sensitive instruments [1]. As a by-product quantum measurement theory, a topic quite isolated from the frontier of physics some decades ago, has been revived and new experiments have been proposed for its better understanding. Models of quantum measurement theory comparable to the outcomes of the experiments are therefore welcome. Here we describe some of the results obtained using one of these models, originally developed in [2] (see also [3] for a complete account of the approach). The goal of this paper is twofold: to discuss specific examples of systems in which quantum measurement theory plays a crucial role and to show that they are all described within the *unified* context of the model we use. For reasons of space we do not deal here with a comparison of our approach to the other ones described elsewhere in these Proceedings [4]. Moreover, we remand the interested reader to the mentioned references of our papers for more details.

A MODEL FOR IMPULSIVE MEASUREMENTS

In classical mechanics the effect of a sequence of measurements on the subsequent dynamics is negligible. In quantum mechanics instead the execution of a measurement, either continuous or stroboscopic, influences the system. In this last case it is therefore important to include the effect of an actual measurement on the dynamics of the observed system. This has been obtained by Mensky [2, 3] through the path-integral approach (other similar approaches have been proposed in [4]). In this framework the propagator of a system described by a Lagrangian $\mathcal{L}(x(t), \dot{x}(t), t)$, and undergoing a continuous measurement of its position between the times 0 and τ, with result $a(t)$ and instrumental uncertainty Δa, is written as a weighted path-integral:

$$K_{[a]}(x'', \tau; x', 0) = \int_{x(0)\equiv x'}^{x(\tau)\equiv x''} \mathcal{D}[x(t)] \exp\left\{\frac{i}{\hbar}\int_0^\tau \mathcal{L}(x(t), \dot{x}(t), t)dt\right\} w_{[a]}[x], \quad (1)$$

where

$$w_{[a]}[x] = \exp\left\{-\frac{1}{2\Delta a^2 \tau}\int_0^\tau [x(t) - a(t)]^2 dt\right\}. \quad (2)$$

The most natural way to represent an impulsive measurement of position at time 0 is as limit of a continuous one for infinitesimal time intervals. In this approximation,

$$\lim_{\tau\to 0} K_{[a(t)]}(x'', \tau; x', 0) = K_a(x'', x') = e^{-\frac{(x'-a)^2}{2\Delta a^2}} K(x'', 0; x', 0) \quad (3)$$

where $a = a(0)$. Let us suppose that $\psi(x, t)$ be the wavefunction of a system subjected to an impulsive measurement of position at time t, with result a. Since $K(x'', 0; x', 0) = \delta(x'' - x')$, it follows

$$\psi(x, t^+) = R_{[\psi]}(a) \int_{-\infty}^{+\infty} dy K_a(x, y)\psi(y, t^-) = R_{[\psi]}(a) \exp\left\{-\frac{(\hat{x} - a)^2}{2\Delta a^2}\right\} \psi(x, t^-). \quad (4)$$

where $R_{[\psi]}(a)$ is a renormalization constant. The effect of the measurement is therefore a Gaussian filtering around the measurement result. This is similar to the usual measurement theory of von Neumann. The latter is simply recovered by choosing for the measurement operator the discontinuous and therefore less realistic form

$$\hat{w}_a^{v.N.} \propto \theta(\hat{x} - [a - \Delta a])\theta([a + \Delta a] - \hat{x}). \quad (5)$$

For a sequence of impulsive measurements the Gaussian reduction (4) is alternatively followed by free evolution periods ΔT (quiescent time). The probability that the Nth measurement gives a result a_N, when the results of the previous $N - 1$ ones are known, is expressed through

$$P_{a_1, a_2, .., a_{N-1}}(a_N) \equiv \frac{\left|\left\langle \psi_{a_1, a_2, .., a_N}(t_N^+)\right| \psi_{a_1, a_2, .., a_N}(t_N^+)\right\rangle\right|^2}{\displaystyle\int_{-\infty}^{+\infty} \left|\left\langle \psi_{a_1, a_2, .., a_N}(t_N^+)\right| \psi_{a_1, a_2, .., a_N}(t_N^+)\right\rangle\right|^2 da_N}. \quad (6)$$

which allows one to evaluate the effective uncertainty defined as

$$\Delta a_{eff}^2 = \frac{2\int (a_N - \tilde{a})^2 \; P_{a_1,a_2,..,a_{N-1}}(a_N) \; da_N}{\int P_{a_1,a_2,..,a_{N-1}}(a_N) \; da_N}. \tag{7}$$

where $\tilde{a}$ is the most probable measurement result. The effective uncertainty Δa_{eff} expresses the spreading of the possible measurement results [5] and is equal to Δa, the instrumental error, in the classical limit. In a quantum regime Δa_{eff} is always larger or equal to Δa. A convenient situation for evaluating the Δa_{eff} is available in the case of impulsive measurements when energy eigenstates and eigenvalues of the system are known. In such a case the evolution of the state during the quiescent time is obtained once an expansion in terms of energy eigenstates of the output of a measurement is made and the numerator of (6) can be rewritten as

$$\left| \left\langle \psi_{a_1,a_2,..,a_N}(t_N^+) \right| \psi_{a_1,a_2,..,a_N}(t_N^+) \right\rangle \Big|^2 = \left(\sum_{m=1}^{\infty} \left| c_m^{(N)} \right|^2 \right)^2 \tag{8}$$

where $c_m^{(N)}$ is the projection coefficient on the mth eigenstate $|m\rangle$ at the Nth measurement. Thus from the energy eigenstates expansion of the initial wavefunction it is possible to derive the effective uncertainty

$$\Delta a_{eff}^2 = \frac{2\int_{-\infty}^{+\infty} (a_N - \tilde{a})^2 \left(\sum_{m=1}^{\infty} \left| \sum_{l=1}^{\infty} B_{ml}^N(a_0, \ldots, a_{N-1}, a_N) c_l^{(0)} \right|^2 \right)^2 da_N}{\int_{-\infty}^{+\infty} \left(\sum_{m=1}^{\infty} \left| \sum_{l=1}^{\infty} B_{ml}^N(a_0, \ldots, a_{N-1}, a_N) c_l^{(0)} \right|^2 \right)^2 da_N}, \tag{9}$$

where

$$c_m^{(N)} \stackrel{\text{def}}{=} \sum_{l=1}^{\infty} B_{ml}^N c_l^{(0)}, \tag{10}$$

$$B_{ml}^N(\Delta T, \Delta a, a_1, a_2, .., a_N) \stackrel{\text{def}}{=} \langle m | \hat{w}_{a_N} \left(\prod_{j=1}^{N} e^{-\frac{i}{\hbar}\hat{H}\Delta T} \hat{w}_{a_{N-j}} \right) | l \rangle. \tag{11}$$

For quantum measurements on harmonic oscillators, having energy levels equally spaced, the optimal quiescent time is half the oscillation period and all can be evaluated without approximations. This is already known for the limit case of instantaneous and infinite accuracy measurements as seen using the canonical approach: the commutator for the position at different times gives [6]

$$[\hat{x}(t + \Delta T), \hat{x}(t)] = \frac{i\hbar}{m\omega} \sin \omega \Delta T \tag{12}$$

which implies that for two instantaneous measurements of position spaced by multiples of half period of oscillation the observable is quantum nondemolition. The path integral formalism allows us to extend such a result to finite accuracy and finite duration measurements, as shown in [5]. It turns out that the effective uncertainty holds values very close to the instrumental uncertainty only for quiescent times which are multiples of half period of the harmonic oscillator. The optimal measurements of position for a harmonic oscillator are also known in literature as quantum nondemolition measurements [1, 6], and are quite important for the detection of small displacements like pulses of gravitational waves of astrophysical origin.

QUANTUM MEASUREMENTS IN SQUIDS

Superconducting Quantum Interferometer Devices (SQuIDs) have been proposed to test macrorealism versus quantum mechanics. A global understanding of quantum mechanics and its relationship to the classical limit requires to extend its validity to the macroscopic world. In this domain the conflict between its structure and the classical sense of physical reality cannot be overcome [7] and experiments aimed to compare the predictions of the two worldviews are crucial. By realism here we mean that "a macroscopic system with two or more macroscopically distinct states available to it will at all times *be* in one or the other of these states" (*macroscopic realism*) as well as that "it is possible, in principle, to determine the state of the system with arbitrarily small perturbation on its subsequent dynamics" (*non-invasive measurability at the macroscopic level*) [8]. This leads to a quantitative test by introducing inequalities between correlation functions of observables of a macroscopic system, a superconducting quantum interferometer device subjected to a sequence of repeated measurements of magnetic flux. The proposal of Leggett and Garg [8] has been criticized due to the limitations given by quantum mechanics to the accuracy obtainable in a set of repeated measurement of the same observable [9]. However, measurement schemes following that path have been proposed both using a set of SQuIDs and two-level mesoscopic systems [10]. We have studied quantitatively optimal strategies for repeated measurements of magnetic flux in bistable potentials which can schematize the SQuID behaviour [11]. It turns out that measurement strategies must be chosen in a particular way to minimize the influence of the previous measurements on the state of the observed system: quantum nondemolition strategies for the measurement of the magnetic flux in a SQuID have to be implemented. The magnetic flux in a SQuID is schematized, if the coupling to the external environment can be neglected, through the effective potential

$$V(\varphi) = -\frac{\mu}{2}\varphi^2 + \frac{\lambda}{4}\varphi^4 \tag{13}$$

where φ is the trapped magnetic flux, a generalized coordinate describing the system, μ and λ are parameters associated to the superconducting circuit. This allows us to describe the system in terms of pure states $\psi(\varphi)$. Since general arguments exist on the fundamental noise introduced by any linear amplifier [12], the problem of the measurement of flux in a superconducting circuit is independent upon the detailed scheme used to detect the quantum state of the SQuID. As already discussed in [5] the optimality of the measurement is dictated by the spectral properties of the system. The asymptotic collapsed wavefunction can be expanded in terms of the eigenstates of the system and an optimal measurement is obtained provided that the quiescent time is commensurable to the characteristic times for the wavefunction reformation

$$T_{ij} \equiv \frac{h}{|E_i - E_j|}, \tag{14}$$

where E_i, E_j are the energy eigenvalues which have maximal projections on the asymptotic state. The two eigenstates which maximally contribute to the wavefunction reformation after tunneling are those corresponding to the first two eigenvalues, whose splitting is related to the tunneling period. The effective error $\Delta\Phi_{eff}$ reached after a series of mesurement pulses has minima close to $\Delta\Phi$ when the quiescient time ΔT is a multiple of the tunnelling period T_{12}. Unless the measurements are repeated with this periodicity noise due to the measurement process is fed into

the system affecting the following measurements. In the case of the potential of Eq. (13) we have numerically evaluated the eigenstates using a selective relaxation algorithm [13]. The predictability of the measurement outcome is affected by the uncertainty due to the previous measurements. One can minimize this by simply using optimal quiescent times which are analytically determined through (14) once the eigenvalues and the eigenstates are given.
The general hypotheses under which we have obtained this result allows us to conclude that a direct experimental test of temporal Bell-type inequalities in principle should remain possible even in a pure quantum mechanical framework, without any classical assumption on the measurement process such as noninvasivity [8]. However, this happens only if the repetition times between consecutive measurements which violates Bell inequalities are compatible with the optimal quiescent times. Otherwise either the inequalities are not violated or quantum noise is introduced in the successive measurement, making it invasive [11]. A preliminary analysis on which we will refer in the future shows that this is the case. The region of violation occurs when the two quiescent times required to have a set of three possible measurements are both of the order of half of the tunnelling period. Unfortunately for such periodicities the effect of the measurement is such that the spreading on the possible results induced by the measurement does not allow one to distinguish between the two wells of the bistable potential. As a consequence violation of temporal Bell inequalities cannot be observed.

QUANTUM MEASUREMENTS IN PENNING TRAPS

Penning traps consist of a combination of static magnetic and electric fields where a single charged particle, for instance an electron, can be stored for long times [14]. A magnetic field confines the motion in a cylinder whose size is determined by the strenght of the magnetic field itself. A further confinement along the axis of the cylinder is obtained through a quadrupole electrostatic field creating a harmonic force parallel to the magnetic field. Thus the motion of the electron is a combination of a harmonic motion along the axis of the cylinder and a circular motion, called cyclotron motion, completely decoupled in the non-relativistic limit. They become coupled, with the spin too, by including relativistic corrections. In this case the Hamiltonian of an electron in a Penning trap is

$$H_{sys} = \hbar\omega_c a_c^+ a_c \left\{ 1 - 1/2\left(\frac{\omega_z}{\omega_c}\right)^2 - \frac{\hbar\omega_c}{2m_0c^2} \right\} - \frac{(\hbar\omega_c)^2}{2m_0c^2}(a_c^+ a_c)^2 +$$
$$+\frac{p_z^2}{2m_0} + \frac{m_0\omega_z^2 z^2}{2} - \frac{1}{2m_0c^2}\left(\frac{p_z^2}{2m_0}\right)^2 + \frac{\hbar\omega_z\sigma_z}{4}\left[g - \frac{\hbar\omega_c}{m_0c^2}\left(1 + \frac{g-1}{2}\left(\frac{\omega_z}{\omega_c}\right)^2\right)\right] -$$
$$-\frac{(\hbar\omega_c)^2}{2m_0c^2}\left[1 + \frac{g-1}{2}\left(\frac{\omega_z}{\omega_c}\right)^2\right]\sigma_z a_c^+ a_c - \frac{p_z^2}{2m_0^2c^2}\hbar\omega_c a_c^+ a_c - g\frac{p_z^2}{8m_0^2c^2}\hbar\omega_c\sigma_z \quad (15)$$

where a_c and a_c^+ are the annihilation and creation operators associated to the cyclotron motion, z, p_z are position and momentum for the axial motion and σ_z is the projection of the spin on the z-axis. The interactions among the three degrees of freedom have small coupling constants proportional to c^{-2}.

Recently it has been proposed to measure the energy of the cyclotron motion of an electron in a Penning trap in a quantum nondemolition way [15]. A measurement of the energy associated to the cyclotron motion is obtained through the measurement of the axial motion of the trapped particle provided that the spin is preassigned. A detailed analysis of the dynamics of such a measurement is still missing, and the model discussed here can be applied to this situation. We note two peculiarities of such an application. Firstly, one is dealing with the case of non-quadratic terms in the Hamiltonian and numerical analysis is mandatory. Secondly, this is an example of an *indirect measurement*, in which the informations obtained on one degree of freedom allow one to measure some other quantities related to another degree of freedom coupled to the first (in this particular case the cyclotron energy through the direct measurement of spin and axial motion of the electron). The simplest prototype of systems in which indirect measurements are defined could consist of two coupled harmonic oscillators with only one subjected to measurement. In the particular case of position measurements the corresponding Hamiltonian could be written as

$$H_{eff} = \frac{p_1^2}{2m_1} + \frac{m_1\omega_1^2 x_1^2}{2} + \gamma x_1 x_2 + \frac{p_2^2}{2m_2} + \frac{m_2\omega_2^2 x_2^2}{2} + \frac{i\hbar}{\Delta a_1^2 \tau}[x_1(t) - a_1(t)]^2 \quad (16)$$

where the two oscillators are linearly coupled with strenght γ, and the measurement term directly affects only the first oscillator. It is easy to realize that, due to the coupling, a restriction of the paths in the coordinate space of the first oscillator will induce a restriction of the paths also in the second oscillator, therefore defining an indirect effective uncertainty Δa_{2eff}. A perturbative evaluation of such a quantity in the case of Eq. (16) is ongoing and we will refer on it in the future.
The interest of the above problem goes beyond the Penning traps application, and involves topics such as gravitational wave antennae coupled to a transducer [16] and quantum measurements of energy in microwave cavities [17].

We acknowledge M. B. Mensky for collaboration and stimulating discussions.

References

[1] V. B. Braginsky and F. Ya. Khalili, *Quantum Measurement*, K. S. Thorne editor (Cambridge University Press, Cambridge, 1992), and references cited therein.

[2] M. B. Mensky, Phys. Rev. D **20**, 384 (1979); Zh. Eksp. Teor. Fiz. **77**, 1326 (1979); [Sov. Phys. JETP **50**, 667 (1979)].

[3] M. B. Mensky, *Continuous Quantum Measurements and Path-Integrals* (IOP Publishers, Bristol and Philadelphia, 1993).

[4] F. Ya Khalili, *Vestnik Mosk. Univers.*, ser. 3 **22**, 37 (1981); A. Barchielli, L. Lanz and G. M. Prosperi, *Nuovo Cimento B* **72**, 79 (1982); C. M. Caves, *Phys. Rev. D* **33**, 1643 (1986); *Phys. Rev. D* **35**, 1815 (1987); C. M. Caves and G. J. Milburn, *Phys. Rev. A* **36**, 5543 (1987); V. P. Belavkin, *Phys. Lett. A* **140**, 355 (1989).

[5] M. B. Mensky, R. Onofrio, and C. Presilla, Phys. Rev. Lett. **70**, 2828 (1993).

[6] C. M. Caves, K. S. Thorne, R. W. Drever, V. Sandberg, and M. Zimmermann, Rev. Mod. Phys. **52**, 341 (1980).

[7] J. S. Bell, see for instance *Speakable and unspeakable in quantum mechanics: collected papers in quantum mechanics*, (Cambridge University Press, Cambridge, 1987), in particular pp. 14-21.

[8] A. J. Leggett and A. Garg, Phys. Rev. Lett. **54**, 857 (1985).

[9] L. E. Ballentine, Phys. Rev. Lett. **59**, 1493 (1987); A. Peres, Phys. Rev. Lett. **61**, 2019 (1988).

[10] C. D. Tesche, Phys. Rev. Lett. **64**, 2358 (1990).

[11] T. Calarco and R. Onofrio, *Optimal measurements of magnetic flux in superconducting circuits and macroscopic quantum mechanics*, submitted to Phys. Rev. Lett. (1994).

[12] H. Heffner, Proc. IRE **50**, 1604 (1962); H. A. Haus and J. A. Mullen, Phys. Rev. **128**, 2407 (1980).

[13] C. Presilla and U. Tambini, *A selective relaxation method for numerical solution of Schrödinger problems*, submitted to Phys. Rev. Lett. (1994).

[14] L. S. Brown and G. Gabrielse, Rev. Mod. Phys. **58**, 233 (1986).

[15] I. Marzoli and P. Tombesi, Europhys. Lett. **24**, 515 (1993).

[16] E. Majorana, N. Pergola, P. Puppo, P. Rapagnani, and F. Ricci, Phys. Lett. A **180**, 43 (1993).

[17] M. Brune, S. Haroche, J. M. Raimond, L. Davidovich, and N. Zagury, Phys. Rev. A **45**, 5193 (1992).

QUANTUM MEASUREMENTS, PHASE TRANSITIONS AND SPONTANEOUS SYMMETRY BREAKING

S.N.Mayburov

Lebedev Inst. of Physics
Leninsky Prospect 53
Moscow, Russia, 117924
E-mail: Mayburov@SCI.FIAN.MSK.SU

Abstract

It is shown that if quantum measurement results in a phase transition then due to orthogonality of Fock spaces of different vacuums the final state can't be distinguished from the mixture. Examples for transitions in ferromagnetic and vapour–liquid systems are considered. The nonlocality interpretation is proposed in which EPR-Bohm and other quantum nonlocal correlations described as the spontaneous symmetry breaking of system state.

1 INTRODUCTION

There are many open questions now in the quantum measurement theory [1-3]. In this paper we want to show that some of them can be cleared basing on the analogy between the quantum measurement and the spontaneous symmetry breaking (SSB) mechanism of the nonperturbative quantum field theory (NQFT) [4,5]. Their possible connection was pointed out first by Neeman [6] . We'll show that this breaking can be related to the phase transition in the macroscopic detector media initiated by its interaction with the measured quantum particle. If the measured particle belongs to the EPR-Bohm pair in this case the global symmetry of the quantum system state will be broken.

2 MEASUREMENTS AND PHASE TRANSITIONS

Besides gauge field theories the phenomena of SSB are well known for the phase transitions in macroscopic bodies. Some of these transitions are applied for the detection of the elementary particles, for example, liquid-vapour in Wilson and bubble chambers. In the standard quantum mechanics their interaction should result in the entanglement of the quantum states of particle g and macroscopic detector S during the measurement of the observable l of g.

$$|g_0\rangle\,|S_0\rangle \rightarrow \sum C_i\,|l_i\rangle\,|S_i\rangle \tag{1}$$

So in principle the interference between different $|l_i\rangle$, $|s_j\rangle$ states can be observed analogously to the case of the entangled two photon states [7].

First we regard qualitatively the interaction of relativistic particle g coordinate in Wilson chamber filled by saturated vapour of some liquid. We take $|g\rangle$ is a superposition of two flight paths $|x_1\rangle$ inside and $|x_2\rangle$ outside of chamber. If g passes through the vapour it can transfer high momentum to one of the atomic electrons, which will leave the chamber practically without secondary interactions. In the interaction point will be left positive ion at rest which will attract electrostatically surrounding atoms due to their multipole moments. If the vapour pressure is higher than saturation pressure it will result in formation of the liquid drop which can grow into size unrestrictedly, corresponding to phase transition of the first kind [8]. It's impossible to solve directly evolution equations for particle-detector state. Current theory can only predict the direction of the system evolution and parameters of phase transition from the study of the minimum of free energy or thermodynamic potential for two phases system [8]. The formation of the new phase is checked by the approximation of the surface free energy to its normal macroscopic value. It occurs when the drop radius $R \gg 10^{-6}$ cm, which corresponds to $10^{4\div 6}$ atoms in it. Further growth of R is unimportant and we can suppose that it stops due to approximation to phase equilibrium.

Interference presence is stipulated strictly by nondiagonal terms of the operator $\hat{M} = \hat{M}_A \cdot \hat{M}_S$, where $\hat{M}_A$ and $\hat{M}_S$ are particle and detector observables measurements. So we must consider terms of $\bar{M} = Tr(\rho_f \hat{M})$ of the type $\langle x_1|\hat{M}_A|x_2\rangle \cdot \langle L_I|\hat{M}_S|V_J\rangle$. But due to orthogonality of Fock spaces of different phases M_S matrix elements could be exactly zeroes [8]. It occurs for any linear operator M_S which could be written as a sum of finite products of annihilation-production operators of one phase excitations a_i, a_i^+ (quasiparticles) and so can't connect states from different phases. For the vapour the Fock ground state is normal vacuum and for the liquid the new vacuum is its ground state and quasiparticles are phonons.

Now we regard the phase transitions of second kind where some approximate calculations are possible for ferromagnetic media [8]. Again to solve evolution equations for phase transitions exactly is impossible, but we'll make some estimates for the particular simple model. We consider the charged particle's state which is the superposition of two parallel paths $|x_{1,2}\rangle$ pass along opposite sides of small ferromagnetic sample FS which initial magnetisation is negligible. It can be if:

a) the sample is in paramagnetic state at $T > T_c(P)$, and at the instant by the change of pressure P on FS it transfers into ferromagnetic phase.

b) for soft magnetic materials for $T \ll T_c$, when many domain fields equalise each other. Hamiltonian of FS in external (particle's) field $\vec{H}$ is

$$\hat{H}_F = -\sum b_{ij}\vec{S}_i\vec{S}_j - M\sum \vec{H}\vec{S}_j \tag{2}$$

where b_{ij} is atomic exchange integral. The change of spin system S^2 and S_z is allowed due to interaction with the crystal lattice orbital momentum which is omitted in (2). We'll take particle's magnetic field to be nearly constant in the FS volume and directed along Z axis. For the coherent paths superposition the field is operator

$$\vec{H} = \frac{q[\vec{P} \cdot \vec{R}_{1,2}]}{r^3_{1,2}}, \tag{3}$$

where q is the charge, $\vec{p}$-momentum operator of particle, $\vec{r}_{1,2}$ – radius vectors from particle two paths to FS centre. Also it's supposed that time of flight when $\vec{H}$ acts on FS $\tau_F \gg \tau_T$ – phase transition time which for ferromagnetic is of the order 10^{-3} sec [8]. Under this conditions it can be easily shown that for both examples a) and b) the minimum of energy for hamiltonian H_F density achieved in the entangled state

$$|\Psi_f\rangle = \frac{1}{\sqrt{2}}(\alpha_1|x_1\rangle|U_F\rangle + \alpha_2|x_2\rangle|D_F\rangle) \tag{4}$$

where $\alpha_{1,2}$ probability amplitudes for x_1, x_2, $|U_F\rangle = |u_1\rangle \cdot |u_2\rangle ...$ the FS spin quantum state polarized completely along Z and $|D_F\rangle$ – in opposite direction. $|U_F\rangle$ and $|D_F\rangle$ are two different vacuum states of ferromagnetic which excitations (magnons) are also orthogonal, which we'll denote as $|k_u\rangle$, $|k_d\rangle$.

Alike the previous example any nondiagonal terms are infinitely small for the product of finite number of a_k, a^+_k and $S + A$ interference is unobservable in this case. For example that's true for the magnetisation $\vec{I}$ which define the FS induction. More exactly the system evolve to the minimum of free energy F which differ from Hamiltonian density by magnon excitation terms depending of the temperature. It was shown, that at $T < .5T_c$ their relative weight is of the order 10^{-2} relative to vacuum state weight, as illustrated by Bloch law for the magnetisation [8, 9]. It shouldn't change our conclusions, and the system should stay in nearly coherent state in this temperatures range.

Preliminary NQFT path integral calculations for the SSB in the infinite spin fields confirm the described picture. In the low temperature limit superposition of the small external fields drives the spin field with $SU(2)$ invariant Langragian into the entangled superposition of two ground states analogous to Ψ_f of (4). It was shown that the same breakdown occurs also for the restricted field volume or finite N changing only magnon effective mass and minimal field value required for the SSB initiation [5].

Note that the obtained final state practically coincide with the one of Koleman–Hepp (KH) model [10]. It considers interaction of test particle with ensemble E_S of $N < \infty$ polarised free spins. It was shown that in this case for the observable $\hat{Q} = \prod^N(\sigma_{xi})$ nondiagonal matrix element A_Q is to be large.

$$A_Q = \langle U_F| \hat{Q} | D_F\rangle \sim \left|\langle u_i| \sigma_{zi}| u_i\rangle\right|^N.$$

To measure Q each particle of E_S should be send to Stern-Goerlach magnet where each S_X is to be measured separately. Situation is different for FS states for which no measurement on FS as whole body can't define Q value. Note that $\hat{H}_F$ is only electron exchange part of the full lattice Hamiltonian $\hat{H}_{em}$. If try to measure $\hat{Q}$ by use of Stern-Goerlach magnet before it we must dismount the FS lattice into atoms not disturbing spins collective state. Obviously it seems to be impossible to perform. Even if a single atom is removed from the lattice by the projectile scattering or some externed field it'll result in magnon and other excitations of the lattice and distort the Fock space basis [9]. Both these factors will distort the measurement and reduce A_Q absolute value which as calculations show after $\sim \alpha_k N$ kicks will become nearly zero ($\alpha_k < 1$). From other point even if it can be performed coherently when only $N < 10^{5\div 6}$ is left in the lattice it becomes paramagnetic due to additional electron momentum acquired in small volume [11]. So further spins measurement loses any sense.

This measurability restriction for individual particles seems quite natural for the secondarily quantised states originating from the collective interactions. It seems that the all observables which can be really measured for FS for example the magnetisation $\vec{I} = \sum^{N} \vec{\sigma}_i$ behave as classic stochastic values. It means that our set of FS observables is described by C^* algebra which introduces restrictions of measurability analogous to classic meter pointer observables [10]. This change in comparison with K–H model in N finite case is explained by the presence of the electromagnetic field which have infinite number degrees of freedom. Note that obtained superselection of FS eigenstates is stipulated by full lattice Hamiltonian which can contain other spin dependent parts beside included in $\hat{H}_F$.

3 EPR CORRELATIONS

In this chapter we'll study possible connections of state vector symmetry breaking with some space-time symmetries alike to the gauge field theories [6]. In the previous section the value of SSB parameter – for example the direction of magnetisation have purely stochastic behavior. In quantum mechanical formalism if two particles have interacted previously then some of their measured observables are correlated. These correlations defined by the reduction postulate (RP) formulation, which conserve some relations between particles parameters stipulated by initial interaction. For the EPR-Bohm singlet two fermion system the final state after both particles S_Z measurements is

$$\Psi_f = \frac{1}{\sqrt{2}}\left(|S^1_u\rangle|u\rangle_1|d\rangle_2|S^2_d\rangle - |S^1_d\rangle|d\rangle_1|u\rangle^2|S^2_4\rangle\right) \tag{5}$$

where $S^1_u, \ldots$ are the states of detectors D_1, D_2 for u, d states of particles.

For the simplicity only strongest two particle correlations are studied here, but we expect that all qualitative results for weak correlations will be principally the same. For the discrete observables the maximal correlated states have the form :

$$\psi_0 = |N\rangle = \sum_k a_k|B_k\rangle|C_k\rangle \tag{6}$$

For the continuous observables arbitrary hydrogenlike potential systems $M = A_1 + A_2$ which can be free or binded to some point are studied. Their wave function can be written in general:

$$|M\rangle = \psi(r_1, r_2) = \chi(R)\phi(r_{12}) \tag{7}$$

where R_1, r_2 coordinate vectors of particles A_1, A_2, $r_{12} = r_1 - r_2$, $R = (m_1 r_1 + m_2 r_2)/(m_1 + m_2)$ system center of mass coordinate. For free M system $\chi(R) = c \cdot exp(ipR)$, where p is M total momentum, and M is uniformly distributed inside the large volume H^3 with dimensions $\sim H$. We also assume M have finite size, characterized by $R_M = 0.5 * \overline{|r_1 - r_2|}$ and $\varphi(r) \Rightarrow 0$ at $r \gg R_M$.

R and r_{12} are M generalized coordinates (GC) and in their terms $|M\rangle$ factorised into wave function χ of the system M as the whole and ϕ describing relation between its parts. If idealized detectors without distributing another particle measure on the instant r_1 and r_2 then their correlations due to potential binding are to be found. In principle such r_1, r_2 correlations can be detected if M is unstable and decays in flight at the moment t inside and A_1, A_2 have very low velocities. At $t_1 = t + dt$ we insert in H^3 detectors D_1, D_2 and measure r_1, r_2. The wave packets smearing is small for small dt, so we find $|r_1 - r_2| \ll H$ and $R = .5(r_1 - r_2)$. This correlation can be regarded as induced by R GC reduction to the same value both in r_1 and r_2 measurement, when $\phi(r_{12}) = \delta(r_{12})$ for H scale.

This example demonstrate in a more straightforward way that strong nonlocal correlations are connected with the factitious dependence of wave function on one GC which is reduced to the same unique value in both measurements, despite possible large distance between their points. Described set-up is in fact the alternative realisation of coordinate-momentum EPR experiment.

For EPR-Bohm singlet pair its spin state vector can be rewritten through the GC, $T_z = S_{z_1} - S_{z_2}$, $J_z = S_{z_1} + S_{z_2}$

$$|V_{12}\rangle = 1/\sqrt{2}\left(|T_z = 1\rangle - |T_z = -1\rangle\right)\delta(J_Z) \qquad (8)$$

and in fact is the function of one variable describing system V_{12} electron spin orientation in space for the $J_z = 0$, $J^2 = 0$ which are also GC. Obviously this approach agrees with Bell theorem, because it conserve amplitudes structure, despite the transformation of observables. For example of spin $\frac{1}{2}$ pair if the angle between z_1 and z_2 is α calculations for amplitudes of T_z, J_z give the expectation value for projection on z_1, z_2 to be $E(z_1, z_2) = -\cos\alpha$ [4].

This formalism leads to the new interpretation of quantum nonlocality which derived from the supposed hierarchy of systems space-time (and spins) observables. We assume to regard on equal base macroscopic world W and any free quantum systems M constituting the same set of independent systems S. The main assumption is that in the nature there are really fundamental quantum observables which describe relations between such independent systems or this system internal coordinates. Such observable logically is to be unique and unambiguous for each system despite its uncertainty and consequently to be reduced to the same value being independently measured in the separate space points. Individual particles observables are derivatives of such primary system GC which result in their distant correlations. For M system if r_1 or r_2 are measured the reduction of R occurs and due to it r_1, r_2 correlations are observed as shows above analysis. The same approach is applicable for spin space where EPR-Bohm singlet state can be regarded as independent of external world W and T describe their relative orientation. In classical physics any choice of system GC is equivalent, but in quantum mechanics the preferable GC seems to be revealed by the reduction of the nonlocal correlated states. Proposed interpretation implies that the origin of stochastic values in the measurement can't be ascribed to the fluctuations in the local interaction with the measuring device, but rather have fundamental character related to the quantum systems space-time properties and described as the breakdown of system state global symmetry.

It's well known that the standard RP prescription which implies the state vector reduction process is to occur instantaneously in all space formally isn't covariant [1]. It was shown that the covariant RP which agree with QFT causality can be formulated for the observer dependent wave functions $\Psi_O(t, x_O, x_1, d_1 x_2, d_2 ...)$ where x_O and t are observer position and his time axis and x_I, d_I particle's coordinates and measured observables [12]. It settles that after the measurement in x_1 ,t the reduced wave function Ψ_O‘ will exist inside x_1, t_1 future light cone and initial Ψ_O outside of it. For the EPR-Bohm correlated states it means that observable S_{2Z} which value can be predicted from the measurement of the S_{1Z} in the point x_1 stay uncertain for some time in the point x_2. Proposed prescription is obviously covariant as the light cone transforms into itself and satisfy the relativistic locality which demands that no influence, in our case state reduction can spread faster than light. Such prescription corresponds with the idea of state vector representing the information about the system available at a given point.

Now we consider correlated state $\psi_O(t, x_O, D)$ evolution in the approach which use formalism of GC D_I when d_1 , d_2 are their functions. In this case if we try to introduce the observer dependent wave functions of system GC the simpliest covariant prescription is that reduced function Ψ_O‘(D_I) will exist inside two light cones starting in x_1, t and x_2, t and $\Psi_O(D_I)$ stays outside of it. In this approach space like separated measurements in x_1 and x_2 can be regarded as independent of each other in a sense that the fulfilment of the measurement in x_1 can't influence the result of the measurement in x_2 and vice versa. This independence is a consequence of covariance and symmetry of the measurement in x_1 and x_2 under Lorentz transformations in this approach, in which one event can become the past or the future relative to another without change of the measurement result. In fact in the GC formalism this independence follows from the idea that the reduction of single GC D leads to the unique reduced value. In relativistic framework the conclusion that value of GC measured in x_1 doesn't depend if the measurement in x_2 was performed or not acquire more significance. So this interpretation is based on single assumption that GC of isolated quantum system are fundamental variables of the theory and are always reduced to the unique value being measured several times on different parts of system. In the relativistic framework it means that certain GC stays locally uncertain after the measurement in x_1 light cone. In other words it suppose that quantum uncertainty behave alike the transferred information and obey causality.

These results suppose that the correlation between the two measuring devices which measure parameters of EPR pair is explained by the space-time properties of system state which symmetry is broken by this measurement. It reflected by the fact that as the SSB parameter for two devices in formula (6) can be taken the same observable $T_z = S_{z_1} - S_{z_2}$. So in this case two separated phase transitions in these services will evolve in correlation defined by this parameter.

We conclude that SSB phenomena in the form of phase transitions can be applied for the description of the quantum measurement processes. We can't assert that in our model RP in some form is unnecessary and SSB performs the 'dicing' in a single quantum event [6]. At least RP can be formulated in this model in much weaker and self-consistent form: if the short-time interaction with measured microparticle results in SSB of detector state and if for all measurable observables their nondiagonal matrix elements are negligible then the reduction of initial state vector to the mixture occurs. The GC formalism of nonlocal correlations evidence that quantum measurement processes are stipulated by quantum systems space-time properties, in particular separate systems relations or their parts internal relations described by GC observables. This system GC are initially uncertain relative to macroscopic frame. Being measured on one particle from M system in x_1 it also can stay uncertain for some time for other particles of M.

References

[1] Y. Aharonov, D.Z. Albert, Phys. Rev. D 24, 359 (1981)

[2] B. D'Espagnat, Found. Phys. 20, 1329 (1990)

[3] J.B. Kramer, Rev. Mod. Phys. 58, 647 (1986)

[4] T.P. Cheng, L.F. Li, "Gauge Theory of Elementary Particles" (Clarendon Press, Oxford, 1984)

[5] E. Celeghini, E. Graziano, G. Vitiello, Phys. Lett. A145, 1, (1990)

[6] Y. Neeman in "Microscopic Reality and Quantum Formalism" (Kluwer Academic Publishers, London, 1988) Vol. 1, p. 145; in Proc. Symp. on Foundations of Modern Physics (World Sci., Singapore, 1993) p. 289

[7] D.M. Greenberger et al., Am. J. Phys. 58, 1131 (1990)

[8] L.D. Landau, E.M. Lifshitz, "Statistical Physics" (Pergamon Press, London, 1958)

[9] J.M. Ziman, "Principles of the theory of solids", Univ. Press, Cambridge, 1964.

[10] K. Hepp, Helv. Phys. Acta 45, 237 (1972); J.S. Bedl, Ibid. 48, 93 (1975)

[11] S. Shtrikman, D. Treves in 'Magnetism' (ed. G.T. Rado, H. Suhl), vol. 3, Acad. Press, N.Y., 1963, p. 395
S.V. Vonsovsky, 'Magnetism', Nauka, Moscow, 1971, p. 800

[12] J. Finkelstein, Found. Phys. Lett. 6, 383 (1992)

A QUANTUM LOGIC DESCRIPTION OF SOME IDEAL MEASUREMENTS

Sylvia Pulmannová

Mathematical Institute
Slovak Academy of Sciences
Štefánikova 49
Sk-814 73 Bratislava, Slovakia
e-mail: pulmann@mau.savba.sk

Abstract

Some ideality conditions of quantum measurements of ordinary observables are studied in the frame of quantum logics with special properties. Relations between different ideality conditions are shown.

A classical experiment is described by a triple $(\Omega, \mathcal{F}, \mathcal{P})$, where $\mathcal{F}$ is a σ-algebra of subsets of a nonempty set Ω, and P is a probability measure on $\mathcal{F}$. $\mathcal{F}$ is interpreted as the set of all random events of the experiment.

A generalization of the classical probability theory is given in the quantum logic approach. The Kolmogorovian triple $(\Omega, \mathcal{F}, \mathcal{P})$ is replaced by a couple (L, m), where L is a quantum logic, which is interpreted as the set of all quantum random events. Most usually, L is assumed to be an orthomodular σ-lattice or poset, and m is a state on L. (See [V, PP].)

From the algebraic point of view, distributivity which holds in $\mathcal{F}$ is replaced by a weaker property - orthomodularity.

A *σ-orthomodular poset* is a partially ordered set L with 0 and 1 as the smallest and the greatest element, respectively, endowed with a unary operation - orthocomplementation, such that: (i) $a'' = (a')' = a$; (ii) $a \leq b \Rightarrow b' \leq a'$; (iii) $a \vee a' = 1$; (iv) $(a_i)_{i \in N}$, $a_i \leq a_j'$, $i \neq j \Rightarrow \bigvee_{i \in N} a_i \in L$; (v) $a \leq b \Rightarrow b = a \vee (a' \wedge b)$.

Condition (iv) with at most countable sets N expresses the σ-completeness of L. If (iv) holds only for finite sets N, then L is an orthomodular poset.

Condition (v) is the orthomodular law. We say that $a, b \in L$ are *orthogonal* ($a \perp b$) if $a \leq b'$.

A *state* on L is a mapping $m : L \to [0,1]$ such that $m(1) = 1$ and $a \perp b \Rightarrow m(a \vee b) = m(a) + m(b)$. A state m is σ-additive (completely additive) if $m(\bigvee_{i\in I} a_i) = \sum_{i\in I} m(a_i)$ for any countable (arbitrary) set I of mutually orthogonal elements of L such that $\bigvee_{\in I} a_i$ exists in L.

An *observable* on L is σ-homomorphism $X : \mathcal{F} \to L$, where $(\Omega, \mathcal{F})$ is a measurable space, which is called the *value space* of X. An observable X is *discrete* if there is a countable set $\{\omega_i\}_{i\in N}$ such that $X(\{\omega_i\}_{i\in N}) = 1$. Most usually, $(\Omega, \mathcal{F}) = (R^n, \mathcal{B}(R^n))$, $\mathcal{B}(R^n)$ is Borel σ-algebra. A *spectrum* of such an observable is the smallest closet set $C \subset R^n$ such that $X(C) = 1$. An observable with value space $(R, \mathcal{B}(R))$ is called *real* observable. A real observable with the spectrum contained in $\{0,1\}$ is a *proposition* observable. (See [V, PP]).

In a measurement process, a quantum system is coupled with a measuring device, which is usually assumed to be a classical system. To the description of such a coupled system, a Boolean power seems to be an appropriate tool. (See [P1, P2].) Let L be a logic, B a complete Boolean algebra. A *Boolean power* $L[B]$ is the set of all functions

$$L[B] := \{f : L \to B : f(a_1) \wedge f(a_2) = 0_B, a_1 \neq a_2; \bigvee_{a\in L} f(a) = 1_B\}.$$

A Boolean power endowed with appropriately defined partial order and orthocomplementation, is an orthomodular poset (or lattice if L is a lattice), not necessarily σ-complete. (See [Gr] for the general definition and properties of Boolean powers, and [P1] for properties of Boolean powers of orthomodular posets.)

For example, if $L = \mathcal{P}(\mathcal{A}_1)$, $B = \mathcal{P}(\mathcal{A}_2)$, where $\mathcal{P}(\mathcal{A}_i)$ $(i = 1,2)$ are projection lattices of von Neumann algebras $\mathcal{A}_i$; and $\mathcal{A}_2$ is abelian, then

$$L[B] = \{\sum_i p_i \otimes b_i : \; p_i \in \mathcal{P}(\mathcal{A}_1); \; b_i \in \mathcal{P}(\mathcal{A}_2), \sum_i b_i = I_2\},$$

where I_2 is the identity in $\mathcal{A}_2$. (See [P2].)

If the Boolean algebra B is discrete, the Boolean power $L[B]$ is isomorphic to the direct product of copies of L indexed by atoms of B.

There are complete embeddings $\lambda : L \to L[B]$, $\beta : B \to L[B]$. Moreover, to any state m on L and every completely additive state μ on B there exists a so-called *product state* $m \otimes \mu$ on $L[B]$ such that $m \otimes \mu(\lambda(a) \wedge \beta(b)) = m(a)\mu(b)$, $a \in L$, $b \in B$.

Assume that a quantum system is described by a logic L and a measuring apparatus by a complete Boolean algebra B, and that the joint system is described by the Boolean power $L[B]$. Let $\mathcal{P}_L$ be the set of physical (σ-additive) states on L, and $\mathcal{P}_B$ be the set of physical states (completely additive) on B. As physical states of $L[B]$, we will consider σ- convex combinations of product states.

Let an observable X with the value space $(\Omega, \mathcal{F})$ be fixed. (We consider only "ordinary" observables - an analogue of PV-measures, although an analogue of POV-measures can also be defined on quantum logics, [PP].) Let $X_{\mathcal{A}}$ be an observable on the apparatus described by B with value space $(\Omega_{\mathcal{A}}, \mathcal{F}_{\mathcal{A}})$. Let $f : \Omega \to \Omega_{\mathcal{A}}$ be a measurable function (so-called *pointer function*). In analogy with [BLM], a *measurement* of the observable X is a five-tuple

$$\mathcal{M} = (B, X_{\mathcal{A}}, m_{\mathcal{A}}, f, V),$$

where $m_{\mathcal{A}}$ is an initial state of the apparatus, and $V : \mathcal{P} \to \mathcal{P}$ is a convexity preserving transformation of the state space of the coupled system, which describes the measuring interaction. The five-tuple $\mathcal{M}$ is classified as a measurement of X if the following equality holds:

$$m(X(F)) = V(m \otimes m_{\mathcal{A}})(\beta(X_{\mathcal{A}} \circ f^{-1}(F)))$$

for all $m \in \mathcal{P}_L$, and all $F \in \mathcal{F}$. The convexity preserving transformation V can be extended to the cone $K(\mathcal{P}) := \{\alpha s : \alpha \geq 0, \ s \in \mathcal{P}\}$ by homogeneity, and by linearity to $K(\mathcal{P}) - K(\mathcal{P})$.

Let L be an orthomodular poset. Recall that $a, b \in L$ are *compatible* (aCb) if they are contained in the same Boolean subalgebra of L. The set of all elements of L which are compatible with any other element of L is the *centre* of L. If m is a state and c is a central element of L, then the mapping $s_c : L \to [0,1]$, $s_c(a) = s(a \wedge c)$, is an additive measure on L. If $L[B]$ is a Boolean power, then the set $\beta[B]$ is contained in the centre of $L[B]$ ([P1]).

For every measurement $\mathcal{M}$, an *instrument* $I_{\mathcal{M}}$ is defined by

$$I_{\mathcal{M}}(F)(m) = V(m \otimes m_{\mathcal{A}})_{\beta(X_{\mathcal{A}} \circ f^{-1}(F))} \circ \lambda.$$

$I_{\mathcal{M}}(F)(m)$ is interpreted as the final nonnormalized state of the system, provided that the measurement lead to an outcome in F. From the definitionof $I_{\mathcal{M}}$ and properties of $\mathcal{P}$, it follows

$$\{F_i\}_{i \in N} \subset \mathcal{F}, \ F_i \cap F_j = \emptyset, \ i \neq j \ \Rightarrow \ I_{\mathcal{M}}(\bigcup_{i \in N} F_i)(m) = \sum_{i \in N} I_{\mathcal{M}}(F_i)(m)$$

for every at most countable set N. We note that in some approaches, the notion of an instrument was introduced independently of a measurement, see e.g. [D, L1, L2].

Before introducing some important classes of measurements (or instruments, equivalently), we will define some special properties of quantum logics. In what follows, L denotes a quantum logic.

Let m be a state on L, $b \in L$ such that $m(b) \neq 0$. A function $p_m(a|b) : L \to [0,1]$ satisfying

(i) $p_m(a|b)$ is a state on L and $m(b|b) = 1$;

(ii) $a, b \in L, \ aCb \ \Rightarrow \ p_m(a|b) = \frac{m(a \wedge b)}{m(b)}$

is called a *conditional state* (with respect to m and b).

We say that L *admits conditional states* with respect to a set of states S if for every $m \in S$ and $b \in L$ with $m(b) \neq 0$ there is a unique conditional state $p_m(.|b)$.

Von Neumann algebras with no I_2-factor as direct summand are examples of logics with conditional states with respect to the set of all states [CZ].

A set M of states on L is said to be *rich* if

$$a, b \in L, \ a \not\leq b \ \Rightarrow \ \exists m \in M : m(a) = 1, m(b) < 1.$$

A logic L with a rich convex set of states M is called *u-spectral* if for every convex functional $f : M \to [0,1]$ there is a unique observable $x : \mathcal{B}(R) \to L$ such that $f(m) = m(x)$, where $m(x) = \int \lambda m(x(d\lambda))$ is the expectation of x in the state m, for all $m \in M$. Von Neumann algebras yield an example of u-spectral logics [R].

A measurement $\mathcal{M}$ or the corresponding instrument I is called (cf. [LBM])

- *repeatable*, if $I(E)(I(F)(m))(1) = I(E \cap F)(m)(1)$ $(E, F \in \mathcal{F}, \ m \in \mathcal{P}_{\mathcal{L}})$;

- *first kind*, if $m(x(E)) = I(\Omega)(m)(x(E))$;
- *ideal*, if $a \in C(X)$, $m(a) = 1 \Rightarrow I(\Omega)(m)(a) = 1$; where $C(X)$ denotes the set of all $a \in L$ such that $aCX(E)$ for all $E \in \mathcal{F}$.

Assume that X is a discrete observable with the spectrum $\{\omega_i\}_i$, $X(\{\omega_i\}) = b_i$. A measurement $\mathcal{M}$ of X is

- *d-ideal*, if $m(b_i) = 1 \Rightarrow I(\{\omega_i\})(m) = m \ \forall m \in \mathcal{P}_L, \ \forall i$ ([LBM]);
- *c-ideal*, if $a \in C(X) \Rightarrow I(\{\omega_i\})(m)(a) = m(b_i \wedge a) \ \forall m \in \mathcal{P}_L, \ \forall i$ ([P2]);
- *Lüders* if $I(\{\omega_i\})(m)(a) = m(b_i)p_m(a|b_i)$ for all $m \in \mathcal{P}_L$ and all i on a logic with conditional states with respect to $\mathcal{P}_L$ ([P2]).

In [P2], the following relations between different types of measurements have been proved (we note that if we consider measurements on u-spectral logics with rich convex set M of states, we assume $\mathcal{P}_L = M$, and if we consider measurements on logics with conditional states with respect to a set S of states, we assume $\mathcal{P}_L = S$): (i) On any logic, a repeatable measurement is of the first kind. (ii) On u-spectral logics, a first kind measurement or an ideal measurement is repeatable. (iii) On a logic with conditional states a discrete measurement is c-ideal iff it is Lüders, and then it is repeatable.

Proposition 1 *On a logic with conditional states, a Lüders measurement is d-ideal.*

Proof. For a Lüders instrument I,

$$m(b_i) = 1, aCb_i \Rightarrow I(\{\omega_i\})(m)(a) = m(b_i)p_m(a|b_i) = m(a \wedge b_i) = m(a).$$

Since the conditional state $p_m(.|b_i)$ is uniquely defined by its values on $C(b_i)$ (where $C(b_i) = \{a \in L : aCb_i\}$), we have $I(\{\omega_i\})(m) = p_m(.|b_i) = m$. □

Proposition 2 *On a u-spectral logic, an instrument I is ideal iff*

$$a \in C(X) \Rightarrow I(\Omega)(m)(a) = m(a)$$

for all $m \in \mathcal{P}_L$.

Proof. Ideality of I implies that for every $m \in \mathcal{P}_L$,

$$a \in C(X), m(a) = 1 \Rightarrow m(I^*(\Omega)(a)) = 1,$$

where I^* is the dual instrument which can be defined on a u-spectral logic (see [P2] for more details. We recall that $I(E)(m)(a) = m(I^*(E)(a))$ ($a \in L$) for every $m \in \mathcal{P}_L$ and every $E \in \mathcal{F}$, where $I^*(E)(a)$ is a real observable on L with spectrum in $[0,1]$.) Since $\mathcal{P}_L$ is rich, this yields $a \leq I^*(\Omega)(a)(\{1\})$. Similarly, $a' \leq I^*(\Omega)(a')(\{1\}) = I^*(\Omega)(a)(\{0\})$. The latter equality follows from

$$I(\Omega)(m)(a) + I(\Omega)(m)(a') = I(\Omega)(m)(1) = 1,$$

where the last equality follows from ideality of I. This yields

$$1 = a \vee a' \leq I^*(\Omega)(a)(\{1\}) \vee I^*(\Omega)(a)(\{0\}),$$

hence the spectrum of $I^*(\Omega)(a)$ is contained in $\{0,1\}$, i.e., $I^*(\Omega)(a)$ is a proposition observable. Moreover, $I^*(\Omega)(a') = I^*(\Omega)(a)'$ (we identify the proposition observable $I^*(\Omega)(a)$ with the element $I^*(\Omega)(a)(\{1\})$ of L). From $a \leq I^*(\Omega)(a)$, $a' \leq I^*(\Omega)(a')$ we get $I^*(\Omega)(a) = a$. The converse statement is immediate. □

Proposition 3 *On a u-spectral logic, a d-ideal measurement is ideal.*

Proof. d-ideality of I implies that

$$m(b_i) = 1 \Rightarrow m(I^*(\{\omega_i\})(b_i)) = 1,$$

hence $b_i \leq I^*(\{\omega_i\})(b_i)(\{1\})$. This implies that, for every m,

$$m(b_i) \leq m(I^*(\{\omega_i\})(b_i)(\{1\})) \leq m(I^*(\{\omega_i\})(b_i)) \leq m(I^*(\Omega)(b_i)).$$

Summing over i on both sides we easily derive that for every i,

$$b_i = I^*(\{\omega_i\})(b_i)(\{1\}) = I^*(\{\omega_i\})(b_i) = I^*(\Omega)(b_i).$$

Now let $a \in C(X)$. Then $m(a \wedge b_i) = 1$ implies $m(b_i) = 1$, hence by d-ideality, $I(\{\omega_i\})(m) = m$, and hence $m(I^*(\{\omega_i\})(a \wedge b_i))(\{1\})) = 1$. This yields $a \wedge b_i \leq I^*(\{\omega_i\})(a \wedge b_i)(\{1\})$. Using similar technique as above, we obtain

$$a = I^*(\{\omega_i\})(a) = I^*(\Omega)(a),$$

and this yields the ideality condition. □

Theorem 1 *On u-spectral logics with conditional states and for discrete measurement, the following conditions are equivalent: (i) ideal; (ii) d-ideal; (iii) c-ideal; (iv) Lüders.*

Proof. By Proposition 1, (*iv*) $\Rightarrow$ (*ii*). By Proposition 3, (*ii*) $\Rightarrow$ (*i*). By [P2], (*iv*) $\Leftrightarrow$ (*iii*). It remains to prove (*i*) $\Rightarrow$ (*iv*). Let I be an ideal instrument for a discrete observable X. For any j and aCb_j we have

$$I(\{b_j\})(m)(a) = I(\{b_j\})(m)(a \wedge b_j) = m(a \wedge b_j).$$

The first equality holds owing the fact that an ideal measurement is repeatable. The second equality follows by Proposition 2 and $a \wedge b_j \in C(X)$. This yields $I(\{b_j\})(m)(a) = p_m(a|b_j)m(b_j)$ for any $a \in C(b_j)$. Owing to uniqueness of conditional states, it follows the latter relation holds for any $a \in L$. This proves that I is Lüders. □

We note that the equivalence (*i*) $\Leftrightarrow$ (*ii*) $\Leftrightarrow$ (*iv*) in the traditional approach was proved in [LBM]; the equivalence (*i*) $\Leftrightarrow$ (*iv*) in von Neumann algebras approach was proved in [L2]. Our results are a generalization of the latter ones.

References

[BLM] P. Busch, P. Lahti and P. Mittelstaedt, Quantum Theory of Measurement. Springer-Verlag, Berlin, 1991

[CZ] G. Cassinelli and N. Zanghi, Conditional probabilities in quantum mechanics. *Nuovo Cimento*, **B 73**, 237–245, 1983

[D] B. Davies, Quantum Theory of Open Systems. Academic Press, New York, 1976

[Gr] G. Grätzer, Universal Algebras. Van Nostrand, Princeton, NJ, 1968

[L1] A. Luczak, Instruments on von Neumann algebras. *Preprint*

[L2] A. Luczak, On ideal measurements and their corresponding instruments on von Neumann algebras. *Preprint*

[LBM] P. Lahti, P. Busch and P. Mittelstaedt, Some important classes of measurements and their information gain. *J. Math. Phys.*, **32**, 2770–2775, 1991

[P1] S. Pulmannová, Boolean powers and quantum measurements. *Rep. Math. Phys.*, **32**, 235–250, 1993

[P2] S. Pulmannová, Quantum measurements and quantum logics. *J. Math. Phys.*, **35**, 1555–1572, 1994

[PP] P. Pták and S. Pulmannová, Orthomodular Structures as Quantum Logics. Kluwer, Dordrecht, 1991

[R] G. Rüttimann, Detectable properties and spectral quantum logics. in *Interpretations and Foundations of Quantum Theory.*, Proceedings of a conference held in Marburg, 28-30 May 1979 (Bibliographisches Institut, Mannheim, 1981)

[V] V. Varadarajan, Geometry of Quantum Theory, Springer-Verlag, New York, 1985

ON THE QUANTUM EVOLUTION OF CHAOTIC SYSTEMS AFFECTED BY REPEATED FREQUENT MEASUREMENT

B. Kaulakys

Institute of Theoretical Physics and Astronomy,
A. Goštauto 12, 2600 Vilnius, Lithuania
E-mail: kaulakys@itpa.fi.lt

Abstract

We investigate the effect of repeated measurement for quantum dynamics of the suppressed systems which classical counterparts exhibit chaos. The essential feature of such systems is the quantum localization phenomena strongly limiting motion in the energy space. Repeated frequent measurement of suppressed systems results to the delocalization. Time evolution of the observed chaotic systems becomes close to the classical frequently broken diffusion-like process described by rate equations for the probabilities rather than for amplitudes.

1 INTRODUCTION

Evolution of quantum systems, when they are not being observed, are described by the Schrödinger equation. Measurements change abruptly the state of the system and project it to the eigenstate of the measured observation. Repeated frequent measurement can inhibit or even prevent the quantum dynamics. This is the essence of the quantum Zeno effect or the quantum watched pot [1-3].

The quantum Zeno effect means that time evolution of the system from an eigenstate of the measured observable into a superposition of eigenstates is inhibited by the measurement. It has been shown in a recent experiment [3], that a variation of the quantum Zeno effect in a three level system, originally proposed by Cook [2], can be realized. Fierichs and Shenzle [4] have obtained the outcome of the

Quantum Communications and Measurement
Edited by V. P. Belavkin *et al.*, Plenum Press, New York, 1995

experiment [3] on the basis of the standard three-level Bloch equations without the *ad hoc* collapse of the wave function.

Aharonov and Vardi [5] showed that frequent measurement can not only stop the dynamics of a system, but also may induce a time evolution of a system: a dense sequence of measurements along a presumed path induces a dynamics of the system close to this arbitrary chosen trajectory (see, also [6]).

One of the specific features of the systems mentioned above is that they consist only of the few (two or three) quantum states and are purely quantum. The purpose of this report is to consider the influence of the repeated frequent measurement on the evolution of the multilevel quasiclassical systems which classical counterparts exhibit chaos. It has been established that the chaotic dynamics of such, e. g. strongly driven by a periodic external field non-linear systems, is suppressed of the quantum interference effect and results to the quantum localization of the classical dynamics in the energy space of system [7,8]. Thus, the quantum localization phenomena strongly limits the quantum motion.

Here we investigate the effect of the repeated measurement for quantum dynamics of suppressed systems. Our investigation is based on the mapping equations of motion for systems with one degree of freedom perturbed by a periodic force.

2 QUANTUM MAPS

From the standpoint of an understanding of manifestation of the classical chaos in the quantum description of systems, the region of large quantum numbers is of greatest interest. Here we can use the quasiclassical approximation and convenient variables are the angle ϑ and the action I. Transition from the classical to the quantum (quasiclassical) description can be undertaken replacing I by the operator $\hat{I} = -i\frac{\partial}{\partial\vartheta}$ [9,10]. (We use the system of units with $\hbar = m = e = 1$). One of the simplest systems in which dynamic chaos and quantum localization of states can be obtained is a system with one degree of freedom described by Hamiltonian $H_0(I)$ and perturbed by periodic kicks:

$$H(I,\vartheta,t) = H_0(I) + K\cos\vartheta\sum_j \delta(t - jT). \tag{1}$$

Here T and K are the period and parameter of the perturbation. The intrinsic frequency of the unperturbed system is $\Omega = dH_0/dI$. In particular for a linear oscillator we have $H_0 = \Omega I$. For $H_0 = I^2/2$ we have the widely investigated rotator [7,8].

Integration of the Hamilton equations of motion over one perturbation period leads to the classical map for the action and angle [9-11]:

$$\left\{ \begin{array}{l} I_{j+1} = I_j + K\sin\vartheta_j \\ \vartheta_{j+1} = \vartheta_j + \Omega(I_{j+1})T. \end{array} \right. \tag{2}$$

For derivation of quantum equations of motion we expand the state function $\psi(\vartheta,t)$ of the system in quasiclassical eigenfunctions, $\varphi_n(\vartheta) = e^{in\vartheta}/\sqrt{(2\pi)}$, of the Hamiltonian H_0,

$$\psi(\vartheta,t) = (2\pi)^{-1/2}\sum_n a_n(t)e^{in\vartheta}. \tag{3}$$

Integrating the Schrödinger equation over a period T, we obtain the following maps for the amplitudes [10]

$$a_n(t_{j+1}) = e^{-iH_0(n)T} \sum_m a_m(t_j) J_{m-n}(K), \quad t_j = jT \tag{4}$$

where $J_m(K)$ is the Bessel function.

The form (4) of the map for the quantum dynamics is rather common: similar maps may be derived, e. g., for the atom in a microwave field [10,12].

3 DYNAMICS

Classical dynamics of the system described by map (2) in the case of global distinct stochasticity is diffusion-like with the diffusion coefficient in the I space resulting from (2),

$$B(I) = K^2/4T. \tag{5}$$

From equations (4) we have the transitions probabilities $P_{n,m}$ between n and m states during the period T:

$$P_{n,m} = J^2_{m-n}(K). \tag{6}$$

Using the expression $\sum_n n^2 J^2_n(K) = K^2/2$ and approximation of uncorrelated transitions we may formally evaluate the local quantum diffusion coefficient in the n space (see also [13,14]):

$$B(n) = \frac{1}{2T} \sum_m (m-n)^2 J^2_{m-n}(K) = \frac{K^2}{4T}. \tag{7}$$

Therefore, the local quantum diffusion coefficient coincides with the classical one (5).

However, it turns out that such a quantum diffusion takes place only for some finite time t^* after which an essential decrease of the diffusion rate is observed. Such a behaviour of the quantum systems in the region of strong classical chaos was called "the quantum suppression of classical chaos". This phenomenon turns out to be typical for models (1) with the nonlinear Hamiltonians $H_0(I)$ and other quantum systems. Thus, the diffusion coefficient (7) derived in the approximation of uncorrelated transitions (6) does not describe the true quantum diffusion in the energy space. The quantum interference effect is essential and results to the quantitatively different dynamics. This is due to the pseudorandom nature of the phases $H_0(n)T$ in equation (4) as functions of n (but not j). Replacing $\exp[-iH_0(n)T]$ in (4) by $\exp[-i2\pi g(n)]$, where $g(n)$ is a sequence of random numbers that are uniformly distributed in the interval [0,1] we observe the quantum localization as well [10]. However, the essential point is the independence of the phases $H_0(n)T$ on the step of iteration or time t_j. For the random phases as functions of iteration steps we observe the unlimited motion in the n space.

4 INFLUENCE OF MEASUREMENTS

Each measurement of the energy projects the system into one of the energy eigenstates with definite n. Therefore, if we make a measurement before the next kick we will find the system in the states φ_m with the appropriate probabilities

$P_m = |a_m|^2$. After the measurement the phase of the amplitude $a_m(t_j)$ is random. Thus, if there are measurements of the system's state before or after each kick, we have from equations (4) the system of rate equations for probabilities

$$P_n(t_{j+1}) = \sum_m J^2_{m-n}(K) P_m(t_j). \tag{8}$$

The rate equations (8) result to the unlimited diffusion-like motion in the n space with the diffusion coefficient (7). Therefore, the frequent measurement of the quantum system described by map (4), which exhibits the quantum suppression of classical chaos, results to the diffusion-like motion, similar to the classical dynamics. The analogous conclusion may be drawn and for other systems, as well.

CONCLUSIONS

Repeated frequent measurement of the simple two- or three-state system inhibit or even prevent the quantum dynamics of the system. The similar measurement of the multilevel quasiclassical system with quantum suppression of classical chaos results to the dynamics described by rate equations for probabilities rather than for amplitudes and causes the delocalization of the states superposition. The quantum dynamics of such chaotic systems affected by repeated frequent measurement resembles the classical diffusion-like motion.

It should be noted that the same effect may be derived without the *ad hoc* collapse hypothesis but from the quantum theory of irreversible processes. Even the simplest detector follows irreversible dynamics due to the coupling to the multitude of vacuum modes. The quantum system attains irreversibility due to the interaction with the detector and the phases of the amplitudes or the off-diagonal matrix elements of the density matrix decay steadily (see for analogy [4]). Measurement of such a type may also result to the disappearance of the effect of the quantum suppression. Therefore, the quantum evolution of chaotic system interacting with the environment or detector is more classical-like than the evolution of the idealized isolated system.

Acknowledgement

The research described in this publication was made possible in part by Grant No. LAA000 from the International Science Foundation.

References

[1] B. Misra and E. C. G. Sudarshan, J. Math. Phys. **18**, 756 (1977).

[2] R. J. Cook, Phys. Sripta **T21**, 49 (1988).

[3] W. M. Itano, D. J. Heinzen, J. J. Bollinger and D. J. Wineland, Phys. Rev. A **41**, 2295 (1990).

[4] V. Frerichs and A. Schenzle, Phys. Rev. A **44**, 1962 (1991).

[5] Y. Aharonov and M. Vardi, Phys. Rev. D **21**, 2235 (1980).

[6] T. P. Altenmüller and A. Schenzle, Phys. Rev. A **48**, 70 (1993).

[7] G. Casati, B. V. Chirikov, D. L. Shepelyansky and I. Guarneri, Phys. Rep. **154**, 77 (1987).

[8] F. M. Izrailev, Phys. Rep. **196**, 299 (1990).

[9] G. M. Zaslavskii, Stochastic Behaviour of Dynamical Systems, Moscow: Nauka (1984).

[10] V. G. Gontis and B. P. Kaulakys, Liet. Fiz. Rink. **28**, 671 (1988) [Sov. Phys.-Collec. **28**(6) 1 (1988)].

[11] A. J. Lichtenberg and M. A. Lieberman, Regular and Stochastic Motion, New York: Springer (1983).

[12] R. V. Jensen, S. M. Susskind and M. M. Sanders, Phys. Rep. **201**, 1 (1991)

[13] V. Gontis and B. Kaulakys, J. Phys. B **20**, 5051 (1987).

[14] B. Kaulakys, V. Gontis, G. Hermann and A. Scharmann, Phys. Lett. **A159**, 261 (1991).

Part III.
Quantum Jumps, Diffusion & Localization

A HAMILTONIAN SOLUTION TO QUANTUM COLLAPSE, STATE DIFFUSION AND SPONTANEOUS LOCALIZATION

V.P.Belavkin[1] and O. Melsheimer[2]

[1]University of Nottingham, NG7 2RD, UK
[2]Fachbereich Physik, Renthof 7,
Marburg, W-3550, Germany

Abstract

The dynamical solution [1] to the quantum measurement problem is applied to construct a Hamiltonian model for quantum jumps and spontaneous localizations. We give an explicit stochastic Hamiltonian of discontinuous unitary evolution for a system of quantum particles that interacts singularly with "bubbles" which admit a continual counting observation. This model allows to watch a quantum particle trajectory like in cloud chamber by unsharp localisations of the positions of spontaneously scattered bubbles. Thus, the continuous reduction and spontaneous localization theory is obtained from a Hamiltonian singular interaction as a result of quantum filtering, i.e., a sequential time continuous conditioning of an input quantum process by the output measurement data. We show that in the case of indistinguishable particles the a posteriori dynamics is mixing, giving rise to an irreversible Boltzmann-type reduction equation even for an individual trajectory. The latter coincides with the nonstochastic Schrödinger equation only in the mean field approximation, whereas the central limit yields Gaussian mixing fluctuations described by a quantum state reduction equation of diffusive type.

Keywords :
Quantum jumps, continuous reduction, state diffusion, quantum trajectories, spontaneous localization.

INTRODUCTION: A solution to the collapse problem

The quantum measurement theory based on the ordinary von Neumann reduction postulate [7] applies neither to instantaneous observations with continuous spectra nor to continual (continuous in time) measurements. Although such phenomena can be described in the more general framework of Ludwig's Davies-Lewies operational approach [2]– [6] there is a particular interest in describing quantum measurements by concrete Hamiltonian models from which the operational description can be derived by an averaging procedure. A unitary scattering model of such kind for instantaneous unsharp measurement of particle localization was given in [1]. Let us consider the singular interaction Hamiltonian

$$h_x(t) = x\delta(t)\frac{\hbar}{i}\frac{\mathrm{d}}{\mathrm{d}\lambda}, \quad \delta(t) = \begin{cases} \infty\,, & t = 0 \\ 0\,, & t \neq 0 \end{cases} \tag{0.1}$$

with Dirac δ-function, producing the translation operator

$$s_x = \exp\left\{-\frac{i}{\hbar}\int_{-\infty}^{\infty} h_x(t)\mathrm{d}t\right\} = e^{-x\mathrm{d}/\mathrm{d}\lambda} \tag{0.2}$$

at time $t = 0$ of the measurement. Here x is the position of the particle and λ is the pointer position given by the q-coordinate of a quantum meter. The particle scattering operator $S = \{s_x\}$ applied to the (generalized) position eigen-kets $|x\rangle$ as $S|x\rangle = |x\rangle s_x$ does not affect the position x of the particle but changes the meter coordinate q to

$$y = s_x^\dagger q s_x = x + q\,, \tag{0.3}$$

which implies that the initial wave function $\psi_0(x, \lambda)$ of the system "particle plus meter" is transformed into

$$\psi(x, \lambda) = s_x\psi_0(x, \lambda) = \psi_0(x, \lambda - x)\,. \tag{0.4}$$

If in the initial state the particle and the meter were not coherent, $\psi_0(x, \lambda) = \eta(x)f_0(\lambda)$, and the wave function f_0 of the meter was fixed, then one can obtain the unitary transformation $S : \psi_0 \mapsto s_\bullet\psi_0$ via a family $\{F(\lambda)\}$ of reduction transformations $F(\lambda) : \eta \mapsto f_\bullet(\lambda)\eta$ for the prior particle vector-states η. Specifically, Eq. (0.4) can be defined as

$$\psi(x, \lambda) = f_0(\lambda - x)\eta(x) \equiv f_x(\lambda)\eta(x)\,. \tag{0.5}$$

The linear nonunitary operators $F(\lambda) = s_\bullet f_0(\lambda)$ act on $|x\rangle$ as the multiplication $F(\lambda)|x\rangle = |x\rangle f_x(\lambda)$ by $f_x(\lambda) = f_0(\lambda - x) = s_x f_0(\lambda)$ and would give a sharp localization of any particle wave function $\eta(x)$ at the point $x = \lambda$ of the pointer position provided that the wave function $f_0(\lambda)$ could be initially localized at $\lambda = 0$. But there are no such sharply localized quantum states for the continuous pointer, and the best that one can do is to take an approximately sharp wave packet $f_0(\lambda)$. Say, f_0 is the Gaussian packet rescaled to the standard form

$$f_0(\lambda) = \exp\left\{-\frac{\pi}{2}\lambda^2\right\}. \tag{0.6}$$

This results in the unsharp localization $[F(\lambda)\eta](x) = \psi(x, \lambda)$,

$$\psi(x, \lambda) = \exp\left\{-\frac{\pi}{2}(x - \lambda)^2\right\}\eta(x) \tag{0.7}$$

of any wave function $\eta(x)$ about the observed value λ of the pointer. The localized wave function $\psi(x,\lambda)$ defines for each measurement result λ the posterior particle vector-state

$$\eta_\lambda = c_\lambda^{-1}\psi(\lambda)\,, \quad |c_\lambda|^2 = \|\psi(\lambda)\|^2 \tag{0.8}$$

up to the normalization

$$\|\psi(\lambda)\|^2 = \int |\psi(x,\lambda)|^2 \mathrm{d}x = \mathrm{p}(\lambda) \tag{0.9}$$

to the probability density

$$\mathrm{p}(\lambda) = \int |f_0(\lambda - x)|^2 |\eta(x)|^2 \mathrm{d}x\,. \tag{0.10}$$

For commuting operators x and y, this is equivalent to the classical measurement model $y = x + q$ for unsharp measurement of an unknown signal x via the sharp measurement of the signal plus Gaussian noise q with given probability density $|f_0(\lambda)|^2$ of the random values $q = \lambda$.

These arguments illustrate how to interpret the reduction model involving the continuous spectrum of a quantum measurement as a Hamiltonian interaction model with nondemolition observation for a quantum object using the measurement of the pointer coordinate of the quantum meter. Since von Neumann introduced this approach, they were used by numerous other authors [8, 9] for the derivation of a generalized reduction $\eta \mapsto F(\lambda)\eta$ that would replace the von Neumann postulate $\eta \mapsto E(\lambda)\eta$ given by orthoprojections $\{E(\lambda)\}$ in the case of discrete values λ.

As extended to nondemolition observations continual in time [10]– [16], this idea constitute an essence of the quantum filtering method for the derivation of nonlinear stochastic wave equations describing the quantum dynamics under the observation. Since a particular type of such equations has been taken as a postulate in the phenomenological theory of continuous reduction and spontaneous localization [17]– [21], the question arises whether it is possible to obtain this equation from an appropriate Schrödinger equation. Here we shall show how this can be done by second quantisation of the singular interaction Hamiltonian considered by von Neumann obtaining a stochastic model of continual nondemolition observation for the position of a quantum particle by counting some other quanta.

First we show in the Sec.1 that even the projection postulate can be derived in the framework of this approach with the suggested Hamiltonian interaction and a proper nondemolition observation with a discrete spectrum and sharp initial state of the meter. Then we define a single-kick stochastic wave equation for the reduced state-vector, corresponding to the unsharp $f_0(\lambda)$.

In the Sec.2 we develop the differential treatment of discontinuous unitary evolution in terms of generalized Schrödinger equation corresponding to the scatterings at given or randomly distributed time instants. Then we show how it can be reduced to the many-kick stochastic wave equation, describing spontaneous localization of a quantum particle under the continual observation of its trajectory in a bubble chamber.

The quantum system of many similar interacting particles in a bubble chamber is treated in the Sec.3. We prove that the reduced dynamics of the particles is mixing under the continual observation of positions of the scattered bubbles. It can not be described by any reduction equation for the posterior wave function but can be described by an irreversible stochastic equation for the density matrix of the particles.

Finally we obtain in the Sec.4 the macroscopic and diffusion limits of the generalized Schrödinger and reduction equations under the limit of weak coupling constant and higher frequency of the interaction with the apparatus. In the macroscopic limit these two equations coincide with the ordinary Schrödinger equation, corresponding to the mean field approximation, while in the diffusion approximation, which describes the fluctuations, they lead to the essentially different kind of stochastic wave equations [24, 25].

These results are presented by the equations (1.10), (2.10), (3.10) and (4.10).

1 HAMILTONIAN MODEL FOR A GENERALIZED REDUCTION

Let $\mathcal{H}$ be a Hilbert space called the state space of a particle, and let R be a selfadjoint operator in $\mathcal{H}$ with either integer or continuous spectrum $\mathbf{Z}$ or $\mathbf{R}$. Let $\kappa > 0$ be a scaling parameter. One can regard the scaled operator κR as the position x of the particle in $\mathbf{R}$ or in the lattice $\kappa\mathbf{Z}$ if it is quantised so that κR is given in the x-representation, the operator acts as the multiplication $\kappa R|x\rangle = |x\rangle\lambda(x)$ by $\lambda(x) = \kappa\lfloor x/\kappa\rfloor$, where $\lfloor x\rfloor \in \mathbf{Z}$ denotes the integer part of x.

A quantum meter with continuous ($\Lambda = \mathbf{R}$) or discrete ($\Lambda = \varepsilon\mathbf{Z}$) pointer scale is described by the Hilbert space $L^2(\Lambda)$ of complex-valued functions $f : \Lambda \to \mathbf{C}$ square integrable in the sense that $\|f\|^2 = \int |f(y)|^2 \mathrm{d}y < \infty$. Here $f(y) = \sum_{l\in\mathbf{Z}} \delta_l^\varepsilon(y) f(\varepsilon l)\varepsilon$ is the isometric interpolation of $f(\varepsilon l)$, $l \in \mathbf{Z}$ in the discrete case, given by

$$\delta_l^\varepsilon(y) = \int_0^{1/\varepsilon} e^{2\pi i k(\varepsilon l - y)} \mathrm{d}k\,, \quad y \in \mathbf{R}\,,$$

such that $\sum_{y\in\Lambda} |f(y)|^2\varepsilon = \|f\|^2$.

Consider a moving particle with Hamiltonian H in $\mathcal{H}$. Its singular evolution corresponding to the position measurement at time $t = 0$ is described in the product space $\mathcal{H}_1 = \mathcal{H} \otimes L^2(\Lambda)$ by time-dependent Hamiltonian

$$H_1(t) = H \otimes \mathbf{1} + \kappa R \otimes \delta(t) P\,, \tag{1.1}$$

in the interaction picture corresponding to the free translation evolution of the meter in $L^2(\Lambda)$. Here $\mathbf{1}$ is the identity operator in $L^2(\Lambda)$, defining the particle Hamiltonian $H_0 = H \otimes \mathbf{1}$ in $\mathcal{H}_1$, and P is the momentum operator of the meter $P = \hbar\partial/i\partial y$, having the bound spectrum $2\pi\hbar[0, 1/\varepsilon)$ for $\Lambda = \varepsilon\mathbf{Z}$. This generates the shift operator (0.2) in $L^2(\Lambda)$ with $x \in \Lambda$ and the scattering operator

$$S_t = \exp\Big(-\frac{i}{\hbar}\kappa R \otimes P_t\Big) = \begin{cases} S\,, & t > 0 \\ I\,, & t \leq 0 \end{cases} \tag{1.2}$$

in $\mathcal{H}_1$, where $P_t = 1_t P$, $1_t = 1$ if $t > 0$, and $1_t = 0$ otherwise.

The singular time dependence of the Hamiltonian (1.1) makes it impossible to define the Schrödinger equation $i\hbar \mathrm{d}\psi/\mathrm{d}t = H_1(t)\psi(t)$ at $t = 0$ in the usual sense. But one can define a discontinuous unitary evolution $U_1(t) : \mathcal{H}_1 \to \mathcal{H}_1$ for some $t_0 < 0$ as the single-jump unitary process

$$U_1(t) = \exp\Big(-\frac{i}{\hbar}\int_{t_0}^t H_1(s)\mathrm{d}s\Big) = e^{iH_0(t_0 - t)/\hbar} S_t$$

provided that $[R, H] = 0$. If R and H do not commute, the usual method which a physicists would follow at facing this difficulty is to consider the free evolution $\psi_-(t)$

for $t < 0$ and $\psi_+(t)$ for $t > 0$, supplemented by the boundary condition $\psi_+(0) = S\psi_-(0)$. Although this method is hardly applicable in the case of spontaneous interaction at a random instant $t_1 > t_0$, let us show that the discontinuous unitary evolution corresponding to the fixed $t_1 = 0$ can be defined as the fundamental solution $\psi(t) = U_1(t)\psi_0$ of a generalized Schrödinger equation. Such regularized equation can be written in terms of the forward differentials

$$\mathrm{d}\psi(t) = \psi(t + \mathrm{d}t) - \psi(t), \quad \mathrm{d}1_t = 1_{t+\mathrm{d}t} - 1_t\,,$$

as the Ito integral of

$$\mathrm{d}\psi(t) + \frac{i}{\hbar} H_0\psi(t)\mathrm{d}t = (S - I)\psi(t)\mathrm{d}1_t\,, \quad \psi(t_0) = \psi_0\,. \tag{1.3}$$

For every initial $\psi_0 \in \mathcal{H}$ the equation (1.3) has a unique solution at $t_0 \le 0$; this solution is given by the unitary operator $U_1(t) = U_0(t - t_0)S_t(-t_0)$, where $U_0(t) = \exp\{-iH_0t/\hbar\}$, $S_t(r) = U_0^\dagger(r)S_tU_0(r)$.

To prove this let us rewrite the generalized Schrödinger equation in the integral form

$$\psi(t) = e^{-iH_0t/\hbar}\left(e^{iH_0t_0/\hbar}\psi_0 + \int_{t_0}^{t} e^{iH_0r/\hbar}(S - I)\psi(r)\mathrm{d}1_r\right) \tag{1.4}$$

(the equivalence of (1.4) to (1.3) can be shown by straightforward differentiation). We can write $\psi(0) = U_0(-t_0)\psi_0$ for the solution (1.4) when $t = 0$ since $\int_{t_0}^{0} \varphi(r)\mathrm{d}1_r = 0$ if $t_0 \le 0$; hence $\psi(t)$ can be rewritten as

$$\psi(t) = U_0(t)(\psi(0) + (S - I)1_t\psi(0)) = U_0(t)S_t\psi(0)\,.$$

It follows from $\int_{t_0}^{t} \varphi(r)\mathrm{d}1_r = 1_t\varphi(0)$ for any function φ and any $t_0 < 0$, and $(S - I)1_t = S_t - I$ by the definition of the scattering operator (1.2). Thus, $U_1(t)$ is the unitary operator

$$U_0(t)S_tU_0(-t_0) = U_0(t - t_0)S_t(-t_0).$$

This gives the solution to equation (1.3) for $t_0 = 0$ as well, since $\psi(t) = U_0(t)S_t\psi(0)$ is equal to ψ_0 for $t = 0$.

The rescaled pointer $q_t = 1_tq$ switched on at the instant $t = 0$ of the scattering is described in the space $\mathcal{H} \otimes L^2(\Lambda)$ by the operator $Q_t = \kappa^{-1}I \otimes q_t$, having in the Heisenberg picture $Y_t = U_1^\dagger(t)Q_tU_1(t)$ the constant form $Y_t = Y(-t_0)$ for all $t > 0$. Here $Y(r) = S^\dagger(r)QS(r)$ taken at $r = -t_0$, $t_0 \le 0$ is the shifted operator

$$Y(r) = R(r) \otimes \mathbf{1} + \frac{1}{\kappa} I \otimes q\,, \tag{1.5}$$

where $R(r) = U^\dagger(r)RU(r)$ and $S(r) = \exp(-\kappa R(r) \otimes iP/\hbar)$. Obviously the observables $\{Y_t\}$ are self-nondemolition in the sense of their joint measurability. It follows from the trivial commutativity condition

$$[Y_s, Y_t] = 0\,, \quad \forall\, s, t\,. \tag{1.6}$$

And they are nondemolition with respect to an arbitrary particle operator $X_t(r) = U_1(t)^\dagger(X \otimes \mathbf{1})U_1(t)$ in the Heisenberg picture at $t_0 = -r$ in the sense of their predictability [10, 14]

$$[X_s, Y_t] = 0\,, \quad \forall\, s \ge t, \tag{1.7}$$

indeed, $[X_s, Y_s] = 0$ and $Y_s = Y_t$ for s, $t > 0$ since

$$U_1^\dagger(s)Q_tU_1(s) = S^\dagger(-t_0)Q_tS(-t_0)\,, \quad \forall\, s \ge t\,.$$

Let us fix a state vector $f_0 \in L^2(\Lambda)$, $\|f_0\| = 1$, given by a wave function $f_0(y)$ on Λ localized at $y = 0$. Let $|y)$ denote the (generalized) eigen function in the spectral representation $q = \int y|y)(y|\mathrm{d}\lambda$ of the meter coordinate $(qf)(y) = yf(y)$, where $\mathrm{d}\lambda = \mathrm{d}y$ if $y \in \mathbf{R}$ and $\mathrm{d}\lambda = \varepsilon$ if $y \in \varepsilon\mathbf{Z}$. This yields the localizing transformations $F_t(y) = (y|S_t f_0$ in the form

$$F_t(y) = f_0(yI - \kappa R_t) = \begin{cases} F(y)\,, & t > 0 \\ f_0(y)I\,, & t \leq 0 \end{cases} \tag{1.8}$$

where $F(y) = f_0(yI - \kappa R)$. The reduction transformations $\eta \mapsto \psi(t, y)$, defined on the particle space $\mathcal{H}$ by the formula

$$\psi(t, y) = U(t - t_0)F_t(-t_0, y)\eta\,, \quad y \in \Lambda,$$

with $U(t) = \exp(-iHt/\hbar)$ and $F_t(r) = U^\dagger(r)F_tU(r)$ reproduce the unitary evolution $U_1(t)$ in the y-representation on $\eta \otimes f_0 \in \mathcal{H}_1$ similarly to Eq. (0.4) and (0.5), namely,

$$U(t - t_0)F_t(-t_0, y)\eta = U(t)(y|S_t f_0 U(-t_0)\eta = (y|U_1(t)(\eta \otimes f_0)\,.$$

The operator $F(r, y) = f_0(yI - \kappa R(r))$ at $r = -t_0$ with a given initial vector-state η of the particle before the scattering ($t_0 \leq 0$) define the probability measure

$$\mu(\Delta) = \int_\Delta \|F(r, y)\eta\|^2 \mathrm{d}\lambda = \langle \eta, \Pi_t(\Delta)\eta\rangle\,, \quad \Delta \subseteq \Lambda\,,$$

for the statistics of the nondemolition measurement of κR via the observation of the pointer position $y \in \Delta$ after the scattering. It is given by a positive operator-valued measure $\Pi_t(\Delta) = \int_\Delta |F_t(y, -t_0)|^2 \mathrm{d}\lambda$ at $t > 0$. (Before the scattering it defines the initial probability measure

$$\mu_0(\Delta) = \langle \eta, \Pi_t(\Delta)\eta\rangle = \int_\Delta |f_0(y)|^2 \mathrm{d}\lambda\,, \quad t \leq 0$$

which is independent of η.)

In the case of discrete scale the eigen functions $|y)$ are normalisable: $(y|y) = 1/\varepsilon$, and one can take a sharply localized $f_0 = \varepsilon^{d/2}|0)$, given by the eigen function $f_0(y) = e(y)/\varepsilon^{d/2}$, where $e(y) = 1$ if $y = 0$, $e(y) = 0$ if $y \neq 0$. By renormalising the operator $F(y)$ as $E(y) = \varepsilon^{d/2}F(y)$ with step $\varepsilon = \kappa$ of the position quantisation, one obtain the orthogonal projections

$$e(\kappa R - yI) = \int_{x:\lfloor x/\kappa \rfloor = y/\kappa} |x\rangle\langle x|\mathrm{d}x = E(y)\,, \quad y \in \kappa\mathbf{Z}$$

of $\kappa R = \int \lambda(x)|x\rangle\langle x|\mathrm{d}x = \sum_{y \in \kappa\mathbf{Z}} yE(y)$ corresponding to eigen values $\lambda(x) = \kappa\lfloor x/\kappa \rfloor$. Thus, $\Pi_t(\Delta)$, $\Delta \subseteq \Lambda$ is the spectral measure $\sum_{y \in \Delta} E(t_0, y)$ of the quantised position $\kappa R(-t_0)$ of the particle for a $t > 0$ given by the eigen orthoprojections $E(r, y) = U^\dagger(r)E(y)U(r)$ of the operator $R(r)$, corresponding to the rescaled pointer integer values y/κ. Thus, the projection reduction postulate has been deduced from the Hamiltonian interaction (1.1) and the nondemolition measurement for the sharply localized initial state f_0. But there is no continuous limit as $\kappa \to 0$ of such sharp reduction with nontrivial $e \neq 0$, since $\|e\|^2 = \int |e(y)|^2 \mathrm{d}\lambda = \kappa \to 0$ and the sharp function $e(y) = \delta_0^\kappa(y)\kappa$ on $\Lambda = \kappa\mathbf{Z}$ disappears as an element of the state space $L^2(\mathbf{R})$ of the continuous meter.

In the continuous case, one can take only an approximately sharp $f_0 \in L^2(\mathbf{R})$, say $f_0(y) = \delta_0^\varepsilon(y)\varepsilon$, $y \in \mathbf{R}$, and renormalise the operator (1.8) as $G_t(y) = f_0(yI -$

$\kappa R_t)/f_0(y)$ if $f_0(y) \neq 0$, as in the Gaussian case (0.6). They define $G_t(y)$ as the identity operator for $t \leq 0$ or $r > 0$, whereas for $t > 0$ and $r \leq 0$ the operator

$$G(y) = \langle y|Sf_0 = (y|G\,, \quad G = f_0^{-1}Sf_0\,, \tag{1.9}$$

say, of the Gaussian form

$$G(y) = \exp\{\pi\kappa R(yI - \frac{1}{2}\kappa R)\}$$

given by the generalized eigen functions $|y\rangle = |y)/f_0(y)$ for the spectral representation $q = \int y|y\rangle\langle y|\mathrm{d}\mu_0$ with respect to the initial probability measure $\mathrm{d}\mu_0 = |f_0(y)|^2\mathrm{d}\lambda$ with density $|f_0(y)|^2 = \exp\{-\pi y^2\}$ in the Gaussian case.

The corresponding propagators

$$T(t,y) = U(t-t_0)G_t(-t_0,y), \quad y \in \Lambda,\ t_0 \leq 0\,,$$

where $G_t(r) - U^\dagger(r)G_tU(r)$, define the operator–valued measure $\Pi_t(\Delta)$ as

$$\Pi_t(\Delta) = \int_\Delta T^\dagger(t,y)T(t,y)\mathrm{d}\mu_0 = \int_\Delta |G_t(-t_0,y)|^2\mathrm{d}\mu_0$$

so that the output probability measure μ is absolutely continuous with respect to μ_0 for any $\eta \in \mathcal{H}$: $\mu(\Delta) = \int_\Delta \|G(-t_0,y)\eta\|^2\mathrm{d}\mu_0$. Hence, the reduced state vector $\chi(t,y) = T(t,y)\eta$ is normalized to 1 as a stochastic vector process $\chi(t) : y \mapsto \chi(t,y) \in \mathcal{H}$ in the mean square sense with respect to the input probability measure μ_0,

$$\|\chi(t)\|_0^2 = \int \langle\chi(t,y),\chi(t,y)\rangle\mathrm{d}\mu_0 = \langle\eta,\eta\rangle = 1\,.$$

This model of nondemolition observation with continuous data $y \in \mathbf{R}$ also applies to unsharp measurement of operator R with discrete spectrum. In contrast to sharp measurement, unsharp measurement is not sensitive to the continuous spectrum limit as $\kappa \to 0$ of $R = \lfloor x/\kappa \rfloor$, corresponding to the replacement of κR by $x \in \mathbf{R}$.

For any initial $t_0 \leq 0$ the stochastic vector process $\chi(t) = T(t)\eta$ satisfies the single-kick equation

$$\mathrm{d}\chi(t) + \frac{i}{\hbar}H\chi(t)\mathrm{d}t = \mathrm{d}1_t[G-I]\chi(t)\,, \quad \chi(t_0) = \eta \tag{1.10}$$

generated by the random differential $\mathrm{d}1_t[G-I](y) = (G(y)-I)\mathrm{d}1_t$ on $y \in \Lambda$ with respect to the initial probability measure μ_0. This simplest reduction equation is written in terms of the forward differentials $\mathrm{d}\chi(t,y) = \chi(t+\mathrm{d}t,y) - \chi(t,y)$, that is, is understood in the sense of Ito.

Indeed, by representing G_t as $G_t(y) = (G(y)-I)1_t + I = I + 1_t[G-I](y)$, one can describe $\chi(t) = U(t-t_0)G_t(-t_0)\eta$ by the integral equation

$$\begin{aligned}\chi(t) &= U(t-t_0)\eta + U(t)1_t[G-I]U(-t_0)\eta = \\ &= e^{-iHt/\hbar}\Big(e^{iHt_0/\hbar}\eta + \int_{t_0}^t e^{iHr/\hbar}\mathrm{d}1_r[G-I]\chi(r)\Big)\,.\end{aligned}$$

But this equation is equivalent to the differential equation (1.10), a fact that can be proved by straightforward differentiation taking into account the Ito multiplication table

$$(\mathrm{d}t)^2 = 0\,, \quad \mathrm{d}t\mathrm{d}1_t = 0 = \mathrm{d}1_t\mathrm{d}t\,, \quad (\mathrm{d}1_t)^2 = \mathrm{d}1_t\,.$$

Similarly, one can obtain the simplest nonlinear stochastic equation for the normalized reduced state vector $\chi_y(t) = \chi(t, y)/\|\chi(t, y)\|$:

$$\mathrm{d}\chi_y(t) + \frac{i}{\hbar} H \chi_y(t)\mathrm{d}t = (G_y(t) - I)\chi_y(t)\mathrm{d}1_t\,, \quad \chi_y(0) = \eta$$

where $G_y(t) = G(y)/\|G(y)\chi_y(t)\|$, $\eta \in \mathcal{H}$. This equation is an equivalent differential form the nonlinear integral stochastic equation

$$\begin{aligned}\chi_y(t) &= e^{-iHt/\hbar}\Big(e^{iHt_0/\hbar}\eta + \int_{t_0}^{t} e^{iHr/\hbar}(G_y(r) - I)\chi_y(r)\mathrm{d}1_r\Big) = \\ &= U(t - t_0)\eta + U(t)(G_y - I)1_t\chi_y(0) = U(t)G_t(y)\chi_y(0)/\|G_t(y)\chi_y(0)\|\,.\end{aligned}$$

This yields $\chi_y(t) = T(t, y)\eta/\|T(t, y)\eta\|$ for a $t_0 \leq 0$ because of $\|G_t(y)\eta\| = \|T(t, y)\eta\|$.

Note that the random state vector $\chi_y(t)$ is obtained by conditioning with respect to the output (rather than input) probability measure $\mathrm{d}\mu = \|\chi(t, y)\|^2\mathrm{d}\mu_0$.

2 SPONTANEOUS LOCALIZATION OF A SINGLE PARTICLE

Let us consider a spontaneous process of scattering interactions (1.1) of a quantum particle at random time instants $t_n > 0$, $t_1 < t_2 < \ldots$, with a renewable meter in an apparatus of the cloud chamber type with bubbles serving as the meter. We consider the increasing sequences $(t_1, t_2, \ldots)$ as countable subsets $\tau \subset \mathbf{R}_+$ such that $\tau_t = \tau \cap [0, t)$ is finite for any $t \geq 0$ in accordance with the finiteness of the number of scattered bubbles on the finite observation interval $[0, t)$. The set of all such infinite τ will be denoted by Γ_∞, and Γ is the inductive limit $\cup\Gamma_t$ as $t \to \infty$ of $\Gamma_t = \{\tau_t | \tau \in \Gamma_\infty\}$, which is the disjoint union $\Gamma_t = \sum_{n=0}^{\infty} \Gamma_t(n)$ of n-simplice $\Gamma_t(n) = \{t_1 < \ldots < t_n\} \subset [0, t)^n$.

The measurement apparatus is assumed to be a quantum system of infinitely many bubbles each of which is identical to the single meter described in the previous section. The coordinate q_n, of a bubble labeled by the scattering number $n \in \mathbf{N}$ shows the position $\lambda_n \in \Lambda$ of the pointer at time $t_n \in \tau$.

The corresponding Hamiltonian of the moving particle is given by the series

$$H(t, \tau) = H_0 + \kappa R \otimes \sum_{n=1}^{\infty} \delta(t - t_n)P(n) \tag{2.1}$$

having at most two nonzero terms if $t \in \tau$. Here $H_0 = H \otimes \mathbf{1}^{\otimes\infty}$ is the Hamiltonian describing the time evolution on the intervals between the scatterings $t \in \tau$ and $P(n)$ is the momentum of the nth scattered bubble, given as the operator $P(n) = -i\hbar\mathrm{d}/\mathrm{d}\lambda_n$.

The generalized Schrödinger equation corresponding to the Hamiltonian (1.1) can be written for fixed $\tau \in \Gamma_\infty$ by analogy with the single-kick case

$$\mathrm{d}\psi(t) + \frac{i}{\hbar} H_0\psi(t)\mathrm{d}t = (S(n_t) - I)\psi(t)\mathrm{d}n_t, \quad \psi(0, \tau) = \psi_0\,. \tag{2.2}$$

Here $S(n) = \exp\{-\frac{i}{\hbar}\kappa R \otimes P(n)\}$ and $n_t(\tau) = |\tau_t|$ is the numerical process that gives the cardinality $|\tau_t| = \sum_{r\in\tau} 1_{t-r}$ of the localized subset $\tau_t = \{t_n < t\}$, so that $\mathrm{d}n_t(\tau)$ is equal to 1 for $t \in \tau$, and zero otherwise.

The solution to this equation is uniquely determined for every $\tau \in \Gamma_\infty$ by the initial state ψ_0 of the system. Namely,

$$\psi(t,\tau) = U(t,\tau)\psi_0 ,$$

where $U(t,\tau) = U_0(t)V_t^\dagger(\tau)$, $V_t^\dagger(\tau)$ is the chronological product $\prod_{r\in\tau}^{\leftarrow} S_t(r) = S(t_{n_t})\dots S(t_1)$, and

$$V_t(\tau) = S_t^\dagger(t_1)S_t^\dagger(t_2)\dots = \Big(\prod_{r\in\tau}^{\leftarrow} S_t(r)\Big)^\dagger . \tag{2.3}$$

Here $S_t(t_n) = U_0^\dagger(t_n)S_t(n)U_0(t_n)$ for $t_n < t$, where $S(n) = \exp\{-\frac{i}{\hbar}\kappa R\otimes P(n)\}$, and $S_t(t_n) = I$ if $t_n \geq t$ so that the infinite product (2.3) contains only a finite number $n_t = \sum_{r\in\tau} 1_{t-r}$ of factors different from the identity operator I.

Recall that the differential equation (2.2) is equivalent to the integral equation given by the recurrence relation

$$\psi(t,\tau) = e^{-iH_0t/\hbar}\Big(\psi_0 + \sum_{r\in\tau}^{r<t} e^{iH_0r/\hbar}(S(n_r)-I)\psi(r,\tau)\Big) \tag{2.4}$$

for every $\tau \in \Gamma_\infty$. Hence, $\psi(t,\tau) = U_0(t)V_t^\dagger(\tau)\psi_0$, where $U_0(t) = e^{-iH_0t/\hbar}$ and $V_t(\tau)$ is a solution to the operator equation

$$V_t(\tau) = I + \sum_{r\in\tau}^{r<t} V_r(t)(S_t^\dagger(r,\tau)-I) , \quad V_0(\tau) = I ,$$

where $S^\dagger(t,\tau) = U_0(t)^\dagger S(n_t(\tau))^\dagger U_0(t)$. But this equation has a unique solution (2.2), which can be written as the binomial sum

$$[L_t(t_1)+I][L_t(t_2)+I]\dots = \sum_{\sigma\subseteq\tau_t} L(s_1,\tau)\dots L(s_n,\tau)$$

in terms of $\sigma = \{s_1,\dots,s_n\}$, $s_1 < \dots < s_n$, $n \leq n_t$, $L_t(r) = S_t^\dagger(r) - I$ ($= 0$ if $r \geq t$) and $L(r,\tau) = S^\dagger(r,\tau) - I$. Indeed, this sum contains I as the null product corresponding to $\sigma = \emptyset$, and the sum of the other terms is equal to

$$\begin{aligned} V_t(\tau) - I &= \sum_{r\in\tau}^{r<t}\sum_{\sigma\subseteq\tau_r} L(s_1,\tau)\dots L(s_m,\tau)L(r,\tau) = \\ &= \sum_{r\in\tau}^{r<t} V_r(\tau)L(r,\tau) = \sum_{r\in\tau}^{r<t} V_r(\tau)(S^\dagger(r,\tau)-I) , \end{aligned}$$

where $m \leq n_t - 1$.

Note that the generalized differential equation (2.2) depending on $\tau \in \Gamma_\infty$ via $n_t = n_t(\tau)$ is not necessarily stochastic as long as we have not fixed a probability distribution for the instants $\tau = (t_1, t_2, \dots)$ of the singular interactions. In order to obtain a continuous (at least in the mean) dynamics for such a quantum jump process it is necessary to assume the interactions are spontaneous with a continuous probability distribution of random instants τ. One can assume that the number process $n_t(\tau)$ is stochastic, given by the Poisson law $\pi_0(\mathrm{d}\tau)$ on Γ_∞ presented as the projective limit as $t \to \infty$ of the probability measures

$$\pi_0(\mathrm{d}\tau_t) = e^{-\nu t}\nu^{|\tau_t|}\mathrm{d}\tau_t , \quad \nu > 0 . \tag{2.5}$$

Here $\tau_t = \tau$ is a finite time-ordered sequence $\tau(n) = (t_1, \ldots, t_n) \in \Gamma_t$ with $n = n_t$, $\mathrm{d}\tau_t = \prod_{k=1}^{n_t} \mathrm{d}t_k$ is the measure on Γ_t given by the sum of product measures $\mathrm{d}t_1, \ldots, \mathrm{d}t_n = \mathrm{d}\tau(n)$ on the simplices $\Gamma_t(n)$, $\mathrm{d}\tau(0) = 1$ on $\Gamma_t(0) = \{\emptyset\}$ such that

$$\int_{\Gamma_t} \nu^{|\tau|} \mathrm{d}\tau := \sum_{n=0}^{\infty} \nu^n \int \cdots \int_{0 \le t_1 < \ldots < t_n < t} \mathrm{d}t_1 \ldots \mathrm{d}t_n = e^{\nu t}.$$

Note that any other numerical process can be described by a positive density function $f(\tau)$ with respect to the Poissonian measure, that is, has the form $f(\tau)\pi(\mathrm{d}\tau)$.

Let us fix an initial state $\varphi_0 = f_0^\infty$ of the bubbles as the infinite product $f_0^\infty = \otimes_{k=1}^\infty f_k$ of the identical state vectors $f_k = f_0$ of the bubbles given by a normalized element $f_0 \in L^2(\Lambda)$. This defines the solutions $\psi(t, \tau)$ of the stochastic equation (2.2) with the initial data $\psi_0 = \eta \otimes \varphi_0$ given by state vectors $\eta \in \mathcal{H}$ in the particle space $\mathcal{H}$ as the state vectors in the product space $\mathcal{H}_\infty = \mathcal{H} \otimes \mathcal{E}$, where $\mathcal{E}$ is the Hilbert space generated by the infinite-product functions $\varphi(v) = \prod_{k=1}^\infty f_k(\lambda_k)$ with the equal elements $f_k = f_0$ for almost all k. We suppose, as in Sec.1, that $\varphi_0(v) \neq 0$ for almost all $v = (\lambda_1, \lambda_2, \ldots)$ such that one can identify the space $\mathcal{E}$ with the space $L_0^2(\Lambda^\infty)$ of all square-integrable functions $f = \varphi/\varphi_0$ with respect to the product measure $\mu_0^\infty(\mathrm{d}\boldsymbol{\lambda}) = \mu_0(\mathrm{d}\lambda_1) \cdot \mu_0(\mathrm{d}\lambda_2) \cdots$ on the space $\Lambda^\infty = \Lambda \times \Lambda \times \ldots$ of the sequences $\boldsymbol{\lambda} = (\lambda_1, \lambda_2, \ldots)$: $\|f\|_0^2 = \int |f(\boldsymbol{\lambda})|^2 \mu_0^\infty(\mathrm{d}\boldsymbol{\lambda}) = \|\varphi\|^2$. The generalized product-vectors $|\boldsymbol{\lambda}\rangle = |\lambda_1\rangle \otimes |\lambda_2\rangle \otimes \ldots$ of this space are defined by tensor-product of the δ-functions $\langle\lambda'|\lambda\rangle$, respectively to μ_0 such that $\int |\boldsymbol{\lambda}\rangle\langle\boldsymbol{\lambda}|\mu_0^\infty(\mathrm{d}\boldsymbol{\lambda}) = \mathbf{1}^{\otimes\infty}$.

Consider the sequence $(q_1, q_2, \ldots)$ of coordinates q_n, of the scattered bubbles at the time instants $\{t_1, t_2, \ldots\}$. The commuting operators q_n, $n \in \mathbf{N}$, described in $\mathcal{E}$ by the multiplications $q_n|\boldsymbol{\lambda}\rangle = |\boldsymbol{\lambda}\rangle\lambda_n$, are assumed to be measured at the random time instants t_n, $n \in \mathbf{N}$. The point trajectories of such measurements are given by the sequences $= (y_1, y_2, \ldots)$ of pairs $y_n = (t_n, \lambda_n)$ with $t_1 < t_2, \ldots$ and $\lambda_n \in \Lambda$, identified with countable subsets $\{y_1, y_2, \ldots\} \subset \mathbf{R}_+ \times \Lambda$. As elements $v = (\tau, \boldsymbol{\lambda})$ of the Cartesian product $\Upsilon_\infty = \Gamma_\infty \times \Lambda^\infty$, they have the probability distribution $\mathrm{P}_0(\mathrm{d}v) = \pi_0(\mathrm{d}\tau)\mu_0^\infty(\mathrm{d}\boldsymbol{\lambda})$, where Λ^∞ is the space of all sequences $\boldsymbol{\lambda} \in \Lambda^\infty$, $\lambda_n \in \Lambda$, equipped with the probability product-measure $\mu_0^\infty(\mathrm{d}\boldsymbol{\lambda})$.

The measurement data of the observable process up to a given time instant $t > 0$ is described by a finite sequence $v_t = (y_1, \ldots, y_n)$ with $n = n_t(v)$ given by the numerical process $n_t(\tau)$ for the component τ of v.

Let us introduce the counting distribution $n_t(\Delta) = |v_t \cap (\mathbf{R}_+ \times \Delta)|$ as the number $n_t(\Delta, v)$ of scatterings in the time-space region $[0, t) \times \Delta$ and define the counting integral $\int_0^\infty \int_\Lambda L(r, \lambda)\mathrm{d}n_r(\mathrm{d}\lambda)$ over $y \in \mathbf{R}_+ \times \Lambda$ as the series

$$n[L](v) = \sum_{y \in v} L(y), \quad \forall\, v \in \Upsilon_\infty. \tag{2.6}$$

Having fixed an integer-valued distribution $n_t(\Delta) \in \{0, 1, \ldots\}$ as a function of $t \ge 0$ and of measurable sets $\Delta \subseteq \Lambda$, one can obtain the corresponding trajectory v as a sequence of the counts of the jumps of $n_t(\Delta)$ in the time-space $\mathbf{R}_+ \times \Lambda$.

Given an initial state vector in $\mathcal{H}_\infty$ of the form $\psi_0 = \eta \otimes \varphi_0$ with fixed $\varphi_0 = f_0^\infty$, one can define a nonunitary stochastic evolution $\eta \mapsto T(t, v)\eta$ by setting

$$T(t, v) = \langle\boldsymbol{\lambda}|U(t, \tau)\varphi_0, \quad v = (\tau, \boldsymbol{\lambda}),$$

which reproduces the unitary evolution $U(t, \tau) = U_0(t)V_t^\dagger(\tau)$ defined by (2.2). This can also be written as $T(t, v) = U(t)F_t^\dagger(v)$, since $U_0(t) = U(t) \otimes \mathbf{1}^{\otimes\infty}$ commutes with the (generalized) eigen-bras $\langle\boldsymbol{\lambda}| = \langle\lambda_1, \lambda_2, \ldots|$ of the bubble coordinates $(q_1, q_2, \ldots)$:

$\langle\boldsymbol{\lambda}|U_0(t) = U(t)\langle\boldsymbol{\lambda}|$. The reduction transformations $F_t(v)$, $v \in \Upsilon_\infty$, are given by the chronological products

$$F_t(v) = G_t^\dagger(y_1)G_t^\dagger(y_2)\ldots \equiv \overrightarrow{\prod_{y\in v}} G_t^\dagger(y) \tag{2.7}$$

of $G_t(t_n, \lambda_n) = U^\dagger(t_n)G(\lambda_n)U(t_n)$ for $t_n < t$, where $G(\lambda) = \langle\lambda|Sf_0$, owing to the product form (2.3) of the unitary transformations $V_t(\tau)$, $\tau \in \Gamma$, and $\langle\boldsymbol{\lambda}| = \otimes_{k=1}^\infty\langle\lambda_k|$, $\varphi_0(\boldsymbol{\lambda}) = \prod_{k=1}^\infty f_0(\lambda_k)$, $\langle\boldsymbol{\lambda}|\varphi_0 \equiv 1$ for $\boldsymbol{\lambda} \in \Lambda^\infty$. The stochastic operator (2.7) defined by the single–point reductions

$$G_t(r,\lambda) = \langle\lambda|S_t(r)f_0 = \begin{cases} G(r,\lambda)\,, & r < t \\ I\,, & r \geq t \end{cases} \tag{2.8}$$

is normalized with respect to the initial probability $\mathrm{P}_0(\mathrm{d}v)$, $\Pi_{\Upsilon_\infty}[I](t) = I$, where

$$\Pi_A[X](t) = \int_A T^\dagger(t,v)XT(t,v)\mathrm{P}_0(\mathrm{d}v) = \int_A F_t(v)XF_t^\dagger(v)\mathrm{P}_0(\mathrm{d}v)\,, \tag{2.9}$$

is a continual operational-valued measure [4]– [6] defined on measurable sets $A \subseteq \Upsilon_\infty$ of the point trajectories $v_t = \{(r,\lambda) \in v|r < t\}$ given by the operations $\Phi_t(v) : X \mapsto F_t(v)XF_t^\dagger(v)$ for particle operators $X : \mathcal{H} \to \mathcal{H}$.

The positive operator-valued measure $\Pi_t(A) = \Pi_A[I](t)$ gives the statistics

$$\mathrm{P}_t(A) = \int_A \|\langle\boldsymbol{\lambda}|U(t,\tau)\psi_0\|^2\mu_0^\infty(\mathrm{d}\boldsymbol{\lambda})\pi_0(\mathrm{d}\tau_t)$$

for the continual observation with respect to an arbitrary initial wave-function $\psi_0 = \eta \otimes f_0^\infty$, $\eta \in \mathcal{H}$ in the form

$$\mathrm{P}_t(A) = \langle\eta, \Pi_t(A)\eta\rangle\,.$$

The output probability measure $\mathrm{P}(A)$, $A \subseteq \Upsilon_\infty$, is defined by the marginales $\mathrm{P}_t(A)$, $A \subseteq \Upsilon_t$, as $t \to \infty$.

The reduced wave function $\chi(t,v) = T(t,v)\eta$ is normalized

$$\|\chi(t)\|^2 = \int \|\chi(t,v)\|^2\mathrm{P}_0(\mathrm{d}v) = 1$$

as a stochastic vector process $\chi(t) : \Upsilon_\infty \to \mathcal{H}$ with respect to the initial probability distribution P_0 on Υ_∞.

It satisfies the stochastic wave equation

$$\mathrm{d}\chi(t) + \frac{i}{\hbar}H\chi(t)\mathrm{d}t = \mathrm{d}n_t[G-I]\chi(t)\,, \quad \chi(0) = \eta\,, \tag{2.10}$$

expressed in terms of the random differential $\mathrm{d}n_t[G-I](v) = (G(\lambda_{n_t}(v)) - 1)\mathrm{d}n_t(v)$, $n_t(v) = n_t(\tau)$ for the point distribution $n_t[L] = \int_\Lambda L(\lambda)n_t(\mathrm{d}\lambda) = n[L_t]$ over $\lambda \in \Lambda$ defined in (2.6) for $L_t(r,\lambda) = 1_t(r)L(\lambda)$.

To prove Eq.(2.10), discovered for the first time in [14], we rewrite it in the following integral form

$$\chi(t) = e^{-iHt/\hbar}\Big(\eta + \int_0^t\int_\Lambda e^{iHr/\hbar}(G(\lambda) - I)\chi(r)\mathrm{d}n_r(\mathrm{d}\lambda)\Big),$$

given for each $v \in \Upsilon_\infty$ by the finite sum

$$n[1_t U^\dagger(r)(G-I)\chi](v) = \sum_{(r,\lambda)\in v}^{r<t} U^\dagger(r)(G(\lambda)-I)\chi(r)\,.$$

We express the solution to this equation in the form $\chi(t,v) = U(t)F_t^\dagger(v)\eta$ via the solution (2.7) to the recursion equation

$$F_t(v) = I + \sum_{(r,\lambda)\in v}^{r<t} F_r(v)(G(r,\lambda)-I)\,, \quad F_0(v) = I,$$

with $G(t,\lambda) = U^\dagger(t)G(\lambda)U(t)$, as was done for the unitary case.

Let us also write on F the nonlinear equation

$$\mathrm{d}\chi_v(t) + \frac{i}{\hbar}H\chi_v(t)\mathrm{d}t = (G_v(t)-I)\chi_v(t)\mathrm{d}n_t(v)\,, \quad \chi_v(0) = \eta,$$

with $G_v(t) = G(\lambda_{n_t(v)})/\|G(\lambda_{n_t(v)})\chi_v(t)\|$. Its solutions define the normalized reduction $\chi_v(t) = \chi(t,v)/\|\chi(t,v)\|$ for continual counting measurements as a stochastic vector process $\chi_v(t) \in \mathcal{H}$ with respect to the output probability measure P of the point process $t \mapsto v_t$. This can easily be obtained, as in [15] by applying the Ito multiplication table

$$(\mathrm{d}t)^2 = 0\,, \quad \mathrm{d}t\mathrm{d}n_t = 0 = \mathrm{d}n_t\mathrm{d}t\,, \quad (\mathrm{d}n_t)^2 = \mathrm{d}n_t\,.$$

3 MIXING REDUCTION FOR MANY IDENTICAL PARTICLES

We now consider M identical particles interacting independently with the bubbles in accordance with the scattering term in the Hamiltonian (2.1). The spontaneous process of scatterings is described by the time-ordered sequences of pairs (k_n, t_n), $t_1 < t_2 < \ldots$, where $k_n \in \{1,\ldots,M\}$ is the number of the particle labeled by the scattering number $n \in \mathbf{N}$ at time instant $t_n > 0$. We have excluded possibility of two or more scatterings of the bubbles at the same instant of time, as was done for a single particle in Sec.2. The sequence (k_1,t_1), (k_2,t_2), ... of the scatterings can be represented by the occupational subsets $\tau_k = \{t_n \in \tau | k_n = k\}$ of the time set $\tau = \{t_1, t_2, \ldots\}$, which are disjoint, $\tau_k \cap \tau_l = \emptyset$ if $k \neq l$ since the scatterings for different particles are independent. We shall consider the M-tuples $\tau_\bullet = (\tau_1,\ldots,\tau_M)$ of these countable subsets $\tau_k \subset \mathbf{R}_+$ as elements $\tau_\bullet \in \Gamma_\infty^M$ of the Cartesian M- product of Γ_∞, given by the partition $\tau = \sqcup\tau_k := \cup\tau_k$, $\tau_k \cap \tau_l = \emptyset$ if $k \neq l$ of a $\tau \in \Gamma_\infty$.

The Hamiltonian of the interacting particles for fixed $\tau_\bullet = \{(k_1,t_1),$ $(k_2,t_2),\ldots\}$ reads

$$H(t,\tau_\bullet) = H_0^M + \kappa\sum_{n=1}^{\infty} R(k_n)) \otimes \delta(t-t_n)P(n)\,. \tag{3.1}$$

Here $H_0^M = H^M \otimes \mathbf{1}$ is the Hamiltonian of the particles describing the time evolution on the intervals between the scatterings with the bubbles:

$$H^M = \sum_{k=1}^{M} H(k) + \sum_{k=1}^{M}\sum_{l>k}^{M} W(k,l)\,,$$

where $H(k) = I^{\otimes(k-1)} \otimes H \otimes I^{\otimes(M-k)}$ is the Hamiltonian of the kth particle and $W(k,l)$ is the interaction potential in $\mathcal{H}^{\otimes M}$ of the kth and lth particle, $1 \leq k < l \leq M$.

Let $\mathcal{H}^M$ denote the M-particle Hilbert space, which is an invariant subspace $\mathcal{H}^M \subseteq \mathcal{H}^{\otimes M}$ of symmetric (bosons) or antisymmetric (fermions) M-tensors $\eta^M \in \mathcal{H}^{\otimes M}$ generated by the product-vectors $\otimes_{k=1}^M \eta_k \in \mathcal{H}^{\otimes M}$ with $\eta_k \in \mathcal{H}$. The correspondent Ito-Schrödinger equation for the stochastic state vector $\psi^M(t) : \tau_\bullet \mapsto \psi(t, \tau_\bullet)$ with values $\psi(t, \tau_\bullet) \in \mathcal{H}_\infty^M$ in the product space $\mathcal{H}_\infty^M = \mathcal{H}^M \otimes \mathcal{E}$ of the M-particle space $\mathcal{H}^M$ by $\mathcal{E} = \lim_{n\to\infty} L^2(\Lambda^n)$ reads

$$\mathrm{d}\psi^M(t) + \frac{i}{\hbar} H_0^M \psi^M(t)\mathrm{d}t = (S(k_t, n_t) - I)\psi^M(t)\mathrm{d}n_t \,. \tag{3.2}$$

Here $S(k,n) = \exp\{-\frac{i}{\hbar}\, \kappa R(k) \otimes P(n)\}$, $R(k) = I^{\otimes(k-1)} \otimes R \otimes I^{\otimes(M-k)}$,

$$k_t(\tau_\bullet) = \sum_{k=1}^{M} k 1_{\tau_k}(t)\,, \quad \text{where } 1_\tau(t) - \begin{cases} 1, & t \in \tau \\ 0, & t \notin \tau \end{cases}$$

is the random number $k_t : \Gamma_\infty^M \to \{1, \ldots, M\}$, labeling a particle by k at any instant $t \in \tau_k$ of its collision with a bubble labeled by $n_t(\tau) = \sum_{k=1}^M n_{k,t} = |\tau \cap [0,t)|$, where $n_{k,t} = |\tau_k \cap [0,t)|$, $\tau = \cup \tau_k$.

The solutions $\psi(t, \tau_\bullet) = U(t, \tau_\bullet)\psi_0^M$, $\psi_0^M \in \mathcal{H}_\infty^M$, of equation (3.2) can be written as $U(t, \tau_\bullet) = U_0^M(t) V_t^\dagger(\tau_\bullet)$ in terms of the finite chronological product

$$V_t(\tau_\bullet) = S_t^\dagger(k_1, t_1) S_t^\dagger(k_2, t_2) \ldots, \qquad \tau_\bullet \in \Gamma_\infty^M\,, \tag{3.3}$$

where $S_t(k, t_n) = I$ if $t_n \geq t$, $S_t(k, t_n) = U_0^{M\dagger}(t_n) S(k,n) U_0^M(t_n)$ if $t_n < t$, and $U_0^M(t) = \exp\{-\frac{i}{\hbar}\, H_0^M(t)\}$. The proof is exactly the same as for the case of a single particle $M = 1$.

Let $\omega = (w_1, w_2, \ldots)$ denote a chronologically ordered sequence of triples $w_n = (k_n, t_n, \lambda_n)$ and Ω the space of all such sequences with $\{t_1, t_2, \ldots\} \in \Gamma_\infty$. Every sequence $\omega \in \Omega$ can be represented as a pair $\omega = (\tau_\bullet, \boldsymbol{\lambda})$, where $\tau_\bullet = (\tau_1, \ldots, \tau_M)$ is a partition of the corresponding sequence $\tau = \{t_1, t_2, \ldots\}$ and $\boldsymbol{\lambda} = (\lambda_1, \lambda_2, \ldots)$, so that Ω can be identified with the product $\Gamma_\infty^M \times \Upsilon_\infty$. The space Ω is equipped with the probability measure $\mathrm{P}_0(\mathrm{d}\omega) = \pi_0(\mathrm{d}\tau_\bullet)\mu_0^\infty(\mathrm{d}\boldsymbol{\lambda})$, where $\pi_0(\mathrm{d}\tau_1, \ldots, \mathrm{d}\tau_M) = \prod_{k=1}^M \pi_0(\mathrm{d}\tau_k)$ is the product of the identical Poisson measures (2.5), in accordance with the independence of the spontaneous interactions of each particle with the bubbles.

Given an initial state vector $\psi^M = \eta^M \otimes \varphi_0$, where $\varphi_0 = f_0^\infty$, one can easily prove that the nonunitary stochastic evolution

$$T(t, \omega) = \langle \boldsymbol{\lambda} | U(t, \tau_\bullet)\varphi_0\,, \quad \omega = (\tau_\bullet, \boldsymbol{\lambda})$$

is also a finite chronological product

$$T(t, \omega) = U^M(t) F_t^\dagger(\omega)\,, \quad F_t(\omega) := G_t^\dagger(w_1) G_t^\dagger(w_2) \ldots .$$

Here $U^M(t) = \exp\{-\frac{i}{\hbar}\, H^M t\}$, $G_t(k, t_n, \lambda) = I$ for $t_n \geq t$, and $G_t(k, t_n, \lambda) = U^{M\dagger}(t_n) G(k, \lambda) U^M(t_n)$, $t_n < t$, is defined by the reduced scattering operator

$$G(k, \lambda) = I^{\otimes(k-1)} \otimes G(\lambda) \otimes I^{\otimes(M-k)}\,, \quad G(\lambda) = \langle \lambda | S f_0 \tag{3.4}$$

for $f_0 \in L^2(\Lambda)$, $\|f_0\| = 1$, applied to the kth particle only in $\mathcal{H}^M$. The stochastic operator $T(t)$ defines the solutions $\chi^M(t,\omega) = T(t,\omega)\eta^M$ to the Ito differential equation

$$\mathrm{d}\chi^M(t) + \frac{i}{\hbar}H^M\chi^M(t)\mathrm{d}t = \sum_{k=1}^{M} \mathrm{d}n_{k,t}[G(k) - I]\chi^M(t)\,, \quad \chi^M(0) = \eta^M$$

for the stochastic vector states $\chi^M(t) : \Omega \to \mathcal{H}^M$ of the M-particle system, corresponding to an initial $\eta^M \in \mathcal{H}^M$. The right–hand side of this equation is written as the forward increment of the point integral

$$n[L_t] = \sum_{k=1}^{M} \int_0^t \int_\Lambda L(k,r,\lambda)\mathrm{d}n_{k,r}(\mathrm{d}\lambda)$$

given by the stochastic distribution $n[L](\omega) = \sum_{w\in\omega} L(w)$ for $L_t(k,r,\lambda) = 1_t(r)L(k,r,\lambda)$. The vector $\chi^M(t,\omega) \in \mathcal{H}^M$ as well as $\chi(t,\upsilon)$ in (2.10), is no longer normalized ($\|\chi^M(t,\omega)\| \neq 1$) for an initial state-vector $\eta^M \in \mathcal{H}^M$, $\|\eta^M\| = 1$, but it is normalized with respect to the probability measure P_0 on Ω in the mean square sense. But, in contrast to $\chi(t,\upsilon)$, $\chi^M(t,\omega)$ is not yet the reduced description of the M-particle system under the observation of the scattering process $\upsilon_t = \{(r,\lambda) \in \upsilon | r \le t\}$, given by the registration of the pointer positions $\lambda_n \in \Lambda$ at random time instants t_n.

The reduced dynamics corresponding to the observation is described by a stochastic operational process $X \mapsto \Theta[X](t)$,

$$\Theta[X](t,\upsilon) = \Phi_t[U^{M\dagger}(t)XU^M(t)](\upsilon) \tag{3.5}$$

for the M-particle operators $X : \mathcal{H}^M \to \mathcal{H}^M$ given by the conditional expectation

$$\Phi_t[X](\tau,\boldsymbol{\lambda}) = \frac{1}{M^{|\tau_t|}} \sum_{\sqcup\sigma_k=\tau} F_t(\sigma_\bullet,\boldsymbol{\lambda})XF_t^\dagger(\sigma_\bullet,\boldsymbol{\lambda}) \tag{3.6}$$

where the sum is taken over all partitions $\sigma_\bullet = (\sigma_1,\ldots,\sigma_M)$ of a finite subset $\tau_t = \tau \cap [0,t)$. This averaging is due to the impossibility to detect the individuality of the identical particles producing the indistinguishable effects on the bubbles by measuring the scatterings of the bubbles.

To prove Eq. (3.6), we need to compare the correlations of $F_t(\omega)XF_t^\dagger(\omega)$ and of an arbitrary functional $g(\upsilon_t)$ of the observable point process υ_t with the correlations of (3.6) and of $g(\upsilon_t)$. But by applying the well known formula [22] one can easily find

$$\int_{\Gamma_t^M} x(\sigma_1,\ldots,\sigma_M)\prod_{k=1}^{M}\mathrm{d}\sigma_k = \int_{\Gamma_t} \sum_{\sigma_k:\sqcup\sigma_k=\sigma} x(\sigma_1,\ldots,\sigma_M)\mathrm{d}\sigma$$

for the multiple point integration that these correlations with respect to the probability measure P_0 on Ω given by the Poisson law (2.5) simply coincide:

$$\begin{aligned}
&\int_{\Gamma_t^M} \pi_t(\mathrm{d}\sigma_\bullet)\int_{\Lambda^\infty} g(\sqcup_{k=1}^M\sigma_k,\boldsymbol{\lambda})F(\sigma_\bullet,\boldsymbol{\lambda})XF_t^\dagger(\sigma_\bullet,\boldsymbol{\lambda})\mu_0^\infty(\mathrm{d}\boldsymbol{\lambda})\\
&= \int_{\Gamma_t^M} \langle g(\sqcup_{k=1}^M\sigma_k), X(\sigma_1,\ldots,\sigma_M)\rangle_0 \prod_{k=1}^{M} e^{-\nu t}\nu^{|\sigma_k|}\mathrm{d}\sigma_k\\
&= \int_{\Gamma_t} \langle g(\sigma), \sum_{\sigma_k:\sqcup\sigma_k=\sigma} X(\sigma_1,\ldots,\sigma_M)\rangle_0 e^{-M\nu t}\nu^{|\sigma|}\mathrm{d}\sigma\\
&= \int_{\Gamma_t} \pi_t^M(\mathrm{d}\sigma)\int_{\Lambda^\infty} g(\sigma,\boldsymbol{\lambda})\frac{1}{M^{|\sigma|}}\sum_{\tau_k:\sqcup\tau_k=\sigma} F(\omega)XF^\dagger(\omega)\mu_0^\infty(\mathrm{d}\boldsymbol{\lambda})\,.
\end{aligned}$$

Here $\langle .,. \rangle_0$ is the abbreviation for the inner product in $\mathcal{E}$ of the test function $v \mapsto g(\tau, \boldsymbol{\lambda})$ with fixed $\tau \in \Gamma_t$ and the operator function $X_t(\tau_\bullet, \boldsymbol{\lambda}) = F_t(\tau_\bullet, \boldsymbol{\lambda}) X F_t^\dagger(\tau_\bullet, \boldsymbol{\lambda})$ with fixed $\tau_\bullet \in \Gamma_t^M$. The probability measure

$$\pi_0^M(\mathrm{d}\tau_t) = \sum_{\sqcup \sigma_k = \tau_t} \prod_{k=1}^M \pi_0(\mathrm{d}\sigma_k) = e^{-M\nu t} |M\nu|^{|\tau_t|} \mathrm{d}\tau_t \tag{3.7}$$

on Γ_t has the intensity $M\nu$. It is induced by the measure $\pi_0(\mathrm{d}\tau_\bullet)$ on Γ_∞^M with respect to the particle identification map $\tau_\bullet \in \Gamma_\infty^M \mapsto \tau = \sqcup_{k=1} \tau_k$, defining the observable data $v = (\tau, \boldsymbol{\lambda})$ by the stochastic map $\omega = (\tau_\bullet, \boldsymbol{\lambda}) \mapsto v \in \Upsilon_\infty$ on $\omega \in \Omega$.

By the coincidence of the correlations proved above, the stochastic operator (3.5) is indeed the conditional expectation of the stochastic operators $X_t(\omega)$ with respect to the observable process v_t.

In contrast to the pure operations $X \mapsto X_t(\omega)$ the reduction operation $X \mapsto \Phi_t[X](v)$, preserves the symmetry of the M-particle operators X with respect to particle permutation. It is the least mixing operation which preserves the indistinguishability of the particles respectively to the observations of the bubble scatterings, corresponding to the complete nondemolition measurement of the particles.

Indeed, the reduction operation (3.6) can be simply written as the finite iteration

$$\begin{aligned} \frac{1}{M^n} \sum_{k_1,\ldots,k_n=1}^M & G_t^\dagger(k_1, y_1) \ldots G_t^\dagger(k_n, y_n) X G_t(k_n, y_n) \ldots G_t(k_1, y_1) \\ & = \Psi_t[\ldots \Psi_t[X](y_1) \ldots](y_n) = \Phi_t[X](y_1, \ldots, y_n) \end{aligned}$$

with $n = |\tau_t|$ single mixing reductions

$$\Psi[X](y) = \frac{1}{M} \sum_{k=1}^M G^\dagger(k, y) X G(k, y) \,. \tag{3.8}$$

Given as the arithmetic mean value of the permutations for the pure operations $X \mapsto G_t^\dagger(k, y) X G_t(k, y)$, corresponding to the identical operators (3.4), the reductions $X \mapsto \Psi_t[X](y)$ are permutationally symmetric, and are not mixing only if the pure operations if they do not break this symmetry.

The derived mixing property of the reduced stochastic dynamics $t \mapsto \rho[X](t, v)$ derived above for the corresponding statistical states

$$\rho^M[X](t, v) = \langle \eta^M, \Theta[X](t, v) \eta^M \rangle = \mathrm{Tr}\{X \varrho^M(t, v)\} \tag{3.9}$$

gives an increase in the entropy

$$\sigma^M(t, v) = -\mathrm{Tr}\{\varrho^M(t, v) \ln \varrho^M(t, v)\}$$

for an ensemble of identical particles even under the condition of complete nondemolition observation. According to (3.9), the reduced density operators $\varrho^M(t, v)$ for the system of M identical particles gives the probability density

$$\mathrm{p}^M(t, v_t) = \mathrm{Tr}\{\varrho^M(t, v_t)\} = \mathrm{P}^M(\mathrm{d}v_t)/\mathrm{P}_0^M(\mathrm{d}v_t)$$

of the output process v_t. Here $\mathrm{P}_0^M = \pi^M \otimes \mu_0^\infty$ is the probability measure on $\Upsilon_\infty = \Gamma_\infty \times \Lambda^\infty$ defined by the Poisson measure (3.7). This means that the a posteriori density operator $\varrho^M(t)$ is defined as a stochastic positive trace class operator normalized in the mean sense

$$\|\varrho^M(t)\|_1 = \int \mathrm{Tr}\{\varrho^M(t, v)\} \mathrm{P}_0^M(\mathrm{d}v) = 1 \,.$$

The density $\varrho^M(t)$ satisfies the stochastic operator equation

$$\mathrm{d}\varrho^M(t)+\frac{i}{\hbar}[H,\varrho^M(t)]\mathrm{d}t=\mathrm{d}n_t\Big[\frac{1}{M}\sum_{k=1}^{M}G(k)\varrho^M(t)G^\dagger(k)-\varrho^M(t)\Big], \tag{3.10}$$

which has a unique solution for every initial condition $\varrho^M(0,v)=\varrho_0^M$ given by the density operator ϱ_0^M for the M-particle states $\varrho_0^M[X]=\mathrm{Tr}\{X\varrho_0^M\}$.

Let derive the differential equation (3.10) in the equivalent integral form

$$\rho^M[X](t)=\rho^M[X(t)](t)+\int_0^t\int_\Lambda\rho^M[(\Psi[X(t-r)](\lambda))-X(t-r)](r)\mathrm{d}n_t(\mathrm{d}\lambda)$$

for an $\rho_0^M[X]=\langle\eta^M,X\eta^M\rangle$, $\eta^M\in\mathcal{H}^M$, where $X(t)=U^M(t)^\dagger XU^M(t)$,

$$U^M(t)=e^{-iH^Mt/\hbar},\quad \Psi[X](\lambda)=\frac{1}{M}\sum_{k=1}^{M}G^\dagger(k,\lambda)XG(k,\lambda)\,.$$

Taking into account the fact that the stochastic integral (2.6) in this equation is simply a finite sum for every $v\in\Upsilon_\infty$ and t, one can write it as recursive operator equation

$$\Phi_t[X](v)=X+\sum_{(r,y)\in v}^{r<t}\Phi_r[\Psi[X](r,\lambda)-X](v)\,,$$

for a stochastic operation Φ_t, defining the solutions to Eq. (3.10), as in (3.9), in terms of the composition (3.5).

But such recurrence has the unique solution $\Phi_t(v)=\Psi_t(y_1)\circ\Psi_t(y_2)\circ\ldots$, defined for a $v=(\tau,\boldsymbol{\lambda})\in\Upsilon_\infty$ as the chronological composition of the maps $\Psi_t(r,\lambda):X\mapsto\Psi[X](r,\lambda)$ if $r<t$ and $\Psi_t[X](r,\lambda)=X$ if $r\geq t$. This solution can be found by the iterations

$$\begin{aligned}\Phi_t(v)&=\mathrm{I}+\sum_{(r,\lambda)\in v}^{r<t}\Phi_r(v)\circ\Lambda(r,\lambda)\\&=\mathrm{I}+\sum_{(r,\lambda)\in v}^{r<t}\Big(\mathrm{I}+\sum_{(s,\lambda)\in v}^{s<r}\Phi_s(v)\circ\Lambda(s,\lambda)\Big)\circ\Lambda(r,\lambda)\\&=\ldots=\sum_{\sigma\subseteq v_t}\Lambda(z_1)\circ\ldots\circ\Lambda(z_n)\,,\end{aligned}$$

where $\Lambda(y)=\Psi(y)-\mathrm{I}$, $\Phi_s(v)$ is the identical map $\mathrm{I}:X\mapsto X$ if $s=t_1$ and $\sum_{\sigma\subseteq v_t}\Lambda(z_1)\circ\ldots\circ\Lambda(z_n)=\Psi_t(y_1)\circ\Psi_t(y_2)\circ\ldots$ in terms of $\sigma=\{z_1,\ldots,z_n\}$, $z=(s,\lambda)$, $s_1<\ldots<s_n$, $n\leq n_t$.

Let us also write the nonlinear stochastic equation

$$\mathrm{d}\varrho_v^M(t)+\frac{i}{\hbar}[H,\varrho_v^M(t)]\mathrm{d}t=\varrho_v^M(t)\circ(\Psi_v(t)-\mathrm{I})\mathrm{d}n_t(v)\,,$$

where $\Psi_v(t)=\Psi(\lambda_{n_t(v)})/\mathrm{Tr}\{E(\lambda_{n_t(v)})\rho\}$,

$$\varrho\circ\Psi(\lambda)=\frac{1}{M}\sum_{k=1}^{M}G(k,\lambda)\varrho G^\dagger(k,\lambda)\,,\quad E(\lambda)=\frac{1}{M}\sum_{k=1}^{M}G^\dagger(k,\lambda)G(k,\lambda))$$

for the normalized density operator

$$\varrho_v^M(t)=\varrho^M(t,v)/\mathrm{p}^M(t,v)\,.$$

This describes the conditional expectations $\rho_v^M[X](t)=\mathrm{Tr}\{X\varrho_v^M(t)\}$ of the M-particle operators with respect to the output probability measure

$$\mathrm{P}^M(\mathrm{d}v)=\mathrm{p}^M(t,v)\mathrm{P}_0^M(\mathrm{d}v)\,,$$

where $\mathrm{p}^M(t,v)=\mathrm{Tr}\{\varrho^M(t,v)\}$.

4 MACROSCOPIC AND CONTINUOUS REDUCTION LIMITS

We now consider the mean field approximation of the measurement apparatus fixing its total effect $\nu\kappa = -\gamma$ given by the mean number ν of scattered bubbles per second and an interaction constant κ coupling each bubble to a particle in the Hamiltonian (2.1). We look for the limits of the unitary and reduced evolutions (2.2) and (2.10) as $\nu \to \infty$ and $\kappa \to 0$ such that γ is a real constant. To perform these limits we need the expansions

$$\begin{aligned} S(n) &= I \otimes \mathbf{1} - i\frac{\kappa}{\hbar} R \otimes P(n) - \left(\frac{1}{2}\right)\left(\frac{\kappa}{\hbar}\right)^2 (R \otimes P(n))^2 + \dots \\ G(\lambda) &= I - \kappa \frac{f_0'(\lambda)}{f_0(\lambda)} R + \frac{1}{2}\kappa^2 R \frac{f_0''(\lambda)}{f_0(\lambda)} R + \dots \end{aligned} \quad (4.1)$$

of the scattering operator $S(n) = \exp\{-\frac{i}{\hbar}\kappa R \otimes P(n)\}$ and the reduced operator $G(\lambda) = f_0(\lambda I - \kappa R)/f_0(\lambda)$ with respect to the coupling constant κ. The first term of the expansion for $S(n)$ is disappearing in the r.h.s. of Eq.(2.2) while the second and third terms are appearing as the differentials of the operator-valued stochastic integrals

$$\hat{n}_t[P] = \int_0^t P(n_r)\mathrm{d}n_r\,, \quad \hat{n}_t[P^2] = \int_0^t P(n_r)^2 \mathrm{d}n_r\,.$$

The corresponding terms

$$n_t\left[\frac{f_0'}{f_0}\right] = \int_0^t \int_\Lambda \frac{f_0'(\lambda)}{f_0(\lambda)} \mathrm{d}n_r(\mathrm{d}\lambda)\,, \quad n_t\left[\frac{f_0''}{f_0}\right] = \int_0^t \int_\Lambda \frac{f_0''(\lambda)}{f_0(\lambda)} \mathrm{d}n_r(\mathrm{d}\lambda)$$

in the right–hand side. of eq.(2.10) can also be written as the integrals $\hat{n}_t[L] = \int_0^t L(n_r)\mathrm{d}n_r$ with values in operator functions of $\tau \in \Gamma_\infty$

$$\hat{n}_t[L](\tau) = \sum_{n=1}^{n_t(\tau)} L(n)\,, \quad L(n) = \mathbf{1}^{\otimes(n-1)} \otimes L \otimes \mathbf{1}^{\otimes\infty}\,, \quad (4.2)$$

where $L = [l(\lambda)]$ is one of the multiplication operators

$$L' = [f_0'(\lambda)/f_0(\lambda)]\,, \quad L'' = [f_0''(\lambda)/f_0(\lambda)]\,.$$

Here $f_0'(\lambda)$ denotes the derivative $\partial f_0(\lambda) = \partial f_0(\lambda)/\partial q$ and $f''(\lambda)$ denotes $\partial\partial f(\lambda)$. The stochastic integral $\hat{n}_t[L](\tau)$ corresponding to the pointwise multiplication $L : f \mapsto lf$ of $f \in L^2(\Lambda)$ by a function l of the bubble coordinate λ acts in $\mathcal{E}$ as the multiplication operator

$$[\hat{n}_t[L](\tau)\varphi](\boldsymbol{\lambda}) = \sum_{n=1}^{n_t(\tau)} l(\lambda_n)\varphi(\boldsymbol{\lambda}) \equiv n_t[l](\tau, \boldsymbol{\lambda})\varphi(\boldsymbol{\lambda})\,.$$

Hence, the main terms on the right–hand side of Eq.(2.2) and (2.10) for $\kappa \to 0$ are given by the renormalised stochastic integrals

$$\hat{\lambda}(t) = \frac{1}{\nu t}\int_0^t L(n_r)\mathrm{d}n_r = \frac{1}{\nu t}\hat{n}_t[L] \quad (4.3)$$

of the operator-valued stochastic functions $L(t, \tau) = L(n_t(\tau))$ with respect to the numerical process $n_t(\tau)$ that has the Poisson probability distribution (2.5) on Γ_∞.

To pass to the large number limit $\nu \to \infty$ in (4.3) for an arbitrary operator L in $L^2(\Lambda)$, we need to use the quantum stochastic representation [23] of the integral (4.3) in the Fock space $\mathcal{F}$ over $L^2(\mathbf{R}_+ \times \Lambda)$. The space $\mathcal{F}$ can be defined as the $L^2(\Upsilon)$-space of all square integrable functions $\varphi : \Upsilon \to \mathbf{C}$, $\|\varphi\|^2 = \int_\Upsilon |\varphi(\upsilon)|^2 \lambda(\mathrm{d}\upsilon) < \infty$ of time ordered finite sequences $\upsilon = (y_1, \ldots, y_n)$, $y = (t, \lambda)$ identified with subsets $\upsilon \subset \mathbf{R}_+ \times \Lambda$ of cardinality $|\upsilon| = 0, 1, 2, \ldots$. The measure $\lambda(\mathrm{d}\upsilon)$ on the union $\Upsilon = \sum_{n=0}^{\infty} \Upsilon(n)$ of the disjoint subsets $\Upsilon(n) = \{\upsilon \in \Upsilon : |\upsilon| = n\}$ is given as the sum

$$\lambda(A) = \sum_{n=0}^{\infty} \lambda(\Upsilon(n) \cap A)$$

of the product $\lambda(\mathrm{d}\upsilon) = \prod_{y \in \upsilon} \mathrm{d}y$ of measures $\mathrm{d}y = \mathrm{d}t\mathrm{d}\lambda$ on $\mathbf{R}_+ \times \Lambda$ such that

$$\|\varphi\|^2 = \sum_{n=0}^{\infty} \int\!\!\int_{0 \le t_1 < \ldots \le t_n < \infty} |\varphi(y_1, \ldots, y_n)|^2 \prod_{i=1}^{n} \mathrm{d}y_i \,.$$

Let define $N_t[L]$ define the numerical integral in $\mathcal{F}$ by the action of the operator (4.2)in each Fock component $\mathcal{F}(\tau) = L^2(\Lambda^{|\tau|})$ for all $\tau \in \Gamma$:

$$[N_t[L]\varphi](\tau) = \sum_{n=1}^{n_t(\tau)} L(n)\varphi(\tau) = \hat{n}_t[L](\tau)\varphi(\tau) \,. \qquad (4.4)$$

Here $\varphi(\tau)$ is the function $\varphi(\tau, \boldsymbol{\lambda}) = \varphi(\upsilon)$ of $\boldsymbol{\lambda} \in \Lambda^{|\tau)}$, corresponding to a $\varphi \in L^2(\Upsilon)$ with a fixed time component of $\upsilon = (\tau, \boldsymbol{\lambda}) \in \Upsilon$.

In order to obtain the initial probability measure $\mathrm{P}_0(\mathrm{d}\upsilon) = \pi(\mathrm{d}\tau)\mu_0^\infty(\mathrm{d}\boldsymbol{\lambda})$ on Υ_∞ induced by an initial Fock vector $\varphi_0 \in L^2(\Upsilon)$, we need an isomorphic transformation of (4.4)

$$\hat{N}_t[L] = N_t[L] + \sqrt{\nu}(A_t[f_0^\dagger L] + A_t^\dagger[L f_0]) + \nu t f_0^\dagger L f_0 \qquad (4.5)$$

which can be locally performed by a unitary transformation

$$\hat{N}_t[L] = U_s^\dagger N_t[L] U_s \,, \quad U_s = \exp\{\sqrt{\nu}(A_s^\dagger[f_0] - A_s[f_0^\dagger])\}$$

for every $t < s$. Here $A_t^\dagger[f]$ and $A_t[f^\dagger]$ are the creation and annihilation integrals of $f \in L^2(\Lambda)$, $f^\dagger \in L^2(\Lambda)^*$, given by the operators

$$\begin{aligned} \left[A_t^\dagger[f]\varphi\right](\upsilon) &= \sum_{y \in \upsilon_t} f(\lambda)\varphi(\upsilon \backslash y) \\ \left[A_t[f^\dagger]\varphi\right](\upsilon) &= \int_{([0,t) \times \Lambda} f(\lambda)^* \varphi(\upsilon \sqcup y)\mathrm{d}y \end{aligned}$$

in the Fock space $L^2(\Upsilon)$, where $\upsilon \backslash y$ means the sequence $\upsilon \in \Upsilon$ with deleted $y = (r, \lambda)$, $r < t$, and $\upsilon \sqcup y$ means the sequence $\upsilon \in \Upsilon$ with an additional element $y \notin \upsilon$. The characteristic functional of the stochastic operators $\hat{n}_t[L]$ with respect to the initial state-vector $f_0^\infty \in \mathcal{E}$ and the Poisson probability measure (2.5) is now given simply by the vacuum expectation

$$\int_{\Gamma^\infty} (f^\infty, e^{i\hat{n}_t[L]} f^\infty)\pi_0(\mathrm{d}\tau) = \langle \delta_\phi, e^{i\hat{N}_t[L]}\delta_\phi \rangle \,,$$

where $\delta_\phi(\upsilon) = 1$ if $\upsilon = \emptyset$; otherwise, $\delta_\phi(\upsilon) = 0$.

The corresponding representation $\hat{l}(t) = \frac{1}{\nu t} \hat{N}_t[L]$ for (4.3) helps us immediately obtain the quantum large number limit

$$\lim_{\nu \to \infty} \frac{1}{\nu t} \hat{N}_t[L] = f_0^\dagger L f_0 \hat{1}$$

as the mean value $l_0 = (f_0, Lf_0) \equiv f_0^\dagger L f_0$ of a single-bubble operator with respect to an initial wave packet $f_0 \in L^2(\Lambda)$. This gives the following macroscopic limit

$$\mathrm{d}\psi(t) + \frac{i}{\hbar} H_0\psi(t)\mathrm{d}t = \frac{i}{\hbar}\gamma(R \otimes p_0\hat{1})\psi(t)\mathrm{d}t$$

of the generalized Schrödinger equation (2.2) which turns out to be a nonsingular one with an additional potential $-\gamma p_0 R$ corresponding to the mean momentum $p_0 = (f_0, Pf_0)$ of a bubble in the initial state f_0. As one could expect, the mean field dynamics preserves the product structure $\psi(t, v) = \eta(t)\varphi_0(v)$ of an initial product-vector $\psi_0 = \eta \otimes \varphi_0$ being trivial on the Fock component $\varphi_0 \in \mathcal{F}$ because of $\mathcal{H}_0 = H \otimes \hat{1}$. But unexpectedly (compare with [20]) the macroscopic limit

$$\mathrm{d}\chi(t) + \frac{i}{\hbar} H\chi(t)\mathrm{d}t = \frac{i}{\hbar}\gamma p_0 R\chi(t)\mathrm{d}t \tag{4.6}$$

of the reduction equation (2.10) corresponds to the same unitary dynamics $\eta(t) = U(t)\eta = \chi(t)$ of the particle state-vector if $\chi(0) = \eta$, because of

$$\frac{1}{\nu t} n_t\left[\frac{f_0'}{f_0}\right] \to (f_0, L'f_0) = \int f_0(\lambda)^* f_0'(\lambda)\mathrm{d}\lambda = \frac{i}{\hbar} p_0 \,.$$

The macroscopic limits for the M-particle system also give essentially the same continuous unitary evolutions in the large space $\mathcal{H}^M \otimes \mathcal{F}$ and in the reduced space $\mathcal{H}^M$. To get this correspondence, one has only to replace the measure (2.5) on Γ_∞ for Eq.(3.2) by the product measure $\pi_0^{\otimes M}$ on Γ_∞^M and for Eq.(3.10) by the induced measure (3.7) on Γ_∞, taking $M\nu$ instead of ν in (4.5). This means that the mixing property of the reduced equation (3.10) vanishes in the mean field approximation for the bubble system.

Let us now pay attention to the fluctuations with respect to the obtained large number limits. Such fluctuations might appear for $-\kappa = \gamma/\nu \to 0$ in the large time scale $t \sim 1/\kappa$. We can get these fluctuations without rescaling the time t if we assume that $p_0 = 0$ and $-\kappa = \gamma/\sqrt{\nu}$, so that we have to take into account also the κ^2-terms in (4.1).

It follows from the Fock space representation (4.5) that the quantum central limit

$$\lim_{\nu\to\infty} \frac{l}{\sqrt{\nu}} \hat{N}_t[L] = A_t[f_0^\dagger L] + A_t^\dagger[Lf_0]$$

exists for any single-bubble operator L with zero mean value $(f_0, Lf_0) = 0$. We first apply this central limit theorem to the right–hand side in (2.2):

$$\mathrm{d}\psi(t) + K_0\psi(t)\mathrm{d}t = \frac{i}{\hbar}\gamma(R \otimes \mathrm{d}\hat{u}_t)\psi(t) \,. \tag{4.7}$$

Here $K_0 = K \otimes \hat{1}$, $K = \frac{i}{\hbar} H + \frac{1}{2}\left(\frac{\gamma}{\hbar}\right)^2 R\sigma^2 R$,

$$\hat{u}_t = A_t[f_0^\dagger P] + A_t^\dagger[Pf_0] = 2\Re A_t^\dagger[Pf_0]$$

is the Fock space representation of the Wiener process u_t with the dispersion $\sigma^2 = \hbar^2(\partial f_0, \partial f_0)$, defined by the quantum stochastic multiplication formula [22, 23]

$$\mathrm{d}\hat{u}\mathrm{d}\hat{u} = \mathrm{d}A_t[f_0^\dagger P]\mathrm{d}A_t^\dagger[Pf_0] = f_0^\dagger PPf_0\mathrm{d}t \,.$$

The central limit equation (4.7) for the unitary evolution of the coupled system turns out to be a stochastic Schrödinger-Ito equation of diffusive type driven by the Wiener process u_t. The same conclusion obviously holds for the M-particle system driven by M independent Wiener processes $\hat{u}_t(k)$ identical to $\hat{u}_t$, $k = 1, \ldots, M$:

$$\mathrm{d}\psi^M(t) + K_0^M \psi(t)\mathrm{d}t = \frac{i}{\hbar}\gamma\Big(\sum_{k=1}^{M} R(k) \otimes \mathrm{d}\hat{u}_t(k)\Big)\psi^M(t)\,,$$

where $K^M = \frac{i}{\hbar} H^M + \frac{1}{2}\left(\frac{\gamma}{\hbar}\right)^2 \sum_{k=1}^M R(k)\sigma^2 R(k)$, $K_0^M = K^M \otimes \hat{1}$.

The application of the central limit theorem to the r.h.s. of the reduction equation (2.10) yields an essentially different type of the stochastic evolution,

$$\mathrm{d}\chi(t) + K\chi(t)\mathrm{d}t = \gamma R \chi(t)\mathrm{d}\hat{v}_t\,, \tag{4.8}$$

originally derived in [22] by quantum calculus method. Here v_t is a complex Wiener vector-process, with the Fock-space representation

$$\hat{v}_t = A_t[f_0^\dagger L'] + A_t^\dagger[L' f_0] = \Re A_t^\dagger[(w_0 + \bar{w}_0) f_0] + i\Im A_t^\dagger[(w_0 - \bar{w}_0) f_0] \tag{4.9}$$

given by the complex osmotic velocity $w_0(\lambda) = \partial \ln f_0(\lambda)$ of a single bubble, and the operator K is essentially the same as in (4.7). The linear stochastic equation (4.8) has a unique solution $\chi(t, v) = T(t, v)\eta$ for a given $\chi(0, v) = \eta \in \mathcal{H}$ which is not normalized $\|\chi(t,v)\| \neq 0$ for every Wiener trajectory $t \mapsto v_t$ but is normalized in the mean square sense $\int \|\chi(t,v)\|^2 \mathrm{P}_0(\mathrm{d}v) = 1$ with respect to the Gaussian probability measure P_0 of $v = \{v_t | t > 0\}$. The measure P_0 is defined by the zero mean values of v_t and by the table

$$\begin{aligned}
\mathrm{d}\hat{v}\mathrm{d}\hat{v} &= \mathrm{d}A_t[f_0^\dagger L']\mathrm{d}A_t^\dagger[L' f_0] = f_0^\dagger L' L' f_0 \mathrm{d}t \\
\mathrm{d}\hat{v}^*\mathrm{d}\hat{v} &= \mathrm{d}A_t[f_0^\dagger L'^\dagger]\mathrm{d}A_t^\dagger[L' f_0] = f_0^\dagger \hat{L}'^\dagger L' f_0 \mathrm{d}t
\end{aligned}$$

of commuting multiplication

$$\mathrm{d}\hat{v}\mathrm{d}\hat{v}^* = \mathrm{d}\hat{v}^*\mathrm{d}\hat{v}\,.$$

But the reduction noise $\hat{v}_t$ obtained in the Fock space representation does not commute with the real Wiener process $\hat{u}_t = \hat{u}_t^*$ in (4.7),

$$\begin{aligned}
\mathrm{d}\hat{v}\mathrm{d}\hat{u} &= \mathrm{d}A_t[f_0^\dagger]\mathrm{d}A_t^\dagger[P f_0] = f_0^\dagger L' P f_0 \mathrm{d}t \\
\mathrm{d}\hat{u}\mathrm{d}\hat{v} &= \mathrm{d}A_t[f_0^\dagger]\mathrm{d}A_t^\dagger[L' f_0] = f_0^\dagger P L' f_0 \mathrm{d}t
\end{aligned}$$

if $g(\lambda) \equiv \partial^2 \ln f_0(\lambda) \neq 0$, since

$$[\mathrm{d}\hat{v}, \mathrm{d}\hat{u}] = f_0^\dagger [L', P] f_0 \mathrm{d}t = \frac{\hbar}{i}(f_0, g f_0)\mathrm{d}t\,.$$

In the same way one can obtain the continuous reduction equation

$$\begin{aligned}
&\mathrm{d}\varrho^M(t) + (K\varrho^M(t) + \varrho^M(t)K^\dagger)\mathrm{d}t = \\
&\quad \left(\frac{\gamma}{\hbar}\right)^2 \sum_{k=1}^{M} R(k)\sigma^2 \varrho^M(t) R(k)\mathrm{d}t + \gamma(\mathrm{d}w_t R\varrho^M(t) + \varrho^M(t) R \mathrm{d}w_t^*)
\end{aligned} \tag{4.10}$$

where $R = \frac{1}{M}\sum_{k=1}^M R(k)$ and the operator $K = K^M$ is the same as in the equation for the unitary evolution of the M-particle system coupled with the bubbles. The derived stochastic equation for the M-particle density operator $\varrho^M(t, w)$ normalized in the mean is driven by the complex Wiener process $w_t = v_t^M$ having the same multiplication table as the process $\sqrt{M}v_t$ in (4.8). The diffusive type equation (4.10), (3.10) also has the mixing property. It was also derived in [6] by operational method.

CONCLUSION

The projection postulate for the reduction process follows as a result of conditioning by the results of a nondemolition measurement [1] in the Hamiltonian model for the particle-meter interaction. This model gives also all unsharp reductions for the measurements with continuous spectrum.

The singular interaction, corresponding to the collision model of scatterings can be treated in quantum mechanics also in terms of Schrödinger equation in the generalized sense. The spontaneous localization [20, 21] of the particle under the continual observation is explained as the result of the nondemolition countings of the scattered bubbles in a measurement apparatus like cloud chamber.

The spontaneous localization of a system of the identical particles in a bubble chamber is mixing due to the indistinguishability of the particles via the measurements of the bubbles. It is described in terms of the filtering equation for the density matrix of the particles, driven by the Poisson process of the total scatterings of the bubbles. This equation was derived in [6] within the operational approach to the quantum continual measurements.

The macroscopic limit of the spontaneous localization equation is described in terms of the Schrödinger one, corresponding to the mean field approximation. The central limit of the spontaneous localization equation proves the diffusive reduction equation, which appeared in [18, 19] for some cases of R and H. Our theory shows the origin of the Wiener processes generating the Ito diffusion equations for quantum states [24, 25], and also derives a diffusive mixing equation for the reduced density matrix of the system of particles under the continual observation.

The quantum jump model, based on Poisson distributed impulsive measurements, is not new, see for example [26, 27]. On the statistical (a priori) level it is described by the master equation which also results from the quantum diffusion models. However the treatment of these processes on the individual (a posteriori) level gains different types of quantum dynamics which can be described only in terms of the stochastic differential equations. The first such treatment of the counting measurements and the a posteriori dynamics was suggested in [14, 15] in terms of quantum stochastic calculus method [23]. The present work puts this method on the solid mathematical basis by treatment of δ-function couplings with the classical Ito calculus applied to the quantum jumps and spontaneous localizations.

References

[1] Belavkin V. P. Nondemolotion principle of quantum measurement theory. *Foundations of Physics*, 24(5):685–714, 1994.

[2] G. Ludwig. *Commun.Math.Phys.*, 9:1, 1968.

[3] E.B. Davies and J. Lewis. *Commun.Math.Phys.*, 17:239–260, 1970.

[4] A. Barchielli, L. Lanz, and G.M. Prosperi. *Nuovo Cimento*, 72B(79), 1982.

[5] Belavkin V. P. Theory of control of observable quantum systems. *Automatica and Remote Control*, 44(2):178–188, 1983.

[6] Belavkin V.P. and A Barchielli. Measurements continuous in time and posteriori states in quantum mechanics. *J. Phys. A. Math. Gen.*, 24:1495–1514, 1991.

[7] J. von Neumann. *Mathematical Foundations of Quantum Mechanics.* Princeton University Press, 1955.

[8] A. Peres. *Am.J.Phys.*, 52:644, 1984.

[9] L.E. Balentine. *Int.J.Theor.Phys.*, 27:211–218, 1987.

[10] Belavkin V.P. A new wave equation for a continuous non-demolition measurement. *Phys Letters A*, 140(78):355–358, 1989.

[11] Belavkin V.P. and P Staszewski. A quantum particle undergoing continuous observation. *Phys Letters A*, (140):359–362, 1989.

[12] Belavkin V.P. A posterior Schrödinger equation for continuous non- demolition measurement. *J of Math Phys*, 31(12):2930–2934, 1990.

[13] Belavkin V.P. and Staszewski P. Nondemolition observation of a free quantum particle. *Phys. Rev. A.*, 45(3):1347–1356, 1992.

[14] Belavkin V.P. A continuous counting observation and posterior quantum dynamics. *J Phys A Math Gen*, (22):L 1109–L 1114, 1989.

[15] Belavkin V.P. A stochastic posterior Schrödinger equation for counting nondemolition measurement. *Letters in Math Phys*, (20):85–89, 1990.

[16] Belavkin V.P. and Staszewski P. A continuous observation of photon emission. *Rep. on Math. Phys.*, 29(2):213–225, 1991.

[17] P. Pearle. *Phys.Rev.*, D29:235, 1984.

[18] N. Gisin. *J.Phys.A.*, 19:205–210, 1986. Math.Gen.

[19] L. Diósi. *Phys.Rev.*, A40:1165–1174, 1988.

[20] Ghirardi G.C., A. Rimini, and T. Weber. *Phys. Rev.*, 34D:470, 1986.

[21] Ghirardi G.C., Pearle P., and Rimini A. Markov processes in Hilbert space and continuous spontaneous localization of systems of indentical particles. *Phys. Rev. A.*, 42:78–89, 1990.

[22] Belavkin V. P. Chaotic states and stochastic integrations in quantum systems. *Usp. Math Nauk (Russian Math Surveys)*, (1), 1992.

[23] Hudson R.L. and Parthasarathy K.R. Quantum Ito's formula and stochastic evolution. *Comm. Math. Phys.*, 93:301–323, 1984.

[24] Gisin N. and Percival I.C. Quantum state diffusion. *J.Phys. A*, 25:5677, 1992.

[25] Garraway B.M. and Knight P.L. Phase space picture of quantum measurement; localization and jumps. *Phys. Rev. Lett.*, Submitted, 1993.

[26] Milburn G. *Phys. Rev. A*, 36:744, 1987.

[27] Milburn G. and Gagen. *Phys Rev. A*, 46:1578, 1992.

EVENT-ENHANCED FORMALISM OF QUANTUM THEORY OR COLUMBUS SOLUTION TO THE QUANTUM MEASUREMENT PROBLEM

Ph. Blanchard[1] and A. Jadczyk[2]

[1]Faculty of Physics and BiBoS
University of Bielefeld Universitätstr. 25
D-33615 Bielefeld
[2]Inst. of Theor. Physics
University of Wrocław Pl. Maxa Borna 9
PL-50 204 Wrocław

1 INTRODUCTION

Quantum Mechanics has proved to be tremendously powerful, practical, and successful in the description of the micro–world of elementary particles, atoms and molecules. There seems to be no limit to the versality of the Schrödinger equation and to the power of Quantum Theory as an incredibly accurate computational tool for the physicist, chemist, and biologist. The progress made in the last 70 years has really been a matter of sharpening the quantum mechanical mathematical formalism rather than of our understanding of it. As Quantum Mechanics amassed success after success only a few physicists remained fascinated by the fundamental problems that remained unsolved. The proposed solutions to the quantum measurement problem by e.g. von Neumann and Wigner – are no solution at all. They merely shift the focus from one unsolved problem to another. On the other hand the predictions for the outcomes of measurements performed on statistical ensembles of physical systems are excellent. What is however completely missing in the standard interpretation is an explanation of experimental facts i.e. a description of the actual individual time series of events of the experiment. That an enhancement of Quantum Theory allowing the description of single systems is necessary is nowadays clear. Indeed advances in technology make fundamental experiments on quantum

Quantum Communications and Measurement
Edited by V. P. Belavkin *et al.*, Plenum Press, New York, 1995

systems possible. These experiments give us series of events for which there are definitely no place in the original, standard version of quantum mechanics, since each event is classical, discrete and irreversible. In recent papers [1–8] we provided a definite meaning to the concepts of experiment and event in the framework of mathematically consistent models describing the information transfer between classical event–space and quantum systems. We emphasize that for us the adjective 'classical' has to be understood in the following sense: to each particular experimental situation corresponds a class of classical events revealing us the Heisenberg transition from the possible to the actual and these events obey the rules of classical logic of Aristotle and Boole. The World of the Potential is governed by quantum logic and has to account for the World of Actual, whose logic is classical. We accept both and we try to see what we gain this way. It appears that working with so enhanced formalism of quantum theory we gain a lot. [1] We proposed mathematical and physical rules to describe

- the two kinds of evolution of quantum systems namely continuous and stochastic
- the flow of information from quantum systems to the classical event–space
- the control of quantum states and processes by classical parameters.

In our event-enhanced formalism the quantum system Σ_q is coupled to a classical space Σ_c – where events do happen – and a measurement is nothing else but a particular coupling between Σ_q and Σ_c; in a simplest case – via a completely positive semigroup $\alpha_t = e^{tL}$ in such a way that information can be transferred from Σ_q to Σ_c. We consider the total system $\Sigma_{tot} = \Sigma_q \times \Sigma_c$ and the behaviour associated to the total algebra of observables

$$\mathcal{A}_{tot} = \mathcal{A}_q \otimes \mathcal{A}_c = \mathcal{L}(\mathcal{H}_q) \otimes C(X_c)$$

where X_c is the classical event–space and $\mathcal{H}_q$ the Hilbert space associated to the quantum system. The "classical" pure state corresponds to a point in X_c and the coordinates of this point corresponds exactly to the properties of Σ_c. There is no such correspondence in $\mathcal{H}_q$. The quantum state is a unique sort of entity. Time evolution of ensembles of coupled system, prepared by the same algorithm, is described by a Liouville equation in the Hilbert space $\mathcal{H}_t$ of the total system with $\mathcal{H}_t = \mathcal{H}_q \otimes L^2(X_c)$. Individual quantum systems and the classical degrees of freedom are described by pairs of pure states: a pure state of the classical system Σ_c and a pure state of the quantum system Σ_q. Time evolution of this pair is derived from the Liouville equation. It is a piecewise deterministic random Markov process. At random times distributed according to a specific inhomogeneous Poisson process jumps occur. There are jumps of the quantum state vectors and also at the same time jumps of the states of Σ_c. These second jumps we can "see" (to measure a quantity we must "look at it") and these classical events can be recorded if necessary. Sometimes these can be jumps of a "pointer" discrete positions, sometimes jumps in the pointer's velocity. Knowing this PD-process one can answer many (perhaps even all) kinds of questions about time correlations of the events as well as simulate numerically the possible histories of individual systems. What we achieved in this

[1] Referring in the title to Christopher Columbus seems to us appropriate for two reasons. First he is known mainly as the discoverer of the New World despite the fact that he was certainly not the first European to land on the coast of the Americas. But his achievement is distinguished from the earlier adventures by its consequences. Second, according to legend, in addition to being a famous navigator, Columbus also knew a thing or two about the phenomenological solution to an otherwise unsolvable problem – namely that of balancing an egg on its end on a flat table. His solution, however, albeit simple and effective, was hardly acceptable to anybody else.

way is the maximum of what can be achieved without introducing hidden variables, which is more than original orthodox interpretation gives. That is why we call our approach the "Event–Enhanced Quantum Theory". Within this framework there need not be cat paradoxes anymore – cats are allowed to behave as cats; we cannot predict individual events as they are random, but we can simulate the observations of individual systems. Our formalism is therefore closely linked to practical matters and is of relevance to today's experimental Quantum Physics.

In Section 2 we will briefly describe the mathematical and physical ingredients of our class of phenomenological models. The range of applications is rather wide as will be shown in Section 3 and 4 with a discussion of Bohm's version of the EPR experiment and the discussion of a general cloud chamber model. In a particular case this model explains the extension of Quantum Theory formulated by Ghirardi, Rimini and Weber in 1986. The GRW theory is based on a process of amplification called spontaneous localization [9]. Born's interpretation of the wave function can be derived and has not to be postulated. EPR connection between a pair of distant quantum entities that have previously interacted in some way seems to involve a strange communication called by Einstein a "spooky action at a distance". We discuss this problem briefly in Section 3. Section 5 deals with some other applications and concluding remarks.

2 COUPLING QUANTUM SYSTEM TO THE CLASSICAL EVENT–SPACE

For a long time the theory of measurements in Quantum Mechanics, elaborated by Bohr, Heisenberg and von Neumann in the 1930-s has been considered as an esoteric subject of little relevance for real physics. But in the 1980-s the technology has made possible to transform "Gedankenexperimente" of the 1930-s into real experiments. This progress implies that the measurement process in Quantum Theory is now a central tool for physicists testing experimentally by high-sensitivity measuring devices the deeper aspects of Quantum Theory. Quantum mechanical measurement brings together a macroscopic and a quantum system.

Let us briefly describe the mathematical framework we will use. A good deal more can be said and we refer the reader to [1–6]. Our aim is to describe a simple model of a non-trivial transfer of information between a quantum system Σ_q coupled to a classical event–space Σ_c. To concentrate on main ideas rather than on technical details let us describe first a simple situation, namely that of a coupling corresponding to a measurement of a discrete quantum observable. In this case the classical event–space can be finite. As we shall see later in the applications it is possible to handle continuous and infinite dimensional generalization of this framework. To the quantum system there corresponds a Hilbert space $\mathcal{H}_q$. In $\mathcal{H}_q$ we consider a family of orthonormal projectors $e_i = e_i^* = e_i^2$, $(i = 1, ..., n)$, $\sum_{i=1}^n e_i = 1$, associated to an observable $A = \sum_{i=1}^n \lambda_i e_i$ of the quantum mechanical system. The space of classical events is supposed to have m distinct pure states, and it is convenient to take $m \geq n$, otherwise some information about the quantum system can be lost. The algebra $\mathcal{A}_c$ of classical observables is in this case nothing else than $\mathcal{A}_c = \mathbf{C}^m$. The set of classical statistical states coincides with the space of probability measures on X_c. Using the notation $X_c = \{s_0, ..., s_{m-1})$, a classical state is therefore an m-tuple $p = (p_0, ..., p_{m-1}), p_\alpha \geq 0$, $\sum_{\alpha=0}^{m-1} p_\alpha = 1$. The state s_0 plays in some cases a distinguished role and can be viewed as the neutral initial state of a pointer. The algebra of observables of the total system $\mathcal{A}_{tot}$ is given by $\mathcal{A}_{tot} = \mathcal{A}_c \otimes L(\mathcal{H}_q) = \mathbf{C}^m \otimes L(\mathcal{H}_q) = \bigoplus_{\alpha=0}^{m-1} L(\mathcal{H}_q)$, and it is convenient to realize $\mathcal{A}_{tot}$ as an algebra of operators on an auxiliary Hilbert space $\mathcal{H}_{tot} = \mathcal{H}_q \otimes \mathbf{C}^m = \bigoplus_{\alpha=0}^{m-1} \mathcal{H}_q$. $\mathcal{A}_{tot}$ is then isomorphic to the algebra of block diagonal $m \times m$ matrices $A = \mathrm{diag}(a_0, a_1, ..., a_{m-1})$ with $a_\alpha \in L(\mathcal{H}_q)$. States on $\mathcal{A}_{tot}$ are represented by block diagonal matrices $\rho = \mathrm{diag}(\rho_0, \rho_1, ..., \rho_{m-1})$, where the ρ_α are positive trace class operators in $L(\mathcal{H}_q)$ satisfying moreover Σ_α Tr $(\rho_\alpha) = 1$. By taking partial traces each state ρ projects onto an effective quantum state $\pi_q(\rho)$ and an effective classical state $\pi_c(\rho)$ given respectively by $\hat{\rho} \equiv \pi_q(\rho) = \sum_\alpha \rho_\alpha$, $\pi_c(\rho) = (\mathrm{Tr}\ \rho_0, \mathrm{Tr}\ \rho_1, ..., \mathrm{Tr}\ \rho_{m-1})$. Let us consider dynamics. A nontrivial coupling between both systems is impossible without a dissipative term. The time evolution of the total system is given (in the simplest case that we consider) by a semigroup $\alpha^t = e^{tL}$ of completely positive maps of $\mathcal{A}_{tot}$ – preserving hermiticity, identity and positivity – with L of the Lindblad form

$$L(A) = i[H, A] + \sum_{i=1}^{n} (V_i^* A V_i - \frac{1}{2}\{V_i^* V_i, A\}). \tag{1}$$

There is a simple method of constructing appropriate couplings. In order to couple Σ_q to Σ_c in such a way that the coupling will correspond to measurement of the given quantum observable $A = \sum_{i=1}^{n} \lambda_i e_i$, the V_i are chosen as tensor products $V_i = \sqrt{\lambda} e_i \otimes \phi_i$, where ϕ_i act as transformations on classical (pure) states. Denoting $\rho(t) = \alpha_t(\rho(0))$, the time evolution of the states is given by the dual Liouville equation

$$\dot{\rho}(t) = -i[H, \rho(t)] + \sum_{i=1}^{n} (V_i \rho(t) V_i^* - \frac{1}{2}\{V_i^* V_i, \rho(t)\}), \tag{2}$$

where in general H and the V_i may depend explicitly on time (in fact, H can also carry an index: $H \to H_\alpha$).

In [1] we propose a simple, purely dissipative Liouville operator (i.e. we put $H = 0$) that describes an interaction of $\sum_q$ and $\sum_c$, for which $m = n + 1$ and $V_i = e_i \otimes \phi_i$, where ϕ_i is the flip transformation of X_c transposing the neutral state s_0 with s_i. We show that the Liouville equation can be solved explicitly for any initial state $\rho(0)$ of the total system. The quantum probabilities are after switching on of the interaction, mirrored by the state of the classical system. Moreover we show that the effective quantum state $\hat{\rho}(t) = \pi_q(\rho(t))$ of the quantum subsystem tends for $t \to +\infty$ to a limit which coincides with the standard von Neumann-Lüders quantum measurement postulate. The model can be easily generalized (cf. Refs. [4–8]) to include measurements of fuzzy or noncommuting observables (in fact, in the cloud chamber model, discussed in Section 4, we are measuring fuzzy position of a quantum particle).

The total system $\Sigma_t = \Sigma_q \otimes \Sigma_c$ is open. Thus one can try to understand its dynamical behaviour as an effective evolution of a subsystem of unitarily evolving larger quantum systems. Although mathematically possible and studied by many authors (for a recent discussion related to quantum measurements cf. Ref. [10]) – such an enlargement is non unique and our understanding of the problem of "minimal" extensions in the case of mixed, classical+quantum, systems is still rather poor. Moreover, it is not clear at all what practical gain can be achieved this way. Therefore we prefer to extend the prevailing paradigm and learn as much as possible how to deal directly with open systems.

A complete general theory of dissipative couplings of quantum systems to classical ones does not yet exist. The best we can do is to study a lot of examples. In the following sections two characteristic situations will be presented. See Refs. [2–6] for more examples. For every example we have considered, a piecewise deterministic random process has been constructed that takes place on the state of pure states of the total system and which reproduces the Liouville equation of the total system by averaging. A theory of piecewise deterministic processes (PDP) is described in a recent book by M.H. Davis [11]. Examples of processes of this type but without a non-trivial evolution on the classical space were also discussed in a physical context [12–15]. Piecewise deterministic Markov processes enjoy deterministic dynamics punctuated by random jumps generated by a Markov stochastic structure. The model introduced by Davis contains virtually all nondiffusion models of applied probability. A PDP is determined by its local characteristics:

i) A vector field X which determines a flow Φ on the state space S

$$\frac{d}{dt}f(\Phi(t,x)) = Xf(\Phi(t,x)), \qquad \Phi(0,x) = x,$$

ii) A jump rate function λ,

iii) A transition probability matrix $\mathcal{Q}$.

From these characteristics a right-continuous sample path x_t of the process $\{x_t\}_{t\geq 0}$ starting at x may be constructed as follows. Define $x_t = \Phi(t,x)$ for $0 \leq t < t_1^*$, where t_1^* is the realization of the first jump time t_1 with generalized exponential distribution determined by $P_x[t_1 > t] = \exp\left(-\int_0^t \lambda(\Phi(s,x))ds\right)$. We may now restart the process at $x_{t_1^*}$ according to the same recipe and proceeding recursively we obtain a sequence of jump times $t_1^*, t_2^*, \ldots$ between which X_t follows the integral curve of X. The space S itself is a disjoint union of smooth manifolds S_ι, and jumps happen between different S_ι (the vector field is also parametrized by the index ι). Let us denote now by α the parameter characterizing the point in the classical event–space (it corresponds to ι above). Each observable A of the total system defines now a function $f_A(\psi,\alpha)$ on the space $S = \{(\psi,\alpha)\}$ of pure states of the total system $\Sigma_t = \Sigma_q \otimes \Sigma_c$

$$f_A(\psi,\alpha) = <\psi, A_\alpha \psi> \ .$$

In [3–8] we showed that the time evolution for observables can be written in a Davis form

$$\begin{aligned}\frac{d}{dt}f_A(\psi,\alpha) &= [(X_H + X_D)f](\psi,\alpha) \\ &\quad + \lambda(\psi,\alpha)\sum_\beta \int Q(\psi,\alpha;d\phi,\beta)[f_A(\phi,\beta) - f_A(\psi,\alpha)]\end{aligned}$$

where Q is the transition probability matrix of the PDP. In Section 3 and 4 we will describe processes associated to semigroups of the above type.

3 BOHM'S VERSION OF THE EPR EXPERIMENT

The anti-realism of the Copenhagen interpretation of Quantum Mechanics was met head – on by the very nice thought experiment proposed 1935 by Einstein,

Podolsky and Rosen [16], today commonly called EPR. The argument proceeds by characterizing and using the key notions of Completeness, Reality and Locality. The most perspicuous Bohm version of the EPR argument considers a system which decays into a pair of particles, which travel in opposite directions along the x axis. Ignoring all but spin, each particle, call them L and R for left and right, is associated with its own two dimensional Hilbert space $\mathbf{C}^2$. The spin of the system is zero to start with and this is supposed to be conserved. Thus if L has spin $+1$ in the z direction then R must have spin -1 in the same direction. In the singlet state the system is represented by the state $\psi_{LR} = (|+L> \otimes |-R> + |-L> \otimes |+R>)/\sqrt{2}$ where $|+L>$ and $|-L>$ are the spin eigenstates for the L particle and $|+R>$ $|-R>$ the eigenstates for the R particle. If we measure the spin of the L particle, and if we know that the total spin is conserved, we then know the state of R. While it might be concerned that the spin measurement on the L particle may have disturbed it the same cannot be said of the R particle which should be unaffected by the measurement. In other words we are in a position to predict with probability one the state of the R particle and since we could not have influenced it (Locality) it follows that the spin of the R particle exists independently of measurement (Reality), which implies that Quantum Mechanics, not being able to predict the result with certainty, does not completely describe the whole of reality (Completeness).

Let us describe an EPR-type set up. The quantum Hilbert space $\mathcal{H}_q$ is $\mathbf{C}^2 \otimes \mathbf{C}^2$ (we have two particles). For the classical Hilbert space that accomodates events we choose $\mathbf{C}^3 \times \mathbf{C}^3$ (we have detectors on the left and right).

Let us first remark that the statistics (Bose, Fermi,...) of the two particles that originate from a common source does not play a role here, since in the experimental situation EPR considered the two particles fly apart to the left and right ends of the laboratory. We define now four properties to be measured
E_1: Is spin up in the z-direction at the left end?, $E_2 = E_1^{\perp} = \mathbf{1} - E_1$

E_1': Is spin up in the n-direction at the right end?, $E_2' = E_1'^{\perp} = \mathbf{1} - E_1'$
It is clear that $[E_i, E_j'] = 0, \quad i,j = 1,2$.
In the Hilbert space $\mathcal{H}_t = \mathbf{C}^2 \otimes \mathbf{C}^2 \otimes \mathbf{C}^3 \otimes \mathbf{C}^3$ of the total system we introduce now four operators $V_i, V_j' \quad i,j = 1,2$ defining the Lindblad generator

$$V_i = \sqrt{\lambda}\; E_i \otimes A_i \otimes \mathbf{1}, \qquad V_i' = \sqrt{\lambda'}\; E_i' \otimes \mathbf{1} \otimes A_i,$$

where $(A_i)_{\alpha\beta} = \delta_{i\,\alpha}\delta_{0\,\beta}$, for $i = 1,2\,, \alpha, \beta = 0,1,2$. Assuming quantum Hamiltonian H, it is now easy to describe the associated PD-process. We get the following table of transitions, there rates, probabilities and jumps:

$$\begin{array}{llll}
(0,0) \to (1,0) & \lambda + \lambda' & \frac{\lambda}{\lambda+\lambda'}\|E_1\psi_t\|^2 & \psi_t \to \frac{E_1\psi_t}{\|E_1\psi_t\|} \\
(0,0) \to (2,0) & \lambda + \lambda' & \frac{\lambda}{\lambda+\lambda'}\|E_2\psi_t\|^2 & \psi_t \to \frac{E_2\psi_t}{\|E_2\psi_t\|} \\
(0,0) \to (0,i) & \lambda + \lambda' & \frac{\lambda'}{\lambda+\lambda'}\|E_i'\psi_t\|^2 & \psi_t \to \frac{E_i'\psi_t}{\|E_i'\psi_t\|} \\
(0,1) \to (i,1) & \lambda & \frac{\lambda}{\lambda+\lambda'}\|E_i\psi_t\|^2 & \psi_t \to \frac{E_i\psi_t}{\|E_i\psi_t\|} \\
(1,0) \to (1,i) & \lambda' & \frac{\lambda'}{\lambda+\lambda'}\|E_i'\psi_t\|^2 & \psi_t \to \frac{E_i'\psi_t}{\|E_i'\psi_t\|}
\end{array}$$

where $\psi_t = U(t)\psi = \exp(-iHt)\psi$.

We may ask now the question whether the statistics of events on the right depends on measurements performed on the left. To answer this question let us compute the probability that the pointer on the right will jump from 0 to 1 during the time interval $(0,t)$. This can happen in three ways:

1. as a direct transition $(0,0) \mapsto (0,1)$
2. as a composite transition $(0,0) \mapsto (1,0) \mapsto (1,1)$
3. as a composite transition $(0.0) \mapsto (2,0) \mapsto (2,1)$

Transition 1 happens in the time interval $(0,t)$ with probability

$$p_1 = \frac{\lambda'}{\lambda+\lambda'}||E_1'\psi_t||^2 \times \left(1 - e^{-(\lambda+\lambda')t}\right). \tag{3}$$

The composite transition 2 happens with probability

$$\begin{aligned} p_2 &= \int_0^t \frac{\lambda}{\lambda+\lambda'}||E_1U(s)\psi||^2 e^{-(\lambda+\lambda')s}(\lambda+\lambda')ds \\ &\times||E_1'U(t-s)\frac{E_1U(s)\psi}{||E_1U(s)\psi||}||^2 \times \left(1-e^{-\lambda'(t-s)}\right) \\ &= \int_0^t \lambda e^{-(\lambda+\lambda')s}ds \times ||E_1'U(t-s)E_1U(s)\psi||^2 \times \left(1-e^{-\lambda'(t-s)}\right) \end{aligned}$$

and a similar formula with E_1 replaced with E_2 gives p_3. If now

$$[U(s)^\star E_iU(s), U(s')^\star E_j'U(s')] = 0 \qquad s, s' \leq t \tag{4}$$

then a straightforward computation gives the result:

$$p_1 + p_2 + p_3 = ||E_1'U(t)\psi||^2 \left(1-e^{-\lambda' t}\right). \tag{5}$$

It follows that as long as the usual locality assumptions (4) are satisfied, the event statistics seen on the right does not depend on what is measured on the left, and whether anything is measured there at all. We stress that this observation alone should not be used to conclude that superluminal signalling using EPR is impossible – this for the very reason that we were considering a particular and simplified model. What we proved is only that superluminal communicators must necessarily use more refined methods than the one considered above.

4 CLOUD CHAMBER MODEL AND GRW SPONTANEOUS LOCALIZATION THEORY

Our aim is now to account for the tracks that quantum particles leave in cloud chambers. Physically a cloud chamber is a highly complex system. To describe the response of the chamber to a quantum particle it is sufficient to assume that we have to do with a collection of two state systems able to change their state when a particle passes nearby a sensitive center. Let us sketch the model proposed in [7], [8].

Let us consider the space $E = \mathbf{R}^3$ as filled with a continuous medium which can be at each point $a \in E$ in one of two states: "on" represented by $\binom{1}{0}$ and "off" represented by $\binom{0}{1}$. The set of all possible states of the system is then 2^E. But we will be only interested with a continuum of states namely the "vacuum" and states which differ from the vacuum only in a finite number of points. As vacuum let us choose the state "off" everywhere. The space of classical events can be identified with the space of finite subsets of E from which it follows that the total system

$\Sigma_{tot} = \Sigma_q \otimes \Sigma_c$ is described by families $\{\rho_\Gamma\}_{\Gamma \subset E}$, Γ finite subset of E. For each $a \in E$ let g_a be a Hermitian bounded operator which represents heuristically the sensitivity of the counter located at a. We can think of g_a as a gaussian function $g_a(x)$ centered at $x = a$. We denote

$$\int_E g_a^2(x)da = \Lambda(x). \tag{6}$$

The quantum mechanical Hilbert space is then $\mathcal{H}_q = L^2(\mathbf{R}^3, dx)$. Each state ρ of the total system can be, formally, written as $\rho = \sum_{\Gamma \in \mathcal{S}} \rho_\Gamma \otimes \epsilon_\Gamma$, where, for $\Gamma \in \mathcal{S}$,

$$\epsilon_\Gamma = \prod \otimes_{a \in E} \begin{pmatrix} \chi_\Gamma(a) & 0 \\ 0 & 1 - \chi_\Gamma(a)) \end{pmatrix}, \tag{7}$$

and where χ_Γ stands for the characteristic function of Γ. The Lindblad coupling is now choosen in the following way

$$L_{int}(\rho) \equiv \int_{\mathbf{R}^3} da[V_a \rho V_a - \frac{1}{2}\{V_a^2, \rho\}] \tag{8}$$

where $V_a = g_a \otimes \tau_a, \tau_a$ denoting the flip at the point $a \in \mathbf{R}^3$. Let us introduce the following notation: $a(\Gamma)$ represents the state Γ with the counter at position a flipped, i.e. $a(\Gamma) = (\Gamma \setminus \{a\}) \cup \{\{a\} \setminus \Gamma\}$. The Liouville equation is given by $\dot{\rho} = -i[H, \rho] + L_{int}(\rho)$. But using the following identity in Eq. (8)

$$V_a \rho V_a = \sum_\Gamma g_a \rho_\Gamma g_a \otimes \epsilon_{a(\Gamma)} = \sum_\Gamma g_a \rho_{a(\Gamma)} g_a \otimes \epsilon_a \tag{9}$$

we can write

$$\dot{\rho}_\Gamma = -i[H, \rho_\Gamma] + \int_{\mathbf{R}^3} da g_a \rho_{a(\Gamma)} g_a - \frac{1}{2}\{\Lambda, \rho_\Gamma\} \ . \tag{10}$$

Summing up over Γ we get for the effective quantum state $\hat{\rho} = \sum_\Gamma \rho_\Gamma$

$$\dot{\hat{\rho}} = -i[H, \hat{\rho}] - \int_{\mathbf{R}^3} da g_a \ \hat{\rho} \ g_a - \frac{1}{2}\{\Lambda, \hat{\rho}\} \ . \tag{11}$$

Let us emphasize that the time derivative of $\hat{\rho}$ depends only on $\hat{\rho}$. Moreover (11) is exactly of the type discussed in connection with the spontaneous localization model of Ghirardi, Rimini and Weber [9], the difference being that GRW considered only the constant rate case, and were simply not interested in the classical traces of particles.

We can also construct the associated PD Markov process. We get for time evolution observables the same equation as in (10, except for the sign in front of the Hamiltonian. By taking expectation values we obtain a Davis generator corresponding to rate function $\lambda(\psi) = (\psi, \Lambda\psi)$, and probability kernel Q with non-zero elements of Q given by

$$Q(\psi, \Gamma; d\psi', a(\Gamma)) = \frac{||g_a\psi||^2}{\lambda(\psi)} \delta(\psi' - \frac{g_a\psi}{||g_a\psi||}) d\psi'. \tag{12}$$

Time evolution between jumps is given by:

$$\psi_t = \frac{\exp\left(-iHt - \frac{\Lambda t}{2}\right) \psi_0}{|| \exp\left(-iHt - \frac{\Lambda t}{2}\right) \psi_0 ||} \ . \tag{13}$$

The PD process can be described as follows: $\psi \in L^2(\mathbf{R}^3, dx)$ develops according to the above formula until at time t_1 jump occurs. The jump consists of a pair: (classical event, quantum jump). The classical medium jumps at a with probability density

$$p(a; \psi_{t_1}) = \|g_a \psi_{t_1}\|^2 / \lambda(\psi_{t_1}) \ , \tag{14}$$

(flip of the detector) while the quantum part of the jump is jump of the Hilbert space vector ψ_{t_1} to $g_a\psi_{t_1}/\|g_a\psi_{t_1}\|$ and the process starts again. The random jump time t_1 is governed by the inhomogeneous Poisson process with rate function $\lambda(\psi_t)$. If the medium is homogeneous, then $\lambda(\psi) = \text{const} = \lambda$, and we obtain for quantum jumps the GRW model. More complete discussion can be found in Refs. [7,8]. [2]

Derivation of Born's interpretation Let us consider now the idealized case of a homogeneous medium of particle detectors that are coupled to the particle only for a short time interval $(t, t+\Delta t)$, $\Delta t \to 0$ with intensity λ, so that $\lambda\Delta t \to \infty$. Let us also assume that the detectors are strictly point-like that is, that $g_a^2(x) \to \lambda\delta(x-a)$. In that case the formula (14), giving the probability density of firing the detector at a, becomes $p(a;\psi) = \|g_a\psi\|^2/\lambda = |\psi(a)|^2$ and we recover the Born interpretation of the wave function. The argument above goes as well for the case of a particle with spin.

Remark 1 For free particles in a homogeneous medium, i.e. $H = -\frac{\hbar^2}{2m}\Delta$, and for gaussian wave packets, straight lines are the most probable one. Indeed starting with a moving gaussian wave packet ψ_t then the probability of a registration of the particle at a is $\|f_a\psi_{t_1}\|^2/\lambda$ which is maximum if a coincides with the center of the gaussian.

Remark 2 For different values of the parameters we obtain different situations. Choosing $\lambda \simeq 10^{-9}$ years and a "universal" medium we obtain exactly the spontaneous localization model á la GRW. But we can also obtain standard nice particle tracks. The behaviour depends essentially on the relation between the two time scales: the one given by the energy spectrum and the other provided by the jump rate function.

Remark 3 In [8] the above model is discussed in more details introducing a multiparticle cloud chamber model. For a homogeneous medium one gets, for the effective statistical state of the quantum system, exactly the same equation as in Ref. [18]. For N particles the localization effect is proportional to the number of particles. The rate of jump, even in a homogeneous medium, is no more constant and the formulas given there provide the framework for a numerical simulation.

Remark 4 Several authors tried to explain track formation by a pure Hamiltonian theory (for recent attempts cf. Ref [19,20]). We know of no successful derivation that leads to a clear law relating rate of detections to geometry of detector locations and shape of wave function. The law we derive could have been in principle obtained already by E.B. Davies [21].

[2] We could translate our simple algorithm of track formation into the language of stochastic differential equations and filtering theory (cf. e.g. Ref. [17]), but that would serve no useful purpose at all – as all questions of interest can be answered using PDP algorithm, either by analytic computation, or by numerical simulation as in Monte Carlo Wave Function method of Refs. [12–15].

CONCLUDING REMARKS

We have seen that the word "measurement" instead of being banned, as suggested by J. Bell [22,23] can be given a precise and acceptable meaning: an appropriate CP coupling between the quantum system and a classical event-system, where information about the quantum states is transfered to the classical recording device by a continuous family of CP maps of the total system. It is fundamental to note that a transfer of information from Σ_q to Σ_c is impossible by any automorphic evolution. For a discussion of this fact see [6] and also Landsmann [24] and the no-go theorem by Ozawa [25]. In the framework we propose, all probabilistic properties of Quantum Mechanics – as e.g. Born's interpretation of the scalar product as a probability amplitude – can be derived thanks to the PD Markov process.

Our approach is in some respects similar in spirit to the idea of Quantum Stochastic Processes of E.B. Davies [26], especially to his class of "transition processes". The main difference being that he took the space of events as a primary object, without recognizing usefulness of introducing a classical system whose states can parametrize quantum dynamics and jump rates, and whose changes of states constitute events.

References

[1] Blanchard, Ph., and Jadczyk, A.: *Phys. Lett.* **A 175** (1993), 157-164

[2] Blanchard, Ph., and Jadczyk, A.: *Phys. Lett.* **A 183** (1993), 272-276

[3] Blanchard, Ph., and Jadczyk, A.: "Classical and quantum intertwine", in *Symposium on the Foundations of Modern Physics 1993*, Eds. P. Busch et al., World Scientific (1993)

[4] Blanchard, Ph., and Jadczyk, A.: "From quantum probabilities to classical facts", in *Advances in Dynamical Systems and Quantum Physics*, Capri, May 1993, Ed. R. Figari, World Scientific (1994), hep – th 9311090

[5] Blanchard, Ph., and Jadczyk, A.: "How and When Quantum Phenomena Become Real", in Proc. Third Max Born Symp. "Stochasticity and Quantum Chaos", Sobotka, Eds. Z. Haba et al., Kluwer Publ. (1994)

[6] Jadczyk, A.: "Topics in Quantum Dynamics", in Proc. I Caribbean Spring School of Math. and Theor. Physics, Saint François, Guadeloupe, June 1993, Eds. M. Dubois-Violette and R. Coquereaux, World Scientific (1994); BiBoS 635 / 5 / 94 and CPT - 94 / P. 3022, hep - th 9406204

[7] Jadczyk A.: "Particle Tracks, Events and Quantum Theory", RIMS Preprint N. 989, Kyoto August 94 , hep-th 9407157

[8] Jadczyk, A.: "On Quantum Jumps, Events and Spontaneous Localization Models", ESI-Wien Preprint (1994)

[9] Ghirardi, G.C., Rimini, A., and Weber, T.: "An Attempt at a Unified Description of Microscopic and Macroscopic Systems", in *Fundamental Aspects of Quantum Theory*, Como 1985, Eds. V. Gorini and A. Frigerio, NATO ASI Series B 144, Plenum Press, New York 1986, 57–64

[10] Belavkin, V.P. : *Found. Phys* **24** (1994), pp. 685–714

[11] Davis, M. H. A. : *Markov models and optimization*, Monographs on Statistics and Applied Probability, Chapman and Hall, London 1993

[12] Carmichael, H.: "An open systems approach to quantum optics", Lecture Notes in Physics m18 Springer Verlag, 1993.

[13] Dalibard, J., Castin, Y., and Mølmer, K.:*Phys. Rev. Lett.* **68** (1992) 580–583.

[14] Dum, R., Zoller, P., and Ritsch, H.: *Phys. Rev.* **A 45** (1992) 4879–4887.

[15] Gardiner, C.W., Parkins, A.S., and Zoller, P.: *Phys. Rev.* **A 46** (1992) 4363–4381.

[16] Einstein, A., Podolsky, B. and Rosen, N. : *Phys. Rev.* **47** (1935), 777–780

[17] Barchielli, A., Belavkin, V.P. : *J. Phys.* **A 24** (1991), 1495–1514

[18] Ghirardi, G.C., Pearle, P., Rimini, W. : *Phys. Rev.* **A 42** (1990), 78–89

[19] Wan, K.K. and Harrison, F.E. : *Found. Phys.* **24** (1994), 831

[20] Belavkin, V. and Melsheimer, O. : "A Hamiltonian theory for continuous reduction and spontaneous localization", Preprint Centro Vito Volterra N. 139, Roma May 93 (see also the present volume)

[21] Davies, E.B. :*J. Funct. Anal.* **6** (1970), 318–346

[22] Bell, J.: "Against measurement", in *Sixty-Two Years of Uncertainty*, Proc. NATO Adv. Study Inst., Erice, Ed. Arthur I. Miller, NATO ASI Series B vol. 226, Plenum Press, New York 1990

[23] Bell, J.: "Towards an exact quantum mechanics", in *Themes in Contemporary Physics II.* , Deser, S., and Finkelstein, R.J., Ed., World Scientific, Singapore 1989

[24] Landsman, N. P.: *Int. J. Mod. Phys.* **A6** (1991), 5349–5371

[25] Ozawa, M.: *Progr. Theor. Phys.* **88** (1992), 1051–1064

[26] Davies, E.B. :q *Commun. Math. Phys.* **15** (1969), 277-304; **19** (1970), 83-105; **22** (1971), 51–70

DISSIPATION AND REDUCTION OF SUPERCONDUCTING STATES DUE TO SPONTANEOUS LOCALIZATION[1]

M. Buffa[1], O. Nicrosini[2], and A. Rimini[1]

[1]Università di Pavia,
Dipartimento di Fisica Nucleare e Teorica,
Pavia, Italy.
[2]CERN – TH Division, Geneva,
Switzerland, on leave from INFN,
Sezione di Pavia.

Abstract

The problem of quantum measurement is briefly recalled with particular reference to theories of reduction as a real physical process. The spontaneous localization models for systems containing identical particles are described and the choice of the model when several kinds of particles are present is discussed. Dissipation and reduction effects induced by spontaneous localization in superconducting systems are analyzed. It is found that dissipation effects are negligibly small and that reduction times are long and far beyond any hope of testability with present–day technology.

Keywords :
quantum measurement, spontaneous localization, superconductivity, macroscopic quantum coherence.

1 INTRODUCTION

The basic principle of evolution of quantum mechanics, the Schrödinger equation, does not include the stochastic features that emerge during quantum measurements and are incorporated, in the ordinary presentation, into the reduction

[1]Supported in part by the Ministero dell'Università e della Ricerca Scientifica e Tecnologica and by the Istituto Nazionale di Fisica Nucleare.

principle. In other words, the Schrödinger equation does not describe the particularity of the world we experience. One possible answer to this problem is to introduce a stochastic modification to the Schrödinger equation, so that reduction is described as a real physical process affecting the wave function. Then the wave function is put in correspondence to the result of the measurement, as in the ordinary presentation based on the reduction principle. The alteration must be ruled by definite, universally valid, precise assumptions. It must have practically unobservable consequences in all ordinary situations and, nevertheless, it must rapidly reduce superpositions of macroscopically distinguishable states. A theory satisfying these requirements was proposed some years ago with the name of quantum mechanics with spontaneous localization [1, 2], and important improvements were introduced later [3–6]. A consequence of the theory in all its forms is that the internal structure of macroscopic objects is unaffected by the modification carried out on the Schrödinger equation, an essential ingredient of the proof being that the internal wave function of the macroscopic object be sharply localized with respect to a parameter of the theory whose order of magnitude is 10^{-5} cm. This condition is well satisfied in normal states but is violated in superconducting states, so that this case needs separate consideration.

A qualitative estimate of the effects of spontaneous localization on superconducting states has been given by Rae [7] in the framework of the theory in its original form. In this form, which is much more intuitive than the later developments, the theory assumes that each particle i of a system of distinguishable particles undergoes at random times, with mean frequency λ^i, a spontaneous localization process within regions of linear dimensions of the order of 10^{-5} cm. The space distribution of the localization regions is in accordance with the quantum probability rule. The frequency λ^i is assumed to be of the order of $10^{-16}\,\mathrm{s}^{-1}$, so that practically nothing happens to a microscopic system, consisting of a small number of particles. On the other hand, it can be shown [1, 2] that, in the case of a macroscopic object, linear superpositions of macroscopically distant positions are forbidden, except for very short times. Furthermore, it is clear that the internal structure of the macroscopic object is unaffected by the localization process, provided the microscopic constituents are already localized within regions much smaller than 10^{-5} cm because of the Schrödinger dynamics. As already noted, this condition is not satisfied in the case of superconducting states. Rae argues as follows. The wave function of the superconducting state can be written as

$$\psi = \psi_{\boldsymbol{k}_1}\psi_{\boldsymbol{k}_2}\cdots\psi_{\boldsymbol{k}_C}\exp[iS(\boldsymbol{r})], \tag{1.1}$$

where $\psi_{\boldsymbol{k}_j}$ represents the wave function of a Cooper pair with electron wave vectors $\pm\boldsymbol{k}_j$ and $S(\boldsymbol{r})$ is the macroscopic phase associated with the supercurrent. Then the maximum consequence of the localization of an electron would be the break of one of the Cooper pairs. The overall rate of the process can be written $N\lambda$, where Rae uses for N the number of paired electrons $\approx 10^{20}\,\mathrm{cm}^{-3}$ and for λ the value $10^{-15}\,\mathrm{s}^{-1}$ (somewhat larger than the value referred to above). In this way one finds for the rate of destruction of Cooper pairs $10^5\,\mathrm{cm}^{-3}\,\mathrm{s}^{-1}$, which corresponds to a decay in the supercurrent well below the experimental detection limit. Moreover Rae observes that the above argument does not take into account the possibility of recreation of Cooper pairs, which would lower the observability of the localization process even further.

Another problem considered by Rae is that of the superposition of the states corresponding to two different supercurrents, which possibly takes place in superconducting devices such as SQUIDs. The two different supercurrents are typically separated by an intensity of the order of 10^{-7}A and, in this sense, they are macroscopically different. The experimental proof of the existence of such a superposition, as originally proposed by Leggett [8], could provide evidence for the existence of macroscopic quantum coherence. The wave function of the superposition can be written

$$\psi = \psi_{\boldsymbol{k}_1}\psi_{\boldsymbol{k}_2}\cdots\psi_{\boldsymbol{k}_C}\{\exp[iS_1(\boldsymbol{r})] + \exp[iS_2(\boldsymbol{r})]\}, \tag{1.2}$$

where the phases $S_1(\boldsymbol{r})$ and $S_2(\boldsymbol{r})$ correspond to the two different supercurrents. Rae argues that the two terms in eq. (1.2) are not located in different places, so that neither of them is selected by the localization process. The only result is again the break up of one of the Cooper pairs, giving rise simply to some negligible additional dissipative effect.

The spontaneous localization theory sketched in this section cannot be applied to systems of identical particles, because the proper symmetry or antisymmetry of the wave function is spoiled by the envisaged process (on the other hand, the wave functions used by Rae either are not properly antisymmetrized). Another unpleasant feature of the old theory is that the stochastic process takes place discontinuously in time. Both these inconveniences can be cured independently. It is the purpose of this work to present the results of calculations performed within the most recent and satisfactory theory. In the next section we describe the continuous spontaneous localization theory for identical particles. In section 3 we apply to superconducting states the theory of section 2. In section 4 we report the results of analytical and numerical calculations.

2 SPONTANEOUS LOCALIZATION OF A SYSTEM WITH IDENTICAL PARTICLES

The spontaneous localization of a system consisting of a single kind of identical particles can be described [4, 6] through the Itô stochastic equation

$$\mathrm{d}|\psi\rangle = \left[-\frac{i}{\hbar}\hat{H}\mathrm{d}t + \sqrt{\gamma}\int \mathrm{d}^3\boldsymbol{x}\hat{N}_\psi(\boldsymbol{x})\mathrm{d}B(\boldsymbol{x}) - \tfrac{1}{2}\gamma\int \mathrm{d}^3\boldsymbol{x}\left(\hat{N}_\psi(\boldsymbol{x})\right)^2\mathrm{d}t\right]|\psi\rangle, \tag{2.1}$$

where $B(\boldsymbol{x},t)$ is a set of Wiener processes (we drop for shortness the argument t) such that

$$\overline{\mathrm{d}B(\boldsymbol{x})} = 0, \qquad \overline{\mathrm{d}B(\boldsymbol{x})\mathrm{d}B(\boldsymbol{x}')} = \delta^3(\boldsymbol{x}-\boldsymbol{x}')\mathrm{d}t. \tag{2.2}$$

The nonlinear operators $\hat{N}_\psi(\boldsymbol{x})$ are defined by

$$\hat{N}_\psi(\boldsymbol{x}) = \hat{N}(\boldsymbol{x}) - \langle\psi|\hat{N}(\boldsymbol{x})|\psi\rangle, \tag{2.3}$$

in terms of the mutually commuting Hermitian linear operators $\hat{N}(\boldsymbol{x})$. Eq. (2.1) conserves the norm of $|\psi\rangle$ for any realization of $B(\boldsymbol{x},t)$; $\hat{N}(\boldsymbol{x})$ is a locally averaged number density operator,

$$\hat{N}(\boldsymbol{x}) = \sum\nolimits_s \int \mathrm{d}^3\boldsymbol{x}'\, G(\boldsymbol{x}'-\boldsymbol{x})\, a^\dagger(\boldsymbol{x}',s)a(\boldsymbol{x}',s), \tag{2.4}$$

where the function $G(\boldsymbol{x})$ can conveniently be taken to be of a Gaussian shape

$$G(\boldsymbol{x}) = \left(\frac{\alpha}{2\pi}\right)^{\frac{3}{2}} \exp\left(-\tfrac{1}{2}\alpha\,\boldsymbol{x}^2\right). \tag{2.5}$$

The stochastic process is characterized by the range $1/\sqrt{\alpha}$ of the average (2.4) and by the strength γ. The sizes of these parameters were discussed in ref. [6], and the suggested values, with reference to systems of protons, were

$$1/\sqrt{\alpha} \approx 10^{-5}\text{cm}, \qquad \gamma \approx 10^{-30}\,\text{cm}^3\,\text{s}^{-1}. \tag{2.6}$$

The equation for the statistical operator following from eq. (2.1) is

$$\frac{\mathrm{d}}{\mathrm{d}t}\rho = -\frac{i}{\hbar}\,[\hat{H},\rho] - \tfrac{1}{2}\gamma\int \mathrm{d}^3\boldsymbol{x}\,\left[\hat{N}(\boldsymbol{x}),\left[\hat{N}(\boldsymbol{x}),\rho\right]\right]. \tag{2.7}$$

If eqs. (2.1) or (2.7) are applied to a system consisting of one single particle, the process described is essentially equivalent to a localization of the particle with accuracy $1/\sqrt{\alpha}$ taking place with mean frequency

$$\lambda = \gamma\left(\frac{\alpha}{4\pi}\right)^{\frac{3}{2}} \approx 10^{-17}\,\text{s}^{-1}. \tag{2.8}$$

Equation (2.1) enjoys the same symmetry properties as the corresponding Schrödinger equation, except for time–reversal invariance, of course. This is obvious for space translations and rotations and for time translations. Galilei invariance too is almost evident and can easily be checked [9].

As already said in the introduction, the stochastic process in eq. (2.1) has completely negligible effects on microscopic systems and, provided the structural wave function is sharply localized with respect to $1/\sqrt{\alpha}$, on the internal structure of macroscopic objects. On the other hand, the external motion of macroscopic objects is affected in two different ways. First, a disturbance is introduced in the free motion of a wave packet, giving rise to a sort of dissipative effect. This is characterized by the appearance of a momentum diffusion whose coefficient, for a homogeneous macroscopic particle, under suitable assumptions on its shape and orientation, is given by [6]

$$D = \hbar^2\sqrt{\frac{\alpha}{4\pi}}\,\gamma\left(\frac{N}{V}\right)^2 S_{\perp}, \tag{2.9}$$

where N/V is the number of particles per unit volume and $S_{\perp}$ is the transverse (with respect to the direction to which D refers) section of the particle. The consequences of momentum diffusion are completely negligible [6]. The second effect on the behaviour of a macroscopic object is reduction. If the object is in a superposition of two macroscopically distant positions, the wave function is reduced to one of the two terms. The reduction rate, for a homogeneous macroscopic particle, under suitable assumptions on its shape and orientation, is given by [6]

$$\Gamma_{\mathrm{R}} = L_{\mathrm{out}}\,\gamma\left(\frac{N}{V}\right)^2 S_{\perp}, \tag{2.10}$$

where L_{out} is the length of the object or, if smaller, the separation between the two superposed positions. Reduction is very quick for any macroscopic $S_{\perp}$ and L_{out}.

In the case in which several kinds of identical particles are present, one can formulate a generalization of eq. (2.1) along two possible lines. In the first formulation, one independent stochastic field is assumed for each kind of particles. In the second formulation, a single stochastic field is assumed for all kinds of particles. An indication for selecting one of the possibilities is obtained by considering a microscopic compound system and requiring that the dynamical equation for the motion of the compound system as a whole be again of the type (2.1). The fulfilment of this requirement guarantees the indispensable arbitrariness in the choice of the types of microscopic particles to be considered the elementary constituents of the system. For example, if the system contains nuclei, it shall be immaterial to consider as constituents the nuclei themselves or the protons and neutrons contained in them. It can be shown that the second formulation allows the most simple and natural assumptions concerning the stochastic field for the microscopic compound system and the coupling to it [9].

Since ions are microscopic systems, on the basis of what we have just seen the system in which we are interested can anyhow be considered to be made of ions (i) and electrons (e). Then the equation for the state vector is [9]

$$\mathrm{d}|\psi\rangle = \Big[-\frac{i}{\hbar}\hat{H}\mathrm{d}t + \sqrt{\gamma_0}\sum_k^{(=\mathrm{i,e})} \frac{m_k}{m_0}\int \mathrm{d}^3\boldsymbol{x}\hat{N}_\psi^k(\boldsymbol{x})\mathrm{d}B(\boldsymbol{x}) - \tfrac{1}{2}\gamma_0 \sum_{kk'}^{(=\mathrm{i,e})} \frac{m_k m_{k'}}{m_0^2}\int \mathrm{d}^3\boldsymbol{x}\hat{N}_\psi^k(\boldsymbol{x})\hat{N}_\psi^{k'}(\boldsymbol{x})\mathrm{d}t\Big]|\psi\rangle, \tag{2.11}$$

and that for the statistical operator is

$$\frac{\mathrm{d}}{\mathrm{d}t}\rho = -\frac{i}{\hbar}[\hat{H},\rho] - \tfrac{1}{2}\gamma_0 \sum_{kk'}^{(=\mathrm{i,e})} \frac{m_k m_{k'}}{m_0^2}\int \mathrm{d}^3\boldsymbol{x}\left[\hat{N}^k(\boldsymbol{x}),\left[\hat{N}^{k'}(\boldsymbol{x}),\rho\right]\right], \tag{2.12}$$

where m_0 is a reference mass, for which we take the proton mass. The discussion in ref. [6] can be repeated and suggests the values

$$1/\sqrt{\alpha} \approx 10^{-5}\mathrm{cm}, \qquad \gamma_0 \approx 10^{-30}\,\mathrm{cm}^3\,\mathrm{s}^{-1}. \tag{2.13}$$

We assume that the electron–ion interaction can be incorporated effectively into the electron and ion Hamiltonians, so that the wave function for the whole system can be assumed factorized into an electron and an ion wave function. Moreover we assume that the ions are well localized, with respect to $1/\sqrt{\alpha}$, at the lattice positions. It can be shown that in this case all stochastic terms cancel, except the pure electron ones, so that we get

$$\frac{\mathrm{d}}{\mathrm{d}t}\rho^{\mathrm{i}} = -\frac{i}{\hbar}\left[\hat{H}^{\mathrm{i}},\rho^{\mathrm{i}}\right], \tag{2.14a}$$

$$\frac{\mathrm{d}}{\mathrm{d}t}\rho^{\mathrm{e}} = -\frac{i}{\hbar}\left[\hat{H}^{\mathrm{e}},\rho^{\mathrm{e}}\right] - \tfrac{1}{2}\gamma\int \mathrm{d}^3\boldsymbol{x}\left[\hat{N}^{\mathrm{e}}(\boldsymbol{x}),\left[\hat{N}^{\mathrm{e}}(\boldsymbol{x}),\rho^{\mathrm{e}}\right]\right]. \tag{2.14b}$$

In the following we shall drop the electron index "e" in eq. (2.14b). For γ we shall use the value

$$\gamma = 2.97\ 10^{-37}\,\mathrm{cm}^3\,\mathrm{s}^{-1}, \tag{2.15}$$

corresponding to the value (2.13) of γ_0.

3 DISSIPATION AND REDUCTION EFFECTS IN SUPERCONDUCTING SYSTEMS

Let the statistical operator be initially

$$\rho(0) = |\boldsymbol{j}\rangle\langle\boldsymbol{j}|, \tag{3.1}$$

$|\boldsymbol{j}\rangle$ being a state carrying a supercurrent, in an indefinite superconductor, obtained by performing a Galilei boost on the quasiparticle vacuum $|0\rangle$.

The transition probability from the state (3.1) to an arbitrary orthogonal state is

$$P_{\boldsymbol{j}\to\{l\}}(t) = \sum_{l}^{(\neq \boldsymbol{j})} \langle\, l\,|\rho(t)|\, l\,\rangle, \tag{3.2}$$

where $\rho(t)$ is evolved from (3.1) according to eq. (2.14b). We are interested in the initial value of the transition probability per unit time, Γ. It can be easily shown [9] that Γ is given by

$$\Gamma = \left[\frac{\mathrm{d}}{\mathrm{d}t} P_{\boldsymbol{j}\to\{l\}}(t)\right]_{t=0} = \gamma \int \mathrm{d}^3\boldsymbol{x}\, \left(\langle 0|\hat{N}^2(\boldsymbol{x})|0\rangle - \langle 0|\hat{N}(\boldsymbol{x})|0\rangle^2\right), \tag{3.3}$$

independently of $\boldsymbol{j}$. Γ gives a measure of the dissipation effects due to spontaneous localization in superconducting states. As we shall see in the next section, it will turn out to be very small, leading to the conclusion that dissipation effects are negligible.

Once one has ascertained that superconducting states are not quickly destroyed by the localization process, one can conveniently introduce besides Γ another quantity to measure the size of the effect. The sample is maintained at a fixed, nonzero temperature by the cooling device. This means that the considered excitations, which add to the ones that are anyway present because of the finite temperature, eventually decay, filling a hole by a particle or recreating a pair, and supplying energy to the lattice. Recreation of Cooper pairs was already envisaged by Rae [7]. In this picture, the quantity that really matters is the amount of heat pumped in by the spontaneous localization process, which is to be pumped out by the refrigerator. In calculating this quantity, the role played by the ions cannot be neglected because of their large mass, so that eq. (2.12) has to be used. It has already been noted by Pearle and Squires [10] that the total rate of energy production implied by this equation can be calculated exactly (at least if the potentials contained in $\hat{H}$ are local) and the result is

$$\frac{\mathrm{d}}{\mathrm{d}t}\langle H\rangle = \mathrm{Tr}\,(\dot{\rho} H) = \frac{3\,\hbar^2 \gamma_0\, \alpha^{\frac{5}{2}}}{4\,(4\pi)^{\frac{3}{2}}\, m_0^2} M \tag{3.4}$$

for any state ρ, M being the total mass. Within the total energy production rate the fraction that refers to the center–of–mass motion of a macroscopic object is negligible (see ref. [6]), so that eq. (3.4) gives in practice the rate of increase of the internal energy $\mathrm{d}E/\mathrm{d}t$. We shall discuss the numerical value of this quantity in the next section.

To study the reduction effects, let us mimic the superpositions that take place in a SQUID by the state

$$|\psi\rangle = c_+|+\boldsymbol{j}\rangle + c_-|-\boldsymbol{j}\rangle. \tag{3.5}$$

Since both $|+\boldsymbol{j}\rangle$ and $|-\boldsymbol{j}\rangle$ are eigenstates of $\hat{H}$, it is impossible, as it was immediately obvious, to describe the oscillations taking place in the SQUID in this way. However, there is no reason to believe that the reduction to one term or the other due to the localization process takes place in our indefinite superconductor in a different way than in the SQUID, so that the reduction rate shall be correctly calculated by the equation

$$\frac{\mathrm{d}}{\mathrm{d}t}\rho = -\tfrac{1}{2}\gamma \int \mathrm{d}^3\boldsymbol{x}\, \left[\hat{N}(\boldsymbol{x}), \left[\hat{N}(\boldsymbol{x}), \rho\right]\right]. \tag{3.6}$$

Disregarding dissipation effects amounts to assuming that ρ is a mixture of states of the type (3.5). The reduction process is then described by the damping of the off–diagonal element $\langle -\boldsymbol{j}|\rho|+\boldsymbol{j}\rangle$. From eq. (3.6), taking into account that the terms containing $\langle -\boldsymbol{j}|\hat{N}^2(\boldsymbol{x})|+\boldsymbol{j}\rangle$ or $\langle -\boldsymbol{j}|\hat{N}(\boldsymbol{x})|+\boldsymbol{j}\rangle$ are negligible because the operators $\hat{N}^2(\boldsymbol{x})$ and $\hat{N}(\boldsymbol{x})$ cannot connect in an appreciable way states corresponding to macroscopically different supercurrents [9], one finds

$$\frac{\mathrm{d}}{\mathrm{d}t}\langle -\boldsymbol{j}|\rho|+\boldsymbol{j}\rangle = -\Gamma\langle -\boldsymbol{j}|\rho|+\boldsymbol{j}\rangle, \tag{3.7}$$

where Γ is again given by eq. (3.3). It is seen that dissipation and reduction effects in superconducting states are ruled by the same time rate, so that the two effects are both small or both large. In fact, as we shall see in the next section, they are both small. This situation is in contrast with that of the corresponding effects for the motion as a whole of a macroscopic body, where, as described in section 2, dissipation effects are negligible while reduction effects are important.

4 ANALYTICAL AND NUMERICAL CALCULATIONS

In order to calculate the value of the rate Γ given by eq. (3.3), one needs the expression of $\hat{N}(\boldsymbol{x})$ in terms of quasiparticle operators in the wave–vector representation. Inserting such expression into eq. (3.3) and replacing sums over wave vectors by corresponding integrals, one finds

$$\begin{aligned}\Gamma &= \frac{V\gamma}{(2\pi)^6}\int \mathrm{d}^3\boldsymbol{k}_1\int \mathrm{d}^3\boldsymbol{k}_2 \exp\left[-(\boldsymbol{k}_1-\boldsymbol{k}_2)^2/\alpha\right](u_{k_1}v_{k_2}+v_{k_1}u_{k_2})^2 \\ &= \frac{V\gamma\alpha}{(2\pi)^4}\int_0^\infty \mathrm{d}k_1\int_0^\infty \mathrm{d}k_2\; k_1k_2 \\ &\quad\left(\exp\left[-(k_1-k_2)^2/\alpha\right]-\exp\left[-(k_1+k_2)^2/\alpha\right]\right)(u_{k_1}v_{k_2}+v_{k_1}u_{k_2})^2, \end{aligned} \tag{4.1}$$

where u_k and v_k are the coefficients of the Bogoliubov-Valatin transformation. An exact analytical evaluation of the double integral in the last line of eq. (4.1) is impossible, and even an approximate calculation is very cumbersome because of the different structures in the k_1k_2 plane of the factors. Let us write

$$\Gamma = \Gamma_0(1+C), \tag{4.2}$$

where Γ_0 is Γ in the absence of superconducting effects ($\Delta = 0$); Γ_0 can be calculated exactly and is given by

$$\begin{aligned}\Gamma_0 &= \frac{\gamma\,\alpha^2k_{\mathrm{F}}^2}{(2\pi)^4}V\Big\{1-\tfrac{1}{6}\frac{\alpha}{k_{\mathrm{F}}^2}\left[1-\exp\left(-4k_{\mathrm{F}}^2/\alpha\right)\right] \\ &\quad - \tfrac{1}{3}\exp\left(-4k_{\mathrm{F}}^2/\alpha\right)-\tfrac{2}{3}\sqrt{\pi}\frac{k_{\mathrm{F}}}{\sqrt{\alpha}}\left[\operatorname{erf}\left(2k_{\mathrm{F}}/\sqrt{\alpha}\right)-1\right]\Big\} \\ &= \frac{\gamma\,\alpha^2k_{\mathrm{F}}^2}{(2\pi)^4}V, \end{aligned} \tag{4.3}$$

where k_{F} is the Fermi wave number, the correction neglected in the last line being of the order of 10^{-7}. For the correction C in eq. (4.2) we find

$$C \simeq x_1^2\left\{4\ln(x_1)[\ln(x_1)-1]-\left[\ln^2(x_2)+\ln^2(x_3)-\ln^2(x_4)\right]\Theta(1-x_2)\right\}, \tag{4.4}$$

where

$$x_1 = k_\Delta/\sqrt{2\alpha}, \qquad x_2 = k_D/\sqrt{\alpha}, \qquad x_3 = k_D/k_F, \qquad x_4 = \sqrt{\alpha}/k_F. \tag{4.5}$$

k_Δ and k_D are linked to the superconducting gap Δ and to the Debye energy ϵ_D by

$$\Delta = \hbar^2 k_F k_\Delta/m, \qquad \epsilon_D = \hbar^2 k_F k_D/m. \tag{4.6}$$

The numerical values given by eq. (4.4) were checked against the results of a double numerical integration and found to be in pretty good agreement for wide ranges of values of the parameters. In the case of tin one has

$$k_F = 1.64\ 10^8\ \mathrm{cm}^{-1}, \qquad k_\Delta = 4.51\ 10^3\ \mathrm{cm}^{-1}, \qquad k_D = 1.37\ 10^5\ \mathrm{cm}^{-1}, \tag{4.7}$$

from which

$$x_1 = 3.19\ 10^{-2}, \qquad x_2 = 1.37, \tag{4.8}$$

and

$$C \simeq 6.23\ 10^{-2}, \qquad C_{\mathrm{num}} = 6.49\ 10^{-2}, \tag{4.9}$$

where C is calculated from eq. (4.4) and C_{num} is the "exact" value obtained numerically. As can be seen, pairing does not affect Γ in an important way. Eq. (4.3) gives for tin

$$\Gamma_0 = 5.13\ 10^{-4}\ V\mathrm{cm}^{-3}\,\mathrm{s}^{-1}. \tag{4.10}$$

Eqs. (4.3) or, in the numerical example, (4.10) provide, apart from the correction C, the value of the rate of production of quasiparticle excitations (including particle–hole pairs) due to the spontaneous localization process. When compared to the number of electrons $N \approx 10^{23} V\mathrm{cm}^{-3}$ or to the number of paired electrons $N_p \approx 10^{20} V\mathrm{cm}^{-3}$, such a value shows that dissipation effects due to the process are practically completely negligible for superconducting states, similarly to what happens for the motion as a whole of a macroscopic object.

It is interesting to compare our result with the previous estimate by Rae [7]. This estimate was

$$[\Gamma]_{\mathrm{Rae}} = N\lambda = N\gamma\left(\frac{\alpha}{4\pi}\right)^{\frac{3}{2}}, \tag{4.11}$$

where we have inserted the value (2.8) for the single–particle frequency λ. Comparison with our value (4.3), neglecting C, gives

$$[\Gamma]_{\mathrm{ours}} = \frac{3}{2\sqrt{\pi}}\frac{\sqrt{\alpha}}{k_F}[\Gamma]_{\mathrm{Rae}}, \tag{4.12}$$

which shows that the correct treatment taking into account the indistinguishability of electrons gives a further reduction by a factor of the order of $\sqrt{\alpha}/k_F$.[2]

As suggested in sect. 4, one can measure the size of dissipation effects due to spontaneous localization also by evaluating the rate of increase of internal energy $\mathrm{d}E/\mathrm{d}t$. From eq. (3.4) one gets

$$\frac{\mathrm{d}E}{\mathrm{d}t} = \frac{\mathrm{d}}{\mathrm{d}t}\langle H\rangle \simeq 6\,10^{-14}\,\mathrm{erg\,s}^{-1} M\,\mathrm{g}^{-1}. \tag{4.13}$$

[2]The value used by Rae for λ roughly corresponds to the proton value of γ given in eq. (2.6), which is considerably larger than our value (2.15). Our analysis for systems containing different kinds of particles shows that the value (2.15) is more reliable. On the other hand he used for N the number of paired electrons N_p. This choice does not take into account that particle–hole excitations eventually decay, possibly affecting paired electrons.

To appreciate how small is this small quantity, one can conveniently estimate the corresponding temperature increase, using as an order of magnitude of heat capacity

$$\frac{\mathrm{d}E}{\mathrm{d}T} \approx 3NB. \tag{4.14}$$

One finds

$$\frac{\mathrm{d}T}{\mathrm{d}t} \approx 2\ 10^{-22} A\,{}^\circ\mathrm{K}\ \mathrm{s}^{-1}, \tag{4.15}$$

A being the atomic weight.

As we have seen in the previous section, reduction effects too are ruled by Γ. The reduction time, in our numerical example, amounts to

$$\frac{1}{\Gamma_0} = 1.95\ 10^3\,\frac{\mathrm{cm}^3}{V}\,\mathrm{s}, \tag{4.16}$$

i.e., for a superconductor of centimetric dimensions, to a time of the order of 1 h. Since, with present day technology, it is only a hope to succeed in demonstrating the persistence of superposed macroscopically different supercurrents for times of the order of one SQUID oscillation, say 10^{-6} s, we see that also this way to experimentally test the assumption of spontaneous localization is impracticable.

CONCLUSIONS

The spontaneous localization process described in sect. 2, with the choice (2.13) of the parameters α and γ_0, maintains valid ordinary quantum mechanics for microscopic objects and, provided the structural wave function is sharply localized with respect to $1/\sqrt{\alpha}$, preserves the microscopic structure of macroscopic bodies. As for the motion as a whole of macroscopic objects, dissipative effects are completely negligible, while reduction of superpositions of macroscopically distinguishable states takes place quickly. All that follows easily by adapting the results of ref. [6], obtained in the framework of a model with a single kind of identical particles, to the generalized model of sect. 2.

Superconducting states are a part of the internal wave function which is not necessarily preserved because it is not sharply localized in configuration space. Such states have some macroscopic character, so that they may realize macroscopic quantum coherence [8]. As in the case of the motion of macroscopic objects, the spontaneous localization process gives rise, in superconducting states, to dissipation and reduction effects. From the analysis of sects. 3 and 4, it first follows that dissipation effects are negligible, so that the assumption of the spontaneous localization process is not in contradiction with the existence and persistence of supercurrents, just as it is not in contradiction with the deterministic behaviour of macroscopic objects. Secondly, it turns out that the time of reduction of superpositions of macroscopically different supercurrents is large, of the order of 1 h, contrary to what happens in the case of macroscopically different positions of macroscopic objects. This fact has the consequence that the assumption of the spontaneous localization process cannot be tested experimentally by revealing macroscopic quantum coherence of superconducting states.

A further point needs to be discussed. Independently of their experimental evidence, superpositions of macroscopically different supercurrents (possibly entangled with other degrees of freedom, thereby making it arduous to detect of coherence) are predicted by quantum mechanics, without or with spontaneous localization. The question is whether the existence of such superpositions is in contrast with the program of the spontaneous localization theory of allowing the interpretation of the wave function, when the considered system has a wave function, as the real (and sole) individual property of the system. We think that the detection of macroscopic quantum coherence would be an important achievement, because it would be a significant confirmation of the quantum mechanical edifice. But we do not think that the existence of superpositions of macroscopically different supercurrents is more striking than the existence of, e. g., superpositions of macroscopically distant paths of a neutron in neutron interferometry experiments. If a measurement is performed to ascertain which term in the superposition is there, according to the spontaneous localization assumption reduction takes place quickly at the level of the measuring apparatus or, in the extreme eventuality, as soon as the result is perceived [11]. In all cases, each one of us, human beings and cats, has a reasonably well–defined wave function.

References

[1] G.C. Ghirardi, A. Rimini, and T. Weber, in *Quantum Probability and Applications II*, L. Accardi and W. von Waldenfels, eds., Lecture Notes in Mathematics, vol. 1136 (Springer, Berlin, 1985), p. 223.

[2] G.C. Ghirardi, A. Rimini, and T. Weber, Phys. Rev. D **34** (1986), 470; Phys. Rev. D **36** (1987), 3287.

[3] L. Diósi, Phys. Lett. A **129** (1988), 419.

[4] P. Pearle, Phys. Rev. A **39** (1989), 2277.

[5] N. Gisin, Helv. Phys. Acta **62** (1989), 363.

[6] G.C. Ghirardi, P. Pearle, and A. Rimini, Phys. Rev. A **42** (1990), 78.

[7] A. I. M. Rae, J. Phys. A **23** (1990), L57.

[8] A. J. Leggett, Suppl. Progr. Theor. Phys. **69** (1980), 80.

[9] M. Buffa, O. Nicrosini and A. Rimini, CERN–TH.7323/94, to be submitted to Phys. Rev. A.

[10] P. Pearle and E. Squires, to be published in Phys. Rev. Lett.

[11] F. Aicardi, A. Borsellino, G. C. Ghirardi, and R. Grassi, Found. Phys. Lett. **4** (1991), 109.

PERMANENT STATE REDUCTION: MOTIVATIONS, RESULTS, AND BY-PRODUCTS [1]

Lajos Diósi

KFKI Research Institute for Particle and Nuclear Physics
H-1525 Budapest 114, POB 49, Hungary
E-mail: diosi@rmki.kfki.hu

Abstract

We enlist various motivations underlying the idea that von Neumann collapses occur permanently, spontaneously and even universally. In addition to the conceptual value of permanent reduction theory, we mention its important by-product promoting new quantum Monte-Carlo simulations of irreversible quantum processes, especially in laser optics.

1 INTRODUCTION AND CONTENTS

In the last two decades, serious efforts were concentrated to find the proper mathematical formalism unifying the unitary evolution with permanent collapses. Up to now, things have become well understood at least in the Markovian regime though the universality of permanent reduction has merely been modeled on heuristic grounds. We attempt to classify main contributions that have led to our present-day theory of permanent state reduction. We also emphasize that, as for the fundamental purpose of the project of permanent reduction, we see a spectacular convergence between it, on one hand, and the so-called consistent history concept, on the other hand. This is not very surprising since both projects were initiated to eliminate the fundamental dichotomy of the ordinary quantum theory.

Contents:
1 C-numbers from ψ: Reduction

[1]Talk given at the Int. Symp. on Quantum Communication and Measurement, Nottingham, 10-16 July 1994.

Quantum Communications and Measurement
Edited by V. P. Belavkin *et al.*, Plenum Press, New York, 1995

2 Permanent Reduction—Eq. for $\psi(t)$
3 Permanent Reduction—Eqs. for $\psi(t)$ and $z(t)$
4 Consistent Histories
5 By–Products and Conclusion

2 C-NUMBERS FROM ψ: REDUCTION

However well is quantum mechanics understood its famous dichotomy has remained unresolved after all. The quantum state ψ evolves according to the Schrödinger equation:

$$\dot{\psi} = -iH\psi. \tag{1}$$

The knowledge of ψ in itself does not still provide suitable experimental "c-number" predictions. One has to turn to the von Neumann reduction theory. An apparatus interacts with the system in question and changes the original state of it:

$$\psi \rightarrow \psi_z \tag{2}$$

with probability $p_z = |\langle\psi_z|\psi\rangle|^2$, where z is the c-number recorded by the apparatus. The possible final states $\{\psi_z\}$ form a complete orthogonal system. The reduction process above is nonlinear whereas the initial and final statistical (density) operators are related linearly:

$$\rho \rightarrow \sum_z P_z \rho P_z \tag{3}$$

with projectors $P_z = |\psi_z\rangle\langle\psi_z|$.

Let us see what we have done starting with the general ambition to eliminate the above fundamental dichotomy of the standard quantum mechanics. What have we invented for our original purpose and what for others ?

3 PERMANENT REDUCTION—EQ. FOR $\psi(t)$

Jumps (2) are not acceptable in Nature, let us interpolate them by equivalent processes. This was the main motivation to construct the detailed time evolution equation for $\psi(t)$ interpolating the von Neumann reduction (2). Let us see the basic results in historical order.

Bohm and Bub [1] proposed deterministic evolution equation where randomness were only put into initial conditions. Nonlinear Wiener process was introduced by Pearle, later he found a a clever gambling game analogy, too [2]. Gisin [3] pointed out that the stochastic average of $|\psi(t)\rangle\langle\psi(t)|$, i.e. $\rho(t)$, must all time obey to linear master equation, and he constructed the first such nonlinear Wiener process. Diósi [4] showed how a unique nonlinear Wiener process follows from an arbitrarily given master equation of Lindblad class. And finally, Gisin and Percival [5] presented the Ito equations of the unique Wiener process. This has been known as Quantum State Diffusion (QSD) theory (see Appendix A).

In addition the the first mentioned motivation, it is worthwhile to invoke two further ones.

Environmental Reduction. In everyday experiences, Nature shows spontaneous classicality. Logically, all c-numbers emerge in corresponding von Neumann reductions. Macroobjects are never isolated and their environment acts formally as apparatus [6]. This action is permanent and spontaneous. The macrosystem's density operator $\rho(t)$ is uniquely calculable, at least in principle. The stochastic process for the unraveling $\rho(t)$ into pure state stochastic process $\psi(t)$ can be done in many ways. The QSD theory yields a unique one.

Universal Reduction. On philosophical grounds we can hardly accept that classicality is only due to the particular, e.g., thermal environment. A certain — dynamically not specified — universal "environment" is to be assumed that guarantees the observed generality and universality of permanent reduction. Then, postulating the form of the ensemble evolution equation, the pure state evolution, representing classicality, follows from QSD theory automatically. For the state of art, see Appendix B.

4 PERMANENT REDUCTION—EQS. FOR $\psi(t)$ AND $z(t)$

A typical class of quantum measuring apparatuses acts permanently, i.e., performs unsharp continuous detection. (Also universal permanent reduction can be represented formally as if Nature would apply certain apparatuses performing permanent detection.) The joined statistics of the continuously reduced state $\psi(t)$ and of the classical record $z(t)$ represents both practical and fundamental interests. The task that follows is twofold. Firstly, one has to construct plausible mathematical model of continuous (unsharp) von Neumann reductions and, secondly, the simplest form of the corresponding stochastic equations for $\psi(t)$ and $z(t)$ has to be found.

Let us see a deliberate choice of results. To my knowledge, the idea of restricting Feynman-path integrals for a tube along $z(t)$ is due to Mensky [7]. Barchielli *et al.* [8,9] introduced Gaussian "tubes" and derived generating functionals for the distribution functions of the processes $\psi(t), z(t)$. Caves and Milburn [10] invented the feed-back mechanism in the path-integral formalism. Diósi [11] derived 2 separate Ito-equations: one for $\psi(t)$ and one for $z(t)$. The first one turned out to be the Gisin nonlinear equation [3]. Recently, Diósi, Gisin, Halliwell, and Percival [12] proved that the standard QSD corresponds to permanent reduction onto the (approximate) eigenstates of the Lindblad generators accompanied by a certain feed-back *à la* Caves and Milburn [10]. The classical record $z(t)$ is given by $z(t) = \langle L \rangle + \dot{w}(t)$ where $w(t)$ is the same standard (complex) Wiener process that drives $\psi(t)$ in QSD.

5 CONSISTENT HISTORIES

The motivation for the reinterpretation of quantum mechanics in terms of CH [13–16] is the same that the motivation for the permanent reduction project: both models explain the spontaneous emergence of the classical from the quantum. The CH theory constructs families of repeated (or even continuous) reductions, consistent according to classical logics. It is not at all surprizing that the CH and the permanent reduction theories are closely related. Actually, it can be shown that the samples $\psi(t)$ of QSD are histories [12] and in most cases they are consistent as well. At least in those cases where the closed quantum mechanical system splits into a reservoir and a Markovian subsystem in it, the well-developed Ito-calculus of permanent reduction offers explicit equations for the CH theory, too.

6 BY-PRODUCTS AND CONCLUSION

Even if permanent reduction would not have been successful in clarifying fundamental problems it could have, nevertheless, initiated a number of more practical results. Seemingly, the relevance of nonlinear stochastic wave equations were realized first in the cited works of fundamental purposes in the eighties. Nonlinear stochastic equations, based mostly on independent researches (see, e.g., Refs. [17–20]) have now become extensively used in quantum optics. It is not worthless to note that the full technical apparatus for wave function Monte Carlo simulations [21] in any Markovian system has been granted, e.g., from 1988 [4] or, in terms of QSD, from the nineties. Several good examples could be invoked to show that practical by-products of the researches on permanent reduction might well be fertilizing other fields such as quantum optics and communication.

A. QUANTUM STATE DIFFUSION

Consider a system whose state is subjected to permanent reduction and assume that we know the Lindblad master equation governing the evolution of the density operator:

$$\dot{\rho} \equiv \mathcal{L}\rho = -i[H,\rho] + L\rho L^\dagger - \frac{1}{2}\{L^\dagger L, \rho\} \tag{4}$$

where L denotes the (vector of) Lindblad generator(s). According to the QSD theory, the following Ito stochastic equation belongs to the Liouville superoperator $\mathcal{L}$:

$$\dot{\psi} = \left(\mathcal{L}|\psi\rangle\langle\psi|\right)\psi + \dot{w}^*\left(L - \langle\psi|L|\psi\rangle\right)\psi \tag{5}$$

where $w(t)$ is the (vector of) standard complex Wiener process(es):

$$\dot{w}(t)\dot{w}^*(t') = \delta(t-t') \quad , \qquad \dot{w}(t)\dot{w}(t') = 0 \quad . \tag{6}$$

The QSD process (5) has, among others, two noticeable properties. From the viewpoint of fundamental principles: the above QSD is the only pure state diffusion process free of any additional parameters or functions once $\mathcal{L}$ has been defined. It depends only on $\mathcal{L}$; it is invariant under the redefinition of the Lindblad generator(s) L. From a more practical viewpoint: the above pure state diffusion process put as much of the dynamics as possible into the nonlinear drift term hence the stochasticity is kept minimum. In other words, the nonlinear drift of QSD offers the best nonlinear Schrödinger equation to approximate the short time behaviour of the density operator.

B. UNIVERSAL PERMANENT REDUCTION

A well–known first attempt to model Universal Permanent Reduction is the GRW-theory [22]. Let us illustrate the status of Universal Permanent Reduction project by a model which relates permanent reduction to gravity [23]. A compact definition of the model can be achieved if we specify the Lindblad generators. They are chosen to be proportional to the square root of the gravitational constant G:

$$L_{\mathbf{k}} = \sqrt{\frac{G}{4\pi k}} \sum_m m \exp\left(i\mathbf{k}\mathbf{x}_m\right) \tag{7}$$

where $m, \mathbf{x}_m$ are the masses and coordinate operators, resp., of the objects belonging to our system; summation should be taken for all masses. The vector $L_\mathbf{k}$ is labelled by the wavenumber $\mathbf{k}$. A short distance cutoff must be applied at $k \approx 10^5 cm^{-1}$ [24]. It can be shown that the corresponding QSD equations reduces to the ordinary quantum mechanics for microscopic masses. For more massive objects, permanent reduction become effective and plausible classicality emerges, e.g., in unique position distributions. Gravitational permanent reduction will hopefully find new justification in the frame of quantum cosmology.

This work was supported by the grant OTKA No. 1822/1991. The author thanks the Symposium organisers for the kind and generous invitation.

References

[1] D. Bohm and J. Bub, Rev. Mod. Phys. 38 (1966) 453.

[2] P. Pearle, Found. Phys. 12 (1982) 249.

[3] N. Gisin, Phys. Rev. Lett. 52 (1984) 1657; 53 (1984) 1776; Helv. Phys. Acta 62 (1989) 363.

[4] L. Diósi, J. Phys. A 21 (1988) 2885.

[5] N. Gisin and I.C. Percival, J. Phys. A 25 (1992) 5677; A 26 (1993) 2233; A 26 (1993) 2245; also in: Stochastic Evolution of Quantum States in Open Systems and in Measurement Processes, eds.: L. Diósi and B. Lukács (World Scientific, Singapore, 1994).

[6] H.D. Zeh, Found. Phys. 1 (1970) 69.

[7] M.B. Mensky, Phys. Rev. D 20 (1979) 384.

[8] A. Barchielli, L. Lanz and G.M.Prosperi, Nuovo Cim. 72 B (1982) 79.

[9] A. Barchielli, Nuovo Cim. 74 B (1983) 113.

[10] C.M. Caves and G.J. Milburn, Phys. Rev. A 36 (1987) 5543.

[11] L. Diósi, Phys. Lett. A 129 (1988) 419.

[12] L. Diósi, N. Gisin, J. Halliwell and I.C. Percival, Imperial College preprint, also as gr-qc/9403047.

[13] R.B. Griffiths, J. Stat. Phys. 36 (1984) 219.

[14] R. Omnès, Rev. Mod. Phys. 64 (1992) 339.

[15] M. Gell-Mann and J.B. Hartle, Phys. Rev. D 47 (1993) 3345.

[16] H.F. Dowker and J.J. Halliwell, Phys. Rev. D 46 (1992) 1580.

[17] V.P. Belavkin, Phys. Lett. A (1989) 335.

[18] A. Barchielli and V.P. Belavkin, J. Phys. A (1991) 1495.

[19] V.P. Belavkin and P. Staszewski, Phys. Rev. A 45 (1992) 1347.

[20] M.J. Gagen, H.M. Wiseman and G.J. Milburn, Phys. Rev. A 48 (1993) 132.

[21] J. Dalibard, Y. Castin and K. Mølmer, Phys. Rev. Lett. 68 (1992) 580.

[22] G.C. Ghirardi, A. Rimini and T. Weber, Phys. Rev. D 34 (1986) 470.

[23] L. Diósi, Phys. Rev. A 40 (1989) 1165.

[24] G.C. Ghirardi, R.Grassi and P. Pearle, Phys. Rev. A 42 (1990) 1057.

QUANTUM TRAJECTORIES FOR QUANTUM OPTICAL SYSTEMS

G.J.Milburn, J.K.Breslin and H.M.Wiseman

Department of Physics
University of Queensland
St Lucia 4072 Australia.

1 INTRODUCTION

A quantum trajectory refers to the conditional state of a quantum system, conditioned on a time sequence of measurement records [1–3]. This state is thus dependent on a measurement history. The measurement record is a stochastic process. The source of the randomness is two-fold; it is due to the irreducible quantum fluctuations in the measured system, the source of the information, and it is due to noise added by the measurement apparatus itself. As the stochastic measurement record is accumulated, the conditional state of the measured system will undergo a stochastic evolution. If the measurement record is complete, and if the initial state is pure, the stochastic evolution of the conditional state can be represented by a pure quantum state, otherwise the conditional state is represented by a mixed state. A quantum trajectory thus connects a classical stochastic process, the measurement record, to a quantum stochastic process. We will primarily be concerned with continuous measurement, for which the measured system is always coupled to an environment from which information can be extracted. From this point of view there is nothing particularly new about quantum trajectories. The conditional evolution of a measured system has been considered by many authors [4–8]. Stochastic Schroedinger equations have also been postulated as a means to describe all open quantum systems, not just measured systems [9–11]. In this paper we will primarily be concerned with particular applications of the theory of continuous measurement for quantum optical systems. In the first part of this paper we will use a theory of continuous homodyne detection to derive a master equation for electro-optic feedback.

Quantum Communications and Measurement
Edited by V. P. Belavkin *et al.*, Plenum Press, New York, 1995

There is a close connection between a quantum trajectory and irreversible evolution of an open quantum system. Clearly a measured quantum system is an open system (although not every open system is a measured system). If we do not take account of the measured results, that is, we sum over all possible measurement records, the evolution of the measured system is described by a non-unitary evolution which in general takes pure states to mixed states. This evolution is usually described by a Markov master equation. The measurement record labels environment states. This may be done in many ways. There may be many ways of extracting information about the quantum source from the environment to which it is coupled, and each will give rise to a different quantum trajectory. None the less when the measurement records are summed over, the same master equation must result. A very natural question to ask in this context is; for a given master equation is there an optimal quantum trajectory ? Is there an optimal way of extracting information from the environment that enables the most efficient characterisation of the conditional state of the quantum source ? This is the question we address in the second part of this paper.

As the quantum trajectory approach makes an explicit reference to the measured record, one way to approach the question of optimality is to ask for records that are optimal. To define this, we turn to algorithmic information theory. As discussed below the optimal record is that for which the average algorithmic information of the record is a minimum. That minimum is constrained by the particular master equation which governs the evolution of the quantum source.

Finally, we apply these ideas to a quantum optical system with a rich dynamical structure; the parametrically kicked nonlinear oscillator. This system, in a classical description, exhibits regular and chaotic motion.

2 THE QUANTUM THEORY OF HOMODYNE CURRENT FEEDBACK

We will illustrate the general principles of quantum limited feedback by deriving a master equation to describe the evolution of an optical cavity subject to feedback control. How information is extracted from the cavity determines the minimum noise added by the measurement. We first consider the simplest situation in which the cavity is simply damped through an end mirror. Information about the cavity field is then extracted by measurements on the output field. We will assume the cavity has two outputs at least and use one output to make the measurements required for feedback. Information is extracted from the output field by balanced homodyne detection, the optical analogue of continuous position measurements.

The detected photocurrent at any time depends on the state of the cavity field at that time as well as the entire history of the photocurrent. In a simple-minded theory the state of a system conditioned on a previous measurement is determined by the projection postulate, however we need a much more sophisticated theory of measurement to treat homodyne detection. This theory has previously been given by us [12]. The result is that the current is given by

$$I_c(t) = \eta\langle a + a^\dagger\rangle_c(t) + \sqrt{\eta}\xi(t) \tag{1}$$

where $\xi(t)$ is a delta-correlated white noise term, η is a measure of the efficiency of the detection process and $\langle a + a^\dagger\rangle_c(t)$ is the conditional mean of the quadrature amplitude inside the cavity. Time is measured in units of inverse cavity damping rate. Using the generalised theory of measurement for such processes we can derive a stochastic master equation which gives the evolution of the system as a functional of a particular photocurrent history.

We now assume the feedback works by driving an intracavity process with the photocurrent in Eq.(1). One then derives a modified stochastic master equation which when averaged over all histories gives the evolution of the intracavity field in the presence of feedback. If the effect of feedback on the cavity field is described by the operation

$$[\dot{\rho}_c]_{fb} = I_c(t)\mathcal{K}\rho_c \tag{2}$$

where $\mathcal{K}$ is a Liouville superoperator, then the appropriate master equation is [13]

$$\dot{\rho} = \mathcal{L}\rho + \mathcal{K}(a\rho + \rho a^{\dagger}) + \frac{1}{2\eta}\mathcal{K}^2\rho \tag{3}$$

The first term represents the free dynamics of the cavity and the standard damping term of the field. This term depends on how the information is extracted from the cavity. The second term is a (non-hamiltonian) nonlinear term, while the last term represents the noise which necessarily accompanies feedback. The second and last terms are tied together in just the same way that fluctuations always accompany damping. Thus this equation is an example of the general rule that a given kind of feedback is necessarily accompanied by a particular noise term. A similar result was obtained by Caves and Milburn [6] based on a general model of continuous position measurements.

It is now easy to add other intracavity dynamics, such as a laser, to the master equation. However here we emphasise an important result of Eq.(3). Firstly, adding any term which leads to a linear semiclassical dynamics for the intracavity field amplitude, even if modulated by a feedback current, cannot produce nonclassical light. This immediately follows when one looks at the Glauber P-function which results from the second term in Eq.(3). Furthermore the accompanying noise will not change this situation. Thus we have the general result that homodyne current feedback on to a linear intracavity element cannot produce noise reduction below the quantum limit.

The ultimate explanation for this effect is due to the rather clumsy way we have attempted to extract information from the cavity, ie. by simply letting the field propagate out of the cavity in the usual way and then making measurements on the output. One can do much better by making good measurements directly on the intracavity field. How is one to do this ? The answer is by quantum nondemolition (QND) measurement [14]. Feedback control based on information extracted from the cavity by a QND scheme can produce perfect noise reduction in the output field of the cavity if the gain of the QND measurement is good enough.

Quantum limited electro-optic feedback is quantum measurement theory put to work. The instantaneous state of a system is conditioned by the result of a measurement and the future of the system is altered depending on the result of that measurement. If the measurement is a good one, then feedback via the measured variable can reduce noise below the quantum limit despite the fact that the measurement process itself adds noise into the system.

3 OPTIMAL MEASUREMENT

If a measurement is made of some physical quantity, upon a system in state ρ, the result, a, is a random variable distributed according to

$$P(a) = tr(\hat{F}(a)\rho) \tag{4}$$

where $\hat{F}(a)$ is a positive operator (an effect [15]). Clearly we must have $\sum_a \hat{F}(a) = 1$. (The sum here would be an integral if the results of the measurement were a continuous variable).

The state of the system, conditioned on the result of the measurement is given by

$$\rho^{(a)} = (P(a))^{-1}\,\phi_a\rho \tag{5}$$

The operation ϕ_a must satisfy $tr(\phi_a\rho) = tr(\hat{F}(a)\rho)$. Clearly there are many possible operations which would satisfy this relation. That is to say, there are many measurement schemes which yield the same statistics for the measured quantity. All operations can be written in terms of operators by

$$\phi_a\rho = \hat{\Omega}(a)\rho\hat{\Omega}(a)^\dagger \tag{6}$$

The unconditioned state of the system, that is the system state given that the measurement has taken place but for which the outcome is unknown, is given by

$$\rho' = \sum_a \phi_a\rho \tag{7}$$

In general this transforms pure states to mixed states.

It is well known that there are infinitely many ways to write a general density operator as a sum over pure state projectors. Eq.(7) represents the most natural way to write the state of the system after it has interacted with a known measuring device, however it is not the only way. So let us pose another question. Given that we know the initial state ρ and the final state ρ' what is the 'best' way to decompose this as a sum over conditional states ? Or in more suggestive language, what is the best measurement we can make that corresponds to the transformation in Eq.(7) ? To answer this consider, the average algorithmic information in the record of measurement results. To a very good approximation the average algorithmic information needed to describe the measurement results $\{a\}$, is given by [16]

$$I \approx -\sum_a P(a)\log_2 P(a) \tag{8}$$

where $P(a) = tr(\phi_a\rho)$. The quantity on the right hand side of Eq.(8) is the 'ensemble entropy', $H(\mathcal{A})$ [17], for the particular ensemble decomposition in Eq.(7), (here $\mathcal{A}$ labels the particular ensemble used). We now define an optimal measurement to be that for which the average algorithmic information, needed to specify an outcome, is a minimum. How small can this be ? The answer is provided by an inequality relating the ensemble entropy to the von Neumann entropy of the final state ρ' [17].

$$H(\mathcal{A}) \geq S(\rho') \tag{9}$$

where

$$S(\rho) = -tr(\rho\log_2\rho), \tag{10}$$

is the von Neumann entropy of a state ρ. The von Neumann entropy is the ensemble entropy for the eigen-ensemble of state ρ. The eigen-ensemble is defined to be the ensemble made up of the (orthogonal) eigenstates of the density operator.

How is the optimal measurement related to an arbitrary measurement ? Suppose a measurement is described by the unconditional state

$$\rho' = \sum_{\nu=1}^{s} \hat{A}_\nu \rho \hat{A}_\nu^\dagger \ . \tag{11}$$

and let $\hat{P}_\mu$ be a projector onto the $\mu-$th eigenvector of ρ'. Then it follows that [17]

$$\hat{A}_\nu = \sum_{\mu=1}^{k} N_{\nu\mu} \hat{P}_\mu \tag{12}$$

where the $s \times k$ matrix $N_{\nu\mu}$ satisfies

$$\sum_{\nu=1}^{s} N_{\nu\mu} N_{\nu\mu'}^* = \delta_{\mu\mu'} \tag{13}$$

and k is the rank of ρ'. All measurements are unitarily related to the optimal measurement defined by the projectors $\hat{P}_\mu$. Then, in principle, there is an optimal way of extracting information from the measurement apparatus, consistent with the non-unitary transformation in Eq.(11) for which the average algorithmic information of the record is a minimum. This optimal 'unravelling' of the unconditional state is simply the eigen-ensemble of the unconditional state after the measurement. In reality it may be difficult, if not impossible, to extract the relevant information from the apparatus.

As an example we consider measurements made on a spin-half system. At a fixed time suppose the measured system is coupled to another spin-half system, called the meter. Let the interaction Hamiltonian be

$$H_I = \frac{2\hbar\epsilon}{\tau} \hat{\sigma}_z \hat{S}_z \tag{14}$$

where $\hat{\sigma}_z$ and $\hat{S}_z$ are the $z-$components of the spin operator for the system and apparatus respectively. We assume an impulsive coupling so that the free evolution of the system and apparatus may be neglected over the time of interaction τ. The apparatus is initially prepared in an eigenstate of the $x-$component of spin. Clearly we need to measure the $x-$component of the apparatus spin to obtain information of the system spin. It is easy to see that the unconditional state of the system after the measurement is given by

$$\rho' = \frac{1}{2} \left(U^+ \rho (U^+)^\dagger + U^- \rho (U^-)^\dagger \right) \tag{15}$$

where $U^+ = (U^-)^\dagger = e^{-i\epsilon\hat{\sigma}_z}$. If ρ is the positive eigenstate of σ_x, the von Neumann entropy for the transformed state is

$$S(\rho') = \frac{1}{2}(1 + \cos\epsilon) \log_2(1 + \cos\epsilon) + \frac{1}{2}(1 - \cos\epsilon) \log_2(1 - \cos\epsilon) \tag{16}$$

Taken on its own the transformation in Eq.(15) may be interpreted as describing an equal mixture of systems which have undergone opposite precessions about the

$z-$axis. (Such a perturbation is considered in reference [18] where it is called the turn perturbation). With this interpretation one is led to regard each term in Eq.(15) as describing the conditional state of the system given that the sign of the precession is known. The average information gained on obtaining that knowledge is one bit. Clearly this is greater than the von Neumann entropy in general and thus such an unravelling of the measurement is not optimal.

However by explicit construction the transformation in Eq.(15) may be given a different interpretation in terms of a measurement of the $x-$component of the apparatus spin. There are two possible outcomes to such a measurement (assumed to be perfectly accurate) labelled by up (u) or down (d). We assume for simplicity that the initial system state ρ is a pure state, $|\psi\rangle$. The conditional(unnormalised) states of the system, given the results of the measurement, are

$$|\tilde{\psi_u}\rangle = \cos(\epsilon\hat{\sigma}_z)|\psi\rangle \tag{17}$$

$$|\tilde{\psi_d}\rangle = \sin(\epsilon\hat{\sigma}_z)|\psi\rangle \tag{18}$$

The probabilities for each outcome are

$$p_u = \langle\psi|\cos^2(\hat{\sigma}_z)|\psi\rangle \tag{19}$$

$$p_d = \langle\psi|\sin^2(\hat{\sigma}_z)|\psi\rangle \tag{20}$$

For example, suppose $|\psi\rangle$ is the up-eigenstate of the $x-$component of spin. Then

$$p_u = \frac{1}{2}(1+\cos\epsilon) \tag{21}$$

$$p_d = \frac{1}{2}(1-\cos\epsilon) \tag{22}$$

Noting that in this case the conditional states are orthogonal, we see that the ensemble defined by this unravelling of the transformation in Eq.[15] is indeed the eigen-ensemble. Thus the average algorithmic information needed to store a record of this measurement result is equal to the von Neumann entropy for the unconditional state, and this readout of the apparatus is optimal.

A number of authors have discussed the case of continuous measurements [5–8]. In such cases the quantum system of interest is always coupled to an environment from which information may be continually extracted. We will follow the presentation of Wiseman [8] which is best suited to the quantum optical applications discussed below.

Let us suppose that in a small time interval dt the result of a measurement is mapped onto the integers $\{0,1\}$. As we shall show, this is a very natural description of measurement in the case that the measured system state obeys a Markov evolution equation. Then following the description in the previous section, the probability of a result $\nu \in \{0,1\}$ is given by

$$P(\nu) = tr\left(\Omega_\nu(dt)\rho(t)\Omega_\nu^\dagger(dt)\right) \tag{23}$$

where

$$\tilde{\rho}(t+dt) = \Omega_1(dt)\rho(t)\Omega_1^\dagger(dt) + \Omega_0(dt)\rho(t)\Omega_0^\dagger(dt) \tag{24}$$

is the unconditional density operator at time $t+dt$.

To proceed, we must specify the nature of the measurement interaction. The simplest way to do this is first to determine the equation governing the irreversible evolution of the measured system, and then choose the measurement operations in Eq.(24) accordingly. A measured system is of course an open system. We will only consider the case where the nature of the system/environment interaction is such as to justify describing the evolution of the system by a Markov master equation [19]. The most general form of the master equation for the density operator of an open quantum system is (we use $\hbar = 1$),

$$\dot{\rho} = -i[H, \rho] + \sum_{\mu} \mathcal{D}[c_\mu]\rho \, , \tag{25}$$

where H is an hermitian operator and $\mathcal{D}$ is a superoperator defined by $\mathcal{D}[c] = \mathcal{J}[c] - \mathcal{A}[c]$ where

$$\mathcal{J}[c]\rho = c\rho c^\dagger \quad ; \mathcal{A}[c]\rho = \frac{1}{2}(c^\dagger c\rho + \rho c^\dagger c) \tag{26}$$

The set of operators c_μ is not unique, and it is precisely this non-uniqueness that we will use to define optimal measurements. For simplicity consider the case of a single output channel so that

$$\dot{\rho} = -i[H, \rho] + \mathcal{D}[c]\rho \tag{27}$$

This master equation is invariant under the transformation

$$c \to c + \alpha \; ; \quad H \to H - i\frac{1}{2}(\alpha^* c - \alpha c^\dagger) \tag{28}$$

It is now possible to assign the measurement operations for the master equation Eq.(27). To begin we consider the case $\alpha = 0$, which we will refer to as the *standard unravelling*. We will further assume perfect measurement (see reference [8] for the generalisation to imperfect measurements). The operations are defined by the operators

$$\Omega_1(dt) = \sqrt{dt}c \tag{29}$$

$$\Omega_0(dt) = 1 - \left(iH + \frac{1}{2}c^\dagger c\right) dt \tag{30}$$

It is easily verified that summing over the measurement results reproduces the master equation, that is

$$\rho(t + dt) = \sum_{\nu=1,0} \Omega_\nu(dt)\rho(dt)\Omega_\nu^\dagger(dt) \tag{31}$$

is equivalent to the master equation Eq.(27). The probabilities for the elementary result is

$$P(1) = \langle c^\dagger c\rangle dt \tag{32}$$

where $\langle A\rangle = tr(\rho(t)A)$.

If we assume the system starts in a pure state, the conditional state of the system at any time must also be pure as we are assuming perfect measurements. Thus the conditional state of the system given a particular result $\nu \in \{0, 1\}$, is the (unnormalised) pure state $|\tilde{\psi}^{(\nu)}(t + dt)\rangle = \Omega_\nu(dt)|\psi_c(t)\rangle$ where the subscript c refers to the conditional state up to time t, that is, this subscript refers to an entire history of measurement records up to time t. In general the measurement thus defined cannot be optimal as the conditional states $|\tilde{\psi}^{(0)}(t+dt)\rangle$, $|\tilde{\psi}^{(1)}(t+dt)\rangle$ need not be orthogonal. We can however, determine how far from optimal they are.

Solve the master equation up to time t, giving the unconditional state $\rho(t)$. This is determined by the master equation alone, and not the particular way in which we extract information from the environment. Given this state we can compute the von Neumann entropy $S = -tr(\rho(t)\log_2\rho(t))$. We can also compute the average algorithmic information I of the measurement records, that is of the binary strings of measurement results. This can be done, in principle, as the probability of a particular history, up to time t, is given by

$$P(\{\nu_1, \nu_2, \ldots, \}; t) = \langle\tilde{\psi}_c(t)|\tilde{\psi}_c(t)\rangle \tag{33}$$

In practice, it is easier to do this by a quantum trajectory simulation. In general, the von Neumann entropy S will be very much less than the average algorithmic information I for the record.

We are now in a position to define a class of optimal measurements using the freedom provided by the transformation Eq.(28). If α is not equal to zero, the measurement operators are

$$\bar{\Omega}_1(dt) = \sqrt{dt}(c+\alpha) \tag{34}$$

$$\bar{\Omega}_0(dt) = 1-(iH+\alpha^*c+\frac{1}{2}c^\dagger c+\frac{1}{2}|\alpha|^2)dt \tag{35}$$

If we are following the conditional evolution so that at some time t the conditional state is $|\psi_c(t)\rangle$ (we assume the initial state was pure), then for an optimal measurement we require that the unnormalised conditional states $\bar{\Omega}_1(dt)|\psi_c(t)\rangle$ and $\bar{\Omega}_0(dt)|\psi_c(t)\rangle$ are orthogonal. One easily sees that this can only occur if we choose α at every infinitesimal time step to be equal to the negative of the conditional mean of c at that time,

$$\alpha = -\langle c\rangle_c\ , \tag{36}$$

then the probabilities for the elementary event is

$$\bar{P}(1) = \left(\langle c^\dagger c\rangle_c - |\langle c\rangle_c|^2\right) dt \tag{37}$$

Of course we will need to know the initial state if we are to know the conditional means at every time. The condition Eq.(36) was first proposed by Diósi in an unrelated context [20]. Comparing Eqs.(37) with Eq.(32), we see that in optimal measurement the elementary event 1 is less likely. In some circumstances it may never occur, in which case the conditional state is simply the solution to the master equation. The measurement record would then be the string of zeros which occurs with unit probability. The average algorithmic information of such a string is zero. Thus this extreme case can only occur if the master equation takes a known initial pure state to a pure state at any time. In general however the average algorithmic information of the record, while less than that for a nonoptimal measurement, will be greater than the von Neumann entropy calculated at the end of the time interval.

4 QUANTUM TRAJECTORIES FOR A CHAOTIC SYSTEM

In this section we illustrate some aspects of quantum trajectories, entropy and information by considering the quantum dynamics of a parametrically kicked nonlinear optical oscillator subject to quantum nondemolition measurement of photon number. This model was first proposed by one of us [21] and in the classical limit exhibits a rich dynamical structure including regions of regular and chaotic behaviour.

The quantum description reflects these two kinds of behaviour in terms of regular or irregular revivals. Apparently the quantum description does not show any sign of chaotic behaviour in terms of a sensitive dependence on initial conditions. However Caves and co-workers have recently challenged this orthodox view [22,23,18], giving a new characterisation of quantum chaos using algorithmic information theory. In this section we first describe the model with quantum counting nondemolition measurements. Recently Barnum et al. [24] have considered similar measurements on this system. We show that the average algorithmic information needed to specify a measurement record, for a system initially localised in a chaotic region of phase space, is much greater than that needed for a system initially localised in the regular region, and very much greater than the von Neumann entropy of the unconditional quantum state.

We consider a single mode field of an optical cavity containing a medium with an intensity dependent refractive index (a Kerr medium). Such a system is modelled as a nonlinear oscillator in which the nonlinear frequency of oscillation is a linear function of the energy. This is an integrable system. To make the dynamics a little more interesting we suppose that periodically the oscillator is subject to a parametric amplification process which acts impulsively, that is on a much shorter time scale than all other dynamical processes. In such a case we model the effect of the parametric process as a periodic sequence of fixed unitary transformations. Physically, such a kick would correspond to pumping a second order nonlinear crystal, inside the optical cavity, with a sequence of very short pulses of light.

The simplest description of the quantum dynamics is in terms of a map for the quantum state, which relates the state from one kick to the next. The map is defined by

$$|\psi'\rangle = \exp\left(\frac{r}{2}((a^\dagger)^2 - a^2)\right) \exp\left(-i\frac{\mu}{2}(a^\dagger)^2 a^2\right) |\psi\rangle \tag{38}$$

where a is the annihilation operator for the cavity field and where r is the strength of the parametric pumping while μ is proportional to the nonlinear refractive index, χ, times the interaction time. The corresponding classical model may be defined by computing the effect of this unitary transformation on a pair of canonically conjugate operators and then replacing these operators by commuting phase-space variables.

In addition to the nonlinear oscillation between kicks we now assume the system is subject to quantum nondemolition measurement of photon number [25]. Unlike ordinary absorptive photon counting, quantum counting does not destroy photons. Physically it can be achieved by a two-mode Kerr interaction with demolition quantum counting on a second mode to determine an intensity dependent phase shift [26]. The effect of a continual nondemoliton count process between kicks is modelled by a master equation for the unconditional state;

$$\dot{\rho} = -i\chi[(a^\dagger)^2 a^2, \rho] - \frac{\gamma}{2}[a^\dagger a, [a^\dagger a, \rho]] \tag{39}$$

where γ is a parameter that determines the signal-to-noise ratio for the measurement. One attractive feature of this master equation is that it leaves the photon number distribution unchanged. (Of course the kicks will change the photon number distribution.).

The master equation defines a *standard unravelling* in terms of the operators

$$\Omega_1(dt) \quad = \quad \sqrt{\gamma dt}\; a^\dagger a \tag{40}$$

$$\Omega_0(dt) = 1 - \left(i\chi(a^\dagger)^2a^2 + \frac{1}{2}\gamma(a^\dagger a)^2\right) dt \tag{41}$$

The measurement results are a Poisson process at the conditional rate $\gamma\langle(a^\dagger a)^2\rangle_c$. A fine-grained history of measurement results would be given in terms of the times of each event on some interval $[0,t)$. A course-grained history would be given in terms of the total count recorded in each of a series of equal time intervals on the interval $[0,t)$. The time interval, for example, could comprise the entire interval between kicks. Indeed the conditional state before each kick only depends on the total count over this interval and is given (unnormalised) in the photon number basis by

$$\langle n|\tilde{\psi}^{(m)}(t+T)\rangle = \frac{(n^2\gamma T)^{m/2}}{\sqrt{m!}} \exp\left(-(i\chi + \frac{\gamma}{2})n^2T\right) \langle n|\psi(t)\rangle \tag{42}$$

where m is the total count recorded in the time T from just after one kick to just before the next. The probability to record m kicks on this interval is given by

$$P(m) = \langle\tilde{\psi}^{(m)}(t+T)|\tilde{\psi}^{(m)}(t+T)\rangle \tag{43}$$

$$= \sum_{n=0}^{\infty} \frac{(n^2\gamma T)^m}{m!} e^{-\gamma n^2 T} |\langle n|\psi(t)\rangle|^2 \tag{44}$$

Eq.(44) provides a particularly easy way to simulate a quantum trajectory corresponding to this course-grained history.

In a number of recent papers Schack, Caves and co workers [16, 18, 22, 23] have considered the sensitivity to perturbation of classical chaotic systems and their quantum analogues. This sensitivity is defined in terms of the rapid increase with time of the information needed to describe the perturbed evolution of a system state. The information required to follow the perturbation is exponentially larger than the usual increase in entropy that results when averaging over the perturbation. In [18] the perturbations are modelled as unitary transformation at fixed time intervals, with a random parameter. This effectively models the perturbation by a stochastic Hamiltonian. Following the effect of the perturbation is very much like following a quantum trajectory. It can only be done if the entire record of perturbation parameters is known. In this section we will also consider the effect of perturbation on a chaotic system but the perturbation results from a quantum back-action in a QND measurement of photon number. The effect of the peturbation on the system can be completely reconstructed so long as the record of measurement results is known. Our objective remains the same as that of Schack and Caves; to determine how much information (in the measurement record) is needed to track the state of the system.

The measurement record comprises strings of positive integers, with each integer being the photon count recorded over the time interval between kicks. The string of measurement results is usually all that is available to the experimentalist. It is not necessarily known in advance that the measurements are ideal (that is of unit efficiency). However we will assume that such is the case. We will also assume that the initial pure state is known. With these assumptions the pure conditional quantum state of the system at every time step may be reconstructed form the measurement record. In the model of Schack et al. [18], all possible perturbation sequences were assumed to be equally probable. This is not the case here. We now describe how the simulation is done.

We first truncate the infinite oscillator Hilbert space to dimension R, chosen sufficiently large to ensure that states are normalized to within a specified tolerance

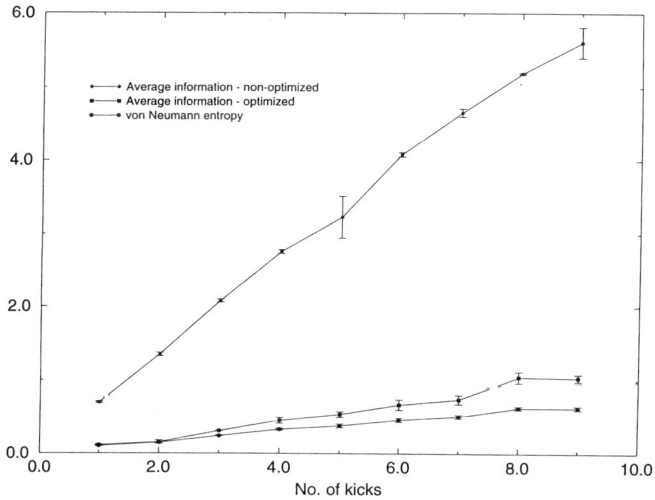

Figure 1: The von Neumann entropy and optimized and nonoptimized average algorithmic information in the measurement record versus kick number for a quantum nondemolition measurement of photon number. The system is initially localized in the regular region In all cases, $g = 1.5$, $\mu = 0.01\pi$, $\gamma = 0.001$, and the total number of trajectories is 6000.

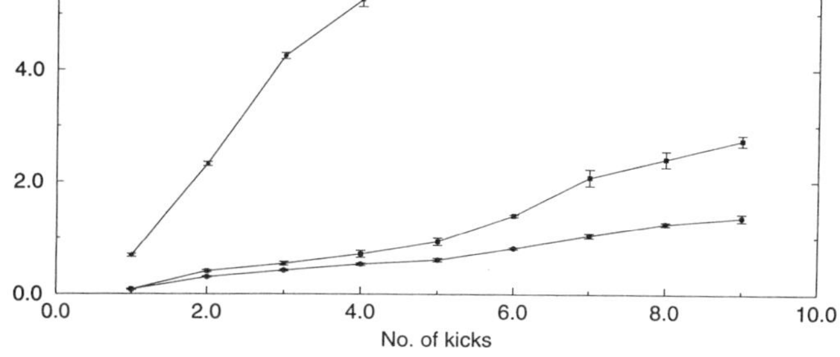

Figure 2: As in Fig. (1), except with the system initially localized in the chaotic region.

for the number of kicks of interest. This depends to a large extent on the initial state [25]. Next, the initial state is chosen to be a coherent state localized in either the regular or the chaotic region of the classical phase portrait. A random positive integer m_1 is then chosen by sampling the distribution in Eq.(44). This number is the count recorded on the first interval, from which the conditional state is determined. If we are interested in computing ensemble average values, for example of the average photon number, the conditional state can then be normalized and the first kick applied. We repeat this procedure for as many kicks as desired. The entire process is then repeated for another trial starting from the same initial state.

The case of optimized measurements however must be done quite differently. It is no longer the case that only the total count over the time between kicks determines the conditional state, due to the fact that the rate operator is continually adjusted as counts are recorded and the conditional mean changes. In this case we divide the interval between kicks into 100 equally spaced intervals. The duration $\Delta t = T/100$ of each interval is sufficiently small that it is most unlikely that more than one count will be recorded. A random number r, uniform on the unit interval, is then selected and compared with the probability $\bar{P}(1)\Delta t$ where $\bar{P}(1)$ is as in Eq.(37) with $c = \sqrt{\gamma}a^\dagger a$. If r is less than this probability, a one is recorded; otherwise the result is zero. Depending on the result of this record, the operator $\Omega_\nu(dt)$ is applied to the state at the start of the interval. The normalized state can then be computed before the next step. As a check on the method we can compute the average photon number for comparison with the results in [25]. In all cases the agreement is good.

To compute the average algorithmic information in the measurement record we proceed as follows. A quantum trajectory is generated by the methods described above and the resulting measurement records stored. While the probability of a measurement record could in principle be computed from the unnormalized conditional state up to the final kick, it is easier to simply count distinct measurement records and compute probabilities by dividing by the number of quantum trajectories simulated. This makes it easy to compute the entropy of the records and assign the average algorithmic information of the measurement records. The result depends on whether the initial state is localized in the regular or chaotic region. In figure (1) we plot the average algorithmic informations and the von Neumann entropy for the regular case. We used three ensembles each with 2000 quantum trajectories to compute these quantities and calculate the error bars. Note that the algorithmic information of the optimized record is much closer to the von Neumann entropy than that of the standard record.

In figure (2) we plot the same quantities for a state initially localized in the chaotic region of phase space. Note first that the rate of increase of the von Neumann entropy is higher than in the regular case. The rate of increase of the information in the optimal measurement record is higher also, and the difference is greater than for the von Neumann entropy. Finally, the rate of increase of the standard record's algorithmic information is also higher in the chaotic case, and the difference is by far the greatest out of the three quantities. The fact that in the chaotic case the average information required to track the state rises much more quickly than the von Neumann entropy is an example of hypersensitivity to perturbation discussed by Schack and Caves [22, 23, 18]. In our case the perturbation is supplied by the quantum back-action due to the measurement. Thus our results provide further indications that the definition of Schack and Caves for quantum and classical chaos in terms of sensitivity to perturbation is very useful.

The reason that the optimal measurement does not realize exactly the von Neumann entropy is, as explained above, that our optimal measurement is only optimal for each infinitesimal time interval for a particular trajectory, while the von Neumann entropy is defined by diagonalizing the density operator at every time. It would seem that it is not in general possible to have quantum trajectories corresponding to an average algorithmic information equal to the von Neumann entropy without modifying the master equation, that is by changing the coupling between the system and the apparatus at every time step. It is however possible that, for a given initial state and finite time interval, there may exist a different unraveling of the original master equation which will have a lower information content in its record, one closer to the von-Neumann entropy. Indeed it is possible to think of simple systems in which there is an obvious measurement scheme which is more optimal, in this sense, than the optimal measurement which we have defined. However, we do not see how such a measurement scheme could be determined in general.

References

[1] H.J.Carmichael, *An Open Systems Approach to Quantum Optics*, Lecture Notes in Physics, m18, (Springer,Berlin, 1993).

[2] H.M.Wiseman and G.J.Milburn, Phys. Rev A **47**, 1652 (1994).

[3] K. Mølmer, ICTP Lectures (1994).

[4] M.D.Srinivas and E.B.Davies, Optica Acta **28**, 981 (1980).

[5] A. Barchielli, Phys. Rev A **34**, 1642 (1986).

[6] C.M.Caves and G.J.Milburn, Phys. Rev A **36**, 5543 (1987).

[7] V.P.Belavkin, J.Phys. A **22**, L1109 (1989).

[8] H.M.Wiseman, Phys. Rev A **49**, 2133 (1994).

[9] N. Gisin, Phys. Rev. Lett., **52**, 1657 (1984).

[10] L.Diósi, J.Phys. A **21**, 2885 (1988).

[11] P.Pearle, Phys. Rev A **39**, 2277 (1989).

[12] H.M.Wiseman and G.J.Milburn, Phys. Rev A **47**, 642 1993

[13] H.M.Wiseman and G.J.Milburn, Phys. Rev. Letts. **70**, 548 (1993)

[14] P.Alsing, G.J.Milburn and D.F.Walls, Phys. Rev. A **37**, 2970 (1988).

[15] K.Kraus, *States, Effects and Operations: Fundamental Notions of Quantum Theory* (Springer, Berlin, 1983).

[16] C.M.Caves, Phys Rev E **47**, 4010 (1993).

[17] L.P.Hughston, R.Jozsa and W.K.Wooters, Phys. Letts. **183**,14 (1993).

[18] R.Schack and C.M.Caves, Phys Rev E, August (1994)

[19] C.W.Gardiner, *Quantum Noise*, (Springer-Verlag, Berlin, 1991).

[20] L.Diósi, Phys. Letts. **114A**, 20 (1986).

[21] G.J.Milburn, Phys. Rev A **41**, 6567 (1990).

[22] R.Schack and C.M.Caves, Phys. Rev. Lett. **69**, 3413 (1992).

[23] R.Schack and C.M.Caves, Phys. Rev. Lett. **71**, 525 (1993).

[24] H. Barnum, R.Schack and C.M.Caves, *preprint.*

[25] B.Wielinga and G.J.Milburn, Phys Rev A, **46**, 762 (1992).

[26] M.J.Gagen and G.J.Milburn, Phys Rev A **45**, 5228 (1992).

ENVIRONMENTAL AND PRIMARY QUANTUM STATE DIFFUSION

Ian C Percival

Department of Physics
Queen Mary and Westfield College,
University of London
Mile End Road, London E1 4NS, England

Abstract

Environmental quantum state diffusion (EQSD) models an indivdual quantum system interacting with its environment. It provides a method for practical computations that is more efficient than the numerical solution of the master equation for the density operator, provides physical insight, and has been widely applied. Primary state diffusion (PSD) is a related alternative quantum theory based on principles that are as simple as those of ordinary quantum theory, but its predictions are different. Compared with other alternative theories, PSD is less difficult to test experimentally. It is based on four principles and three conditions and contains one undetermined universal time constant τ_0. Special relativity is needed for extended systems even when they contain only slowly moving massive particles. Experiments on matter interferometry are proposed which either measure τ_0 or put bounds on it, and which might distinguish between PSD and ordinary quantum mechanics within the next decade. Both EQSD and PSD are based on Langevin-Itô diffusion equations for pure quantum states.

EXPERIMENT

Recent advances in the quantum theory of measurement and of open systems have been stimulated by experiments, for example

E1. Gravitational wave detectors.

Quantum Communications and Measurement
Edited by V. P. Belavkin *et al.*, Plenum Press, New York, 1995

E2. Cold particles in traps.

E3. Interacting photons and atoms in cavities.

E4. Tests of EPR - Bell's Theorem.

E5. Atom interferometry.

E6. Quantum cryptography and communication.

Many of these recent experiments control and detect coherence in individual quantum systems, beyond Bohr's wildest dreams!.

Pure states. In Bohr's time it was reasonable to consider only ensembles of *quantum* systems, although *classical* systems were represented individually. Because of the new experiments this is no longer reasonable.

So now we have *individual* quantum systems represented by state vectors obeying stochastic equations. These representations are necessary for the new experiments, but they are also useful for some old ones, for the dynamics of pure states is easier to compute, even if the equations are stochastic, and also a picture of the evolution of individual systems helps physical insight (Gisin and Percival, 1992a).

TWO MAIN APPROACHES

A1. The usual (Copenhagen) picture is so beautiful and useful, that we should change it as little as possible. *Method:* Base stochasticity on the detection process, eg homodyne and heterodyne detection, giving 'quantum trajectories'. There are new measurement hypotheses to replace those of Born and von Neumann.

A2. Since we now represent individual quantum systems and individual classical systems, for consistency in our physics we should use the same representation for both: a big change from the usual picture. *Methods:* Base stochasticity on interaction of the system of interest with its environment (EQSD) or a fundamental stochasticity of nature (PSD, also GRW and pilot wave theory). There is no independent measurement hypothesis. Measuring apparatus is just one type of environment, and the measurement process just one example of interaction between system and environment.

Failure to distinguish these two approaches in the past has led to some confusion. Both are discussed extensively at this meeting. I shall discuss A2.

EQSD provides a model for localisation and measurement, by adding a diffusion term to the Schrödinger equation. A detailed theory is presented in Gisin and Percival (1992a, 1993a, 1993b) and Percival (1994a), which will be called GP1, GP2, GP3 and Pe2 respectively. The model resolves the problem of wave-particle duality by providing solutions that behave like extended waves or like localised particles, depending on circumstances.

PSD is an alternative quantum theory, which is treated in detail in Percival (1994b), called Pe1.

ENVIRONMENTAL QUANTUM STATE DIFFUSION (EQSD)

John Bell remained to the end of his life a demanding physicist, who said that quantum mechanics was about beables, and not about observables (Bell, 1987a,

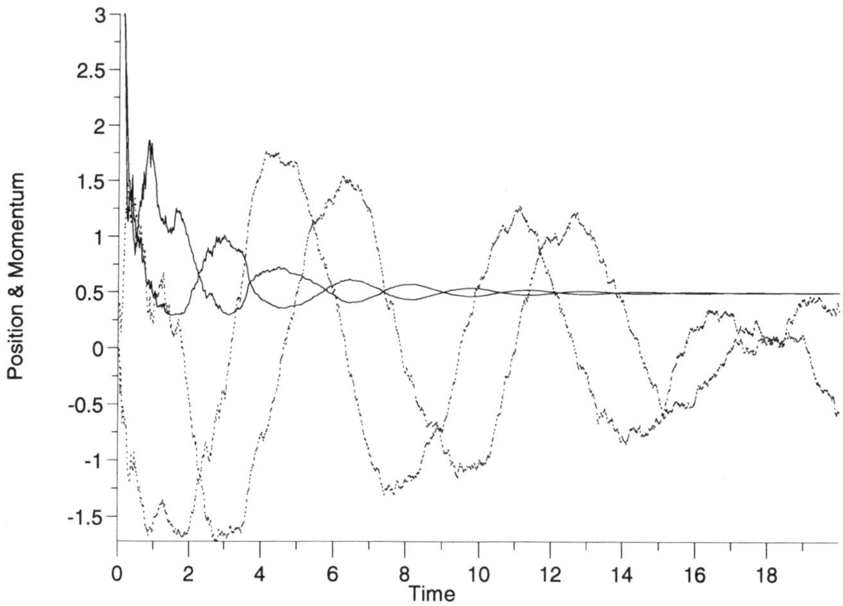

Figure 1: Approach to thermal equilibrium of a harmonic oscillator.

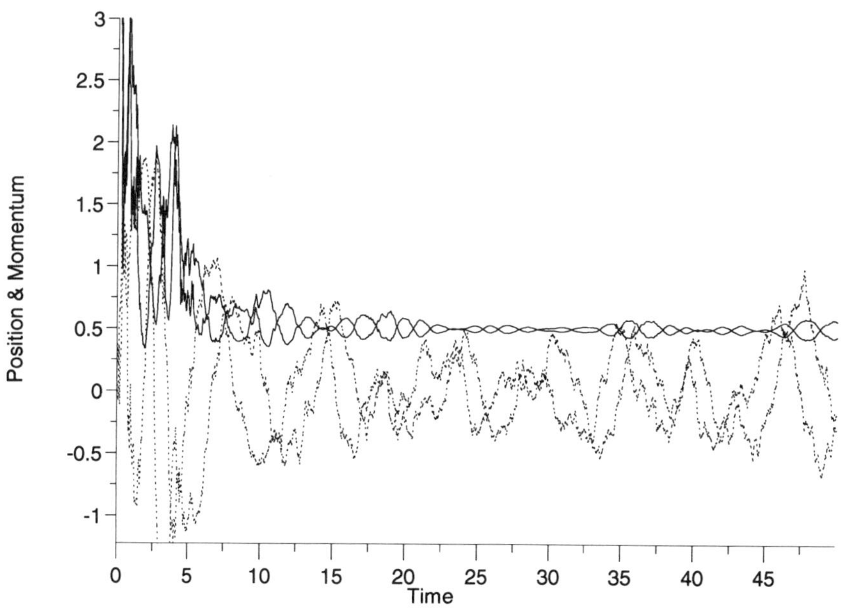

Figure 2: Nonlinear oscillator.

1987b, 1990), and who gave support to complete theories in which indivdual systems are explicitly represented, in particular the pilot wave theory of de Broglie and Bohm (Holland, 1993; Bohm and Hiley, 1993), and the theory of Ghirardi, Rimini and Weber (1986; Bell, 1987a). From Bell's viewpoint these are good theories because they are complete and because they satisfy the following further conditions:

G1. The state and evolution of an individual system are explicitly and unambiguously represented.

G2. There is no arbitrary division between the quantum and classical domains.

G3. There is no arbitrary division between system and environment.

From this viewpoint classical mechanics is a good theory because it satisfies G1 and G3, whereas G2 is irrelevant. But ordinary quantum theory is not. It does not satify G1, because the state vector represents probabilities in ensembles, not individual systems, and it does not satisfy G2 because the localisation of classical variables is incompatible with Schrödinger's equation. However it satisfies G3.

There are now several alternative quantum theories based on quantum state diffusion. Following earlier pioneering work of Bohm and Bub (1966) and Pearle (1976, 1979), Gisin (1984) introduced a simple example of quantum state diffusion with real fluctuations that was generalised by Diósi (1988) and Gisin (1989). The complex Itô form of EQSD was introduced in GP1 and satisfies a condition of unitary invariance (Diósi 1988; Percival 1989). The detailed theory and its applications are described in GP1-3, Pe2. Quantum state diffusion also appears in connection with the theory of continuous quantum measurements as shown, for example, in the many references in Barchielli & Belavkin (1991), Carmichael (1993) and in the presentations of Barchielli and Belavkin at this meeting.

EQSD satisfies G1 and G2, but not G3, since it depends on an arbitrary division betwen system and environment. However it has a unique advantage among quantum state diffusion theories, for it has been shown to be useful, in the sense that it provides a practical numerical method for the solution of problems of open quantum systems, where ordinary quantum theory based on density operators does not. In Gisin, Knight et al (1993) EQSD is used to describe a quantum jump experiment. It is used to describe some nonlinear optical processes in Goetsch and Graham (1993). Garraway & Knight (1994a) have compared EQSD and quantum jump simuations for two-photon processes, and in (1994b) they have compared the phase space picture of EQSD and quantum jumps, showing that the former gives localisation and the latter does not. Spiller in this Volume describes the approach to thermal equilibrium of harmonic oscillators and investigates open angular systems such as quantum capacitors and rotors. Gisin (1993) investigates the Heisenberg picture for EQSD. Further examples are given in GP1-3, Gisin and Percival 1992b, 1993c, Salama & Gisin 1993.

Figure 1 illustrates the approach to equilibrium of a harmonic ocillator with a thermal environment, represented in interaction representation by the operators

$$H = \omega a^\dagger a, \qquad L_1 = \sqrt{\bar{n}\gamma}a^\dagger, \qquad L_2 = \sqrt{(\bar{n}+1)\gamma a}$$

where $\bar{n}$ is the equilibrium mean photon number. The initial state is the pure number state $n = 3$ The two dotted lines represent the mean position $\langle q \rangle$ and momentum $\langle p \rangle$, and the two full lines represent the quantum mean square (QMS) deviations. The decrease of the latter show how initial states tend asymptotically to a coherent state with minimum uncertainty product.

Figure 2 represents the evolution of individual states of a nonlinear oscillator, in which $\frac{1}{10}a^\dagger a^\dagger aa$ is added to the above Hamiltonian. In this case the QMS deviation decreases, but the final states "breath" as the wave packet follows the classical phase space trajectory.

Quantum state diffusion represents the evolution of a quantum system through a unique correspondence between the solutions of a deterministic Bloch or master equation for the ensemble density operator ρ and the solutions of an Itô diffusion equation for the normalised state vector $|\psi\rangle$ of an individual system of the ensemble (GP1). The uniqueness follows from the invariance of the solutions under unitary transformations and also ensures that the diffusion equation and the Bloch equation give the same physics for the ensemble. If the Bloch equation is

$$\dot{\rho} = -\frac{i}{\hbar}[H,\rho] + \sum_m \left(L_m \rho L_m^\dagger - \tfrac{1}{2}L_m^\dagger L_m \rho - \tfrac{1}{2}\rho L_m^\dagger L_m\right), \tag{1}$$

then the corresponding quantum state diffusion equation is a nonlinear stochastic differential equation for the normalised state vector $|\psi\rangle$, whose general differerential Itô form is

$$\begin{aligned} |d\psi\rangle = -\frac{i}{\hbar}H|\psi\rangle dt \; &+ \; \sum_m (\langle L_m^\dagger\rangle L_m - \tfrac{1}{2}L_m^\dagger L_m - \tfrac{1}{2}\langle L_m^\dagger\rangle\langle L_m\rangle)|\psi\rangle dt \\ &+ \; \sum_m (L_m - \langle L_m\rangle)|\psi\rangle d\xi_m, \end{aligned} \tag{2}$$

where H is a Hamiltonian and $\langle L_m\rangle = \langle\psi|L_m|\psi\rangle$ is the quantum expectation of L_m for state $|\psi\rangle$. The density operator is given by the mean over the projectors onto the quantum states of the ensemble, $\rho = \mathrm{M}|\psi\rangle\langle\psi|$, where M represents a mean over the ensemble (GP1). Primary state diffusion theory principally uses the special form of state diffusion equation

$$|d\psi\rangle = (\langle L^\dagger\rangle L - \tfrac{1}{2}L^\dagger L - \tfrac{1}{2}\langle L^\dagger\rangle\langle L\rangle)|\psi\rangle dt + (L - \langle L\rangle)|\psi\rangle d\xi, \tag{3}$$

with $H = 0$ and only one diffusion operator L.

The first sum in (2) represents the 'drift' of the state vector and the second sum the random fluctuations. The $d\xi_m$ are independent complex differential random variables, each of which has normalised independent white noise in its real and imaginary parts, leading to an isotropic Brownian motion or Wiener process in the complex ξ_m-plane, which satisifies the conditions

$$\mathrm{M}d\xi_m = 0 \tag{4a}$$

$$d\xi_n d\xi_m = 0, \qquad d\xi_n^* d\xi_m = \delta_{nm}dt. \tag{4b}$$

The complex Wiener process is normalised to dt, so the independent real and imaginary Wiener processes are each normalised to $dt/2$. This is the same normalisation as in GP2, GP3 and Pe2, but is different from that of GP1.

Lindblad (1976) showed that the standard form (1) for the master equation follows from the linearity of the equations, and from the constant trace and a 'complete positivity' condition on ρ. The correspondence between the master equation for ρ and the diffusion equation for $|\psi\rangle$ ensures that the diffusion equation satisfies corresponding conditions. Gisin (1990) used the correspondence between nonlinear state vector equations and the linear Lindblad form of the master equation as a powerful method to obtain good nonlinear equations for the state vector and to reject bad ones. This Lindblad-Gisin condition appears as one of the principles of primary state diffusion.

Many theorems on localisation have been derived for EQSD (GP2, Pe2), which can also be used for PSD.

PRIMARY STATE DIFFUSION (PSD)

Primary state diffusion can be considered as a development of environmental state diffusion in which the dependence on the arbitrary division is removed. It is introduced as a good and useful theory, which satisfies all the conditions G1-3 and whose differences from ordinary quantum theory might be tested experimentally.

Classical or quantum diffusion is often accompanied by drift, but for sufficiently short time intervals Δt, the diffusion, which is of order $(\Delta t)^{1/2}$, dominates the drift, which is of order Δt. In quantum state diffusion, the regular Schrödinger evolution is of order Δt, so for sufficiently small time intervals the quantum state diffusion dominates the Schrödinger evolution. If we were able to follow a state vector for these small time intervals we would see a state diffusion with a relatively small Schrödinger correction.

All physics is based on the assumption that the evolution of a physical system over longer time intervals arises from the combination of its evolutions over smaller time intervals, so if state diffusion is part of the foundations of quantum mechanics, Schrödinger dynamics should be derived from it. Thus the diffusion is primary and the Schrödinger dynamics is secondary. At best the dynamics of a system should depend on only one operator, to replace, for example, the Hamiltonian operator of ordinary nonrelativistic quantum mechanics. These are the principal features of primary state diffusion theory (PSD).

The state diffusion equations of primary state diffusion theory lead to the whole of ordinary quantum dynamics, localisation dynamics and classical Hamiltonian dynamics .

In the foundations of quantum theory the destruction of coherence, which is necessary to produce localisation or reduction and make the transition from quantum to classical mechanics, must be distinguished from the dissipation of coherence. As part of a fundamental theory, the destruction cannot depend on any division between system and environment.

Schrödinger dynamics dissipates coherence, as when a small system like a molecule interacts with a thermal environment and gets entangled with it. This is represented mathematically by the change in the reduced density operator for the small system. But it cannot and does not destroy coherence, because the entanglement between system and environment is retained if a sufficiently large part of the environment is included as part of the system. And because the Schrödinger dynamics does not destroy coherence, it cannot produce the localisation which is necessary to make the transition from quantum to classical mechanics. Primary state diffusion and related theories *do* destroy coherence, and hence they produce localisation, and for this they require a change in the equation for the state vector. *Decoherence* here means destruction of coherence. For relatively large systems primary state diffusion and the resultant decoherence can be very rapid, and when small systems get entangled with them, their states share in the rapid diffusion, and localise (GP2, GP3, Pe2). So decoherence in the larger system accellerates the decoherence and localisation of the smaller one. This is secondary decoherence.

The primary state diffusion and the consequent primary destruction of coherence in small systems like atoms and nuclei is very slow, but in relatively large systems, like heat baths, it can be very rapid. When small systems get entangled with large systems, their states share in the rapid diffusion, localise and decohere. This is secondary destruction of coherence. So the dissipation of coherence into the larger system accelerates the destruction of coherence in the small system. According to PSD theory, primary destruction of coherence in small systems is too slow to have been seen in experiments that are not specifically designed to detect it. For larger systems like heat baths and measuring apparatus where the destruction of coherence is fast, the coherence that is destroyed is already so dissipated and entangled that it is usually impossible to locate it, as shown in (Percival, 1992). That is why the dynamical processes of primary state diffusion, primary decoherence and primary localisation have yet to be seen experimentally.

Primary state diffusion has the special form (3), with only one diffusion operator, which is denoted K for primary state diffusion, and no explicit Hamiltonian term, so the Langevin-Itô equation for the state vector is correspondingly simple. In the quantum state diffusion theory of GP1-3, the Langevin-Itô state diffusion equations with a single diffusion operator have the form

$$|d\psi\rangle = |\psi_t\rangle dt + |\psi_\xi\rangle d\xi, \tag{5}$$

where $|\psi_t\rangle dt$ is the drift term and $|\psi_\xi\rangle d\xi$ is the diffusion term. In the Langevin-Itô theory, the stochastic fluctuation or noise of the diffusion process is all contained in the complex normalised Wiener white noise fluctuation factor $d\xi$. In primary state diffusion the drift term contains the Schrödinger dynamics.

PRINCIPLES OF PSD

The state diffusion equations are nonlinear, with operators and scalars that may be functions of the state vector $|\psi(t)\rangle$. Where no such dependence is indicated, an operator or scalar is independent of $|\psi(t)\rangle$.

The principles of primary state diffusion are

P1. *State vector.* The state of a system at any time t is represented by a normalised state vector $|\psi(t)\rangle$.

P2. *Operator.* The dynamics of the system is determined by a system diffusion operator K, which is linear.

P3. *State diffusion.* The change of state obeys a nonlinear Langevin-Itô state diffusion equation with normalised complex fluctuations $d\xi$.

P4. *Lindblad-Gisin.* The derived equation for the ensemble density operator

$$\rho = \mathrm{M}|\psi\rangle\langle\psi| \tag{6}$$

satisfies the elementary linear Bloch or master equation

$$\dot{\rho} = K\rho K^\dagger - \tfrac{1}{2}K^\dagger K\rho - \tfrac{1}{2}\rho K^\dagger K. \tag{7}$$

These principles need some explanation.

The first principle P1 applies to classical systems like the solar system, pointers and Brownian particles as well as to quantum systems like electrons and photons. For the classical variables of the classical systems, the states become so localised in configuration and phase space as a result of the state diffusion that they cannot be distinguished from classical points (Pe2).

The operator K in P2 is not selfadjoint. The quantum fluctuations of P3 are independent of spatial coordinates in the nonrelativistic theory but are space dependent in any relativistic theory. Notice that there is no Schrödinger term written into the equation (6) of P4. The operator K determines the Schrödinger dynamics along with the state diffusion

These principles lead (Pe1) to the *basic* PSD equation.

$$|d\psi\rangle = (k(\psi)^* K - \tfrac{1}{2}(K^\dagger K + k(\psi)^* k(\psi)))|\psi\rangle dt + (K - k(\psi))|\psi\rangle d\xi, \tag{8}$$

where

$$k(\psi) = \langle\psi|K|\psi\rangle. \tag{9}$$

There is no explicit regular Schrödinger term in the equation. The first term in the drift has an operator $k^*(\psi)K$, which has a skewadjoint part, and this term generates the Schrödinger dynamics, so the usual dynamics of quantum theory appears as a drift term in the diffusion equation defined by the diffusion operator K alone.

CONDITIONS FOR PSD

The application of these principles to a given system depends on the system operator K. The choice of this operator is severely constrained by the following conditions, which are needed if the theory is to agree with experiment and observation.

C1. *Schrödinger evolution.* Where state diffusion term is negligible, the regular Schrödinger evolution must be given by the skewadjoint part of the drift operator in equation (8), so the quantum state diffusion equation must be of the form

$$|d\psi\rangle = -(i/\hbar)H|\psi\rangle dt + R(\psi)|\psi\rangle dt + K_\Delta(\psi)|\psi\rangle d\xi, \tag{10}$$

where the first term represents the regular Schrödinger evolution, where the second term represents the *residual* drift, where the third term represents the stochastic fluctuations and

$$K_\Delta = K - k(\psi). \tag{11}$$

This condition C1 is a severe constraint on the quantum state diffusion.

C2. *Limited disruption.* The diffusion is not so fast that it disrupts the regular Schrödinger evolution to a degree that violates experimental or observational evidence.

C3. *Fast diffusion.* The rate of diffusion is sufficiently fast to localise classical dynamical variables.

Notice that conditions C2 and C3 are opposing restrictions on PSD theory, one ensuring that the state diffusion is not too fast, and the other that it is not too slow. The conditions can be expressed as bounds on the time constant τ_0 introduced later. The theory could be falsified by evidence of contradictory constraints.

SCHRÖDINGER DYNAMICS FROM DIFFUSION DYNAMICS

By applying condition C1, it is shown in Pe1 that K has the form

$$K = a_1 H + \frac{i}{\hbar a_1} I. \tag{12}$$

By substituting this K into the elementary state diffusion equation and simplifying, the *transformed* state diffusion equation

$$|d\psi\rangle = ((-i/\hbar)H_\Delta dt - \tfrac{1}{2}a_1^2 H_\Delta^2 dt + a_1 H_\Delta d\xi)|\psi\rangle, \qquad \text{(transformed)} \tag{13}$$

is obtained, where

$$H_\Delta = H - \langle\psi|H|\psi\rangle. \tag{14}$$

This equation of nonrelativistic PSD has the same form as the state diffusion equation for the measurement of energy, as shown in GP1. So the principles P1-P4 and the condition C1 lead to a universal intrinsic energy measurement, which Bedford and Wang suggested in 1975. The equation (13) was proposed tentatively by Gisin (1989) as a possible intrinsic state diffusion equation, with real instead of complex noise, and the corresponding density operator equation appears in Milburn (1991).

TIME CONSTANT τ_0 AND CHARACTERISTIC TIMES

These are needed to apply the limited disruption condition C2. Let ΔE be the energy difference between any two energy eigenstates of a quantum system, and δt some short interval of time in which the change of state is small:

$$|(|\psi(t+\delta t)\rangle - |\psi(t)\rangle)| \ll 1. \tag{15}$$

Because the root mean square of $\Delta\xi$ is $(\delta t)^{\frac{1}{2}}$, the transformed PSD equation (13), restricted to the space of these two states, shows that the Hamiltonian drift and the diffusion are equal in the mean when the time interval satisfies

$$\delta t \Delta E/\hbar = (\delta t)^{\frac{1}{2}} a_1 \Delta E \tag{16}$$

or

$$\delta t = \tau_0 = a_1^2 \hbar^2 \qquad a_1 = \tau_0^{1/2}/\hbar. \tag{17}$$

The time interval τ_0 is a universal time constant for primary state diffusion, and a more convenient constant than the constant a_1.

So in PSD there is a universal time interval below which the diffusion term dominates the Hamiltonian term. This is the sufficiently short time interval of the introduction. Experiment shows that τ_0 is a very short time. Paradoxically, to measure τ_0, experiments have to be carried out over relatively *long* times. The constant τ_0 is not a coherence lifetime, but for particular processes the primary coherence lifetime can be obtained from it, and vice versa.

The localisation theorems of [GP1-3, Pe2] then lead to a theory of coherence lifetimes. As the energy is localised, so the state diffusion destroys the coherence between the energy components of a pure state. To make a rough estimate of the lifetime of the coherence, suppose the system is initially prepared in a coherent linear combination

$$|\psi(0)\rangle = \alpha_1|1\rangle + \alpha_2|2\rangle \tag{18}$$

of two energy eigenstates with energies differing by

$$E_2 - E_1 = \Delta E = \hbar\omega \tag{19}$$

and that both coefficients are significant, so that interference effects between the two states could be detected.

If there is no environmental dissipation of coherence, then the localisation in energy becomes significant at a time t when

$$\mathrm{M}\sigma^2(E) < \hbar^2\omega^2/2, \tag{20}$$

which happens by the time

$$T_\mathrm{p} = \frac{1}{\omega^2\tau_0}, \tag{21}$$

which is the primary coherence lifetime. There is an inverse square dependence on the energy separation of the states, corresponding to the same dependence of the rate of density operator diagonalisation in Milburn's theory (1991).

But the coherence can be dissipated earlier by decay, after which the coherence is no longer observable. The environments that are normally considered are sufficiently big for this dissipation of coherence to lead to secondary destruction of coherence, and the lifetime for this will be denoted T_s, the secondary coherence lifetime.

Experiments are not normally designed to measure coherence lifetimes, but nevertheless many do so, and because all relevant experiments to date have been found consistent with ordinary quantum theory

$$T_\mathrm{p} > T_\mathrm{s} \tag{22}$$

and

$$\tau_0 < \frac{1}{\omega^2 T_\mathrm{s}} = \tau_\chi, \tag{23}$$

where τ_χ is a characteristic time for the secondary decoherence, which is different for each experiment, and puts a bound on the universal time constant τ_0. Assuming that the coherence of a linear combination of an initial and final state of the sodium D line is maintained over the radiative lifetime of the excited state, the characteristic time is

$$\tau_\chi = 6.10^{-24}\mathrm{s} > \tau_0. \tag{24}$$

In Pe1 it is shown that it may be possible to achieve much smaller characteristic times, and so much stronger limits on τ_0, now that particles can be kept isolated for long times in traps.

SPECIAL RELATIVITY

Since PSD theory is independent of any choice of boundary between system and environment, it can be applied to the whole universe. This implies that there is a single universal quantum fluctuation $\xi(t)$ which is a function of time alone; the value of this fluctuation is the same for simultaneous events at all points in space.

Simultaneity has no role in special relativity, so the assumption of simultaneous fluctuations has to be abandoned, but relativity also provides a possible solution to the problem. Although there is no relativistic theory of quantum state diffusion, simple relativistic arguments suggest how the difficulties of nonrelativistic PSD theory might be overcome. The single quantum fluctuation $d\xi(t)$ for all points in space is replaced by a fluctuation with coherence length c/Ω for frequency component Ω.

Suppose two systems A and B with centres of mass r_A and r_B are small compared to the distance r between them and suppose that $\Delta t = r/c$.

If Δt is a sufficiently small time interval, then the quantum fluctuations at A and B are almost the same, and if it is sufficiently large, they are almost statistically independent. What is small and what is large depends on the time constants for the effects of the diffusion. For example, if the primary decoherence time for linear combinations of energy eigenstates in A or B is T_p, then the fluctuations at at r_A and r_B have little coherence if

$$\Delta t/T_p - r/(T_p c) \gg 1 \qquad r > L_p = T_p c, \tag{25}$$

where L_p is the *coherence length* for the system. For a given process, diffusion over distances significantly less than the coherence length can be represented to a good approximation by a single quantum fluctuation $d\xi$, whereas over significantly greater distances, it can be represented by independent fluctuations $d\xi_1$ and $d\xi_2$ satisfying conditions (4).

As shown in GP3, an extended de Broglie wave that is subject to independent fluctuations in different regions of space becomes localised to one of those regions, so the waves localise into regions of approximate linear dimension L_p in a time of order T_p.

In the limit $c \to \infty$, there is no localisation, so relativity is necessary to satisfy the fast diffusion condition C3.

MATTER-WAVE INTERFEROMETRY

When a particle of rest mass m has a dispersed de Broglie wave, then by the equivalence principle of mass and energy, the local state is a linear combination of the particle wave and the vacuum, with an energy difference close to the rest energy. So by (21) the primary coherence lifetime is

$$T_p = \frac{\hbar^2}{(mc^2)^2 \tau_0}, \tag{26}$$

where c is the velocity of light. An inverse square dependence on particle mass also appears in the gravitational theory of Ghirardi, Grassi and Rimini (1990).

The best bounds are given by the heaviest particles, so it is worth considering neutrons and atoms. Let A be the atomic mass number of the particle, and u the atomic mass unit. Then the lifetime is

$$T_p = \frac{\hbar^2}{A^2 (\mathrm{u}c^2)^2 \tau_0} = \frac{4.9 \times 10^{-49} \mathrm{s}^2}{A^2 \tau_0}, \tag{27}$$

where relatively small energy differences are neglected in these order of magnitude calculations.

If L is a typical distance between the interfering beams or parts of the matter wave, L/c is the time for a light signal to propagate. Apart from a relatively small geometrical effect which is ignored here, the state diffusion destroys the interference only if this time is longer than the coherence lifetime. So the characteristic time is

$$\tau_\chi = \frac{\hbar^2 c}{A^2(\mathrm{uc}^2)^2 L} = \frac{1.5 \times 10^{-40}\mathrm{m\,s}}{A^2 L}. \tag{28}$$

There is significant primary destruction of coherence when $\tau_\chi < \tau_0$, that is when

$$\begin{aligned} L > \frac{\hbar^2 c}{A^2(\mathrm{uc}^2)^2 \tau_0} &= \frac{1.5 \times 10^{-40}\mathrm{m\,s}}{A^2 \tau_0} \\ &= 2.7\mathrm{km}/A^2 \qquad (\tau_0 = T_{\mathrm{Planck}}). \end{aligned} \tag{29}$$

For τ_0 equal to the Planck time, the oscillations in the interference pattern should be significantly suppressed by primary state diffusion in a cesium beam Mach-Zehnder interferometer of dimension

$$L(\mathrm{Cs}) = 15\mathrm{cm}. \qquad (\tau_0 = T_{\mathrm{Planck}}). \tag{30}$$

The matter-interferometry experiment of Kasevich and Chu (1991), provides a bound $\tau_0 \geq 2000 T_{\mathrm{Planck}}$ on the time constant τ_0.

DISCUSSION OF PSD

In primary state diffusion theory, quantum and classical dynamics follow from quantum state diffusion dynamics. The theory is based on four principles, that a system is represented by an operator, its state by a normalised state vector, the state vector satisfies a Langevin-Itô state diffusion equation, and the resultant density operator for an ensemble must satisfy an equation of elementary Lindblad form. There are three conditions. The first determines the operator, to within an undetermined universal time constant τ_0. The second and third conditions put opposing bounds on τ_0, which represents the boundary between time intervals dominated by diffusion and those dominated by the regular Schrödinger evolution.

The nonrelativisitic theory applies to the decoherence of the internal states of particles, but does not deal satisfactorily with spatial dependence. In particular it does not satisfy the third fast diffusion condition.

Special relativity is an essential component of PSD. Although there is no relativistic theory, simple relativistic arguments suggest how this difficulty may be overcome, by introducing a coherence length c/Ω for components of the quantum fluctuations $d\xi$ of frequency Ω. This leads to sufficiently fast localisation to be consistent with the classical motion of Brownian particles without the need for an observer. In GRW and related theories position localisation is needed to perform this role. The difference is because the dynamical processes of localisation are different in the two theories.

Károlyházy (1966), Hawking (1982), Penrose (1986) and particularly Diósi (1989, 1992) and Ghirardi, Grassi and Rimini (1990) have connected quantum decoherence with quantum gravity (Penrose and Isham, 1986), and this proposal is found to be numerically consistent with PSD, through the possible role of the Planck time.

The experimental consequences of PSD for coherence lifetimes and matter interferometry are independent of the detailed physics of space-time granularity, whose effect is summarised in the diffusion equations and the time constant τ_0. In this way the formidable theoretical difficulties of quantum gravity are avoided.

The nonrelativistic PSD equation derived from the principles P1-P4 and condition C1 is the same as that proposed by Gisin (1989) and is equivalent to the theory obtained independently by Milburn (1991) on different grounds. However, the equation does not satisfy the important fast diffusion condition C3, which is necessary to localise classical particles and to provide a complete theory of measurement, a point made also by Milburn. This flaw is remedied here by introducing the relativistic hypothesis.

An important feature of PSD is the proposed experiments to test the theory and the relativistic hypothesis against ordinary quantum mechanics. According to PSD theory the transition from quantum to classical takes many forms, none of which have yet been sufficiently controllable to expose the quantum-classical boundary. This is because the quantum fluctuations have been swamped by thermal noise or because the necessary experiments have not been attempted. Experiments in optics, beam-foil spectroscopy and the persistence of coherence in the $K^0\overline{K}^0$ system put weak upper bounds on the value of the universal time constant τ_0, but these bounds could be greatly improved by coherence lifetime and matter interferometry experiments designed specifically for that purpose. It may be possible at some time in the future to determine τ_0 even if it is of the same order as the Planck time 5.10^{-44} of quantum gravity, which some have suggested plays a role in the localisation of quantum states. This would provide a new optical window onto quantum gravity. The experiment of Kasevich and Chu (1991) already puts an upper bound on τ_0 of 2000 times the Planck time.

Fortunately the same experiments would be valuable checks on the validity of ordinary quantum mechanics in new domains.

Acknowledgements

It is a pleasure to thank the research groups of P. Knight, P. Kabachnik, J. Mlynek, H. Walther, A. Zeilinger and particularly T. Clark for helpful information and discussion on experiments, the group of J. Briggs for their stimulating hospitality at the University of Freiburg, and the Alexander von Humboldt Foundation for the support that made the research possible.

References

[1] Barchielli, A. & Belavkin, V. P. 1991 *J. Phys. A* **24**, 1495-1514.

[2] Bedford, D. & Wang, D. 1975 *Nuo. Cim. B* **26**, 313-315; 1977 *Nuo. Cim. B* **37**, 55-62.

[3] Bell, J. S. 1987a *Speakable and Unspeakable in Quantum Mechanics* Cambridge: Cambridge University Press.

[4] Bell, J. S. 1987b *Schrödinger, Centenary of a Polymath*, ed Kilmister, C. Cambridge: Cambridge Univ. Press, reprinted in Bell, (1987a).

[5] Bell, J. S. 1990 *Physics World* **3**, 33-40

[6] Bohm, D. & Bub, J. 1966 *Rev. Mod. Phys.*, **38**, 453-469.

[7] Bohm, D. & Hiley, B. J. 1993 *The Undivided Universe*, London and New York: Routledge.

[8] Carmichael, H. J. 1993 *An Open systems Approach to Quantum Optics, Lecture Notes in Physics m18*, Berlin: Springer.

[9] Diósi, L. 1988 *J. Phys. A* **21**, 2885-2898.

[10] Diósi, L. 1989 *Phys. Rev. A* **40**, 1165-1173.

[11] Diósi, L. 1992 In *Quantum Chaos, Quantum Measurement; NATO ASI Series C: Math. Phys. Sci. 357* eds Cvitanovic P., Percival, I. C. & Wirzba A., Dordrecht: Kluwer, 299-304.

[12] Garraway, B. M. & Knight, P. L. 1994a *Phys. Rev. A* **49** 1266-1274; 1994b Dissipitive environments: quantum jumps or phase space localisation?. To appear in *Quantum Optics VI*, Eds Harvey and Walls.

[13] Ghirardi, G.-C., Rimini, A. & Weber, T. 1986 *Phys. Rev. D* **34**, 470-491.

[14] Ghirardi, G.-C., Grassi, R. & Rimini, A.. 1990 *Phys. Rev. A* **42**, 1057-1064.

[15] Gisin, N. 1984 *Phys. Rev. Lett.* **52**, 1657-1660.

[16] Gisin, N. 1989 *Helv. Phys. Acta* **62**, 363-371.

[17] Gisin, N. 1990 *Phys. Lett. A* **143**, 1-2.

[18] Gisin, N. 1993 *J. Mod. Optics* **40**, 2313-2319.

[19] Gisin, N., Knight, P. L., Percival, I. C., Thompson, R. C. & Wilson, D. C. 1993 **167**, 315-318.

[20] Gisin, N. & Percival, I. C. 1992a The quantum state diffusion model applied to open systems. *J. Phys. A* **25**, 5677-5691. Reference GP1.

[21] Gisin, N. & Percival, I. C. 1992b *Phys. Lett. A* **167**, 315-318.

[22] Gisin, N. & Percival, I. C. 1993a Quantum state diffusion, localisation and quantum dispersion entropy. *J. Phys. A* **26**, 2233-2244. Reference GP2.

[23] Gisin, N. & Percival, I. C. 1993b The quantum state diffusion picture of physical processes. *J. Phys. A* **26**, 2245-2260. Reference GP3.

[24] Gisin, N. & Percival, I. C. 1993c *Phys. Lett. A* **175**,144-145.

[25] Goetsch P and Graham R 1993 *Ann. d. Physik* **2** 706.

[26] Hawking, S. 1982 *Commun. Math. Phys.* **87**, 395-415.

[27] Holland, P. R. 1993 *The Quantum Theory of Motion* Cambridge: Cambridge University Press.

[28] Károlyházy, F. 1966 *Nuo. Cim. A* **42**, 390-402.

[29] Kasevich, M., & Chu, S. 1991 *Phys. Rev. Letts.* **67**, 181-184.

[30] Lindblad, G. 1976 *Commun. Math. Phys.* **48**, 119-130.

[31] Milburn, G. J. 1991*Phys. Rev. A* **44**, 5401-5406.

[32] Pearle, P. 1976 *Phys. Rev. D* **13**, 857-868.

[33] Pearle, P. 1979 *Internat. J. Theor. Phys.* **18**, 489-518.

[34] Penrose, R. & Isham, C. J. (eds) 1986 *Quantum Concepts in Space and Time* Oxford: Clarenden, Oxford Science Publications. Several articles give the Planck mass: Shimony p182, Károlyházy et al p109.

[35] Penrose, R. 1986 In Penrose & Isham (1986) 129-146.

[36] Percival, I. C. 1989 Diffusion of Quantum States 2, Preprint QMC DYN 89 4, School of Mathematics, Queen Mary College London.

[37] Percival, I. C. 1992 In *Quantum Chaos, Quantum Measurement. NATO ASI Series C: Math. Phys. Sci. 357* eds Cvitanovic, P., Percival, I. C. & Wirzba, A., Dordrecht: Kluwer 199-204.

[38] Percival, I. C. 1994a Localisation of wide open quantum systems. *J. Phys. A* **27**, 1003-1020. Reference Pe2.

[39] Percival, I.C. 1994b Primary state diffusion. *Proc. Roy. Soc. A* **447**, 189-209. Reference Pe1.

[40] Salama, Y. & Gisin, N. 1993 *Phys. Letts. A* **181**, 269-275.

QUANTUM JUMPS IN MOLECULES EXCITED BY INTENSE LASER FIELD

Boris A. Grishanin and Victor N. Zadkov

Physics Department and International Laser Center,
Moscow State University, Russia

Abstract

Computer simulation results on HCl^+ dissociation in intense ($I \sim 10^{14}$ W/cm^2) laser field in two-level approximation in the model of random quantum jumps are presented. The probabilities of ground and excited states of a molecule at the end of laser pulse are calculated. It is shown that molecule's dissociation dynamics in intense laser field has a quantum stochastic nature. It is argued that the used model can be applied to multiatomic molecules as well.

1 INTRODUCTION

It was shown recently that such coherent effects as quantum interference of wave packets within a molecule [1] and manifestation of squeezed vibrational states [2], which are typical for optics, take place in molecule's dynamics as well. A key feature of molecule's dynamics is a competition between coherent and incoherent dynamics. Here we discuss a transformation of a coherent molecule's dynamics to a chaotic one basing on the two-electronic-level one-dimensional model of a molecule excited by intense laser pulsed field. The results obtained earlier [3] show the evidence of destruction of coherence between the two electronic states and in the wave packets of each states. The approach to simulate the dynamics is based on the dynamics represented by classical movement within a single electronic state and random quantum jumps corresponding to a quantum measuring of a quantum number of electronic states dressed by intense laser field. We applied this approach to computer simulation of ionization-dissociation process in HCl^+ molecule basing

Quantum Communications and Measurement
Edited by V. P. Belavkin *et al.*, Plenum Press, New York, 1995

on the parameters and experimental data presented in Ref. [4]. The calculations results are discussed and effectiveness of our approach is argued.

Using of superintense laser pulses with intensities comparable with the threshold intensity for tunnel ionization of atoms $I_{\rm T} \sim 10^{15}$ W/cm^2 provides new possibilities for controlling photoinduced reactions in molecules. For example, at moderate intensities $I \ll I_{\rm T}$ of a non-resonant laser field electronic excitation of a molecule is insignificant, while at $I \sim I_{\rm T}$ strong excitation of electronic degrees of freedom takes place, despite of non-resonant character of interaction. In this case, dynamics of nuclei is governed simultaneously by the potential energy surfaces of a number of the excited states, and is expected to be essentially different from one typical for resonant excitation by laser fields with moderate intensities.

Molecule's dynamics in strong laser field (10^{12}–10^{14} W/cm^2) could lead to at least two competitive processes, namely photodissociation and photoionization [5–7]. Better understanding of the dynamics of these processes is necessary to achieve control over them, and this is a goal of experimental and theoretical studies [8].

Here we offer an adequate theoretical model for computer simulation of the molecule's photodissociation dynamics. There is a simple molecule, HCl^+, that could be a probe tested both theoretically and experimentally. It was obtained as an intermediate product of ionization of neutral HCl molecule. The experimental results on HCl^+ photodissociation in strong laser field with $I \leq 10^{14}$ W/cm^2 are presented in Ref. [4]. The authors suggested a two-level electronic structure of HCl^+ molecule to describe the dissociation process. The structure of the two electronic levels considered is presented in Fig. 1 for the intensities $I = 0$ and $I = 10^{15}$ W/cm^2. Theoretical studies of such processes are to be based on approximate methods of molecular dynamics within the framework of quantum theory. Simplicity of the approximation used determines its practical efficiency, because the exact solution of the quantum mechanical problem for multiatomic molecules is considered to be impossible. The earlier approaches similar to the Gaussian wave packet approach [9] and others that could be applied to atoms [10] seem to be not appropriate for the case of dissociation dynamics of multiatomic molecules. There have been developed also several very promising approaches [11–13] utilizing semiclassical properties of dynamics within a single electronic level. Some limitations on their validity closely concerned with our specific problem were discussed in Ref. [13]. There are direct indications, especially in case of very strong field, that these powerful approaches are also inappropriate due to the specific chaotic-like behavior of trajectories strongly disturbed by a coupling effect. Calculations of autoionization rate of atoms in IR laser field demonstrate a very strong dependence on the parameters of atomic states [14] due to the well known exponential factor that is typical for tunnel effects [15]. So, we can suggest that the dynamical model we are discussing is an example of a quantum stochastic system. This model differs from the well known Jaynes and Cummings one due to its Hamiltonian, but their quantum space structures are identical. In the latter model one can also reveal a kind of stochasticity [16].

Here, we present a new approach, which combines a quantum description of the dynamics of electronic states of a molecule and classical dynamics within a single electronic state. The quantum transitions between electronic states are described by Quantum Jumps (QJ) model. They are similar to the jumps used to simulate quantum stochastic processes induced by an external noise [17]. Our previous pure quantum calculations of wave packets dynamics for short time distances show that such approach can be effective [3]. Actually, one can expect that electronic dynamics of a molecule in a strong laser field turns to be irregular as a result of strong

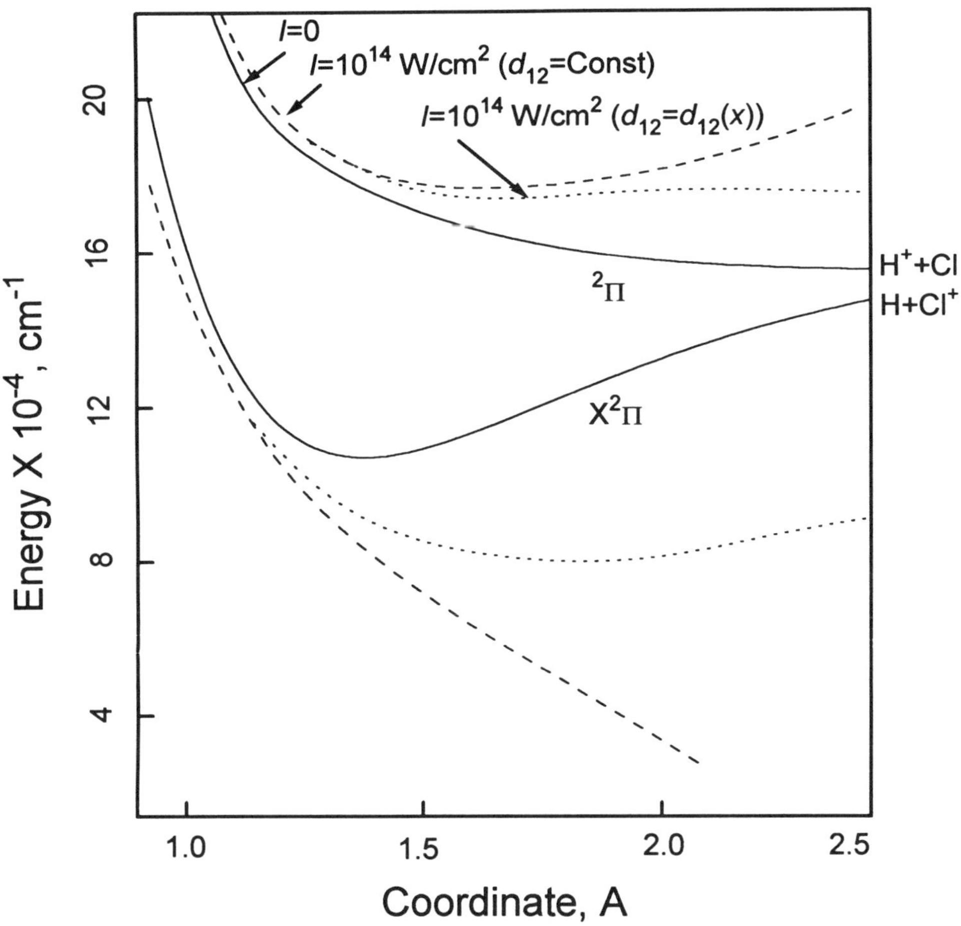

Figure 1: Potential structure of HCl^+ molecule (ground and first excited states) in the presence of external laser field: $I = 0$ (solid line); $I = 10^{14}$ W/cm^2 with (dotted line) and with no (dashed line) spatial dependence of the transition dipole moment d_{12} of the molecule.

modulation of electronic wave function with the nuclear vibrations.

2 QUANTUM JUMPS MODEL

To clarify the QJ model, we consider here a one-dimensional model of a diatomic molecule with two electronic levels. In this model Hamiltonian takes the form

$$\hat{\mathcal{H}} = \frac{\hat{p}^2}{2m} + \hbar\hat{O}(t), \tag{1}$$

where $\hbar\hat{O}(t)$ is the potential energy operator of the molecule in IR field:

$$\hat{O}(t) = \begin{pmatrix} \omega_1(\hat{x},t) & \Omega(\hat{x},t) \\ \Omega(\hat{x},t) & \omega_2(\hat{x},t) \end{pmatrix}. \tag{2}$$

Here $\omega_{1,2}(\hat{x},t) = [V_{1,2}(\hat{x}) + d_{1,2}(\hat{x})E_{\mathrm{L}}(t)]/\hbar$, $\Omega(\hat{x},t) = E_{\mathrm{L}}(t)d_{12}(\hat{x})/\hbar$; $V_{1,2}(\hat{x})$ and $d_{1,2}(\hat{x})$ are the potentials and the dipole moments of the electronic terms, correspondingly; $\hat{p}$ is the momentum operator, m is the reduced mass, and $E_{\mathrm{L}}(t)$ is the electric field strength. Elements of the matrixes $\hat{\mathcal{H}}$, and $\hat{O}(t)$ are functions of the operator $\hat{x}$ of the vibrational coordinate x. The off-diagonal terms in Eq. 2 feature a nonresonant interaction of IR laser field with the electric dipole moment described by the transition matrix element $d_{12}(\hat{x})$.

As far as molecule's dynamics in an external IR laser field is concerned, there are two limits, weak-field and strong-field limits. In case of weak-field limit needless to take into account molecule's electronic dynamics, and one can calculate molecule's dynamics by means of classical Molecular Dynamics (MD) simulation [18, 19]. By contrast, in case of strong-field limit the off-resonant terms in Eq. (2) come into the play and it is necessary to take into account electronic dynamics as well.

The key idea of QJ model is to simulate quantum electronic dynamics of a molecule in strong IR laser field as a sequence of random quantum jumps due to the presence of intense IR field. The molecule's dynamics between the two jumps is treated as a classical one and is calculated by means of classical MD simulation.

Essentially, QJ model is based on the results of some exact quantum calculations we performed earlier of the wave function ψ defined by a time-dependent Schrödinger equation with the Hamiltonian presented by Eq. (1) [3]. The wave function reveals irregular coordinate dependence that looks like a quasi-stochastic one. In terms of molecule's electronic dynamics it means that both electronic terms, lower and upper, take part in the dynamics. These quantum calculations bring us to the idea of substituting the complex electronic dynamics for a sequence of random quantum jumps between the electronic terms (QJ model).

The justification of the QJ model is based on the following representation of the time evolution operator:

$$\hat{U}(t) = \mathrm{T}\prod_{j=1}^{t/\Delta}\sum_{k=1}^{2}\hat{P}_k(t_j)\hat{U}_k(t_j)\hat{P}_k(t_j), \tag{3}$$

where T is the time ordering symbol,

$$\hat{U}_k(\Delta) = \exp\{-(i/\hbar)[\hbar\hat{\omega}_k^{\mathrm{eff}}(t) + (\hat{p}^2/2m)]\Delta\} \tag{4}$$

is the time evolution operator corresponding to the fixed electronic state k and time interval Δ, $\hat{P}_k(t_j)$ are the orthogonal projectors onto the eigen subspaces of matrix

$\hat{O}(t_j)$ and $\omega_k^{\text{eff}}(t_j)$ are the correspondent eigenvalues. The time evolution operator $\hat{U}(t)$ is essentially an averaging of the fluctuations of the momentum $\hat{p}$ over x, due to the dependence of the projectors $\hat{P}_k$ on coordinate $\hat{x}$, for the chosen time interval Δ. This averaging is not essential when Δ is small with respect to the characteristic frequency of fluctuations corresponding to the characteristic value a of a region where essential variations of $\hat{P}_k(x)$ take place, namely when $\hbar\Delta/(ma^2) \ll 1$. As soon as a is expected to be greater than a coherent length this condition is satisfied for $\Delta \ll 1/\omega_0$ where ω_0 is a characteristic vibration frequency. Actually, time averaging is essential on the short time intervals Δ lying between the period of electronic vibrations and that of the IR laser field.

As a next step, we neglect the non-adiabatic terms with $k \neq l$ in the exact expression for the time dependent density matrix

$$\hat{\rho}(t) = \hat{U}(t)\hat{\rho}_0\hat{U}^{-1}(t) = \ldots \sum_{k=1}^{2} \hat{P}_k(t_j)\hat{U}_k(\Delta_j)\ldots\hat{\rho}_0\ldots\sum_{l=1}^{2}\hat{U}_l^{-1}(\Delta_j)\hat{P}_l(t_j)\ldots .$$

The reason for this is that the non-adiabatic terms with $k \neq l$ are not positive in contrast to the diagonal ones with $k = l$. Therefore, in case when the coordinate $\hat{x}(t)$ can be described as a stochastic process with the required correlation properties one can expect that non-adiabatic terms produce no resulting contribution into the total transformation because of the interference of the terms corresponding to different j–th time intervals. It means that we account the evolution of non-diagonal terms as a purely stochastic one thanks to the internal stochasticity of the model. This stochasticity that is due to the strong coupling between the effective potential electronic states affect the vibrational movement along the x coordinate. (It is possible to present a simple analogue of the situation where the stochasticity has an external origin and an exact calculation of the evolution operator is available). After all, we come to the following simplified expression:

$$\hat{\rho}(t) \approx \ldots \sum_{k=1}^{2} \hat{P}_k(t_j)\hat{U}_k(\Delta_j)\ldots\hat{\rho}_0\ldots\hat{U}_k^{-1}(\Delta_j)\hat{P}_k(t_j)\ldots . \tag{5}$$

Basically, this equation means that after any time interval Δ_j the molecule undergoes a self-induced procedure of "quantum measurement" of the electronic state k, that is an instant quantum jump. The combination of terms $\hat{P}_{k_{j+1}}(t_{j+1})$ and $\hat{P}_{k_j}(t_j)$ arising in Eq. (5) after combining multiplications of the adjacent terms corresponding to different time instants t_j, t_{j+1}, where the combinations of different k-values corresponding to t_j, t_{j+1} are presented, describes an instant quantum transition between electronic states, or so called quantum jump. These quantum jumps are possible at any time instant t_j. Introducing for the simplicity wave vectors Ψ_k, so that $\hat{P}_k = \Psi_k\Psi_k^+$, we can easily obtain from Eq. 3 the transition probability

$$P_j = \text{Tr}\hat{P}_{k_{j+1}}(t_{j+1})\hat{P}_{k_j}(t_j) = |\langle\Psi_{k_{j+1}}^+\Psi_{k_j}\rangle|^2, \tag{6}$$

which depends on time intervals Δ_j.

Molecular dynamics between the two successive jumps can be appropriately described by means of classical MD simulation. This dynamics at time intervals Δ_j is governed by the effective potentials $\hbar\hat{\omega}_k^{\text{eff}}(t_j)$. As far as the electronic evolution of a molecule is concerned, it is treated as markovian quantum jumps between the preceding eigen electronic state $\Psi_k(t_j)$ and the following electronic state $\Psi_l(t_{j+1})$.

3 COMPUTER SIMULATION RESULTS

We applied the described above QJ model to a computer simulation of HCl^+ dissociation dynamics in strong IR pulsed laser field, which is governed by Eqs. (1)–(2). To do so we adopt the electronic structure of HCl^+ molecule similar to that one in Ref. [4]. Functions and parameters used in our model are listed below:

$$V_1(x) = D_1(1 - e^{-a_1 x})^2, \quad V_2(x) = D_2 e^{-a_1(x+r_0)} + D_1,$$
$$D_1 = 1.0625 \times 10^4\,\mathrm{cm}^{-1}, \quad D_2 = 5.0 \times 10^6\,\mathrm{cm}^{-1}, \tag{7}$$
$$a_1 = 1.84\text{Å}^{-1}, \quad a_2 = 4.10\text{Å}^{-1}, \quad m = 1.59 \times 10^{-24}\,\mathrm{CGS}; \tag{8}$$

$$d_{1,2}(x) = e^{\mathrm{eff}}_{1,2} x, \quad e^{\mathrm{eff}}_1 = e^{\mathrm{eff}}_2 = 1.486 \times 10^{-10}\,\mathrm{CGS}. \tag{9}$$

In contrast to Ref. [4] we define the spatial dependence of the transition dipole moment of the molecule in the following form:

$$d_{12}(x) = d^0_{12} \times \begin{cases} 1, & x < 0 \\ \exp(-x/r^{\mathrm{eff}}_{12}), & x \geq 0 \end{cases}, \quad x = r - r_0,$$
$$d^0_{12} = 1\,\mathrm{Db}, \quad r_0 = 1.5142\text{Å}, \quad r^{\mathrm{eff}}_{12} = 5.0\text{Å}. \tag{10}$$

Hence, $d_{12}(x)$ goes to zero for large internuclear distances r.

Functions and parameters defining the electromagnetic field are as follows:

$$E(t) = E_{\mathrm{L}} \exp[-(t - t_{\mathrm{c}})^2/\tau_{\mathrm{d}}^2],$$
$$t_{\mathrm{c}} = 4\tau_{\mathrm{d}}, \quad \tau_{\mathrm{d}} = 1.0\,\mathrm{ps}, \quad \omega_{\mathrm{L}} = 1591.5\,\mathrm{cm}^{-1}, \tag{11}$$

where t_c is the time delay of the intensity peak with respect to the starting point.

The main computer simulations results on HCl^+ dissociation dynamics in intense pulsed IR laser field are as follows. We discovered that effects of dressing of the electronic states by the IR field are essential that is in accordance with the predictions in Ref. [4]. It is necessary to stress that if the spatial dependence of the transition dipole moment is not taken into account the lower effective potential curve $\omega_{1\mathrm{eff}}$ corresponds to the repulsion for all distances x and molecule dissociates within this single electronic state with the corresponding products $H+Cl^+$.

The calculated dissociation threshold intensity is about 1.0×10^{14} W/cm^2. This result is not sensitive to the details of the spatial dependence of the transition dipole moment. Nevertheless, this dependence is of a great importance for the dissociation dynamics. Using the approximation (9) of this dependence and the expected parameters we reveal new dynamical features that are in contrast with the calculations presented in Ref. [4] where the spatial dependence of the transition dipole moment was ignored. Namely, the potential curve for the lower dressed state does not correspond to the repulsive one (see Fig. 1), and the dissociation dynamics is also essentially different. Our computer simulation runs show that a number of jumps occur before the molecule undergoes complete dissociation.

Actually, in case of such a simple molecule like HCl^+ these jumps are not exactly the real ones and are used in our QJ model to simulate the exact quantum dynamics when we are not interested in investigation of quantum correlation effects. Nevertheless, the existence of an essential number of the jumps demonstrates an important role of the both electronic states in the dissociation process. Jumps feature the specific jump-like origin of stochasticity deals with a high value of the laser field strength. Intense laser field leads to the electronic excitation of the molecule, plus to stochastization of the dynamics due to the coupling of electronic levels.

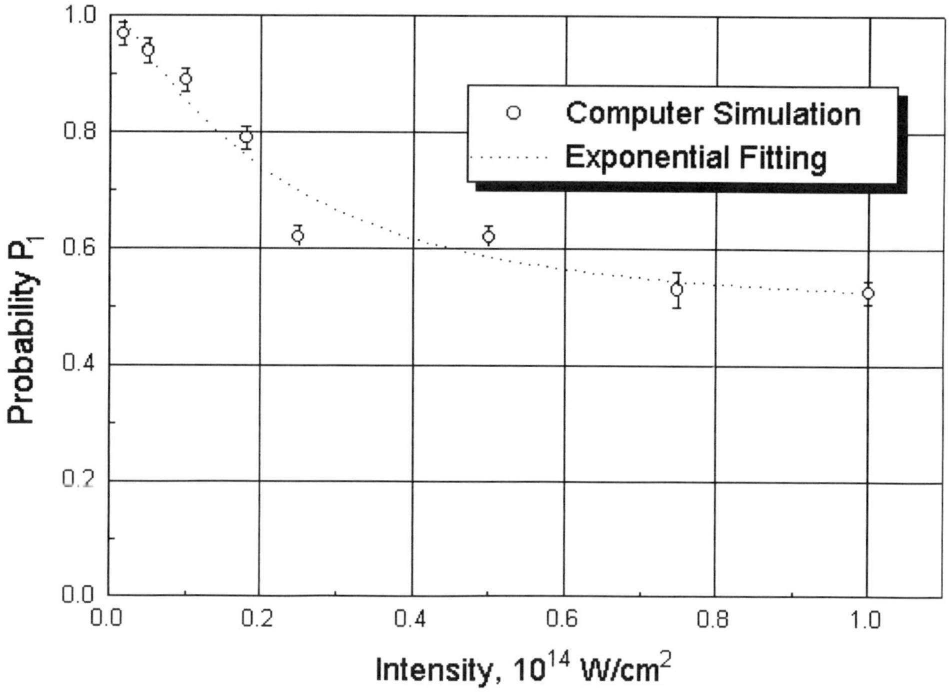

Figure 2: Probability P_1, the probability to find the photodissociation reaction products by the end of laser pulse action at the lower electronic state regardless to that whether the dissociation occurred or not, vs laser field intensity.

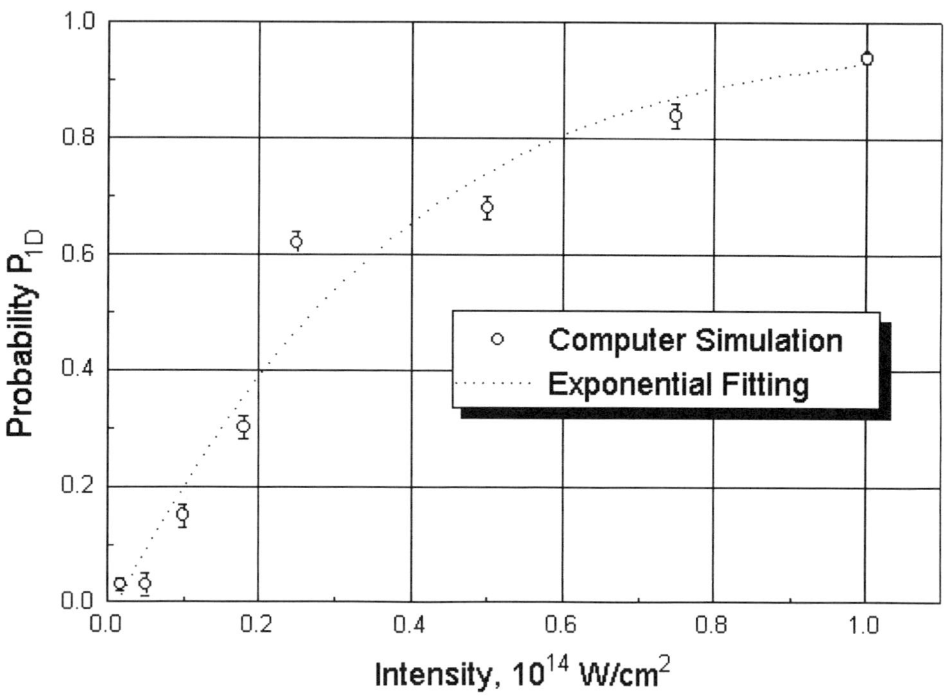

Figure 3: Conditional probability P_{1D}, the probability to find the dissociation products $H+Cl^+$ divided to the probability P_1, vs laser field intensity.

The only quantities that could be measured experimentally at the modern state of experimental techniques are the probabilities to find the different dissociation reaction products by the end of laser field action, namely, the probabilities P_1 and $P_2 = 1 - P_1$ to find the dissociation products at the lower/upper electronic state regardless to that whether the dissociation occurred or not, and the probabilities $P(\mathrm{H} + \mathrm{Cl}^+)$ and $P(\mathrm{H}^+ + \mathrm{Cl})$ to find the dissociation products corresponding to the lower/upper electronic states. To calculate the probability P_1 for the different intensities we performed a great variety of dynamical calculations and analyzed the final states. Fig. 2 shows that $P_1, P_2 \rightarrow 0.5$ if one increases the field intensity up to the value of $I \gg 1 \times 10^{14}$ W/cm^2. The conditional probability of the dissociation $P_{1\mathrm{D}}$, the probability to find the dissociation reaction products H+Cl$^+$ divided to the probability P_1, is also presented in Fig. 3 (probability $P_{2\mathrm{D}}$ for the upper level is equal to 1 for all the intensities).

CONCLUSIONS

The dissociation dynamics of HCl^+ molecule described in frame of the QJ model has several specific features stressed below.

First, it is not simply a dissociation dynamics within a single electronic level. The dissociation process is strongly modified because of the coupling of the electronic levels in intense IR field resulting to much more complicated dynamics. Its quantum nature can be described as a sequence of random quantum jumps between the dressed electronic levels even for the simplest two-level model with a single vibrational coordinate.

It is also possible to apply our calculations to the rigorous quantitative analysis of the experimental data [4]. Actually, the results already obtained may be considered as qualitatively correspondent to the available experimental data.

Finally, our computer simulation results demonstrate that the proposed QJ model yields a proper physical description of a molecule's dissociation dynamics in the presence of a strong IR laser field. The nature of the multilevel dynamics of strongly excited molecules is a stochastic one, and this type of quantum stochasticity is rather different from the ones discussed earlier (see, for example, Ref. [20]). In our view, QJ model seems to be even more effective in the case of multiatomic molecules because the Markov's approach is expected to be more appropriate one to complex systems, plus at the moment there are no other effective methods to simulate multiatomic molecule's dynamics in a strong laser field. In addition, QJ model could be easily run on PC-like computers for diatomic and multiatomic molecules as well.

Acknowledgments

Stimulating discussions with Prof. Stig Stenholm, Dr. Ilkka Tittonen, and Wai Lai (University of Helsinki) are gratefully acknowledged. Authors are grateful for partial support from the International Science Foundation under Grant No. MS9000.

References

[1] B. Garraway, S. Stenholm, Phys. Rev. **A46**, 1413 (1992).

[2] T. J. Dunn, J. N. Sweetser, I. A. Walmsley, C. Radzewicz, Phys. Rev. Lett., **70**, 3388 (1992).

[3] B. A. Grishanin, V. N. Zadkov, In *Quantum Electronics and Laser Science Conference*, 1993 OSA Technical Diegest Series, **3** (Optical Society of America, Washington, DC, 1993), 103.

[4] P. Dietrich, P. B. Corkum, J. Chem. Phys., **97**, 3187 (1992).

[5] M. Saeed, B. Yang, X. Tang, and L. F. DiMauro, Phys. Rev. Lett., **68**, 3519 (1993)

[6] B. Yang, M. Saeed, L. F. DiMauro, et. al., Phys. Rev. **A44**, R1458 (1991).

[7] P. Lambropoulos, Phys. Rev. Lett., **55**, 2141 (1985).

[8] *Coherence Phenomena in Atoms and Molecules in Laser Fields*, A. D. Bandrauk, and S. C. Wallace, Eds.(New York: Plenum, 1992).

[9] R. D. Coalson, M. Karplus, Chem. Phys. Lett., **90**, 301 (1982).

[10] J. K. Liakos, M. Horbatsch, J. Phys. **B**, Atomic and Mol. Opt. Phys. (UK), **24**, 3387 (1991).

[11] M. S. Child, *Molecular Collision Theory* (Academic, New York, 1974).

[12] E. J. Heller, J. Chem. Phys., **65**, 1289 (1976).

[13] S. Stenholm, Phys. Rev. **A47**, 2523 (1993).

[14] Y. Abranyos and M. H. Mittleman, Phys. Rev. **A42**, 4284 (1990).

[15] L. V. Keldysh, Sov. Phys.–JETP, **20**, 1307 (1965).

[16] B. Shore, P. L. Knight, J. Mod. Opt., **40**, 1195 (1993).

[17] K. Molmer, Y. Castin, and J. Dalibard, J. Opt. Soc. Am. **B10**, 524 (1993).

[18] B. A. Grishanin, V. D. Vachev, V. N. Zadkov, Mol. Cryst. Liq. Cryst. Sci. Technol.—Sec. B: Nonlinear Optics, **3**, 375 (1992).

[19] V. D. Vachev, J. G. Frederick, B. A. Grishanin, V. N. Zadkov, N. I. Koroteev, Chem. Phys. Lett., **215**, 306 (1993).

[20] F. Haake, *Quantum Signatures of Chaos* (Springer-Verlag, Berlin, 1991).

THE QUANTUM STATE DIFFUSION OF AN ANGULAR SYSTEM

T.P. Spiller

Physics and Astronomy Subject Group,
School of Mathematical and Physical Sciences,
University of Sussex, Brighton,
Sussex, BN1 9QH, U.K.

Abstract

State evolution, in contrast to that of a density operator, is particularly relevant for open quantum systems which are individually accessible in experiments. Here we apply state diffusion theory to such a system, a planar quantum rotor, or a quantum capacitor. The experimental investigation of ultra-small capacitance devices, where the charging energy of one fundamental charge can well exceed a typical thermal energy, is currently a large and active research area of modern condensed matter physics. We consider one specific example, where the capacitor has additional (driving and tunnelling) Hamiltonian terms, and an environment interaction which describes Ohmic dissipation. For parameter values which render the corresponding classical motion chaotic, we show how evidence for this motion begins to emerge in the quantum treatment. This suggests that the individual system approach plays an important role regarding the classical limit of open quantum evolution.

Keywords :
Quantum state diffusion, tunnelling, chaos.

In this paper quantum state diffusion theory [1-5] is applied to a system possessing an angular symmetry. Quantum state diffusion is an approach through which the behaviour of an individual open quantum system, that is some system of specific interest coupled to an environment, may be modelled. In this approach, the system is described by a state $| \psi >$ whose Schrödinger evolution is modified by the environment. This state evolves according to [1-5]

$$| d\psi >= -iH | \psi > d\tau + \sum_m [L_m - < L_m >_\psi] | \psi > d\xi_m$$
$$+ \sum_m \left[< L_m^\dagger >_\psi L_m - \frac{1}{2} L_m^\dagger L_m - \frac{1}{2} < L_m^\dagger >_\psi < L_m >_\psi \right] | \psi > d\tau. \tag{1}$$

Here H is the Hamiltonian for the system of interest, the operators $\{L_m\}$ represent the action of the environment, and we have set $\hbar = 1$ so that these, and the time τ, are in appropriate dimensionless units. The environment introduces diffusive drift (the second sum in (1)), and fluctuations (the first sum), into the evolution of $| \psi >$. The $d\xi_m$ are independent complex differential random variables, which obey

$$\mathbf{M} d\xi_m = \mathbf{M}(d\xi_m d\xi_n) = 0; \ \mathbf{M}(d\xi_m^* d\xi_n) = \delta_{mn} d\tau \tag{2}$$

where $\mathbf{M}$ denotes the ensemble average.

The ensemble approach, where the statistical behaviour of an ensemble of identical systems is described by a reduced density operator ρ, is perhaps the more conventional and familiar approach to open quantum evolution. From (1), it follows [1] that the density operator $\rho = \mathbf{M} | \psi >< \psi |$ obeys

$$\dot{\rho} = -i[H, \rho] + \sum_m \left(L_m \rho L_m^\dagger - \frac{1}{2} L_m^\dagger L_m \rho - \frac{1}{2} \rho L_m^\dagger L_m \right), \tag{3}$$

which is the well known Bloch form of the master equation for ρ. Quantum state diffusion theory can thus be applied to any system where the ensemble behaviour is describable by (3).

The system we consider here is a planar quantum rotor, or quantum capacitor. Modern technology now permits the fabrication of ultra-small capacitance devices, for which the electrostatic energy of one stored fundamental charge $q^2/2C$ (where $q = e$ for normal metals, or semiconductors and $q = 2e$ for superconductors) is comparable with or in excess of the typical thermal energy kT. (At liquid helium temperatures this means $C \leq 10^{-16}$F. The quantum behaviour of fundamental charges, and the interplay of this with environment effects, is therefore currently a large and active research field[6-10], both theoretical and experimental. Since experiments generally look at one device, there is clearly great value in considering a theoretical approach, like quantum state diffusion, which does likewise, and is thus able to mimic the behaviour seen in individual runs, underpinned by the correct statistical predictions. The example we focus on here is how characteristic classical behaviour, chaotic motion, begins to emerge in the quantum evolution of a single system.

The charge ($Q = Nq$) eigenstates of an isolated capacitor are denoted by $| N >$, for integer N. The quantization of N means that the conjugate co-ordinate, θ, is angular. We define the (unitary) charge raising operator $\zeta^\dagger$ by $\zeta^\dagger | N >=| N+1 >$, so in the θ-representation $\zeta^\dagger = \exp(i\theta)$, and $[N, \zeta^\dagger] = \zeta^\dagger$. We consider a capacitor which is biased by an external displacement charge [6-8] ($Q_x \equiv N_x q$), and which permits the tunnelling of charge [11] between its plates with a dimensionless amplitude $\nu/2$. The dimensionless Hamiltonian is therefore

$$H = \frac{1}{2} (N - N_x)^2 - \frac{\nu}{2} \left(\zeta^\dagger + \zeta \right). \tag{4}$$

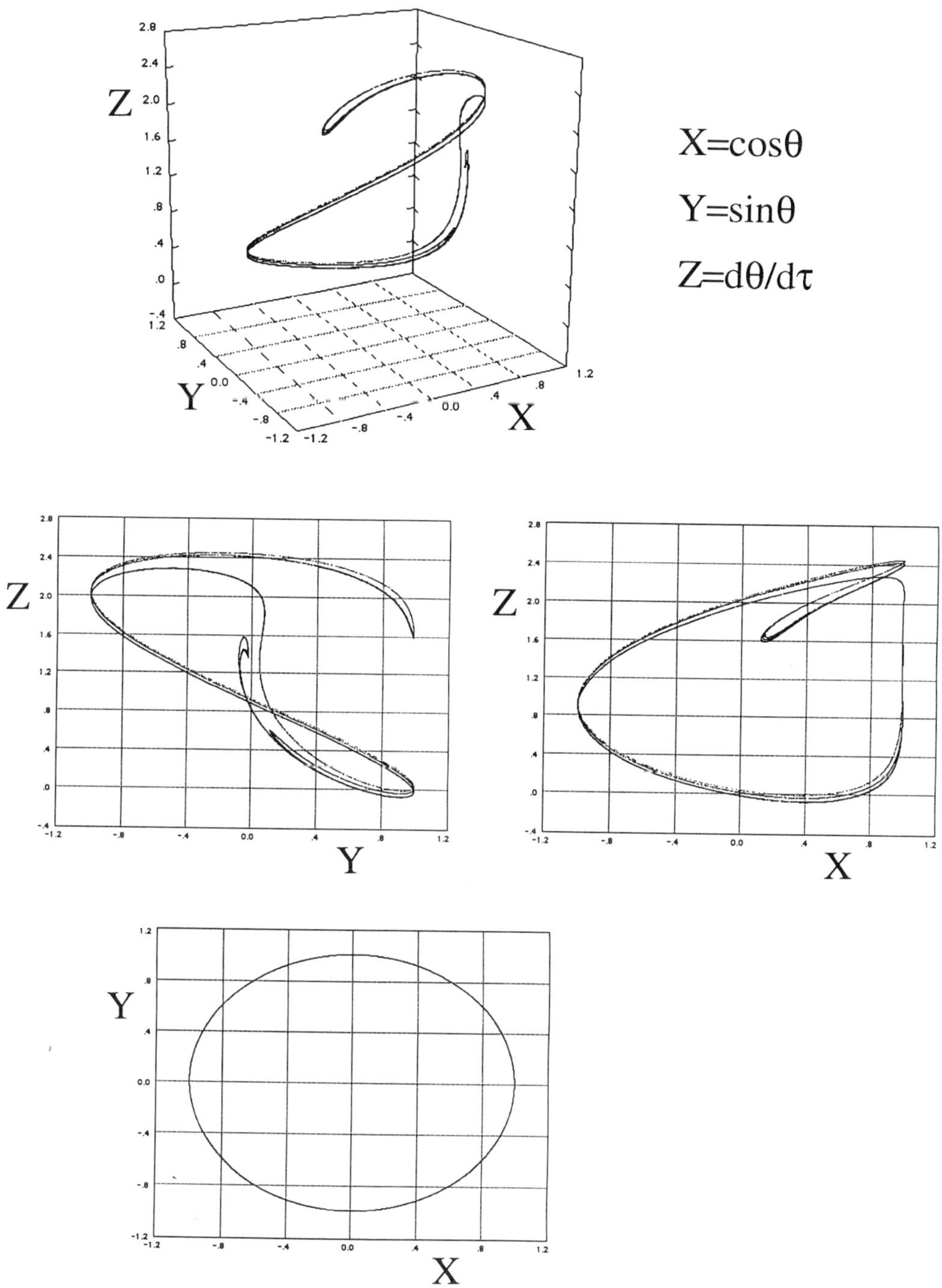

Figure 1: The chaotic Poincare section (25000 points) of motion given by (6), for $\gamma = \frac{1}{2}$, $\omega = \frac{2}{3}$, $\nu = 1$ and $g = \frac{3}{2}$. The transient motion is removed.

In the θ-representation the tunnelling term is the familiar Josephson cosine potential. In the alternative quantum rotor view of this sytem, N is the angular momentum quantum number, N_x is an external rotation, θ is the orientation angle of the rotor in real space, and the cosine potential could, for example, be due to a gravitational field.

We also wish to consider a dissipative force acting on the system, due to its environment. In the capacitor view, this could be due to a resistor which shorts out the charging; in the rotor view it could be due to a surrounding viscous fluid, (which is at rest in the rotating frame). For Ohmic dissipation, which acts to bring $< N >_\psi$ to N_x, and at a rate proportional to their difference, an environment operator L is required which acts to raise N if $N < N_x$, and lower N if $N > N_x$. An operator which does the job is

$$L = \gamma^{\frac{1}{2}} \left[\zeta \theta \left(N - N_x - \epsilon\right) + \zeta^\dagger \left(1 - \theta \left(N - N_x + \epsilon\right)\right) e^{i 2\pi R}\right] \mid N - N_x \mid^{\frac{1}{2}} . \quad (5)$$

Here γ is the decay rate, and the factor $\mid N - N_x \mid^{\frac{1}{2}}$ ensures the proportionality to $(< N >_\psi - N_x)$; the square roots occur because L enters bilinearly in the terms in equations (1) and (3) which are responsible for the dissipation. $\theta(x)$ is the conventional unit step function at the origin in x, and ϵ is a positive infinitesimal. R is a random number in the interval 0 to 1, which ensures that the Hermitian conjugate $L^\dagger$ does not act to produce coherent superpositions of charge states, and thus that complete decay occurs independent of the initial state. The operator L generates the required form of dissipation. For $\nu = 0$ it can be shown that $\mathbf{M}< N >_\psi$ decays exponentially to N_x, at a rate γ.

The classical counterpart of our quantum system can exhibit chaos. For example, with $N_x \equiv -(g/\omega) \sin \omega\tau$, and $\dot{\theta} \equiv N - N_x$, the equation of motion is

$$\ddot{\theta} + \gamma\dot{\theta} + \nu \sin\theta = g \cos \omega\tau, \quad (6)$$

and it is well known that for certain parameter values this damped, driven pendulum is chaotic [12]. An example is given in figure 1. Here we choose to plot the Poincare section of the chaotic motion (sampling once per period of the driving term) on a cylinder (along with its projections). Given the problems with defining an angle operator in quantum mechanics, this makes for an easy comparison with the quantum motion.

It has recently been suggested [13] that the individual system viewpoint is a good approach to adopt, in order to look for the origins of classical chaos in the corresponding quantum evolution. We therefore calculate the state evolution according to (1), for parameters which generate the classical chaos, and sample $\left(Re < \zeta^\dagger >_\psi, Im < \zeta^\dagger >_\psi, < N >_\psi - N_x\right)$ once per period of the drive N_x, in order to construct the quantum analogue of a Poincare section. For the parameters used in figure 1, we find that the results do not show a great deal of structure. However, there is a very good reason for this. The intrinsic quantum noise in the open evolution dominates the motion, and so the Poincare section is an essentially featureless blur of points, of comparable size to the attractor in figure 1.

Since the noise cannot be reduced, the way to overcome this is to amplify the underlying classical motion [13]. The scaling $\gamma \rightarrow \alpha\gamma$, $\nu \rightarrow \alpha^2\nu, g \rightarrow \alpha^2 g$ and $\omega \rightarrow \alpha\omega$ increases the classical $\dot{\theta}$ motion by a factor of α, so the attractor looks the same as figure 1 except for a rescaling of the $\dot{\theta}$ axis. In figure 2 we show the corresponding quantum motion for the scaled parameters with $\alpha = 5$. Now, because the size of the quantum noise relative to that of the underlying classical motion has

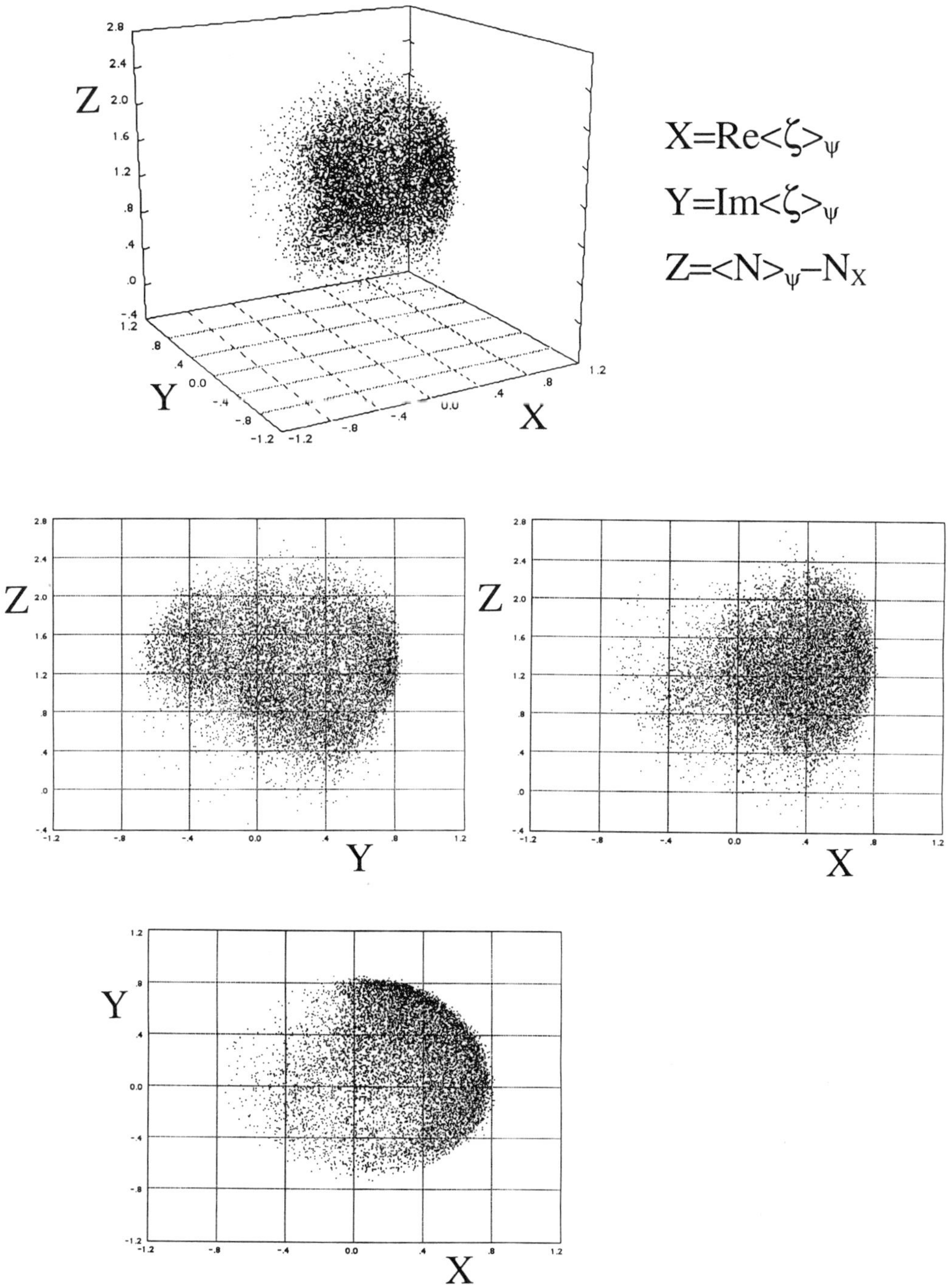

Figure 2: The quantum analogue of figure 1 scaled up by a factor 5, generated from solutions to (1) with $\gamma = \frac{5}{2}$, $\omega = \frac{10}{3}$, $\nu = 25$ and $g = \frac{75}{2}$. The results from a number of independent runs are superimposed to build up the attractor ($\sim$ 12000 points).

changed, evidence for the shape of the classical attractor is now apparent in the quantum motion. Reference 13 gives an example of this for a different system.

Clearly it would be nice to push this further, scaling the parameters so the classical attractor becomes more apparent in the quantum motion, and, indeed, work is in progress on this. However, the computation becomes more difficult due to the increasing size of the parameters and the larger basis of quantum states which is needed. Nevertheless, the data shown here provide sufficient evidence for the case that the individual system approach, quantum state diffusion, has an important role to play in unravelling the classical limit of quantum evolution. It provides a tool for calculating the quantum equivalent of a single dissipative classical chaotic trajectory. Also noteworthy is the fact that the model discussed here is one which is genuinely accessible in experiments, in both its classical and quantum regimes. Indeed, the inference from this work that the system becomes more classical as the damping and the tunnelling are increased, relative to the charging energy, is in keeping with the current view of quantum weak link devices [6-10]. Hopefully, then, quantum state evolution will have a crucial role to play in future work in this field.

Acknowledgements

The author thanks his Quantum Circuits Group colleagues at Sussex for stimulating interactions, and the Royal Society for a University Research Fellowship.

References

[1] N. Gisin, I.C. Percival, J. Phys. A25, 5677 (1992).

[2] N. Gisin, I.C. Percival, Phys. Lett. A167, 315 (1992).

[3] N. Gisin, I.C. Percival, J. Phys. A26, 2233 (1993).

[4] N. Gisin, I.C. Percival, J. Phys. A26, 2245 (1993).

[5] Note that the notation differs in the references 1-4. Here we adopt that of the later papers, 3 and 4.

[6] A. Widom, Y. Srivastava, Phys. Rep. 148, 1 (1987).

[7] G. Schön, A.D. Zaikin, Phys. Rep. 198, 237 (1990).

[8] T.P. Spiller, T.D. Clark, R.J. Prance, A. Widom, Prog. Low Temp. Phys. XIII, 219 (1992).

[9] Physica B152 (1988), Proceedings of the NATO Workshop on Coherence in Superconducting Networks, Delft, December 1987, edited by J.E. Mooji, G. Schön.

[10] 'Single electron tunnelling and mesoscopic devices', Proceedings of the 4th International Conference SQUID'91, Berlin, June 1991, edited by H. Koch, H. Lübbig (Springer Series in Electronics and Photonics 31, Springer-Verlag 1992).

[11] B.D. Josephson, Phys. Lett. 1, 251 (1962).

[12] e.g. G.L. Baker, J.P. Gollub, 'Chaotic dynamics, an introduction', (Cambridge University Press, 1990).

[13] T.P. Spiller, J.F. Ralph, Phys. Lett. A194, 235 (1994).

OBSERVATION OF QUANTUM JUMPS IN SQUID RINGS

R.J. Prance, R. Whiteman, T.D. Clark, J. Diggins,
H. Prance, J.F. Ralph, G. Buckling, G. Colyer[1]
and
C. Vittoria, A. Widom and Y. Srivastava[2]

[1]Physics Department,
University of Sussex,
Brighton, Sussex, BN1 9QH, U.K.
[2]Physics Department, Northeastern
University, Boston,
Massachusetts 02115, U.S.A.

Abstract

In this paper we describe an experimental technique which we have used to probe the low lying energy levels of a quantum mechanical SQUID ring and to observe transitions between these levels.

It is well established that the quantum description of a SQUID ring (a superconducting weak link ring) can be made in terms of the lumped component circuit Hamiltonian[1]

$$H = Q^2/2C + (\Phi - \Phi_x)^2/2\Lambda - \hbar\nu \cos(2\pi\Phi/\Phi_0) \tag{1}$$

where C is the weak link capacitance, Λ is the ring inductance, Φ_x is the magnetic flux applied to the ring, $\hbar\nu/2$ is the matrix element for coherent pair tunnelling through the weak link and $\Phi_0 = h/2e$. In this circuit description the conjugate variables are the total magnetic flux Φ threading the ring and the total displacement flux Q (with units of charge) threading across the weak link capacitor. Thus we may write $[\Phi, Q] = i\hbar^{(1)}$ with $\Delta\Phi\Delta Q \geq \hbar/2$. In the SQUID regime of behaviour $\Delta\Phi < \Phi_0$ and $\Delta Q \geq q$ (=2e, the pair charge).

In the SQUID Hamiltonian (1) there is an electric energy storage term ($Q^2/2C$), a magnetic energy storage term [$(\Phi - \Phi_x)^2/2\Lambda$, with $\Phi = \Phi_x + \Lambda I_s$ for a screening supercurrent I_s circulating around the ring] and a Josephson coupling term[2]

Quantum Communications and Measurement
Edited by V. P. Belavkin *et al.*, Plenum Press, New York, 1995

[$\hbar\nu\cos(2\pi\Phi/\Phi_0)$ in this SQUID ring geometry]. In this representation Φ and Q play the roles of coordinate and momentum, respectively, with $Q \rightarrow -i\hbar\partial/\partial\Phi$. It follows from (1) that C acts as the effective mass of the whole SQUID ring, treated as a single macroscopic quantum object, operating in Φ space in the potential $U(\Phi;\Phi_x) = (\Phi - \Phi_x)^2/2\Lambda - \hbar\nu\cos(2\pi\Phi/\Phi_0)$.

Here, we are concerned with the energy levels of such a quantum mechanical SQUID ring. These may be found by solving the time independent Schrödinger equation[1]

$$H\Psi_\kappa(\Phi;\Phi_x) = E_\kappa(\Phi_x)\Psi_\kappa(\Phi;\Phi_x), \tag{2}$$

where $\kappa = 0$ denotes the ground state, $\kappa = 1$ the first excited state, etc.. The E_κ are all dependent on, and Φ_0-periodic in, Φ_x, as are their derivatives $< I_s(\Phi_x) >_\kappa -$ $-\partial E_\kappa(\Phi_x)/\partial\Phi_x$, the screening current response, and $\chi_\kappa(\Phi_x) = -\Lambda\partial^2 E_\kappa/\partial\Phi_x^2$, the ring magnetic susceptibility. By way of illustration we show in figure 1 the first three SQUID ring energy levels for $\hbar\nu = 0.0400\Phi_0^2/\Lambda$, $\Lambda = 3\times10^{-10}$H and $\hbar\omega_0 = 0.0406\Phi_0^2/\Lambda$, where C is expressed in terms of the (quantized) ΛC oscillation of the SQUID ring [i.e. $\omega_0 = 1/(\Lambda C)^{\frac{1}{2}}$]. In this example $\hbar\nu$ is somewhat larger than the values typically found in our experiments. Nevertheless, it can be seen that each $E_\kappa(\Phi_x)$ displays a distinctive Φ_0-periodic signature. In particular, the $\kappa = 0$ and $\kappa = 1$ levels (or flux bands, as these are commonly termed) have a Φ_x-dependent structure which is maintained irrespective of the values chosen for $\hbar\nu$ and $\hbar\omega_0$. We note that both $< I_s(\Phi_x) >_\kappa$ and $\chi_\kappa(\Phi_x)$ arise as a consequence of the quantum description of (1) and (2). Thus, if there exists an experimental means to use these derivatives to probe the $E_\kappa(\Phi_x)$ of a SQUID ring, no projective quantum measurement will need to be made, i.e. in the sense of measuring current or flux states of the ring. It is apparent that the $\chi_k(\Phi_x)$ provide us with just such an experimental route. These $\chi_k(\Phi_x)$ allow for the introduction of a very simple experimental technique to monitor the $E_\kappa(\Phi_x)$, provided that the SQUID ring can be maintained adiabatically in one of these levels over the time scale required for the experiment. All that is required is that the ring be coupled inductively to a low frequency (usually radio frequency - rf) resonant circuit. This almost invariably takes the form of a parallel inductor (L_T) - capacitor (C_T) resonator, commonly referred to as a tank circuit. The $\chi_\kappa(\Phi_x)$ [and, by inference, the $E_\kappa(\Phi_x)$] can then be followed by plotting the resonant frequency of the coupled system as a function of Φ_x. For weak inductive coupling between the tank circuit coil and the SQUID ring in its κth level, the resonant frequency of the combination is given by[1,3] $f_{r\kappa}(\Phi_x) = f_R/[1 + K^2\chi_\kappa(\Phi_x)]^{\frac{1}{2}}$where f_R is the resonant frequency of the tank circuit alone and K is the flux coupling coefficient between the coil and the ring [$= M_{TS}^2/L_T\Lambda_{eff}$ for coil - ring mutual inductance of M_{TS} and effective coupling inductance $\Lambda_{eff} = \Lambda/2$ for a 2-hole SQUID ring]. It can be seen that by measuring K and following $f_{r\kappa}(\Phi_x)$ we can compute $\chi_\kappa(\Phi_x)$ and hence infer $\mathrm{E}_\kappa(\Phi_x)$, again provided that the SQUID ring can be maintained adiabatically in an energy eigenstate. Obviously, if the ground and first excited state flux bands display characteristic Φ_0-periodic patterns, irrespective of the particular values of $\hbar\nu$ and $\hbar\omega_0$ selected, then so will their derivatives $< I_s(\Phi_x) >_{0,1}$ and $\chi_{0,1}(\Phi_x)$. This allows us a straight forward way of recognizing the ground and excited states of a SQUID ring through observation of the $f_{r\kappa}(\Phi_x)$, themselves a function of $\chi_\kappa(\Phi_x)$.

The question as to whether such adiabaticity can be maintained (so that any decohering effects of the environment can be neglected) must, of course, be answered through experiment. By experience, we have found that this can be achieved by a

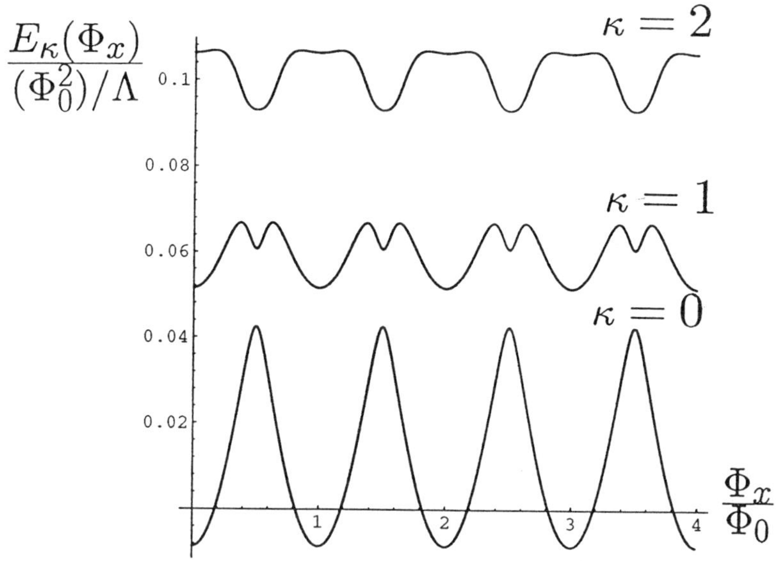

Figure 1: Magnetic flux periodic SQUID ring energy bands calculated from the time-independent Schrödinger equation (4) with $\Lambda = 3 \times 10^{-10}$H, $\hbar\nu = 0.0400\Phi_0^2/\Lambda$, $\hbar\omega_0 = 0.0406\,\Phi_0^2/\Lambda$ with $C = 1.1 \times 10^{-16}$F.

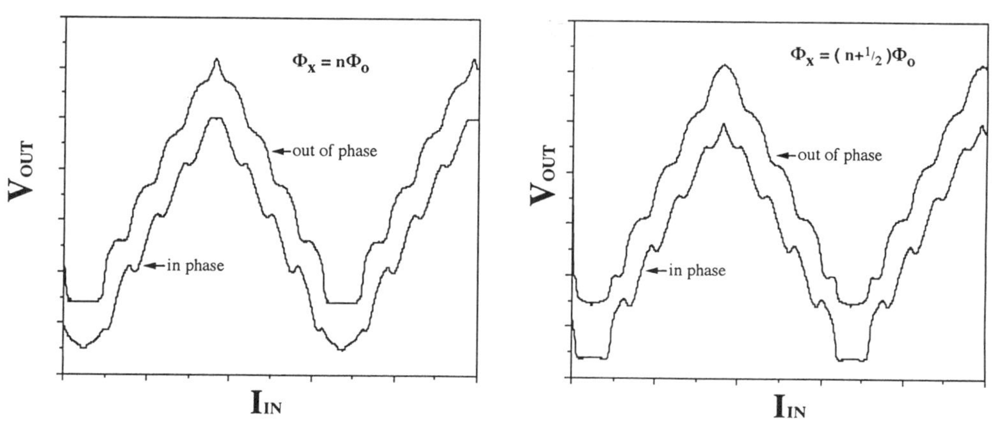

Figure 2: Typical experimental rf ground state dynamical characteristics of a tank circuit - SQUID ring system current driven close to resonance for $\Phi_x = n\Phi_0$ and $(n + \frac{1}{2}\Phi_0)$, n integer, showing the two quadrature components of V_{OUT} (in phase and out of phase).

combination of large energy separations (on the scale of $k_B T$) between SQUID ring energy levels, relatively low circuit frequencies, extensive electromagnetic shielding and ultra low noise and extremely stable rf receiver electronics. The essential requirement in the receiver electronics is a liquid helium cooled first stage GaAsFET preamplifier[3]. With this cryogenically cooled amplifier in place we were constrained for technical reasons to operate the SQUID ring at around 4K. In this temperature range our adiabaticity requirement that the energy level separations be large on the scale of $k_B T$ meant that we had to use SQUID rings with extremely small weak link capacitances, i.e. close to 10^{-16}F. The only SQUID rings available to us with capacitances in this range were of the niobium point contact variety. These took the form of two hole Zimmerman-type SQUID rings[4], the weak link being made mechanically in situ at liquid helium temperatures by contact between a point screw and a pre-oxidised post screw. By this method we could routinely make weak links with capacitances $\leq 10^{-16}$F.[1,5]

In practice the diagnostic technique we used to establish ground state behaviour relied on following the rf dynamics of the SQUID ring, operating as a magnetometer. As a magnetometer the tank circuit is excited by an external rf current source I_{IN} with the SQUID ring set at some value of static Φ_x This current source is triangularly amplitude modulated at audio frequency (typically $\sim$ 1kHz) with a maximum rf peak to peak amplitude of a few Φ_0. In our experiments the voltage response V_{OUT} to the drive (excitation) current is then detected, in our case at a known phase with respect to this current. The ground state rf V_{OUT} versus I_{IN} characteristic consists of a set of voltage features, periodic in I_{IN} , on a linear (ohmic) background. For a quantum SQUID magnetometer the whole characteristic, down to the origin, changes periodically as a function of static Φ_x. With this in mind the point contact was adjusted until the rf V_{OUT} versus I_{IN} characteristic matched the expected form. This could, and was, computed from the full non-linear equation of motion for the coupled tank circuit - SQUID ring system, i.e.[6]

$$C_T \ddot{\phi} + \frac{\dot{\phi}}{R_T} + \frac{\phi}{L_T} = I_{IN}(t) + \mu < I_s(\Phi_x + \mu\phi) >_0 \tag{3}$$

where ϕ is the rf flux in the tank circuit coil, R_T is the tank circuit resistance on resonance, I_{IN} is the drive current (which may include both coherent rf and noise terms), $\mu = M_{TS}/L_T$ and the screening current response $< I_S(\Phi_x + \mu\phi) >$ is for the SQUID ring in its ground state ($\kappa = 0$). Typical ground state dynamical (V_{OUT} versus I_{IN}) characteristics are shown in figure 2 for the two (external) static flux states $\Phi_x = n\Phi_0$ and $(n + \frac{1}{2})\Phi_0$, n integer.

With this diagnostic procedure completed we could then concentrate on the problem of following $f_{r\kappa}(\Phi_x)$. The experimental set up was a modification of that used to extract the rf SQUID magnetometer dynamics displayed in figure 2. For monitoring $f_{r\kappa}(\Phi_x)$ the tank circuit is excited by a small amplitude rf current derived from the tracking generator output of a Rohde and Schwartz spectrum analyzer. This output provided a constant amplitude rf drive over the frequency range of the tank circuit - SQUID ring resonance with the frequency ramped in synchronism with the spectrum analyzer sweep display. Typically, the signal $\mu\phi$ fed to the tank circuit from the spectrum analyzer $\simeq \Phi_0/30$. This is in contrast to the operating condition for observing SQUID magnetometer dynamics, where $\mu\phi \geq \Phi_0$. The combination of small applied signal ($\mu\phi << 1$) and weak coupling ($K^2 \sim 0.01$) ensures that we can invoke the simple relation for $f_{r\kappa}(\Phi_x)$ in terms of $\chi_\kappa(\Phi_x)$.

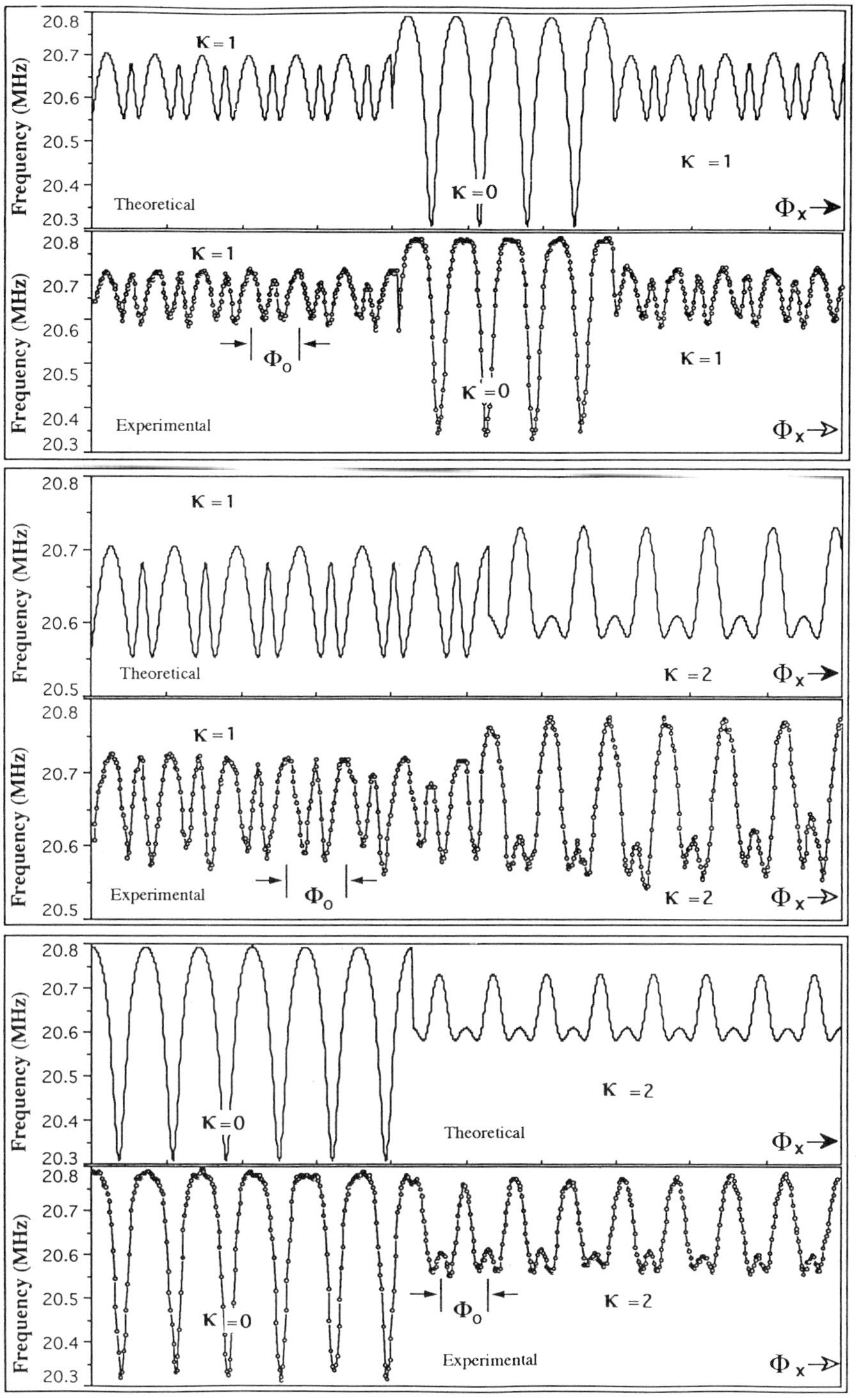

Figure 3: (a) Experimental, noise induced transitions between the $\kappa = 0, 1$ and 2 states for a niobium point contact SQUID ring at an operating temperature of 4.2K and a signal level $\simeq \Phi_0/30$. The times of the noise bursts are shown arrowed.
(b) Theoretical best fits to the data of 3(a) with $\hbar\nu = 0.0162\Phi_0^2/\Lambda$, $\hbar\omega_0 = 0.0406\Phi_0^2/\Lambda$, $\Lambda = 3 \times 10^{-10}$H, $K^2 = 0.0496$ and $f_R = 20.632$MHz.

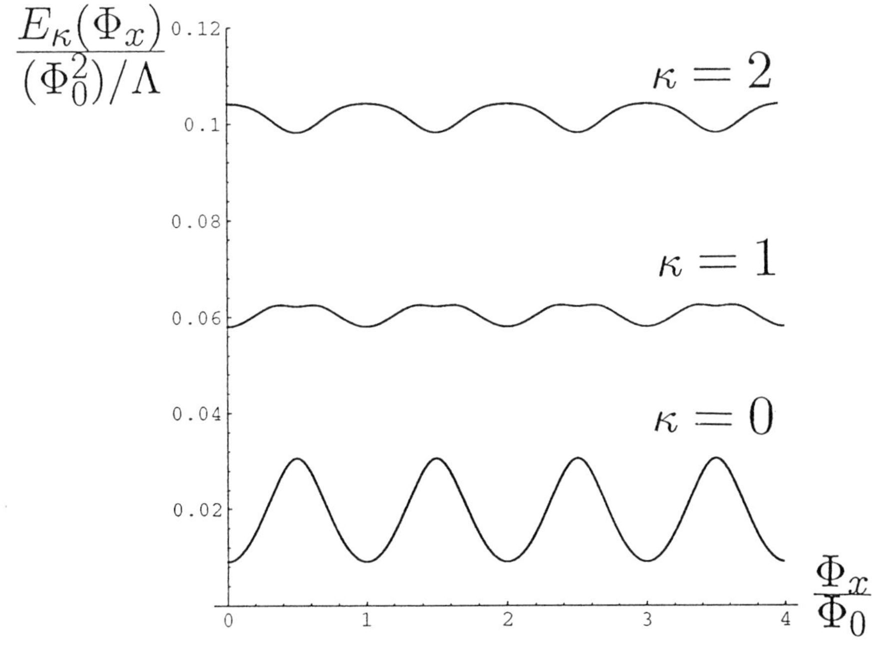

Figure 4: $\kappa = 0, 1$ and 2 flux bands concomitant with the best fit $f_{r\kappa}(\Phi_x)$ patterns of figure 3(b).

In figure 3 we show typical experimental $f_{r\kappa}(\Phi_x)$ data for an ultra small capacitance niobium point contact SQUID ring at 4.2K with $\mu\phi \simeq \Phi_0/30$ and $f_R =$ 20.632MHz. For this experiment the bias flux Φ_x was set at some value, and the resonant frequency of the tank circuit - SQUID ring system was recorded on the spectrum analyzer as an average over three measurements. The flux was then stepped by a small amount ($<< \Phi_0$) and the process was repeated. With this fine scale of discrete shifts, Φ_x was triangularly ramped over a peak to peak amplitude of approximately $10\Phi_0$ at rates close to $0.01\Phi_0$/sec. The ground state (and also excited state) frequency shift data exhibited no hysteresis at the maximum and minimum turning points in $\Phi_x(t)$, which is to be expected for a single quantum object possessing time reversal symmetry. Transitions between the low lying levels of a SQUID ring could be induced by means of extremely broadband (up to THz and greater) noise bursts generated in an external, GaAsFET-based noise source. These bursts were created by making very small changes ($\sim$ few to 10mV) in the gate supply voltage to the GaAsFET. Various transitions between the $\kappa = 0, 1$ and 2 states, as reflected in the different characteristic $f_{r\kappa}(\Phi_x)$ patterns, are shown in figure 3. The difference between the frequency shift signatures for the different bands is quite apparent. It is also apparent that the lifetimes in the two excited states (κ = 1 and 2) are long, i.e. $\sim$ minutes.[7] Clearly, the measured $f_{r\kappa}(\Phi_x)$ patterns of figure 3 constitute a kind of spectroscopy. Our basic problem in this work is, of course, that we are constrained to deal with just one SQUID ring. By definition, we cannot construct two precisely identical (indiscernable) SQUID rings, let alone an ensemble, since these are macroscopic condensed matter objects. We are thus forced to devise a spectroscopy based on one SQUID ring. The frequency shift technique is our solution to the problem.

We have seen that we can progress from flux bands $[E_\kappa(\Phi_x)]$ to frequency shifts $[f_{r\kappa}(\Phi_x)]$ via ring susceptibilities $[\chi_\kappa(\Phi_x)]$. By monitoring $f_{r\kappa}(\Phi_x)$ for several levels, we can reverse this sequence and compute the flux band structure of the SQUID ring from this frequency shift data. The accuracy of this process obviously improves with the number of different $E_\kappa(\Phi_x)$ band patterns observed. In figure 3 we show the results of the best fit computation of $f_{r\kappa}(\Phi_x)$ for the $\kappa = 0, 1$, and 2 bands using the experimental data presented in the same figure. Here, the band transitions are simulated (i.e. patched in) to correspond with experiment. For $\Lambda = 3 \times 10^{-10}$H and $K^2 = 0.0496$ this best fit requires that $\hbar\nu = 0.0162\Phi_0^2/\Lambda$ and $\hbar\omega_0 = 0.0406\,\Phi_0^2/\Lambda$. The first three flux bands corresponding to these values of SQUID ring parameters are shown in figure 4. Experience of calculating $E_\kappa(\Phi_x)$, and their corresponding $f_{r\kappa}(\Phi_x)$, has demonstrated quite clearly that the phase space of $\hbar\nu$ and $\hbar\omega_0$ which can generate best fits to data, such as that shown in figure 3, is extremely limited. Quite generally, correct matching of theory to experimental data restricts the range of choice of these SQUID ring parameters to within a few per cent of the established best fit values.

The match of the experimental data with theoretical fits in figure 3, and the computation of the concomitant flux bands, as shown in figure 4, provide clear proof that ultra small capacitance SQUID rings do indeed behave as single quantum objects. Furthermore, even though the technique is in an early stage of development, we have established a novel spectroscopic method for investigating the energy level structure of these macroscopic objects.

References

[1] H. Prance, T.D. Clark, R.J. Prance, T.P. Spiller, J. Diggins and J.F. Ralph, Nucl. Phys. B. (Proc. Suppl.) 33C, 35 (1993).

[2] B.D. Josephson, Phys. Lett. 1, 251 (1962).

[3] R.J. Prance, T.P. Spiller, T.D. Clark, H. Prance, J. Ralph, A. Widom and Y. Srivastava, Il Nuovo Cimento 106B, 431 (1991).

[4] J. Zimmerman, P. Thieve and J.T. Harding, J. Appl. Phys. 41, 1572 (1970).

[5] H. Prance, R.J. Prance, T.P. Spiller, T.D. Clark, A. Clippingdale, J.F. Ralph, J. Diggins and A. Widom, Phys. Letts. A181, 259 (1993).

[6] J. Diggins, J.F. Ralph, T.P. Spiller, T.D. Clark, H. Prance and R.J. Prance, Phys. Rev. E49, 1854 (1994).

[7] J.F. Ralph, G.J. Colyer, T.D. Clark, T.P. Spiller, R.J. Prance and H. Prance, "A Quantum Electrodynamic Model for Thick Superconducting Rings", to be published in Proc. NATO Advanced Research Workshop on "Mesoscopic Superconductivity" (Physica B), Karlsruhe, May, 1994.

Part IV.
Quantum Channels, Entropy & Information

STATE CHANGE, COMPLEXITY AND FRACTAL IN QUANTUM SYSTEMS

Masanori Ohya

Department of Information Sciences
Science University of Tokyo
Noda City, Chiba 278, Japan

Abstract

Information dynamics was proposed in [1] and it is a synthesis of the dynamics of state change and the theory of complexity. Information dynamics is now developing for uses in several different topics such as quantum physics, fractal theory, quantum information and genetics. Here we discuss entropic complexities in quantum systems and some applications to quantum communication process and Fractal theory.

1 INFORMATION DYNAMICS

Let us first remind our view of information dynamics (ID for short).

An input dynamical system and an output dynamical system are described by $(\mathcal{A}, \Sigma, \alpha(G))$ and $\left(\overline{\mathcal{A}}, \overline{\Sigma}, \overline{\alpha}\left(\overline{G}\right)\right)$, respectively. Here $\mathcal{A}$ is a set of all objects to be observed and Σ is a set of all means getting the observed value for each element A in $\mathcal{A}$, and α describes an inner evolution of the input system on an index set G. We call Σ a "state space" here. Same for the output system $\left(\overline{\mathcal{A}}, \overline{\Sigma}, \overline{\alpha}\right)$. Thus we may say

[Giving a mathematical structure to input and output triples $\equiv$ Having a theory].

For instance, to non-mathematical frame $(\mathcal{A}, \Sigma, \alpha(G))$, a speculation containing classical and quantum systems is a C^*-system, in which we are interested here.

Quantum Communications and Measurement
Edited by V. P. Belavkin *et al.*, Plenum Press, New York, 1995

The dynamics of the state change is given by a channel Λ^* from Σ to $\overline{\Sigma}$ (sometimes $\Lambda^* : \Sigma \to \overline{\Sigma}$ such as $\Lambda^* = \alpha^*$), which is specified by each problem to be studied.

Another emphasis of information dynamics is the complexities associated to the above input and output systems. There exist two complexities in information dynamics, one of which is the complexity $C(\phi)$ associated to a state ϕ of a system itself and another is the transmitted complexity $T(\phi; \Lambda^*)$ accompanied by the state ϕ and a channel Λ^*. These complexities should satisfy the following properties: Let $(\mathcal{A}, \Sigma, \alpha(G))$, $\left(\overline{\mathcal{A}}, \overline{\Sigma}, \overline{\alpha}\left(\overline{G}\right)\right)$ be two mathematically nonspecified systems and $(\mathcal{A}_t, \Sigma_t, \alpha^t(G^t))$ be the compound system and Λ^* be a channel from Σ to $\overline{\Sigma}$. It is important [2,3] to look at the complexities in a subset $\mathcal{S}$ of Σ, in which measurement is performed.

(i) For any state $\varphi \in \mathcal{S} \subset \Sigma$, $C^{\mathcal{S}}(\varphi) \geq 0$ and $T^{\mathcal{S}}(\varphi; \Lambda^*) \geq 0$.

(ii) For any bijection j from $ex\Sigma$, the set of all extreme states, to $ex\Sigma$ (or $\exists$ bijection $j : \Sigma \to \Sigma$)

$$C^{j(\mathcal{S})}(j(\varphi)) = C^{\mathcal{S}}(\varphi)$$
$$T^{j(\mathcal{S})}(j(\varphi); \Lambda^*) = T^{\mathcal{S}}(\varphi; \Lambda^*).$$

(iii) For any $\Phi \in \mathcal{S}_t \subset \Sigma_t$, put $\varphi \equiv \Phi \mid \mathcal{A}$ (i.e., the restriction of Φ to $\mathcal{A}$) $\in \mathcal{S} \equiv \mathcal{S}_t \mid \mathcal{A}$ and $\psi \equiv \Phi \mid \overline{\mathcal{A}} \in \overline{\mathcal{S}} \equiv \mathcal{S}_t \mid \overline{\mathcal{A}}$

$$C^{\mathcal{S}_t}(\Phi) \leq C^{\mathcal{S}}(\varphi) + C^{\overline{\mathcal{S}}}(\psi).$$

Moreover

(iv) $0 \leq T^{\mathcal{S}}(\varphi; \Lambda^*) \leq C^{\mathcal{S}}(\varphi)$.

(v) $T^{\mathcal{S}}(\varphi; id) = C^{\mathcal{S}}(\varphi)$, where "$id$" is an identity channel from Σ to Σ; $id(\varphi) = \varphi,\ \forall \varphi \in \Sigma$.

The information dynamics is described by $(G); \overline{\mathcal{A}}, \Lambda^*, C^{\mathcal{S}}$

$\left(\mathcal{A}, \Sigma, \alpha(G); \overline{\mathcal{A}}, \overline{\Sigma}, \overline{\alpha}\left(\overline{G}\right); \Lambda^*, C^{\mathcal{S}}(\varphi), T^{\mathcal{S}}(\varphi; \Lambda^*)\right)$ and some other functions (relations) R

Information dynamics can be applied to the study of chaos in the following means.

(a) ψ is more chaotic than φ seen from the reference system $\mathcal{S}$ if $C^{\mathcal{S}}(\psi) \geq C^{\mathcal{S}}(\varphi)$.

(b) When φ changes to $\Lambda^*\varphi$, a degree of chaos of this change is given by $C^{\overline{\mathcal{S}}}(\Lambda^*\varphi) - T^{\mathcal{S}}(\varphi\,;\,\Lambda^*)$.

Therefore, for each system of interest, we

(i) mathematically determine $\mathcal{A}, \Sigma, \alpha(G); \overline{\mathcal{A}}, \overline{\Sigma}, \overline{\alpha}\left(\overline{G}\right)$,

(ii) choose Λ^* and R, and

(iii) define $C^{\mathcal{S}}(\varphi)$, $T^{\mathcal{S}}(\varphi;\Lambda^*)$.

Once we set the above (i) $\sim$ (iii) , then our theory is applied to almost all (scientific) systems and we could study the following topics within this framework.

(1) Study of optical communication processes.

(2) Formulation of fractal dimensions of states, and study of complexity for some sequences.

(3) Definition of genetic matrix and construction of phylogenetic tree for evolution of species.

(4) Study of recognition processes.

In section 2, we explain one of the above fundamental objects in general quantum system (GQS); channel Λ^* and its generalization. In section 3, we discuss informational complexities and prove the Kolmogorov-Sinai type theorem with an application to quantum communication. In section 4, we briefly mention how to formulate the ε-entropy and the fractal dimension of a general quantum state in the contents of ID.

2 CHANNEL AND LIFTINGS

Let $(\mathcal{A},\Sigma)$ be an input system and $\left(\overline{\mathcal{A}},\overline{\Sigma}\right)$ be an output system with a mathematical structure to those considered in section 1, where $\mathcal{A}$ is a C^*-algebra or the set all bounded operators $B(\mathcal{H})$ on a Hilbert space $\mathcal{H}$, and Σ is the set of all positive linear functionals on $\mathcal{A}$ or the set of all density operators in $\mathcal{H}$, respectively. We denote the set of all density operators on $\mathcal{H}$ by $\Sigma(\mathcal{H})$. A map $\Lambda^*:\Sigma\to\overline{\Sigma}$ is called a channel. Moreover :

(1) Λ^* is linear channel if Λ^* is affine.

(2) Λ^* is completely positive (CP) channel if its dual map $\Lambda:\overline{\mathcal{A}}\to\mathcal{A}$ satisfies

$$\sum_{i,j=1}^{n} B_i^*\Lambda\left(A_i^*A_j\right)B_j\geq 0$$

for any $\{B_j\}\subset\mathcal{A}$, $\{A_i\}\subset\overline{\mathcal{A}}$ and any $n\in N$.

When some outside effects should be considered in a certain physical process such as noise and an effect of reservoir, it is convenient to extend the system $\mathcal{A}$ to $\mathcal{A}\otimes\mathcal{B}$, where $\mathcal{B}$ describes the outside system. In such cases, the concept of lifting is useful [4]. A lifting from $\mathcal{A}$ to $\mathcal{A}\otimes\mathcal{B}$ is a continuous map

$$\mathcal{E}^*:\Sigma\left(\mathcal{A}\right)\to\Sigma\left(\mathcal{A}\otimes\mathcal{B}\right)$$

where we denote the set of states on $\mathcal{A}$ by $\Sigma\left(\mathcal{A}\right)$ and that on $\mathcal{A}\otimes\mathcal{B}$ by $\Sigma\left(\mathcal{A}\otimes\mathcal{B}\right)$.

(3) A lifting $\mathcal{E}^*$ is linear if it is affine.

(4) A lifting $\mathcal{E}^*$ is nondemolition for a state $\varphi \in \Sigma(\mathcal{A})$ if $\mathcal{E}^*\varphi(A \otimes I) = \varphi(A)$ for any $A \in \mathcal{A}$.

Given a channel $\Lambda^* : \Sigma \to \overline{\Sigma}$ and $\varphi \in \mathcal{S} \subset \Sigma$, take a certain decomposition such that

$$\varphi = \int_{\mathcal{S}} \omega \, d\mu$$

Then

$$\mathcal{E}^*\varphi = \int_{\mathcal{S}} \omega \otimes \Lambda^*\omega \, d\mu \qquad \text{(compound state[3, 5])}$$

is a nondemolition lifting. This compound state (lifting) can be extensively used in the sequel sections.

We take $\mathcal{A} = \overline{\mathcal{A}} = B(\mathcal{H})$, hence $\Sigma = \overline{\Sigma} = \Sigma(\mathcal{H})$, the set of all trace class operators on a Hilbert space $\mathcal{H}$, and let $\mathcal{B}$ be an algebra on another Hilbert space $\mathcal{K}$. Therefore the state space of the total system is described by $\Sigma(\mathcal{H} \otimes \mathcal{K})$. As is discussed in [1,2,4], we have several examples of channels and liftings encountered in physics and engineering.

⟨1⟩ Open System Dynamics :
When two systems $\mathcal{A}$ and $\mathcal{B}$ are interacting through some time evolution U_t (e.g. $U_t = \exp(itH)$, total Hamiltonian H), an initial state $\rho \otimes \sigma$, $\rho \in \Sigma(\mathcal{H})$, $\sigma \in \Sigma(\mathcal{K})$, evolves by U_t such as

$$\mathcal{E}_t^*\rho \equiv U_t(\rho \otimes \sigma) U_t^*$$

which is a lifting of ρ. Moreover a channel is constructed by taking the partial trace of $\mathcal{E}_t^*\rho$ with respect to $\mathcal{K}$, namely

$$\Lambda_t^*\rho \equiv tr_{\mathcal{K}}\mathcal{E}_t^*\rho.$$

Using this lifting, we can construct a Markov chain of Accardi's type [6,7].

⟨2⟩ Isometric Lifting :
Let V be an isometry from $\mathcal{H}$ to $\mathcal{H} \otimes \mathcal{K}$. Then $\mathcal{E}^*\rho \equiv V\rho V^*$, $\rho \in \Sigma(\mathcal{H})$ is a lifting from $\Sigma(\mathcal{H})$ to $\Sigma(\mathcal{H} \otimes \mathcal{K})$. An important example of this isometric lifting comes from the attenuation process of optical communication, that is, the above isometry V is defined as

$$V_{\alpha,\beta}|\theta\rangle = |\alpha\theta\rangle \otimes |\beta\theta\rangle$$

for a coherent state $|\theta\rangle$ with $|\alpha|^2 + |\beta|^2 = 1$. Therefore, the attenuation channel is given by

$$\Lambda_{\alpha,\beta}^*\rho = tr_{\mathcal{K}}\mathcal{E}_{\alpha,\beta}^*\rho \equiv tr_{\mathcal{K}}V_{\alpha,\beta}\rho V_{\alpha,\beta}^*$$

Note that $V_{\alpha,\beta}^* V_{\alpha,\beta} = I$, and

$$\begin{aligned}
\left\langle \theta \left| V_{\alpha,\beta}^* V_{\alpha',\beta'} \right| \theta \right\rangle &= \langle \alpha\theta | \alpha'\theta\rangle \langle \beta\theta | \beta'\theta\rangle \\
&= \exp\left\{-|\alpha - \alpha'|^2 |\theta|^2\right\} \exp\left\{-|\beta - \beta'|^2 |\theta|^2\right\} \\
&= \exp\left\{-\left(|\alpha - \alpha'|^2 + |\beta - \beta'|^2\right) |\theta|^2\right\}
\end{aligned}$$

⟨3⟩ General Quantum Communication Process :
Quantum communication process is described by the following scheme [8].

$$\nu \in \Sigma(\mathcal{K})$$
$$\downarrow$$

$$\Sigma(\mathcal{H}) \ni \rho \longrightarrow \overline{\rho} = \Lambda^* \rho \in \Sigma(\mathcal{H})$$

$$\downarrow$$

$$\text{Loss}$$

$$\begin{array}{ccc} \Sigma(\mathcal{H}) & \xrightarrow{\Lambda^*} & \Sigma(\mathcal{H}) \\ \gamma^* \downarrow & & \uparrow a^* \\ \Sigma(\mathcal{H}\otimes\mathcal{K}) & \underset{\pi^*}{\longrightarrow} & \Sigma(\mathcal{H}\otimes\mathcal{K}) \end{array}$$

The above maps γ^*, a^* are given as

$$\begin{aligned} \gamma^*(\rho) &= \rho \otimes \nu, \quad \rho \in \Sigma(\mathcal{H}) \\ a^*(\theta) &= tr_{\mathcal{K}}\theta, \quad \theta \in \Sigma(\mathcal{H}\otimes\mathcal{K}), \end{aligned}$$

where ν is a noise coming from the outside of the system. The map π^* is a certain channel determined by physical properties of the combined system. Hence the lifting and the channel for the above process are given as

$$\begin{aligned} &\mathcal{E}^*\rho \equiv \pi^*(\rho\otimes\nu) = (\pi^*\circ\gamma^*)(\rho) \\ &\Lambda^*\rho \equiv tr_{\mathcal{K}}\pi^*(\rho\otimes\nu) = (a^*\circ\pi^*\circ\gamma^*)(\rho) \end{aligned}$$

Stinespring Representation Theorem: *Let π^* be a completely positive linear channel and the channel Λ^* be defined by $\langle 3\rangle$.*
(1) There exist the operators $W_k \in B(\mathcal{H})$ such that

$$\Lambda^*\rho = \sum_k W_k\rho W_k^*, \quad \sum_k W_k^* W_k = I.$$

(2) There exists a representation ν on an extended Hilbert space $\overline{\mathcal{H}}$ and $W \in B\left(\mathcal{H},\overline{\mathcal{H}}\right)$ such that

$$\Lambda^*\rho = W\rho W^*\cdot\nu, \quad W^*W = I.$$

For the isometric lifting introduced in the example $\langle 2\rangle$, the above theorem are satisfied. The operators $\{W_k\}$, W are concretely realized in some optical communication processes, which will be discussed elsewhere.

3 COMPLEXITIES IN QUANTUM INFORMATION

Let $(\mathcal{A}, \Sigma)$ be a unital C^*-system and $\mathcal{S}$ be a weak* compact convex subset of Σ. Take a state $\varphi \in \mathcal{S}$ and $M_\varphi(\mathcal{S})$ be the set of all maximal measures μ on $\mathcal{S}$ with the fixed barycenter φ :

$$\varphi = \int_{\mathcal{S}} \omega d\mu,$$

where μ is pseudosupported by the extreme points $ex\mathcal{S}$ of $\mathcal{S}$. Note that $M_\varphi(\mathcal{S}) \neq \phi$ by Choquets theorem and $M_\varphi(\mathcal{S}) = \{\mu\}$ iff $\mathcal{S}$ is a Choquet simplex [9].

The $\mathcal{S}$ (mixing) entropy of φ was defined as [2]

$$S^{\mathcal{S}}(\varphi) = \inf\{H(\mu)\ ;\ \mu \in M_\varphi(\mathcal{S})\}$$

with

$$H(\mu) = \sup\left\{ -\sum_{A_k \in \tilde{A}} \mu(A_k) \log \mu(A_k) \ : \ \tilde{A} \in P(\mathcal{S}) \right\}$$

where $P(\mathcal{S})$ is the set of all finite partitions of $\mathcal{S}$.

The CNT entropy [10] for a subalgebra $\mathcal{N}$ of $\mathcal{A}$ is

$$H_{\varphi}(\mathcal{N}) = \sup\left\{ \sum_k \lambda_k S(\omega_k \mid \mathcal{N}, \varphi \mid \mathcal{N}) \, ; \, \varphi = \sum_k \lambda_k \omega_k \text{ finite decomposition of } \varphi \right\}.$$

where $\varphi \mid \mathcal{N}$ is the restriction of a state φ to $\mathcal{N}$ and $S(\cdot, \cdot)$ is the relative entropy for C^*-algebra [13,14].

The relations between $S^{\mathcal{S}}(\varphi)$ and $H_{\varphi}(\mathcal{N})$ are discussed in [11] and it is shown that $S^{\mathcal{S}}(\varphi)$ distinguishes states more sharply than $H_{\varphi}(\mathcal{N})$.

For a channel $\Lambda^* : \Sigma(\mathcal{A}) \to \Sigma(\mathcal{B})$, the mutual entropy w.r.t. φ and Λ^* is given by [3]

$$I^{\mathcal{S}}(\varphi \, ; \Lambda^*) = \sup\left\{ I_{\mu}^{\mathcal{S}}(\varphi \, ; \Lambda^*) \ ; \ \mu \in M_{\varphi}(\mathcal{S}) \right\}$$

where

$$I_{\mu}^{\mathcal{S}}(\varphi \, ; \, \Lambda^*) = S\left(\Phi_{\mu}^{\mathcal{S}}, \, \Phi_0\right) \equiv \text{Relative entropy of } \Phi_{\mu}^{\mathcal{S}} \text{ w.r.t. } \Phi_0 \quad [12, 13, 14]$$

with

$$\Phi_{\mu}^{\mathcal{S}} = \int_{\mathcal{S}} \omega \otimes \Lambda^* \omega d\mu \quad \text{and} \quad \Phi_{\theta} = \varphi \otimes \Lambda^* \varphi.$$

It corresponds to a transmitted complexity and the complexity $C_I^{\mathcal{S}}(\varphi)$ is given $I^{\mathcal{S}}(\varphi; id)$, where "id" is the identity map from Σ to Σ. Other informational complexities are given as [1]

$$T^{\mathcal{S}}(\varphi \, ; \Lambda^*) = \sup\left\{ \int_{\mathcal{S}} S(\Lambda^* \omega, \Lambda^* \varphi) \, d\mu \, ; \, \mu \in M_{\varphi}(\mathcal{S}) \right\}$$

$$C_T^{\mathcal{S}}(\varphi) = T^{\mathcal{S}}(\varphi \, ; \, id)$$

Moreover the quasimutual entropy introduced in [11] is a transmitted complexity :

$$J^{\mathcal{S}}(\varphi \, ; \Lambda^*) = \sup\left\{ \sum_k \lambda_k S(\Lambda^* \omega_k, \Lambda^* \varphi) \, ; \, \varphi = \sum_k \lambda_k \omega_k \text{ finite decomposition in } \mathcal{S} \right\}$$

Its associated complexity is

$$C_J^{\mathcal{S}}(\varphi) = J^{\mathcal{S}}(\varphi \, ; id) \, .$$

These complexities satisfy the following relations[3,11,12,15]

Theorem : *We have the following inequalities and equalities.*

(1) $0 \le I^{\mathcal{S}}(\varphi \, ; \Lambda^*) \le C_I^{\mathcal{S}}(\varphi), 0 \le T^{\mathcal{S}}(\varphi \, ; \Lambda^*) \le C_T^{\mathcal{S}}(\varphi), 0 \le J^{\mathcal{S}}(\varphi \, ; \Lambda^*) \le C_J^{\mathcal{S}}(\varphi)$.

(2) $0 \le I^{\mathcal{S}}(\varphi \, ; \Lambda^*) \le T^{\mathcal{S}}(\varphi \, ; \Lambda^*) \le J^{\mathcal{S}}(\varphi \, ; \Lambda^*)$

(3) $C_I^{\Sigma}(\varphi) = C_T^{\Sigma}(\varphi) = C_J^{\Sigma}(\varphi) = S^{\Sigma}(\varphi) = H^{\Sigma}(\mathcal{A})$

(4) When $\mathcal{A} = \mathcal{B} = B(\mathcal{H})$, *for any density operator* ρ,

$$0 \le I^{\mathcal{S}}(\rho \, ; \Lambda^*) = T^{\mathcal{S}}(\rho; \Lambda^*) = J^{\mathcal{S}}(\rho \, ; \Lambda^*)$$

Let us consider the information transmission from an initial state ϕ to the final state in various ways, that is, through several channels.

(I) Let us send the information carried by a state $\varphi \in \mathcal{S}$ through some channel $\Lambda_k^* \, (k = 1, \cdots, N)$. Then the transmitted information and complexity are given by

$$I^{\mathcal{S}} (\varphi \, ; \Lambda_N^* \cdots \Lambda_1^*) \quad \text{and} \quad T^{\mathcal{S}} (\varphi \, ; \Lambda_N^* \cdots \Lambda_1^*)$$

In the case of $\Lambda_k^* = \Lambda^*$ for every k, put

$$f(N) = I^{\mathcal{S}} \left(\varphi \, ; \Lambda^{*N} \right) \quad \text{or} \quad T^{\mathcal{S}} \left(\varphi \, ; \Lambda^{*N} \right)$$

Then $0 \leq f(N+1) \leq f(N)$, so that there exists a limit $f(\infty) = \lim_{N \to \infty} f(N)$. The detail discussion for this limit is studied in [16,17] for special channels given by $\langle 1 \rangle$ and $\langle 3 \rangle$ of section 2.

(II) When we send an information through N-repeated channels Λ^{*N} and measure the transmitted information at every site of each channel, then the averaged transmitted information and complexity are given by

$$\frac{1}{N} I^{\mathcal{S}} \left(\varphi \, ; \Lambda^*, \cdots, \Lambda^{*N} \right) \quad \text{and} \quad \frac{1}{N} T^{\mathcal{S}} \left(\varphi \, ; \Lambda^*, \cdots, \Lambda^{*N} \right)$$

Here the above I and T are defined as follows.

$$I_\mu^{\mathcal{S}} \left(\varphi \, ; \Lambda^*, \cdots, \Lambda^{*N} \right) = S \left(\Phi_\mu^{\mathcal{S}} \left(\Lambda^*, \cdots, \Lambda^{*N} \right), \, \Phi_\theta \right)$$

$$I^{\mathcal{S}} \left(\varphi \, ; \Lambda^*, \cdots, \Lambda^{*N} \right) = \sup \left\{ S \left(\Phi_\mu^{\mathcal{S}} \left(\Lambda^*, \cdots, \Lambda^{*N} \right), \, \Phi_\theta \right) \, : \, \mu \in M_{\mathcal{S}} (\varphi) \right\}$$

with the compound states (liftings)

$$\Phi_\mu^{\mathcal{S}} = \int_{\mathcal{S}} \omega \otimes \Lambda^* \omega \otimes \cdots \otimes \Lambda^{*N} \omega d\mu$$

$$\Phi_\theta = \varphi \otimes \Lambda^* \varphi \otimes \cdots \otimes \Lambda^{*N} \varphi$$

and

$$T_\mu^{\mathcal{S}} \left(\varphi \, ; \Lambda^*, \cdots, \Lambda^{*N} \right) = \int_{\mathcal{S}} \sum_{k=1}^{N} S \left(\Lambda^{*k} \omega, \, \Lambda^{*k} \varphi \right) d\mu$$

$$T^{\mathcal{S}} \left(\varphi \, ; \Lambda^*, \cdots, \Lambda^{*N} \right) = \sup \left\{ \int_{\mathcal{S}} \sum_{k=1}^{N} S \left(\Lambda^{*k} \omega, \, \Lambda^{*k} \varphi \right) d\mu \, ; \, \mu \in M_{\mathcal{S}} (\varphi) \right\}$$

Put $g(N) = \frac{1}{N} I^{\mathcal{S}} \left(\varphi \, ; \Lambda^*, \cdots, \Lambda^{*N} \right)$ or $\frac{1}{N} T^{\mathcal{S}} \left(\varphi \, ; \Lambda^*, \cdots, \Lambda^{*N} \right)$. Then there exists $\mu' \in M_\varphi (\mathcal{S})$ such that

$$\frac{1}{N} \left(1 + I_{\mu'}^{\mathcal{S}} \; or \; T_{\mu'}^{\mathcal{S}} \right) \geq g(N) \geq \frac{1}{N} \left(I_\mu^{\mathcal{S}} \; or \; T_\mu^{\mathcal{S}} \right)$$

Therefore it is easily seen that $g(N)$ decreases in N and is lower bounded, which implies the existence of the limit $g(\infty)$ as $N \to \infty$.

(III) Let $\theta_{\mathcal{A}}$ (resp. $\theta_{\mathcal{B}}$) be an automorphism of $\mathcal{A}$ (resp. $\mathcal{B}$) such that $\varphi \circ \theta_{\mathcal{A}} = \varphi$ and Λ be a covariant C.P. map (i.e., $\Lambda \circ \theta_{\mathcal{A}} = \theta_{\mathcal{B}} \circ \Lambda$) from $\mathcal{B}$ to $\mathcal{A}$. Let $\mathcal{A}_k$ (resp. $\mathcal{B}_k$) be a finite subalgebra of $\mathcal{A}$ (resp. $\mathcal{B}$) and α_k (resp. β_k) be a C.P. unital map from $\mathcal{A}_k$ (resp. $\mathcal{B}_k$) to $\mathcal{A}$ (resp. $\mathcal{B}$). Put

$$\alpha^M \equiv (\alpha_1, \, \alpha_2, \, \cdots, \alpha_M)$$

$$\beta_\Lambda^N \equiv (\Lambda \circ \beta_1, \Lambda \circ \beta_2, \cdots, \Lambda \circ \beta_N)$$

Then the compound states for the above α^N and β_Λ^N in $\mu \in M_\varphi(\mathcal{S})$ are

$$\Phi_\mu^{\mathcal{S}}\left(\alpha^M\right) = \int_{\mathcal{S}(\mathcal{A})} \bigotimes_{m=1}^{M} \alpha_m^* \omega d\mu$$

$$\Phi_\mu^{\mathcal{S}}\left(\alpha^M \cup \beta_\Lambda^N\right) = \int_{\mathcal{S}(\mathcal{A})} \bigotimes_{m=1}^{M} \alpha_m^* \omega \bigotimes_{n=1}^{N} \beta_n^* \Lambda^* \omega d\mu$$

Hence we have

$$I_\mu^{\mathcal{S}}\left(\varphi; \alpha^M, \beta_\Lambda^N\right) = S\left(\Phi_\mu^{\mathcal{S}}\left(\alpha^M \cup \beta_\Lambda^N\right), \Phi_\mu^{\mathcal{S}}\left(\alpha^M\right) \otimes \Phi_\mu^{\mathcal{S}}\left(\beta_\Lambda^N\right)\right)$$

$$T_\mu^{\mathcal{S}}\left(\varphi; \alpha^M, \beta_\Lambda^N\right) = \int_{\mathcal{S}(\mathcal{A})} S\left(\bigotimes_{m=1}^{M} \alpha_m^* \omega \bigotimes_{n=1}^{N} \beta_n^* \Lambda^* \omega, \Phi_\mu^{\mathcal{S}}\left(\alpha^M\right) \otimes \Phi_\mu^{\mathcal{S}}\left(\beta_\Lambda^N\right)\right) d\mu$$

$$I^{\mathcal{S}}\left(\varphi; \alpha^M, \beta_\Lambda^N\right) = \sup\left\{I_\mu^{\mathcal{S}}\left(\varphi; \alpha^M, \beta_\Lambda^N\right) ; \mu \in M_\varphi(\mathcal{S})\right\}$$

$$T^{\mathcal{S}}\left(\varphi; \alpha^M, \beta_\Lambda^N\right) = \sup\left\{T_\mu^{\mathcal{S}}\left(\varphi; \alpha^M, \beta_\Lambda^N\right) ; \mu \in M_\varphi(\mathcal{S})\right\}.$$

Proposition: *When* $\max\left\{\|\alpha_n - \overline{\alpha}_n\|, \left\|\Lambda \circ \beta_n - \Lambda \circ \overline{\beta}_n\right\|\right\} < \varepsilon$ *is satisfied,*

$$\left|I^{\mathcal{S}}\left(\varphi; \alpha^M, \beta_\Lambda^N\right) - I^{\mathcal{S}}\left(\varphi; \overline{\alpha}^M, \overline{\beta}^N\right)\right| \leq 12(M+N)\varepsilon\left(\frac{1}{2} + \log\left(1 + \frac{d}{\varepsilon}\right)\right),$$

where $d =$ *max of the dimensions of* $\mathcal{A}_n$ *and* $\mathcal{B}_n$.

Remark:

(1) $I^{\mathcal{S}}(\varphi; \Lambda^*) = I^{\mathcal{S}}(\varphi; id, \Lambda)$

(2) $0 \leq I_\mu^{\mathcal{S}}\left(\varphi; \alpha^M, \beta^N\right) \leq \min\left\{T_\mu^{\mathcal{S}}\left(\varphi ; \alpha^M\right), T_\mu^{\mathcal{S}}\left(\varphi ; \beta_\Lambda^N\right)\right\}$

(3) $T_\mu\left(\varphi ; \alpha^M\right) + T_\mu\left(\varphi ; \beta_\Lambda^N\right) = S_\mu\left(\varphi ; \alpha^M \cup \beta_\Lambda^N\right) + I_\mu\left(\varphi ; \alpha^M, \beta_\Lambda^N\right)$

(4) If $\mathcal{A}_k, \mathcal{A}$ *are abelian* C^**-algebras and each* α_k *equals to the embedding, then our complexities concide with classical K-S entropy:* $T_\mu^{\Sigma}\left(\varphi; \alpha^M\right) = S_\mu^{classical}\left(\bigvee_{m=1}^{M} \tilde{A}_m\right)$, $I_\mu^{\Sigma}\left(\varphi ; \alpha^M, \beta_{id}^N\right) = I_\mu^{classical}\left(\bigvee_{m=1}^{M} \tilde{A}_m, \bigvee_{n=1}^{N} \tilde{B}_n\right)$ *for any finite partitions* $\tilde{A}_m, \tilde{B}_n$ *of a probability space* $(\Omega, \mathcal{F}, \varphi)$.

We are interested in a special case that $\mathcal{A}_k = \mathcal{A}_0$, $\mathcal{B}_k = \mathcal{B}_0$ for all k and $\alpha_k = \theta_{\mathcal{A}}^{k-1} \circ \alpha$, $\beta_k = \theta_{\mathcal{B}}^{k-1} \circ \beta$. Then

$$\tilde{I}^{\mathcal{S}}(\varphi; \Lambda^*, \theta_{\mathcal{A}}, \theta_{\mathcal{B}}, \alpha, \beta) = \limsup_{N \to \infty} \frac{1}{N} I^{\mathcal{S}}\left(\varphi; \alpha^N, \beta_\Lambda^N\right)$$

$$\tilde{I}^{\mathcal{S}}(\varphi; \Lambda^*, \theta_{\mathcal{A}}, \theta_{\mathcal{B}}) = \sup_{\alpha,\beta} \tilde{I}^{\mathcal{S}}(\varphi; \Lambda^*, \theta_{\mathcal{A}}, \theta_{\mathcal{B}}, \alpha, \beta)$$

Theorem : *If there exist* $\alpha'_m : \mathcal{A} \to \mathcal{A}_m$, *and* $\beta'_m : \mathcal{B} \to \mathcal{B}_m$ *such that*

$$\alpha_m \circ \alpha'_m \to id, \ \beta_m \circ \beta'_m \to id \quad as \quad m \to \infty,$$

then

$$\tilde{I}^{\mathcal{S}}(\varphi;\Lambda^*,\theta_{\mathcal{A}},\theta_{\mathcal{B}})=\lim_{m\to\infty}\tilde{I}^{\mathcal{S}}(\varphi;\Lambda^*,\theta_{\mathcal{A}},\theta_{\mathcal{B}},\alpha_m,\beta_m)$$

Same limit exists for the mean transmitted complexity T.

This is a Kolmogorov-Sinai type convergence theorem for the mutual entropy and the transmitted complexity. Up to here, the detail discussion of (III) can be seen in [18]. The computation of this limit for an attenuation channel is discussed in [16].

In particular, a quantum extension of classical formulation for information transmission giving a basis of Shannon's coding theorems can be considered in the case that $\mathcal{A}=\bigotimes_{-\infty}^{\infty}\mathcal{A}_0$, $\mathcal{B}=\bigotimes_{-\infty}^{\infty}\mathcal{B}_0$, $\mathcal{S}=\Sigma$ and $\theta_{\mathcal{A}}$, $\theta_{\mathcal{B}}$ are shift operators, both denoted by θ. In this case, the channel capacity is defined as

$$C(\Lambda^*)\equiv\sup\left\{\tilde{I}^{\Sigma}(\varphi;\Lambda^*,\theta)\ ;\ \varphi\in\Sigma\right\}.$$

Using this capacity, we formulate Shannon's coding theorems in fully quantum systems [15].

4 FRACTAL DIMENSION OF STATES

Usual fractal theory mostly treats geometrical sets. It is desirable to extend the fractal theory so as to be applicable to some other objects. For this purpose, we introduce the fractal dimension for general states. First we recall two fractal dimensions of geometrical sets.

⟨Scaling dimension ⟩

We observe a complex set F built from a fundamental pattern. If the number of the patterns observed is $N(1)$ when the scale is very rough, say 1,and the number is $N(r)$ when the scale is r, then we call the dimension defined by

$$d_s(F)=\frac{\log(N(r)/N(1))}{\log(1/r)}$$

the scaling dimension of the set F.

⟨ Capacity dimension ⟩

Let try to cover a set F in the n-dimension Euclidian space R^n by a certain convex set with the diameter ε. If the smallest number of the convex sets needed to conver the set F is $N(\varepsilon)$, then we call the dimension given by

$$d_c(F)=\lim_{\varepsilon\to 0}\frac{\log N(\varepsilon)}{\log(1/\varepsilon)}$$

the capacity dimension (or the ε-entropy dimension) of the set F.

These two fractal dimensions become equal for almost all sets in which we can compute these dimensions.

The ε-entropy is extensively studied by Kolmogorov [19] and his ε-entropy is for a probability measure, which gives us an idea to define the ε-entropy for a general state.

Kolmogorov introduced the notion of ε-entropy in probability space $(\Omega,\mathcal{F},\mu)$. His formulation is as follows: For two random variables $f,g\in M(\Omega)$, the mutual entropy $I(f,g)$ is defined by the joint probability measure $\mu_{f,g}$ and the direct product measure $\mu_f\otimes\mu_g$ such that

$$I(f,g)=S(\mu_{f,g},\mu_f\otimes\mu_g)$$

The ε-entropy for a random variable f is given by

$$S(f,\varepsilon) \equiv \inf\{I(f,g)\,; g \in M_d(f,\varepsilon)\}$$

where $M_d(f,\varepsilon) \equiv \left\{g \in M(\Omega)\,;\, \sqrt{\int_\Omega d(f,g)^2\, d\mu} < \varepsilon\right\}$ with the distance $d(f,g)$ between f and g.

We introduced the ε-entropy of a general quantum state ϕ in [3,20], and an application of ε-entropy was discussed in [21]. Here we a bit generalize it. Let $\mathcal{C}$ be the set of some channels physically interested and define two sets;

$$\mathcal{C}_1(\Lambda^*;\varphi) = \{\Gamma^* \in \mathcal{C}; \Gamma^*\varphi = \Lambda^*\varphi\},$$

$$\mathcal{C}_2(\varphi;\varepsilon) = \{\Gamma^* \in \mathcal{C}\,;\, ||\varphi - \Gamma^*\varphi|| \le \varepsilon\},$$

Then the ε-entropy of a state of ϕ w.r.t. $\mathcal{S}$ is defined by using the transmitted complexity $T^{\mathcal{S}}(\varphi;\, \Lambda^*)$ as

$$S_{\mathcal{C}}^{\mathcal{S}}(\varphi;\varepsilon) = \inf\left\{J_{\mathcal{C}}^{\mathcal{S}}(\varphi;\Lambda^*)\,;\Lambda^* \in \mathcal{C}_2(\varphi;\varepsilon)\right\},$$

where

$$J_{\mathcal{C}}^{\mathcal{S}}(\varphi;\Lambda^*) = \sup\left\{T^{\mathcal{S}}(\varphi;\Gamma^*)\,;\Gamma^* \in \mathcal{C}_1(\Lambda^*;\varphi)\right\}.$$

When $\mathcal{S} = \Sigma$ and $\mathcal{C}$ is the set of all channels on Σ, the ε-entropy is denoted by $S(\varphi;\varepsilon)$ for simplicity. When T equals to the mutual entropy I, the ε-entropy is called a Kolmogorov type.

Theorem :

(1) If $\mathcal{A} = C(\Omega)$, the set of all continuous functions on $\Omega, \varphi = \mu_f$ with a random variable f and T is a classical mutual entropy, then $S(\varphi;\varepsilon) = S(f;\varepsilon)$ of Kolmogorov.

(2) If T is the mutual entropy I in GQS, then $S(\varphi)$ is a complexity and $S(\varphi;\varepsilon)$ is a transmitted complexity.

The capacity dimension of a state φ w.r.t. $\mathcal{S}$ and $\mathcal{C}$ is defined by

$$d_{\mathcal{C}}^{\mathcal{S}}(\varphi) \equiv \lim_{\varepsilon\to\infty} d_{\mathcal{C}}^{\mathcal{S}}(\varphi;\varepsilon),$$

where

$$d_{\mathcal{C}}^{\mathcal{S}}(\varphi;\varepsilon) = \frac{S_{\mathcal{C}}^{\mathcal{S}}(\varphi;\varepsilon)}{\log(1/\varepsilon)}.$$

The above $d_{\mathcal{C}}^{\mathcal{S}}(\varphi;\varepsilon)$ is called the capacity dimension of ε-order. The information dimension of a state φ for $\mathcal{S}$ of ε order is defined by

$$d_{I,\mathcal{C}}^{\mathcal{S}}(\varphi;\varepsilon) = \frac{S_{\mathcal{C}}^{\mathcal{S}}(\varphi;\varepsilon)}{I_{\mathcal{C}}(\varepsilon)},$$

where $I_{\mathcal{C}}(\varepsilon)$ is a certain normalization constant function in the limit $\varepsilon \to 0$: $\lim_{\varepsilon\to 0} I_{\mathcal{C}}(\varepsilon) = S_{\mathcal{C}}^{\mathcal{S}}(\varphi)$.

The fractal dimensions above are applied to genome sequences, spin systems and so on.

Acknowledgments

This study is partially supported by Ciba-Geigy Foundation (Japan) for the Promotion of Science.

References

[1] Ohya, M. : "Information dynamics and its application to optical communication processes", Lecture Notes in Physics 378, Springer, 1991, pp. 81–92.

[2] Ohya, M. : "Entropy change in linear response dynamics", Nuovo Cimento, 91B, 1986, pp. 25–30.

[3] Ohya, M. : "Some aspects of quantum information theory and the applications to irreversible processes", Rep. Math. Phys., 27, 1989, pp. 19–47.

[4] Accardi, L. and Ohya, M. : "Compound channels, transition expectations and liftings", to appear in J. Multivariate Analysis.

[5] Ohya, M : "Note on quantum probability", L. Nuovo Cimento, 38, 1983, pp. 402–406.

[6] Accardi L. : "Noncommutative Markov chains" In : international School of Mathematical Physics, Camerino, 1974, 268–295.

[7] Accardi L., Ohya M. and Suyari H. : "Computation of mutual entropy in quantum Markov chains", preprint in CENTRO MATEMATICO V.VOLTERRA.

[8] Ohya, M : "On compound state and mutual information in quantum information theory", IEEE Trans. Information Theory, 1983, 29, pp. 770–777.

[9] Choquet G. : Lecture Analysis I, II, III, Benjamin, New York, 1969.

[10] Connes, A. Narnhofer, H. and Thirring, W. : "Dynamical entropy of C*-algebras and von Neumann algebras", Commun. Math. Phys. 112, 1987, pp. 691–719.

[11] Muraki, N., Ohya, M. and Petz, D. : "Entropies of general quantum systems", Open Systems and Information Dynamics, 1(No.1), 1992, pp. 43-56.

[12] Ohya M. and Petz D. : "Quantum Entropy and its Use", Springer-Verlag 1993.

[13] Araki, H. : "Relative entropy for states of von Neumann algebras Publ. RIMS Kyoto Univ. 11, 1976, pp. 809–833.

[14] Uhlmann, A. : "Relative entropy and the Wigner-Yanase-Dyson-Lieb concavity in an interpolation theory", Commun. Math. Phys. **54**, 1977, pp. 21–32.

[15] Ohya, M. : in preparation.

[16] Ohya, M. and Watanabe, N. : "A mathematical study of information transmission in quantum communication processes", in this volume.

[17] Accardi L., Ohya M. and Suyari H. : "Mutual entropy in quantum Markov chains", in this volume.

[18] Muraki, N. and Ohya, M. : preprint.

[19] Kolmogorov, A.N. : "Theory of transmission of information", Amer. Math. Soc. Translation, Ser.2, 33, 1963, pp. 291-.

[20] Ohya, M. : "Fractal dimensions of states", Quantum Probability and Related Topics VI, 1991, pp. 359–369.

[21] Akashi, S. : "A relation between Kolmogorov-Prokhorov's condition and Ohya's fractal dimensions", IEEE Transaction on Inf. Theory., 38, 1992, pp.1567–1570.

INFORMATION IN M-ARY QUANTUM OPTICAL COMMUNICATIONS: AN INEQUALITY PROVIDING AN UPPER LIMIT

A. Vourdas

Department of Electrical
Engineering and Electronics
University of Liverpool
P.O. Box 147 Liverpool L69 3BX

Abstract

It is shown that

$$I(\rho_i; p_i) \leq \sum_{i=1}^{M} p_i I(\rho_i) + I_0 \tag{1}$$

where $I(\rho_i; p_i)$ is the mutual information transmitted in an M-ary quantum optical communications system that uses the signal ρ_i for the letter i and transmits this letter with prior probability p_i; $I((\rho_i)$ is the information of the density matrix ρ_i; and I_0 is an information related to the use of signals with different average number of photons, for different letters.

1 INTRODUCTION

Quantum fluctuations become important at angular frequencies (which are much greater than the temperature T (in units $h = c = k_B = 1$). For "reasonable" temperatures quantum fluctuations at optical frequencies are significant. The aim of quantum optical communications is to take these quantum fluctuations into account; and to exploit quantum phenomena with no classical analogue for novel ways of communication (e.g. quantum interference, quantum correlations, Einstein-Podolsky-Rosen experiment, etc.)

In previous work [1] we have studied the information $I(\rho)$ of a signal described by a density matrix ρ, defined as the difference between the entropy of ρ and the entropy of a thermal signal with an equal average number of photons. This is

Quantum Communications and Measurement
Edited by V. P. Belavkin *et al.*, Plenum Press, New York, 1995

a fundamental quantity that gives the maximum information that this signal can transmit. It is fundamental in the sense that it is independent of any specific quantum communication system that uses this signal. In [2] we have considered binary communication systems and calculated the transmitted (mutual) information $I(\rho_i; p_i)$ that depends on the signals ρ_i and the prior probabilities p_i. Intuitively we expect that the $I(\rho_i; p_i)$ might be related to the $\sum p_i I(\rho_i)$. We prove inthis paper that $I_0 + \sum p_i I(\rho_i)$ is an upper bound for $I(\rho_i; p_i)$; where I_0 is an information related to the use of signals with different average number of photons, for different letters.

In section 2 we define the information of a signal ρ and discuss briefly the implications of a measurement Λ. In section 3 we introduce the information $I(\rho_i; p_i)$ that is transmitted in M-ary quantum optical communication systems, and prove the inequality mentioned in the abstract. Two special cases are discussed which clarify the physical significance of I_0. We conclude in section 4 with a discussion of our results.

2 INFORMATION (NEGENTROPY) OF A QUANTUM SIGNAL DESCRIBED BY A DENSITY MATRIX ρ

Let ρ be a density matrix and n the average number of photons

$$n = Tr(\rho a^+ a) \tag{2}$$

It is known that among density matrices with a fixed average number of photons, the thermal one has the maximum entropy

$$\rho_{th}(n) = \sum_{N=0}^{\infty} \frac{n^N}{(1+n)^{N+1}} |N\rangle\langle N| \tag{3}$$

$$S_{th}(n) = \ln(1+n) + n\ln(1+n^{-1}) \tag{4}$$

In the spirit of the negentropy principle of Brillouin, we define the quantity

$$I(\rho) = S_{th}(n) - S(\rho) \tag{5}$$

$$S(\rho) = -Tr\rho\ln\rho \tag{6}$$

which characterises the information content of the density matrix ρ. It should be emphasised that this quantity gives the maximum information that a beam described by the density matrix ρ can transmit. In [1] we have calculated numerically $I(\rho)$ for squeezed states, number eigenstates and mixtures of these signals with thermal noise.

We next consider a measurement (Hermitian operator) Λ with eigenvalues λ_N

$$\Lambda = \sum \lambda_N \pi_N \tag{7}$$

$$\pi_N \pi_M = \delta_{NM} \pi_N \tag{8}$$

A system with density matrix ρ will be described after the measurement Λ by the

$$\rho_\Lambda = \sum_N \pi_N \rho \pi_N \tag{9}$$

We define the following measurement dependent information [3]

$$I(\rho; \Lambda) = S_{th}(n) - S(\rho; \Lambda) \tag{10}$$

$$S(\rho; \Lambda) = -Tr\rho_\Lambda \ln\rho_\Lambda \tag{11}$$

and using the known inequality

$$S(\rho;\Lambda) \geq S(\rho) \tag{12}$$

we prove

$$I(\rho;\Lambda) \leq I(\rho) \tag{13}$$

3 INFORMATION IN M-ARY QUANTUM OPTICAL COMMUNICATION SYSTEMS

We consider an M-ary quantum optical communication system i.e. a system that uses an alphabet with M letters that we call $1, 2, \ldots M$. We call p_i the probability (relative frequency) with which the letter i is transmitted.

$$0 \leq p_i \leq 1; \quad i = 1, \ldots M \tag{14}$$

$$\sum_{i=1}^{M} p_i - 1$$

The p_i are assumed to be known. We call ρ_i the density matrix of the signal corresponding to the letter i and define the transmitted information per letter as

$$I(\rho_i; p_i) = S\left(\sum_i p_i \rho_i\right) - \sum_i p_i S(\rho_i) \tag{15}$$

The property of concavity of the entropy ensures that this quantity is non-negative. Numerical evaluation of this quantity for certain examples has been presented in [2]. The information (14) is measurement-independent. The measurement process is expected to reduce the transmitted information and refs [4] have proved this rigorously.The quantity $I(\rho_i; p_i)$ depends on all signals (i and all probabilities p_i. It would be nice if we could express it as a sum of the information $I(\rho_i)$ (equ.(4)) of the various signals. The following inequality clarifies the relationship between the quantities (4), (14).

$$I(\rho_i; p_i) \leq I_0 + \sum_i p_i I(\rho_i) \tag{16}$$

$$I_0 = S_{th}\left(\sum_i p_i n_i\right) - \sum p_i S_{th}(n_i) \tag{17}$$

$$n_i = Tr(a^+ a \rho_i) \tag{18}$$

The right hand side of equ. (15) is the information of each signal multiplied with the probability p_i with which this signal is transmitted; plus an extra term I_0. Since the $S_{th}(n)$ is a concave function, I_0 is related to the use of signals with different average number of photons for different letters. In order to appreciate this, we consider two special cases. In the first. all the density matrices ρ_i have the same average number of photons and we easily see that $I_0 = 0$. in this case equ. (15) becomes:

$$n_1 = n_2 \ldots = n_M \rightarrow I(\rho_i; p_i) \leq \sum_i p_i I(\rho_i) \tag{19}$$

. In the second special case, we consider a binary system ($M = 2$), which uses two thermal signals for the transmission of two letters.

$$\rho_1 = \rho_{th}(n_1) \tag{20}$$

$$\rho_2 = \rho_{th}(n_2) \tag{21}$$

In this special case $I(\rho_1) = I(\rho_2) = 0$ and equ.(15) gives:

$$I(\rho_i; p_i) \leq I_0 \tag{22}$$
$$I_0 = S_{th}(p_1 n_1 + p_2 n_2) - p_1 S_{th}(n_1) - p_2 S_{th}(n_2) \tag{23}$$

We have considered the case $p_1 = p_2 = 0.5, n_1 = 1$ and calculated numerically the quantities $I(\rho_i, p_i)$, I_0 as a function of n_2. We have also calculated the probability of error p_e, which is known [5] to be given by the formula

$$p_e = p_1 \left[1 - \sum \tau_k H(\tau_k)\right] \tag{24}$$

where τ_k are the eigenvalues of the operator

$$\left[\rho_1 - \left(p_2 p_1^{-1}\right) \rho_2\right] |\tau_k\rangle = \tau_k |\tau_k\rangle \tag{25}$$

and $H(\tau_k)$ is the step function which is equal to one for positive values of τ_k, and equal to zero for negative values. The infinite summation needed for the calculation of the entropies has been truncated at some value which is much larger than the average number of photons in the signals. The results are presented in table 1 where it is seen that as the difference between the average number of photons in the two signals increases, the transmitted information increases and the probability of error decreases.

These two special cases clarify the physical significance of I_0 as information related to the use of signals with different average number of photons for different letters. The right hand side of equ.(15) contains the informations $I(\rho_i)$ of the various signals with weight the prior probability p_i; plus the information I_0.

For completeness we also give another upper limit of the quantity $I(\rho_i; p_i)$. Using the Lanford-Robinson inequality [6] we easily prove

$$I(\rho_i; p_i) \leq -\sum p_i \ln p_i \leq \ln M \tag{26}$$

4 DISCUSSION

The main result of this paper is the inequality of equ.(15). It states that an upper bound for the information $I(\rho_i, p_i)$ contains the sum of all $p_i I(\rho_i)$ plus the information I_0. $I(\rho_i)$ expresses how much less entropy the ρ_i has in comparison with a thermal density matrix with an equal average number of photons. I_0 is related to the fact that signals with different average number of photons are used for different letters. The result is useful for an easy estimate of $I(\rho_i, p_i)$.

n_1	n_2	$I(\rho_i, p_i)$(nats)	I_0(nats)	p_e
1	0	0.21	0.26	0.25
1	1	0.00	0.00	0.50
1	2	0.03	0.03	0.40
1	3	0.08	0.09	0.34
1	4	0.12	0.14	0.30
1	5	0.16	0.20	0.27
1	6	0.19	0.25	0.24
1	7	0.21	0.30	0.22
1	8	0.23	0.34	0.21
1	9	0.24	0.38	0.19
1	10	0.25	0.42	0.18

References

[1] A. Vourdas, Phys. Rev. A37, 3890 (1988); Phys. Rev. A39, 206 (1989)

[2] A. Vourdas, IEEE Trans. Info. Theo. IT-36, 208 (1990); J. Mod. Opt. 37, 1661 (1990)

[3] R. Balian, M. Veneroni, N. Balazs, Europhys. Let. 1, 1 (1986)

[4] A.S. Holevo, Problemy Peredachi Inform. 9(3), 3 (1973) H.P. Yuen, M. Ozawa, Phys. Rev. Let. 70, 363 (1993) M.J.W. Hall, M.J. O'Rourke, Quantum Opt. 5, 161 (1993)

[5] C.W. Helstrom, "Quantum detection and estimation theory" (Academic, London, 1976).

[6] A. Wehrl, Rev. Mod. Phys. 50, 221 (1978).

NON-COMMUTATIVE EXTENSION OF THE INFORMATION GEOMETRY

H. Hasegawa

Department of Electronics Engineering,
Fukui University,
3-9-1 Bunkyo, Fukui 910, Japan

Abstract

It is shown that the differential-geometrical formulation of statistics concerning the structure of a smooth manifold in the parameter space of classical probabilities, $S = \{p(\,\cdot\,, \theta)\}$, which was constructed by Amari can be extended to the non-commutative framework, at least, within finite dimensional algebras where the classical probabilities $p(\,\cdot\,, \theta)$ are replaced by density matrices, $\rho(\theta)$. A brief outline of the framework is presented with the emphasis on elucidating the remarkable dualistic structure of the so-called α-families of ρ's, specifically, in the case of $|\alpha| = 1$ (i.e. the dualistic structure between $\log \rho$ and ρ).

1 INTRODUCTION

This is a brief outline of an investigation to construct a non-commutative version of the differential-geometrical formulation of statistics developed by Amari (in his Lecture Notes in Statistics **28** 1985 [1]) which is known by the name of *Information Geometry.* The first attempt to this end by the present author dealing with the *α-divergence*, $D_\alpha(\rho, \sigma)$, was published in a recent issue of Reports on Mathematical Physics [2], and hence the author wishes to present its outgrowth which emphasizes that the most beautiful part of Amari's theory, namely, the dual structure of the smooth manifold of certain probability functionals can be exactly preserved in such a non-commutative formulation. A full account requires many pages which must be postponed to other publications.

Quantum Communications and Measurement
Edited by V. P. Belavkin *et al.*, Plenum Press, New York, 1995

2 DECOMPOSITION THEOREM FOR A NEIGHBORHOOD OF THE DENSITY MATRIX

We investigate a smooth, C^∞, manifold in the parameter space $\Theta(\theta^1, \theta^2, \ldots, \theta^n)$ of a finite dimensional C^*-algebra $\mathcal{B}(H^N)$, in particular, that of all the density matrices $\mathcal{S} \subset \mathcal{B}(H^N)$ as a statistical model of our object, which may be regarded as a non-commutative extension of the *multinomial distribution*, one of Amari's examples for such models[1].

Our starting point is the fact that for any density matrix $\rho \in \mathcal{S}$ there exists a presentation of a topological neighborhood $\delta\rho(\equiv \rho(\theta + d\theta) - \rho(\theta))$ up to first order in $d\theta$) in terms of the commutant $\mathcal{C}(\rho) = (X \in \mathcal{B}(H^N);\ [X, \rho] = 0)$ and an antihermitian matrix Δ such that

$$\delta\rho = \delta^c\rho + [\rho, \Delta] \qquad \delta^c\rho \in \mathcal{C}(\rho), \qquad \sqrt{-1}\Delta \in \mathcal{B}^s(H^N) \tag{1}$$

where δ, δ^c and $[\,\cdot\,, \Delta]$ have derivation properties (i.e. linearity and Leibnitz rule of differentiation on products) as superoperators on $\mathcal{B}(H^N)$: in particular, for any smooth function $F(\rho) \in \mathcal{C}(\rho)$ of ρ together with (1), one can write

$$\delta F(\rho) = F'(\rho)\delta^c\rho + [F(\rho), \Delta], \qquad F'(\rho) \in \mathcal{C}(\rho). \tag{2}$$

(In [2], we used the notation d, d_c and Ω instead of δ, δ^c , and Δ, respectively.

The new symbols better indicate that they are superoperators, i.e. operations on algebras and have other advantages.) Note that the representation (1) is unique in spite of the non-uniqueness of Δ.

We may remark that, whenever we say commutant $\mathcal{C}(\rho)$ of ρ, it means the subalgebra of $\mathcal{B}(H^N)$ of all continuous functions of ρ which commutes with ρ at a *fixed set of parameter values* θ in Θ; a change of the θ-values causes a non-commutative portion with ρ, and that expressions (1) and (2) claim the representation of such a non-commutativity to be always expressible in terms of the single operation of commutation $[\,\cdot\,, \Delta]$, to lowest order.

Thus, the antihermitian element Δ in (1) and (2) must vary also with a change in the parameter space associated with the variation of ρ, and its form can be deduced in an elementary manner as follows. Let ρ be written as a unitary transform $\rho = \mathcal{U}\rho_D\mathcal{U}^+$ of the diagonal ρ_D (i.e. the Schatten decomposition of ρ), then by requiring that ρ_D is always diagonal so that $[\rho_D, \delta\rho_D] = 0$,

$$\begin{aligned}\delta\rho &= \mathcal{U}\delta\rho_D\mathcal{U}^+ + \mathcal{U}[\mathcal{U}^+\delta\mathcal{U}, \rho_D]\mathcal{U}^+ \\ &= \delta^c\rho + [\rho, \mathcal{U}\delta\mathcal{U}^+]\end{aligned}$$

from which $\delta^c\rho$ and Δ can de identified by

$$\delta^c\rho = \mathcal{U}\delta\rho_D\mathcal{U}^+ \in \mathcal{C}(\rho) \tag{3}$$

$$\Delta = \mathcal{U}\delta\mathcal{U}^+ = -\delta\mathcal{U}\mathcal{U}^+ = -(\mathcal{U}\delta\mathcal{U}^+)^+. \tag{4}$$

This will also assure the validity of expression (2) almost intuitively. But we give here another proof of this fact by a use of the contour integration method applied to non-commuting linear operators [3].

For any analytic function $F(z)$ of z on the complex z-plane,

$$F(\rho) = \frac{1}{2\pi i}\int_C \frac{F(z)}{z - \rho}dz \tag{5}$$

where the contour C in the z-plane is chosen to encircle all the eigenvalues of ρ (lying on the real axis; $[0,1]$ because ρ is a density matrix) and is assumed to lie in a simply-connected domain where $F(z)$ is assumed to be holomorphic. By a use of expanding the resolvent for $\rho+\delta\rho$ to the first order in $\delta\rho$ i.e.

$$\frac{1}{z-(\rho+\delta\rho)} = \frac{1}{z-\rho} + \frac{1}{z-\rho}\delta\rho\frac{1}{z-\rho},$$

$$F(\rho+\delta\rho) - F(\rho) = \frac{1}{2\pi i}\int_C \frac{1}{z-\rho}\delta\rho\frac{1}{z-\rho}F(z)dz,$$

we substitute expression (1) into the above contour integral to obtain

$$\begin{aligned}
\delta F(\rho) &= \frac{1}{2\pi i}\int_C \delta^c\rho\frac{F(z)}{(z-\rho)^2}dz \qquad ([\delta^c\rho,\rho]=0)\\
&+ \frac{1}{2\pi i}\int_C\{\frac{1}{z-\rho}(\rho\Delta-\Delta\rho)\frac{1}{z-\rho}\}F(z)dz\\
&= F'(\rho)\delta^c\rho + \frac{1}{2\pi i}\int_C[\frac{F(z)}{z-\rho},\Delta]dz \qquad (F'(z)=\frac{dF}{dz})\\
&= F'(\rho)\delta^c\rho + [F(\rho),\Delta]. \qquad (6)
\end{aligned}$$

Such represents the method of proof used frequently in the present work.

3 TANGENT SPACE AND α-REPRESENTATION

The fact that the manifold $\mathcal{S}$ of our statistical model is n-dimensional in the parameter space is shown by the existence of a local homeomorphism of each open set of $\mathcal{S}$ containing a point $\theta(\theta^1,\ldots,\theta^n)$ into R^n[1]. The integer n can be identified with the dimensionality of the tangent space T_θ at θ as for the classical case where a tangent vector $A\in T_\theta$ is expressed as

$$A = A^i\partial_i \qquad \text{i.e.} \quad Af = A^i\frac{\partial f}{\partial\theta^i} \qquad \text{for any} \quad f\in S.$$

Thus for the present non-commutative case

$$\begin{aligned}
A = A^i\delta_i &= A^i(\delta_i^c + [\cdot\;\; \Delta_i]), &\qquad \Delta_i\in\mathcal{B}(H^N), \quad \text{i.e.}\\
A^i\delta_i F &= A^i(\delta_i^c F + [F,\Delta_i]), &\qquad \delta_i^c F\in\mathcal{C}(\rho),
\end{aligned}$$

where A^i's are c-number functions of θ, and $F\in\mathcal{B}(H^N)$. Since $\delta_i^c F$'s are commutative with each other, they can be understood just as the classical case so that they may be denoted by $\partial_i F$ (i.e. partial derivatives with respect to θ^i's). By the explicit construction of Δ in (4), one has therefore

$$AF = A^i(\partial_i F + [F,\Delta_i]); \qquad \partial_i F\in\mathcal{C}(\rho), \quad \text{and} \quad \sqrt{-1}\Delta_i\in\mathcal{B}^s(H^N). \qquad (7)$$

If only the commutative parts $\partial_i F$'s are considered, the model becomes identical with Amari's multinomial example for which the dimension n is restricted to $n\le N$. For the most general non-commutative case, the dimensionality must be restricted such that

$$\begin{aligned}
n &\le N + \frac{1}{2}N(N-1) &\qquad \text{for real orthogonal } \mathcal{U}, \qquad \text{and}\\
&\le N + N(N-1) &\qquad \text{for general unitary } \mathcal{U}.
\end{aligned}$$

The above upper bound of the dimensionality must be reduced by one, when the normalization of the probability is taken into account (i.e. $\int p(x)dx = 1$ and $\mathrm{Tr}\rho = 1$). Consequently, we obtain

$$\langle \partial_i \log p \rangle = 0 \qquad \text{in the classical case [1];}$$

We also have:

$$\langle \delta_i \log \rho \rangle = Tr\rho\delta_i \log \rho = 0. \tag{8}$$

This is because $\mathrm{Tr}\rho[\log \rho, \Delta_i] = 0$ for $F(z) = \log z$ in (5).

Let $L^\alpha(\rho)$ be a one-parameter family of functions of ρ defined by

$$\begin{aligned} L^\alpha(\rho) &= \tfrac{2}{1-\alpha}\rho^{\frac{1-\alpha}{2}} && \alpha \neq 1, \quad \text{and} \\ &= \log \rho && \alpha = 1. \end{aligned} \tag{9}$$

Then, it is possible to introduce the α-representation of the tangent space T_θ along the same lines as the classical formulation:

$$T_\theta^{(\alpha)} = \{A^{(\alpha)} \in \mathcal{B}(H^N) | A^{(\alpha)} = A^i \delta_i L^\alpha(\rho)\},$$

and

$$\begin{aligned} \delta_i L^\alpha(\rho) &= \rho^{\frac{1-\alpha}{2}} \partial_i \log \rho + \tfrac{2}{1-\alpha}[\rho^{\frac{1-\alpha}{2}}, \Delta_i] && \alpha \neq 1 \\ &= \partial_i \log \rho + [\log \rho, \Delta_i] && \alpha = 1. \end{aligned} \tag{10}$$

As one of the most important ingredients for elucidating the dual structure of the geometry, we introduce a *bilinear form* between $T_\theta^{(\alpha)}$ and $T_\theta^{(-\alpha)}$ as follows.

$$\langle A^{(\alpha)}, B^{(-\alpha)} \rangle = TrA^{(\alpha)}B^{(-\alpha)} = A^i B^j Tr\delta_i L^\alpha(\rho)\delta_j L^{-\alpha}(\rho),$$

where

$$\begin{aligned} &Tr\delta_i L^\alpha(\rho)\delta_j L^{-\alpha}(\rho) \qquad (= \langle \delta_i L^\alpha(\rho), \delta_j L^{-\alpha}(\rho) \rangle) \\ &= Tr\rho\partial_i \log \rho \partial_j \log \rho \; + \; \frac{4}{1-\alpha^2} Tr[\rho^{\frac{1-\alpha}{2}}, \Delta_i][\rho^{\frac{1+\alpha}{2}}, \Delta_j], \qquad \alpha \neq \pm 1 \\ &\qquad \text{or} \qquad + \; Tr[\log \rho, \Delta_i][\rho, \Delta_j], \qquad \alpha = \pm 1. \end{aligned} \tag{11}$$

Remark. $Tr[\rho^{\frac{1-\alpha}{2}}, \Delta_i][\rho^{\frac{1+\alpha}{2}}, \Delta_j] = Tr(\rho^{\frac{1-\alpha}{2}} \Delta_i \rho^{\frac{1+\alpha}{2}} \Delta_j - \rho\Delta_i\Delta_j) + term(i \rightleftharpoons j)$ which shows that it is invariant under the interchange $i \rightleftharpoons j$ and hence $\alpha \rightleftharpoons -\alpha$.

Let the c-number vector fields A^i's and B^j's be confined to real fields. Then, by the above remark

$$\langle A^{(\alpha)}, B^{(-\alpha)} \rangle = \langle B^{(-\alpha)}, A^{(\alpha)} \rangle = \langle A^{(-\alpha)}, B^{(\alpha)} \rangle, \qquad \text{and}$$

in particular

$$\langle \delta_i L^\alpha(\rho), \delta_j L^{-\alpha}(\rho) \rangle = \langle \delta_j L^{-\alpha}(\rho), \delta_i L^\alpha(\rho) \rangle = \langle \delta_j L^\alpha(\rho), \delta_i L^{-\alpha}(\rho) \rangle. \tag{12}$$

4 INFORMATION METRIC TENSOR

An introduction of a bilinear form into the tangent space of a smooth manifold means a writing of the Riemannian metric in terms of a symmetric tensor, called the metric tensor, and when it is applied to a statistical model S it is to be identified with the (Fisher) information tensor [1]. From expression (11), for the present non-commutative statistical model, therefore,

$$\begin{aligned} g_{ij}^{(\alpha)} &= Tr\delta_i L^\alpha(\rho)\delta_j L^{-\alpha}(\rho) \\ = Tr\rho\partial_i \log\rho\partial_j \log\rho &+ \frac{4}{1-\alpha^2}Tr[\rho^{\frac{1-\alpha}{2}},\Delta_i][\rho^{\frac{1+\alpha}{2}},\Delta_j], \qquad |\alpha| \neq 1 \\ \text{or} \qquad &+ Tr[\log\rho,\Delta_i][\rho,\Delta_j], \qquad |\alpha| = 1. \end{aligned} \tag{13}$$

with symmetries

$$g_{ij}^{(\alpha)} = g_{ji}^{(\alpha)} = g_{ij}^{(-\alpha)}. \tag{14}$$

This is the first significant deviation from the classical theory, namely, the information tensor is not independent of the parameter α, as is the case in the commutative formulation which is manifest in the first part of the above expression.

5 α-CONNECTION, TORSION AND CURVATURE

First , we present two formulas. From the definitions,

$$\delta_i L^\alpha(\rho) = \partial_i L^\alpha(\rho) + [L^\alpha(\rho),\Delta_i]$$

$$\delta_i\delta_j L^\alpha = \partial_i\partial_j L^\alpha + [\partial_i L^\alpha,\Delta_j] + [\partial_j L^\alpha,\Delta_i] + [L^\alpha,\partial_i\Delta_j] + [[L^\alpha,\Delta_j],\Delta_i]$$

and with the explicit expression for $L^\alpha(\rho)$ in (9),

$$\partial_i L^\alpha(\rho) = \rho^{\frac{1-\alpha}{2}}\partial_i \log\rho$$

$$\partial_i\partial_j L^\alpha(\rho) = \rho^{\frac{1-\alpha}{2}}\{\partial_i\partial_j\log\rho + \frac{1-\alpha}{2}(\partial_i\log\rho)(\partial_j\log\rho)\},$$

we obtain

$$\langle\delta_i\delta_j L^\alpha(\rho), L^{-\alpha}(\rho)\rangle = -\langle\delta_j L^\alpha(\rho),\delta_i L^{-\alpha}(\rho)\rangle = -g_{ji}^{(\alpha)}, \tag{15}$$

and

$$\begin{aligned} \langle\delta_j\delta_i L^\alpha(\rho),\delta_k L^{-\alpha}(\rho)\rangle &= Tr\rho\{\partial_i\partial_j\log\rho + \frac{1-\alpha}{2}(\partial_i\log\rho)(\partial_j\log\rho)\}\partial_k\log\rho \\ &+ Tr[[L^\alpha(\rho),\Delta_i],\Delta_j][L^{-\alpha}(\rho),\Delta_k]. \end{aligned} \tag{16}$$

Expression (16) can be identified with the coefficient of the affine α-connection $\Gamma_{ijk}^{(\alpha)}$. (See [1],) An important point of this connection is that it is not symmetric with respect to the first two indices i and j: $\Gamma_{jik}^{(\alpha)} \neq \Gamma_{ijk}^{(\alpha)}$, which is a consequence of the non-commutativity of the operations of δ_i and δ_j such that

$$\delta_i\delta_j - \delta_j\delta_i = [\cdot\ \Sigma_{ij}]$$

where

$$\Sigma_{ij} = \partial_i\Delta_j - \partial_j\Delta_i - [\Delta_i,\Delta_j] \tag{17}$$

which satisfies $\mathrm{Tr}\rho[L^{(\alpha)}(\rho), \Sigma_{ij}] = 0$ but in general $\Sigma_{ij} \neq 0$. Thus we obtain the coefficient of the α-connection and its associated torsion tensor:

$$\begin{aligned}\Gamma^{(\alpha)}_{ijk} = Tr\rho\{\partial_i\partial_j \log\rho \ &+ \ \frac{1-\alpha}{2}(\partial_i \log\rho)(\partial_j \log\rho)\}\partial_k \log\rho \\ &+ \ Tr[[L^{\alpha}(\rho), \Delta_i], \Delta_j][L^{-\alpha}(\rho), \Delta_k] \end{aligned} \tag{18}$$

$$\begin{aligned} S^{(\alpha)}_{ijk} = \Gamma^{(\alpha)}_{ijk} - \Gamma^{(\alpha)}_{jik} &= Tr[L^{\alpha}(\rho), [\Delta_i, \Delta_j]][L^{-\alpha}(\rho), \Delta_k] \\ &= Tr[L^{\alpha}(\rho), \Sigma_{ji}][L^{-\alpha}(\rho), \Delta_k] \end{aligned} \tag{19}$$

(Remark: $\partial_i\Delta_j - \partial_j\Delta_i \in \mathcal{C}(\rho)$)

The presence of nonvanishing torsion tensor $S^{(\alpha)}_{ijk} = -S^{(\alpha)}_{jik}$ is the second significant deviation from the classical theory [1].

Another important notion in differential geometry, namely, the Riemann-Christoffel curvature tensor, in particular, that associated with the α-connection (18) will be discussed elsewhere. There, the important issue is that the vanishing of $R^{(\alpha)}_{ijkm}$ implies the existence of the affine coordinates and vice versa so that the α-family of the statistical model can be characterized.

6 α-DIVERGENCE AND α-FAMILY

As we have seen already, the information metric tensor g_{ij} introduced via the bilinear form between the two mutually dual tangent spaces $T^{(\alpha)}$ and $T^{(-\alpha)}$ is not independent of the real parameter α as it is in the classical Fisher form: that is, the tensor depends on α explicitly in its non-commutative components in (11). This gives rise to a new question about the existence of a scalar functional for two arbitrary density matrices ρ and σ from which the metric tensor can be derived when these ρ and σ are infinitesimally distant with each other. The answer can be provided from our previous study [2].

We define the α-divergence for two density matrices ρ and σ in $\mathcal{B}(H^N)$by the expression

$$D_\alpha(\rho, \sigma) = \frac{4}{1-\alpha^2} Tr\{(1 - \sigma^{\frac{1+\alpha}{2}}\rho^{-\frac{1+\alpha}{2}})\rho\}, \qquad \alpha \in R, \tag{20}$$

with $Tr\rho = Tr\sigma = 1$. It is well defined for $\alpha \neq \pm1$, but for the special values of ±1, it can also be defined by requiring continuity in α

$$lim_{\alpha\to1} D_\alpha(\rho, \sigma) = Tr\sigma(\log\sigma - \log\rho), \quad \text{and} \tag{21}$$

$$lim_{\alpha\to-1} D_\alpha(\rho, \sigma) = Tr\rho(\log\rho - \log\sigma). \tag{22}$$

Thus, $D_\alpha(\rho,\sigma)$ is defined as a continuous function of the real variable α in $(-\infty,\infty)$, and satisfies the α-duality $D_\alpha(\rho,\sigma) = D_{-\alpha}(\sigma,\rho)$ and the strict positivity $D_\alpha(\rho,\sigma) \geq 0$ with equality only if $\rho = \sigma$, which can be proved from the c-number real, convex function $f_\alpha(u) = \frac{4}{1-\alpha^2}(1 - u^{\frac{1+\alpha}{2}})$ for $u > 0$ with the property $f_\alpha(1) = 0$ and $f''_\alpha(1) = 1$ [4]. But in order for $D_\alpha(\rho,\sigma)$ to be jointly convex with respect to ρ and σ, the parameter α must be restricted to lie in $[-3,3]$ (see [2]).

Our answer to the information metric tensor associated with the α-divergence can be stated in the following

Theorem 1 *Let the density matrix σ be infinitesimally distant from ρ in the α-divergence $D_\alpha(\rho,\sigma)$ such that*

$$\sigma = \rho + \delta\rho, \qquad \delta\rho = d\theta^i \delta_i \rho.$$

Then,

$$D_\alpha(\rho, \rho+\delta\rho) = D_\alpha(\rho+\delta\rho, , \rho) = \frac{1}{2} d\theta^i d\theta^j g_{ij}^{(\alpha)} + O(d\theta^3)$$

where

$$g_{ij}^{(\alpha)} = \langle \delta_i L^\alpha(\rho), \delta_j L^{-\alpha}(\rho) \rangle$$

which is more explicitly given by (11), satisfying the symmetry $g_{ji}^{(\alpha)} = g_{ij}^{(\alpha)} = g_{ij}^{(-\alpha)}$.

Remark. The α-dependent, non-commuting part of $g_{ij}^{(\alpha)}$ i.e.

$$\frac{4}{1-\alpha^2} \mathrm{Tr}[\rho^{\frac{1-\alpha}{2}}, \Delta_i][\rho^{\frac{1+\alpha}{2}}, \Delta_j]$$

in (11) originates from Wigner-Yanase-Dyson-Lieb convex function [5] under the condition $|\alpha| \leq 3$.

The proof of the above theorem given in [2] utilizes the contour integration method by which $D_\alpha(\rho, \rho+\delta\rho)$ is expanded up to the second order, $(d\theta)^2$, on setting $F(\rho) = \rho^{\frac{1+\alpha}{2}}$ and $G(\rho) = \rho^{\frac{1-\alpha}{2}}$, showing

$$D_\alpha(\rho, \rho+\delta\rho)^{(1)} = \frac{-4}{1-\alpha^2} Tr\{\delta^{(1)} F(\rho) \times G(\rho)\} = 0, \quad \text{and}$$

$$D_\alpha(\rho, \rho+\delta\rho)^{(2)} = \frac{-4}{1-\alpha^2} Tr\{\delta^{(2)} F(\rho) \times G(\rho)\} = \frac{1}{2} d\theta^i d\theta^j g_{ij}^{(\alpha)},$$

where

$$\delta^{(1)} F(\rho) = \frac{1}{2\pi i} \int_C \frac{1}{z-\rho} \delta\rho \frac{F(z)}{z-\rho} dz \tag{23}$$

$$\delta^{(2)} F(\rho) = \frac{1}{2\pi i} \int_C \frac{1}{z-\rho} \delta\rho \frac{1}{z-\rho} \delta\rho \frac{F(z)}{z-\rho} dz. \tag{24}$$

Therefore, its validity is quite general (except that the density matrix has to be assumed to be invertible): in particular, it does not require the *flatness* of the manifold $\mathcal{S}$ that implies the existence of the affine coordinate system.

An α-family of the statistical model is defined, following [1], by

$$L^\alpha(\rho(\theta)) = \theta^i C_i + k(\theta) C_{n+1}, \qquad C_{i=1,2,\ldots,n+1} \in \mathcal{B}(H^N) \tag{25}$$

which may be put into a system with flatness. The treatment is given in [1] within the classical (commutative) framework. Its non-commutative extension is not certain at this moment except the special case $\alpha = 1$ ($L^1(\rho) = \log\rho$) and $\alpha = -1$ ($L^{-1}(\rho) = \rho$), known as the *exponential family* and the *mixture family*, respectively. We discuss these special families in detail in the rest of this report.

7 DUAL COORDINATE SYSTEM FOR $\alpha = \pm 1$ FLAT FAMILY

It was elucidated by Amari in the classical framework [1] that further fertile geometrical structures of duality (Pythagorian theorem, α-projection and the related extremum property etc.) exist in the statistical manifold S, provided that it is equipped with an α-affine coordinate system for which $\Gamma^{(\alpha)}_{ijk} = 0$. It is our interest all the way to clarify them in the non-commutative formulation. This is still not yet done, one of main obstacles being that the present non-commutative manifold is generally not torsion-free. Nevertheless, it can be shown by a direct construction that the case $\alpha = \pm 1$ (the exponential family $\alpha = 1$, the mixture family $\alpha = -1$) can be treated fully in parallel with the classical framework.

To this end, let us define a *biorthogonal* basis of the tangent space T_θ such that

$$\langle \delta_i L^\alpha(\rho), \delta^j L^{-\alpha}(\rho)\rangle = \delta_i^j \qquad \text{(Kronecker's } \delta\text{)}, \tag{26}$$

and

$$A(\in T_\theta) = A^i\delta_i = A_i\delta^j,$$

where

$$\delta^j F = \partial^j F + [F, \Delta^j] = g^{jk}\delta_k F, \qquad (g^{jk}) = (g_{ij})^{-1}, \tag{27}$$

under the assumption that the Riemannian metric tensor is non-degenerate (then, the biorthogonal basis surely exists). We further assume the existence of a new coordinate system $\{\eta_k\}$ such that

$$\eta_k = \eta_k(\theta), \qquad \text{or} \qquad \theta^i = \theta^i(\eta) \tag{28}$$

according to which

$$\delta_i = \frac{\partial \eta_k}{\partial \theta^i}\delta^k \qquad \text{and} \qquad \delta^j = \frac{\partial \theta^j}{\partial \eta_k}\delta_k \tag{29}$$

This implies $\langle \delta^j, \delta^k\rangle = g^{jk}$ together with $\langle \delta_j, \delta_k\rangle = g_{ij}$.

The pair of coordinate systems θ and η is said to be a *dual coordinate system*. One of basic theorems proved by Amari which is useful here is

Theorem 2 *A necessary and sufficient condition for a dual coordinate system (θ, η) to exist in the smooth manifold S is that at least one of the metric tensor $g_{ij}(\theta)$ or $g^{ij}(\eta)$ can be given in terms of a scalar function $\psi(\theta)$ or $\phi(\eta)$ (called the potential function) by*

$$g_{ij}(\theta) = \frac{\partial^2 \psi}{\partial \theta^i \partial \theta^j}, \qquad \textit{or} \qquad g^{ij}(\eta) = \frac{\partial^2 \phi}{\partial \eta_i \partial \eta_j}, \tag{30}$$

and once this is satisfied, the other is automatically satisfied by the Legendre transformation

$$\eta_i = \frac{\partial \psi}{\partial \theta^i} \qquad \textit{or} \qquad \theta^i = \frac{\partial \phi}{\partial \eta_i} \qquad \textit{with the identity} \quad \psi(\theta) + \phi(\eta) - \theta^i\eta_i = 0. \tag{31}$$

On the above basis we now present our main result for $\alpha = \pm 1$ in the following

Theorem 3 *Let $\rho^{(1)}(\theta)$ denote the family of density matrices, and $\rho^{(-1)}(\eta)$ another family, specified respectively by*

$$\rho^{(1)}(\theta) = \exp\{\theta^i C_i - \psi(\theta)\} \qquad C_i \in \mathcal{B}^s(H^N) \qquad \textit{(exponential family)} \tag{32}$$

$$\rho^{(-1)}(\eta) = \sum_{i=1}^{n} \eta_i \bar{C}^i + \eta_{n+1}\bar{C}^{n+1} \qquad \bar{C}^i \in \mathcal{B}^s(H^N) \qquad \textit{(mixture family)}, \tag{33}$$

where

$$\psi(\theta) = \log(Tre^{\theta^i C_i}) \quad ensures \quad Tr\rho^{(1)}(\theta) = 1 \quad and \tag{34}$$

$$Tr\bar{C}^i = 1, \quad i = 1, 2, \ldots, n+1 \quad and \quad \eta_{n+1} = 1 - \sum_{i=1}^{n} \eta_i \quad ensures \quad Tr\rho^{(-1)}(\eta) = 1. \tag{35}$$

Then, $\{\theta^i\}$ is a $+1$ affine coordinate for the $\rho^{(1)}(\theta)$ family and $\{\eta^i\}$ a -1 affine coordinate for the $\rho^{(-1)}(\eta)$family, respectively. Suppose that either one of the two families is given. Then, the other exists through a coordinate transformation between them, thus $\eta_i = \eta_i(\theta)$ or $\theta^i = \theta^i(\eta)$, by which (θ, η) becomes a dual coordinate system of a single family of density matrices, $\rho^{(1)}(\theta) = \rho^{(-1)}(\eta)$.

In order to show the proof of Th.3 most concisely, several preliminary matters are prepared.

Preliminary 1. Supplement to (34) and (35)

For the exponential family defined by (32), one has

$$\partial_i \psi(\theta) = \frac{1}{Tr(e^{\theta^j C_j})} Tr(e^{\theta^j C_j} C_i) = \langle C_i \rangle \tag{36}$$

$$\begin{aligned} \partial_i \partial_j \psi(\theta) &= \int_0^1 \langle \rho^{-\lambda} C_i \rho^{\lambda} C_j \rangle d\lambda - \langle C_i \rangle \langle C_j \rangle \\ &= \int_0^1 \langle \rho^{-\lambda} (C_i - \langle C_i \rangle) \rho^{\lambda} (C_j - \langle C_j \rangle) \rangle d\lambda. \end{aligned} \tag{37}$$

This is known as Kubo's perturbation formula to second order for the free energy of a canonical system [6]. For the mixture family defined by (33), on the other hand, an interesting quantity whose expansion to order of $(d\theta)^2$ should be established is the *entropy* $H(\rho^{(-1)}) = H(\rho) = -Tr\rho\log\rho$, and one has

$$-\partial^i H(\rho) = Tr(\bar{C}^i - \bar{C}^{n+1}) \log \rho \tag{38}$$

$$-\partial^i \partial^j H(\rho) = Tr\rho^{-1} (\bar{C}^i - \bar{C}^{n+1})^0 (\bar{C}^j - \bar{C}^{n+1})^0 + Tr[\rho, \Delta^i][\log \rho, \Delta^j] \tag{39}$$

where $(\bar{C}^i - \bar{C}^{n+1})^0$ is a projection of the operator $\bar{C}^i - \bar{C}^{n+1}$ onto the commutant $\mathcal{C}(\rho)$. The derivation of this formula can be made by using the contour integration method applied to $F(z) = z \log z$ in (5). It is noted that expression (39) may be put into another form which retains a parallelism with (37):

$$-\partial^i \partial^j H(\rho) = \int_0^1 \langle \rho^{-\lambda} (\delta^i \log \rho) \rho^{\lambda} \delta^j \log \rho \rangle d\lambda.$$

Preliminary 2. Decomposition formulas

For the exponential family

$$\begin{aligned} \delta_i \rho^{(1)} &= \rho^{(1)} (C_i^0 - \langle C_i \rangle) + [\rho^{(1)}, \Delta_i] \\ \delta_i \log \rho^{(1)} &= C_i^0 - \langle C_i \rangle + [\log \rho^{(1)}, \Delta_i] \end{aligned} \tag{40}$$

where $C_i^0 = proj(C_i)|_{\mathcal{C}(\rho^{(1)})} \in \mathcal{C}(\rho^{(1)})$ which satisfies $\langle C_i^0 \rangle = \langle C_i \rangle$, but from (32) and (36)

$$\begin{aligned} \delta_i \log \rho^{(1)} = C_i - \partial_i \psi = C_i - \langle C_i \rangle \qquad &\text{and hence} \\ [\log \rho^{(1)}, \Delta_i] = C_i - C_i^0. & \end{aligned} \tag{41}$$

Also, the expression $\delta_i \log \rho^{(1)} = C_i - \partial_i \psi$ enables us to obtain

$$\delta_i \delta_j \log \rho^{(1)} = -\partial_i \partial_j \psi \qquad \text{so that} \qquad \langle \delta_i \delta_j \log \rho^{(1)}, \delta_k \rho^{(1)} \rangle = 0$$

which implies $\Gamma^{(1)}_{ijk} = 0$ i.e. θ's are $+1$ affine coordinates.

For the mixture family

$$\begin{aligned} \delta^i \rho^{(-1)} &= (\bar{C}^i - \bar{C}^{n+1})^0 + [\rho^{(-1)}, \Delta^i] \\ \delta^i \log \rho^{(-1)} &= \frac{1}{\rho}(\bar{C}^i - \bar{C}^{n+1})^0 + [\log \rho^{(-1)}, \Delta^i], \end{aligned} \tag{42}$$

but from (33),

$$\delta^i \rho^{(-1)} = \bar{C}^i - \bar{C}^{n+1} \qquad \text{and hence}$$

$$\delta^i \delta^j \rho^{(-1)} = 0 \qquad (\langle \delta^i \delta^j \rho^{(-1)}, \delta^k \log \rho^{(-1)} \rangle = 0)$$

which implies $\Gamma^{(-1)ijk} = 0$ i.e. η's are -1 affine coordinates.

Preliminary 3. Information metric tensor

For $\alpha = +1$, from (40),

$$g^{(1)}_{ij} = \langle (C^0_i - \langle C_i \rangle)(C^0_j - \langle C_j \rangle) \rangle + Tr[\rho, \Delta_i][\log \rho, \Delta_j]$$

but since

$$Tr[\rho, \Delta_i][\log \rho, \Delta_j] = \int_0^1 \langle \rho^{-\lambda}[\log \rho, \Delta_i] \rho^\lambda [\log \rho, \Delta_j] \rangle d\lambda$$

and by using (41) we get

$$\begin{aligned} g^{(1)}_{ij} &= \langle (C^0_i - \langle C_i \rangle)(C^0_j - \langle C_j \rangle) \rangle + \int_0^1 \langle \rho^{-\lambda}(C_i - C^0_i) \rho^\lambda C_j - C^0_j) \rangle d\lambda \\ &= \int_0^1 \langle \rho^{-\lambda} C_i \rho^\lambda C_j \rangle d\lambda - \langle C_i \rangle \langle C_j \rangle \qquad (\langle C_i C^0_j \rangle = \langle C^0_i C^0_j \rangle \quad \text{etc}) \\ &= \partial_i \partial_j \psi \qquad \text{given by (37).} \end{aligned} \tag{43}$$

For $\alpha = -1$, from (42) we get

$$\begin{aligned} g^{(-1)}_{ij} &= Tr \rho^{-1} (\bar{C}^i - \bar{C}^{n+1})^0 (\bar{C}^j - \bar{C}^{n+1})^0 + Tr[\rho, \Delta^i][\log \rho, \Delta^j] \\ &= -\partial^i \partial^j H(\rho) \qquad \text{given by (39).} \end{aligned} \tag{44}$$

Proof of Theorem 3. It is now possible to assure the validity: If the starting family is the exponential family (32), then, the result (43) shows that its metric tensor is expressed as $\partial_i \partial_j \psi$ so that this $\psi(\theta)$ is just the potential function and by putting $\langle C_i \rangle = \eta_i$ in (36), (θ, η) becomes a dual coordinate through the Legendre transformation with the identity between $\psi(\theta)$ and $\phi(\eta) = -H(\rho(\theta(\eta))) = -\psi(\theta) + \theta^i \eta_i$. Similarly, if the starting family is the mixture (33), the result (44) shows that the corresponding metric tensor is expressed as $\partial^i \partial^j \phi(\eta), \phi(\eta) = -H(\rho(\eta))$, the potential function for this family, and by putting $\theta^i = Tr(\bar{C}^i - \bar{C}^{n+1}) \log \rho$ in (38), (η, θ) becomes a dual coordinate through the same Legendre transformation as above. This concludes the present report.

References

[1] S. Amari, *Differential Geometrical Method in Statistics: Lecture Notes in Statistics* Vol 28 (Springer 1985).

[2] H.Hasegawa, *Rep. Math. Phys.* **33** (1993),87.

[3] T. Kato, *Perturbation Theory for Linear Operators* Springer Verlag New York Inc. 1966.

[4] D. Petz, *Rep. Math. Phys.* **11**(1986),57.

[5] E. H. Lieb, *Adv. Math.* **11**(1973),267.

[6] R. Kubo, *J. Phys. Soc. Japan* **12**(1957),570.

INFORMATION GAIN IN QUANTUM COMMUNICATION CHANNELS

G. M. D'Ariano, C. Macchiavello, and M. G. A. Paris

Dipartimento di Fisica "A. Volta",
via Bassi 6, I27100 Pavia, Italy

1 INTRODUCTION

In quantum communications the performances of an amplifier depend on the scheme of the channel in which the device is inserted. For example, both gain and noise figure of the amplifier depend on the kind of coding at the transmitter, and detection at the receiver. [1] In an optimized ideal channel, detection and coding are ideal, and both are "matched" on the same observable; the alphabet probability is optimized in order to satisfy the physical constraints on the line. In this case, the channel capacity is already achieved, and there is no need of amplification. However, when the channel is nonideal—either because of quantum mismatch between transmitter and receiver, because of losses along the line, or as a result of nonunit quantum efficiency at detectors—an appropriate amplification can improve the transmitted information, ideally achieving the channel capacity for infinite gain.

In this paper we analyze the effect of an amplifier insertion in a quantum nonideal channel, for coherent and quadrature/squeezed channels. Nonefficient heterodyne and homodyne detections are described by means of the Wigner and marginal-Wigner POM's [2–4] respectively, whereas the quantum evolution along the communication line is given in terms of CP-maps, [5,6] whose infinitesimal versions are "master" and "Langevin" equations. We show that for a large class of linear amplifiers (attenuators) these equations have differential representations on the POM's in form of Fokker-Planck or Ornstein-Uhlenbeck equations, with Gaussian Green-function solutions. This allows a simple description of the device in terms of drift and diffusion coefficients.

Within the above approach we show that an ideal amplifier matched with an ideal POM has a pure-drift differential representation, corresponding to a POM rescaling which leaves mutual information invariant. For nonunit quantum efficiency, an amplifier matched with ideal detection can improve the transmitted information, eventually recovering the channel capacity for infinite gains: this can be signalled by the fact that the diffusion coefficient becomes negative and, correspondingly, the noise figure becomes lower than unit. When mismatched, an amplifier can be re-matched with the detector either by "squeezing" the idler, or by means of measurement-and-feedback controls, with both measurement and feedback matched with detection.

2 BASICS OF QUANTUM COMMUNICATIONS

In a quantum communication channel information transmission between source and user is described as follows. [7] The source consists of a coder which maps letters θ from an alphabet Θ into a set of density operators ρ_θ on the Hilbert space $\mathcal{H}$ of the dynamical system supporting communication. The alphabet Θ (either a discrete set, a subset of R^d, or a mixed discrete-continuous set) is distributed according to an "a priori" probability $dP(\theta)$, which also determines the mixture of parametrized states

$$\bar{\rho} \doteq \int dP(\theta)\rho_\theta \,. \tag{1}$$

The set Θ, the map ρ_θ and the probability $dP(\theta)$ will be globally refereed to as "coding", and denoted by $\mathcal{C} = \{dP(\theta), \rho_\theta, \theta \in \Theta\}$. We will consider two main types of coding: i) the quadrature/squeezed coding $\mathcal{C}_x$, where the parametrized states ρ_λ ($\lambda \in R$) are squeezed states for the quadrature $X = \frac{1}{2}(a+a^\dagger)$ with average $\langle X\rangle = \lambda$; ii) the coherent coding $\mathcal{C}_\alpha$, which uses coherent states $\rho_\alpha = |\alpha\rangle\langle\alpha|$, $\alpha \in C$. Any device inserted in the communication line produces a Schrödinger evolution $\rho \to A_S(\rho)$ of the density matrix ρ carrying information. Thus the series "coder+device" is equivalent to a new coder with coding $\mathcal{C} \circ \mathcal{A} = \{dP(\theta), A_S(\rho_\theta), \theta \in \Theta\}$, where $\mathcal{A} \doteq \{\rho \to A_S(\rho)\}$ denotes the device map.

At the end of the communication line the user determines the transmitted letter θ as the result of a quantum measurement on the system. This is conveniently described by means of a probability-operator-valued measure [2–4] (POM) $d\mu(\zeta)$ on $\mathcal{H}$ for the outcome $\zeta \in R^{d'}$, namely a resolution of identity $\int d\mu(\zeta) = 1$ with $d\mu(\zeta) \geq 0$ [we use notation $d\mu(\zeta) \equiv \mu(d\zeta)$]. The POM's generalize the customary orthogonal-projector-valued measures associated to quantum observables. When the measurement is performed on the system state ρ the output probability distribution is given by

$$dP[\rho](\zeta) \doteq \mathrm{Tr}[\rho d\mu(\zeta)] \,. \tag{2}$$

Each detection apparatus is described by a POM. [8] In the following we will denote the detection by $\mathcal{D} = \{d\mu(\zeta), \zeta \in R^{d'}\}$. We will only consider homodyne detection $\mathcal{D}_x = \{d\mu(x), x \in R\}$ and heterodyne detection $\mathcal{D}_\alpha = \{d^2\mu(\alpha,\bar{\alpha}), \alpha \in C\}$: the explicit form of the POM for such detectors will be given in the next section. If the detector is preceded by a device described by the Heisenberg evolution $d\mu \to A_H(d\mu)$, the series "device+detector" is equivalent to a new detector with $\mathcal{A} \circ \mathcal{D} = \{A_H(d\mu(\zeta)), \zeta \in R^d\}$. This equivalence should be compared with that "coder+device$\simeq$coder", and reflects duality $A_H \equiv A_S^\vee$ under trace (2).

In terms of the detector POM the conditional probability of receiving ζ is given by Eq. (2) with $\rho \equiv A_S(\rho_\theta)$. The amount of information transfer between the coder and the detector is described by the "mutual information" $I(\theta;\zeta)$ between the random variables θ and ζ. Here, also in consideration of the symmetry property $I(\theta;\zeta) = I(\zeta;\theta)$ we prefer to label information with the maps of coding, detection, and other devices along the line. Then the mutual information is given by

$$I[\mathcal{C} \circ \mathcal{A} \circ \mathcal{D}] = \int dP(\theta) \int dP[\rho_\theta](\zeta) \log \frac{dP[\rho_\theta]}{dP[\bar{\rho}]}(\zeta) \,, \tag{3}$$

where $dP[\rho_\theta]/dP[\bar{\rho}]$ is the Radon-Nikodym derivative of $P[\rho_\theta]$ with respect to $P[\bar{\rho}]$. The maximum of I over $dP(\theta)$ (with $\mathcal{A}$ the identical map) is the "capacity" C of the information channel; the global maximum obtained by varying also $\mathcal{C}$ and $\mathcal{D}$ is the "ultimate quantum capacity" of the system. Using the Holevo-Ozawa-Yuen information bound, [7,3] it can be proved that the ultimate quantum capacity is achieved by number coding/detection, with $\bar{\rho}$ as the thermal state for photon number $\bar{n}$ corresponding to the average power physically allowed along the line. In the following we are interested only in coherent and quadrature/squeezed channels. For such cases the channel capacities are given by

$$C[\mathcal{C}_x \circ \mathcal{D}_x] = \log(1 + 2\bar{n}) \,, \qquad C[\mathcal{C}_\alpha \circ \mathcal{D}_\alpha] = \log(1 + \bar{n}) \,. \tag{4}$$

For the quadrature channel the capacity is achieved by squeezed states $|\lambda; r\rangle$, with squeezing parameter $e^{2r} = 1 + 2\bar{n}$ and zero-average Gaussian superposition $dP(\lambda)$ with variance $\sigma^2 = \bar{n} - \sinh^2 r$. For the coherent channel again the probability $dP(\alpha, \bar{\alpha})$ is Gaussian, and corresponds to a thermal $\bar{\rho}$ with $\bar{n}$ average photons.

3 HOMODYNE AND HETERODYNE DETECTION WITH NON-UNIT QUANTUM EFFICIENCY

The photon-count distribution $P_\eta(n)$ for a phototube cathode small with respect to radiation coherence length is given by the Bernoulli convolution

$$P_\eta(n) = \sum_{m=n}^{\infty} \binom{m}{n} \eta^n (1-\eta)^{m-n} \langle m|\rho|m\rangle,$$

with η denoting the quantum efficiency of the detector. In other words, the detector is equivalent to an ideal detector ($\eta = 1$) preceded by a beam splitter with transmissivity η. Using this simple rule it is easy to evaluate the POM's of homodyne and heterodyne detectors equipped with nonideal photocounters. [9] For a homodyne detector (with strong local oscillator and two equal photodetectors of efficiency η) the POM is simply the Gaussian convolution

$$d\mu_\eta(x) = \int \frac{dx'}{\sqrt{\pi(1-\eta)/2\eta}} \exp\left[-\frac{2\eta}{1-\eta}(x-x')^2\right] |x'\rangle\langle x'| \,, \tag{5}$$

where $|x\rangle$ are eigenstates of the quadrature $a_\phi = \frac{1}{2}(a^\dagger e^{i\phi} + a e^{-i\phi})$ at phase ϕ with respect to the local oscillator (in the following we will consider either $\phi = 0$, or $\phi = \pi/2$, for simplicity). For $\eta = 1$ one recovers the POM $d\mu(x) = dx|x\rangle\langle x|$ of the ideal homodyne detector. For an heterodyne detector with photodetector efficiency

η (independent on frequency in the range between signal and image frequencies), the POM is again a Gaussian convolution of the ideal POM, namely

$$d^2\mu_\eta(\alpha,\bar\alpha) = \frac{d^2\alpha}{\pi}\int \frac{d^2\alpha'}{\pi(1-\eta)/\eta}\exp\left[-\frac{\eta}{1-\eta}|\alpha-\alpha'|^2\right]|\alpha'\rangle\langle\alpha'| \,. \tag{6}$$

It is convenient to write both POM's (5) and (6) in terms of the Wigner POM

$$d^2 w_s(\alpha,\bar\alpha) \doteq \frac{d^2\alpha}{\pi}\int \frac{d^2\beta}{\pi} e^{\alpha\bar\beta-\bar\alpha\beta+\frac{1}{2}s|\beta|^2} D(\beta)\,, \tag{7}$$

where $D(\beta) = \exp(\beta a^\dagger - \bar\beta a)$ is the displacement operator. The probability associated to the POM (7) is the Wigner function $W_s(\alpha,\bar\alpha)$ for ordering parameter s. It is nonnegative for $s \leq -1$. Comparing Eqs. (6) and (7) it is easy to show that $d^2\mu_\eta(\alpha,\bar\alpha) \equiv d^2 w_{1-2\eta^{-1}}(\alpha,\bar\alpha)$, whereas the homodyne POM can be written as marginal POM of $d^2 w_s$, namely

$$d\mu_s(x) = \int \frac{dy}{\pi}\frac{d^2 w_s(x,y)}{dxdy}\,, \qquad s = 1-\eta^{-1}\,. \tag{8}$$

In Eq. (8) we used the change of variables $\alpha = x + iy$. Notice that all instrumental POM are nonnegative for quantum efficiency $0 \leq \eta \leq 1$.

4 AMPLIFICATION AND LOSS

4.1 CP-maps

An amplifier is an open system [5] where the amplified mode of radiation—the "signal" mode—resonantly interacts with other modes (parametric amplifier) or with matter degrees of freedom (active medium amplifier). "Pump" modes provide the necessary energy for amplification, whereas "idler" modes account for resonance condition and phase-matching. Also a beam splitter or a lossy cavity/fiber are open systems, with the signal mode gradually lost into external modes. The dynamical evolution of the density matrix ρ of an open system is described by a map of the form

$$\rho_t = A_S(\rho) \doteq \mathrm{Tr}_P\left[U_t\,\rho_P\otimes\rho\,U_t^\dagger\right]\,, \tag{9}$$

where U_t is a unitary operator, and ρ_P is the density matrix for the "probe" (=idler+pump). Corresponding to the Schrödinger-picture (9) one has the Heisenberg evolution of POM's $d\mu$, which is defined through duality $A_H = A_S^\vee$ under trace (2). The Heisenberg map A_H is a unit-preserving normal completely positive (CP) map, [6] and transforms POM's into POM's. The Schrödinger map evolves density matrices, preserving positivity, trace and convex linear combinations. In terms of the unitary evolution U_t the CP-map A_H is written as follows

$$d\mu_t = A_H(d\mu) = \mathrm{Tr}_P\left[\rho_P\otimes\hat 1\,U_t^\dagger\,\hat 1\otimes d\mu\,U_t\right]\,. \tag{10}$$

The normal unit preserving CP-maps (and their dual) admit the Stinespring representation [5]

$$A_H(d\mu) = \sum_k V_k^\dagger d\mu V_k\,, \qquad A_S(\rho) = \sum_k V_k\rho V_k^\dagger\,, \qquad \sum_k V_k^\dagger V_k = 1\,. \tag{11}$$

4.2 Master and Langevin equations

The infinitesimal versions of Eqs. (9) and (10) are referred to as "master" and "Langevin" equations, respectively. Lindblad proved that the most general form of master equation is [11]

$$d\rho = dA_S(\rho) = \sum_k D[A_k]\rho\,, \qquad D[A]\rho \doteq A\rho A^\dagger - \frac{1}{2}\left\{A^\dagger A, \rho\right\}\,, \tag{12}$$

$\{\,,\,\}$ denoting the anticommutator, and A_k being (complex) operators. The Langevin equation corresponding to Eq. (12) is [12]

$$d\,d\mu = dA_H(d\mu) = \sum_k D^\vee[A_k]d\mu\,, \qquad D^\vee[A]d\mu \doteq A^\dagger d\mu A - \frac{1}{2}\left\{A^\dagger A, d\mu\right\}\,. \tag{13}$$

4.3 Fokker-Planck equations

When a differential representation of the Langevin equation is available, one obtains a Fokker-Planck equation (FPE). The explicit form of the FPE generally depends on the analytic form of the POM under consideration, and coincides with the differential equation for the probability distribution. For the Wigner functions $W_s(\alpha,\bar\alpha)$ a differential representation is available for all superoperators that are polynomial functions of a and $a^\dagger$. In Table 1 the representation of right and left multiplication of ρ by a and $a^\dagger$ are reported along with the composition rules for obtaining the representation of any monomial.

Superoperator	Wigner representation	Superoperator	Wigner representation
$L[a] \doteq a\cdot$	$\alpha - \frac{1}{2}(s-1)\partial_{\bar\alpha}$	$R[a] \doteq \cdot a$	$\alpha - \frac{1}{2}(s+1)\partial_{\bar\alpha}$
$L[a^\dagger] \doteq a^\dagger\cdot$	$\bar\alpha - \frac{1}{2}(s+1)\partial_\alpha$	$R[a^\dagger] \doteq \cdot a^\dagger$	$\bar\alpha - \frac{1}{2}(s-1)\partial_\alpha$
$L[O_1O_2] = L[O_1]L[O_2]$		$R[O_1O_2] = R[O_2]R[O_1]$	

Table 1: Left and right superoperators and corresponding Wigner differential representations.

For the homodyne POM the FPE [in this case referred to as Ornstein-Uhlenbeck equation (OUE)] can be obtained from the Wigner representation using marginal integration (8) and dropping out boundary terms for the exponential decay of W_s at infinity. As an example, the OU representation of $D[a]$ can be obtained as follows

$$\begin{aligned}\int_{-\infty}^{\infty} dy \int_{-\infty}^{\infty}\int_{-\infty}^{\infty} \frac{dv\,du}{4\pi^2} e^{\frac{s}{8}(u^2+v^2)}\mathrm{Tr}\left[D[a]e^{iu(X-x)+iv(Y-y)}\right] &= \\ \int_{-\infty}^{\infty}\frac{dy}{\pi}\left[\frac{1}{2}(\partial_x x + \partial_y y) + \frac{1-s}{8}(\partial^2_{xx} + \partial^2_{yy})\right] W_s(x,y) &= \\ \left(\frac{1}{2}\partial_x x + \frac{1-s}{8}\partial^2_{xx}\right)\int_{-\infty}^{\infty}\frac{dy}{\pi} W_s(x,y)\,,&\end{aligned} \tag{14}$$

where $a = X + iY$ is the quadrature decomposition of a. Notice that there are Wigner differential operators which admit no marginal form [for example, the operator $\alpha\partial_\alpha = \frac{1}{2}(x+iy)(\partial_x - i\partial_y)$ has terms linear in y having no marginal representation]. On the other hand, the OU differential operators are marginal representations of many different Wigner operators: hence, the OU representation is neither injective, nor defined on the whole superoperator algebra, whereas the Wigner representation is one-to-one. In Table 2 the representation of the superoperators used within this paper are reported.

Superoperator	Wigner	Marginal-X
$D[a]$	$\frac{1}{2}(\partial_\alpha \alpha + \partial_{\bar\alpha}\bar\alpha) + \frac{1-s}{2}\partial^2_{\alpha\bar\alpha}$	$\frac{1}{2}\partial_x x + \frac{1-s}{8}\partial^2_{xx}$
$D[a^\dagger]$	$-\frac{1}{2}(\partial_\alpha \alpha + \partial_{\bar\alpha}\bar\alpha) + \frac{1+s}{2}\partial^2_{\alpha\bar\alpha}$	$-\frac{1}{2}\partial_x x + \frac{1+s}{8}\partial^2_{xx}$
$[a^2, \cdot]$	$2\alpha\partial_{\bar\alpha} - s\partial^2_{\bar\alpha\bar\alpha}$	—
$[a^{\dagger 2}, \cdot]$	$-2\bar\alpha\partial_\alpha + s\partial^2_{\alpha\alpha}$	—
$[a, [a, \cdot]]$	$\partial^2_{\bar\alpha\bar\alpha}$	$\frac{1}{4}\partial^2_{xx}$
$[a^\dagger, [a^\dagger, \cdot]]$	$\partial^2_{\alpha\alpha}$	$\frac{1}{4}\partial^2_{xx}$
$D[a_\phi]$	$\frac{1}{4}\partial^2_{\alpha\bar\alpha} - \frac{1}{8}(e^{2i\phi}\partial^2_{\alpha\alpha} + e^{-2i\phi}\partial^2_{\bar\alpha\bar\alpha})$	$\frac{1}{8}\sin^2\phi\partial^2_{xx}$
$[a^2 - a^{\dagger 2}, \cdot]$	$2(\alpha\partial_{\bar\alpha} + \bar\alpha\partial_\alpha) - s(\partial^2_{\alpha\alpha} + \partial^2_{\bar\alpha\bar\alpha})$	$2\partial_x x - \frac{1}{2}s\partial^2_{xx}$
$i[Y, a \cdot + \cdot a^\dagger]$	$\frac{1-s}{4}\partial^2_{\bar\alpha\bar\alpha} + \frac{1-s}{4}\partial^2_{\alpha\alpha} + \frac{1-s}{2}\partial^2_{\alpha\bar\alpha} + \frac{1}{2}(\alpha\partial_{\bar\alpha} + \bar\alpha\partial_\alpha + \partial_\alpha\alpha + \partial_{\bar\alpha}\bar\alpha)$	$\partial_x x + \frac{1-s}{4}\partial^2_{xx}$
$i[X, -ia \cdot + \cdot ia^\dagger]$	$\frac{1-s}{4}\partial^2_{\bar\alpha\bar\alpha} + \frac{1-s}{4}\partial^2_{\alpha\alpha} + \frac{s-1}{2}\partial^2_{\alpha\bar\alpha} + \frac{1}{2}(\alpha\partial_{\bar\alpha} + \bar\alpha\partial_\alpha - \partial_\alpha\alpha - \partial_{\bar\alpha}\bar\alpha)$	0
$i[Y, a^\dagger \cdot + \cdot a]$	$-\frac{1+s}{4}\partial^2_{\bar\alpha\bar\alpha} - \frac{1+s}{4}\partial^2_{\alpha\alpha} - \frac{1+s}{2}\partial^2_{\alpha\bar\alpha} + \frac{1}{2}(\alpha\partial_{\bar\alpha} + \bar\alpha\partial_\alpha + \partial_\alpha\alpha + \partial_{\bar\alpha}\bar\alpha)$	$\partial_x x - \frac{1+s}{4}\partial^2_{xx}$
$i[X, ia^\dagger \cdot - \cdot ia]$	$-\frac{1+s}{4}\partial^2_{\bar\alpha\bar\alpha} - \frac{1+s}{4}\partial^2_{\alpha\alpha} + \frac{1+s}{2}\partial^2_{\alpha\bar\alpha} + \frac{1}{2}(\alpha\partial_{\bar\alpha} + \bar\alpha\partial_\alpha - \partial_\alpha\alpha - \partial_{\bar\alpha}\bar\alpha)$	0
$i[Y, [a^\dagger, \cdot]]$	$-\frac{1}{2}(\partial^2_{\alpha\alpha} + \partial^2_{\alpha\bar\alpha})$	$-\frac{1}{4}\partial^2_{xx}$
$i[X, [ia^\dagger, \cdot]]$	$\frac{1}{2}(-\partial^2_{\alpha\alpha} + \partial^2_{\alpha\bar\alpha})$	0
$i[Y, [a, \cdot]]$	$\frac{1}{2}(\partial^2_{\bar\alpha\bar\alpha} + \partial^2_{\alpha\bar\alpha})$	$\frac{1}{4}\partial^2_{xx}$
$i[X, [-ia, \cdot]]$	$\frac{1}{2}(\partial^2_{\bar\alpha\bar\alpha} - \partial^2_{\alpha\bar\alpha})$	0

Table 2: Master equation superoperators and corresponding Wigner and marginal-X differential representations

In the following we will consider FPE's of the form

$$\partial_t W_s(\alpha, \bar\alpha; t) = \left[Q(\partial_\alpha \alpha + \partial_{\bar\alpha}\bar\alpha) + 2D_s\partial^2_{\alpha\bar\alpha}\right] W_s(\alpha, \bar\alpha; t) \,, \tag{15}$$

having Gaussian solution

$$\begin{aligned} W_s(\alpha, \bar\alpha; t) &= \frac{1}{\pi\Delta_s^2(t)} \exp\left[-\frac{|\alpha - \alpha_0 e^{-Qt}|^2}{\Delta_s^2(t)}\right] \,, \\ \Delta_s^2(t) &= \frac{D_s}{Q}(1 - e^{-2Qt}) + \Delta_s^2(0)e^{-2Qt} \,. \end{aligned} \tag{16}$$

The corresponding marginal OUE is

$$\partial_t P_s(x; t) = \left[Q\partial_x x + \frac{1}{2}D_s\partial^2_{xx}\right] P_s(x; t) \,, \tag{17}$$

again with Gaussian solution

$$\begin{aligned} P_s(x; t) &= \frac{1}{\sqrt{2\pi d_s^2(t)}} \exp\left[-\frac{(x - x_0 e^{-Qt})^2}{2d_s^2(t)}\right] \,, \\ d_s^2(t) &= \frac{D_s}{2Q}(1 - e^{-2Qt}) + d_s^2(0)e^{-2Qt} \,. \end{aligned} \tag{18}$$

It is clear that Eq. (17) could also be obtained as marginal OUE of FPE's different from (15). Moreover, notice that the diffusion terms in (15) and (17) generally depend on the quantum efficiency of detection through the parameter s. For negative drift $Q < 0$ (amplification) negative diffusion coefficients are admissible, without violating positivity of variances in (16) and (18) during the time evolution. In the following we will see that Eqs. (15) and (17) describe many different linear devices, with Gaussian solutions (16) and (18) modeling both coherent and squeezed coding (in Table 3 the parameters of the probability distributions for these coding are reported).

	Coherent state Wigner repr.	Thermal state Wigner repr.	$dP(\alpha,\bar{\alpha})$ coherent channel
Δ_s^2	$\frac{1}{2}(1-s)$	$\frac{1}{2}(1-s)+\bar{n}$	$\bar{n}$
α_0	α	0	0
	Squeezed state Marginal-X repr.	Thermal super. Marginal-X repr.	$dP(\lambda)$ quadrature channel
d_s^2	$\frac{1}{4}(\frac{1}{1+2\bar{n}}-s)$	$\frac{1}{4}(1-s)+\frac{\bar{n}}{2}$	$\bar{n}(\bar{n}+1)/2\bar{n}+1$
x_0	λ	0	0

Table 3: Variance and average of the Gaussian distributions in Eqs. (16) and (18), corresponding to optimal coding achieving capacities (4).

4.4 Gain, noise figure, and mutual information

Both gain and noise figure of any device inserted in a quantum communication channel generally depend on the whole chain of devices in the line, namely they depend on the overall equivalent detection and coding referred at the output and input of the device, respectively. The gain is defined as the ratio between output and input signals, the "signal" being the amplitude of the modulation $S \doteq \mathrm{Tr}\{\Delta\rho O\}$ of the detected operator O with respect to the reference state ρ_0 ($\Delta\rho = \rho - \rho_0$) [at the beginning of the line the state ρ_0 is usually the vacuum state, but evolution along the line generally leads to nonvacuum ρ_0]. The device is named "linear" when the gain does not depend on the input signal (i.e. on the input state ρ). Notice that, in general, a device can be linear for a particular detection scheme, being nonlinear for other schemes. For homodyne and heterodyne detection the gain is given by

$$g[\Delta\rho, \mathcal{A}\circ\mathcal{D}_x^{(\eta)}] = \frac{\mathrm{Tr}[\Delta\rho A_H(X_\eta)]}{\mathrm{Tr}[\Delta\rho X_\eta]}, \qquad g[\Delta\rho, \mathcal{A}\circ\mathcal{D}_\alpha^{(\eta)}] = \frac{\mathrm{Tr}[\Delta\rho A_H(Z_\eta)]}{\mathrm{Tr}[\Delta\rho Z_\eta]}, \tag{19}$$

where $X_\eta^n = \int_{-\infty}^{\infty} d\mu_\eta(x) x^n$, and similarly $Z_\eta^n = \int d^2\mu_\eta(z,\bar{z}) z^n$. Notice that for heterodyne detection the gain g is complex and carries information on field phase.

Besides the gain, the most significant characteristic of a device is its noise figure R. This represents the degradation of the signal-to-noise ratio (SNR) from the input to the output, and thus it is defined as $R = SNR_{in}/SNR_{out}$. For homodyne detection the noise is just the customary root mean square (r.m.s.) of the detected observable X. For on-off modulation and binary alphabet the noise is mediated on the (equally likely) states ρ and ρ_0. More generally, for an alphabet corresponding to an *a priori* matrix $\bar{\rho}$ as in Eq. (1) one evaluates the average noise N over the alphabet probability, namely

$$N[\mathcal{C}_x \circ \mathcal{D}_x^{(\eta)}] = \int dP(\theta) \langle \Delta X_\eta^2 \rangle_{\rho_\theta} = \langle \Delta X_\eta^2 \rangle_{\bar{\rho}} - \overline{\Delta \langle X_\eta \rangle_{\rho_\theta}^2} \,, \tag{20}$$

where $\langle \Delta O^2 \rangle_\rho = \mathrm{Tr}(\rho O^2) - [\mathrm{Tr}(\rho O)]^2$ and $\overline{\Delta f^2(\theta)} = \int dP(\theta) f^2(\theta) - [\int dP(\theta) f(\theta)]^2$. For linear devices the gain does not depend on θ, and thus the noise figure is simply given by

$$R[\mathcal{C}_x \circ \mathcal{A} \circ \mathcal{D}_x^{(\eta)}] = \frac{N[\mathcal{C}_x \circ \mathcal{A} \circ \mathcal{D}_x^{(\eta)}]}{N[\mathcal{C}_x \circ \mathcal{D}_x^{(\eta)}]} \frac{1}{g^2} \,. \tag{21}$$

For heterodyne detection one has to define the noise of a complex variable, namely the noise of joint detection of two real variables. Hence, generally there are two noises $\frac{1}{2}\left[\langle |Z|_\eta^2 \rangle_{\rho_\theta} - |\langle Z_\eta \rangle_{\rho_\theta}|^2 \pm |\langle \Delta Z_\eta^2 \rangle|\right]$ where $|Z|_\eta^2 = \int d^2\mu_\eta(z,\bar{z})|z|^2$, corresponding to the eigenvalues of the covariance matrix. For "isotropic" devices $\langle \Delta Z_\eta^2 \rangle = 0$ and the noise is simply

$$N[\mathcal{C}_\alpha \circ \mathcal{D}_\alpha^{(\eta)}] = \frac{1}{2}\left[\langle \Delta |Z|_\eta^2 \rangle_{\bar{\rho}} - \overline{\Delta |\langle Z_\eta \rangle_{\rho_\theta}|^2}\right] \,, \tag{22}$$

where $\langle \Delta |O|^2 \rangle_\rho = \mathrm{Tr}(\rho |O|^2) - |\mathrm{Tr}(\rho O)|^2$, $\overline{\Delta |f(\theta)|^2} = \int dP(\theta) |f(\theta)|^2 - |\int dP(\theta) f(\theta)|^2$ and the "modulus operator" has already been defined. For linear devices the figure of noise is then defined similarly to Eq. (21). For devices described by equations (15) and (17) the gain is $g = \exp(-Qt)$, independently on the signal, and for both coherent and homodyne detection. The device is a linear amplifier for negative drift $Q < 0$, and a linear attenuator for $Q > 0$. From Eqs. (20-22) and Table 2 one obtains the noise figures

$$R[\mathcal{C}_x \circ \mathcal{A} \circ \mathcal{D}_x^{(\eta)}] = 1 + \frac{\eta(1+2\bar{n})}{1+2\bar{n}(1-\eta)} \frac{2D_{1-\eta^{-1}}}{Q}(g^{-2}-1) \tag{23}$$

$$R[\mathcal{C}_\alpha \circ \mathcal{A} \circ \mathcal{D}_\alpha^{(\eta)}] = 1 + \eta \frac{D_{1-2\eta^{-1}}}{Q}(g^{-2}-1) \,. \tag{24}$$

For optimized coding and ideal detection the noise figure is bounded as $R \geq 1$. [13] On the other hand, for $\eta < 1$ the noise figure may become lower than unit, signaling that the device (preamplifier) is improving detection performance. Noise figures $R < 1$ are equivalent to negative diffusion coefficients, whereas $R > 1$ corresponds to positive diffusion. An ideal amplifier for ideal detection has $R = 1$, corresponding to pure (negative) drift $-\nabla_\zeta \cdot \zeta$ which gives the POM rescaling $A_H(d\mu(\zeta)) = d\mu(g^{-1}\zeta)$ ($\zeta \in R^{d'}$ is used as in Sect. 2). The POM rescaling leaves the mutual information invariant: therefore, for ideal detection an amplifier can only degrade the mutual information, and improvements are possible only for $\eta < 1$. In the following we will see that in order to improve the mutual information for $\eta < 1$ the amplifier needs to be matched with the detector POM, namely one one should use a quadrature (or phase sensitive) amplifier for homodyne, and a coherent (or phase insensitive) amplifier for heterodyne detection. Here we only give the mutual information after the insertion of a device modeled by Eqs. (15) and (17). For squeezed/quadrature and coherent channels one has

$$I[\mathcal{C}_x \circ \mathcal{A} \circ \mathcal{D}_x^{(\eta)}] = \frac{1}{2}\log\left[1 + \frac{\bar{n}(\bar{n}+1)/2\bar{n}+1}{\frac{D_{1-\eta^{-1}}}{2Q}(g^{-2}-1) + \frac{1}{4}\left(\eta^{-1} - \frac{2\bar{n}}{1+2\bar{n}}\right)}\right] \,. \tag{25}$$

$$I[\mathcal{C}_\alpha \circ \mathcal{A} \circ \mathcal{D}_\alpha^{(\eta)}] = \log\left[1 + \frac{\bar{n}}{\frac{D_{1-2\eta^{-1}}}{Q}(g^{-2}-1) + \eta^{-1}}\right] \,. \tag{26}$$

4.5 Modeling master equations for linear amplifiers

The following master equation models a linear phase insensitive device

$$\partial_t \rho_t = 2\left[AD[a^\dagger] + BD[a]\right]\rho_t \,, \tag{27}$$

with superoperator D defined in Eq. (12). The device is phase insensitive as a consequence of invariance $D[ae^{-i\phi}] = D[a]$. $B > A$ leads to attenuation, whereas $A > B$ produces amplification. For example, with A and B proportional to atomic populations on the upper and lower lasing levels respectively, Eq. (27) describes an active medium amplifier in the linear regime (far from saturation). On the other hand, for $A = \frac{\Gamma}{2}\bar{m}$ and $B = \frac{\Gamma}{2}(\bar{m}+1)$ the same equation describes a field mode damped with photon lifetime Γ^{-1} toward the thermal distribution with $\bar{m}$ average photons. Eq. (27) has the following general CP-map solution [14]

$$\begin{aligned} \rho_t &= \mathrm{Tr}_P[U_t\,\rho\otimes\nu\,U_t^\dagger]\,, \\ U_t &= \begin{cases} \exp[-\arctan\sqrt{e^{\Gamma t}-1}(ab^\dagger - a^\dagger b)] & (B>A)\,, \\ \exp[-\mathrm{arctanh}\sqrt{1-e^{-\Gamma t}}(a^\dagger b^\dagger - ab)] & (A>B)\,, \end{cases} \end{aligned} \tag{28}$$

with $\Gamma/2 = |A-B|$ and ν the thermal state of the idler mode b with average photons $\bar{m} = \min\{A,B\}/|A-B|$ [A and B are non negative, otherwise one would have negative idler photons]. Thus, Eq. (27) also models either parametric amplification with thermal idler, or loss due to frequency conversion. [15] From Table 2 one obtains the Wigner representation of Eq. (27) in form of the FPE (15) with $Q = B - A$ and $2D_s = A+B+s(A-B)$. For ideal heterodyne detection ($s=-1$) the device is ideal for $B=0$ (ideal phase insensitive amplifier). On the contrary, linear attenuation is always nonideal. The diffusion coefficient can be negative for $s = 1-2\eta^{-1} < -1$, and the amplifier can improve both SNR and mutual information when $A-B > \eta A$. The mutual information for the coherent channel is

$$I[\mathcal{C}_\alpha \circ \mathcal{A} \circ \mathcal{D}_\alpha^{(\eta)}] = \log\left[1 + \frac{\bar{n}}{\left(\frac{A}{B-A}+\eta^{-1}\right)(g^{-2}-1)+\eta^{-1}}\right] \tag{29}$$

with $g = \exp[(A-B)t]$: for ideal amplification ($B=0$) the channel capacity is recovered in the limit of infinite gain g.

The phase insensitive amplifier modeled by Eq. (27) is "matched" to heterodyne detection (the field is detected with no preferred phase). The amplifier can be made ideal for the matched detection, and the information loss for $\eta < 1$ is completely recovered in the limit $g \to \infty$. Such information gain for nonefficient detection is described by negative diffusion coefficient D and, correspondingly, by noise figures $R < 1$. On the other hand, when there is a mismatch between amplifying CP-map and detecting POM, it is no longer possible to recover the whole lost information, even for infinite gain. This happens, for example, when amplifier (27) is inserted in a quadrature channel using homodyne detection. In this case the amplifier is described by the OUE (17), and the diffusion coefficient $\frac{1}{2}D_s$ now is evaluated for $s = 1-\eta^{-1}$, as a result of the POM marginal projection (8). One has $D_{1-\eta^{-1}} = A+\frac{1}{2}\eta^{-1}(B-A)$: for $\eta = 1$ $D_0 = \frac{1}{2}(A+B) > 0$, and the amplifier is no longer ideal. For $\eta < 1$ the diffusion coefficient becomes negative when $A - B > 2\eta A$, namely the amplifier is able to improve the SNR only for quantum efficiencies half than the corresponding ones for heterodyne detection. Despite for $\eta < \frac{1}{2}$ the amplifier can reduce the SNR, it can no longer completely recover the lost information. In fact, from Eq. (25) one has

$$I[\mathcal{C}_x \circ \mathcal{A} \circ \mathcal{D}_x^{(\eta)}] = \frac{1}{2} \ln \left[1 + \frac{4\bar{n}(\bar{n}+1)}{(2\bar{n}+1)\frac{2A}{B-A}(g^{-2}-1) + (2\bar{n}+1)\eta^{-1}g^{-2} - 2\bar{n}} \right] \tag{30}$$

which is always lower than the channel capacity in Eq. (4), and leads to $I = \frac{1}{2}\ln(1+2\bar{n})$ for $B = 0$ and $g \to \infty$.

In order to improve the performance of the amplifier for homodyne detection one can "squeeze" the master equation (27) as follows

$$\partial_t \rho_t = 2\left\{AD[a^\dagger] + BD[a]\right\}\rho_t + C^*[a,[a,\rho_t]] + C[a^\dagger,[a^\dagger,\rho_t]] \;, \tag{31}$$

with $|C|^2 \leq AB$. The master equation (31) could be obtained either by "squeezing" the atomic lasing bath, or using transformations (31) with b in a squeezed-thermal state. For a parametric amplifier with the idler in a squeezed-thermal state with N photons in total and $\bar{m}_s$ squeezing photons, one has $A = \frac{\Gamma}{2}(N+1)$, $B = \frac{\Gamma}{2}N$ and $|C|^2 = \frac{\Gamma}{4}[N(N+1) - \bar{m}(2\bar{m}+1)]$, where now $\bar{m}$ denotes the pure thermal photons. The limiting case $|C|^2 = AB$ corresponds to the idler in a squeezed vacuum. From Table 2 one can see that the OUE of the device has the form (17) with the same drift $Q = B - A$ of the unsqueezed amplifier, but with diffusion $D_{1-\eta^{-1}} = A + \frac{1}{2}\eta^{-1}(B-A) + \mathrm{Re}(C)$. For $\mathrm{Re}(C) < 0$ the overall effect of squeezing is a reduction of the diffusion: for $\eta = 1$ and $\bar{m} = 0$ one has $D_0 = \frac{\Gamma}{16}N^{-1} + O(N^{-2})$, namely at infinite squeezing the amplifier becomes ideal for homodyne detection. Also, in this limit for $\eta < 1$ the diffusion coefficient becomes negative $D_{1-\eta^{-1}} = \frac{Q}{2}(\eta^{-1}-1)$ and from Eq. (25) one obtains

$$\begin{aligned} I[\mathcal{C}_x \circ \mathcal{A} \circ \mathcal{D}_x^{(\eta)}] &= \frac{1}{2}\ln\left[1 + \frac{4\bar{n}(\bar{n}+1)}{(2\bar{n}+1)(\eta^{-1}-1)g^{-2}+1}\right] \\ &= \log(1+2\bar{n}) + O(g^{-2}) \;, \end{aligned} \tag{32}$$

recovering the channel capacity.

As a consequence of noninjectivity of marginal integration (8) there are many different ways to achieve ideal quadrature amplification. For example, the popular phase sensitive amplifier is obtained by means of a degenerate parametric amplifier (with strong coherent pump) having effective Hamiltonian of the form

$$H = i\frac{\lambda}{2}(a^2 - a^{\dagger 2}) \;. \tag{33}$$

Again, from Table 2 one can see that Hamiltonian (33) corresponds to an OUE of the form (17) with $Q = \lambda$ and $\frac{1}{2}D_{1-\eta^{-1}} = \frac{Q}{2}(\eta^{-1}-1)$, leading to ideal amplification for $\lambda < 0$, and ideal attenuation for $\lambda > 0$.

Before concluding this subsection, we want to remark that it is not possible to recover lost information when the loss occurs before amplification. For example, let us consider the maps $\mathcal{L}$ and $\mathcal{A}$ both corresponding to Eq. (27), but the former describing a loss ($A = 0$, $B = \Gamma/2$), and the latter an ideal phase insensitive amplifier ($A = t^{-1}\log g$, $B = 0$). Then, the mutual informations for the coherent channel when the amplifier is inserted before and after loss are: $I[\mathcal{C}_\alpha \circ \mathcal{A} \circ \mathcal{L} \circ \mathcal{D}_\alpha^{(\eta)}] = \log\left[1 + \frac{\bar{n}}{1+(1-\eta\gamma)/\eta\gamma g^2}\right]$, $I[\mathcal{C}_\alpha \circ \mathcal{L} \circ \mathcal{A} \circ \mathcal{D}_\alpha^{(\eta)}] = \log\left[1 + \frac{\gamma\bar{n}}{1+(1-\eta)/\eta g^2}\right]$. It is clear that only in the first case the information is recovered, however with an increased power impinged into the line.

4.6 Ideal amplification via feedback

In Ref. [16] it is shown that it is possible to produce squeezing using feedback mediated by intracavity quantum non demolition (QND) measurements. The cavity supports a second mode b, which is coupled to the signal mode a via the interaction $H_I = \chi X_a Y_b$, and which is simultaneously homodyne detected (measurement of X_b) with efficiency ϵ. The evolution due to feedback has the form $\partial_t \rho_t|_{fb} = i\frac{\Lambda}{\epsilon\chi} I(t - 0^+)[Y_a, \rho_t]$, where $I(t - 0^+)$ is the detected photocurrent immediately before the feedback action. When fastly decaying to the vacuum ($\chi^2/\gamma \ll 1$) the mode b can be adiabatically eliminated, and the following master equation for the reduced density matrix of the signal mode is obtained

$$\partial_t \rho_t = L\rho_t + \frac{\Lambda^2}{2h} D[Y]\rho_t + \Gamma D[X]\rho_t + i\Lambda[Y, [X\rho_t + \rho_t X]] \ . \tag{34}$$

In Eq. (34) L is the original Liouvillian for a, $h = \epsilon\Gamma$, $\Gamma = \chi^2/\gamma$, and γ is the damping rate of mode b. From Table 2 one can see that the master equation (34) admits the OU representation (17). The term with $D[Y]$ adds a contribution $\frac{\Lambda^2}{16h}$ to diffusion D_s and could be made arbitrarily small for $h \propto \chi^2 \to \infty$. The term with $D[X]$—describing the QND measurement of X—has no effect on the POM $d\mu_\eta(x)$. Finally, the last term in Eq. (34) adds Λ to the drift, and $\frac{Q}{2}(\eta^{-1} - 1)$ to the diffusion $D_{1-\eta^{-1}}$. One thus concludes that the present intracavity feedback for $\Lambda < 0$ (within the validity limits of the adiabatic approximation) can either convert a loss L into amplification, or improve the original amplifier, adding an ideal term to the OU representation of the Liouvillian L.

The above measurement-and-feedback control has both steps and feedback matched with detection (measurement of X and commutator with Y which drives X). Again, if there is no matching with detection, the device is no longer improved. This happens, for example, if the roles of X and Y are exchanged in the above example. In Ref. [17] other master equations are considered, also from complex-amplitude and heterodyne feedback: using Table 2 one could immediately obtain the corresponding FPE or OUE, checking if feedback is effective in improving the device. The effectiveness of feedback loop is the result of competition between driving terms and positive-diffusion terms due to unavoidable introduction of noise by the measurement step in the quantum-limited feedback loop: as noticed in Ref. [17] the positive diffusion is doubled when the measurement step is achieved by heterodyne detection (the latter having effective quantum efficiency 1/2, due to joint detection of both quadratures. [3])

Acknowledgements

G. M. D'Ariano is indebted with M. Ozawa for illuminating discussions and emails on POM/CP-map approach, and is grateful to him for hospitality in Nagoya University. This work represents part of a preliminary study in the context of a collaboration on "Quantum amplification" with M. Ozawa himself and H. P. Yuen (Northwestern University).

References

[1] H. P. Yuen, Phys. Lett. **113A**, 405, (1986); Opt. Lett. **12**, 789 (1987).

[2] C. W. Helstrom, *Quantum Detection and Estimation Theory* (Academic Press, New York, 1976).

[3] H. P. Yuen, Phys. Lett. **91A**, 101 (1982).

[4] For the description of quantum measurement probability in terms of "probability-operator-valued measure" (POM) see: M. Ozawa, J. Math. Phys. **25**, 79 (1984).

[5] E. B. Davies, *Quantum theory of open systems*, (Academic Press, London, New York, 1976).

[6] For the "completely positive instrument" description of quantum measurements see Ref. [4]

[7] H. P. Yuen and M. Ozawa, Phys. Rev. Lett. **70**, 363 (1993).

[8] Essentially the POM comes from an orthogonal projection-valued measure that includes "probe" degrees of freedom of the apparatus, which are traced out to produce an outcome referring to the system alone.

[9] For evaluation of POM's of ideal homodyne, double homodyne, and heterodyne detectors see: G. M. D'Ariano and M. G. A. Paris, Phys. Rev. A**49**, 3022 (1994); for the nonideal double homodyne detector see: U. Leonhardt and H. Paul, Phys. Rev. A **48**, 4598 (1993).

[10] A. S. Holevo, Probl. Inf. Transm. **9**, 177 (1973).

[11] G. Lindblad, Commun. Math. Phys. **48**, 119 (1976).

[12] Notice that here the Langevin equation has no noise term. However, this is unnecessary as long as the evolution of a POM is concerned.

[13] Actually, the quadrature channel is not perfectly optimized with respect to detection, because quadrature eigenstates have infinite energy. Thus, in principle, it might be possible to get R slightly lower than unit for this case.

[14] G. M. D'Ariano, Phys. Lett. A **187**, 231 (1994).

[15] G. M. D'Ariano and C. Macchiavello, Phys. Rev. **A 48** 3947, (1993).

[16] H. M. Wiseman and G. J. Milburn, Phys. Rev. Lett. **70** 548 (1993); Phys. Rev. A**49** 1350 (1994). For the adiabatic elimination see of the same authors: Phys. Rev. A**47** 642 (1994).

[17] H. M. Wiseman and G. J. Milburn, Phys. Rev. A**49** 4110 (1994).

MUTUAL ENTROPY IN QUANTUM MARKOV CHAINS

Luigi Accardi[1], Masanori Ohya[2] and Hiroki Suyari[2]

[1]Dipartimento di Matematica
Centro Matematico V.Volterra,
Universitá di Roma II, Italy
[2]Department of Information Sciences
Science University of Tokyo, Japan

Abstract

There are several damping phenomena in quantum optics. Such phenomena have been usually explained by open systems. In statistical physics, open system dynamics have been used to study the irreversibility and the approach to equilibrium.

In this paper the dynamical change of the mutual entropy is discussed for a model of quantum Markov chain. The concrete Stinespring-Kraus's expression for this model is obtained and applied to the derivation of the mutual entropy, and some computational results are presented.

1 INTRODUCTION

The purpose of this paper is to show the use of the mutual entropy in quantum Markov chain (QMC for short) for a study of the irreversible process.

In section 2, a well known model in quantum optics is presented. This model is often called "damping, and its dynamical change with a unitary evolution is popular not only in quantum optics but also in statistical physics. In section 3, we review some concepts for the mutual entropy and a generalized QMC, introduced in [O1] and [A1, A2], respectively. We here stand on a concept "lifting [A3] to study the above concepts in a unified manner. The Stinespring-Kraus expression of a completely positive map has mathematically been established, but its concrete

Quantum Communications and Measurement
Edited by V. P. Belavkin *et al.*, Plenum Press, New York, 1995

expression is missing. In section 4, we show its concrete expression for our model. and apply this expression to the computation of the mutual entropy in QMC. In section 5, some computational results are presented.

2 MODEL FOR OPEN SYSTEMS WITH HARMONIC OSCILLATORS

In statistical mechanics, a state of interacting system is changed under some effects from the outside of the system (reservoir), that is, the interaction between two systems is considered and the reduced state after interaction is studied. Mathematical expression of this process is given as follows.

Let a system Σ_0 be described by a Hilbert space $\mathcal{H}_0$, which interacts with an external system Σ_1 described by another Hilbert space $\mathcal{H}_1$, and let the initial states of Σ_0, Σ_1 be $\rho \in \Sigma_0$, $\omega \in \Sigma_1$ (Σ_i is the set of all states on $\mathcal{H}_i$ $(i = 0, 1)$), respectively. Then the combined state $\tilde{\rho} \in \Sigma_0 \otimes \Sigma_1$ after the interaction between the two systems is given by

$$\tilde{\rho} = U_t \left(\rho \otimes \omega\right) U_t^* \tag{2.1}$$

where $U_t = \exp\left(-itH\right)$ with a total Hamiltonian H on $\mathcal{H}_0 \otimes \mathcal{H}_1$. Here we took $\hbar = 1$.

The above total Hamiltonian H for two weakly coupled oscillators in quantum optics is given by [L1]:

$$H = H_0 + H_1 + H_{01} \tag{2.2}$$

$$H_0 = a^* a, \quad H_1 = \sum_j b_j^* b_j, \quad H_{01} = \sum_j \left(\varepsilon_j b_j a^* + \varepsilon_j^* b_j^* a\right) \tag{2.3}$$

where H_0, H_1, H_{01} are the Hamiltonians for the observed system Σ_0, the external system (reservoir) Σ_1 and an interaction between Σ_0 and Σ_1, respectively. a, a^* on $\mathcal{H}_0$ and b_j, b_j^* on $\mathcal{H}_1$ are pairs of annihilation and creation operators, respectively, and ε_j $(j \in N)$ are the coupling constants.

In this paper we assume, for simplicity, that the reservoir has a single mode, that is, the total hamiltonian H is given by

$$H = H_0 + H_1 + H_{01} = a^* a + b^* b + \varepsilon \left(a^* b + a b^*\right) \tag{2.4}$$

In the model, the evolution of the initial state ρ after the interaction is mathematically expressed as

$$\rho \mapsto \bar{\rho} = tr_{\mathcal{H}_1} U_t \left(\rho \otimes \omega\right) U_t^*. \tag{2.5}$$

3 MUTUAL ENTROPY IN QUANTUM MARKOV CHAIN

In this section the mutual entropy is discussed in quantum Markov chain (QMC) of our model. For the sake of the formulation of mutual entropy in QMC, we review some definitions and fundamental results :

Definition 3.1 *[O1] : A completely positive map $\Lambda^* : \Sigma\left(\mathcal{H}\right) \to \Sigma\left(\mathcal{K}\right)$ is called a quantum mechanical channel. Here $\Sigma\left(\mathcal{H}\right)$ is the set of states on a Hilbert space $\mathcal{H}$.*

The quantum mechanical channel $\Lambda^* : \Sigma_0 \to \Sigma_0$ in our model can be written as:

$$\Lambda^* \rho = tr_{\mathcal{H}_1} U_t \left(\rho \otimes \omega\right) U_t^* \tag{3.1}$$

The concept of channel is related to that of "lifting introduced in [A3].

Definition 3.2 *[A3] : Let Σ_0, Σ_1 be the set of all states on $\mathcal{H}_0, \mathcal{H}_1$ respectively. A lifting from Σ_0 to $\Sigma_0 \otimes \Sigma_1$ is a continuous map*

$$\mathcal{E}^* : \Sigma_0 \to \Sigma_0 \otimes \Sigma_1 \tag{3.2}$$

For example, the correspondence $\rho \mapsto \tilde{\rho}$ in (2.1) is a lifting ; $\mathcal{E}^* \cdot = U_t (\cdot \otimes \omega) U_t^*$ is a continuous map from Σ_0 to $\Sigma_0 \otimes \Sigma_1$. The correspondence between a quantum mechanical channel Λ^* and a lifting $\mathcal{E}^*$ is explained in [A3].

Let the Schatten decomposition of $\rho \in \Sigma(\mathcal{H})$ be

$$\rho = \sum_k \lambda_k E_k, \quad \lambda_1 \geq \lambda_2 \geq \cdots \geq \lambda_n \geq \cdots, \quad , E_i \perp E_j \quad (i \neq j) \tag{3.3}$$

where λ_k is an eigenvalue of ρ and E_k is the associated onedimensional projection. The decomposition is not unique unless every λ_k is not degenerated.

The relation between an initial state ρ and the final state $\Lambda^* \rho$ is expressed by the compound states introduced in [O1, O2] such that

$$\sigma_E = \sum_k \lambda_k E_k \otimes \Lambda^* E_k \tag{3.4}$$

The definition of mutual entropy $I(\rho; \Lambda^*)$ is given as follows :

Definition 3.3 *[O1] : Using the above notations (3.3) and (3.4), the mutual entropy $I(\rho; \Lambda^*)$ is defined by*

$$I(\rho; \Lambda^*) = \sup_E \{S(\sigma_E, \sigma_0) \; ; \; E = \{E_k\}\} \tag{3.5}$$

where $\sigma_0 = \rho \otimes \Lambda^* \rho$ and $S(\sigma_E, \sigma_0)$ is the relative entropy defined by

$$S(\sigma_E, \sigma_0) = tr \sigma_E (\log \sigma_E - \log \sigma_0) \tag{3.6}$$

In (3.5), we have to take "sup over all Sahtten decompositions when some eigenvalues are degenerated. We have the following fundamental inequality of Shannon's type.

Theorem 3.4 *[O1] : For an input state ρ and a channel Λ^*, the following inequalities hold.*

$$0 \leq I(\rho; \Lambda^*) \leq \min\{S(\rho), \; S(\Lambda^* \rho)\} \tag{3.7}$$

This theorem means that the information correctly transmitted through a channel Λ^* is always less than the initial information. In other words, the mutual entropy is useful as an efficiency of communication processes. [B1, W1]. Other properties of the mutual entropy are discussed in [O3, O4].

Then, before reviewing QMC, we need the concept of a transition expectation :

Definition 3.5 *[A4] : Let $\mathcal{B}_0, \mathcal{B}_1$ be the algebras of all bounded operators on Hilbert spaces $\mathcal{H}_0, \mathcal{H}_1$ respectively, and let $\mathcal{B}_0 \otimes \mathcal{B}_1$ be a fixed tensor product of $\mathcal{B}_0$ and $\mathcal{B}_1$. A transition expectation from $\mathcal{B}_0 \otimes \mathcal{B}_1$ to $\mathcal{B}_0$ is a completely positive linear map $\mathcal{E} : \mathcal{B}_0 \otimes \mathcal{B}_1 \to \mathcal{B}_0$ satisfying*

$$\mathcal{E}(1 \otimes 1) = 1 \tag{3.8}$$

When $\mathcal{B} = \mathcal{B}_0 = \mathcal{B}_1$ we say that $\mathcal{E}$ is a transition expectation on $\mathcal{B}$.

By using the transition expectations, a generalized QMC is defined as follows :

Definition 3.6 *[A4] : Let $\{\mathcal{E}_n\}_{n\geq 0}$ be a sequence of transition expectations on $\mathcal{B}$. Then there exists a unique completely positive identity preserving map $E_{0]} : \otimes^n \mathcal{B} \to \mathcal{B}$ such that for each integer n and for each $A_0, A_1, \cdots, A_n \in \mathcal{B}$, one has*

$$E_{0]}\left(A_0 \otimes A_1 \otimes \cdots \otimes A_n \otimes 1 \otimes \cdots\right) = \mathcal{E}_0\left(A_0 \otimes \mathcal{E}_1\left(A_1 \otimes \cdots \otimes \mathcal{E}_n\left(A_n \otimes 1\right)\cdots\right)\right) \tag{3.9}$$

Let φ_0 be a state on $\mathcal{B}$ and define a state φ on $\otimes_N \mathcal{B}$ by

$$\varphi = \varphi_0 \circ E_{0]} \tag{3.10}$$

Then φ satisfies

$$\varphi\left(A_0 \otimes A_1 \otimes \cdots \otimes A_n \otimes 1 \otimes \cdots\right) = \varphi_0\left(\mathcal{E}_0\left(A_0 \otimes \mathcal{E}_1\left(A_1 \otimes \cdots \otimes \mathcal{E}_n\left(A_n \otimes 1\right)\cdots\right)\right)\right) \tag{3.11}$$

The state φ, characterized by (3.10) is called the generalized Markov chain associated to the pair $(\varphi_0, \{\mathcal{E}_n\})$. If for each n $\mathcal{E} \equiv \mathcal{E}_n = \mathcal{E}_0$, then the state φ is called homogeneous.

Clearly the dual of a "linear lifting is a transition expectation, therefore to any linear lifting one can associate a QMC in the standard way as explained in [A3]. If the lifting is of convex product type, then we can take an advantage of the special structure to extend the construction of QMC to the case of a not necessarily linear lifting.

As in section 2, take $\Sigma = \Sigma_0 \otimes \Sigma_1$ and $\mathcal{B} = \mathcal{B}_0 \otimes \mathcal{B}_1$. Let φ_0 and φ_0 be states on $\mathcal{B}_0$ and $\mathcal{B}_1$ respectively and let u_0 be the partial shift on $\otimes_{\boldsymbol{N}}(\mathcal{B}_0 \otimes \mathcal{B}_1) \cong \otimes_{\boldsymbol{N}} \mathcal{B} \cong \left(\otimes_{\boldsymbol{N}} \mathcal{B}_0\right) \otimes \left(\otimes_{\boldsymbol{N}} \mathcal{B}_1\right) \equiv \mathcal{A}$, that is, u_0 acts trivially on $\otimes_{\boldsymbol{N}} \mathcal{B}_0$ and does as the usual shift on $\otimes_{\boldsymbol{N}} \mathcal{B}$. Let $U \in \mathcal{B}_0 \otimes \mathcal{B}_1$ be a unitary operator, and define

$$U_n = u_0^{n-1}(U)\ ;\ n \geq 1 \tag{3.12}$$

If $\mathcal{E}_n : \mathcal{B}_0 \otimes \mathcal{B}_1 \to \mathcal{B}_0$ is a transition expectation given by

$$\mathcal{E}_n(x) = tr_{\mathcal{H}_1} U_n^* x U_n \cdot;\ n \geq 1 \tag{3.13}$$

then we can easily construct a QMC according to the above definition 3.6. This class of Markov chains was studied by Kümmerer [K2].

4 A CONCRETE EXAMPLE OF STINESPRING-KRAUS EXPRESSION AND COMPUTATION OF MUTUAL ENTROPY IN OUR MODEL

In this section the Stinespring-Kraus expression is concretely obtained in our model and the mutual entropy is computed [A6].

The Stinespring-Kraus theorem is well known in the following form :

Theorem 4.1 *[K1,S1] : Let $\mathcal{H}$ be a Hilbert space. A linear map $\Lambda : \mathcal{B}(\mathcal{H}) \to \mathcal{B}(\mathcal{H})$ is a completely positive if and only if it has the form*

$$\Lambda(A) = \sum_{i=1}^{n} V_i^* A V_i \qquad A \in \mathcal{A} \tag{4.1}$$

where $V_i : \mathcal{H} \to \mathcal{H}$ are partial isometries and $n \leq \dim(\mathcal{H})$.

We shall give a concrete expression for partial isometries V_i in our model.

From the form (2.4) of the Hamiltonian H of the total system $\mathcal{H}_0 \otimes \mathcal{H}_1$, it is easy to see that

$$[H_0 + H_1,\ H_{01}] = 0. \tag{4.2}$$

By using (4.2) and some calculations [A6], we get the following formula :

$$\Lambda^* \rho = \sum_{\nu=0}^{\infty} T_\nu \left(\rho \otimes \omega\right) T_\nu^* \tag{4.3}$$

where $$T_\nu \equiv \sum_{n=\nu}^{\infty} |n-\nu\rangle \left\langle \varphi_{n-\nu}^{(n)} \right|, \qquad \left\langle \varphi_{n-\nu}^{(n)} \right| \equiv \sum_{\beta=0}^{n} d_{n-\nu,\beta}^{(n)} \langle \beta| \otimes \langle n-\beta| \in \mathcal{K}_n \tag{4.4}$$

$$d_{\alpha,\beta}^{(n)} \equiv \sum_{j=0}^{n} e^{-it\varepsilon\lambda_j^{(n)}} C_\alpha^{(n,j)} \overline{C_\beta^{(n,j)}} \tag{4.5}$$

Since $\left\{ \left| \varphi_{n-\nu}^{(n)} \right\rangle \ ;\ n \geq \nu,\ n,\nu \in N \right\}$ is othogonal shown in [A6], the above T_ν is a partial isometry. Therefore (4.3) and (4.4) give a concrete expression for Stinespring-Kraus's.

These formulations are useful for the computation of several entropies. In the following, we compute the mutual entropy.

From (4.3) and (4.4), we have

$$\Lambda^* \rho = \sum_i \left(\sum_{\substack{l,\nu \\ (\nu-l)+i \geq 0}} \lambda_l^\rho \mu_{(\nu-l)+i}^\omega \left| d_{i,l}^{(\nu+i)} \right|^2 \right) |i\rangle \langle i| \,. \tag{4.6}$$

Here, in the above calculation, we asuume the following Schatten decomposition :

$$\rho = \sum_l \lambda_l^\rho |l\rangle \langle l|, \omega = \sum_\gamma \mu_\gamma^\omega |\gamma\rangle \langle \gamma|. \tag{4.7}$$

Clearly (4.6) is also a Schatten decomposition. Suppose that the decomposition (4.7) of ρ is unique. From (4.6) we find the mutual entropy is

$$I\left(\rho; \Lambda^*\right) = \sum_{k,t} \lambda_k^\rho \sum_{\substack{\nu \\ (\nu-k)+t \geq 0}} \mu_{(\nu-k)+t}^\omega \left| d_{t,k}^{(\nu+t)} \right|^2 \log \frac{\sum\limits_{\substack{\nu_1 \\ (\nu_1-k)+t \geq 0}} \mu_{(\nu_1-k)+t}^\omega \left| d_{t,k}^{(\nu_1+t)} \right|^2}{\sum\limits_{\substack{l,\nu \\ (\nu_2-l)+t \geq 0}} \lambda_l^\rho \mu_{(\nu_2-l)+t}^\omega \left| d_{t,l}^{(\nu_2+t)} \right|^2} \tag{4.8}$$

The typical situations are shown in the followings :

(i) The state $\omega \in \Sigma_1$ is a vacuum state : $\omega = |0\rangle \langle 0|$. Then

$$I\left(\rho; \Lambda^*\right) = \sum_{k,t} \lambda_k^\rho \left| d_{t,k}^{(k)} \right|^2 \log \frac{\left| d_{t,k}^{(k)} \right|^2}{\sum\limits_l \lambda_l^\rho \left| d_{t,l}^{(l)} \right|^2} \tag{4.9}$$

(ii) The state $\omega \in \Sigma_1$ is a Gibbs state : $\omega_\beta = \frac{e^{-\beta H_1}}{tr e^{-\beta H_1}} = \left(1 - e^{-\beta}\right) \sum\limits_n e^{-\beta n} |n\rangle \langle n|$. Then

$$I\left(\rho; \Lambda^*\right) = \left(1 - e^{-\beta}\right) \sum_{k,t} \lambda_k^\rho D_{k,t}^\beta \log \frac{D_{k,t}^\beta}{\sum\limits_l \lambda_l^\rho D_{l,t}^\beta} \tag{4.10}$$

where

$$D_{k,t}^{\beta} = \sum_{\substack{\nu \\ (\nu-k)+t\geq 0}} e^{-\beta((\nu-k)+t)} \left| d_{t,k}^{(\nu+t)} \right|^2 \tag{4.11}$$

5 COMPUTER EXPERIMENTS

Under the above computations we show some numerical results the change of the mutual entropy $I(\rho; \Lambda^*)$ w.r.t. T, the temperature of Gibbs state in the reservoir, and $I(\rho; \Lambda^{*n})$ w.r.t. n, the times of the channel, respectively.

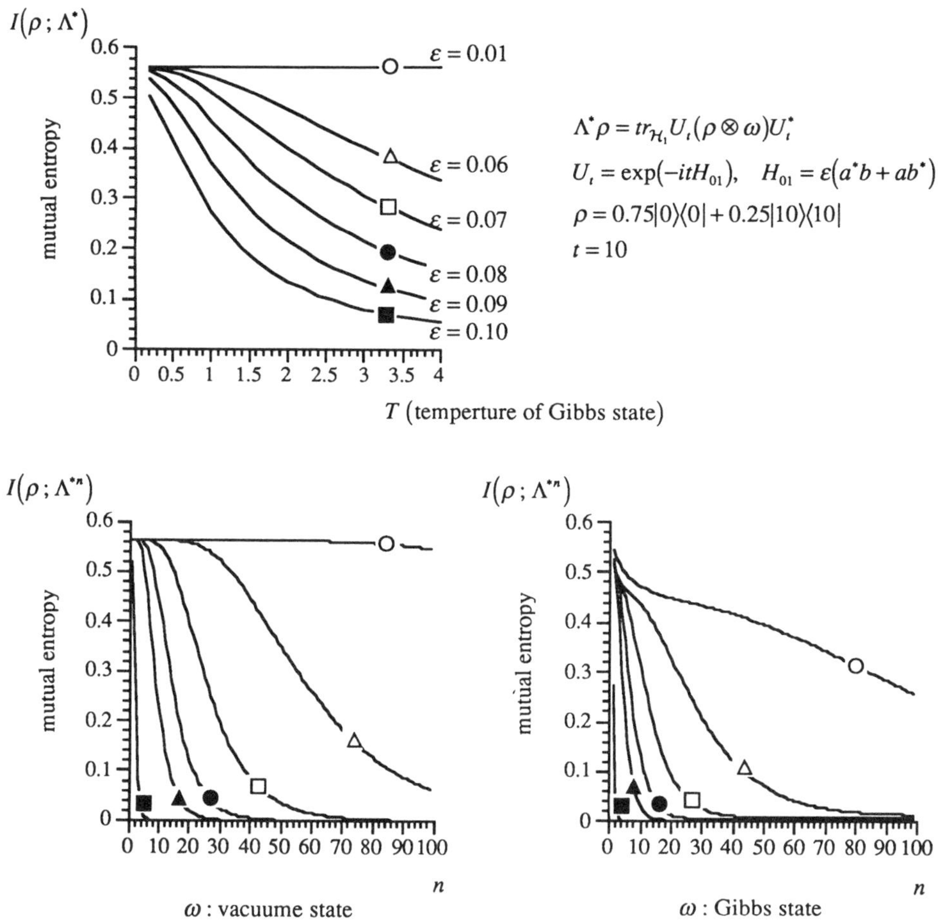

Figure 1.

As expected, the mutual entropy decreases with respect to n, and the Gibbs state plays as a noise and its effect is obviously larger than that of the vacuume state. These numerical experiments show that our model of open system gives a dissipative change of a state, although the interaction Hamiltonian is symmetric in two modes a and b.

6 CONCLUSION

In this paper we first established the Stinespring-Kraus expression of a dissipative evolution of an open system model, and we compute the mutual entropy through a typical model of QMC. Once we have the concrete expression of the mutual entropy, it will be useful to investigate the dynamical change of the mutual entropy in analysis of communication processes and nonequilibrium behavior in statistical mechanics. The detail discussions for the change of mutual entropy will be reported in the forthcoming paper [A7].

References

[A1] Accardi L. ; Noncommutative Markov chains. In : international School of Mathematical Physics, Camerino, 1974, 268-295.

[A2] Accardi L., Frigerio A. ; Markovian cocycles. Proc. Royal Irish Acad. 83A, 1983, 251-263.

[A3] Accardi L. and Ohya M. ; Compound channels, transition expectations and liftings, to appear in J. Multivariate Analysis.

[A4] Accardi L. ; An outline of Quantum Probability, unpublished manuscript, 1990.

[A5] Accardi L. ; Foundation of Quantum Probability, to be published.

[A6] Accardi L., Ohya M. and Suyari H. ; Computation of mutual entropy in quantum Markov chains, preprint in CENTRO MATEMATICO V.VOLTERRA.

[A7] Accardi L., Ohya M. and Suyari H. ; Dynamical change of mutual entropy in quantum Markov chains, in preparation.

[B1] Bendjaballah C., Hirota O. and Ohya M. ; Optical Communication Theory, to be published in Springer-Verlag.

[K1] Kraus K. ; States, effects and operations, Springer LNP N.190, 1983.

[K2] Kümmerer B. and Schroder W. ; in Quantum probability and applications to the quantum theory of irreversible processes, Springer LNM 1055, 1982.

[L1] Louisell, W.H. ; Quantum Statistical Properties of Radiation, John Wiley and Sons, New York, 1973.

[O1] Ohya M.; On Compound State and Mutual Information in Quantum Information Theory. IEEE Transactions of Information Theory IT-29, 1983, 770–774.

[O2] Ohya M. ; Note on Quantum Probability. Letter al Nuovo Cimento 38, 1983, 402-404.

[O3] Ohya M. ; Some Aspects of Quantum Information Theory and Their Applications to Irreversible Processes, Reports on Mathematical Physics, 27, 1989, 19–47.

[O4] Ohya M. and Petz D. ; Quantum Entropy and its Use, Springer-Verlag 1993.

[S1] Stinespring W.F. ; Positive functions on C^*-algebras, Proc. Amer. Math. Soc. 6, 1965, 211-216.

[W1] Watanabe N. ; Efficiency of optical modulations for photon number states, Quantum Probability and related Topics 6, 1991, 489-498.

HAMILTONIAN MODEL OF A TRANSPUTER QUANTUM AUTOMATON

V.A. Yatsenko

V.M. Glushkov Institute of Cybernetics, Kiev, Ukraine

Abstract

The paper represents the Hamiltonian models of the classical and quantum-mechanical transputer cellular automata. The observability, controllability and minimal realizability theorems are formulated for the cases in which such automata are taken into account. The cellular automaton based on quantum oscillators is presented as the example. The hypothesis, according to which the quantum filter proposed by V. Belavkin could be simulated by our cellular automaton, is discussed. A number of applications are described.

Keywords :

Cellular automata, Hamiltonian system, controllability, observability, minimal realization, differential geometry, reconfigurable cell.

1. We describe the cellular automaton (CA) in which every cell of which is considered as the controllable finite-dimensional Hamiltonian system having the form

$$\begin{aligned} \partial_t x^k &= \mathcal{F}_H^k(x^k, a^k, u^k), \quad x^k \in M^k, x^k(0) = x_0^k, \\ \partial_t y_j^k &= -H_j^k(x^k, a^k), \quad j = 1, \cdots, m \quad k = 1, \cdots, p \\ u^k &= (u_1^k, \cdots, u_m^k) \in \Omega \subset R^m, \end{aligned} \tag{1}$$

where $H^k(x^k, a^k, u^k)$ is an arbitrary analytical function of states x^k; of cell interaction parameters a^k and of reconfigurating influences u^k; k is a cell number; $a^k = F^k(x^l); l \neq k$. If there is the Hamiltonian

$$H^k(x^k, a^k, u^k) = H_0^k(x^k, a^k) - \sum_{j=1}^{m} u_j^k H_j^k(x^k, a^k),$$

then we have the system of equations

$$
\begin{aligned}
\dot{x}^k &= g^k_{H_0}(x^k, a^k) + \sum_{j=1}^{m} u^k_j g^k_{H_j}(x^k, a^k),\ x^k(0) = x^k_0,\ x^k \in (M^{2n}, \omega), \\
y^k_j &= -\frac{\partial H^k}{\partial u^k_j}(x^k, u^k) \quad j = 1, \cdots, m, \quad k = 1, \cdots, p, \\
u^k &= (u^k_1, \cdots, u^k_m) \in \Omega \subset R^n
\end{aligned} \tag{2}
$$

Here M is a simplectic manifold with a simpletic form ω and Hamiltonian vectorfields $g^k_{H_j}, j = 1, \cdots, m$; $\Omega \subset R^m$ is a control value domain that contains a point 0; $u^k_j \in \mathcal{U}^k_j$ are the control vector components that belong to specified classes $\mathcal{U}^k$ of admissible functions. Consider the control u^k as a generalized external macroscopic influence.

2. Assume that an elementary cell contained in some CA is described by Euler-Lagrange equation with external forces; a cell is a nondissipative system with n freedom degrees $q^k_1, \cdots, q^k_m$ and a Lagrangian $L^k(q^k_1, \cdots, q^k_m, \dot{q}^k_1, \cdots, \dot{q}^k_m)$. Then, the CA dynamics is defined by the equation

$$
\frac{d}{dt}\left(\frac{\partial L^k}{\partial \dot{q}^k}\right) - \frac{\partial L^k}{\partial q^k} = F^k_i, \quad i = 1, \cdots, n \tag{3}
$$

where $F^k = (F^k_1, \cdots, F^k_n)$ is a vector of generalized external forces. Let some of the vector components be zero and suppose that the rest of them provide the reconfiguration properties possessed by the CA, and here the cell Lagrangian depends explicitly on the vector parameter the current value of which is determined by the neighbouring cell state. Then the representation

$$
\begin{aligned}
\dot{q}^k_k &= \frac{\partial H^k_0}{\partial p^k_i}, \quad (i = 1, \cdots, n), \\
\dot{p}^k_i &= -\frac{\partial H^k_0}{\partial q^k_i} + u^k_i, \quad (j = 1, \cdots, m), \\
\dot{p}^k_i &= \frac{\partial H^k_0}{\partial q^k_i}, \quad (i = 1, \cdots, n),
\end{aligned} \tag{4}
$$

where $p^k_i = \frac{\partial L^k}{\partial \dot{q}^k}$ and H^k_0 is an internal Hamiltonian for the k-th cell, is true. We have the system

$$
\dot{x}^k = g^k_{H_0}(x^k) + \sum_{j=1}^{m} u^k_j g^k_{H_j}(x^k),\ x^k(0) = x^k_0,\ x^k \in M^k, \tag{5}
$$

at the arbitrary coordinates x^k for a smooth manifold M^k ; M^k is a simplectic manifold with a simplectic form ω^k; $g^k_H(x^k)$ and $(j = 0, 1, \cdots, m)$ are Hamiltonian vectorfields meeting the relation $\omega^k(\mathcal{F}_{H_j} \cdot) = -dH^k_j$. According to Darboux theorem, there exist such canonical coordinates (p^k, q^k) that

$$
\omega^k = \sum_{i=1}^{m} dp^k_i \wedge dq^k_i. \tag{6}
$$

from the point of view of locality. If it is assumed that an interaction Hamiltonian H^k_j is equal to $q_j (j = 1, \cdots, m)$, then system of equations (4) is derived. A CA cell 'state-output' map may be represented by the expression

$$
y^k_j = H^k_j(x^k) \quad (j = 1, \cdots, m), \tag{7}
$$

i.e., an output signal is specified by a disturbance $u_1^k F_H^k, \cdots, u_m^k F_H^k$ for a natural input, and this circumstance, in its turn, is the reason why the energy $H_1^k, \cdots, H_m^k$ values are changed. Introduce the following definitions.

Definition 1. A controlled CA model

$$\begin{aligned} &\partial_t x^k = G^k(x^k, u^k) \quad x^k(0) = x_0^k \quad x \in M^k \\ &y_j^k = h_j^k(x^k, u^k) \quad j = 1, \cdots, r \quad u^k = (u_1^k, \cdots, u_m^k) \in \Omega \subset \mathbf{R}^n \end{aligned} \tag{8}$$

is called the Hamiltonian model if it admits the representation in the form of the system

$$\partial_t x^k = \mathcal{F}_{H^k}^k(x^k, u^k), \quad x^k \in M^k,\ x^k(0) = x_0^k, \tag{9}$$

where $H^k(x^k, u^k)$ is the arbitrary analytical function of the k-th cell state x^k and of the control parameters u^k. Let $H^k(x^k, u^k)$ be of the form

$$H_0^k(x^k) - \sum_{j=1}^{m} u_j^k H_j^k(x^k); \tag{10}$$

then system (2) is obtained. CA cell outputs are specified here by the expression

$$y_j^k = -\frac{\partial H^k}{\partial u_j^k}(x^k, u^k) \quad (j = 1, \cdots, m) \tag{11}$$

Since the state space M^k for system (5) is simplectic, then it is possible to yield Poisson bracket

$$\{F^k, G^k\} = \sum_{j=1}^{n}\left(\frac{\partial F^k}{\partial p_i^k}\frac{\partial G^k}{\partial q_i^k} - \frac{\partial F^k}{\partial q_i^k}\frac{\partial G^k}{\partial p_i^k}\right), \quad F^k, G^k : M^k \to \mathbf{R} \tag{12}$$

Definition 2. Let L^k is Lie algebra for Hamiltonian vectorfields of some k-th cell. The linear span for the functions $f^r(H_j^k)$, where $f^k \in L^k$, is termed k-cell observation space $\mathcal{H}^k$.

Since $X_F^k(G^k) = \{F^k, G^k\}$ and $[g_F^k, g_G^k] = g_{\{F,G\}}^k$, then $\mathcal{H}$ is defined by the functions

$$\{F_1^k, \{F_2^k, \{\cdots\{F_k^k, H_k^k\}\cdots\}\}\}, \tag{13}$$

for CA (5,7) and here F_r^k, $r = 1, \cdots, k$ is equal to $H_i^k, i = 0, 1, \cdots, m$.

If the theorems considered for the case with nonlinear system controllability and observability [1] are applied for a CA, then the following results are obtained.

Proposition 1. $\mathcal{H}^k$ is an ideal formed by $H_1^k, \cdots, H_m^k$ in Lie algebra (under Poisson bracket) generated, in its turn, by $H_0^k, \cdots, H_m^k$.

Proposition 2. Let (5,7) be a Hamiltonian CA model. Then:

a) The CA states are strongly accessible and the CA is weakly observable if $dim\, d\mathcal{H}^k(x) = dim\, M^k, \forall x^k \in M^k$ or: the CA is quasiminimal.

b) The CA is strongly accessible and observable if $dim\, d\mathcal{H}^k(x^k) = dim\, M^k, \forall x^k \in M^k$ and $\mathcal{H}^k$ allows to distinguish M^k points, or: the CA is minimal.

A nonminimal CA can be reduced to a quasiminimal type under the same 'input-output' map. If a CA is Hamiltonian, then the Hamiltonian system is yielded again as the result of the transformation procedure.

It was mentioned earlier that the quasiminimal CAs with one and the same 'input-output' map are locally diffeomorphic. In the Hamiltonian case, map equivalence means the smooth simplectomorphism, i.e.: there are the local canonical transformation and the energy equivalence which are accurate to the constants (for both CAs). Therefore, the 'input-output' map determines not only the CA state, but also the canonical CA state structure and the internal energy. This approach can be extended to general-type CAs (2) and here the observation space $\mathcal{H}$ is used, i.e., $\mathcal{H}$ is the linear space that exists for the functions (x, u) and includes $\partial H_1/\partial u, \cdots, \partial H_m/\partial u$ and invariants. Calculations of Poisson brackets of Hamiltonians, Hamiltonian derivatives, etc. forms Lie algebra.

When the well-known accessibility and observability theorems mentioned in [2] are considered for the CAs cases, they are formulated as follows.

Theorem 1 *Let (5,7) be strongly accessible and observable by a CA. Then a CA is minimal if and only if vector field Lie algebra is self-adjoint.*

Theorem 2 *A strongly accessible and observable CA is self-adjoint (and, therefore, Hamiltonian), if cell output variations $\delta_1 y^k$, $\delta_2 y^k$ are determined on a compact $[0,T]$ and the equality*

$$\int^{\infty} [\delta_1^T u^k(t)\delta_2 y^k(t) \; - \; \delta_2^T u^k(t)\delta_1 y^k(t)] = 0.$$

is true for an arbitrary piecewise constant u_j^k and for any two cell input variations $\delta_1 u^k$, $\delta_2 y^k$.

Besides controllability and observability duality, the Hamiltonian CAs possess a number of other useful features. In particular, the Hamiltonian CA theory is sometimes essential for the quantum models of adaptive computational medium.

3. We proceed from the conventional H-Hamiltonian quantization scheme according to which generalized coordinates $q_1, \cdots, q_n$ and momenta $p_1, \cdots, p_n$ are quantized first of all and then H is expressed with respect to these quantum values. Pursuant to the concepts stated in [3,4], when a function set on M is quantized, such quantization means that Hermitian operator f that acts on a complex Hilbert space is brought into correspondence with each function contained in this set. Besides this, such operators must satisfy some commutative relations which, in their turn, correspond to Poisson bracket of the classical functions. Assume $M = R^{2n}$, and let $q_1, \cdots, q_n$ and $p_1, \cdots, p_n$ be the variables that must be quantized, and then the following commutation relations are true:

$$[\hat{q}_i, \hat{q}_j] = 0, \quad [\hat{p}_i, \hat{p}_j] = 0, \quad [\hat{q}_i, \hat{p}_j] = i\hbar\delta_{ij}\hat{I} \tag{14}$$

Let us reduce the value $\hbar$ to 1 in order to simplify our further discussion. Since a skew-Hermitian operator $-i\hat{f}$ corresponds to every Hermitian operator $\hat{f}$, then relations (14) mean that the map $\hat{f} \to -i\hat{f}$ is the morphism of Lie algebra, which contains Poisson bracket and which is built up on the basis of the classical functions $q_1, \cdots, q_n \;\; p_1, \cdots, p_n, 1$ into the skew-Hermitian operator Lie subalgebra. Here the Poisson bracket [f,g] is mapped into the bracket $-i[\hat{f}, \hat{g}] = i[-i\hat{f}, -i\hat{g}]$.

The traditional scheme according to which the values $q_1, \cdots, q_n$ and $p_1, \cdots, p_n$ are quantized and which satisfies expressions (14) means that the Hilbert space $L^2(R^n, C)$ is introduced, that the an operator $\hat{q}_j$ is assigned to a coordinate q_j (multiplication by q_j) and that the operator $\hat{p}_j = i\frac{\partial}{\partial q_j}$ is assigned to p. The same scheme may also be applied for some simple Hamiltonian CAs.

Example. Consider the harmonic oscillator possessing an external force u^k which, in its turn, may depend on quantum-mechanical observables [5]. The Hamiltonian for such system is specified by the expression

$$H^k(q^k, p^k, u^k) = (p^k)^2/2m + 1/2a^k(q^\ell)(q^k)^2 - u^k(t)q^k \tag{15}$$

The quantization results in the relation

$$i\hbar\frac{\partial\Psi^k}{\partial t} = -\frac{1}{2m}\frac{\partial^2\Psi^k}{\partial(q^k)^2} + \left[\frac{1}{2}v^k(t)a^k(q^\ell)(q^k)^2 - u^k(t)q^k\right]\Psi^k \tag{16}$$

$$(\Psi^k \in L^2(\mathbf{R}, \mathbf{C}).$$

which describes, in particular, the dynamics of a particle contained in a single well (bipotential) and this particle is affected, in its turn, by a homogeneous classical external field, and here a field value and a field direction are the arbitrary time functions. Generally, when the Hamiltonians $H_0, \cdots, H_m$ present in system (2) are transformed to the Hermitian operators present in some complex Hilbert space $\mathcal{H}$, then the quantized system looks like

$$i\frac{\partial\Psi^k}{\partial t} = \left(\hat{H}_0^k - \sum_{j=1}^{m} u_j^k \hat{H}_j^k\right)\Psi^k, \quad \Psi^k \in \mathcal{H} \tag{17}$$

(Schrödinger representation). Here $\hat{H}_0^k$ is the Hamiltonian for an isolated quantum system. The second term contained in this expression, represents the interaction with the external sources through the system of the interaction Hamiltonians $\hat{H}_1, \cdots, \hat{H}_m$.

Note that if there is a complex Hilbert space $\mathcal{H}$, then the imaginary part of the inner Hermitian product $< >$ defines a simplectic form on $\mathcal{H}$. With respect to this simplectic form, the skew-adjoint operators $-i\hat{H}k_j$ are the linear Hamiltonian vectorfields on $\mathcal{H}$. Such fields are the expected observables H_j^k, i.e., they correspond to the quadratic Hamiltonians

$$< \hat{H}_j^k\Psi^k, \Psi^k > = < \Psi^k \mid \hat{H}_j^k \mid \Psi^k >, \quad j = 0, \cdots, m. \tag{18}$$

That is why expression (16) provides the Hamiltonian system

$$\frac{\partial\Psi^k}{\partial t} = -i\hat{H}_0^k\Psi^k + \sum_{j=1}^{m} u_j(-i\hat{H}_j^k)\Psi^k \tag{19}$$

$$y_j^k = < \Psi^k \mid \hat{H}_j^k \mid \Psi^k >, \quad j = 1, \cdots, m \tag{20}$$

with the macroscopic controls $u_j^k, \cdots, u_m^k$ and outputs $y_1^l, \cdots, y_m^k$. The outputs on the infinite-dimensional state space $\mathcal{H}$ are equal to the expected observable $\hat{H}_j^k$ [3]. Note that the essence of the fact that y_j^k is equal to the expected values of $\hat{H}_j^k$ does not arise from the measurement problem. In addition, although $\hat{H}_j^k$ measurement process introduces disturbances into a system, these disturbances propagate along the channels that correspond, in their turn, to the inputs $\hat{u}_j^k$ .Since H^k is the simplectic space, then Poisson bracket for two observable values, namely, $\hat{H}_1^k$ and $\hat{H}_2^k$, is calculated in the following way

$$< \hat{\Psi}^k \mid \hat{H}_1^k \mid \Psi^k >, \quad < \Psi^k \mid \hat{H}_2^k \mid \Psi^k >= - < \Psi^k \mid [\hat{H}_1^k, \hat{H}_2^k] \mid \Psi^k > . \tag{21}$$

"-"arises if the commutators existing for $-i\hat{H}_1^k$ and $-i\hat{H}_2^k$ are taken into account. Therefore, the observation space for a quantum-mechanical CA is specified by the ideal formed generated by the expected values of $\hat{H}_1^k, \cdots, \hat{H}_m^k$ within the Lie algebra generated by the expected values of $\hat{H}_0^k, \cdots, \hat{H}_m^k$ under Poisson bracket. Let us compare this observation space with the observation space existing for system (2).

If the quadratic-linear Hamiltonian are present, then according to Erenfest theorem, these two observation spaces are equal (in the sense that there are the averaged values which satisfy the classical equations). Therefore, the transputer quantum-mechanical CA can never be minimal. Finally, note that CA quantization may be started not with the variables $q_1^k, \cdots, q_n^k$ and $p_1^k, \cdots p_n^k$, but with the observation space existing for system (9,11). If there are the CAs based on harmonic oscillators, the essense of the matter is not changed because the observation space is determined by the linear span of q^k, p^k and 1. When it is necessary to quantize the observation spaces existing for the CAs of different types, then such approach is of interest.

The final remark. It is really possible to quantize the externally and nonpotentially reconfigurable Hamiltonian CAs. But the present paper states that it is possible to quantize the "completed" transputer-type CAs.

References

[1] A.J. Van der Shaft, Controllability and observability for affine nonlinear Hamiltonian systems, IEEE Trans. Aut. Cont., AC -27, pp. 490-492, (1982).

[2] P.E. Crouch, A.J.Van der Shaft, Variational characterization of Hamiltonian systems, IMA J.Math. Cont & Inf., 3, pp.123-130,(1985).

[3] P.A.M. Dirac, *The principles of quantum mechanics*, Oxford: Clarendon Press, (1958).

[4] A.G. Butkovsky, Yu.I. Samoilenko, *Control of quantum-mechanical processes and systems*, Kluwer Academic Publishers, Dordrecht, (1990).

[5] T.J. Tarn, G. Huang, J.W. Clark, Modelling of quantum mechanical control systems, Math. Modelling, 1, pp.109-121,(1990).

CIRCUITRY PROBLEMS FOR ONE QUANTUM MULTIPORT

Sergey A. Smirnov

Glushkov Cybernetics Institute, Kiev
e–mail: smr@d310.icyb.kiev.ua

Abstract

The main purpose of quantum circuitry is to develop effective apparatus for solving the problem of designing quantum networks from quantum devices. Inner circuitry studies operations with only one device. The basic operations of overlapping, joining and earthing of quantum multiport channels are considered, and corresponding reduction rules for recalculation of the scattering matrix are obtained. Dissipative quantum multiports are discussed. We also present the selection rules for a discrete spectrum of a 0-port (quantum accumulator) which is produced from some multiport by the above basic operations.

Keywords :
Coherent circuitry, quantum devices, optical waveguides and resonators, nanoelectronics, scattering matrix.

1 INTRODUCTION

The study of quantum interference devices [1] is motivated by more than scientific interest. The fabrication of nanostructures relies on sophisticated crystal growth and lithographic techniques that develop by means of industrial efforts toward miniaturization . Conventional procedures for determining model parameters and scheme designs are based on an interpretation of classical electrical measurements and give acceptable results for transistors with dimensions greater than about one micrometer. Such procedures may not give reliable results for smaller devices. When dimensions are less than about a micrometer, one also needs device physics and circuitry, based on first principles, to understand problems that arise in making reliable devices and to develop strategies to overcome design limits. Achievement of these goals requires a compromise between the sophistication of solid-state physics and the pragmatic demands of electrical engineering.

On the other hand, progress in integral optics and especially in the quantum theory for optical communications [2] gives rise to problems of circuit designing as the next step after successful investigations of individual devices. For the optical situation it is neccessary to take into consideration all its own peculiarities, but the most important from a physical point of view are the coherent properties of an electromagnetic field. Coherent aspects play a key role in the nanoelectronic situation too. Every variant of the above pragmatic compromise must take them into account.

The notion of a quantum multiport was proposed as a possible approach to the realization of this compromise [3]. The main mathematical characteristic of such a multiport is the scattering matrix. Every element of this matrix is the amplitude of probability for electron transfer from some quantum channel (waveguide) to another across the multiport active zone (resonator). This terminology doubling is based on mathematical isomorphy of two different physical situations. Quantum transport in nanoelectronic structures is considered here as a stationary scattering phenomenon. The stationary Schrödinger equation and Helmholtz equation for an electromagnetic field in an open resonator are equivalent (but dispersion dependences are different). Most of the proposed results were obtained by permissible manipulations with their solutions and may be applied in both situations. That is why the circuitry developed here is not only quantum or nanoelectronic, but coherent first of all.

The scattering matrix of a usual quantum multiport is unitary and symmetric. The second property is the result of the reciprocity theorem, and it was established, in fact, by Büttiker in his remarkable paper [4]. It is necessary to note that Büttiker's argument is sufficient for the optical situation of scattering in an open resonator with linear polarization media. The case of nonlinear polarization is being investigated by the author, and will be described in a special article. Henceforth we are concerned only with the linear situation. Examples, which were studied in [4], are an additional motivation for our definition of a quantum multiport. The composition of quantum multiports generates the problem of a scattering matrix calculation for a composite multiport. It was discussed at the "NANO-II" conference [5], and now we present its solution with minimal comment.

Assume that S_1 and S_2 are scattering matrixes of quantum $(n+m)$- and $(m+k)$-ports correspondingly, and we join their m channels. We use S-matrixes in block form:

$$S_1 = \begin{bmatrix} P_1 & Q_1 \\ Q_1^T & R_1 \end{bmatrix}, \qquad S_2 = \begin{bmatrix} P_2 & Q_2 \\ Q_2^T & R_2 \end{bmatrix}.$$

The scattering matrix of the resulting quantum $(n+k)$-port is:

$$S = \begin{bmatrix} P_1 + Q_1 P_2 (I - R_1 P_2)^{-1} Q_1^T & Q_1 (I + (I - R_1 P_2)^{-1} R_1) Q_2 \\ Q_2^T (I - R_1 P_2)^{-1} Q_1^T & R_2 + (I - R_1 P_2)^{-1} R_1 Q_2 \end{bmatrix} \tag{1}$$

where I is the $m \times m$ unit matrix.

Availability of this composition rule allows us to formulate and solve the synthesis problems for optical or nanoelectronic circuits. It is neccessary to note that calculations by expression (1) require only the usual operations of simple matrix algebra. Thus, we may consider S-matrix formalism as the foundation of engineering level coherent circuitry. But the composition of a multiport is not a unique circuitry operation, and we start the investigation of other ones.

2 THE OVERLAP OF A QUANTUM CHANNEL

This operation transforms every n-port to $(n-1)$-port. We shall assume that

$$S = \begin{bmatrix} P & q \\ q^T & r \end{bmatrix} \tag{2}$$

is the scattering matrix, of our initial quantum multiport, where r is the reflection amplitude of some channel, and column q consists of amplitudes of transitions to this channel from others. Matrixes S and P are symmetrical in accordance with reciprocity relations. We must derive the expression for scattering matrix $\tilde{S}$ of an $(n-1)$-port appearing after the overlap of this channel.

Our approach to the solution of this problem is based on the following reasoning. Let us consider the eigenfunction of Hamiltonian of the Schrödinger equation for an $(n-1)$-port. Wave function values on the overlap line (or point) are equal to zero. Its values are equal to zero everywhere in this channel. But the last property is not necessary for the solution of the boundary value problem, and we shall attempt to find the input amplitude in this channel (when it is open), that totally reflects from the active zone, and we have a standing wave in it. Thus, we conclude that input and output amplitudes are different only by sign.

Let (2) be the S-matrix of an initial n-port, c and c' be $(n-1)$ vectors of input and output amplitudes in open channels, and c_n be the input amplitude in the overlapping channel. Our above reasoning allows us to write the matrix equation

$$\begin{bmatrix} P & q \\ q^T & r \end{bmatrix} \begin{bmatrix} c \\ c_n \end{bmatrix} = \begin{bmatrix} c_1 \\ -c_n \end{bmatrix},$$

and to solve it:

$$\begin{aligned} c_n &= (1+r)^{-1}(-q^T c), \\ c' &= Pc - q\frac{1}{1+r}q^T c = \tilde{S}c. \end{aligned}$$

We obtain the scattering matrix of the resulting $(n-1)$-port:

$$\tilde{S} = P - q^T(1+r)^{-1}q.$$

The structure of this expression is conserved in a general situation. If we overlap m channels of some quantum multiport with scattering matrix

$$S = \begin{bmatrix} P & Q \\ Q^T & R \end{bmatrix}, \tag{3}$$

in block form where the $m \times m$ block R corresponds with these channels, then the procedure described above gives the following result:

$$\tilde{S} = P - Q^T(I+R)^{-1}Q. \tag{4}$$

I here is the $m \times m$ matrix unit.

3 THE "EARTHING" OPERATION

Earlier we considered only quantum multiports, which are described by unitary symmetrical scattering matrixes. From a physical point of view they correspond to the case of elastic scattering. But our nondissipative multiports are obviously deficient for circuitry requirements. The main part of its solution is based on the use of irreversible elements. Now we consider the simplest method of defining a quantum dissipative multiport. It is important that we obtain it by using a regular procedure within the scope of inner circuitry.

Consider the quantum n-port described by scattering matrix (2). Assume that we joined the last channel with an absolutely absorbing reservoir. Thus we have a quantum $(n-1)$-port with "dissipative" scattering matrix $\tilde{S} = P$. Its characteristic property is

$$\| P\Psi \| \leq \| \Psi \|,$$

where $\| \cdots \|$ is the norm in Hermit space of $(n-1)$ dimension, and $\det \tilde{S} \leq 1$. It is necessary to note, that reciprocity relations are concerned:

$$P^T = P,$$

and formula (1) is as true for dissipative initial multiports as for nondissipative.

Another way of looking at such multiports is more formal. Our "inelastic" channel is not required to be spatial, and may be considered as a physical one, in accordance with quantum-mechanical inelastic scattering theory [6].

4 THE JOINING OF TWO QUANTUM CHANNELS

The result of this inner circuitry operation is an $(n-2)$-port. We assume expression (3) for an S-matrix of an initial n-port, but with R now a 2×2 matrix. The joining of the last two channels allows us to consider the output amplitude of one as the input amplitude of another and vice versa. On the basis of this geometrical argument we can write the following equation:

$$\begin{bmatrix} P & Q \\ Q^T & R \end{bmatrix} \begin{bmatrix} c \\ c_{n-1} \\ c_n \end{bmatrix} = \begin{bmatrix} c' \\ c_n \\ c_{n-1} \end{bmatrix} = \begin{bmatrix} c' \\ Ja \end{bmatrix} = S \begin{bmatrix} c \\ a \end{bmatrix},$$

where

$$J = \begin{bmatrix} 0 & 1 \\ 1 & 0 \end{bmatrix}.$$

Let us solve these equations:

$$\begin{aligned} c' &= Pc + Qa \\ Ja &= Q^T c + Ra \\ a &= (J-R)^{-1} Q^T c \\ c' &= Pc + Q(J-R)^{-1} Qc, \end{aligned}$$

and we have:

$$\tilde{S} = P + Q(J-R)^{-1} Q. \tag{5}$$

In the case when an m pair of channels joins this formula is also correct, but with a corresponding ordering of channels, and $2m \times 2m$ matrixes R and J.

5 SPECTRUM OF ACTIVE ZONE, 0-PORTS

We have described all the basic operations of quantum multiport inner circuitry. But there is one case when the formulas obtained are insufficient. All proposed methods are based on the assumption that the resulting multiports have one or more channels .

Now we consider a situation when all channels of an n-port are overlapped. The simplest case of a two-port was investigated in [2]. The result of this operation is an active zone of our n-port, or closed resonator. Its spectral property depends on an S-matrix of initial quantum multiport. We carry out the calculation in the same way as in section 2. Let c be a vector whose components are input amplitudes of our multiport, and c' consists of output ones. For every energy value E we have $c' = Sc$.

Consider the case when $Sc = -c$, when the scattering is turned over for some input vector c. In accordance with the above discussion we can treat this wave function as the solution of an eigenvalue problem with zero boundary conditions. Thus we obtain the selection rule for a desired spectrum: it contains all values of energy, that scattering matrix $S(E)$ has -1 as an eigenvalue. It is usual to remember, that eigenvalues of every unitary matrix are unit modulo. Here is an expression of the selection rule convenient for calculations:

$$\det(S(E) + I) = 0. \tag{6}$$

Another opportunity to produce a 0-port is connected with the joining of a pair of channels of quantum $2n$-port. Assuming that we join channels in accordance with the rule $(i \leftrightarrow n + i)$, where $1 \leq i \leq n$, we may write the scattering matrix in the form:

$$S = \begin{bmatrix} S_1 & Q \\ Q^T & S_2 \end{bmatrix}$$

where Q is an $n \times n$ matrix.

$$S \begin{bmatrix} c_1 \\ c_2 \end{bmatrix} = \begin{bmatrix} c_2 \\ c_1 \end{bmatrix} = \begin{bmatrix} 0 & E \\ E & 0 \end{bmatrix} \begin{bmatrix} c_1 \\ c_2 \end{bmatrix} = J \begin{bmatrix} c_1 \\ c_2 \end{bmatrix}$$

Some energy value E is an eigenvalue for our resonator, when $S(E)$ "turns over" (inverts) the states in connecting channels for some vectors c_1 and c_2 . We obtain the selection rule:

$$\det(JS(E) - I) = 0. \tag{7}$$

It is useful to note that the selection rules in the two situations considered are expressions of one geometrical condition: the unit eigenvalue of some matrix. In the case of overlapped channels this matrix is $-S(E)$, and in the case of joined channels it is $JS(E)$. We omit the simple checking of this consequence.

And now we can consider the general case. Assume, that we transform some quantum n-port into a 0-port by operations of joining and overlapping. Our selection rule (7) conserves, but matrix J corresponds to the "geometry" of closing operations. If we overlap channel number i, then we write $(J)_{ii} = -1$. If we join k and m channels, we write $(J)_{km} = (J)_{mk} = 1$. Other matrix elements of J are equal to zero. So, we have described the method of construction of matrix J, and we see, that (7) is true for every case.

CONCLUSION

We have presented a general consideration of quantum circuitry inner problems. There are only two fundamental inner circuitry operations: overlapping and joining, and they have been completely explored. It is clear that more complex multiport transformations may be represented as the sequence of the simplest ones. The third operation considered has some peculiarity. The earthing allows us to extend the notion of quantum multiport on dissipative nanodevices. Their practical importance is determined by the irreversibility of physical transformations, used for information processing. Availability of dissipative multiports in a "quantum hardware development kit" allows us the hope of using many solutions from classical logical design in principally new situations.

Quantum 0-port may be considered as the result of execution for all possible basic operations on some quantum multiport. Formulas of scattering matrix recalculations give (4) and (5). The result is not dependent on the sequence of these operations, but determines the family of possible 0-ports. It is easy to estimate the number of these family members. If the initial multiport has n channels, there are no more than $n!$ different 0-ports, corresponding to it. The main purpose of quantum 0-port is to represent accumulator elements in theory. Selection rules (6) and (7), obtained in this paper, allow us to calculate the spectrum of such elements.

From a mathematical point of view our results were produced from investigation of eigenvalue problem solutions geometry for the Sturm–Liouville operator.

The proposed reduction rules (4), (5), selection rule (7), jointly with the composition law (1) are the complete set of operations for quantum scheme design. They may be considered as an effective basis of coherent circuitry calculating apparatus, suitable for engineering requirements.

References

[1] F. Capasso (Ed.) *Physics of Quantum Electron Devices*: Springer Series in Electronics: v.28, Berlin: 1990.

[2] C. Bendjaballah, O. Hirota, S. Reynaud (Eds.) *Quantum Aspects of Optical Communications*: LNT 378, Springer, Berlin: 1991.

[3] S.A. Smirnov, *"Theory of quantum two-ports"*, Avtomatika, 1993, N.1, p.83-90, (in Russian).

[4] M. Büttiker, *"Symmetry of Electrical Conduction"*, IBM J. Res. Develop., v.32, p.317-334 (1988).

[5] S.A. Smirnov, *"Composition problems of quantum multiports"*, NANO-II, Second International Conference in Nanometer Scale Science and Technology, Moscow 1993, Abstracts, p.57.

[6] L.D. Landay and E.M. Lifshitz, *Quantum Mechanics (Non-Relativistic Theory)*: Pergamon Press, Oxford: 1977.

A MATHEMATICAL STUDY OF INFORMATION TRANSMISSION IN QUANTUM COMMUNICATION PROCESSES

Masanori Ohya and Noboru Watanabe

Department of Information Sciences
Science University of Tokyo
Noda City, Chiba 278, Japan

Abstract

In quantum communication theory, an input signal is represented by the quantum state. The input state changes under the influence of noise and loss associated with a channel. The attenuation process is a model of quantum channel describing an optical communication process. When an input state changes to an output state through a channel, the amount of information carried from the input state to the output state is represented by the quantum mutual entropy (information). The quantum communication theory has been studied by various researchers [4-7,11,13,15,16,18].

In §1, we briefly review a quantum channel for attenuation processes, and we explain a quantum entropy and a quantum mutual entropy in quantum communication systems in §2. By using the general framework defined in [16], we discuss the convergence (speed) of the mutual entropy and mean mutual entropy for attenuation process in §3.

1 QUANTUM COMMUNICATION CHANNEL

Let $(B(\mathcal{H}_1), \Sigma(\mathcal{H}_1))$, $(B(\mathcal{H}_2), \Sigma(\mathcal{H}_2))$ be the input and output systems, respectively, where $B(\mathcal{H}_k)$ is the set of all bounded linear operators on a separable Hilbert space $\mathcal{H}$ and $\Sigma(\mathcal{H})$ is the set of all density operators on $\mathcal{H}$. A channel in the quantum communication theory is defined [1,10,11,13] by a mapping Λ^* from $\Sigma(\mathcal{H}_1)$ to $\Sigma(\mathcal{H}_2)$ satisfying that the dual map Λ from $B(\mathcal{H}_2)$ to $B(\mathcal{H}_1)$ is a completely positive unital map [8,15].

Quantum Communications and Measurement
Edited by V. P. Belavkin *et al.*, Plenum Press, New York, 1995

Let $\mathcal{K}_1$ and $\mathcal{K}_2$ be two Hilbert spaces representing the influence of noise and loss, respectively and let $\xi \in \Sigma(\mathcal{K}_1)$ be a state of the noise system. The optical channel for an attenuation process is defined by

$$\Lambda^*(\rho) = \mathrm{tr}_{\mathcal{K}_2} V(\rho \otimes \xi) V^*, \tag{1.1}$$

where

$$V(x_n^{(1)} \otimes y_0^{(1)}) = \sum_{j=0}^{n} C_j^n x_j^{(2)} \otimes y_{n-j}^{(2)}, \quad C_j^n = \sqrt{\frac{n!}{(n-j)!j!} \eta^j (1-\eta)^{n-j}} \tag{1.2}$$

and η is the transmission ratio for the channel. This channel has be used to study the optical communication process, for instance, to compare with some modulations [12,18,19] and to calculate the error probability [13,14].

2 QUANTUM ENTROPY AND QUANTUM MUTUAL ENTROPY

For an input state of the density operator $\rho \in \Sigma(\mathcal{H}_1)$, the von Neumann entropy is defined by [9,15]

$$S(\rho) = -\mathrm{tr}\rho \log \rho, \tag{2.1}$$

which implies the amount of information having the input state ρ.

Ohya formulated the quantum mutual entropy in [11] with respect to an input state and a channel using the compound states σ_E and σ_0 such as

$$\sigma_E = \sum_n \lambda_n E_n \otimes \Lambda^* E_n, \quad \sigma_0 = \rho \otimes \Lambda^* \rho. \tag{2.2}$$

The relation between these compound states and the liftings in quantum Markov chain is studied in [1,2,3]. The mutual entropy with respect to an input state ρ and a communication channel Λ^* is defined in [11] as

$$I(\rho; \Lambda^*) \equiv \sup\{\mathrm{tr}\sigma_E(\log \sigma_E - \log \sigma_0); E = \{E_n\}\}, \tag{2.3}$$

where $E = \{E_n\}$ is the Schatten decomposition [15] of ρ.

3 CONVERGENCE OF QUANTUM MUTUAL ENTROPY AND QUANTUM MEAN MUTUAL ENTROPY FOR CHANNELING TRANSFORMATION

In this section, we consider the convergence of quantum mutual entropy and quantum mean mutual entropy for an attenuation process based on [16].

3.1 Convergence of quantum mutual entropy for an attenuation process

In optical communication processes, a optical fiber is widely used. In order to calculate the quantum mutual entropy for optical fiber channel, we must know the construction of optical fiber. The optical fiber has the character of uniformly structure, so that it is sufficient to investigate the optical fiber with a fixed length L for calculating the quantum mutual entropy with respect to the optical fiber channel with $N \times L$ length. Let Λ^* be the quantum channel with a fixed length L.

The quantum channel $(\Lambda^*)^N$ with $N \times L$ length is obtained by N step's channeling transformations of Λ^*. Moreover, the optical fiber has the character of low noise, therefore we assume that the optical fiber channel Λ^* is the attenuation channel with the vacuum noise.

Let $\mathcal{H}_0$ be a Hilbert space of an input system and $\mathcal{H}_N$ $(N = 1, 2, \cdots)$ be a Hilbert space of an output system after N step's channeling transformations, respectively. The mutual entropy $\mathrm{I}(\rho^{(0)}; (\Lambda^*)^N)$ with respect to an input state $\rho^{(0)}$ and a channel $(\Lambda^*)^N$ means the amount of information correctly transmitted from the input state to the output state $(\Lambda^*)^N \rho^{(0)}$ after N step's channeling translations. For simplicity, we put $\mathcal{H}_0 = \mathcal{H}_N = \mathcal{H}$ $(N = 1, 2, \cdots)$ and use the attenuation channel Λ^*. After N step's channeling transformations, the input state $\rho^{(0)} \in \Sigma(\mathcal{H}_0)$ changes to the output state of the N-th output system, which is expressed as $(\Lambda^*)^N \rho^{(0)} : \Sigma(\mathcal{H}_0) \to \Sigma(\mathcal{H}_N)$. For any input state $\rho^{(0)} \in \Sigma(\mathcal{H}_0)$, two compound states σ_E^N and σ_0^N are given by

$$\begin{aligned}
\sigma_E^N &= \sum_{n_N=0} \sum_{n_0=n_N} \lambda_{n_0}^{(0)} \mu_{n_N,n_0}^{(N)} E_{n_0}^{(0)} \otimes E_{n_N}^{(N)}, \\
\sigma_0^N &= \sum_{n_0=0} \sum_{n'_N=0} \sum_{n'_0=n'_N} \lambda_{n_0}^{(0)} \lambda_{n'_0}^{(0)} \mu_{n'_N,n'_0}^{(N)} E_{n_0}^{(0)} \otimes E_{n'_N}^{(N)},
\end{aligned}$$

where $\lambda_{n_N}^{(N)} = \sum_{n_{N-1}=n_N} \cdots \sum_{n_0=n_1} \lambda_{n_0}^{(0)} |C_{n_1}^{n_0}|^2 \cdots |C_{n_N}^{n_{N-1}}|^2$, $E_{n_i}^{(i)} = |x_{n_i}^{(i)} >< x_{n_i}^{(i)}|$, $\{x_{n_i}^{(i)}\}$ is the CONS of $\mathcal{H}_i$ and $\mu_{n_N,n_0}^N = \sum_{n_{N-1}=n_N}^{n_0} \cdots \sum_{n_1=n_2}^{n_0} |C_{n_1}^{n_0}|^2 \cdots |C_{n_N}^{n_{N-1}}|^2$. Then we have the quantum mutual entropy as follows:

$$\mathrm{I}(\rho^{(0)}; (\Lambda^*)^N) = \sup\{\sum_{n_0=0} \lambda_{n_0}^{(0)} \mathrm{S}((\Lambda^*)^N(E_{n_0}^{(0)}); (\Lambda^*)^N \rho^{(0)}); E^{(0)} = \{E_{n_0}^{(0)}\}\} \tag{3.1}$$

Lemma 3.1 *For any input state $\rho^{(0)}$ and any channel Λ^*, if $N \leq M$ then*

$$I(\rho^{(0)}; (\Lambda^*)^N) \geq I(\rho^{(0)}; (\Lambda^*)^M). \tag{3.2}$$

For any input state $\rho^{(0)} \in \Sigma(\mathcal{H}_0)$ and the attenuation channel $(\Lambda^*)^N$, if the Schatten decomposition of the input state $\rho^{(0)}$ is represented as $\rho^{(0)} = \sum_{n_0=0} \lambda_{n_0}^{(0)} E_{n_0}^{(0)}$ then the above relative entropy $\mathrm{S}(\sigma_E^N; \sigma_0^N)$ is expressed by

$$\mathrm{S}(\sigma_E^N; \sigma_0^N) = \sum_{n_N=0} \sum_{n_0=n_N} \lambda_{n_0}^{(0)} \mu_{n_N,n_0}^{(N)} \log\Big(\frac{\mu_{n_N,n_0}^{(N)}}{\sum_{n_0"=n_N} \lambda_{n_0"}^{(0)} \mu_{n_N,n_0"}^{(N)}}\Big). \tag{3.3}$$

We calculate the mutual entropy for the input photon number state $\rho^{(0)}$ and the attenuation channel Λ^* and consider the convergence of mutual entropy $\mathrm{I}(\rho^{(0)}; (\Lambda^*)^N)$ with respect to N.

For simplicity, we take the input state $\rho^{(0)}$ as $\rho^{(0),M} = \lambda_0^{(0)} E_0^{(0)} + (1 - \lambda_0^{(0)}) E_M^{(0)}$. where $E_0^{(0)}$ is a vacuume state and $E_M^{(0)}$ is M photon number state. We have the following results:

(1) For $M = 2$, $N = 100$, $\lambda_0^{(0)} = \frac{3}{10}$ and $\eta \in \{0.999, 0.99, 0.98, 0.97, 0.95, 0.9, 0.6\}$, we get (see Figure 1). From Fig 1, we conclude that (i) the mutual entropy decreases to very small number as N is getting to large number and (ii) the convergence of $\mathrm{I}(\rho^{(0)}; (\Lambda^*)^N)$ depends on the transmission ratio η of channel, namely, if η is closer to 0 then the convergence becomes faster and if η is closer to 1 then the convergence becomes slower.

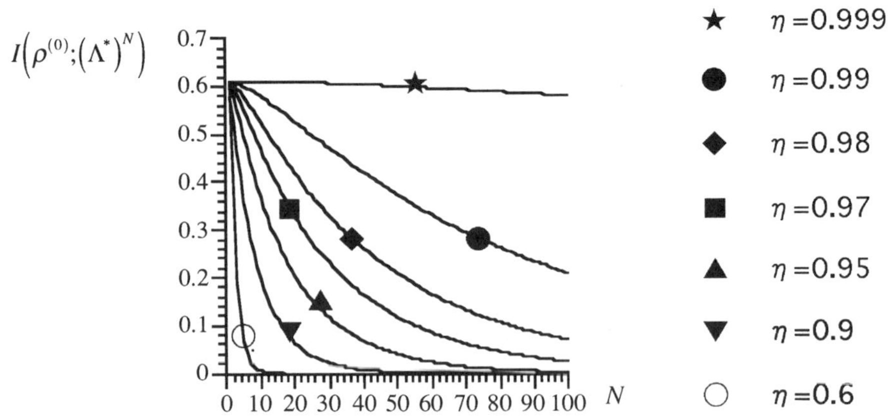

Fig.1 Convergence of mutual entropy w.r.t. η

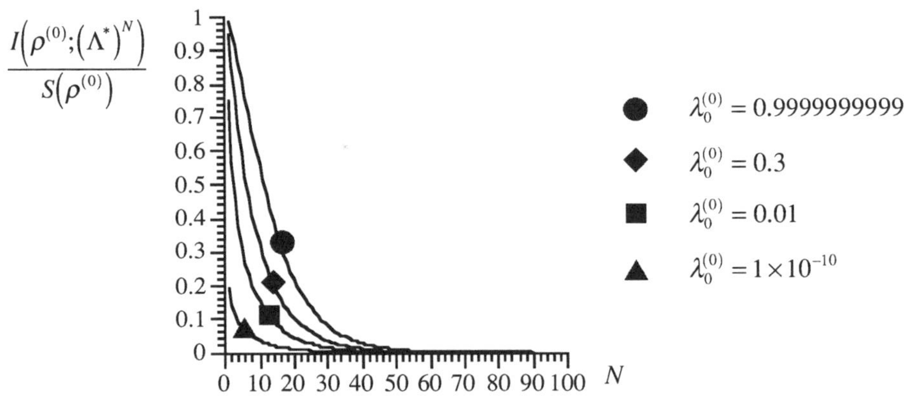

Fig2. Convergence of mutual entropy ratio w.r.t. $\lambda_0^{(0)}$

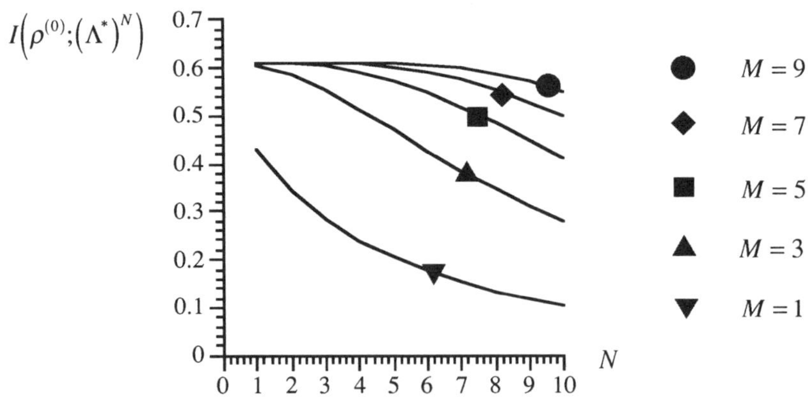

Fig.3 Convergence of mutual entropy w.r.t. M

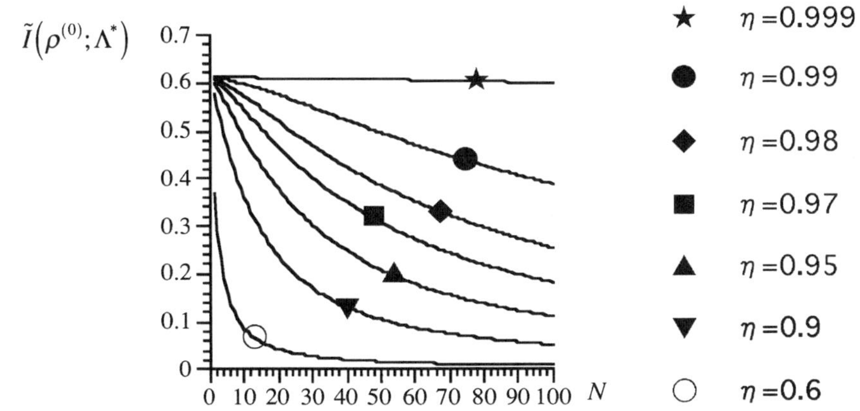

Fig.4 Convergence of mean mutual entropy w.r.t. η

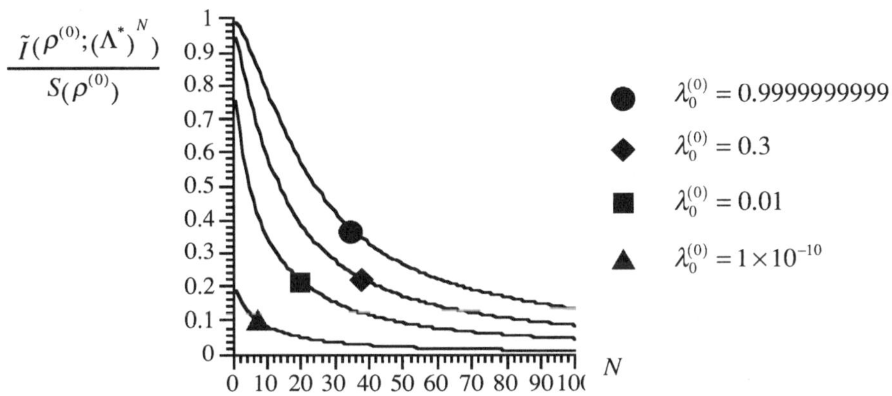

Fig.5 Convergence of mean mutual entropy ratio w.r.t. $\lambda_0^{(0)}$

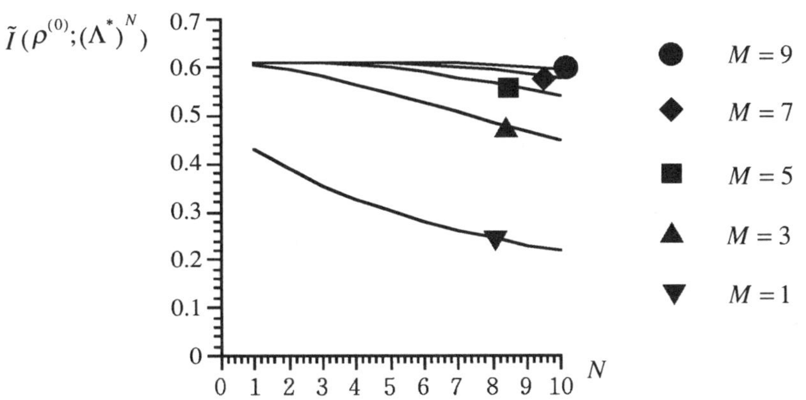

Fig.6 Convergence of mean mutual entropy w.r.t. M

In optical fiber channel, if the transmission ratio η is improved to be very close to 1, then the input information is almost correctly transmitted to the output system even at long distance.

(2) For $M = 2$, $N = 100$, $\lambda_0^{(0)} \in \{0.9999999999, 0.3, 0.01, 1 \times 10^{-10}\}$ and $\eta = \frac{9}{10}$, we have (see Figure 2). From Fig 2, we conclude that the convergence of the mutual entropy ratio $\mathrm{I}(\rho^{(0)}; (\Lambda^*)^N) / \mathrm{S}(\rho^{(0)})$ depends on the probability $\lambda_0^{(0)}$ of the input state, namely, if $\lambda_0^{(0)}$ is closer to 0 then the convergence becomes faster and if $\lambda_0^{(0)}$ is closer to 1 then the convergence becomes slower.

(3) For $M \in \{1, 3, 5, 7, 9\}$, $N = 10$, $\lambda_0^{(0)} = \frac{3}{10}$ and $\eta = \frac{9}{10}$, we get (see Figure 3). From Fig 3, we conclude that the convergence of $\mathrm{I}(\rho^{(0)}; (\Lambda^*)^N)$ depends on the photon number M of the input state, namely, as the input photon number M decreases, the convergence becomes faster and as M increases, the convergence becomes slower.

3.2 Convergence of quantum mean mutual entropy for an attenuation process

The quantum mutual entropy $\mathrm{I}(\rho^{(0)}; (\Lambda^*)^N)$ represents the amount of information correctly translated to the output system at the N times of unit distance from the input system. When we do not know in which distance the receiver exists from the transmitter, we estimate the amount of information sending from the transmitter to the receiver by the mean mutual entropy [16]:

$$\widetilde{\mathrm{I}}(\rho^{(0)}; (\Lambda^*)^N) = \frac{1}{N} \sup\{\sum_{n_0=0} \lambda_{n_0}^{(0)} \sum_{k=1}^{N} \mathrm{S}((\Lambda^*)^k(E_{n_0}^{(0)}), (\Lambda^*)^k \rho^{(0)}); E^{(0)} = \{E_{n_0}^{(0)}\}\} \quad (3.4)$$

where two compound states $\widetilde{\sigma}_E^N$ and $\widetilde{\sigma}_0^N$ are given by

$$\widetilde{\sigma}_E^N = \sum_{n_0=0} \lambda_{n_0}^{(0)} \bigotimes_{k=0}^{N} (\Lambda^*)^k E_{n_0}^{(k)}, \quad \widetilde{\sigma}_0^N = \bigotimes_{k=0}^{N} (\Lambda^*)^k \rho^{(k)} \quad (3.5)$$

Lemma 3.2 *For any input state $\rho^{(0)}$, any channel Λ^* and any N, we have*

$$I(\rho^{(0)}; (\Lambda^*)^N) \leq \widetilde{I}(\rho^{(0)}; (\Lambda^*)^N). \quad (3.6)$$

For any input state $\rho^{(0)} \in \Sigma(\mathcal{H}_0)$ and the attenuation channel $(\Lambda^*)^N$, if the Schatten decomposition of the input state $\rho^{(0)}$ is represented as $\rho^{(0)} = \sum_{n_0=0} \lambda_{n_0}^{(0)} E_{n_0}^{(0)}$ then the above relative entropy $\mathrm{S}(\widetilde{\sigma}_E^N; \widetilde{\sigma}_0^N)$ is expressed by

$$\mathrm{S}(\widetilde{\sigma}_E^N; \widetilde{\sigma}_0^N) = \sum_{k=1}^{N} \sum_{n_k=0} \sum_{n_0=n_k} \lambda_{n_0}^{(0)} \mu_{n_k,n_0}^{(k)} \log\Big(\frac{\mu_{n_k,n_0}^{(k)}}{\sum_{n_0"=n_k} \lambda_{n_0"}^{(0)} \mu_{n_k,n_0"}^{(k)}}\Big). \quad (3.7)$$

We calculate the the mean mutual entropy for the input photon number state $\rho^{(0)}$ and the attenuation channel Λ^* and consider the convergence of mean mutual entropy $\widetilde{\mathrm{I}}(\rho^{(0)}; (\Lambda^*)^N)$ with respect to N .

For simplicity, we take the input state $\rho^{(0)}$ as $\rho^{(0),M} = \lambda_0^{(0)} E_0^{(0)} + (1 - \lambda_0^{(0)}) E_M^{(0)}$. where $E_0^{(0)}$ is a vacuume state and $E_M^{(0)}$ is M photon number state. We have the following results:

(1) For $M = 2$, $N = 100$, $\lambda_0^{(0)} = \frac{3}{10}$ and $\eta \in \{0.999, 0.99, 0.98, 0.97, 0.95, 0.9, 0.6\}$, we get (see Figure 4).

(2) For $M = 2$, $N = 100$, $\lambda_0^{(0)} \in \{0.9999999999, 0.3, 0.01, 1 \times 10^{-10}\}$ and $\eta = \frac{9}{10}$, we have (see Figure 5).

(3) For $M \in \{1, 3, 5, 7, 9\}$, $N = 10$, $\lambda_0^{(0)} = \frac{3}{10}$ and $\eta = \frac{9}{10}$, we get (see Figure 6).

The gradation of the mean mutual entropy is a bit different from that of §3.1. The convergence of the mean mutual entropy is slower than that of the mutual entropy. Other behavior of the change of the mutual entropy is similar as that of Fig.1, 2 and 3.

The calculation of mean mutual entropy of K-S type in optical modulated processes is shown in the following paper [17].

References

[1] Accardi, L., Ohya, M. : "Compound channels, transition expectations and liftings", to appear in J. Multivariate Analysis.

[2] Accardi L. (1974): "Noncommutative Markov chains" In : international School of Mathematical Physics, Camerino, pp. 268–295.

[3] Accardi L., Ohya M. and Suyari H. : "Mutual entropy in quantum Markov chains", in this volume.

[4] Hirota, O (1985): "Optical Communication Theory" (in Japanese), Morikita Publishing.

[5] Hirota, O (1992): "Squeezed Light", Elsevier.

[6] Holevo, A.S. (1977): "Problems in the mathematical theory of quantum communication channels", Rep. Math. Phys. **12**, pp. 273–278.

[7] Ingarden, R.S. and Urbanik, K. (1962): "Quantum information thermodynamics", Acta Phys. Polon. **21**, pp. 281–304.

[8] Lindblad, G. (1975): "Completely positive maps and entropy inequalities, Commun. Math. Phys. **40**, pp. 147–151.

[9] Neumann, J. von (1932): " Die Mathematischen Grundlagen der Quantenmechanik " , Springer-Berlin.

[10] Ohya, M. (1981): "Quantum ergodic channels in operator algebras", J. Math. Anal. Appl. **84**, pp. 318–327.

[11] Ohya, M (1983): "On compound state and mutual information in quantum information theory", IEEE Trans. Information Theory **29**, pp. 770–777.

[12] Ohya, M., Watanabe, N., (1987): "Efficiency of modulations in an optical communication process", Proc. International Conf. Opt. Commun. Theory, pp. 40–43.

[13] Ohya, M. (1989): "Some aspects of quantum information theory and their applications to irreversible processes", Rep. Math. Phys. **27**, pp. 19–47.

[14] Ohya, M., Suyari, H. (1991): "Rigorous derivation of error probability in coherent optical communication", in *Quantum Aspects of Optical Communication* ed. by C. Benjaballah, O. Hirota, S. Reynaud (Lecture Notes in Physics **378**, Springer), pp. 203–212.

[15] Ohya, M., Petz, D. (1993): "Quantum Entropy and its Use", Springer-Verlag.

[16] Ohya M. : "State change, complexity and fractal in quantum systems", in this volume.

[17] Ohya, M., Watanabe, N. : "A calculation of mean mutual entropy of K-S type in optical modulated processes", in preparation.

[18] Watanabe, N. (1991): "Efficiency of optical modulations with coherent states", in *Quantum Aspects of Optical Communication* ed. by C. Benjaballah, O. Hirota, S. Reynaud (Lecture Notes in Physics **378**, Springer), pp.350–360.

Part V.
Quantum Detection, Estimation & Filtering

QUANTUM FILTERING OF MARKOV SIGNALS WITH WHITE QUANTUM NOISE[1]

Belavkin V.P.

Mathematics Department,
University of Nottingham,
NG7 2RD, UK

Abstract

Time-continuous non-anticipating processes of nondemolition measurements in quantum systems are described. In particular, the notion of physically realisable quantum filter is introduced and the problem of its optimisation to obtain the best a posteriori quantum state is considered. The fact that the optimal filtering of a quantum Markovian Gaussian signal with the Gaussian white quantum noise can be described by a coherent Markovian linear filter corresponding to quantum Gaussian state diffusion is proved. As an example, the problem of optimal measurement of complex amplitude for a quantum Markovian oscillator, loaded on a quantum wave communication line, is considered.

INTRODUCTION

At present, the dynamical problem of optimal reception of optical signals with quantum noise, taking into account the experimental limitations caused by the quantum nature of electromagnetic waves, is becoming actual. An attempt to develop the multistep variant of quantum measurement theory and decisions, (see review [1] and its references) applying it to the problems of dynamical filtering in discrete time, was made in [2]. However, because of the very restrictive definition of the quantum measurement process as a set of independent quantum measurements,

[1]Translated from Russian by R.V. Belavkin. The original was published in: Radiotechnika i Electronika, 25 (1980), pp. 1445-1453.

even the simplest Gaussian problem for quantum linear filtering was not solved with the exception of one special case.

Here we show while solving the filtering problems for a quantum diffusion, that these difficulties can be avoided if a natural wider class of sequential compatible quantum measurements is considered assuming their dependence not only on time, but also on the results of preceding measurements. An appropriate quantum device, that realises the adapted in such way measurements, is naturally called the nonanticipating (causal) filter with a memory. The optimisation of independent quantum measurements in the narrow class of filters without memory, is a problem with additional limitations leading to a poorer quality of filtration. As is proved below, the optimal filtering in the case of a Markovian Gaussian signal with the white background quantum noise is given by a coherent quantum filter with a Markovian memory. The next optimal quantum measurement depends on a result of the present measurement and is independent of previous measurement results in this case. Such quantum filter can be realised as an indirect coherent measurement by adding vacuum quantum noise and by processing of the measurement results with a classical linear filter of Kalman-Bucy type [3]. A nonlinear (and quasi linear) generalisation of this result for the non-Gaussian case, which corresponds to the classical nonlinear filtering theory of Stratonovitch [4], is also of interest.

Let us first consider a physical model of quantum open system with output channel, in which the problem of optimal quantum filtering arises, and give its solution, and then we shall get down to a precise formulation and derive this solution.

1 OPTIMAL OBSERVATION OF QUANTUM OSCILLATOR AT THE OUTPUT OF WAVELINE

Let us consider an electromagnetic oscillator with frequency Ω loaded on a couple of communication lines, which are described by wave conductivity G and resistance R respectively, as shown in the figure ($\Omega = 1/\sqrt{LC}$, $G = \sqrt{c_1/l_1}$, $R = \sqrt{l_2/c_2}$, and propagation velocities of the waves $v_i = 1/\sqrt{l_i c_i}$, $i = 1, 2$ are supposed to be equal $v_1 = v_2 = v$). We assume that the lines are homogeneous and that the measuring

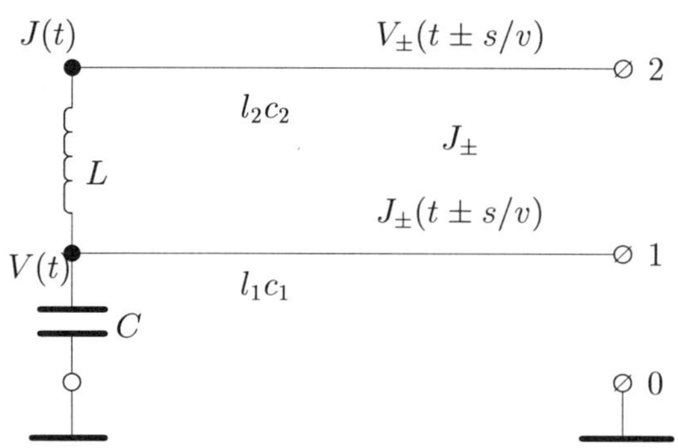

device which is set at their output $(0-1)$, $(0-2)$ is ideally conjugated with them, and the system is entirely described by the dynamic variables of voltage $V(t)$ and current $J(t)$ on the contour (L, C), and by the running waves

$$J_{\pm}(t \pm s/v) = J_1(s,t) \pm GV_1(s,t)\,, \quad V_{\pm}(t \pm s/v) = V_2(s,t) \pm RJ_2(s,t)$$

of current and voltage in the first and second lines respectively. These waves, with the boundary conditions $V_1(0,t) = V(t)$, $J_2(0,t) = J(t)$, satisfy the following equations

$$\begin{aligned} C\dot{V}(t) + GV(t) - J(t) = J_+(t)\,, &\qquad L\dot{J}(t) + RJ(t) + V(t) = V_+(t)\,, \\ J_-(t) = J_+(t) - 2GV(t)\,, &\qquad V_-(t) = V_+(t) - 2RJ(t)\,. \end{aligned}$$

Assuming, for simplicity, that the following conjugation condition $R/L = G/C \equiv \gamma$ is true, the above system of equations can be written in the one-dimension form

$$\dot{x}(t) + \left(i\Omega + \frac{1}{2}\gamma\right)x(t) = x_+(t)\,, \quad x_-(t) = x_+(t) - \gamma x(t)$$

using the complex amplitudes

$$\begin{aligned} x(t) &= \left(\sqrt{C}V(t) + i\sqrt{L}J(t)\right)/\sqrt{2}\,, \\ x_\pm(t) &= \left(J_\pm(t)/\sqrt{C} + iV_\pm(t)/\sqrt{L}\right)/\sqrt{2}\,. \end{aligned}$$

We now move towards the stochastic description of the corresponding quantum system. The complex amplitudes $x(t)$ should be replaced by the corresponding operators $\check{x}(t)$ in the Heisenberg picture, which satisfy the canonical commutation relations

$$[\check{x}(t), \check{x}(t)^*] \equiv \check{x}(t)\check{x}(t)^* - \check{x}(t)^*\check{x}(t) = \hbar\check{1}\,, \tag{1}$$

where $\check{x}(t)^*$ are adjoint operators, and $\check{1}$ is the unit operator in the space of quantum-mechanical states of this system ($\hbar$ is the Planck constant, $\check{x}(t)/\sqrt{\hbar} = \check{a}(t)$ is the annihilation operator for oscillator quanta at time t). By virtue of linearity the corresponding quantum equation for $\check{x}(t)$ has the same form as the one for $x(t)$, where the amplitudes of the propagating wave x_+ have to be replaced by the operators $\check{x}_+(t)$ with the commutation relations ensuring constancy of the commutator (1). Obviously, the last is true if we propose

$$[\check{x}_\pm(t), \check{x}_\pm(t')] = 0\,, \quad [\check{x}_\pm(t), \check{x}_\pm(t')^*] = \gamma\hbar\delta(t - t')\,, \tag{2}$$

where the commutators for the output wave $\check{x}_-(t)$ do not contradict its linear relation with $\check{x}_+(t)$ and $\check{x}(t)$, as it is assumed that $\check{x}_+(t)$ commutates both with $x(t)$ and with $x(t)^*$. As shown in [6], the above equations and commutation relations can be obtained by a quantisation of an open oscillator in a "rotating waves" representation using the narrow-band approximation. Introducing the notations

$$\check{v}(t) = \check{x}_+(t) = -\hat{v}(t)\,, \quad \check{y}(t) = \check{x}_-(t - s/v) = -\hat{v}(t) \tag{3}$$

for the forward and backward waves at the input and output of the line of length s respectively, let us write these equations in the standard form

$$\frac{\mathrm{d}}{\mathrm{d}t}\check{x}(t) + \alpha\check{x}(t) = \check{v}(t)\,, \quad \hat{y}(t + s/v) = \gamma\check{x}(t) + \hat{v}(t)\,, \tag{4}$$

where $\alpha = i\Omega + \gamma/2$. The first equation, rewritten in terms of the stochastic differentials as

$$\mathrm{d}\check{x}(t) + \alpha\check{x}(t)\mathrm{d}t = \mathrm{d}\check{v}_t,$$

where $\check{v}_t = \int_0^t \check{v}(t)\mathrm{d}t$, can be easily integrated. The solution

$$\check{x}(t) = e^{-\alpha t}\check{x}(0) + \int_0^t e^{\alpha(r-t)}\mathrm{d}\check{v}_r$$

does not depend on the output wave $\hat{y}(t')$ with $t' < t + s/v$, i.e. it commutes with $\hat{y}(t')$ as well as with $\hat{y}(t')^*$. This expresses the ideal conjugation between the output measurement device and the transmission lines, and will be called the nondemolition condition. Together with the hypothesis about the Markovian property of $\check{x}(t)$, which is fulfilled under the narrow-band approximation for the equilibrium state of the measurement device sending the noise wave $\check{v}(t)$ into the line, it significantly simplifies the optimisation problem of continuous measurement in the quantum system described.

Let us now consider the problem of the indirect nondemolition observation of the amplitude operator $\check{x}(t)$ of the quantum oscillator, mentioned above. One would like to obtain an optimal estimate $x(t)$ of this amplitude being received during this observation on the output of the transmission line. The mean square optimisation problem for the complex estimate $x(t)$, as a result of quantum nondemolition measurement, is to minimize the mean square error

$$\left\langle \left(\check{x}(t) - x(t)\right)^* \left(\check{x}(t) - x(t)\right) \right\rangle . \tag{5}$$

The continuous in time quantum nondemolition measurement is described by any stochastic non-anticipating transformation of the quantum stochastic process $\hat{y}$ into the classical complex one x. It is given at a fixed $t > 0$ by a stochastic functional of the output operators $\hat{y}(t')^*, \hat{y}(t'), t' \leq t$, represented as a normal operator $x(t)$, which commutes also with the conjugate operators $x(t')^*, x(t')$ at $t' \geq t$ and so at all t'. Note that the estimating operator $x(t)$ does not necessary commute with all estimated operators $\check{x}(t', \check{x}(t')^*$, although due to the nondemolition it commutes with them at $t' \geq t - s/v$. The problem of finding the classical observable process $x(t)$, which is the closest in the mean square sense (5) to the non-observed quantum stochastic process $\check{x}(t)$, is a quantum filtering problem.

Let, for example, the initial quantum oscillator state be a circular-Gaussian with zero mathematical expectation $\langle \check{x}(0) \rangle = 0$ and average number of quanta $\Sigma \geq 0$:

$$\langle \check{x}(0)^2 \rangle = 0\,, \quad \langle \check{x}(0)^* \check{x}(0) \rangle = \hbar\Sigma\,, \tag{6}$$

and $\check{v}(t)$ be a quantum, circular-Gaussian, white noise: $\langle \check{v}(t) \rangle = 0$,

$$\langle \check{v}(t)\check{v}(t') \rangle = 0\,, \quad \langle \check{v}(t)^* \check{v}(t') \rangle = \hbar\nu\gamma\delta(t - t')\,, \tag{7}$$

where $\nu = (\exp\{\hbar\Omega/kT\} - 1)^{-1}$ is an average number of quanta corresponding to an equilibrium state with temperature T. As will be shown below, the optimal estimate $x(t)$ of the operator $\check{x}(t)$ minimizing the criterion (5), is given by a nonorthogonal positive operatorvalued measure, which describes the coherent quasimeasurement of an integral nonanticipating transformation of the output process $\hat{y}$. With the delay s/v being neglected if $s \ll vt$, it can be realized by the measurement of the complex random process $x(t)$, that is described by the Kalman-Bucy filter,

$$\frac{\mathrm{d}}{\mathrm{d}t}x(t) + \alpha x(t) = \kappa(t)\Big(y(t) - \gamma x(t)\Big)\,, \quad y(t) = \hat{y}(t) + \hat{w}(t)\,, \tag{8}$$

which is usually written in the stochastic differential form

$$\mathrm{d}x(t) + \alpha x(t)\mathrm{d}t = \kappa(t)\Big(\mathrm{d}y_t - \gamma x(t)\mathrm{d}t\Big)\,,$$

with $x(0) = 0$ and $y_t = \int_0^t y(t)\mathrm{d}t$. In this equation $\kappa(t) = (\Sigma(t) - \nu)/(1+\nu)$, $\Sigma(t)$ is the solution of the Riccati equation

$$\frac{\mathrm{d}}{\mathrm{d}t}\Sigma(t) = \big(\gamma/(1+\nu)\big)\left(\nu - \Sigma(t)\right)\left(1 + \Sigma(t)\right), \tag{9}$$

with $\Sigma(0) = \Sigma_0$, and $\hat{w}(t)$ is the complex amplitude operator with the commutators

$$[\hat{w}(t), \hat{w}(t')] = 0\,, \quad [\hat{w}(t), \hat{w}(t')^*] = -\hbar\gamma\delta(t - t')$$

and correlations

$$\langle \hat{w}(t)\hat{w}(t')\rangle = 0\,, \quad \langle \hat{w}(t)^*\hat{w}(t')\rangle = \hbar\gamma\delta(t - t')\,, \tag{10}$$

making the right part of the equation (8) a classical commutative δ-correlated complex process. Such an indirect measurement is realised, for example, during the heterodyning [7], when $\hat{w}(t)$ plays the part of reference wave. The error (5) of filtration in this case has a minimal value $\hbar\Sigma(t)$, which is not trivial in contrast to the classical case even under zero temperature T, when $\sigma = 0$. It corresponds to adding into the measurement channel a trivial quantum noise, that is statistically equivalent to a classical white noise of intensity $\gamma\hbar > 0$. This conclusion is also valid for the classical oscillator ($\Sigma_0 \gg 1$) being observed by a quantum wave line. And only in the infinite temperature case $T \to \infty$ when $\sigma \gg 1$, taking into account the quantum nature of the system does not produce any substantial increase of the filtration error compared with the one given by the classical theory.

2 CAUSAL QUANTUM MEASUREMENT PROCESSES AND FILTERS

As far as the sequences $x(t_1)\ldots x(t_2)$ of results of a time measurement process define a function $x(t) \in X$ under continuous observation, the quantum measurement process must be statistically described by the multitime probability distributions $\mathbf{P}(\mathrm{d}x(t_1)\ldots\mathrm{d}x(t_n))$, $t_1 < \ldots < t_n$, $n = 1, 2, \ldots$ defining the probability measure $\mathbf{P}\left(\mathrm{d}\{x(t)\}\right)$ on the space of the trajectories $\{x(t)\}$. Such a measure must be represented in the linear form with respect to the density operator $\hat{\rho}$ as

$$\mathbf{P}\left(\mathrm{d}\{x(t)\}\right) = \mathrm{tr}\,\hat{\rho}\Pi\left(\mathrm{d}\{x(t)\}\right) \tag{11}$$

Here tr means the trace, $\Pi\left(\mathrm{d}\{x(t)\}\right) \geq 0$ is the operator-valued probability measure on the space of realisations $\{x(t)\}$, that is normalised to the identity operator $\int \Pi\left(\mathrm{d}\{x(t)\}\right) = \hat{1}$. This measure defines, in particular, the multitime operator-valued measures $\Pi(\mathrm{d}x(t_1)\ldots\mathrm{d}x(t_n))$, $t_1 < \ldots < t_n$, $n = 1, 2, \ldots$ with similar positivity and normalisation properties, which describe the probabilities of discrete observations as the linear functions of $\hat{\rho}$ according to the formula (11).

Every decomposition of the identity $\hat{1} = \int \Pi(\mathrm{d}\{x(t)\})$ into the positive operators $\Pi(\mathrm{d}\{x(t)\})$ corresponds, according to Naymark's theorem [8], to a measurement of some compatible set of operators, which are observed in an "extended quantum system". But only those measures Π correspond to the measuring processes physically realisable in real time, which satisfy the causality condition $\Pi^t(\mathrm{d}x^t) \in \mathcal{B}^t$, where Π^t is the measure Π on the space of "reduced" realisations $x^t = \{x(s)\}_{s\leq t}$, and $\{\mathcal{B}^t\}, t \geq 0$ is a nondecreasing family of operator algebras $\mathcal{B}^s \subseteq \mathcal{B}^t$, defining the nondemolition observables up to the time instants t. Typically $\mathcal{B}^t$ is generated by the subfamily $\hat{b}^t = \{\hat{b}(s) : s \leq t\}$ of some operator family $\{\hat{b}(t)\}, t \geq 0$, describing an output quantum stochastic process $\hat{b}(t)$ as in the considered example when $\hat{b}(t) = \hat{y}(t)$. But in general $\mathcal{B}^t$ must satisfy only the commutativity condition $\mathcal{B}^t \subseteq \mathcal{A}_t'$ for each t with respect to the algebra $\mathcal{A}_t$, generated by the present and future $\check{x}_t = \{\check{x}(s) : s \geq t\}$ of an estimated quantum stochastic process $\{\check{x}(t)\}$. Moreover, we shall presuppose the existence of the conditional operator-valued measures $\Pi_s^t(x^s, \mathrm{d}x_s^t)$, $x_s^t = \{x(t')\}_{t' \in (s,t)}$, describing by the formula (11) the non-anticipating processes of observation from any moment s till $t > s$ with the fixed preceding results $x^s = \{x(t)\}_{t \leq s}$. They are defined by the compatibility condition

$$\Pi_{t_0}^{t_1}(x^{t_0}, \mathrm{d}x_{t_0}^{t_1})\Pi_{t_1}^{t_2}(x^{t_1}, \mathrm{d}x_{t_1}^{t_2}) = \Pi_{t_0}^{t_2}(x^{t_0}, \mathrm{d}x_{t_0}^{t_2}) \tag{12}$$

for all $t_0 < t_1 < t_2$, where the Hermitian positive operators $\Pi_s^t(x^s, \mathrm{d}x_s^t)^* = \Pi_s^t(x^s, \mathrm{d}x_s^t)$ have values in the relative commutants $\mathcal{B}_s^t$ of the output subalgebras $\mathcal{B}^s$ with respect to $\mathcal{B}^t$. Such a compatible family $\{\Pi_s^t\}_{s<t}$ of the conditional measures $\Pi_s^t(x^s, \mathrm{d}x_s^t) \in \mathcal{B}_s^t$, normalised to the identity operator $\hat{1}$, will be called a *non-anticipating quantum filter*. The measures Π_s^t, which are independent of x^s, correspond to filters without memory. The Markovian filters described by the conditional measures Π_s^t, which depend only on the last preceding value $x(s)$, are the simplest filters with memory. In addition, the transition operator-valued measure $\Pi_s^t(x(s), \mathrm{d}x(t))$, defines according to (12) the multitime conditional measures

$$\begin{aligned}\Pi(x(t_0), \mathrm{d}x(t_1)\ldots \mathrm{d}x(t_n)) &= \\ &= \Pi_{t_0}^{t_1}(x(t_0), \mathrm{d}x(t_1))\ldots\Pi_{t_{n-1}}^{t_n}(x(t_{n-1}), \mathrm{d}x(t_n)),\end{aligned}$$

satisfying the operator-valued Smolukhovsky equation

$$\int_{X_{t_1}} \Pi_{t_0}^{t_1}(x(t_0), \mathrm{d}x(t_1))\Pi_{t_1}^{t_2}(x(t_1), \mathrm{d}x(t_2)) = \Pi_{t_0}^{t_2}(x(t_0), \mathrm{d}x(t_2)). \tag{13}$$

In the case of a linear space X, for example a complex n-dimensional space C^n, it is convenient to describe the Markovian filters by the transition characteristic operator-functions

$$\Phi_s^r(\mathbf{x}(s), \mathbf{u}(t)) = \int e^{i(\mathbf{u}(t)^+\mathbf{x}(t)+\mathbf{x}(t)^+\mathbf{u}(t))}\Pi_s^r(\mathbf{x}(s), \mathrm{d}\mathbf{x}(t)), \tag{14}$$

where $\mathbf{u}$, $\mathbf{x}$ are n-dimensional complex columns, and $\mathbf{x}^+$, $\mathbf{u}^+$ are their conjugate rows. In this case the normalisation condition for Π_s^t gives $\Phi_s^t(\mathbf{x}(s), 0) = \hat{1}$, and the condition of positive semidefiniteness of $\Pi_s^t \geq 0$ implies positive semidefiniteness of the operator-matrix $[\Phi(\mathbf{x}(s), \mathbf{u}_k - \mathbf{u}_l)] \geq 0$, where $\{\mathbf{u}_i, i = 1, 2, \ldots\}$ is any finite collection of vectors $\mathbf{u}_i \in \mathrm{C}^n$. The operators Φ_s^t commute on the non overlapping time intervals, and the Markovian condition (12) for them can be written in the following differential operator form

$$\Phi_{t_1}^{t_2}\left(-i\frac{\partial}{\partial \mathbf{u}(t_1)}, \mathbf{u}(t_2)\right)\Phi_{t_0}^{t_1}(\mathbf{x}(t_0), \mathbf{u}(t_1))|_{u(t_1)=0} = \Phi_{t_0}^{t_2}(\mathbf{x}(t_0), \mathbf{u}(t_2)). \tag{15}$$

It is not difficult to prove the opposite fact: for every family of operator-valued functions $\{\Phi_s^t, s < t\}$, which are continuous at the point $u(t) = 0$, and satisfy the above normalisation, positiveness and conditions (15), the corresponding Markovian filter must exist.

As an example of a Markovian filter we can take a linear coherent filter, that is defined by the characteristic operator-function

$$\Phi_s^t(\mathbf{x}(s),\mathbf{u}(t)) = e^{i\mathbf{u}(t)^+\widehat{\mathbf{x}}(t)} e^{i\widehat{\mathbf{x}}(t)^+\mathbf{u}(t)}, \tag{16}$$

where $\widehat{\mathbf{x}}(t)$ is vector-column composed of the operators $\hat{x}_1(t),\ldots,\hat{x}_n(t)$; $\widehat{\mathbf{x}}(t)^+$ is vector-row composed of the adjoint operators $\hat{x}_1(t)^*,\ldots,\hat{x}_n(t)^*$. They can be found by solving the linear quantum stochastic differential equation

$$\frac{\mathrm{d}}{\mathrm{d}t}\widehat{\mathbf{x}}(t) + B(t)\widehat{\mathbf{x}}(t) = K(t)\widehat{\mathbf{b}}(t), \qquad \widehat{\mathbf{x}}(s) = \mathbf{x}(s); \tag{17}$$

where $B(t) = \{\beta_{kl}(t)\}$, $K(t) = \{\kappa_{kl}(t)\}$ are complex matrices of the size $n \times n$ and $n \times m$ respectively, which may continuously depend on time, and $\widehat{\mathbf{b}}(t)$ is an m -dimensional column composed of the annihilation operators $\hat{b}_1(t),\ldots,\hat{b}_m(t)$ of the quantum output process, satisfying the following canonical commutation relations

$$\left[\hat{b}_k(t),\hat{b}_l(t')\right] = 0, \quad \left[\hat{b}_k,(t),\hat{b}_l^*(t')\right] = \delta_{kl}\delta(t-t').$$

It is not hard to verify by integrating the equation (17) that the characteristic operator-function (16) satisfies all the above mentioned properties, including (15). However, it becomes obvious, if we consider that the operators $\widehat{\mathbf{x}}(r)$ are linear transformations from the annihilation $\widehat{\mathbf{b}}(t)$ ones, so that (16) is the characteristic operator-function of the operator-valued Dirac δ-measure

$$\Pi_s^t\left(\mathbf{x}(s),\mathrm{d}\mathbf{x}(t)\right) = \mathcal{N}_-\{\delta\left(\widehat{\mathbf{x}}(t),\mathrm{d}\mathbf{x}(t)\right)\},$$

which corresponds to the anti normal order "$\mathcal{N}_-$", when the creation operators $\widehat{\mathbf{b}}^*(t)$ act first. It is known well, that such measures are generated by the coherent projectors and Lebesgue measure on the complex space C^n. For example, for the case of a fixed $\mathbf{x}(0) = \mathbf{x}_0$ the measure $\Pi(t,\mathrm{d}\mathbf{x}) \equiv \Pi_0^t(\mathbf{x}_0,\mathrm{d}\mathbf{x})$ has the form

$$\Pi(t,\mathrm{d}\mathbf{x}) = |t,\mathbf{x}\rangle\langle t,\mathbf{x}|\mathrm{d}\lambda(t,\mathbf{x}), \tag{18}$$

where $|t,\mathbf{x}\rangle$ are the normalised to unit right eigen vectors

$$\widehat{\mathbf{x}}(t)|t,\mathbf{x}\rangle = \mathbf{x}|t,\mathbf{x}\rangle, \quad \mathbf{x}\in\mathrm{C}^n \tag{19}$$

for the operators $\widehat{\mathbf{x}}(t)$, which are defined by the equation (17) with the initial condition $\widehat{\mathbf{x}}(0) = \mathbf{x}_0$. The element $\mathrm{d}\lambda$ of the volume in C^n, normalising (18) to unity $\hat{1}$, is given by

$$\mathrm{d}\lambda(t,\mathbf{x}) = \Pi_{i=1}^n \mathrm{d}\Re x_i \mathrm{d}\Im x_i / \det \pi C(t),$$

where $C(t)$ is a matrix of the commutators $[\hat{x}_k(t),\hat{x}_l(t)^*] = (C(t))_{kl}$ corresponding to $C(0) = 0$.

As a physical realisation of the above linear coherent quantum Markovian filter one can consider the optimal indirect measurement [9] of the operators $\{\widehat{\mathbf{x}}(t)\}$, realised by a precise measurement of the Markovian process $\mathbf{x}(t)$, described by the linear classical stochastic differential equation

$$\frac{\mathrm{d}}{\mathrm{d}t}\mathbf{x}(t) + B(t)\mathbf{x}(t) = K(t)\mathbf{b}(t), \quad \mathbf{b}(t) = \widehat{\mathbf{b}}(t) + \widehat{\mathbf{c}}(t). \tag{20}$$

Here $\widehat{\mathbf{c}}(t)$ is independent of $\widehat{\mathbf{b}}(t)$ quantum "vacuum" vector noise with the commutators

$$[\hat{c}_k(t),\hat{c}_l(t')] = 0, \quad [\hat{c}_k(t),\hat{c}_l(t')^*] = -\delta_{kl}\delta(t-t'),$$

which make the components of the sum $\mathbf{b}(t)$ commuting with the components of $\mathbf{b(t)}^*$. Moreover, $\langle\hat{b}(t)\rangle = 0$,

$$\langle\hat{c}_k(t)\hat{c}_l(t')\rangle = 0\,, \quad \langle\hat{c}_k(t)^*\check{c}_l(t')\rangle = \delta_{kl}\delta(t-t')\,,$$

and so $\langle\hat{c}_k(t)\hat{c}_l(t')^*\rangle = 0$. Therefore the momenta of the process $\mathbf{x}(t)$ coincide with the anti normal momenta for $\widehat{\mathbf{x}}(t)$, in particular $\langle\widehat{\mathbf{x}}(t)\rangle = \langle\mathbf{x}(t)\rangle$,

$$\langle\hat{x}_k(t)\hat{x}_l(t)\rangle = \langle x_k(t)x_l(t)\rangle\,, \quad \langle\hat{x}_k(t)\hat{x}_l(t)^*\rangle = \langle x_k(t)x_l(t)^*\rangle\,.$$

3 QUANTUM GAUSSIAN STATE DIFFUSION AND FILTERING

Let $\check{\mathbf{x}}\,(t)$ be an n-dimensional quantum diffuse process with the zero initial expectation $\langle\check{\mathbf{x}}(0)\rangle = 0$ and the initial correlations

$$\langle\check{x}_k(0)\check{x}_l(0)\rangle = 0\,, \qquad \langle\check{x}_k(0)^*\check{x}_l(0)\rangle = (R_0)_{lk}, \tag{21}$$

where R_0 is a given positive-definite matrix. The process is defined by the stochastic equation

$$\frac{\mathrm{d}}{\mathrm{d}t}\check{\mathbf{x}}(t) + A(t)\check{\mathbf{x}}(t) = J(t)\check{\mathbf{a}}(t), \tag{22}$$

where $A(t)$, $J(t)$ are matrices of size $n\times n$ and $n\times m$ respectively, and $\check{\mathbf{a}}(t)$ is a white complex noise with the trivial expectation $\langle\check{\mathbf{a}}(t)\rangle = 0$ and the nonzero normal correlative matrix $Q(t)$:

$$\langle\check{a}_k(t)\check{a}_l(t')\rangle = 0\,, \quad \langle\check{a}_k(t)^*\check{a}_l(t')\rangle = (Q(t)_{lk}\delta(t-t'). \tag{23}$$

Let this process be observed as the output of the quantum linear channel

$$\widehat{\mathbf{b}}(t) = F(t)\check{\mathbf{x}}(t) + \widehat{\mathbf{a}}(t), \tag{24}$$

where $F(t)$ is an $m\times n$ matrix, and $\widehat{\mathbf{a}}(t)$ is an m-dimensional quantum white noise with the canonical commutators

$$[\hat{a}_k(t),\hat{a}_l(t')] = 0\,, \quad [\hat{a}_k(t),\hat{a}_l(t')^*] = \delta_{kl}\delta(t-t')\,, \tag{25}$$

and with zero expectation $\langle\widehat{\mathbf{a}}(t)\rangle = 0$ and a nonzero normal correlation matrix $N(t)$:

$$\langle\hat{a}_k(t)\hat{a}_l(t')\rangle = 0\,, \quad \langle\hat{a}_k(t)^*\hat{a}_l(t')\rangle = (N(t))_{lk}\delta(t-t'). \tag{26}$$

We assume the pair $(\check{\mathbf{a}}(t),\widehat{\mathbf{a}}(t))$ to be independent of $\check{\mathbf{x}}(0)^*$, but it can have the nonzero matrix $T(t)$ of the mutual normal correlations in

$$\langle\check{a}_k(t)\hat{a}_l(t')\rangle = 0\,, \quad \langle\check{a}_k(t)^*\hat{a}_l(t')\rangle = (T(t))_{kl}\delta(t-t'). \tag{27}$$

This is, for example, because of the non-zero commutation relations

$$[\check{a}_k(t),\hat{a}_l(t')] = 0\,, \quad [\check{a}_k(t),\hat{a}_l(t')^*] = (D(t))_{kl}\delta(t-t'), \tag{28}$$

which are necessary for the commutativity of $\widehat{\mathbf{y}}(t')$ with $\check{\mathbf{x}}(t)^*$ for $t\geq t'$ if $\widehat{\mathbf{y}}$ is a nondemolition process with respect to a quantum process $\check{\mathbf{x}}(t)$. If $\check{\mathbf{x}}$ satisfies the commutation relations

$$[\check{x}_k(t),\check{x}_l(t)] = 0\,, \qquad [\check{x}_k(t),\check{x}_l(t)^*] = (C_0)_{kl}, \tag{29}$$

the matrix D is defined by the nondemolition property as a solution to the equation $JD + C_0F^+ = 0$. (It occurs in the example considered in the section 1.) Note, however, that the nonzero commutation relations (2) here are not essential for the nondemolition property

$$[\check{x}_k(t), \hat{y}_l(t')] = 0\,, \quad [\check{x}_k(t), \hat{y}_l(t')^*] = 0$$

for all $t \geq t'$, as far as the process $\check{\mathbf{x}}(t)$ can be classical: $C_0 = 0$.

The problem of optimal nonstationary quantum filtering consist of finding the physically realizable measurement process at the output of the quantum channel, the results of which, denoted by $\mathbf{x}(t)$, minimize the mean of the normal squared error

$$\|\check{\mathbf{x}}(t) - \mathbf{x}(t)\|^2 = \sum_{i=1}^{n}(\check{x}_i(t) - x_i(t))^*(\check{x}_i(t) - x_i(t)). \tag{30}$$

Theorem 1 *Let the initial quantum vector $\check{\mathbf{x}}(0)$ be Gaussian with zero mathematical expectation and the correlations (21); let the pair $(\check{\mathbf{a}}(t), \hat{\mathbf{a}}(t))$ be Gaussian with zero expectations also and with correlations (23), (26), (27), and $\hat{\mathbf{a}}(t)$ also satisfy the canonical commutation relations (25). Then the optimal quantum filter by the criterion (30) for the quantum Markovian Gaussian process $\check{\mathbf{x}}(t)$, which is defined by the linear stochastic equation (22), is the Markovian coherent filter (16), described by the linear stochastic equation (17), where $\mathbf{x}(0) = 0$,*

$$B = A + KF, \quad K = (PF^+ + JT^+)(N + I)^{-1}. \tag{31}$$

Here I is the identity $m \times m$-matrix, and $P = P(t)$ is the a posteriori correlation matrix, that is given by the solution of the Riccati equation

$$\frac{\mathrm{d}}{\mathrm{d}t}P + AP + PA^+ + (PF^+ + JT^+)(N + I)^{-1}(FP + TJ) = JQJ^+ \tag{32}$$

with the initial condition $P(0) = R_0$. The trace of this matrix defines the minimal expectation value of the error (30). (The dependence of all the matrices in (31), (32) on t is omitted for brevity.)

PROOF. Given the fact that the criterion for the point filtering (30) depends on the value $\mathbf{x}(t)$ of the observation $\{\mathbf{x}(t)\}$ only at the last instant t, the average value of the criterion is defined only by the single-time operator-valued measure $\Pi(t, \mathrm{d}\mathbf{x}(t))$. Therefore, one can try to find the solution $\Pi^0(t, \mathrm{d}\mathbf{x}(t))$ of the statistical problem [10] of the optimal observation for each t. If it proves to be a single-time operator-valued measure for some physically realisable measurement process $\Pi^0(\mathrm{d}\{\mathbf{x}(t)\})$ in the sense of the section 2, then the problem of optimal filtering will be solved.

Using the stochastic calculus for normal ordered quantum Gaussian variables, which was developed in [10] one can obtain the following representation for the average of the error (30) as a function of the operator-valued measure $\Pi(t, \mathrm{d}\mathbf{x})$:

$$\langle\|\check{\mathbf{x}}(t) - \mathbf{x}(t)\|^2\rangle = \langle\|\check{\mathbf{x}}(t) - \mathbf{x}(t)\|^2\,\Pi(t, \mathrm{d}\mathbf{x})\rangle = \sum_{i=1}^{n}(P(t))_{ii} + \int \mathrm{Tr}\,\widehat{R}(t, \mathbf{x})\Pi(t, \mathrm{d}\mathbf{x}). \tag{33}$$

Here $P(t)$ is a matrix, that satisfies equation (32), $R(t, \mathbf{x})$ is the operator-valued function

$$\widehat{R}(t, \mathbf{x}) = \sum_{i=1}^{n}(\widehat{x}_i(t) - x_i)^*\widehat{\rho}(t)(\widehat{x}_i(t) - x_i), \tag{34}$$

$\widehat{\mathbf{x}}(t)$ are operators, satisfying equation (17) with the initial condition $\widehat{\mathbf{x}}(0) = 0$ and the matrices (31), and $\widehat{\rho}(t)$ is a Gaussian density operator for the output process $\widehat{\mathbf{x}}(t)$. It can be represented in the form of the normally ordered expression

$$\widehat{\rho}(t) = \det(G(R-P)^{-1})\mathcal{N}\left[\exp\left\{-\sum_{k,l}\widehat{x}_k^*(t)(R-P)_{kl}^{-1}\widehat{x}_l(t)\right\}\right],$$

where $G(t)$, $R(t)$ are the solutions of the equations

$$\frac{\mathrm{d}}{\mathrm{d}t}G + BG + GB^+ = KK^+, \quad \frac{\mathrm{d}}{\mathrm{d}t}R + AR + RA^+ = JQJ^+$$

with the initial conditions $G(0) = 0$ and $R(0) = R_0$ respectively.

The second element of the right-hand side in (33) is not negative because of the nonnegative definiteness of the operator (34) and is equal to zero on the coherent measure (18) by equation (19) for $\mathbf{x}_0 = 0$. Hence, it is clear that the coherent measurement, described by this measure, is optimal for all t. Therefore, the coherent observation described by the characteristic operators (16) with the initial condition $\widehat{\mathbf{x}}(0) = 0$ is the optimal measurement process. Theorem is proved.

Note that the a posteriori state for the quantum diffusion $\check{\mathbf{x}}(t)$ is described by the Gaussian stochastic density operator

$$\check{\rho}(t) = \det(G(P+G)^{-1})\mathcal{N}\left[\exp\left\{-\sum(\check{x}_k(t) - x_k(t))^*(P+G)_{kl}^{-1}(\check{x}_l - x_l(t))\right\}\right]$$

given by the solution of the diffusion equation (20) with $\mathbf{x}(0) = 0$.

As example, let us consider the n-dimensional quantum oscillator (22), that has $\check{\mathbf{a}}(t) = \widehat{\mathbf{a}}(t)$, $JJ^+ = \hbar(A + A^+)$, such that the commutators

$$[\check{x}_k, \check{x}_l] = 0, \qquad [\check{x}_k, \check{x}_l^*] = \hbar\delta_{kl} \tag{35}$$

are preserved. The matching m-dimensional quantum communication line is described by the output wave

$$\widehat{\mathbf{y}}(t) = G(t)\check{\mathbf{x}}(t) - J(t)\widehat{\mathbf{a}}(t). \tag{36}$$

This wave commutes for $G = A + A^+$ with $\check{\mathbf{x}}(t')^*$, $t' \geq t$. In the former dimensionless units this line is described by the wave (24), where $F = -J^+/\hbar$, so that $\widehat{\mathbf{y}}(t) = \hbar F^+\widehat{\mathbf{b}}(t)$. Let the correlation matrix N of the noise $\widehat{\mathbf{a}}(t)$ be proportional to the identity matrix: $N = \nu I$. Then the equation (32) for the dimensionless a posteriori correlation matrix $S = P/\hbar$ will have the form

$$\dot{S} = i[S, H] - \frac{1}{2(\nu+1)}\left((S+I)G(S-\nu I) + (S-\nu I)G(S+I)\right), \tag{37}$$

where $H = (A - A^+)/2i$, and $K(t) = (\nu I - S)J/(1+\nu)$. In particular, for the one-dimensional case $m = n = 1$, putting $S = \Sigma$, $G = \gamma$, we obtain (9). It is interestingly to point out, that the equation (37) has a simple stationary solution $S = \nu I$, for which the optimal filtering does not require any observation, because $K(t) = 0$. The condition of stationary state $R_0 = \hbar\nu I$ in this case is a condition of thermal equilibrium for the initial statistical state of the oscillator and equilibrium state of the communication line, and it corresponds to equality of their temperatures.

References

[1] V.P. Belavkin, Zarubeznaja Radioelectronica, Soviet Radio, 1975, 5, 3.

[2] J.S. Baras, R.O. Harger, Y.H. Park IEEE Trans., 1976, IT-22, 59.

[3] R.E. Kalman, R.S. Bucy, Trans. ASMSE, ser. D, J. Basic Eng., 1961, 83, 95.

[4] R.L. Stratonovitch, Conditional Markovian Processes and their Application to the Optimal Control Theory, Moscow Univ. Publ., 1966.

[5] H.A. Haus. Proc. I.E.E.E., 1970, 58, 1599.

[6] M. Lax, Phys. Rev., 1965, 145, 110.

[7] S.D. Personick, IEEE Trans., 1971, IT-17, 240.

[8] M.A. Naymark, Reports of Sc. Ac. of the USSR, Math. series, 1943, 41, 359.

[9] V.P. Belavkin, Radiotechnics and Electronics, 1972, 17 , 12, 2533.

[10] V.P. Belavkin, Problems of Control and Information Theory, 1975, 4 (33), 241.

COVARIANT POV-MEASURES FOR W^*-DYNAMICAL SYSTEMS

Thomas Breuer

Peterhouse, University of
Cambridge, CB2 1RD, UK

Abstract

POV-measures on homogeneous spaces are defined with respect to automorphic group representations on von Neumann algebras. We prove some existence results for covariant POV-measures: If the automorphic representation of G is integrable in the sense of Connes and Takesaki, then there are covariant POV-measures on all homogeneous spaces G/H. The existence of a covariant POV-measure on G/H, together with the compactness of H, implies integrability of the group representation.

1 INTRODUCTION

POV-measures, as introduced by Davies and Lewis [7] and [16], proved very useful in mathematical as well as in experimental physics. (For some applications see Holevo [10].) Covariant POV measures on groups, or more generally on homogeneous spaces, are an important tool for the description of coherent states [20] and of quantisation procedures. Traditionally covariant POV-measures on homogeneous spaces G/H have been defined with respect to unitary representations of G [1]. Amann [3] proposed to consider POV-measures on G with values in a von Neumann algebra $\mathcal{M} \subset \mathcal{B}(\mathcal{H})$ which are covariant with respect to representations of G as automorphism group of $\mathcal{M}$.

The use of arbitrary von Neumann algebras $\mathcal{M} \subset \mathcal{B}(\mathcal{H})$ instead of $\mathcal{B}(\mathcal{H})$ has the advantage that it allows for the description of classical systems, and of systems with both quantum and classical properties. This is of particular importance in quantisation theory, where superselection rules arise from inequivalent representations of some kinematical group [12, 15].

Quantum Communications and Measurement
Edited by V. P. Belavkin *et al.*, Plenum Press, New York, 1995

The aim of this paper is to investigate the existence of covariant POV-measures on homogeneous spaces with values in a von Neumann algebra. This generalises some of the work reviewed in [1] insofar as we make reference to general von Neumann algebras and automorphic group representations instead of $\mathcal{B}(\mathcal{H})$ and unitary group representations. The results of [3] are generalised in the sense that we regard covariant POV-measures not only on G but on arbitrary homogeneous spaces G/H. The main results will be that that integrability of the automorphic representation is sufficient for the existence of covariant POV-measures, and that the existence of covariant POV-measures together with the compactness of H implies integrability.

2 AUTOMORPHIC GROUP ACTIONS ON VON NEUMANN ALGEBRAS

First we introduce representations of groups as automorphisms of von Neumann algebras, and relate these to unitary group representations. We start with the following

Definition 1 *A* W*-system $(\mathcal{M}, G, \alpha)$ *consists of a W*-algebra* $\mathcal{M}$ *with separable predual, a locally compact, separable group* G*, and a representation* $\alpha : G \to Aut(\mathcal{M})$ *of* G *as a group of automorphisms of* $\mathcal{M}$ *such that*
(i) $\alpha_{g_1 g_2} = \alpha_{g_1}\alpha_{g_2}$ *and*
(ii) for all operators $x \in \mathcal{M}$ *the function* $g \mapsto \alpha_g(x)$ *is* σ*-weakly continuous.*
The action α *of* G *on* $\mathcal{M}$ *is called* ergodic *if the only elements* $x \in \mathcal{M}$ *fulfilling* $\alpha_g(x) = x$ *for all* $g \in G$ *are scalar multiples of the identity,* $x = \lambda\mathbf{1}$.

The assumption that $\mathcal{M}$ has separable predual is equivalent to the fact that $\mathcal{M}$ is isomorphic to a von Neumann algebra on a separable Hilbert space ([19], 3.9.9). Although for some of the mathematical results this assumption is not necessary, it is usually made in physical appications.

EXAMPLE 1: The systems of traditional quantum mechanics can be regarded as W*-systems in the following way. A quantum mechanical system is specified by a unitary ray representation U of a kinematical group G, which is for example the Poincaré group or the Galilei group. Associated to U is a representation α of G as a group of automorphisms of $\mathcal{B}(\mathcal{H})$ defined by

$$\alpha_g(x) := U_g x U_g^*, \qquad g \in G, x \in \mathcal{B}(\mathcal{H}).$$

Then $(\mathcal{B}(\mathcal{H}), G, \alpha)$ is a type I factor W*-system. The choice of a type I factor for $\mathcal{M}$ is a consequence of von Neumann's irreducibility postulate. On $\mathcal{B}(\mathcal{H})$, the irreducibility of U is equivalent to the ergodicity of α [5].

Conversely, every W^*-system, where $\mathcal{M}$ is a factor of type I, can be brought into the form $(\mathcal{B}(\mathcal{H}), G, U \cdot U^*)$. This is due to the fact that all automorphisms α of a type I factor are inner: they are induced by a unitary operator $U \in \mathcal{M}$ by $\alpha(x) = UxU^*$.

EXAMPLE 2: Systems of classical mechanics can also be described by W^*-systems. If Ω is the phase space of the system, take as algebra of observables $L^\infty(\Omega)$, the functions on Ω which are bounded almost everywhere with respect to the Liouville measure. Ω, regarded as the additive group of boosts and translations, acts on $L^\infty(\Omega)$ by

$$(\mathrm{Ad}\lambda(\omega')f)(\omega) := f(\omega - \omega').$$

As these examples show, W^*-systems generalise quantum mechanical and classical systems. They can also describe systems which are partially classical and partially quantum mechanical.

A unitary ray representation U of G leads to an automorphism of $\mathcal{B}(\mathcal{H})$. Thus every system of covariance in the traditional sense can be regarded as a system of covariance with respect to the W^*-system $(\mathcal{B}(\mathcal{H}), G, U \cdot U^*)$. But the converse in not true: an automorphic representation of G on $\mathcal{M}$ cannot always be replaced by a unitary ray representation of G on $\mathcal{H}$.

Let me expand on this for a moment. An automorphic group representation α cannot be replaced with a unitary ray representation if there is some automorphism α_g which is not spatial. (An automorphism α_g is called spatial if it is of the form $\alpha_g = U_g \cdot U_g^*$ for some unitary U_g, U_g not necessarily in $\mathcal{M}$.) I do not know of any necessary and sufficient condition for an automorphism to be spatial. So I will just quote some partial results. If $\mathcal{M}$ is a factor of types I or III, any automorphism is spatial. But if $\mathcal{M}$ is a factor of type II_∞ with commutant of type II_1, then there are examples of automorphisms which are not spatial, even if $\mathcal{H}$ is separable ([13], 13.4.3)). It is possible to give a necessary and sufficient condition when an automorphism of a type II_∞ factor with commutant of type II_1 is spatial ([13], 9.6.33)). If $\mathcal{M}$ is not a factor, then one can guarantee that an automorphism is spatial if the central decomposition of $\mathcal{M}$ does not contain any factor of type II and if the automorphism leaves every central element fixed. This last condition is a severe restriction in physical applications. It can be dropped if for example the commutant of $\mathcal{M}$ is abelian ([8], III.3.2, Corollary to Proposition 2), or if the commutant is of the form $C \otimes F$, where C is an abelian von Neumann algebra and F is an irreducible factor ([8], III.3.2, Proposition 2), or if $\mathcal{M}$ is of type III and operates on a separable Hilbert space ([8], III.8.6, Corollary 8).

Even if all α_g are spatial it is not trivial to replace α by a unitary ray representation. The main difficulty is that for general Neumann algebras $\mathcal{M}$ the continuity requirement (ii) in Definition 1 cannot be formulated for unitary ray representations. Only in the special case that $\mathcal{M}$ has separable predual and that all automorphisms α_g are inner (i.e. of the form $\alpha_g = U_g \cdot U_g^*$ for some unitary operator U_g *in* $\mathcal{M}$), one can replace the automorphic representation of G by a weakly measurable unitary ray representation. But for many systems of interest the automorphisms α_g are not inner. For example, any automorphism which does not leave all central elements fixed cannot be inner.

3 INTEGRABILITY OF AUTOMORPHIC GROUP REPRESENTATIONS ON VON NEUMANN ALGEBRAS

Now I will introduce the notion of integrability for an automorphic group representation on a von Neumann algebra. This definition is due to Connes and Takesaki [6] and generalises the notion of square-integrability for unitary ray representations of a group on a Hilbert space.

Let $(\mathcal{M}, G, \alpha)$ be a W^*-system. The restriction of the left Haar measure dg to a compact subset K of G is finite. For each fixed $x \in \mathcal{M}$ there is exactly one $y \in \mathcal{M}$ such that $\rho(y) = \int_K \rho(\alpha_g(x))dg$ for all $\rho \in \mathcal{M}_*$. The integral $\int_K \alpha_g(x)dg$ is said to exist in the weak sense and one writes $y = \int_K \alpha_g(x)dg$. The Haar measure on G is not finite if G is not compact. One says that the integral $\int_G \alpha_g(x)dg$ exists if the set

$$\{\int_K \alpha_g(x)dg : K \subset G, K \,\text{compact}\}$$

is bounded in $\mathcal{M}$. In this case one writes $\int_G \alpha_g(x)dg < \infty$ and defines $\int_G \alpha_g(x)dg := \sup_K \int_K \alpha_g(x)dg$, where the supremum is taken over all compact subsets of G.

Definition 2 *A W^*-system $(\mathcal{M}, G, \alpha)$ is called* integrable *if the set*

$$n_\alpha := \{y \in \mathcal{M} : \int_G \alpha_g(y^*y)dg < \infty\}$$

is σ-weakly dense in $\mathcal{M}$.

If $\mathcal{M}$ is the von Neumann algebra $\mathcal{B}(\mathcal{H})$ of all bounded operators on a Hilbert space, then every automorphic representation α can be written as

$$\alpha_g(x) := U_g x U_g^*, \qquad x \in \mathcal{M},$$

where U is a weakly measurable unitary ray representation of G on $\mathcal{H}$.

Definition 3 *A Borel measurable unitary ray representation U of G on a Hilbert space $\mathcal{H}$ is* square-integrable *if there exist at least two non-zero vectors $\xi, \eta \in \mathcal{H}$ such that*

$$\int_G |\langle U_g \eta | \xi \rangle|^2 dg$$

exists.

Note that if G is not unimodular, there are always some pairs $\xi', \eta' \in \mathcal{H}$ such that $\langle U_g \eta' | \xi' \rangle$ is not a square integrable function on G [9]. According to ([4], theorem 2), a unitary ray representation U is square integrable if and only if the automorphic representation $U \cdot U^*$ is integrable on $\mathcal{B}(\mathcal{H})$.

Assume that we have two W^*-systems $(\mathcal{M}, G, \alpha)$ and $(\mathcal{N}, G, \beta)$ satisfying $\mathcal{M} \subset \mathcal{N}$ and $\alpha_g(x) = \beta_g(x)$ for all $x \in \mathcal{M}$. It is clear that the integrability of $(\mathcal{N}, G, \beta)$ implies the integrability of $(\mathcal{M}, G, \alpha)$, but not vice versa. In particluar, if an automorphic representation α on $\mathcal{M} \subset \mathcal{B}(\mathcal{H})$ can be replaced with a unitary ray representation U, the integrability of α does not imply the square integrability of U.

4 THE EXISTENCE OF COVARIANT POV-MEASURES

With the help of W^*-systems it is possible to generalise the notion of a covariant POV-measure.

Definition 4 *Let $(\mathcal{M}, G, \alpha)$ be a W^*-system. A POV-measure a on G/H with values in $\mathcal{M}$ is said to be covariant with respect to the automorphic representation α of G on $\mathcal{M}$ if*

$$\alpha_g(a(\Delta)) = a(g\Delta), \qquad \Delta \in \Sigma(X).$$

In this case (a, α) is called a generalised transitive system of covariance (a, α) based on G/H.

Traditonally systems of covariance are defined as POV-measures which are covariant with respect to some unitary ray representation. The present definition is more general than the traditional one because it refers to arbitrary W^*-systems, and not just to those of the form $(\mathcal{B}(\mathcal{H}), G, U \cdot U^*)$.

Now I will turn to some results on the existence of covariant POV-measures on G/H with values in $\mathcal{M}$. The results presented here are straightforward generalisations of the existence theorems of Amann [3] and Homma [11] for covariant POV-measures on G.

Proposition 1 *Let $(\mathcal{M}, G, \alpha)$ be a W^*-system, and let H be a closed subgroup of G. If $(\mathcal{M}, G, \alpha)$ is integrable, then there exists a covariant POV-measure on G/H with values in $\mathcal{M}$.*

PROOF: To each $f \in L^\infty(G/H)$ associate a function $i(f) \in L^\infty(G)$ by

$$i(f)(s) := f(sH), \qquad s \in G.$$

According to [11], the integrability of $(\mathcal{M}, G, \alpha)$ implies the existence of a covariant embedding O of $(L^\infty(G), G, \mathrm{Ad}\lambda)$ into $(\mathcal{M}, G, \alpha)$. (For a definition of covariant embeddings see e.g. [3, 4].) Define a map $O' : L^\infty(G/H) \to \mathcal{M}$ by

$$O'(f) := O(i(f)), \qquad f \in L^\infty(G/H).$$

Normalisation, linearity, and positivity of O' are obviously implied by O having these properties. The covariance of O' follows from

$$\begin{aligned} O'(\mathrm{Ad}\lambda(g)f)(s) &= O(i(\mathrm{Ad}\lambda(g)f))(s) = f(g^{-1}sH) = \\ &= O(\mathrm{Ad}\lambda(g)i(f))(s) = \alpha_g(O(i(f))(s) = \alpha_g(O'(f))(s). \end{aligned}$$

A covariant POV-measure a on G/H is defined by taking $a(\Delta) := O'(\chi_\Delta)$ for any Borel subset of G/H. Since $\mathrm{Ad}\lambda(g)\chi_\Delta = \chi_{g\Delta}$, the covariance of O', i.e. $\alpha_g(O'(f)) = O'(\mathrm{Ad}\lambda(g)f)$, implies the covariance of a, i.e. $\alpha_g(a(\Delta)) = a(g\Delta)$. QED.

In general the existence of a covariant POV-measure on G/H does not imply integrability of $(\mathcal{M}, G, \alpha)$. But if H is a *compact* closed subgroup of G, then this implication holds.

Proposition 2 *Let $(\mathcal{M}, G, \alpha)$ be a W^*-system, and let H be a closed subgroup of G. If there exists a covariant POV-measure on G/H with values in $\mathcal{M}$ and if H is compact, then $(\mathcal{M}, G, \alpha)$ is integrable.*

PROOF: Since H is compact, the right Haar measure dh on H can be normalised, $\int_H dh = 1$. To every function $f \in L^\infty(G)$ associate a function $\overline{f}$ in $L^\infty(G/H)$ by

$$\overline{f}(gH) := \int_H f(gh)dh.$$

By assumption there exists a covariant POV-measure a on G/H with values in $\mathcal{M}$. By the spectral theorem, a can be extended to a covariant embedding O' of $(L^\infty(G/H), G, \mathrm{Ad}\lambda)$ into $(\mathcal{M}, G, \alpha)$. Define a map $O : L^\infty(G) \to \mathcal{M}$ by

$$O(f) := O'(\overline{f}), \qquad f \in L^\infty(G).$$

O is a covariant embedding of $(L^\infty(G), G, \mathrm{Ad}\lambda)$ into $(\mathcal{M}, G, \alpha)$: From $\overline{\mathrm{Ad}\lambda(g)f} = \mathrm{Ad}\lambda(g)\overline{f}$ and from covariance of O' we have

$$O(\mathrm{Ad}\lambda(g)f) = O'(\overline{\mathrm{Ad}\lambda(g)f}) = O'(\mathrm{Ad}\lambda(g)\overline{f}) = \alpha_g(O'(\overline{f}) = \alpha_g(O(f)).$$

Normalisation of the Haar measure dh and of O' imply $O(1) = O'(\int_H 1dh) = O'(1) = \mathbf{1}$. Positivity and linearity of O also follow from the fact that O' has these properties.

By Homma [11] the existence of a covariant embedding $O : L^\infty(G) \to \mathcal{M}$ implies that $(\mathcal{M}, G, \alpha)$ is integrable. QED.

5 OUTLOOK

Ali, Antoine, and Gazeau [2] introduced a weaker form of square integrability for unitary group representations. They defined square integrablity with respect to a given subgroup H and a cross-section $\sigma : G/H \to G$. Square integrability in the traditional sense implies square integrability with respect to all subgroups and cross-sections. But square integrability with respect to some subgroup and some crosssection does not imply square integrability in the tradtional sense. They showed that square integrability with respect to a subgroup H and a cross-section σ is equivalent to the existence of a quasi-covariant POV-measure on G/H.

In this spirit, it should be possible to introduce a notion of integrability of an automorphic representation with respect to a subgroup and a cross-section. One might conjecture that integrability of α with respect to a subgroup H and a cross-section σ is equivalent to the existence of a (quasi)covariant embedding of $L^\infty(G/H)$ into $(\mathcal{M}, G, \alpha)$.

It is perhaps worthwhile to sketch the connection between covariant POV-measures and coherent states. Assume that one is given a pure normal state ψ on type I von Neumann algebra $\mathcal{M}$. ψ serves as fiducial state and defines an atomic projector $p = |\psi\rangle\langle\psi|$ in $\mathcal{M}$. If the representation α is integrable, the integral $\int_G \alpha_g(p)dg$ exists in $\mathcal{M}$ in the weak sense and is fixed under all α_g. If α is additionally ergodic, the integral is a scalar multiple of the identity operator in $\mathcal{M}$. By appropriate normalisation of ψ we can achieve

$$\int_G \alpha_g(p)dg = \mathbf{1}.$$

Denote by H the subgroup of those $g \in G$ such that $\alpha_g(p) = p$. Then $\alpha_{gh}(p) = \alpha_{gh'}(p) =: \alpha_{gH}(p)$. By choosing an appropriately normalised measure μ on G/H one can therefore achieve

$$\int_{G/H} \alpha_{gH}(p)d\mu(gH) = \mathbf{1}.$$

This exactly the completeness condition for coherent states ([14], section I.2). Thus $\{\alpha_{gH}(p)\}$ is a set of coherent states. A covariant POV-measure on G/H with values in $\mathcal{M}$ is defined by

$$a(\Delta) := \int_\Delta \alpha_{gH}(p)d\mu(gH).$$

This way of constructing coherent states only works if $\mathcal{M}$ is a type I von Neumann algebra. If this is not the case we do not have any pure normal states ψ is $\mathcal{M}$. For a more general picture it would be necessary to define and integrate the action of G on mixed states. This will not be attempted here.

Scutaru [21], Moscovici and Verona [17], and Amann [4] studied the relevance of square-integrability for coherent states defined with respect to unitary (ray) representations of G. The present results of the relation between integrability of α on $\mathcal{M}$ and the existence of covaraint POV-measures should be seen in this context.

References

[1] Ali S. T. (1985), Riv. Nuovo Cim. **8** (11), 1-128

[2] Ali S. T., J.-P. Antoine, J.-P. Gazeau (1991a), Ann. Inst. H. Poincaré **55**, 829-855, 857-890

[3] Amann A. (1986a), Fortschr. Phys. **34**, 167-215

[4] Amann A. (1986b), J. Math. Phys. **27**, 2282

[5] Berberian S. K. (1974), *Lectures in Functional Analysis and Operator Theory*, New York: Springer

[6] Connes A., Takesaki M. (1977), Tôhôku Math. J. **29**, 473-575

[7] Davies E. B., Lewis J. T. (1970), Comm Math. Phys. **17**, 239

[8] Dixmier J. (1957), *Les algèbres d'opérateurs dans l'éspace Hilbertien (Algèbres de von Neumann)*, Paris: Gauthier-Villars

[9] Duflo M., Moore C. C. (1976), J. Funct. Anal. **21**, 209-243

[10] Holevo A. S. (1982), *Probabilistic and Statistical Aspects of Quantum Theory*, Amsterdam: North Holland,

[11] Homma M. (1988), Yokohama Math. J. **36**, 105

[12] Isham C. J. (1983), "Topological and global aspects of quantum theory", in B. S. Dewitt, R. Stora (eds.): *Relativity, Groups, and Topology 2.* Les Houches Summer School 1983, Amsterdam: North Holland, 1984

[13] Kadison R.V., Ringrose (1986), *Fundamentals of Operator Algebras*, Vols. I, II, New York: Academic Press

[14] Klauder J. R., B. S. Skagerstam (1984), *Coherent States. Applications in Physics and Mathematical Physics*, Singapore: World Scientific

[15] Landsman N. P. (1990), *Reviews in Mathematical Physics* **2** (1), 45-72, 73-104

[16] Ludwig G., Comm. Math. Phys. bf 4, 331-348; **9**, 1-12; **26**, 78-86

[17] Moscovici H., Verona A. (1978), *Ann. Inst. H. Poincaré* **29**, 139

[18] Nakagami Y., Takesaki M. (1979), *Duality for Crossed Products of von Neumann Algebras*, Lecture Notes in Mathematics **731**, Berlin: Springer

[19] Pedersen G. K. (1979), *C^*-algebras and their Automorphism Groups*, London: Academic Press

[20] Perelomov A. M. (1972), Comm. Math. Phys. **26**, 222

[21] Scutaru H. (1977), Lett. Math. Phys. **2**, 101

ON AN EFFECTIVENESS OF QUANTUM MINI-MAX FORMULA IN QUANTUM COMMUNICATION

Masao Osaki and Osamu Hirota

Center for the study of
production development engineering
Tamagawa University Research Institute
Tamagawa University, Machida-city,
Tokyo 194 (Japan)

1 INTRODUCTION

The quantum detection theory is much attractive in practical design problems of quantum communication systems because it can predict an ultimate performance of optical communication systems[1]. Recently many researchers have tried to apply the quantum Bayes formula to practical optical communication systems. In some case, unfortunately, the quantum Bayes formula can not provide analytical solutions because of the difficulties to solve the nonlinear equations for unknown variables. However, Hirota[2] showed the fact that the quantum mini-max formula may provide a simplification of the calculation process and this was confirmed for binary pure state signals, orthogonal signal for pure state, and others[3]. Especially, for ternary symmetric signal system, it is quite difficult to obtain an analytical solution in the case of Bayes. But the case of mini-max is solvable. Although the analytical solutions of the optimum detection operators are purely mathematical descriptions of the quantum detection process, they will provide the important information to the implementation problem designing a practical system with the minimum error probability, so-called "the optimum quantum receiver" [4,5].

In this paper, we confirm the effectiveness of the quantum mini-max formula compared with the quantum Bayes one. The theoretical investigations are tried in both formulae for binary phase shift keyed (BPSK), ternary symmetric, and 3 phase shift keyed (3PSK) coherent state signal sets. When the analytical solution

Quantum Communications and Measurement
Edited by V. P. Belavkin *et al.*, Plenum Press, New York, 1995

can not be obtained from the formulae, the numerical analysis is also carried out if it is possible. By means of these analyses, the correctness and effectiveness of the quantum mini-max formula are confirmed. Especially in the case of ternary symmetric signal set, we can show the precise error probability by the quantum mini-max formula including the former one given by Hirota[2]. This paper consists of 4 chapters. Chapter 2 is devoted to the brief summary of the quantum detection theory, where fundamental formulae of the quantum Bayes and mini-max strategies are given. In Chapter 3, the identification of the error probabilities by the mini-max and the worst ones by the Bayes is confirmed for BPSK, ternary symmetric, and 3PSK signal sets. The optimum detection operators are also given for every signal set by the quantum mini-max, and for BPSK signal set by the quantum Bayes formula. Chapter 4 is the conclusion of this paper.

2 QUANTUM DETECTION THEORY

There are some strategies in the quantum detection theory standing on the similar concept of the classical detection (decision) theory. That is, the quantum Bayes, Neyman-Pearson, and mini-max strategies are considered[1, 2]. In this paper, the quantum Bayes and mini-max are investigated because the quantum Bayes strategy gives the minimum risk when the signal prior-probabilities are known at a detection side. On the other hand, when the signal prior-probabilities are unknown at a detection side, the quantum mini-max strategy is the most useful since it guarantees the minimum worst risk for all signal prior-probabilities. This minimum worst risk is equal to the minimum risk achieved by the quantum Bayes with the least favorable signal prior-probabilities. Furthermore the quantum mini-max can simplify the derivation process of a solution munch more compared with the case of the quantum Bayes.

Since we discuss the application of the quantum detection theory to communication systems, the risk function to be optimized is the error probability defined as follows:

$$Pe = 1 - \sum_{i=1}^{M} \xi_i \mathrm{Tr} \hat{\rho}_i \hat{\Pi}_i, \tag{1}$$

where $\hat{\rho}_i$, $\hat{\Pi}_i$, and ξ_i are the density operator for a signal quantum state, the detection operator, and a prior-probability of signal $\hat{\rho}_i$ respectively, and M is the number of signals. The quantum detection corresponds to a separation of a signal space which is a subspace of Hilbert space spanned by signal quantum states, $\{\hat{\rho}_i\}$. Therefore, the detection operators, $\left\{\hat{\Pi}_i\right\}$, is the resolution of identity.

$$\sum_{i=1}^{M} \hat{\Pi}_i = \hat{I}, \tag{2}$$

Since we assume that signal quantum states are pure states for practical calculation, the next lemma is much useful.

Lemma (by Kennedy) *when the signal quantum states are linearly independent, the optimum p.o.m. for the error probability is indeed projection-valued.*

Namely, detection operators can be explained by measurement states, $|\omega_i\rangle$, as follows:

$$\hat{\Pi}_i = |\omega_i\rangle\langle\omega_i|, \quad \text{and} \quad \langle\omega_i|\omega_j\rangle = \delta_{ij}, \tag{3}$$

so that signal quantum states, $\{|\psi_i\rangle\}$, can be defined by measurement states with variables, $x_{ji} \equiv \langle\omega_j|\psi_i\rangle$.

$$|\psi_i\rangle = \sum_{j=1}^{M} x_{ji}|\omega_j\rangle. \tag{4}$$

Since this equation can be represented in the matrix form as

$$\begin{bmatrix} |\psi_1\rangle \\ \vdots \\ |\psi_M\rangle \end{bmatrix} = \begin{bmatrix} x_{11} & \cdots & x_{M1} \\ \vdots & \ddots & \cdots \\ x_{1M} & \cdots & x_{MM} \end{bmatrix} \begin{bmatrix} |\omega_1\rangle \\ \vdots \\ |\omega_M\rangle \end{bmatrix}, \tag{5}$$

the measurement states can be explained as follows:

$$\begin{bmatrix} |\omega_1\rangle \\ \vdots \\ |\omega_M\rangle \end{bmatrix} = [x_{ji}]^{-1} \begin{bmatrix} |\psi_1\rangle \\ \vdots \\ |\psi_M\rangle \end{bmatrix}. \tag{6}$$

And the error probability is also explained with $\{x_{ji}\}$

$$Pe = 1 - \sum_{i=1}^{M} \xi_i |x_{ii}|^2. \tag{7}$$

Therefore we need to find the matrix of $\{x_{ji}\}$ to minimize the error probability and to determine the optimum measurement states. The rules to find the optimum matrix, $[x_{ji}]$, are explained in the following.

2.1 Quantum Bayes Strategy

The quantum Bayes strategy is the minimization of the error probability with respect to the detection operators.

$$\min_{\hat{\Pi}} Pe = \min_{\hat{\Pi}} \left\{ 1 - \sum_{i=1}^{M} \xi_i \mathrm{Tr}\hat{\rho}_i\hat{\Pi}_i \right\}. \tag{8}$$

The necessary and sufficient conditions for the optimum detection operators become the next formula[1].

$$\hat{\Pi}_j[\xi_j\hat{\rho}_j - \xi_i\hat{\rho}_i]\hat{\Pi}_i = 0, \quad \forall i, j, \tag{9}$$

$$\hat{\Gamma} - \xi_i\hat{\rho}_i \geq 0, \quad \forall i, \tag{10}$$

where

$$\hat{\Gamma} = \sum_{i=1}^{M} \xi_i\hat{\rho}_i\hat{\Pi}_i = \sum_{i=1}^{M} \xi_i\hat{\Pi}_i\hat{\rho}_i. \tag{11}$$

By means of Kennedy's Lemma, Eq.(9) can be represented by $\{x_{ji}\}$ as follows:

$$\xi_k x_{kk} x_{mk}^* - \xi_m x_{km} x_{mm}^* = 0. \tag{12}$$

Since the inner products of the signal quantum states are related with $\{x_{ji}\}$ as

$$\kappa_{ij} = \langle\psi_i|\psi_j\rangle = \sum_{k=1}^{M} \langle\psi_i|\omega_k\rangle\langle\omega_k|\psi_j\rangle = \sum_{k=1}^{M} x_{kj}x_{ki}^*, \tag{13}$$

M^2 equations are prepared by Eqs.(12) and (13) for M^2 unknown variables, $\{x_{ji}\}$. Therefore all of $\{x_{ji}\}$ will be solved in principle.

2.2 Quantum Mini-max Strategy

The quantum mini-max strategy is also the application of the classical one, where the error probability is maximized with respect to the prior-probability, $\{\xi_i\}$, and minimized by choosing the detection operators, $\left\{\hat{\Pi}_i\right\}$.

$$\min_{\hat{\Pi}} \max_{\xi} Pe = \max_{\xi} \min_{\hat{\Pi}} Pe. \tag{14}$$

The mathematical foundation of the quantum mini-max detection rule such as complete class theorem, existence theorem, strict determinateness was clarified by Holevo[6] and Ozawa[7] taking into account the following items of classical theory:

1) Any mini-max rule is Bayes relative to the least favorable prior-probability.

2) A mini-max rule has a smaller maximum average risk than any other.

Hirota gave the necessary and sufficient conditions of the quantum mini-max detection operator by taking into account the next item,

3) A Bayes decision rule whose conditional risk is constant is a mini-max rule.

This provides the following mini-max condition to the quantum Bayes formula given in Eqs.(9) and (10).

$$\mathrm{Tr}\hat{\rho}_i\hat{\Pi}_i = \mathrm{Tr}\hat{\rho}_j\hat{\Pi}_j, \quad \forall i,j. \tag{15}$$

By use of the relation, $x_{ji} \equiv \langle\omega_j|\psi_i\rangle$, Eq.(15) is represented as follows:

$$|x_{ii}|^2 = |x_{jj}|^2, \quad \forall i,j. \tag{16}$$

This condition makes much easier to solve the nonlinear equations given by Eqs.(12) and (13) as we will see in the following.

3 QUANTUM MINI-MAX STRATEGY FOR SEVERAL SIGNAL SETS

The applications of the quantum detection theory to the practical signal sets have been carried out by many researchers. It has been revealed that the quantum Bayes formula is too complex to obtain the analytical solutions except for a few quantum signal sets while the quantum mini-max can more. In the following, we confirm the correctness and effectiveness of the quantum mini-max formula for BPSK, ternary symmetric, and 3PSK coherent state signal sets.

3.1 Binary Phase Shift Keyed Coherent State Signal

In the case of binary phase shift keyed (BPSK) coherent state signal set, signal quantum states are represented as $[|\psi_1\rangle, |\psi_2\rangle] = [|\alpha\rangle, |-\alpha\rangle]$. Then, based on the quantum Bayes formula, Eqs.(12) and (13) become as follows:

$$\left\{\begin{array}{rcl} x_{11}^2 + x_{21}^2 & = & 1, \\ x_{12}^2 + x_{22}^2 & = & 1, \\ x_{11}x_{12} + x_{21}x_{22} & = & \kappa, \\ \xi_1 x_{21} x_{11} & = & \xi_2 x_{12} x_{22}, \end{array}\right. \tag{17}$$

where κ is an inner product defined as $\kappa = \langle\alpha| - \alpha\rangle = \exp[-2N_s]$ and $N_s = |\alpha|^2$. Unfortunately, these equations for $\{x_{ji}\}$ are too difficult to derive the solutions even if the equal signal prior-probabilities are assumed. However, by use of eigenvalue equations explained in [1], $\{x_{ji}\}$ can be determined analytically while the calculation process is much tedious and requires skillful algorithm. As a result, the error probability becomes

$$Pe = \frac{1 - \sqrt{1 - 4\xi_1\xi_2\kappa^2}}{2}, \tag{18}$$

and the optimum detection operators are obtained as follows:

$$\left\{\begin{array}{lcl} \hat{\Pi}_1 & = & \frac{1}{4R(1-\kappa^2)}\left[\{2R - 2\lambda\kappa^2 + (1+\lambda)\}\,|\alpha\rangle\langle\alpha| \right. \\ & & \quad + \{2R + 2\kappa^2 - (1+\lambda)\}\,|-\alpha\rangle\langle-\alpha| \\ & & \quad \left. -\kappa(2R + 1 - \lambda)(|\alpha\rangle\langle-\alpha| + |-\alpha\rangle\langle\alpha|)\right], \\ \hat{\Pi}_2 & = & \frac{1}{4R(1-\kappa^2)}\left[\{2R + 2\lambda\kappa^2 - (1+\lambda)\}\,|\alpha\rangle\langle\alpha| \right. \\ & & \quad + \{2R - 2\kappa^2 + (1+\lambda)\}\,|-\alpha\rangle\langle-\alpha| \\ & & \quad \left. -\kappa(2R - 1 + \lambda)(|\alpha\rangle\langle-\alpha| + |-\alpha\rangle\langle\alpha|)\right], \end{array}\right. \tag{19}$$

where

$$R = \sqrt{\left(\frac{1+\lambda}{2}\right)^2 - \lambda\kappa^2}, \qquad \lambda = \xi_1/\xi_2. \tag{20}$$

On the other hand, the quantum mini-max formula adds the mini-max condition to the quantum Bayes formula. Then the Eq.(17) is reduced to

$$a^2 + b^2 = 1, \qquad 2ab = \kappa, \tag{21}$$

where $a^2 \equiv x_{11}^2 = x_{22}^2$ and $b \equiv x_{12} = x_{21}$. Therefore we can obtain the solutions of $\{x_{ji}\}$ by elementary algebra, where the signal prior-probabilities for signal '1' and '2' are equal. The error probability results in as follows:

$$Pe = \frac{1 - \sqrt{1 - \kappa^2}}{2}. \tag{22}$$

The optimum detection operators are also derived easily as follows:

$$\left\{\begin{array}{lcl} \hat{\Pi}_1 & = & \frac{1}{2(1-\kappa^2)}\left\{(1 + \sqrt{1-\kappa^2})|\alpha\rangle\langle\alpha| + (1 - \sqrt{1-\kappa^2})|-\alpha\rangle\langle-\alpha| \right. \\ & & \quad \left. -\kappa(|\alpha\rangle\langle-\alpha| + |-\alpha\rangle\langle\alpha|)\right\}, \\ \hat{\Pi}_2 & = & \frac{1}{2(1-\kappa^2)}\left\{(1 - \sqrt{1-\kappa^2})|\alpha\rangle\langle\alpha| + (1 + \sqrt{1-\kappa^2})|-\alpha\rangle\langle-\alpha| \right. \\ & & \quad \left. -\kappa(|\alpha\rangle\langle-\alpha| + |-\alpha\rangle\langle\alpha|)\right\}. \end{array}\right. \tag{23}$$

Evidently the error probability in Eq.(22) is equal to Eq.(18) for $\xi_1 = \xi_2$ and this is the worst case of the Bayes. The optimum detection operators in Eq.(23) are also the same with those by the quantum Bayes in Eq.(19) for $\xi_1 = \xi_2$. Therefore, the correctness of the quantum mini-max formula is confirmed. The optimum detection operators in Eq.(23) will give the good information to understand the practical optimum quantum receiver by Dolinar[4].

3.2 Ternary Symmetric Coherent State Signal

In the case of the ternary symmetric coherent state signal set, signal quantum states are represented by $[|\psi_1\rangle, |\psi_2\rangle, |\psi_3\rangle] = [|0\rangle, |\alpha\rangle, |-\alpha\rangle]$. Because of the symmetry of the inner product matrix and on the assumption that the prior-probabilities for

$|\psi_2\rangle, |\psi_3\rangle$ are equal, the matrix of $\{x_{ji}\}$ can be given with the reduced number of variables as follows:[1]

$$[x_{ji}] = \begin{bmatrix} a & d & d \\ b & c & e \\ b & e & c \end{bmatrix}. \tag{24}$$

The nonlinear equations for these variables by the quantum Bayes formula are too difficult to be solved analytically but the equations from Eq.(13) can be simplified to the representations by 'b'.

$$\begin{aligned} a &= \left\{2\kappa b + (1-\kappa^2)\sqrt{1+\kappa^4-2b^2}\right\}/(1+\kappa^4), \\ c &= \left\{\sqrt{1+\kappa^4-2b^2}+\sqrt{1-\kappa^4}\right\}/2, \\ d &= \left\{\kappa\sqrt{1+\kappa^4-2b^2}-b(1-\kappa^2)\right\}/(1+\kappa^4), \\ e &= \left\{\sqrt{1+\kappa^4-2b^2}-\sqrt{1-\kappa^4}\right\}/2, \end{aligned} \tag{25}$$

where κ is defined as $\kappa = \langle 0|\alpha\rangle = \langle 0|-\alpha\rangle = \exp[-N_s/2]$. Substituting these relations into Eq.(12), the solution of 'b' must satisfy the next equation.

$$0 = \xi_k x_{kk} x_{mk} - \xi_m x_{km} x_{mm} = \xi_1 ab - \xi_2 cd \equiv f(b). \tag{26}$$

By means of numerical analysis, the solution of 'b' can be obtained with respect to the prior-probability of signal 1, ξ_1. Applying the obtained 'b' into Eq.(25) and then into Eq.(7), the numerical error probability is derived for each ξ_1.

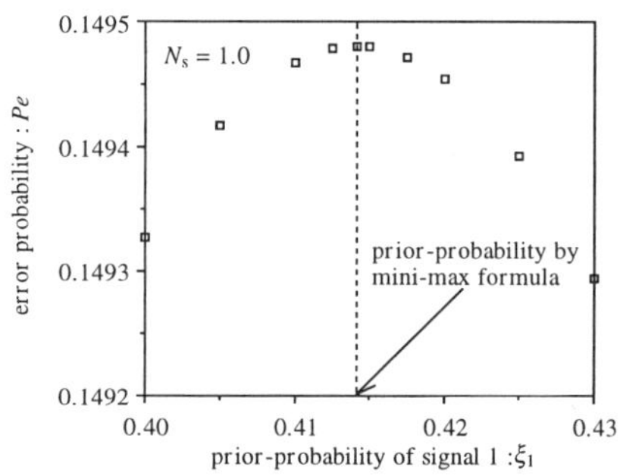

Figure 1: Identification of the solution by mini-max and the worst one by Bayes for signal prior-probability.

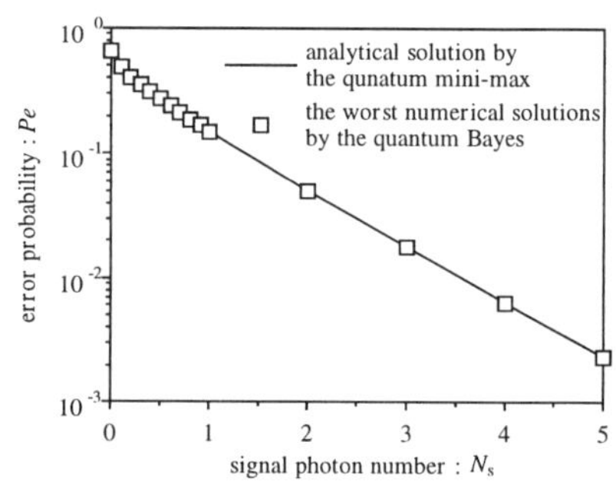

Figure 2: Identification of the solutions by mini-max and the worst ones by Bayes for each signal photon number.

On the other hand, when the quantum mini-max formula is applied to this signal set, we can put $|a|^2 = |c|^2$ in Eq.(25) so that the analytical solution for 'b' is obtained. While there are some variations of result derived from the mini-max formula, we adopt one satisfying the mini-max condition, $|a|^2 = |c|^2$, as the accurate solution. Then the analytical error probability by quantum mini-max formula is given as follows:

$$\begin{aligned} Pe &= 1 - \tfrac{1}{2}\Big\{1 - \tfrac{1}{2A}\left(B^2 - 2AC \pm B\sqrt{B^2-4AC}\right) \\ &\quad + \sqrt{1-\kappa^8 - \tfrac{1-\kappa^4}{A^2}\left(B^2-2AC \pm B\sqrt{B^2-4AC}\right)}\Big\}, \end{aligned} \tag{27}$$

where

$$\left\{\begin{aligned} A &= -2\left\{(\kappa^4+2\kappa^2-1)^2 + 8\kappa^2\right\}, \\ B &= 8\kappa\sqrt{(1-\kappa^8)(1+\kappa^4)}, \\ C &= (1+\kappa^4)\left\{(\kappa^4+2\kappa^2-1)^2 - (1-\kappa^8)\right\}, \end{aligned}\right. \tag{28}$$

formula \ signal set		BPSK	Ternary	3PSK
Bayes	error probability	○	△	◎ *
	detection operator	○	×	◎ *
mini-max	error probability	◎	◎	◎
	detection operator	◎	◎	◎

◎ : analytical solution with elementary algebra
○ : analytical solution with new algorism
△ : numerical solution
× : difficult
* : equal signal prior probabilities

Table 1: Results of application of the quantum Bayes and mini-max formulae

and '+' of '±' is valid for $N_s < \ln[1+\sqrt{2}]$ and '−' of it is valid for $N_s > \ln[1+\sqrt{2}]$ as the condition, $|a|^2 = |c|^2$. In comparison between the solutions by Bayes and mini-max, it is confirmed that the error probability by mini-max is equal to the worst one by Bayes with respect to the signal prior-probability (Fig.1) and to the signal photon number (Fig.2). Therefore we can say that the quantum mini-max formula is correct. The analytical optimum detection operators for this signal set are also obtained by the quantum mini-max formula (see Appendix).
Namely, in the case of ternary symmetric signal set, the mini-max formula can give the analytical solutions of the error probability and the optimum detection operators while the Bayes formula can not. This is the effectiveness of the quantum mini-max formula.

3.3 3 Phase Shift Keyed Coherent State Signal

In the case of 3 phase shift keyed (3PSK) signal set, $[|\psi_1\rangle, |\psi_2\rangle, |\psi_3\rangle] = [|\alpha\rangle, |\alpha e^{2i\pi/3}\rangle, |\alpha e^{-2i\pi/3}\rangle]$, the quantum Bayes formula in Eqs.(12) and (13) can not be solved. Fortunately, assuming the equal signal prior-probabilities and using the eigenvalue equations similar to the case of BPSK, the error probability and the optimum detection operators can be derived analytically (see Appendix). However, these solutions are just equal to those obtained by the quantum mini-max formula since the condition of equal signal prior-probabilities is also the mini-max condition.

CONCLUSION

The correctness and effectiveness of the quantum mini-max formula were confirmed in the cases of BPSK, ternary symmetric, and 3PSK coherent state signal sets. Namely, it was confirmed that the error probabilities obtained by the quantum mini-max formula are equal to the worst ones by the quantum Bayes. This is the correctness. The effectiveness is also verified as shown in table 1. The quantum mini-max formula is much useful to obtain the analytical solutions for these signal sets without the special skill level. The analytically obtained optimum detection operators provide the good information to the implementation problem of the optimum quantum receivers.

Acknowledgment. We wish to thank Dr. K.Yamazaki, Mr. T.S.Usuda in our laboratory and other members of Society of Optical Communication Theory of Japan for their informative advices and discussions. This work was supported by Grants of Ministry of Education and Toray Science and Tech. Foundation. One of authors is supported by Foundation for C & C Promotion.

References

[1] C.W.Helstrom : *Quantum detection and estimation theory* Academic Press, 1976.

[2] O.Hirota and S.Ikehara : The Trans. the IECE Japan. **E65**, 627, 1982.

[3] M.Sekiguchi : Thesis of Master of Science, Keio University, (in Japanese), 1990.

[4] S.J.Dolinar : M.I.T. Res. Lab. Electron. Quart. Prog. Rep., No.111, 115, 1973.

[5] K.Yamazaki : Tech. Rep. IEICE of Japan, **IT94-17**, (in Japanese), 1994.

[6] A.S.Holevo : Proc. of the Steklov Inst. of Mathematics **124**, 1978.

[7] M.Ozawa : Research Reports on Inform. Sciences, Tokyo Inst. Tech. **A-74**, 1980.

APPENDIX: THE OPTIMUM DETECTION OPERATORS FOR 3 PSK AND TERNARY SYMMETRIC SIGNALS

The optimum detection operators for 3-dimensional signals, $[|\psi_1\rangle, |\psi_2\rangle, |\psi_3\rangle]$, can be represent as follows:

$$\begin{aligned} \hat{\Pi}_i = \; & \tfrac{1}{d}\{m_{i,11}|\psi_1\rangle\langle\psi_1| + m_{i,12}|\psi_1\rangle\langle\psi_2| + m_{i,13}|\psi_1\rangle\langle\psi_3| \\ & + m_{i,21}|\psi_2\rangle\langle\psi_1| + m_{i,22}|\psi_2\rangle\langle\psi_2| + m_{i,23}|\psi_2\rangle\langle\psi_3| \\ & + m_{i,31}|\psi_3\rangle\langle\psi_1| + m_{i,32}|\psi_3\rangle\langle\psi_2| + m_{i,33}|\psi_3\rangle\langle\psi_3|\}. \end{aligned} \tag{A1}$$

For 3-PSK signal set, the elements in the above are represented as

$$\left\{ \begin{aligned} d &= 18S^2I^2 \\ m_{1,11} &= m_{2,22} = m_{3,33} = G^2 \\ m_{1,12} &= m_{1,13}{}^* = m_{1,21}{}^* = m_{1,31} = m_{2,12} = m_{2,23} \\ &= m_{2,21}{}^* = m_{2,32}{}^* = m_{3,13}{}^* = m_{3,23} = m_{3,31} = m_{3,32}{}^* = GH \\ m_{1,22} &= m_{1,33} = m_{2,11} = m_{2,33} = m_{3,11} = m_{3,22} = |H|^2 \\ m_{1,23} &= m_{1,32}{}^* = m_{2,13}{}^* = m_{2,31} = m_{3,12} = m_{3,21}{}^* = H^{*2} \end{aligned} \right. \tag{A2}$$

where

$$\begin{aligned} G &= \sqrt{2}J(S+\sqrt{2}I) \\ H &= J(I-\sqrt{2}S) + 3\mathrm{i}K_\mathrm{s}\sqrt{1+2K_\mathrm{c}} \\ I &= \sqrt{(1+2K_\mathrm{c})}J, \\ J &= \sqrt{1+S-K_\mathrm{c}}, \\ S &= \sqrt{(1-K_\mathrm{c})^2 - 3K_\mathrm{s}^2}, \end{aligned} \tag{A3}$$

$$\begin{aligned} K_\mathrm{c} &= k\cos\left[\tfrac{\sqrt{3}}{2}N_\mathrm{s}\right], \\ K_\mathrm{s} &= k\sin\left[\tfrac{\sqrt{3}}{2}N_\mathrm{s}\right], \\ k &= \exp\left[-\tfrac{3}{2}N_\mathrm{s}\right]. \end{aligned} \tag{A4}$$

For the ternary symmetric signal set, the elements in Eq.(A1) become as

$$
\begin{cases}
d &= 4(1-\kappa^2)^2(1+\kappa^4)^2(1-\kappa^4), \\
m_{1,11} &= 4(1-\kappa^4)(1+\kappa^4)^2F^2, \\
m_{1,12} &= m_{1,13} = m_{1,21} = m_{1,31} = 4(1-\kappa^4)(1+\kappa^4)F\left\{b(1-\kappa^2)-\kappa F\right\}, \\
m_{1,22} &= m_{1,23} = m_{1,32} = m_{1,33} = 4(1-\kappa^4)\{b(1-\kappa^2)-\kappa F\}^2, \\
m_{2,11} &= m_{3,11} = D^2, \\
m_{2,12} &= m_{2,21} = m_{3,13} = m_{3,31} = -DE_+, \\
m_{2,13} &= m_{2,31} = m_{3,12} = m_{3,21} = DE_-, \\
m_{2,23} &= m_{2,32} = m_{3,23} = m_{3,32} = -E_+E_-, \\
m_{2,22} &= m_{3,33} = {E_+}^2, \\
m_{2,33} &= m_{3,22} = {E_-}^2,
\end{cases}
\tag{A5}
$$

where

$$
b = \begin{cases}
\left(-B+\sqrt{B^2-4AC}\right)/2 & \text{for } N_\mathrm{s} < \ln\left[1+\sqrt{2}\right], \\
\left(-B-\sqrt{B^2-4AC}\right)/2 & \text{for } N_\mathrm{s} > \ln\left[1+\sqrt{2}\right],
\end{cases}
\tag{A6}
$$

$$
\begin{cases}
A &= -2\left\{(\kappa^4+2\kappa^2-1)^2+8\kappa^2\right\}, \\
B &= 8\kappa\sqrt{(1-\kappa^8)(1+\kappa^4)}, \\
C &= (1+\kappa^4)\left\{(\kappa^4+2\kappa^2-1)^2-(1-\kappa^8)\right\},
\end{cases}
\tag{A7}
$$

and

$$
\begin{aligned}
D &= 2b(1+\kappa^4)\sqrt{1-\kappa^4}, \\
E_\pm &= \left[(1-\kappa^2)(1+\kappa^4) \pm \sqrt{1-\kappa^4}\left\{2b\kappa+(1-\kappa^2)F\right\}\right], \\
F &= \sqrt{1-2b^2+\kappa^4}.
\end{aligned}
\tag{A8}
$$

PROBABILITIES FOR OBSERVING MIXED QUANTUM STATES GIVEN LIMITED PRIOR INFORMATION

Matthew J. Donald

The Cavendish Laboratory,
Madingley Road, Cambridge CB3 0HE,
Great Britain.

Abstract

The original development of the formalism of quantum mechanics involved the study of isolated quantum systems in pure states. Such systems fail to capture important aspects of the warm, wet, and noisy physical world which can better be modelled by quantum statistical mechanics and local quantum field theory using mixed states of continuous systems. In this context, we need to be able to compute quantum probabilities given only partial information. Specifically, suppose that $\mathcal{B}$ is a set of operators. This set need not be a von Neumann algebra. Simple axioms are proposed which allow us to identify a function which can be interpreted as the probability, per unit trial of the information specified by $\mathcal{B}$, of observing the (mixed) state of the world restricted to $\mathcal{B}$ to be σ when we are given ρ – the restriction to $\mathcal{B}$ of a prior state. This probability generalizes the idea of a mixed state (ρ) as being a sum of terms (σ) weighted by probabilities. The unique function satisfying the axioms can be defined in terms of the relative entropy. The analogous inference problem in classical probability would be a situation where we have some information about the prior distribution, but not enough to determine it uniquely. In such a situation in quantum theory, because only what we observe should be taken to be specified, it is not appropriate to assume the existence of a fixed, definite, unknown prior state, beyond the set $\mathcal{B}$ about which we have information. The theory was developed for the purposes of a fairly radical attack on the interpretation of quantum theory, involving many-worlds ideas and the abstract characterization of observers as finite information-processing structures, but deals with quantum inference problems of broad generality.

Keywords :
quantum measurement theory, relative entropy.

Quantum mechanics started with the study of simple isolated systems like an electron and a proton in empty space. Such systems can be described by pure states — by wavefunctions. When we turn to non-isolated systems, however, in quantum statistical mechanics, or many body theory, or quantum optics, or local quantum field theory, description by wavefunctions is often inappropriate. Density matrix descriptions are called for not least because our observations of complex systems are always incomplete.

In this paper, I shall present a mathematical tool for dealing with incomplete information in quantum systems. I shall start by commenting briefly on the foundational questions which haunt any serious consideration of ignorance in quantum mechanics. I have expressed my views on these matters more fully in [1–3], in which I attempt to start from first principles and give a complete quantum measurement theory, but the mathematics given here does not depend on the details of those papers, and it is my hope that it might be of more general use. Indeed, it is that hope which motivates this attempt to bring the mathematics to a wider audience.

In spite of the importance of infinite-dimensional spaces, I shall leave aside some mathematical technicalities by assuming for most of this paper that we are working with a finite-dimensional Hilbert space. This assumption is merely for convenience of exposition and may ignored by those who normally ignore mathematical details. Such details are to be found in [2, 4, and 5].

The formalism will be introduced here through some elementary examples which are sufficient to demonstrate the essential ideas. These examples are given in terms of probability distributions which correspond to quantum states on sets of mutually commuting operators. Following the examples, a set of simple axioms is given which allows the generalization to arbitrary states and sets of operators. Finally, the full definition and some of its properties are presented in a theorem which also states the extension to infinite-dimensions.

According to theory, time propagation in a closed quantum system is governed by a deterministic evolution defined by unitary maps of the form $e^{-itH/\hbar}$. This evolution is well understood and experimentally confirmed in many circumstances. Nevertheless, the fundamental problem of the interpretation of quantum theory remains. This is that such Hamiltonian evolution does not seem sufficient to describe the world that we see. From time to time, the quantum state of the world appears to change abruptly. These changes are referred to as "collapses".

Orthodox wave mechanics, describes "collapse" as the replacement of a wavefunction by an eigenfunction of some "measured" operator. However, there are several problems with this idea. In a real experiment it is often hard to see how the operator being measured can be unambiguously defined. For example, in a bubble chamber experiment, are we measuring bubble positions or particle positions? For practical purposes, it doesn't make much difference, but, if "collapse" is a genuine physical process then it ought to be precisely definable. I have suggested in [1–3] that this problem can be dealt with working at the level of the observer and by characterizing an observer as an abstract processor of information. Observers however, are complex, localized, thermal systems and so should be described by density matrices rather than simply by pure states. By working with local states, it is also possible to deal with the problem that an instanteous collapse of a wavefunction for the whole universe would be manifestly non-relativistic.

Suppose then that we are given a density matrix ρ. We want to develop a formalism to meaure the a priori probability of a " generalized state collapse" taking ρ to some other state σ. The physical interpretation I propose, for such a collapse, is that, although ρ may be the "true" state, it is not a state we are capable of seeing. It is not possible for us, because of our natures, to see a macroscopic system in a mixture of macroscopically distinct states. We cannot see a cat except as something which is either dead or alive. σ — the state collapsed to — is one of the states which we are capable of seeing.

A weaker interpretation just says that measurement in quantum mechanics involves state change. We have information, which is far from complete, about the state before the measurement. We also have information about the possible states after the measurement. For example, we know that, as long as the experiment works, our macroscopic measuring device is going to give us a definite result. At this general level, we want to be able to compute probabilities.

Our preliminary goal then, is a function which we shall denote by $\text{app}(\sigma|\rho)$ which we can interpret as the probability of being able to mistake the state of the world for σ, despite the fact that it is actually ρ.

At the simplest level, a density matrix ρ can be expanded, in terms of eigenstates, into a sum $\rho = \sum_{m=1}^{M} r_m |\psi_m><\psi_m|$ and the number r_m is interpreted as the probability of observing state $\sigma_m = |\psi_m><\psi_m|$ given the prior state ρ. The first question in generalizing this idea is to ask for the probability of observing a general state σ. The second question asks whether it is appropriate to assume that our observations are sufficient to give us complete knowledge of either ρ or σ.

The second question goes to the heart of this contribution. In general, we can have very limited information about a complex system. But this is quantum mechanics! Nothing which is not observed should be taken for granted. For example, the EPR experiment may be interpreted as telling us that we must not assume that the spin of an electron is determined except through the complete circumstances of its observation. What one sees, depends on how one looks.

Suppose then that we have knowledge of only a few expectation values for ρ. We want to know the probability of observing different values for those expectations. In a classical inference problem, we would assume that there was a definite fixed unknown prior distribution. In quantum theory, because only what we observe should be taken to be definite, this assumption is inappropriate.

It is possible to express the idea of an incompletely known state either in terms of sets of states or in terms of restrictions of states. I find the latter to be conceptually simpler. The former is slightly more general but can be easily accomodated in the present framework [2].

Definition.

1) Let $\mathcal{H}$ be a Hilbert space and $\mathcal{B}(\mathcal{H})$ be the set of all bounded operators on $\mathcal{H}$. Then a state ρ on $\mathcal{B}(\mathcal{H})$ is defined to be a density matrix. Such a state can be considered to be a complex-valued function on $\mathcal{B}(\mathcal{H})$. Thus, if $\rho = \sum_{m=1}^{M} r_m|\psi_m><\psi_m|$ and $B \in \mathcal{B}(\mathcal{H})$ then $\rho(B) = \sum_{m=1}^{M} r_m <\psi_m|B|\psi_m>$.

2) If $\mathcal{B} \subset \mathcal{B}(\mathcal{H})$ then a state ρ on $\mathcal{B}$ will be the restriction to $\mathcal{B}$ of a state on $\mathcal{B}(\mathcal{H})$. In other words ρ is a complex-valued function which takes the form $\rho(B) = \rho'(B)$ for some density matrix ρ'.

Definition 2 may seem pernickety, but the central point of the present work is that if we have only have information identifying a state on $\mathcal{B}$ then we should not identify it with any particular possible density matrix ρ', but rather with the set of all possible extensions. In other words, ρ should be identified with $\{\rho' : \rho'|_{\mathcal{B}} = \rho\}$.

Classical example 1. Many of the conceptual issues in the present theory can be understood by considering how the theory applies to inference problems in classical probability theory. Such problems arise when we consider a set $\mathcal{B}$ of commuting operators. A state on $\mathcal{B}$ is then a set of expectations compatible with some probability distribution corresponding to a state on an Abelian von Neumann algebra containing $\mathcal{B}$. For example, when $\mathcal{B}$ takes the form $\{A_n = \sum_{m=1}^{M} a_{nm} P_m : n = 1, \ldots, N\}$ for a sequence $(P_m)_{m=1}^{M}$ of commuting projections and some given matrix (a_{nm}), then a state ρ on $\mathcal{B}$ is determined by given values $\rho(A_n) : n = 1, \ldots, N$ and

$$\begin{aligned}\rho \;&= \{(r_m)_{m=1}^{M} : 0 \leq r_m \leq 1, \textstyle\sum_{m=1}^{M} r_m = 1, \text{ and} \\ &\qquad \textstyle\sum_{m=1}^{M} a_{nm} r_m = \rho(A_n) \text{ for } n = 1, \ldots, N\}.\end{aligned}$$

In general, we shall work with an arbitrary set of operators $\mathcal{B}$. The specification of $\mathcal{B}$ may well be the most difficult task in using the present theory for a practical quantum inference problem. $\mathcal{B}$ should be a set of operators about which the observer has direct information. In general, this will be a set of observables for the macroscopic experimental device rather than for the microscopic system being investigated. ρ will be the restriction to $\mathcal{B}$ of the initial quantum state.

With $\mathcal{B}$, we now have another fundamental element in our interpretation and our goal must be revised. The aim now is to define a function $\text{app}_{\mathcal{B}}(\sigma|\rho)$ which is to be interpreted as the probability per unit trial of the information in $\mathcal{B}$ of being able to mistake the state of the world on $\mathcal{B}$ for σ, despite the fact that it is actually ρ.

A von Neumann algebra represents a complete subsystem. Specifying a state on a von Neumann algebra, corresponds on the classical level to the complete specification of a probability distribution. Having a state on a non-algebra corresponds to having one of an unknown range of probability distributions.

Classical example 2. Suppose that $\mathcal{B} = \mathcal{Z}$ - a finite Abelian von Neumann algebra generated by a sequence $(P_m)_{m=1}^{M}$ of commuting projections. States ρ and σ on $\mathcal{Z}$ correspond to probability distributions $(r_m)_{m=1}^{M}$ and $(s_m)_{m=1}^{M}$ on $\{1, \ldots, M\}$. If the state of the world on $\mathcal{Z}$ is actually ρ then, when we choose a sequence of points from $\{1, \ldots, M\}$, the probability of getting m is given by r_m. We would mistake this state for σ, if in N trials we found that, for each m, we got m roughly $s_m N$ times. The probability (A) of such a result can be explicitly calculated using the multinomial distribution. Indeed, if each $s_m N$ is an integer, then

$$A = N! \prod_{m=1}^{M} \frac{r_m^{s_m N}}{(s_m N)!}.$$

As $N \to \infty$, $\log A$ is asymptotic to

$$N\{\sum_{m=1}^{M} (-s_m \log s_m + s_m \log r_m)\}.$$

This suggests that it would be appropriate for app to satisfy

$$\text{app}_{\mathcal{Z}}(\sigma|\rho) = \exp\{\sum_{m=1}^{M} (-s_m \log s_m + s_m \log r_m)\}.$$

Classical example 3. In a specific case of example 1, we take $\mathcal{B} = \{P = P_1 + P_2, Q = P_1 + P_3\}$ where $(P_m)_{m=1}^{4}$ are orthogonal projections such that $\sum_{m=1}^{4} P_m = 1$. A state ρ on $\mathcal{B}$ is a set of probability distributions on $\{1,\ldots,4\}$ with $\text{Prob}\{1,2\} = \rho(P)$ and $\text{Prob}\{1,3\} = \rho(Q)$ determined:

$$\rho = \{(r_m)_{m=1}^{4} : 0 \leq r_m \leq 1, \sum_{m=1}^{4} r_m = 1 \text{ and } r_1 + r_2 = \rho(P), r_1 + r_3 = \rho(Q)\}.$$

It is possible to make a complete computation of $\text{app}_{\mathcal{B}}$ when $\mathcal{B}$ has the form given in this example [5]. It would be interesting to find a physical situation where such a set was relevant. More typical, physically, might be a case where we have non-commuting projections P and Q and know $\rho(P)$, $\rho(Q)$, and $\rho(P \wedge Q)$, but not $\rho(P^m Q^n)$ for $m, n > 1$.

Classical example 4. In the situation of example 3, knowledge of ρ on $\mathcal{B}$ does not yield a complete prior distribution $(r_m)_{m=1}^{4}$. In example 2, we considered observing a sequence of N trials of the set $\{1,\ldots,M\}$. Each trial can be thought of as an evaluation of the M random variables (X^m) on $\{1,\ldots,M\}$, where X^m is the characteristic function of $\{m\}$. In example 3, however, we only have access at a trial to two random variables: X^P — the characteristic function of $\{1,2\}$ and X^Q — the characteristic function of $\{1,3\}$. All that is known about the distribution of these variables is that $E(X^P) = \rho(P)$ and that $E(X^Q) = \rho(Q)$. We want to find a value for the probability per trial that some distribution compatible with ρ gives values compatible with σ. If we make a choice of $(r_m)_{m=1}^{4}$, then an appropriate value would be given by the asymptotic probability per trial that, under the distribution $(r_m)_{m=1}^{4}$, the set $\{1,2\}$ is visited with relative frequency $\sigma(P)$ and the set $\{1,3\}$ is visited with relative frequency $\sigma(Q)$. A mathematical expression for this value is given by

$$V = \lim_{\epsilon \to 0} \lim_{N \to \infty} \Big(\text{Prob}(|\frac{1}{N}(\sum_{n=1}^{N} X_n^P) - \sigma(P)| \leq \epsilon, |\frac{1}{N}(\sum_{n=1}^{N} X_n^Q) - \sigma(Q)| \leq \epsilon)\Big)^{1/N}$$

where Prob is defined by $(r_m)_{m=1}^{4}$, and $(X_n^P)_{n=1}^{N}$ and $(X_n^Q)_{n=1}^{N}$ are the relevant sequences of independent identically distributed random variables with distributions determined by Prob. According to Sanov, [6, thm. 2],

$$V = \sup\{\exp[\sum_{m=1}^{4} -s_m \log(s_m/r_m)] : \quad 0 \leq s_m \leq 1, \textstyle\sum_{m=1}^{4} s_m = 1 \text{ and}$$
$$s_1 + s_2 = \sigma(P), s_1 + s_3 = \sigma(Q)\}.$$

The definition I am proposing suggests that, in this situation, given no other information, it is appropriate to choose $(r_m)_{m=1}^{4}$ so as to maximize this value. This suggestion is definitely a quantum mechanical one. It is appropriate because what we do not measure has no influence, or even, no reality. We see what we are capable of seeing, in the form which our apparatus has been set up to allow us to see. The situation pulls out one of the states which is possible for us with probabilitities determined only by the information available to us.

These examples have demonstrated the concepts at issue. The generalization to states on arbitrary sets of operators will now be given by the following set of axioms. The first three axioms are entirely self-explanatory. It should be noted that the arguments for the other axioms are merely arguments — they are not proofs. A much more extensive discussion is given in [2], but the reader of this paper may be more impressed just by the simplicity of the axioms and by the fact that they are consistent. Ultimately, the axioms are justified if they provide a definition which is useful and compatible with observation.

Axiom 1. $0 \le \mathrm{app}_{\mathcal{B}}(\sigma|\rho) \le 1$.

Axiom 2. $\mathrm{app}_{\mathcal{B}}(\sigma|\rho) = 1$ if and only if $\sigma = \rho$.

Axiom 3. Let $U \in \mathcal{B}(\mathcal{H})$ be unitary and define $\tau : \mathcal{B}(\mathcal{H}) \to \mathcal{B}(\mathcal{H})$ by $\tau(B) = UBU^*$. Then

$$\mathrm{app}_{\mathcal{B}}(\sigma|\rho) = \mathrm{app}_{\tau^{-1}(\mathcal{B})}(\sigma \circ \tau|\rho \circ \tau).$$

Axiom 4. Suppose that $\rho = p_1\rho_1 + p_2\rho_2$ where $p_1, p_2 \in [0,1]$ with $p_1 + p_2 = 1$. Supppose that there exists a projection $P \in \mathcal{B}$ with $\rho_1(P) = 1$ and $\rho_2(P) = 0$ then $\mathrm{app}_{\mathcal{B}}(\rho_1|\rho) = p_1$.

In this case, ρ is a mixture of the distinct states ρ_1 and ρ_2 so axiom 4 means that app is a generalization of the idea that a mixed state can be written as a sum of terms weighted by probabilities.

Property 4. Suppose that $\rho = p_1\rho_1 + p_2\rho_2$ where $p_1, p_2 \in [0,1]$ with $p_1 + p_2 = 1$. Supppose that there exists a projection $P \in \mathcal{B}$ with $\rho_1(P) \ge 1 - \epsilon$ and $\rho_2(P) \le \epsilon$ for some $\epsilon \in [0, \frac{1}{2}]$. Then $p_1 \le \mathrm{app}_{\mathcal{B}}(\rho_1|\rho) \le p_1 - 3\epsilon \log \epsilon$.

This is a significant improvement on axiom 4.

Quantum measurement theory has long been concerned with providing models of "state decomposition with negligible inference effects". Recently, for example, this has been done under the banner of "environment-induced decoherence" [7]. Such models demonstrate the validity of the sort of state decomposition proposed in property 4 at the macroscopic level with p_1 playing the role of a physically significant and experimentally measurable probability. They can be taken as providing justification and experimental confirmation for the theory proposed here. The central problem in measurement theory is to justify the coarse-graining involved. Here that coarse-graining is represented by the restriction of attention to the set $\mathcal{B}$. An appropriate fundamental choice for $\mathcal{B}$ is justified in [3].

Property 5. Let $\mathcal{B}_1 \subset \mathcal{B}_2$ and σ_2, ρ_2 be extensions to $\mathcal{B}_2$ of states σ_1, ρ_1 on $\mathcal{B}_1$. Then $\mathrm{app}_{\mathcal{B}_2}(\sigma_2|\rho_2) \le \mathrm{app}_{\mathcal{B}_1}(\sigma_1|\rho_1)$.

This is justified in terms of the proposed interpretation of $\mathrm{app}_{\mathcal{B}}(\sigma|\rho)$ because an observer who can draw finer distinctions is less likely to make mistakes.

Axiom 5. Let $\mathcal{B}_1 \subset \mathcal{B}_2$. Then

$$\mathrm{app}_{\mathcal{B}_1}(\sigma_1|\rho_1) = \sup\{\mathrm{app}_{\mathcal{B}_2}(\sigma_2|\rho_2) : \sigma_2|_{\mathcal{B}_1} = \sigma_1, \rho_2|_{\mathcal{B}_1} = \rho_1\}.$$

This improvement on property 5 is justified because although we know nothing else about the extension from $\mathcal{B}_1$ to $\mathcal{B}_2$, we do know for certain, as a consequence of definition 2, that σ_1 and ρ_1 are restrictions of states on $\mathcal{B}_2$.

By axiom 5 and definition 2, app on an arbitrary set $\mathcal{B}$ is determined by app on $\mathcal{B}(\mathcal{H})$.

Property 6. Let σ_1, σ_2 and ρ be states on a set $\mathcal{B}$. Let $p_1, p_2 \in [0,1]$ with $p_1 + p_2 = 1$. Let $\sigma = p_1\sigma_1 + p_2\sigma_2$. Then

$$\mathrm{app}_{\mathcal{B}}(\sigma|\rho) \ge \mathrm{app}_{\mathcal{B}}(\sigma_1|\rho)^{p_1}\mathrm{app}_{\mathcal{B}}(\sigma_2|\rho)^{p_2}.$$

This is plausible because one of the ways for an observer to observe the state σ is to observe σ_1 a fraction p_1 of the time and σ_2 a fraction p_2 of the time.

Axiom 6. Let σ_1, σ_2 and ρ be states on a finite-dimensional von Neumann algebra $\mathcal{A}$. Let $p_1, p_2 \in [0,1]$ with $p_1 + p_2 = 1$. Let $\sigma = p_1\sigma_1 + p_2\sigma_2$. Then

$$\Lambda = \text{app}_{\mathcal{A}}(\sigma_1|\rho)^{p_1}\text{app}_{\mathcal{A}}(\sigma_2|\rho)^{p_2}/\text{app}_{\mathcal{A}}(\sigma|\rho)$$

is independent of ρ.

Consider sequences of trials of the information in $\mathcal{A}$ performed on a system which is in the state ρ. Suppose that all the results from all the sequences are compatible with the state σ. Some of these sequences may have the property that the results are compatible with the state σ_1 a fraction p_1 of the time and with the state σ_2 a fraction p_2 of the time. Because $\mathcal{A}$ is a von Neumann algebra and all the states involved are completely specified, Λ is a measure of the relative probability of this property. Whether a sequence of trials compatible with σ has the property in question or not depends only on σ. On a non-algebra, however, it is possible for Λ to depend on ρ because ρ is not a complete specification of the situation being tested by a trial.

In the classical context of examples 1–4, this argument is a correct factual statement. In the wider context of non-commutative probability theory, it tells us not just about the definition of app, but also about the meaning of a "trial" and of a state being "completely specified". It is, perhaps, remarkable that it is possible to extend such an argument to the non-commutative situation.

Theorem. Let $\mathcal{H}$ be a finite-dimensional Hilbert space and $\mathcal{B} \subset \mathcal{B}(\mathcal{H})$. Then there is a unique function $\text{app}_{\mathcal{B}}(\sigma|\rho)$ satisfying Axioms 1–6. $\text{app}_{\mathcal{B}}(\sigma|\rho)$ also satisfies properties 4, 5, and 6. When $\mathcal{B}$ is $\mathcal{B}(\mathcal{H})$ — the set of all bounded operators — then

$$\text{app}_{\mathcal{B}(\mathcal{H})}(\sigma|\rho) = \exp(\text{tr}(-\sigma\log\sigma + \sigma\log\rho)).$$

For general $\mathcal{B} \subset \mathcal{B}(\mathcal{H})$, $\text{app}_{\mathcal{B}}(\sigma|\rho)$ is then defined by axiom 5. If $\mathcal{B}$ is a von Neuman algebra then $\text{app}_{\mathcal{B}}(\sigma|\rho)$ is the exponential of the relative entropy of σ with respect to ρ. The formula given in example 2 is correct as is the value proposed for example 4.

There is also a unique function $\text{app}_{\mathcal{B}}(\sigma|\rho)$ on infinite-dimensional systems. Axioms 1 - 5 and properties 4, 5, and 6 are satisfied as stated, except that the supremum in axiom 5 has to be taken over all states (non-normal as well as normal). Axiom 6 holds for normal states on an injective von Neumann algebra. Uniqueness is achieved by requiring that $\text{app}_{\mathcal{B}}(\sigma|\rho)$ be the minimal w* upper semicontinuous function having these properties.

References

[1] M.J. Donald, "Quantum theory and the brain". *Proc. R. Soc. Lond. A* **427**, 43–93 (1990).

[2] M.J. Donald, "A priori probability and localized observers". *Found. Phys.* **22**, 1111–1172 (1992).

[3] M.J. Donald, "A mathematical characterization of the physical structure of observers". *Found. Phys.* (to appear).

[4] M.J. Donald, "On the relative entropy". *Commun. Math. Phys.* **105**, 13–34 (1986).

[5] M.J. Donald, "Further results on the relative entropy". *Math. Proc. Camb. Phil. Soc.* **101**, 363–373 (1987).

[6] I.N. Sanov, "On the probability of large deviations of random variables". *Mat. Sbornik* **42**, 11–44 (1957) and *Selected Translations in Math. Stat. and Prob.* **1**, 213–244 (1961).

[7] W.H. Zurek, "Decoherence and the transition from quantum to classical". *Physics Today* 36–44 (October, 1991).

AN EXAMPLE OF A RECEIVED QUANTUM STATE CONTROLLER BY OPTICAL KERR EFFECT

Tsuyoshi Sasaki-Usuda and Osamu Hirota

Research division for quantum
communication science,
Tamagawa University, 6-1-1,
Tamagawagakuen, Machida-city,
Tokyo 194 (Japan).

1 INTRODUCTION

According to recent development of optical technology, performance of optical communication system almost achieves the theoretical limitation predicted by classical communication theory, which is so called standard quantum limit (SQL). The quantum communication theory [1-3] predicts a new possibility to overcome the SQL by controlling quantum noise which is subject to quantum nature of light.

In general, properties of quantum noise depend upon both quantum state of a light and measurement process. As method to reduce effects of quantum noise, two possibilities were proposed and discussed. One is to find the optimum measurement process and the other is to choose or transform a quantum state to reduce the quantum noise. The former concept was given by C. W. Helstrom [1]. He predicted not only an excellent optical receiver but the optimum quantum receiver which has the minimum error probability with fixed quantum state signal, and showed new limitation of optical system. We call this limitation 'Helstrom bound'. However, it is not clear what physical process corresponds to the optimum quantum receiver and this gave rise to a realization problem. The latter concept was first proposed by H. P. Yuen, which is so called quantum state control [4, 5]. The study of quantum state control was started from a use of nonclassical light, such as a squeezed state, as a transmission signal and a certain advantage for a conventional system was shown [5, 6]. Unfortunately, since the advantage is degraded by an energy loss

Quantum Communications and Measurement
Edited by V. P. Belavkin *et al.*, Plenum Press, New York, 1995

in transmission channels, we can expect only little advantage in applying to long distant communications [7]. To cope with this problem, Hirota and his group proposed a concept of received quantum state control [8-10]. It says we can obtain an improvement of a system performance by merely inserting received quantum state controller before quantum measurement process in a conventional system.

We are interested in a physical example of this system, in particular an example for digital communication system. So far some examples of received quantum state controller have been proposed [11-14], but few people understand principles of these systems and believe validity of received quantum state control. Because these systems contain conditional isometric processes [15,16] which are kinds of non-unitary processes and seems to be beyond the scope of conventional quantum mechanics. We need some new ideas in modern quantum mechanics if we mention about such systems.

In this paper, we will discuss a possibility of realization of received quantum state control by a unitary transformation to input quantum states in order to give an easy construction. We consider optical Kerr effect as an example of received quantum state controller. Because optical Kerr effect drastically improves a signal-to-noise ratio (SNR) for a quadrature-phase amplitude as shown in the previous paper [17] and this makes us think that optical Kerr effect may be useful as a received quantum state controller. However improvement of SNR does not necessarily mean that of error probability and we must check their properties individually. Properties of optical Kerr effect have been analyzed quantum mechanically by some people (see for example [18]) and our analysis is based on these theories. We show that our example overcomes the SQL. Furthermore, we discuss some points about the improvement of error probability.

2 CONCEPT OF RECEIVED QUANTUM STATE CONTROL

First, let us survey the concept of received quantum state control. An quantum communication system is called received quantum state control system if it contains at the receiver part a quantum state controller by which the SQL is overcome. The term 'SQL' is used in various ways in literatures. We takes the following definition [19] since our main interest is to construct a digital communication system with lower error probability.

Definition. *The standard quantum limit (SQL) is the minimum error probability when we detect quantum state signals with measuring apparatus described by spectral decomposition of the signal observable.*

The above definition means the SQL is an error probability when the optimum decision by classical communication theory is applied to received signals with quantum noise arising by a measurement of a quantity which has the information of the signal.

In received quantum state control system, coherent state lights are employed as transmission signals since coherent states keep the same quantum nature even if they pass through the transmission line with energy loss [7]. Thus the received signals are also coherent states. The received signals are transformed to another quantum states by received quantum state controller and transformed signals are detected by a conventional quantum measurement which is realized as an optical receiver. A general model of received quantum state control system for digital communication system is given in Fig.1.

What we will discuss here is the realization by a unitary transformation to input quantum states. However, it seems impossible to overcome the SQL by unitary processes which are familiar for us, e.g., displacement, rotation, and squeezing processes. There must be a certain condition for overcoming the SQL. In this statement, we have a following lemma [19].

Lemma. *The unitary operator must be generated by a non-commuting observable to the signal observable if the process overcomes the SQL for the signal observable.*

This lemma was proved by Hirota [19]. One can easily understand that this lemma only gives a necessary condition but indicates a direction to construct a physical example of received quantum state control. We will give an example to overcome the SQL by a unitary transformation in the next section.

3 REALIZATION OF RECEIVED QUANTUM STATE CONTROL BY OPTICAL KERR EFFECT

In this section, we will show that an optical Kerr medium which is a kind of unitary transformer can be a received quantum state controller. Our main interest is to show a physical example of a unitary transformer for received quantum state control in order to make clear the usefulness of this construction method. Here we concentrate the case of system with binary phase-shift-keying (BPSK) and coherent detection. Suppose that the binary signals are coherent state lights with phases 0 and π for signals '0' and '1', respectively. Received signals can be described as $|\alpha_{\rm in}^{(i)}>$, $i = 0, 1$, $\alpha_{\rm in}^{(0)} = \alpha_{\rm in}, \alpha_{\rm in}^{(1)} = -\alpha_{\rm in}$. Binary phase shift keyed signals can be regarded to convey signals by one quadrature component of optical wave. Describing it by $\hat{X}_\phi$, corresponding SQL is gotten by the measurement of $\hat{X}_\phi$ plus optimum decision. Since spectral decomposition of $\hat{X}_\phi$ is formally written as $\hat{X}_\phi = \int x_\phi|x_\phi><x_\phi|dx_\phi$ with eigen values x_ϕ and eigen states $|x_\phi>$ of it, the SQL is expressed as

$$P_{\rm e}(SQL) = \min_{x_{\rm th},\, \phi}\left\{\xi_0 \int_{-\infty}^{x_{\rm th}} \left|<x_\phi|\alpha_{\rm in}^{(0)}>\right|^2 dx_\phi + \xi_1 \int_{x_{\rm th}}^{\infty} \left|<x_\phi|\alpha_{\rm in}^{(1)}>\right|^2 dx_\phi\right\}, \tag{1}$$

where ξ_i are prior probabilities for $|\alpha_{\rm in}^{(i)}>$. This process is approximately realized by well-known homodyne detection [20] and Bayes' decision rule and therefore this agrees with the result of classical communication theory. Since the unitary operator for optical Kerr effect is generated by a number operator which is a non-commuting observable to the signal observable $\hat{X}_\phi$, optical Kerr effect satisfies the condition of lemma in section 2. Therefore it can be a candidate of received quantum state controller. In the following, we first explain some of the results of optical Kerr effect and then we show a configuration and error performance of the system in applying this phenomenon.

3.1 Optical Kerr effect

Optical Kerr effect is one of nonlinear optical phenomena of the third-order and its quantum mechanical analyses were given by some people. Let us survey some of these results of optical Kerr effect in Schrödinger and Heisenberg pictures. In the notation of reference [18], a quantum state for an output of optical Kerr medium is

$$|\psi_{\rm k}>=\hat{U}_{\rm k}(L)|\psi_{\rm in}>, \tag{2}$$

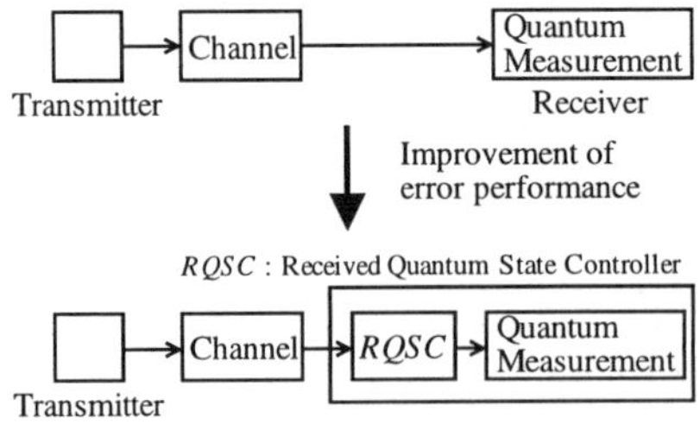

Figure 1: Concept of received quantum state control.

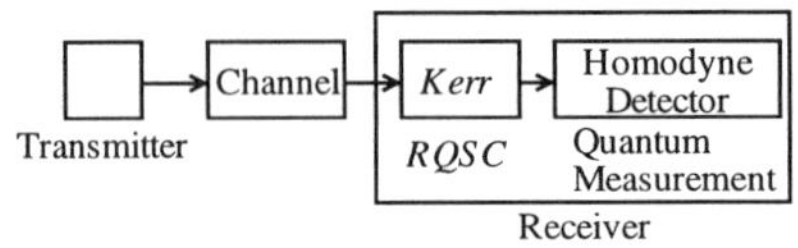

Figure 2: Schematic diagram of proposed received quantum state control system.

where $|\psi_{\rm in}>$ is an input quantum state. The unitary operator $\hat{U}_{\rm k}(L)$ describing the unitary evolution through the Kerr medium of length L is written as

$$\hat{U}_{\rm k}(L)=\exp\left[\frac{\rm i}{2}\gamma\hat{n}(\hat{n}-1)\right], \tag{3}$$

where $\hat{n}=\hat{a}^{\dagger}\hat{a}$, $\gamma=\frac{2\chi L}{v}$, v is a velocity of light in the Kerr medium, and χ is an anharmonicity parameter and is proportional to the third-order nonlinear susceptibility $\chi^{(3)}$ or the nonlinear refractive index n_2 of the Kerr medium. Complex amplitude operator after the evolution by the Kerr medium is

$$\hat{b}=\hat{U}_{\rm k}^{\dagger}(L)\hat{a}\hat{U}_{\rm k}(L)={\rm e}^{{\rm i}\gamma\hat{a}^{\dagger}\hat{a}}\hat{a}={\rm e}^{{\rm i}\gamma\hat{n}}\hat{a}, \tag{4}$$

where $\hat{a}$ is a complex amplitude operator for input and $\hat{n}$ is a photon number operator and is constant of the motion. Eq.(4) implies that a phase of an incident light rotates proportional to the number of photons by optical Kerr effect. This property is well known as a photon number-dependent phase rotation of optical Kerr effect. If the input light has a photon fluctuation, a phase fluctuation arises by the fluctuation. This is sometimes referred as self-phase modulation effect. But as special feature, the photon number statistics is unaffected.

3.2 Proposed system

Fig.2 shows the proposed system. The system configuration is similar to conventional system with BPSK and homodyne detection and the modulation scheme is just the same as mentioned at the beginning of this section. Only difference is optical Kerr medium inserted before homodyne detector. We can describe received signals as mentioned above. The output states of optical Kerr medium are described as $\hat{U}_{\rm k}(L)|\alpha_{\rm in}^{(i)}>$. The input and output signals of optical Kerr medium are described by phase space representation as shown in Fig.3.

The output states from the optical Kerr medium are detected by homodyne detector which corresponds to a measurement of quadrature-phase amplitude $\hat{X}_{\phi}$. Here $\hat{X}_{\phi}\equiv\frac{1}{2}(\hat{a}{\rm e}^{-{\rm i}\phi}+\hat{a}^{\dagger}{\rm e}^{{\rm i}\phi})$ and ϕ corresponds to a phase of local oscillator in homodyne detector. We call it a detection phase.

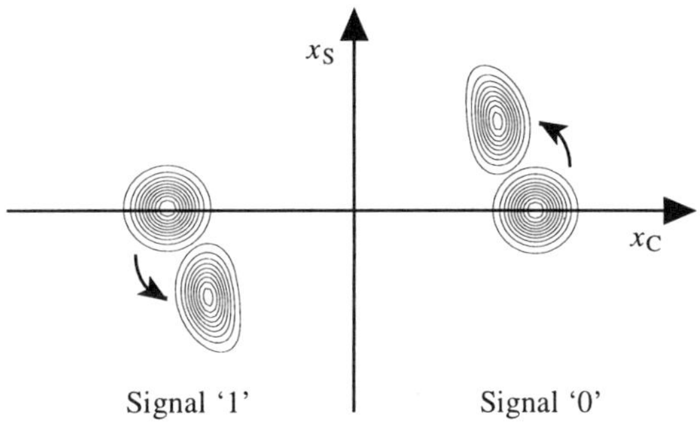

Figure 3: Phase space representation of input and output states of optical Kerr medium.

3.3 Error performance

Now we show the error performance of the system. Quantum noise distribution in the homodyne detector is

$$p(x_\phi|i) = \mathrm{Tr}\hat{\rho}_{\mathrm{out}}^{(i)}\hat{\Pi}(x_\phi) = \mathrm{Tr}\hat{U}_{\mathrm{k}}(L)\hat{\rho}_{\mathrm{in}}^{(i)}\hat{U}_{\mathrm{k}}^{\dagger}(L)\hat{\Pi}(x_\phi) = \left|< x_\phi|\hat{U}_{\mathrm{k}}(L)|\alpha_{\mathrm{in}}^{(i)} >\right|^2, \quad (5)$$

where $\hat{\rho}_{\mathrm{in}}^{(i)} = |\alpha_{\mathrm{in}}^{(i)} >< \alpha_{\mathrm{in}}^{(i)}|$ is a density operator of the input state of optical Kerr medium, $\hat{\Pi}(x_\phi) = |x_\phi >< x_\phi|$ is a projection operator describing measurement of $\hat{X}_\phi$. Since

$$|\alpha_{\mathrm{in}}^{(i)} >= \sum_n C_n^{(i)}|n >; \quad C_n^{(i)} =< n|\alpha_{\mathrm{in}}^{(i)} >= \frac{\alpha_{\mathrm{in}}^{(i)n}}{\sqrt{n!}} \exp\left[-\frac{1}{2}\left|\alpha_{\mathrm{in}}^{(i)}\right|^2\right], \quad (6)$$

$p(x_\phi|i)$ reduces to

$$\begin{aligned} p(x_\phi|i) &= \left|< x_\phi|\exp\left[\frac{\mathrm{i}}{2}\gamma\hat{n}(\hat{n}-1)\right]\left(\sum_n C_n^{(i)}|n >\right)\right|^2 \\ &= \left|\sum_n C_n^{(i)} \exp\left[\frac{\mathrm{i}}{2}\gamma n(n-1)\right] < x_\phi|n >\right|^2, \end{aligned} \quad (7)$$

$$< x_\phi|n >= \left(\frac{1}{2^n n!}\sqrt{\frac{2}{\pi}}\right)^{1/2} H_n\left[\sqrt{2}x_\phi\right] \cdot \mathrm{e}^{-x_\phi{}^2} \cdot \mathrm{e}^{-\mathrm{i}n\phi} \cdot \mathrm{e}^{\mathrm{i}\theta_0}, \quad (8)$$

where θ_0 is an arbitrary phase factor. It follows that the error probability is

$$P_{\mathrm{e}} = \frac{1}{2}\left\{P_{\mathrm{e}}(1|0) + P_{\mathrm{e}}(0|1)\right\}, \quad (9)$$

where $P_{\mathrm{e}}(1|0) = \int_{-\infty}^{x_{\mathrm{th}}} p(x_\phi|0)dx_\phi$ and $P_{\mathrm{e}}(0|1) = \int_{x_{\mathrm{th}}}^{\infty} p(x_\phi|1)dx_\phi$.
Here we assumed that prior probabilities for signals '0' and '1' are equal. The threshold value x_{th} should be zero, so that $P_{\mathrm{e}}(1|0) = P_{\mathrm{e}}(0|1)$ since the quantum noise distributions for the output of optical Kerr medium are symmetric with respect to the origin as shown in Fig.3. We show in Fig.4 an example of the error probability with respect to detection phase in the case of $\alpha_{\mathrm{in}} = 1$ and $\gamma = 0.3$.

We can see from Fig.4 that there is a region of ϕ in which the error probability is improved in comparison with the SQL. This means that the SQL can be overcome by optical Kerr effect and therefore it can be a received quantum state controller. And this is indeed just the result we find. We also see from Fig.4 that there is a detection phase for which the error probability is minimized with fixed $\alpha_{\rm in}$ and γ. Let it be the optimum detection phase $\phi_{\rm opt}$. This property of error probability is the same as that of SNR but it does not necessarily follow that the phase of maximizing SNR is equal to $\phi_{\rm opt}$ [17]. Fig.5 shows the error probability in the case of using $\phi_{\rm opt}$ with respect to nonlinearity γ. From Fig.5, there is a γ which minimize the error probability with fixed $\alpha_{\rm in}$. This is also the same property as the case of SNR [17]. Let it be $\gamma_{\rm opt}$. Although $\gamma_{\rm opt}$ is not equal to the nonlinearity maximizing SNR, it is easily found that they have almost the same values.

Finally, Fig.6 shows relative error probability with respect to $\alpha_{\rm in}$ under the condition of $\phi = \phi_{\rm opt}$ and $\gamma = \gamma_{\rm opt}$, which is defined as $\tilde{P}_{\rm e} \equiv \frac{P_{\rm e}}{P_{\rm e}(SQL)}$.

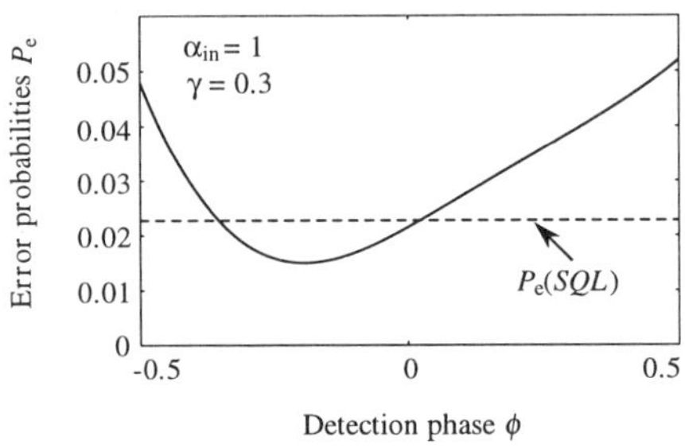

Figure 4: Error probability of proposed system with respect to detection phase ϕ.

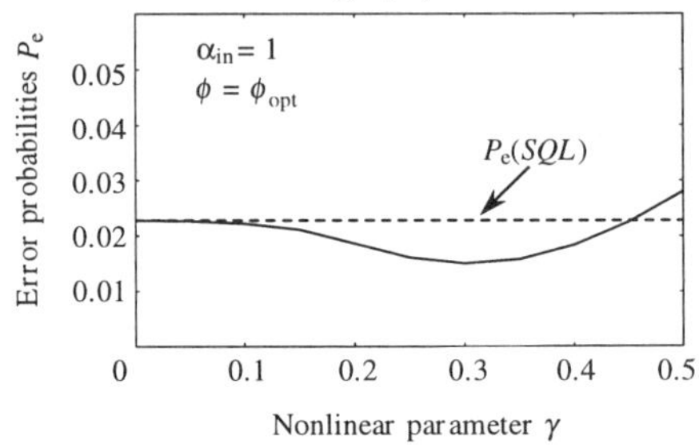

Figure 5: Error probability of proposed system with respect to nonlinear parameter γ, with optimum detection phase.

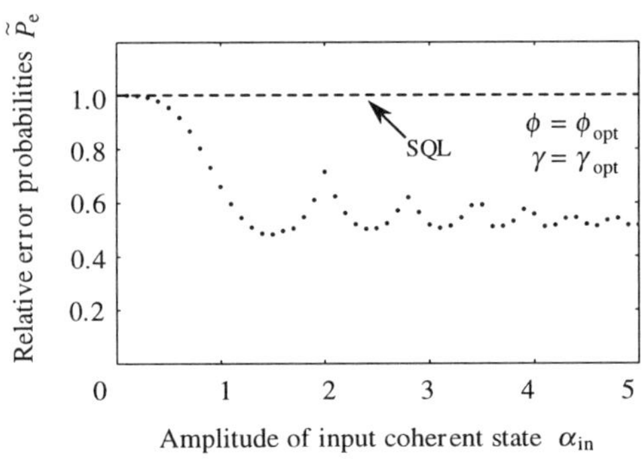

Figure 6: Relative error probability of proposed system with respect to amplitude of input state, with optimum detection phase and optimum nonlinearity.

As shown in Fig.6, the relative error probability is not necessarily lower as $\alpha_{\rm in}$ larger. This property is greatly different from the case of SNR, in which the improvement is drastically higher as $\alpha_{\rm in}$ larger. In addition, degree of the improvement in error probability is not so remarkable one in the case with range of parameters using here. We only mention here the possibility of applying a unitary process and more detailed properties of error performance including the case of applying squeezer will be presented another time.

3.4 Discussion

According to the above result, the error probability can be improved by optical Kerr effect. Here we would like to discuss the reason why it is improved. The improvement seems mysterious since optical Kerr effect is a unitary process which keeps inner products of different quantum signals. The reason of improvement may be quantum effect of the transformation. Because there is no classical transformation to improve the classical optimum receiver which is limited by classical noises and the SQL, the limitation in classical sense, can be overcome by our system. If there is something difference between transformations by classical and quantum Kerr effects, the SQL may be overcome by the Kerr effect. So let us first explain the classical Kerr effect which gives a deterministic transformation in phase space and is described by the following rotation matrix.

$$\begin{pmatrix} X_C' \\ X_S' \end{pmatrix} = \begin{pmatrix} \cos\left[\gamma|\alpha(0)|^2\right] & -\sin\left[\gamma|\alpha(0)|^2\right] \\ \sin\left[\gamma|\alpha(0)|^2\right] & \cos\left[\gamma|\alpha(0)|^2\right] \end{pmatrix} \begin{pmatrix} X_C \\ X_S \end{pmatrix}, \tag{10}$$

where $\alpha(0) = X_C + iX_S$ and $\alpha(L) = X_C' + iX_S'$ are the complex amplitudes for the input and output, respectively.

Corresponding quantum theory is an evolution of complex amplitude operator in Heisenberg picture. Let $\hat{X}_C$ and $\hat{X}_S$ be quadrature-phase amplitude operators for the input and $\hat{X}_C'$ and $\hat{X}_S'$ those for the output, that is, $\hat{a} = \hat{X}_C + i\hat{X}_S$ and $\hat{b} = \hat{X}_C' + i\hat{X}_S'$. Using Eq.(4), $\hat{X}_C'$ and $\hat{X}_S'$ are represented by $\hat{X}_C$ and $\hat{X}_S$ as

$$\begin{aligned} \hat{X}_C' &= \frac{1}{2}\left(\hat{b} + \hat{b}^\dagger\right) = \frac{1}{2}\left(e^{i\gamma\hat{n}}\hat{a} + e^{-i\gamma\hat{n}}\hat{a}^\dagger + \left[\hat{a}^\dagger, e^{-i\gamma\hat{n}}\right]\right) \\ &= \cos\left[\gamma\hat{n}\right]\hat{X}_C - \sin\left[\gamma\hat{n}\right]\hat{X}_S + \frac{1}{2}\left[\hat{a}^\dagger, e^{-i\gamma\hat{n}}\right], \end{aligned} \tag{11}$$

$$\begin{aligned} \hat{X}_S' &= \frac{1}{2i}\left(\hat{b} - \hat{b}^\dagger\right) = \frac{1}{2i}\left(e^{i\gamma\hat{n}}\hat{a} - e^{-i\gamma\hat{n}}\hat{a}^\dagger - \left[\hat{a}^\dagger, e^{-i\gamma\hat{n}}\right]\right) \\ &= \sin\left[\gamma\hat{n}\right]\hat{X}_C + \cos\left[\gamma\hat{n}\right]\hat{X}_S + \frac{i}{2}\left[\hat{a}^\dagger, e^{-i\gamma\hat{n}}\right]. \end{aligned} \tag{12}$$

Since commutator $\left[\hat{a}^\dagger, e^{-i\gamma\hat{n}}\right]$ is in general not equal to zero, the above transformation of quadrature-phase amplitude operators should be represented by a matrix as

$$\begin{pmatrix} \hat{X}_C' \\ \hat{X}_S' \end{pmatrix} = \begin{pmatrix} A & B \\ C & D \end{pmatrix} \begin{pmatrix} \hat{X}_C \\ \hat{X}_S \end{pmatrix}, \tag{13}$$

$$A = \frac{1}{2}(1 + e^{i\gamma})\left\{\cos\left[\gamma\hat{n}\right] - i\tan\left(\frac{\gamma}{2}\right)\sin\left[\gamma\hat{n}\right]\right\}, \tag{14a}$$

$$B = -\frac{1}{2}(1 + e^{i\gamma})\left\{\sin\left[\gamma\hat{n}\right] + i\tan\left(\frac{\gamma}{2}\right)\cos\left[\gamma\hat{n}\right]\right\}, \tag{14b}$$

$$C = \sin\left[\gamma\hat{n}\right] - \frac{1}{2}\sin\gamma\left\{\cos\left[\gamma\hat{n}\right] - i\tan\left(\frac{\gamma}{2}\right)\sin\left[\gamma\hat{n}\right]\right\}, \tag{14c}$$

$$D = \cos\left[\gamma\hat{n}\right] + \frac{1}{2}\sin\gamma\left\{\sin\left[\gamma\hat{n}\right] + i\tan\left(\frac{\gamma}{2}\right)\cos\left[\gamma\hat{n}\right]\right\}. \tag{14d}$$

This shows the transformation matrix in quantum theory is different from that in classical theory and can not be expressed as a simple rotation. Therefore, in optical Kerr effect, the performance of quantum noise is not necessarily explained as analogy of classical theory. That is, the quantum noise distribution we analyzed above may be different from the noise distribution of the output of optical Kerr effect when the input of it is treated as a classical signal with additive noise having the same property as quantum noise of coherent state. This may arise in the case that the transformation matrix in Eq.(13) has an operator as an element, which is noncommutative to $\hat{X}_{\mathrm{C}}$ or $\hat{X}_{\mathrm{S}}$. While if the matrix is constructed only by c-numbers, quantum noise distribution for the output is just the same as the case of classical noise as A. K. Ekert and P. L. Knight showed in their semiclassical approach [21]. Hence, the transformation matrix must have an operator element in order that the process provides a 'pure quantum mechanical effect' which means we can not analyze it by a semiclassical approach. Optical Kerr effect is an example of such process and it overcomes the SQL. More detailed condition and physical interpretation to overcome the SQL will be given in a subsequent paper [22].

Our system is limited by Helstrom bound since the optical Kerr effect is a unitary process. We need a conditional isometric device as a received quantum state controller if we also desired to overcome the Helstrom bound.

CONCLUSIONS

We have shown that an optical receiver system with optical Kerr medium as a received quantum state controller can overcome the SQL for binary phase-shift-keyed system. Our result illustrates a possibility of realization of received quantum state control by a unitary process. Further studies concerning the realization of the receivers, such as conditions for overcoming the SQL and physical interpretations of the improvement, are in progress [19, 22]. Surely our system is limited by the Helstrom bound, so that we need a received quantum state control system with a conditional isometric process [15, 16] in order to get further improvement.

Acknowledgment

The authors would like to thank Drs. K. Yamazaki and M. Osaki (Tamagawa University) for their discussions. T. S. Usuda is supported by Foundation for C & C Promotion.

References

[1] C. W. Helstrom, *Quantum detection and estimation theory*, (Academic Press, New York, 1976).

[2] A. S. Holevo, *Probabilistic and statistical aspect of quantum theory*, (North-Holland, Amsterdam, 1982).

[3] O. Hirota, *Optical communication theory*, (Morikita Pub. Co., Tokyo, 1985).

[4] H. P. Yuen, Phys. Rev., **A13**, 2226, (1976).

[5] O. Hirota and S. Ikehara, Trans. IECE of Japan, **E61**, 273, (1978).

[6] J. H. Shapiro, et. al., IEEE Trans. Inf. Theory, **IT-25**, 179, (1979).

[7] O. Hirota ed., *Squeezed light*, (Elsevier, Amsterdam, 1992).

[8] O. Hirota, et. al., Trans. IEICE of Japan, **E70**, 801, (1987).

[9] O. Hirota, Opt. Commun., **67**, 204, (1988).

[10] O. Hirota, in LNP-378 (Springer-Verlag, Berlin, 1991), pp.223-230.

[11] S. Bun, et. al., Trans. IEICE of Japan, **E73**, 1647, (1990).

[12] H. Koyano, et. al., in LNP-378 (Springer-Verlag, Berlin, 1991), pp.249-258.

[13] T. Sasaki and O. Hirota, Trans. IEICE of Japan, **E75-B**, 514, (1992).

[14] M. Osaki, O. Hirota, and I. Ojima, J. Mod. Opt., **39**, 1555, (1992).

[15] O. Hirota, Phys. Lett. **A155**, 343, (1991).

[16] O. Hirota, in Abstracts of Papers of QCM'94, (1994).

[17] T. S. Usuda, T. Satoh, K. Yamazaki, and O. Hirota, J. Mod. Opt., in press.

[18] M. Kitagawa and Y. Yamamoto, Phys. Rev., **A34**, 3974, (1986).

[19] O. Hirota, Tech. Group Rep. in IEICE of Japan, **93**, No.531, pp.25-30, (1994).

[20] H. P. Yuen and V. W. S. Chan, Opt. Lett., **8**, 177, (1983).

[21] A. K. Ekert and P. L. Knight, Phys. Rev., **A43**, 3934, (1991).

[22] M. Sasaki, T. S. Usuda, and O. Hirota, submitted.

LONG TIME BEHAVIOR OF THE SOLUTIONS OF THE BELAVKIN QUANTUM FILTERING EQUATION

Vassili N. Kolokol'tsov[1,2,3]

[1]Alexander von Humboldt fellow
[2]Fakultät für Mathematik,
Ruhr universität Bochum (Germany)
[3]On leave of absence from Applied Math.
Dep., MIEM, Moscow (Russia)

Abstract

Long time evolution of the operator means is investigated for continuously observed quantum systems governed by the Belavkin quantum filtering equation. The coefficient of the quality of measurement is introduced and studied. This coefficient describes numerically the watch dog effect and the noncertainty principle for continuously observed quantum systems under nondirect nondemolition measurement of diffusion type. Moreover, for the case of a free quantum particle with continuously observed coordinate, the behavior of the solutions of the Belavkin equation is completely investigated. It turns out, in particular, that these solutions have extraordinary property that can be interpreted as a sort of confinement. Namely, as $t \to \infty$, any linear combination of Gaussian wave packets will tend to a single one for almost all realizations of innovating Wiener process. In other words, several initial particles will stick together at large times, i.e. they can not be identified in the process of measurement. As a consequence, it follows that the dispersion of coordinate will always tend to the same limit, as $t \to \infty$, not depending on initial data. In conclusion, the large time behavior of controlled continuously observed quantum systems are discussed.

1 INTRODUCTION

The nondemolition measurement principle for continuously observed quantum system was formulated in [Bel1] as a commutativity condition $[X(t), Y(s)] = 0 \, \forall t \geq$

s for Heisenberg operators of the particle $X(t)$ and the measured operators $Y(t)$ of the output field. This principle implies the singularity of the interaction between the observed quantum system and the measurement apparatus. It was shown in [Bel2] on the example of the continuous observation of the coordinate of a quantum particle in electromagnetic field. The adequate models for description of such interaction were given by the method of quantum stochastic calculus proposed by Hudson and Parthasarathy [HP]. It was shown in [Bel3] that in the framework of this method one can proposed two different types of continuous nondemolition measurement: measurements of counting and of diffusion types. For both these types V.P.Belavkin has developed the theory of quantum filtering (which naturally generalizes the classical filtering theory for stochastic processes) and has written the quantum filtering equation [Bel3] that describes the aposterior dynamics of quantum system under continuous nondemolition measurement. In this paper we study only observations of diffusion type. In this case the general Belavkin filtering equation has the form

$$d\chi + (\frac{i}{h}H + \frac{1}{2}R^{\star}R)\chi\, dt = R\chi\, dQ, \tag{1.1}$$

where χ is the unknown aposterior (non-normalized) wave function of the given continuously observed quantum system in a Hilbert space $\mathcal{H}$, h is the Planck constant, the selfadjoint operator $H = H^{\star}$ in $\mathcal{H}$ is the Hamiltonian of a free (non-observed) quantum system, the operator R in $\mathcal{H}$ stands for the observed physical value, Q is the standard (input) white noise (Wiener process). Moreover, the Belavkin theorem states that the output (measured) diffusion process $\hat{Q}$ has the density $\|\chi\|^2$ with respect to the standard Wiener measure P of the input process Q, i.e. the mean $E_{\hat{Q}}$ of some functional $f(\chi)$ of the state χ over all realization of the output process $\hat{Q}$ has the form

$$E_{\hat{Q}}f(\chi) = E_Q(f(\chi)\|\chi\|^2), \tag{1.2}$$

where E_Q denote the mean with respect to the Wiener measure P of the process Q. In particular, averaging the mean value

$$\langle A\rangle = \langle A\rangle_{\varphi} = (A\varphi, \varphi) = (A\chi, \chi)/(\chi, \chi) \tag{1.3}$$

of some operator A in $\mathcal{H}$ on the normalized wave function $\varphi = \chi/\|\chi\|$ we have

$$E_{\hat{Q}}\langle A\rangle = E_{\hat{Q}}(A\varphi, \varphi) = E_Q(A\varphi, \varphi)(\chi, \chi) = E_Q(A\chi, \chi). \tag{1.4}$$

As in the classical filtering theory, in the quantum filtering theory it is natural and useful to consider the so called innovating process W, defined by the stochastic equation

$$dW = dQ - 2\langle R\rangle dt. \tag{1.5}$$

The measures defined by Q and W are connected by the famous Girsanov formula. Namely, equation (1.2) implies the equation for the density $\|\chi\|^2$ of the output process $\hat{Q}$:

$$d\|\chi\|^2 = 2(R\chi, \chi)\, dQ = 2\langle R\rangle \|\chi\|^2\, dQ.$$

Therefore, $\|\chi\|^2$ is a positive supermartingale and local martingale. The Girsanov theorem states that if for all times t

$$E_Q\|\chi\|^2 = 1, \tag{1.6}$$

i.e. the density $\|\chi\|^2$ is a positive martingale, then the innovating process W is a Wiener martingale with respect to the measure $\tilde{P}$ that has the density $\|\chi\|^2$ with respect to the measure P. Therefore, denoting by E_W the mean with respect to the measure $\tilde{P}$ of the Wiener martingale W, we conclude that the mean (1.2) is equal to

$$E_{\hat{Q}}f(\chi) = E_Q(f(\chi)\|\chi\|^2) = E_W f(\chi). \tag{1.7}$$

Note, that there exist many sufficient conditions for equation (1.8) to be fulfilled. For instance, (1.6) holds, if

$$E_Q \exp\{\epsilon\langle R\rangle\} \leq C \tag{1.8}$$

for all t and some positive constants ϵ and C. In terms of the normalized wave function φ and the innovating process W equation (1.1) takes the form [Bel3]:

$$d\varphi+\left[\frac{i}{h}(H - h\langle ReR\rangle ImR) + \frac{1}{2}(R - \langle ReR\rangle)^\star(R - \langle ReR\rangle)\right]\varphi\, dt = (R-\langle ReR\rangle)\varphi\, d\omega, \tag{1.9}$$

where symmetric operators $ReR = 1/2(R + R^\star)$ and $ImR = 1/2(R - R^\star)$ are the real and imaginary parts of R. Let us stress that due to the equation $E_{\hat{Q}} = E_W$, the consideration of the innovating process is equivalent to the consideration of the output process, and consequently, from the point of view of the theory of measurement, all results should be formulated in terms of W (so, the process Q plays only auxiliary role).

For most natural physical examples, operator R is selfadjoint and (1.1) and (1.9) reduce respectively to the stochastic equations

$$d\chi + (\frac{i}{h}H + \frac{1}{2}R^2)\chi\, dt = R\chi\, dQ \tag{1.10}$$

and

$$d\varphi + \left(\frac{i}{h}H + \frac{1}{2}(R - \langle R\rangle)^2\right)\varphi\, dt = (R - \langle R\rangle)\varphi\, d\omega. \tag{1.11}$$

It was shown in [Bel3,BS,BK] that in the case $\mathcal{H} = L^2(\mathcal{R}), R = x$ and $H = -\Delta$ or $H = -\Delta + x^2$, i.e. when measuring a coordinate of the free quantum particle or of the quantum oscillator, the set of Gaussian wave packets is invariant with respect to dynamics (1.1) and moreover, for them the dynamics of the dispersion $D_R = \langle R^2 - \langle R\rangle^2\rangle$ is deterministic and D_R has a finite limit as $t \to \infty$ not depending on the initial Gaussian packet. In particular, this fact gives an explanation in the framework of quantum stochastic calculus of the so called watchdog effect (that is an aspect of the quantum Zeno paradox [MS]), i.e. it shows that the Gaussian wave packet of a free quantum particle is not spreading when the particle is continuously observed. Clear, that the limit of the dispersion of the measured physical value is an average measure of noncertainty with which the coordinate of a free quantum particle or of a quantum oscillator can be measured.

In general situation the dynamics of the dispersion $D_R = D_R(\varphi(t,[W]))$ is stochastic. Thus it is natural to pose the question if there exists a limit when $t \to \infty$ of the average dispersion $E_W(D_R)$. If this limit exists and does not depend on initial data, then its value

$$\kappa(H, R) = \lim_{t\to\infty} E_W D_R(\varphi(t,[W])) \tag{1.12}$$

is an invariant of a pair of operators (one of them being selfadjoint), which was introduced in [Kol2] and which estimates the quality of the measurement of the value R, when the free dynamics is defined by the Hamiltonian H. In particular, the inequality $\kappa(H,R) < \infty$ stands for the watch dog effect, and the inequality $\kappa(H,R) > 0$ stands for the noncertainty principle. Constant (1.12) will be called the coefficient of the quality of measurement (CQM). Moreover, sometimes it is possible to prove the existence of the pointwise limit. Thus, when the pointwise limit of the dispersion exists and does not depend on initial data, one can define the pointwise version of the coefficient of the quality of measurement as

$$\kappa(H,R) = \lim_{t\to\infty} D_R(\varphi(t,[W])) \quad \text{for a. a. } W. \tag{1.13}$$

On the other hand , sometimes one can prove only more weak convergent results. In these situations, one can , for instance, put liminf instead of lim in (1.12) or include in definition of the coefficient of measurement the averaging in time that gives

$$\kappa(H,R) = \lim_{t\to\infty} \frac{1}{t} \int_0^t E_W D_R(\varphi(\tau,[W])\, d\tau. \tag{1.14}$$

We describe here two situations, where the CQM (1.12) and respectively (1.13) exist and can be calculated and the behavior of other operator means for large times can be completely studied. In Sect. 4 the qualitative behavior of the solutions of the controlled quantum filtering equation is discussed on the base of approach from [Kol4,Kol5].

2 FINITE DIMENSIONAL SYSTEMS

Here the case of finite dimensional Hilbert space $\mathcal{H}$ is studied and thus the state space for dynamical system (1.9) is the finite dimensional projective space. Proofs can be obtains either by means of the qualitative theory of diffusion processes [AKR, IK, Kl] or using the theory of Lyapunov exponents for a product of random matrices [GR]. They are given in [Kol2], as well as some generalizations, including the study of controlled dynamics. Note also that in some models of mobilevision there are quite similar finite dimensional problems [Ju]. We restrict ourselves to the case, when the operator R is a one-dimensional projector. From physical point of view, it stands for the measurement of the probability for the system to be in a fixed state. Let $R^\perp$ be the orthogonal projector on the kernel of R, let $\lambda_0, ..., \lambda_n$ be the eigenvalues of the operator $RHR + R^\perp H R^\perp$, (λ_0 being the eigenvalue of the one-dimensional operator RHR), and let v_j be the corresponding eigenvectors. Suppose that $(Hv_0, v_j) \neq 0$ for all $j = 1, ..., n$ (condition of irreducibility of the Lie algebra generated by operators H and R). Then one can easily prove that the degenerate diffusion defined by equation (1.11) has unique invariant measure μ on CP^n.

Theorem 1. If $n \geq 3$ and the numbers $(\lambda_0 - \lambda_j)^2$ are pair wise different (in the case $n = 2$, if $\lambda_0 \neq \lambda_1$), then the support of the invariant measure μ coincides with the whole state space CP^n, and it has a smooth density with respect to the Lebesgue measure. Moreover, the means $E_W f(\varphi)$ of any continuous function on the solutions of (1.9) tend to the integral of this function with respect to μ. In particular, CQM (1.12) exists.

Theorem 2. If the operator $RHR + R^\perp H R^\perp$ is proportional to the unit operator, then the invariant measure μ has a one-dimensional support and can be calculated explicitly, CQM (1.12) exists and is equal to

$$\frac{1}{4}\int_0^\pi \sin^2\theta\, p(\theta)\, d\theta \left[\int_0^\pi p(\theta)\, d\theta\right]^{-1}, \tag{2.1}$$

where

$$p(\theta) = e^{B \cot \theta} \sin^{-3} \theta \int_0^\theta \sin t e^{-B \cot t} \, dt, \tag{2.2}$$

and $B = 4\|R^\perp H v_0\|$.

3 FREE QUANTUM PARTICLE WITH CONTINUOUSLY OBSERVED COORDINATE

This section is devoted to the investigation of the Belavkin equation

$$d\chi + \frac{1}{2}(-ih\Delta + x^2)\chi \, dt = x\chi \, dQ, \tag{3.1}$$

which is a particular case of (1.10) with $\mathcal{H} = L_2(\mathcal{R})$, $H = -h^2\Delta/2$ and $R = \hat{x}$ being the operator of the multiplication on x. We consider one-dimensional x for simplicity, though all results and proofs hold for any dimensions. As it was already mentioned, the solutions of this equation were studied in details in [Bel3, BS, BK] for initial functions having a Gaussian form

$$g^\omega_{q,p} = \exp\{-\frac{\omega}{2}(x-q)^2 + \frac{i}{h}px\}. \tag{3.2}$$

The set of Gaussian functions is invariant with respect to evolution (1.21) and moreover, coefficient ω in (3.2) is not stochastic (it does not depend on W) and can be calculated explicitly, which gives

$$\lim_{t\to\infty} \omega(t) = \alpha = h^{-1/2}(1-i).$$

Let us note here that the particular case (3.1) of equation (1.10) was discovered independently by L. Diosi in [D1, D2], where also the stationary solutions with the dispersion α were constructed. Since dispersions D_R and $D_{\hat{p}}$ of coordinate and momentum for functions (3.2) depend only on ω, we get automatically the existence of finite limits of these dispersions for Gaussian initial data. For general initial data only some remarks were made in [BK] (but without proof) concerning limit properties of solutions at large times. Here we give complete qualitative description of the large time behavior of the solutions of (1.21) with general initial data, as well as the existence and uniqueness theorem for the Cauchy problem. Proofs are given in [Kol3]. First step is the investigation of the evolution of finite linear combinations of Gaussian packets. It turns out that any solution of these form will necessary tend to a single Guassian packet (whose mean coordinate and momentum satisfy the Newton system for a free classical particle), as $t \to \infty$ and for a. a. realization of the innovating process W. This property of solutions can be interpreted as a sort of confinement. Namely, several particles (several Gaussian packets) will be stick together at large times, i.e. they can not be identified in the process of measurement. Next, we construct the Green function for the Cauchy problem (3.1), which implies a correctness theorem for this problem and the possibility to generalize previous results on general initial functions proving, in particular, the existence of the pointwise CQM (1.13) in the considered situation.

Therefore, let us start with solutions of equation (3.1) of the form

$$\chi = \sum c_j g^\omega_{q_j,p_j} = \sum_{j=1}^k c_j \exp\{-\frac{\omega}{2}(x-q_j)^2 + \frac{i}{h}p_j x\} \tag{3.3}$$

with some initial values $c_j^0, q_j^0, p_j^0, \omega^0$ with $c_j^0 \neq 0, Re\omega_0 > 0$.

Theorem 3. Let the initial function for equation (3.1) has form (3.3) and let

$$q = \langle R\rangle = \frac{(R\chi, \chi)}{(\chi, \chi)}, \quad p = \langle \hat{p}\rangle = \frac{(\hat{p}\chi, \chi)}{(\chi, \chi)}$$

be the mean values of the operators $R = \hat{x}$, $\hat{p} = \frac{h}{i}\frac{\partial}{\partial x}$ respectively on functions (2.15), and let us denote by $\tilde{O}(e^{-\sqrt{h}t})$ any function that is equal to $O(e^{-\epsilon t})$ for any $\epsilon < \sqrt{h}$. Then

(i) Sufficient condition (1.6) for the Girsanov theorem is fulfilled and thus equation (1.7) holds;

(ii) mean values q_j, p_j of the coordinate and momentum of the Gaussian packets included in (3.3) approach one to another with exponential speed, more precisely, the differences $p_j - p_m$ and $q_j - q_m$ are not stochastic (they do not depend on the process Q) and

$$q_j - q_m = \tilde{O}(e^{-\sqrt{h}t}), \quad p_j - p_m = \tilde{O}(e^{-\sqrt{h}t}) \tag{3.4}$$

as $t \to \infty$;

(iii) for all j

$$\begin{cases} \lim_{t\to\infty} E_W(q - q_j) = \lim_{t\to\infty} E_Q((q - q_j)\|\chi\|) = 0, \\ \lim_{t\to\infty} E_W(p - p_j) = \lim_{t\to\infty} E_Q((p - p_j)\|\chi\|) = 0, \end{cases} \tag{3.5}$$

and

$$E_W q = t\tilde{p} + \tilde{q} + O(e^{-\sqrt{h}t}), \quad E_W p = \tilde{p} + O(e^{-\sqrt{h}t}), \tag{3.6}$$

as $t \to \infty$, where $\tilde{q}, \tilde{p}$ are the initial values of q and p respectively;

(iv) in the set of all trajectories of the innovating process W there exists the set Ω of the full measure such that for all $W \in \Omega$ there exist finite and nonvanishing limits of all relations $d_{jm} = c_j/c_m$ such that for each m

$$\sum_{j=1}^{k} \lim_{t\to\infty} d_{jm} \neq 0; \tag{3.7}$$

(v) for all $W \in \Omega$

$$(a) \lim_{t\to\infty} q_j - q = 0, \quad \lim_{t\to\infty} p_j - p = 0,$$

or, more precisely,

$$q_j - q = \tilde{O}(e^{-\sqrt{h}t}), \quad p_j - p = \tilde{O}(e^{-\sqrt{h}t}),$$

as $t \to \infty$, for all $j = 1, ..., k$ and some functions $\tilde{O}(e^{-\sqrt{h}t})$ depending on W;

$$(b) \begin{cases} q = \tilde{q}(W) + \tilde{p}(W)t + \sqrt{h} + h\int_0^t W(s)\,ds + \tilde{O}(e^{-\sqrt{h}t}) \\ p = \tilde{p}(W) + hW + \tilde{O}(e^{-\sqrt{h}t}) \end{cases} \tag{3.8}$$

for some $\tilde{q}(W)$ and $\tilde{p}(W)$;

$$(c) \lim_{t\to\infty} D_R = \sqrt{h}/2, \quad \lim_{t\to\infty} D_{\hat{p}} = h^{3/2}; \tag{3.9}$$

(vi) for all $W \in \Omega$

$$\lim_{t\to\infty} \left\| \frac{\chi}{\sum_{j=1}^k c_j} - g_{q,p}^{\alpha} \right\|^2 = 0, \tag{3.10}$$

i.e. the large times behavior of the solutions of (3.1) for initial data (3.3) is the same as for a single Gaussian packet.

Now we turn to general initial data.

Theorem 4. The solution $G(t,x,\xi)$ of equation (3.1) with initial data

$$G(0,x,\xi)=\delta(x-\xi),$$

(i. e. the Green function of the Cauchy problem for equation (3.1)) exists and has the form

$$G=C_G\exp\{-\frac{\omega_G}{2}(x^2+\xi^2)+\beta_G x\xi+a_G x+b_G\xi+\gamma_G\}, \tag{3.11}$$

where coefficients C_G,ω_G,β_G are deterministic (they do not depend on the Wiener process W) and are equal to

$$\omega_G=\alpha\coth\frac{2t}{\alpha},\quad \beta_G=\alpha(\sinh\frac{2t}{\alpha})^{-1},\quad C_G=\left(\frac{2\pi}{\alpha}\sinh\frac{2t}{\alpha}\right)^{-1/2},$$

and other coefficients are

$$a_G=\left(\sinh\frac{2t}{\alpha}\right)^{-1}\int_0^t\sinh\frac{2\tau}{\alpha}\,dQ(\tau),$$

$$b_G=ih\alpha\int_0^t\frac{a(\tau)}{\sinh(2\tau/\alpha)}\,d\tau,\quad \gamma_G=\frac{ih}{2}\int_0^t a^2(\tau)\,d\tau.$$

This theorem implies a correctness theorem for equation (3.1) [Kol3], as well as the fact that for all $p\in[1,\infty]$, all $t>0$, and a.a. W the solving stochastic operator $\hat{G}_t:\chi_0\mapsto\chi_t[W]$ can be uniquely prolonged by continuity to the linear continuous operator $\hat{G}_t:L_p\mapsto L_p$, and moreover, for any $\chi_0\in L_p$, $p\in[1,\infty]$, and any $t>0$ there exists $\epsilon=\epsilon(t_0,p)>0$ such that for all $t>t_0$ the function $\hat{G}_t\chi_0$ can be represented in the form

$$\chi=e^{-x^2/\epsilon}\int c(p)\,e^{ipx/h}\,dp, \tag{3.12}$$

where c belongs to the Schwarz space S. Therefore, when studying the large times behavior of solutions of equation (3.1) one can confine himself to initial functions of form (3.12), which are a continuous linear combination of Gaussian functions with fast decreasing coefficients. For example, it is easy to prove that for any initial function of form (3.12) with $c\in L_2$, $\epsilon>0$, the norm of the solution $\|\chi\|^2$ of equation (3.1) is a positive martingale (and therefore (1.7) holds). Let us note, however, that the family of operators $\hat{G}_t$ in $L_2(\mathcal{R})$ is not bounded for a.a. innovating trajectories W. Now, we can generalize Theorem 3 on the case of other initial data.

Theorem 5. Let an initial data for equation (3.1) has form (3.12), where the function c belongs to the Schwarz space S. Then for the solution χ formulas (3.6), (3.8), (3.9) hold, as well as the integral variant of formula (3.10):

$$\lim_{t\to\infty}\left\|\chi(t)(\int c(\alpha,t)\,d\alpha)^{-1}-g^{\alpha}_{q,p}\right\|^2=0. \tag{3.13}$$

In particular, CQM (1.13) exists and is equal to $\sqrt{h}/2$.

4 CONTROLLED SYSTEMS UNDER CONTINUOUS MEASUREMENT

On a simple two dimensional controlled model we describe here the qualitative behavior of the optimal income function on large periods of planning. Let us consider the pair of selfadjoint 2×2- matrices $H = (h_{kl})$ and $R = diag(r_0, r_1)$, where $h_{12} = u_1 + iu_2$ and $u_1, u_2 \in [-A, A]$ are the controlled parameters. For simplicity, we confine ourselves to the case of controls $u(t)$ that are constants on the intervals of the length τ with a fixed τ. To describe a Markov process on the state space $\mathcal{C}P^1$ one needs to define integral and terminal incomes by some operators Q and G. Then the optimal income from the period of time $[t, T]$ with $T - t = N\tau$ is given by the formula $B^N\langle G\rangle$, where B is the Belmann operator

$$BS(\varphi) = \max_u E_W \left(\int_0^\tau \langle Q\rangle(\varphi(s, u, [W]))\, ds + S(\varphi(\tau, u, [W])) \right),$$

and $\varphi(\tau, u, [W])$ is the solution of the Belavkin equation.

Theorem 6. For N large enough there exists unique solution (λ, S) of the equation

$$B^N S = \lambda N + S$$

(λ and S do not depend on N) and for all operators G there exists the limit

$$\lim_{N\to\infty} \frac{1}{N} B^N\langle G\rangle = \lambda.$$

Proof is just the application of the methods from [Kol4], [Kol5].

Acknowledgments

I am very thankful to Prof. V.P. Belavkin, who inspired me with problems of quantum measurements, and to Prof. S. Albeverio for his hospitality in Ruhr Universität Bochum, as well as to S. Albeverio, V.P. Belavkin, L. Diosi, P.Staszewski, and A. Ponosov for useful discussions. I am much indebted to the Alexander von Humboldt Stiftung for financial support.

References

[AKR] J. Azema, M.Kaplan-Duflo, D.Reves. Recurrence fine des processes de Markov. Ann.Inst.H.Poincare B, **2** (1966), 185-220.

[Bel1] V.P. Belavkin. A new wave equation for a continuous nondemolition measurement. Phys.Let.A, **140**(1989), 355-358.

[Bel2] V.P.Belavkin. The reconstruction theorem for quantum stochastic process. Teor.i Mat.Fis. **62**(1985), 409-431. Engl. transl. in Theoret.Math.Phys.

[Bel3] V.P.Belavkin. Nondemolition measurements, nonlinear filtering and dynamic programming of quantum stochastic processes. In: Modelling and Control of Systems, Proc. Bellman Continuous Workshop, Sophia-Antipolis 1988, Lect.Notes in Contr. and Inform. Sci., **121**(1988), 245-265.

[BK] V.P.Belavkin, V.N.Kolokoltsov. Quasy-classical asymptotics of quantum stochastic equations. Teor. i Mat. Fis. **89**(1991), 163-178. Engl. transl. in Theor. Math. Phys.

[BS] V.P.Belavkin, P.Staszewski . A stochastic solution of Zeno paradox for quantum Brownian motion. Prepr. **28**(1990). Centro Mat.V.Volterra, Univ.degli studi di Roma 2, published in Phys. Rev. A **45:3** (1992), 1347-1356.

[D1] L. Diosi. Continuous quantum measurement and Ito formalism. Phys. Lett. A **129**(1988), 419-423.

[D2] L. Diosi. Localized solution of a simple nonlinear quantum Langevin equation. Phys. Lett. A **132**(1988), 233-236.

[GR] Y. Guivarch, A. Raugi. Products of random matrices: convergence theorems. Contemp. Math. **50**(1986), 31-54.

[HP] R.L.Hudson, K.R.Parthasarathy. Quantum Ito's formula and stochastic evolution. Comm.Math.Phys. **93**(1984), 301-323.

[IK] K.Ichihara, H.Kunita. A classification of the Second Order Degenerate Elliptic Operator and its Probabilistic Characterization. Z.Wahrsch. Verw.Gebiete **30**(1974), 235-254.

[Ju] D.Juriev. Quantum projective field theory: Quantum projective analogs of the Euler formulas. Teor. i Mat. Fis. **92**(1992), 172-175. Engl.transl. in Theor.Math.Phys.

[Kl] W.Klieman. Recurrence and invariant measures for degenerate diffusion. Ann. Prob. **15**(1987), 690-707.

[Kol1] V.N.Kolokoltsov. Application of the quasyclassical methods to the investigation of the Belavkin quantum filtering equation. Mat. Zametki **50**(1991) 153-156. Transl. in Math. Notes.

[Kol2] V.N.Kolokoltsov. Long time behavior of continuously observed and controlled quantum systems (a study of the Belavkin quantum filtering equation). Preprint **204** (December 1993), Ruhr Universität Bochum, SFB 237. To be published in Quantum Prob. Rel. Top.

[Kol 3] V.N. Kolokoltsov. Scattering theory for the Belavkin equation describing a quantum particle with continuously observed coordinate. Preprint **228** (Juni 1994), Ruhr Universität Bochum, SFB 237.

[Kol4] V.N. Kolokoltsov. Stochastic Bellman equation as a nonlinear equation in Maslov's space. Perturbation theory. Dokl. Akad. Nauk **323** (1992), 223-228. Engl. transl. in Sov. Math. Dokl.

[Kol5] V.N. Kolokoltsov On linear, additive and homogeneous operators in idempotent analysis. Adv. in Sov. Math. **13**: Idempotent Analysis, ed. V.P. Maslov, S.N. Samborski (1992).

[MS] B.Misra, E.C.Sudarshan. The Zeno's paradox in quantum theory. J.Math.Phys. **18**(1977), 756-763.

OPTIMAL QUANTUM MEASUREMENTS FOR TWO PURE AND MIXED STATES

L.B. Levitin

Boston University,
College of Engineering
44 Cummington St. Boston,
MA 02215, USA

Abstract

An explicit formula is obtained for the entropy defect and the (maximum) information for an ensemble of two pure quantum states; an optimal basis is found, that is, an optimal measurement procedure which enables one to obtain the maximum information. Some results are also presented for the case of two mixed states, described by second-order density matrices (for example, spin polarization matrices). It is shown that in the case of two states the optimal measurement is a direct von Neumann measurement performed in the subspace of the two states.

1 INTRODUCTION

The concepts of entropy defect and the information for an ensemble of quantum systems have been first introduced in 1969 [1–4]. If there is probability p_i ($i = 1, 2, \ldots$) for a quantum system to be in a state described by a density matrix $\boldsymbol{\rho}^{(i)}$, the entropy defect of the system is defined to be

$$I_0 = - \text{ Tr } \boldsymbol{\rho} \ln \boldsymbol{\rho} \; + \; \sum_i p_i \text{ Tr } \boldsymbol{\rho}^{(i)} \ln \boldsymbol{\rho}^{(i)} \ , \tag{1}$$

where

$$\boldsymbol{\rho} = \sum_i p_i \boldsymbol{\rho}^{(i)} \tag{2}$$

is the a priori density matrix of the ensemble.

Quantum Communications and Measurement
Edited by V. P. Belavkin *et al.*, Plenum Press, New York, 1995

The entropy defect represents the average decrease in the entropy of the system when a macrostate $\boldsymbol{\rho}^{(i)}$ is specified compared with the situation when only the mixture of the states $\boldsymbol{\rho}$ is known. The information contained in a quantum system about its macrostate is defined to be

$$I = \sup_L I_L = \sup_L \left\{ -\sum_n \rho_{nn} \ln \rho_{nn} + \sum_i \sum_n p_i \rho_{nn}^{(i)} \ln \rho_{nn}^{(i)} \right\} \tag{3}$$

where L is a complete set of variables of the system [5]; $\rho_{nn}, \rho_{nn}^{(i)}$ are the diagonal elements of the density matrix, relative to a basis in which all operators belonging to L are diagonal. (For brevity, we shall refer to this basis as to "basis L"). Thus, the information is defined as the maximum information obtainable by a direct quantum measurement (in von Neumann's sense [6]). Each direct measurement corresponds to an orthogonal resolution of identity in the Hilbert space H of the system.

As shown in [1–4], the quantity I plays the same role in information transmission by quantum agents (under von Neumann measurement procedures) as Shannon's information for classical systems.

An important remark should be made concerning the so called indirect measurements. By "indirect measurements" we mean quantum-mechanical (von Neumann) measurements applied to a physical system one of whose subsystems is a system carrying information (e.g., [7–10, 28, 29]). In mathematical terms, an indirect measurement can be described by means of a nonorthogonal resolution of the identity in the state space of the subsystem and, therefore, often referred to as a probability-operator-valued measurement (POM). Definition of information as the supremum J over all possible (in general, nonortohogonal) resolutions of identity in H seems to be, therefore, more general. Indeed, examples have been given in [11, 14] in which the optimal indirect measurement yields information exceeding the bound (3), i.e. where $J > I$. However, it was shown in 1981 [12, 3, 13] that in the case of an infinite-dimensional separable Hilbert space always $J = I$. Thus, if H is infinite-dimensional we can use definition (3) without any reservations.

In the classical case, entropy defect and information are always equal which constitutes the entropy defect principle (1966, [15]) of classical physics. However, in the realm of quantum physics, these two quantities split apart, and the equality turns into inequality (first published in 1969 [1–4]):

$$I \leq I_0, \tag{4}$$

where equality holds if and only if all the matrices $\boldsymbol{\rho}^{(i)}$ commute. Inequality (4) expresses the quantum-mechanical entropy defect principle. In the quasiclassical limit, all operators commute [5], and information coincides again with entropy defect.

The importance of the entropy defect principle goes far beyond applications to channel capacity problems. In fact, the principle is a cornerstone of a consistent physical information theory [3, 15]. It has been used for thermodynamic characterization of information channels [20, 21]. Both quantities, I_0 and I, have been shown to have a direct thermodynamic meaning in phenomena related to the Gibbs paradox and to the Second Law of thermodynamics [22–24].

Finding explicit expressions for I and for the optimal basis (i.e. the optimal set of projection operators) is, apparently, a very difficult problem. Up to now it has been solved for a few special cases only. In the early work by Belavkin and Stratonovich [16, 17], the optimal indirect measurement has been found for the Gaussian ensemble of partially coherent (so called "Gaussian") states of electromagnetic field oscillators. Davies [14] has found the solution for the special case of group covariant sets of equiprobable states. He also showed that if the state space H is of dimension d, the optimal measurement can be performed in a space of dimension $n \leq d^2$. In [18] a complete classification of POMs has been given. Recently Josza, Robb, and Wootters [19] established a remarkable lower bound on the information I which depends on the a priori density matrix $\boldsymbol{\rho}$ only. Interesting results on quantum coding have been obtained in [26, 27].

In this paper explicit expressions will be found for the information I and for an optimal basis L in the simple case of two pure states. We shall also present some results for the case of two mixed states described by second-order matrices (such as spin polarization matrices).

2 ENTROPY DEFECT AND INFORMATION FOR AN ENSEMBLE OF TWO QUANTUM STATES

Consider the case in which a quantum system has a probability p of being in a state $\boldsymbol{\rho}^{(1)}$ and a probability $1-p$ of being in a state $\boldsymbol{\rho}^{(2)}$. Suppose that these are pure states, i.e., their density matrices $\boldsymbol{\rho}^{(1)}$, $\boldsymbol{\rho}^{(2)}$ are of the form

$$\rho_{nn'}^{(1)} = \psi_n^{(1)}\bar{\psi}_{n'}^{(1)} \quad ; \qquad \rho_{nn'}^{(2)} = \psi_n^{(2)}\bar{\psi}_{n'}^{(2)} \tag{5}$$

where $\psi^{(1)}$, $\psi^{(2)}$ are wave functions (state vectors). (The bar denotes complex conjugation.) The following theorem holds (1973, [22, 23]).

Theorem. The information I_L for an ensemble of two pure states, occuring with probabilities $p_1 = p$, $p_2 = 1-p$, achieves its maximum $I_L = I$ in a basis $L = \{\varphi_i\}$ $(i = 1, 2, \ldots)$ such that two basis vectors (say φ_1 and φ_2) lie in the plane spanned by the state vectors $\psi^{(1)}$, $\psi^{(2)}$, and the following condition holds:

$$\boldsymbol{\sigma} = p\boldsymbol{\rho}^{(1)} - (1-p)\boldsymbol{\rho}^{(2)} \quad \text{is a diagonal matrix.}$$

This result means that in the optimal basis L only two components of the vectors $\psi^{(1)}$ and $\psi^{(2)}$ are different from zero (say $\psi_1^{(i)}$, $\psi_2^{(i)}$), and moreover,

$$p\psi_1^{(1)}\bar{\psi}_2^{(1)} = (1-p)\psi_1^{(2)}\bar{\psi}_2^{(2)} \tag{6}$$

In terms of density matrices, this means that in the optimal basis

$$\begin{array}{ll} \rho_{nn'}^{(1)} = \rho_{nn'}^{(2)} = 0 & \text{for } n > 2, \quad n' > 2 \\ p\rho_{nn'}^{(1)} = (1-p)\rho_{nn'} & \text{for } n \neq n' \end{array} \tag{7}$$

Since the phase of a wave function (a constant factor of the form $e^{i\alpha}$) may be chosen arbitrarily (it does not affect the form of the density matrix), condition (6) implies that the phases of states $\psi^{(1)}$, $\psi^{(2)}$ and the basis vectors φ_1, φ_2 can be chosen in such a way that all the components of $\psi^{(1)}, \psi^{(2)}$ in this basis are real. Condition (6) may then be given a simple geometric meaning. Let (x_1, x_2) be the Cartesian coordinate system formed by the basis vectors φ_1, φ_2. Then the endpoints of the vectors $\sqrt{p}\psi^{(1)}$ and $\sqrt{1-p}\psi^{(2)}$ must lie on a hyperbola $x_1 x_2 = \text{const}$ (see Fig. 1).

Proof. Let I_L achieve its maximum at a basis L_0. Denote the elements of the density matrices in this basis by $\rho_{nn'}^{(i)}$. We shall assume that at least one of the matrices has a nonzero off-diagonal element; otherwise, the matrices commute and, as shown in [1–4], $I_L = I$ for a basis in which they are both diagonal. Suppose, for example, that

$$\rho_{12}^{(1)} = \bar{\rho}_{21}^{(1)} \neq 0 \tag{8}$$

Let $\tilde{L}$ be some other basis. Denote the elements of the density matrices in $\tilde{L}$ by $\tilde{\rho}^{(i)}_{mm'}$. Transformation from the new basis to the old one is performed by a unitary matrix $\|u_{nm}\|$. Hence:

$$\tilde{\rho}_{mm'} = \sum_n \sum_{n'} \bar{u}_{nm} \rho_{nn'} u_{n'm'} \tag{9}$$

Choose the matrix $\|u_{nm}\|$ as follows:

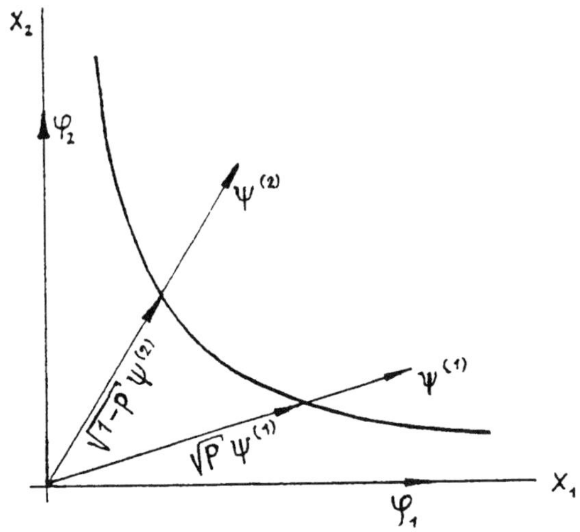

Figure 1: Two nonorthogonal pure quantum states and optimal basis.

$$u_{11} = u_{22} = \sqrt{1-\varepsilon^2} \quad ; \; u_{12} = \vartheta\varepsilon \quad ; \; u_{21} = -\bar{\vartheta}\varepsilon$$
$$u_{nm} = \delta_{nm} \quad (n > 2, m > 2) \tag{10}$$

where ϑ is an arbitrary complex number with modulus unity, and $\varepsilon > 0$ is a small real number.

We now introduce auxiliary matrices $\boldsymbol{\sigma}^{(1)} = p\boldsymbol{\rho}^{(1)}, \boldsymbol{\sigma}^{(2)} = (1-p)\boldsymbol{\rho}^{(2)}$. Like $\boldsymbol{\rho}^{(i)}$, these matrices are Hermitian and nonnegative definite.

Using (9) and (10), we obtain the following expression for the increment to I_L upon the passage from the basis L_0 to $\tilde{L}$:

$$\begin{aligned} I_{\tilde{L}} - I_{L_0} = & \left\{ \left[\bar{\vartheta}\sigma_{12}^{(1)} + \vartheta\sigma_{21}^{(1)}\right] \ln \frac{\sigma_{22}^{(1)}}{\sigma_{11}^{(1)}} + \left[\bar{\vartheta}\sigma_{12}^{(2)} + \vartheta\sigma_{21}^{(2)}\right] \ln \frac{\sigma_{22}^{(2)}}{\sigma_{11}^{(2)}} - \right. \\ & - \left. \left[\bar{\vartheta}\left(\sigma_{12}^{(1)} + \sigma_{12}^{(2)}\right) + \vartheta\left(\sigma_{21}^{(1)} + \sigma_{21}^{(2)}\right)\right] \ln \frac{\sigma_{22}^{(1)} + \sigma_{22}^{(2)}}{\sigma_{11}^{(1)} + \sigma_{11}^{(2)}} \right\} \varepsilon + \\ & + \left\{ \left(\sigma_{22}^{(1)} + \sigma_{22}^{(2)} - \sigma_{11}^{(1)} - \sigma_{11}^{(2)}\right) \ln \frac{\sigma_{22}^{(1)} + \sigma_{22}^{(2)}}{\sigma_{11}^{(1)} + \sigma_{11}^{(2)}} - \right. \\ & - \left(\sigma_{22}^{(1)} - \sigma_{11}^{(1)}\right) \ln \frac{\sigma_{22}^{(1)}}{\sigma_{11}^{(1)}} - \left(\sigma_{22}^{(2)} - \sigma_{11}^{(2)}\right) \ln \frac{\sigma_{22}^{(2)}}{\sigma_{11}^{(2)}} - \\ & - \frac{1}{2} \left[\left(\frac{1}{\sigma_{11}^{(1)} + \sigma_{11}^{(2)}} + \frac{1}{\sigma_{22}^{(1)} + \sigma_{22}^{(2)}} \right) \left(\bar{\vartheta}\left(\sigma_{12}^{(1)} + \sigma_{12}^{(2)}\right) + \right. \right. \\ & \left. + \vartheta\left(\sigma_{21}^{(1)} + \sigma_{21}^{(2)}\right) \right)^2 - \left(\frac{1}{\sigma_{11}^{(1)}} + \frac{1}{\sigma_{22}^{(1)}} \right) \left(\bar{\vartheta}\sigma_{12}^{(1)} + \vartheta\sigma_{21}^{(1)} \right)^2 - \end{aligned}$$

$$- \left(\frac{1}{\sigma_{11}^{(2)}} + \frac{1}{\sigma_{22}^{(2)}} \right) \left(\bar{\vartheta} \sigma_{12}^{(2)} + \vartheta \sigma_{21}^{(2)} \right)^2 \Big] \Big\} \varepsilon^2 + O\left(\varepsilon^3\right) \tag{11}$$

A necessary condition for I_L to achieve a maximum at the basis L_0 is that the term linear in ε vanish and the quadratic term be negative. When is this possible ?

Since ϑ is arbitrary, it may be chosen so that

$$\bar{\vartheta} \sigma_{12}^{(2)} + \vartheta \sigma_{21}^{(2)} = 0 \tag{12}$$

Then the coefficient of ε will vanish if at least one of the following equalities holds:

$$\bar{\vartheta} \sigma_{12}^{(1)} + \vartheta \sigma_{21}^{(1)} = 0 \tag{13}$$

$$\frac{\sigma_{11}^{(1)}}{\sigma_{11}^{(2)}} = \frac{\sigma_{22}^{(1)}}{\sigma_{22}^{(2)}} \tag{14}$$

If (12) and (14) hold, the coefficient ε^2 is nonnegative, vanishing only when $\sigma_{22}^{(2)} = \sigma_{11}^{(2)} - 0$. But in the latter case the information does not change when we go from L_0 to $\tilde{L}$, and, hence, it is possible to choose the initial basis L_0 in such a way that $\sigma_{12}^{(1)} = 0$, which, however, contradicts assumption (8). Thus, if $\sigma_{12}^{(1)} \neq 0$ in the optimal basis, then also $\sigma_{12}^{(2)} \neq 0$. It then follows from the simultaneous validity of (12) and (13) that

$$\sigma_{12}^{(2)} = k \sigma_{12}^{(1)} \tag{15}$$

where $k \neq 0$ is a real number.

Now, using the fact that the states under consideration are pure, we have

$$\sigma_{nn'}^{(i)} = c_n^{(i)} \bar{c}_{n'}^{(i)} \tag{16}$$

It now follows from condition (15) and from the vanishing of the coefficient of ε that

$$(1+k) \ln \frac{1+|k|z}{z+|k|} + (1-k) \ln z = 0 \tag{17}$$

where

$$z = \frac{|c_1^{(1)}||c_2^{(2)}|}{|c_2^{(1)}||c_1^{(2)}|} > 0 \quad \text{(according to (8)) and} \quad z \neq 1$$

(otherwise equality (14) would hold). It is readily shown that when $z > 0$, $z \neq 1$ and $k \neq 0$ equality (17) holds only if $k = 1$. Thus, a necessary condition for I_L to reach its maximum is equality (6):

$$p \psi_1^{(1)} \bar{\psi}_2^{(1)} = (1-p) \psi_1^{(2)} \bar{\psi}_2^{(2)}$$

Thus, in the case that assumption (8) is valid, i.e., when the states $\psi^{(1)}$ and $\psi^{(2)}$ are not identical and not orthogonal, then the wave function $\psi^{(1)}$ in the optimal basis has at least two components $\psi_1^{(1)} \neq 0$, $\psi_2^{(1)} \neq 0$, and equality (6) holds, whence it follows that $\psi_1^{(2)} \neq 0$, $\psi_2^{(2)} \neq 0$.

Now suppose that, in the optimal basis, at least one of the wave functions has some other nonzero component: $\psi_n^{(1)} \neq 0$ or $\psi_n^{(2)} \neq 0 \quad (n \neq 1, 2)$. Arguments similar to those just presented show that all such components must satisfy the following equalities:

$$\begin{aligned} p \psi_1^{(1)} \bar{\psi}_n^{(1)} &= (1-p) \psi_1^{(2)} \bar{\psi}_n^{(2)} \\ p \psi_2^{(1)} \bar{\psi}_n^{(1)} &= (1-p) \psi_2^{(2)} \bar{\psi}_n^{(2)} \end{aligned} \tag{18}$$

Combining (6) and (18), we obtain

$$\frac{\psi_2^{(1)}}{\psi_1^{(1)}} = \frac{\psi_2^{(2)}}{\psi_1^{(2)}} \quad ; \quad \frac{\psi_n^{(1)}}{\psi_1^{(1)}} = \frac{\psi_n^{(2)}}{\psi_1^{(2)}} \tag{19}$$

Thus, all components of the wave functions $\psi^{(1)}$ and $\psi^{(2)}$ must be proportional and (because of normalization) equal, so that states $\psi^{(1)}$ and $\psi^{(2)}$ are identical. It follows that, in the case of nonidentical and nonorthogonal states, only two components, $\psi_1^{(i)}$ and $\psi_2^{(i)}$, differ from zero in the optimal basis, and these two components satisfy condition (6). The negativity of the coefficient of ε^2 in (11) now becomes a sufficient condition for I_L to be maximum. We shall prove that this is always the case.

Note that when (6) is true the second group of terms (in brackets) in this coefficient is always nonnegative; it is a maximum for ϑ such that

$$\bar{\vartheta}\sigma_{12}^{(1)} + \vartheta\sigma_{21}^{(1)} = 2|\sigma_{12}^{(1)}| = 2p|\psi_1^{(1)}||\psi_2^{(1)}| \tag{20}$$

Using (6) and (20), a few manipulations yield the following expression for the coefficient of ε^2:

$$f(z) = (1-z)\ln z + \frac{2(1-z)^2}{1+z}$$

where z is the same quantity as in (17). It is easy to show that if $z \neq 1$ then $f(z) < 0$. This completes the proof of the theorem.

Now we calculate I_0 and I. To calculate I_0 we must find the eigenvalues of the matrix $\boldsymbol{\rho} = p\boldsymbol{\rho}^{(1)} + (1-p)\boldsymbol{\rho}^{(2)}$. For pure states, the eigenvalues depend only on the probability p and on the only joint invariant of $\boldsymbol{\rho}^{(1)}$ and $\boldsymbol{\rho}^{(2)}$ – the trace r of their product:

$$\mathrm{Tr}\, \boldsymbol{\rho}^{(1)}\boldsymbol{\rho}^{(2)} = r = |s|^2$$

where $s = \langle\psi^{(1)}, \psi^{(2)}\rangle$ is the scalar product of the wave vectors $\psi^{(1)}$ and $\psi^{(2)}$. The eigenvalues are:

$$\lambda_{1,2} = \frac{1}{2}\left(1 \pm \sqrt{1-4p(1-p)(1-r)}\right)$$

Hence, the entropy defect is

$$\begin{aligned} I_0 &= -\mathrm{Tr}\, \boldsymbol{\rho}\ln\boldsymbol{\rho} = \ln 2 \; - \\ &-\frac{1}{2}\Big[\left(1+\sqrt{1-4p(1-p)(1-r)}\right)\ln\left(1+\sqrt{1-4p(1-p)(1-r)}\right) + \\ &+\left(1-\sqrt{1-4p(1-p)(1-r)}\right)\ln\left(1-\sqrt{1-4p(1-p)(1-r)}\right)\Big] \end{aligned} \tag{21}$$

The explicit form of the optimal basis L_0 is readily found (it is unique up to permutation of the basis vectors and multiplication by arbitrary phase factors).

To calculate the information I, however, it will suffice to find the diagonal elements of the density matrices in this basis:

$$\begin{aligned} \rho_{11}^{(1)} &= 1-\rho_{22}^{(1)} = \frac{1}{2}\left(1+\frac{1-2(1-p)r}{\sqrt{1-4p(1-p)r}}\right) \\ \rho_{11}^{(2)} &= 1-\rho_{22}^{(2)} = \frac{1}{2}\left(1-\frac{1-2pr}{\sqrt{1-4p(1-p)r}}\right) \end{aligned} \tag{22}$$

Hence,

$$\begin{aligned}
I &= \frac{1}{2}\Big\{p\Big[\Big(\sqrt{1-4p(1-p)r}+1-2(1-p)r\Big)\cdot \\
&\ln\Big(\sqrt{1-4p(1-p)r}+1-2(1-p)r\Big) + \\
&+\Big(\sqrt{1-4p(1-p)r}-1+2(1-p)r\Big)\cdot \\
&\ln\Big(\sqrt{1-4p(1-p)r}-1+2(1-p)r\Big)\Big] + (1-p)\ \cdot \\
&\Big[\Big(\sqrt{1-4p(1-p)r}-1+2pr\Big)\ln\Big(\sqrt{1-4p(1-p)r}-1+2pr\Big) + \\
&+\Big(\sqrt{1-4p(1-p)r}+1-2pr\Big)\ln\Big(\sqrt{1-4p(1-p)r}+1-2pr\Big)\Big] - \\
&-\Big(\sqrt{1-4p(1-p)r}+1-2p\Big)\ln\Big(\sqrt{1-4p(1-p)r}+1-2p\Big) - \\
&-\Big(\sqrt{1-4p(1-p)r}-1+2p\Big)\ln\Big(\sqrt{1-4p(1-p)r}-1+2p\Big)\Big\}\cdot \\
&(1-4p(1-p)r)^{-\frac{1}{2}} \qquad (23)
\end{aligned}$$

In the general case, the formulae for I_0 and I are very cumbersome. We shall compare them for the case $p = \frac{1}{2}$. It is convenient to introduce the angle α between the vectors $\psi^{(1)}$ and $\psi^{(2)}$; then $r = \cos^2\alpha$. In this case the vectors $\psi^{(1)}$ and $\psi^{(2)}$ are symmetric about the bisectrix of the angle between the basis vectors φ_1, φ_2:

$$I_0 = \ln 2 - \frac{1}{2}(1+\cos\alpha) + (1-\cos\alpha)\ln(1-\cos\alpha) \qquad (24)$$

$$I = \frac{1}{2}(1+\sin\alpha)\ln(1+\sin\alpha) + (1-\sin\alpha)\ln(1-\sin\alpha) \qquad (25)$$

The following interesting relationship is noteworthy:

$$I\left(\frac{\pi}{2}-\alpha\right) = \ln 2 - I_0(\alpha)$$

The difference $I_0 - I$ achieves its maximum at $\alpha = \frac{\pi}{4}$. In that case $I_0 + I = \ln 2$ and $I \approx \frac{2}{3}I_0$. Plots of I_0 and I for the case $p = \frac{1}{2}$ are shown in Fig. 2.

Return now to the general case of two mixed states described by second-order density matrices (for example, different spin states). The relationship (15) is valid in this case too, but now k is the root of a complicated transcendental equation and depends in invariants of the matrices $\boldsymbol{\rho}^{(1)}$ and $\boldsymbol{\rho}^{(2)}$: their determinants d_1 and d_2 and the trace r of their product. Even for second-order matrices we could not derive general explicit expressions for I and the optimal basis L_0. Only when $d_1 = d_2 = d$ and $p = \frac{1}{2}$ it can be shown that $k = 1$ and the condition (7) is satisfied. In that case we obtain:

$$\begin{aligned}
I_0 &= \frac{1}{2}\Big[\Big(1+\sqrt{1-4d}\Big)\ln\Big(1+\sqrt{1-4d}\Big) + \\
&+\Big(1-\sqrt{1-4d}\Big)\ln\Big(1-\sqrt{1-4d}\Big) - \\
&-\Big(1+\sqrt{r-2d}\Big)\ln\Big(1+\sqrt{r-2d}\Big) - \\
&-\Big(1-\sqrt{r-2d}\Big)\ln\Big(1-\sqrt{r-2d}\Big)\Big] \qquad (26)
\end{aligned}$$

$$I = \frac{1}{2}\Big[\left(1+\sqrt{1-r-2d}\right)\ln\left(1+\sqrt{1-r-2d}\right) + + \left(1-\sqrt{1-r-2d}\right)\ln\left(1+\sqrt{1-r-2d}\right)\Big] \quad (27)$$

For fixed r, the values of I_0 and I decrease with increasing d.

It is noteworthy that in the case of two pure states the condition for maximum information coincides with the condition for optimal detection, guaranteeing a minimum mean error probability [10, 25]. However, this is not true in the general case of two mixed states.

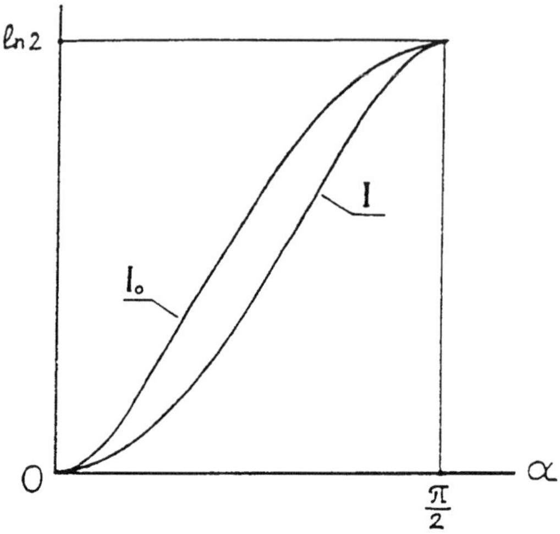

Figure 2: Entropy defect and information for two pure quantum states, $p = \frac{1}{2}$. The point $(\frac{\ln 2}{2}, \frac{\pi}{4})$ is the center of symmetry of the graph.

CONCLUSION

The results of Section 2 provide one of few examples of explicit expressions for information and optimal basis in the cases of states described by noncommuting density matrices. It is remarkable, that in the case considered, the optimal procedure is a direct measurement, i.e., a measurement in the space spanned by the state vectors used as signals. There are grounds for the conjecture that direct measurement is optimal whenever the number of states (signals) does not exceed the dimensionality of the state space.

References

[1] L.B. Levitin, "On the Quantum Measure of Information". *Proc. of the 4th Conference on Information and Coding Theory, Sec. II*, Tashkent, 1969 (English translation: in this book), pp. 111-115.

[2] L.B. Levitin, "Amount of Information and the Quantum-Mechanical Irreversibility of Measurement". *Proc. of the 2nd Intern. Symp. on Info. Theory*, Moscow-Yerevan, 1971, p. 144.

[3] L.B. Levitin, "Information Theory for Quantum Systems". *Information, Complexity and Control in Quantum Physics.* A. Blaquiere, S. Diner, and G. Lochak, eds., Springer-Verlag, Wien-New York, 1987, pp. 15-47.

[4] L.B. Levitin, "Physical Information Theory. Part II: Quantum Systems". *Proc. of the Workshop on Physics of Computation; PhysComp '92*, IEEE Computer Society Press, Dallas, 1993, pp. 215-219.

[5] L.D. Landau and E.M. Lifshitz, *Quantum Mechanics.* Pergamon Press, Oxford, 1965.

[6] J. von Neumann, *Mathematische Grundlagen der Quantenmechanik.* Springer Verlag, Berlin, 1932.

[7] E. Arthurs and J.L. Kelly, "On the Simultaneous Measurements of a Pair of Conjugate Observables". *Bell System Tech. J.*, **44** (1965), pp. 725-729.

[8] L.B. Levitin, and V.V. Mityugov, "Reception of Coherent Signals by Splitting of the Beam". *1st Conference of Problems of Information Transmission by Laser Radiation*, Kiev, 1968.

[9] C.W. Helstrom, J.W.S. Liu, and J.P. Gordon, "Quantum-Mechanical Communication Theory". *Proc. IEEE* **58**(10), (1970), p. 1587.

[10] C.W. Helstrom, *Quantum Detection and Estimation Theory*, Academic Press, New York, 1976.

[11] A.S. Holevo, "Informational Aspects of Quantum Measurements". *Prob. Info. Transm.*, **9**(3) (1973), pp. 31-42.

[12] L.B. Levitin, "Direct and Indirect Quantum Measurements Yield Equal Maximum Information". *1981 IEEE Intern. Symp. on Information Theory*, Santa Monica, CA, USA, 1981.

[13] L.B. Levitin, "Information in Direct and Indirect Quantum Measurements". *Proc. of the Workshop on Physics of Computation: PhysComp '92*, IEEE Computer Society Press, Dallas, 1993, pp. 220-222.

[14] E.B. Davies, "Information and Quantum Measurement". *IEEE Trans. on Info. Theory*, IT-**24**, (1979), pp. 596-599.

[15] D.S. Lebedev, L.B. Levitin, "Information Transmission by Electromagnetic Field". *Inform. Control*, **9**(1), (1966), pp. 1-22.

[16] V.P. Belavkin and R.L. Stratonovich, "On Optimization of Quasimeasurement of Quantum Observables on Information Criteria". *Proc. of 3rd Intern. Symp. on Info. Theory*, Tallin, 1973, pp. 14-18.

[17] V.P. Belavkin and R.L. Stratonovich, "On Optimization of Processing of Quantum Signals by Information Criterion". *Radio Eng. Electron. Physics*, **18**(9), (1973), pp. 1839-1844.

[18] L.P. Hughston, R. Josza, and W.K. Wootters, "A Complete Classification of Quantum Ensembles Having a Given Density Matrix". *Phys. Letters A*, **183** (1993), pp. 14-18.

[19] R. Josza, D. Robb, and W.K. Wootters, "Lower Bound for Accessible Information in Quantum Mechanics". *Phys. Rev.*, **49**(2) (1994), pp. 668-677.

[20] L.B. Levitin, "Ideal Physical Information Channel". *Prob. Info. Transm.*, **1**(1) (1965), pp. 91-93.

[21] L.B. Levitin, "A Thermodynamic Characterization of Ideal Physical Information Channels". *J. of Information and Optimization Sciences*, **2** (1981), pp. 259-266.

[22] L.B. Levitin, "Quantum Amount of Information and Gibbs' Paradox". *3rd Intern. Symp. on Info. Theory*, Tallin, 1973.

[23] L.B. Levitin, "Quantum Amount of Information and Maximum Work". *Proc. of the 13th IUPAP Conf. on Stat. Physics*, D. Cabib, D.G. Kuper, and I. Riess, eds., A. Hilger, Bristol, 1978.

[24] L.B. Levitin, "Gibbs Paradox and Equivalence Relation Between Quantum Information and Work". *Proc. of the Workshop on Physics of Computation: PhysComp '92*, IEEE Computer Society Press, 1993, pp. 223-226.

[25] C.W. Helstrom, "Detection Theory and Quantum Mechanics". *Information and Control*, **10**, (1967), p. 254 .

[26] B. Schumacher, "Quantum Coding". *Phys. Rev. A* (to appear).

[27] R. Josza, B. Schumacher, "A New Proof of the Quantum Noiseless Coding Theorem". *J. of Modern Optics* (to appear).

[28] A. Peres, *Quantum Theory: Concepts and Methods.* Kluwer Academic, Dordrecht, 1993.

[29] A. Peres, "Neumark's Theorem and Quantum Inseparability". *Found. Phys.*, **20** (1990), p. 1441.

DIFFERENTIAL GEOMETRICAL ASPECTS OF QUANTUM STATE ESTIMATION AND RELATIVE ENTROPY

Hiroshi Nagaoka

Graduate school of Information Systems,
University of Electro–Communications,
1-5-1 Chofugaoka, Chofu-shi,
Tokyo 182, Japan.

In this article, we treat the space S of finite-dimensional positive density operators (quantum states) in a differential geometrical viewpoint. We suppose that a generalized covariance for arbitrary two observables (Hermitian operators) is specified at each state in S, which includes the symmetrized inner product and the Bogoliubov inner product as special (but important) cases, and introduce a triplet structure $(g, \nabla^{(\mathrm{e})}, \nabla^{(\mathrm{m})})$ on S via the specified covariance, where g is a Riemannian metric and $\nabla^{(\mathrm{e})}$ and $\nabla^{(\mathrm{m})}$ are affine connections. The structure $(g, \nabla^{(\mathrm{e})}, \nabla^{(\mathrm{m})})$ is regarded as a quantum analogue of the triplet of Fisher metric, exponential connection and mixture connection on a space of probability densities introduced in the information geometry by S. Amari ([1]). Some aspects relating to the quantum state estimation and the relative entropy are treated in terms of the differential geometry, where the theory of dual connections developed by Nagaoka and Amari ([4] [1]) plays an essential role.

Since this article is a summary, see [3] or forthcoming papers for more detailed discussions and the proofs.

GEOMETRICAL STRUCTURES ON A MANIFOLD OF QUANTUM STATES

Let $\mathcal{H}$ be a finite-dimensional Hilbert space, $\mathcal{A} \stackrel{\text{def}}{=} \{A \mid A = A^*\}$ be the totality of Hermitian operators on $\mathcal{H}$ and $S \stackrel{\text{def}}{=} \{\rho \mid \rho = \rho^* > 0, \mathrm{Tr}\, \rho = 1\}$ be the totality of positive density operators on $\mathcal{H}$. Since S is an open subset of the affine space

$\mathcal{A}_1 \stackrel{\text{def}}{=} \{A \mid A \in \mathcal{A}, \text{Tr } A = 1\}$, we can naturally regard S as a smooth manifold equipped with a flat affine connection, which is called the *mixture connection* or the *m-connection*. We denote the covariant derivative of this connection by $\nabla^{(\mathrm{m})}$: $\mathcal{X}(S) \times \mathcal{X}(S) \to \mathcal{X}(S)$, where $\mathcal{X}(S)$ is the totality of smooth vector fields on S, and identify it with the connection itself.

Suppose that we are given a family $\{\langle\cdot,\cdot\rangle_\rho\}_{\rho\in S}$ of inner products on $\mathcal{A}$, where $\langle A, B\rangle_\rho \in \mathbf{R}$ is specified smoothly by ρ for $\forall A, \forall B \in \mathcal{A}$. We further assume that $\langle A, 1\rangle_\rho = \langle A\rangle_\rho \stackrel{\text{def}}{=} \text{Tr } (\rho A)$ for $\forall A \in \mathcal{A}$. Such a family $\{\langle\cdot,\cdot\rangle_\rho\}_{\rho\in S}$ is called a *generalized covariance*. Two important examples are the *symmetrized inner product*

$$\langle A, B\rangle_\rho = \frac{1}{2}\text{Tr } (\rho AB + \rho BA)$$

and the *Bogoliubov inner product* (also called the *Kubo-Mori inner product* or the *canonical correlation*)

$$\langle A, B\rangle_\rho = \int_0^1 \text{Tr } (\rho^\lambda A \rho^{1-\lambda} B) d\lambda.$$

For an arbitrary tangent vector $D \in T_\rho(S)$, where $T_\rho(S)$ denotes the tangent space of S at ρ, there uniquely exists an operator $D^{(\mathrm{e})} \in \mathcal{A}$ such that $D\langle A\rangle = \langle D^{(\mathrm{e})}, A\rangle_\rho$ for $\forall A \in \mathcal{A}$, where $D\langle A\rangle$ is the derivative of the function $\langle A\rangle : \rho \mapsto \langle A\rangle_\rho$ by D. Then a linear isomorphism $D \mapsto D'$ from $T_\rho(S)$ to $T_{\rho'}(S)$ is determined by $D'^{(\mathrm{e})} = D^{(\mathrm{e})} - \langle D^{(\mathrm{e})}\rangle_{\rho'}$. This isomorphism yields another affine connection on S, which we call the *exponential connection* or the *e-connection* and denote by $\nabla^{(\mathrm{e})}$. This connection is curvature free, but is not flat unless the torsion tensor vanishes.

Define the inner product g_ρ on $T_\rho(S)$ by $g_\rho(D_1, D_2) \stackrel{\text{def}}{=} \langle D_1^{(\mathrm{e})}, D_2^{(\mathrm{e})}\rangle_\rho$. Then $g = \{g_\rho\}_{\rho\in S}$ forms a Riemannian metric on S, which we call the *Fisher metric.*

DUALITY OF E,M-CONNECTIONS

In general, if a Riemannian metric g and two affine connections ∇ and ∇^* on a manifold M satisfy

$$Xg(Y, Z) = g(\nabla_X Y, Z) + g(Y, \nabla^*_X Z)$$

for $\forall X, \forall Y, \forall Z \in \mathcal{X}(M)$, we say that ∇ and ∇^* are mutually *dual* (or *conjugate*) w.r.t. g ([4] [1]). It is shown that the e-connection and the m-connection are dual w.r.t. the Fisher metric.

If (M, g, ∇, ∇^*) is *dually flat* in the sense that both ∇ and ∇^* are flat, there uniquely exists ([4] [1]) a function $K : M \times M \to \boldsymbol{R}$ such that $K(\rho, \sigma) > 0$ and $K(\rho, \rho) = 0$ for $\forall\rho \neq \forall\sigma$ and that

$$\frac{\partial^2}{\partial\xi_1^i\partial\xi_1^j} K(\rho_1, \rho_2) = g_{ij}(\rho_1), \quad \forall\rho_1, \forall\rho_2 \in M,$$

or equivalently

$$\frac{\partial^2}{\partial\zeta_{2i}\partial\zeta_{2j}} K(\rho_1, \rho_2) = g^{ij}(\rho_2), \quad \forall\rho_1, \forall\rho_2 \in M,$$

where (ξ^i) ((ζ_i), resp.) is an arbitrary ∇-affine (∇^*-affine, resp.) coordinate system and g_{ij} (g^{ij}, resp.) are the corresponding components of the metric g. The function

K is called the *divergence* of the dually flat space (M, g, ∇, ∇^*) and is shown to satisfy a generalized Pythagorean property ([4] [1]). The Kullback infromation divergence (classical relative entropy) is derived as such a divergence in the classical information geometry. In the quantum case we have the following.

Theorem 1 *For the Bogoliubov product,* $(S, g, \nabla^{(\mathrm{m})}, \nabla^{(\mathrm{e})})$ *is dually flat, and its divergence is the quantum relative entropy* $K(\rho, \sigma) = \mathrm{Tr}\,[\rho(\log \rho - \log \sigma)]$.

TORSION OF E-CONNECTION

For $\forall \rho \in S$, we define the mapping $\Phi_\rho : \mathcal{A} \to \mathcal{A}$ by $\langle A, B \rangle_\rho = \mathrm{Tr}\,(A\Phi_\rho(B))$, $\forall A, \forall B \in \mathcal{A}$. For $\forall A \in \mathcal{A}$ and $\forall D \in T_\rho(S)$, let $\Phi_D(A)$ $(\in \mathcal{A})$ be the derivative of the mapping $\rho \mapsto \Phi_\rho(A)$ by D.

Theorem 2 *Let* $\mathcal{T}^{(\mathrm{e})}$ *be the torsion tensor of the e-connection on* S*. Then we have*

$$\iota_*\{\mathcal{T}^{(\mathrm{e})}(D_1, D_2)\} = \Phi_{D_2}(D_1^{(\mathrm{e})}) - \Phi_{D_1}(D_2^{(\mathrm{e})})$$

for $\forall D_1, \forall D_2 \in T_\rho(S)$*, where* ι *is the immersion* $S \to \mathcal{A}$*.*

Corollary 1 *For the symmetrized product, we have*

$$\iota_*\{\mathcal{T}^{(\mathrm{e})}(D_1, D_2)\} = \frac{1}{4}[[D_1^{(\mathrm{e})}, D_2^{(\mathrm{e})}], \rho],$$

where $[\cdot, \cdot]$ *is the commutator for operators.*

This corollary means that the e-connection for the symmetrized product is not torsion free, and in this case we cannot expect the existence of the divergence such as the relative entropy.

CRAMÉR-RAO INEQUALITY AND E-AUTOPARALLEL SUBMANIFOLDS

Let M be an arbitrary submanifold of S of dimension m $(\leq n)$ and $(\xi^i) = (\xi^1, \ldots, \xi^m)$ be an arbitrary coordinate system of M. In the sequel, we denote the m,e-connections on M by $\nabla_M^{(\mathrm{m})}$, $\nabla_M^{(\mathrm{e})}$, which are defined as the orthogonal projection of the m,e-connections $\nabla^{(\mathrm{m})}$, $\nabla^{(\mathrm{e})}$ onto M via the metric g.

The first theorem in the following is a generalization of the quantum Cramér-Rao inequality based on the symmetric logarithmic derivatives ([2]), which corresponds to the case of the symmetrized inner product. The second theorem describes the condition for the existence of $(F^1, \ldots, F^m)$ achieving the Cramér-Rao bound. For a generalized covariance other than the symmetrized one, however, the meaning of the theorems is not clear in view of the parameter estimation.

Theorem 3 *([5]) If an m-tuple* $F = (F^1, \ldots, F^m) \in \mathcal{A}^m$ *of operators satisfies the unbiasedness condition* $\langle F^i \rangle_\rho = \xi^i(\rho)$ *for* $\forall \rho \in M, \forall i$*, then the matrix* $V(\rho) - G(\rho)^{-1}$ *is nonnegative definite, where* $V(\rho) = [v^{ij}(\rho)]$ *and* $G = [g_{ij}]$ *are the* $m \times m$ *matrices defined by* $v^{ij}(\rho) = \langle F^i - \xi^i(\rho), F^j - \xi^j(\rho) \rangle_\rho$ *and* $g_{ij} = g(\partial_i, \partial_j)$ $(\partial_i \stackrel{\mathrm{def}}{=} \partial/\partial\xi^i)$.

Theorem 4 *For given $(M, (\xi^i))$, there exists an m-tuple $(F^1, \ldots, F^m) \in \mathcal{A}^m$ satisfying the unbiasedness condition in the above theorem and the equality $V(\rho) = G(\rho)^{-1}$ for $\forall \rho \in M$, if and only if M is $\nabla^{(e)}$-autoparallel in S and (ξ^i) is a $\nabla_M^{(m)}$-affine coordinate system of M.*

For the Bogoliubov product, a $\nabla^{(e)}$-autoparallel submanifold M always has a $\nabla_M^{(e)}$-affine coordinate system (ζ_i), and an element of M is parametrized by (ζ_i) as

$$\rho = \exp\left[F^0 + \sum_{i=1}^{m} \zeta_i F^i - \psi(\zeta)\right], \tag{1}$$

where ψ is a real-valued function on $\boldsymbol{R}^m$. It is seen that there is a close analogy between the structure of this form and that of exponential families in the classical case ([1]).

For the symmetrized product, a $\nabla^{(e)}$-autoparallel submanifold M does not generally have a $\nabla_M^{(e)}$-affine coordinate system because of nonvanishing torsion. However, since every 1-dimensional manifold is torsion free, an arbitrary $\nabla^{(e)}$-geodesic M has a $\nabla_M^{(e)}$-affine parameter ζ. We can show that an element of M is parametrized as

$$\rho = \exp\left[\{\zeta F - \psi(\zeta)\}/2\right] H \exp\left[\{\zeta F - \psi(\zeta)\}/2\right]. \tag{2}$$

It is noted in this case that $(M, g, \nabla_M^{(m)}, \nabla_M^{(e)})$ is dually flat and has the divergence. For arbitrary points $\rho, \sigma \in S$, let K be the divergence of the $\nabla^{(e)}$-geodesic connecting these points. Then we have

$$K(\rho, \sigma) = \mathrm{Tr}\,[\rho \log A],$$

where A is the positive operator in $\mathcal{A}$ satisfying $A^{1/2}\sigma A^{1/2} = \rho$, or is explicitly written as

$$A = \left\{\sigma^{-\frac{1}{2}}(\sigma^{\frac{1}{2}}\rho\sigma^{\frac{1}{2}})^{\frac{1}{2}}\sigma^{-\frac{1}{2}}\right\}^2.$$

This K is another quantum analogue of the Kullback divergence, but turns out to be the Kullback divergence itself for the pair of probability distributions on the measurement of the observable A corresponding to the pair of states ρ and σ.

We also note that both (1) and (2) include $\rho = \exp\left[\zeta F - \psi(\zeta)\right]$ as a special case, which is the same form as the Gibbs state in the statistical mechanics.

References

[1] Amari, S.: "Differential-geometrical methods in statistics", Lecture Notes in Statistics 28, Springer-Verlag (1985).

[2] Helstrom, C. W.: "Quantum Detection and Estimation Theory," Academic Press, New York (1976).

[3] Nagaoka, H.: "Differential Geometrical Aspects of Quantum State Estimation and Relative Entropy", METR 94-14. Univ. Tokyo (1994).

[4] Nagaoka, H., Amari, S.: "Differential geometry of smooth families of probability distributions", METR 82-7, Univ. Tokyo (1982).

[5] Petz, D., Toth, G.: "The Bogoliubov inner product in quantum statistics", Lett. Math. Phys., 27, pp.205-216 (1993).

Part VI.
Quantum Optics, Experiments & Simulation

QUANTUM NOISE EATERS

C. Fabre, C. Richy, P. Kurz*,
A. Lambrecht, J.M. Courty, E. Giacobino,
S. Reynaud, A. Heidmann, M. Pinard

Laboratoire Kaster Brossel,
U.P.M.C., Case 74 75252
Paris Cedex 05 - France
*permanent address : University of Konstanz,
Physics Department, Konstanz, Germany

1 INTRODUCTION

"Noise eaters" are optical devices, now commercially available, which are able to reduce the intensity noise of any incoming laser beam. These devices are based on electronic servoloops driving intensity modulators, and it can be shown that they cannot reduce the noise below the standard quantum noise level, or shot noise. This talk will show that one is able to develop "Quantum noise eaters", which are likely to reduce the noise of an incoming beam below the standard quantum noise limit, either on its intensity or on a chosen quadrature component. We will give three examples of such devices, based respectively on laser-cooled atoms, on parametric media, and on mirrors moving under the influence of the radiation pressure force.

2 LASER COOLED ATOMS

Let us consider the optical system sketched in Figure 1a : it consists in a lossless perfect Kerr medium inserted in a one-port optical cavity. Due to the presence of the nonlinear medium, the optical length of the cavity depends on the intracavity field intensity $|E_{cav}|^2$, and therefore the system has a bistable behaviour, which appears for example when one plots $|E_{cav}|^2$ as a function of the input field $|E^{in}|^2$. If

Quantum Communications and Measurement
Edited by V. P. Belavkin *et al.*, Plenum Press, New York, 1995

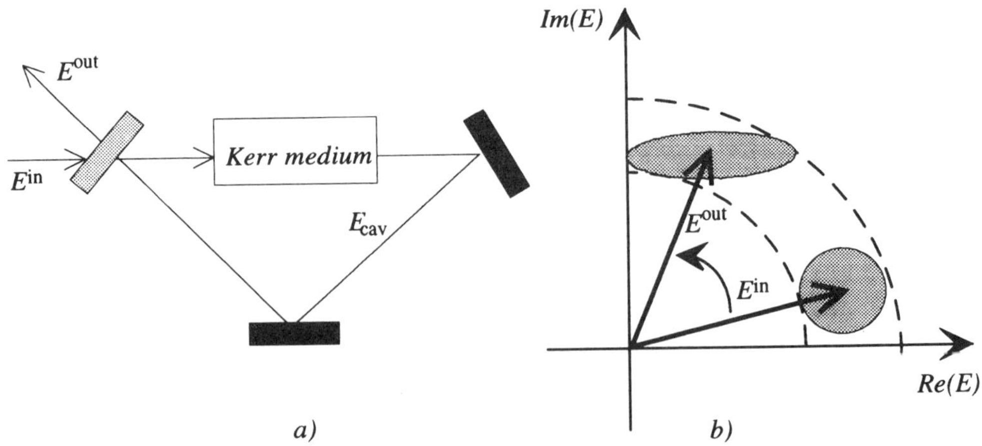

Figure 1.

we assume that the nonlinear medium has no losses and an instantaneous response, then the effect of the cavity on the input is simply a phase shift which is intensity dependent :

$$E^{out} = E^{in} \exp\left(i\Phi\left(|E_{cav}|^2\right)\right) \tag{1}$$

This phase shift is different for the low intensity and for the high intensity fluctuations of the incoming field, assumed to be in a coherent state (Figure 1b). This results in a stretching of the uncertainty area of the input field, and therefore in a squeezing of some phase quadrature for the output fields[(1)].

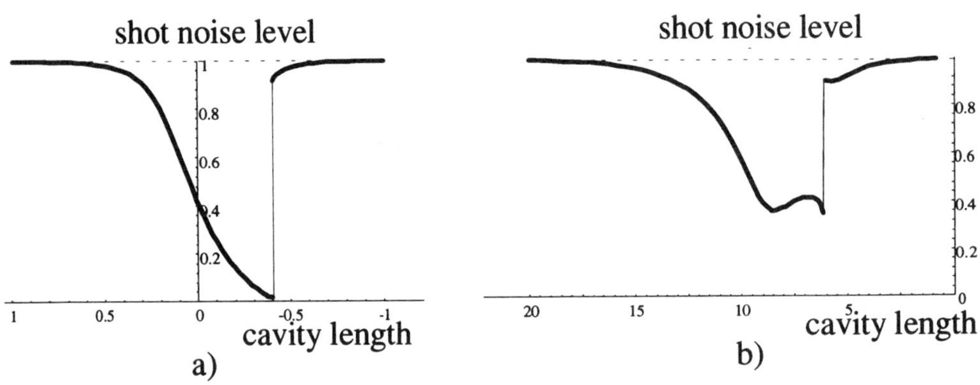

Figure 2.

More precisely, it can be shown that the field reflected by such a nonlinear cavity can be almost perfectly squeezed for noise frequencies smaller than the cavity bandwidth when one approaches the turning points of the bistability curve. As far as the intensity noise is concerned, the same conclusion holds, except for the very low noise frequencies, where the intensity noise is not affected by its passage through the cavity.

A perfect Kerr medium inside an optical cavity is therefore a very good example of a passive quantum noise eater, when the losses of the system are negligible. Unfortunately, such a perfect medium does not exist. Real Kerr media have no instantaneous response, they have also excess noise sources and losses, which reduce the amount and bandwidth of squeezing which can be obtained. Two level atoms inserted in a cavity and irradiated close to resonance[2] can be used as "quantum noise eaters", as shown in Fig. 2 which compares the theoretically predicted amount of squeezing, obtained in similar conditions, by a perfect Kerr medium (Fig. 2a) and by an assembly of two level atoms at rest (Fig. 2b), when one scans the cavity length through a resonance with the input field. Despite the excess noise brought by spontaneous emission, one predicts a large amount of squeezing for moderate input intensities when the atom-field detuning is of the order of 10 times the natural linewidth.

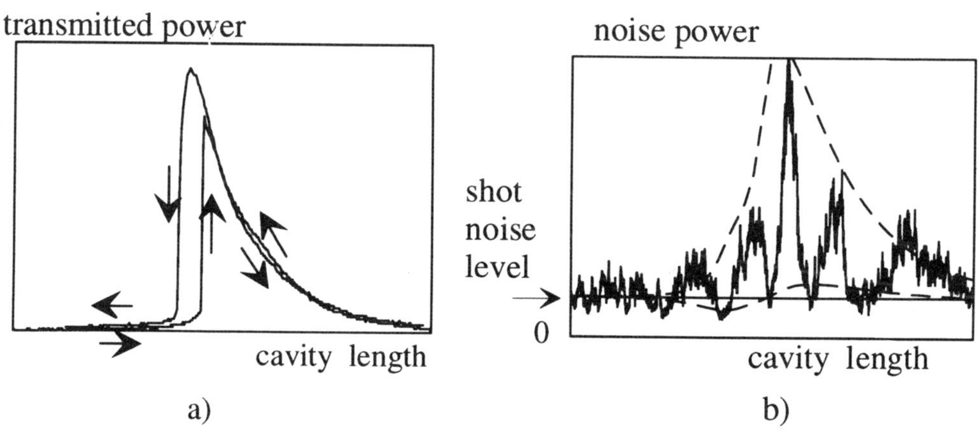

Figure 3.

We are now studying experimentally such a system, using as a nonlinear medium cesium atoms which are cooled and trapped in a magnetooptical trap by auxiliary laser beams. Inserting 10^7 cold atoms in a low finesse optical cavity, we have observed bistability curves for input intensities of the order of $10\mu W$ (Figure 3a). The excess noise due to the cooling and trapping process prevented us from observing squeezing of the output field in a steady state regime. By chopping the trapping beams and the magnetic field, we have recently been able in a preliminary experiment, to measure 30% of quadrature squeezing just after the switching off of the trapping beams[3] (Figure 3b: the oscillations of the noise in the figure are due to a fast scanning of the local oscillator phase, yielding a value of the small and large axes of the noise ellipse on the same recording. The dashed lines give the theoretical variations of the minimum and maximum quadrature noise of the output field).

3 PARAMETRIC $\chi^{(2)}$ CRYSTAL

A three-wave mixing process can also be used as a quantum noise eater : let us consider an optical parametric oscillator, which is simultaneously resonant for the pump, signal and idler fields. If the cavity finesse is large for the generated signal and idler fields, almost all the energy of the input pump field is transferred to the output field, reflected by the nonlinear cavity. Furthermore, if the losses of the system are small, its main effect is, like in the previous case, an intensity dependent phase shift for the pump field, and therefore squeezing is expected to occur on the output pump field.

A detailed theoretical analysis(4) of the triply resonant Optical Parametric Oscillator shows that when the detunings between the cavity modes and the three interacting light modes are slightly different from zero, one can obtain bistable or even unstable behaviours for the mean fields. Like in the Kerr medium case, an important squeezing effect is predicted close to the onset of such regimes.

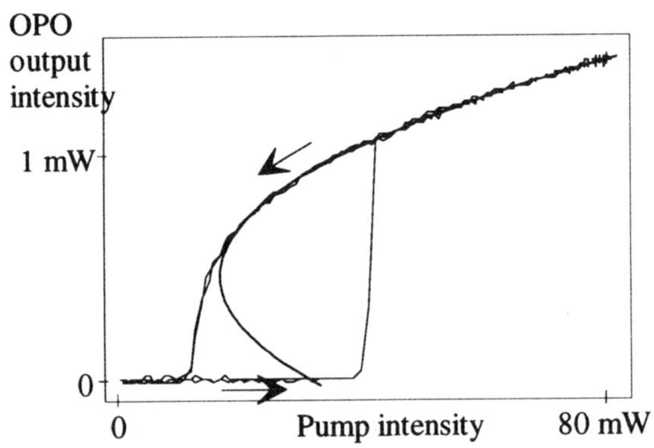

Figure 4.

We have undergone the experimental study of such a device using a nonlinear KTP crystal inserted in a triply resonant "semi-monolithic" cavity (one end face of the nonlinear crystal has a highly reflecting optical coating, so that the number of optical surfaces in the cavity is reduced), which has a finesse of about 60 for the pump field, and 600 for the signal and idler fields (having slightly different frequencies close to $1.06\mu m$). In this system, we have obtained an oscillation theshold of about 1mW of pump intensity(5). As shown in Figure 4, we have also observed a bistable behaviour for pump intensities around 30mW (the full line in Figure 4 corresponds to the theoretically expected variation). The study of the noise properties of the reflected pump beam is now under progress. Such a device will hopefully provide a compact and efficient quantum noise eater for moderate input intensities.

4 MOVABLE MIRROR

When a light beam described by the electric field $E^{in}e^{-i\omega t}$ interacts with a mirror moving with constant velocity v, its amplitude and phase are changed. A simple Lorentz transformation calculation shows that the reflected field is equal to, in the case $v << c$:

$$E^{out} = E^{in} e^{-i\omega\left(1-\frac{2v}{c}\right)t}\left(1 - \frac{2v}{c}\right) \tag{2}$$

There are two different effects related to v : a phase-shift effect, which can be interpreted as a Doppler term for the light beam frequency, and an amplitude effect, which modifies the intensity of the reflected beam. Both effects can be used in noise eating mechanisms, as we will see below.

Furthermore, the light itself can induce the motion of the mirror, because of the radiation pressure force F, which is equal to :

$$F = 2P^{IN}/c\left(1 - \frac{2v}{c}\right) \tag{3}$$

where P^{IN} in the input beam power. We will see that this rather small force ($6 \times 10^{-9} N/W$) may have important effects. Notice the velocity dependent term, which is a friction force giving rise to a radiative damping effect on the mirror motion, with a damping rate Γ_{RAD} equal to :

$$\Gamma_{RAD} = 4P^{IN}/mc^2 \tag{4}$$

This rate is very small for macroscopic mirrors (10^{-14} s^{-1}/W for $m = 4$ mg).

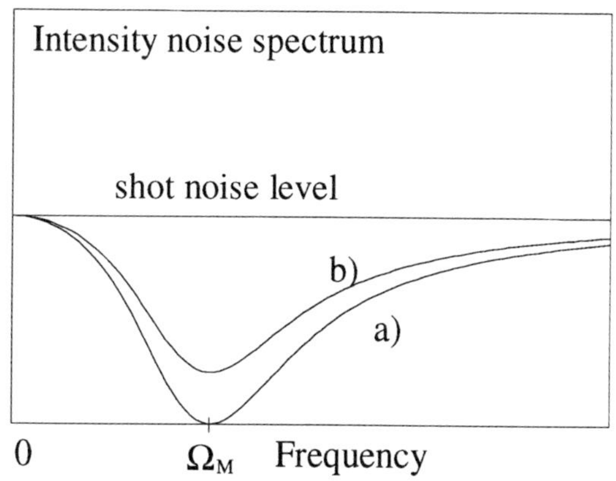

Figure 5.

The velocity dependent amplitude term in (4.1) can be used in a very simple device : it consists in a single mirror submitted to a harmonic restoring force, characterized by a mechanical eigenfrequency Ω_M and a mechanical damping rate Γ_M. When the light reflected by such a mirror has a positive fluctuation, its velocity increases because of an excess of radiation pressure, and therefore the intensity of the reflected field decreases[6] : this yields an intensity controlling mechanism. It is possible from equations (4.1) and (4.2) to calculate the noise spectrum of the flux of reflected photons. It is displayed in figure (5). Curve (5a) gives the spectrum when the only cause of damping for the mirror motion is the radiative friction term

(4.3) : there is a perfect noise reduction around the noise frequency Ω_M. Curve (5b) corresponds to the case $\Gamma_M = 2\Gamma_{RAD}$. In the general case, the minimum noise is given by $S_{MIN} = S_{SQL}\,\Gamma_M/\left(\Gamma_M + \Gamma_{RAD}\right)$, quantity which is close to the standard quantum limit (shot noise) in realistic situations when $\Gamma_M/\Gamma_{RAD} << 1$. So, this very simple and elegant noise eating mechanism cannot be used in real devices.

To enhance the effect, one can use the movable mirror in a one port optical cavity, like for the Kerr or the parametric media of the previous paragraphs. In the steady state regime, and because of the radiation pressure force, the cavity length depends on the intracavity light intensity, and the system behaves as an effective Kerr medium : there is a possibility for some values of the parameters, of a bistable behaviour, which has already been studied experimentally[7]. Assuming that the mirror motion is classical, it is possible to determine the noise characteristics of the reflected field[8], taking into account the input field fluctuations and the Brownian thermal motion of the moving mirror. The intensity noise spectrum is given in figure (6) when the system is operating close to a turning point of the bistability curve. At zero temperature (full line), one predicts a significant squeezing at very low noise frequencies (like in any Kerr media), and around the mechanical frequency Ω_M. At nonzero temperatures (dashed line), the latter is quickly destroyed by thermal fluctuations, while the former subsists for higher values of T, especially in the case when the mechanical quality factor $Q = \Omega_M/\Gamma_M$ is very large.

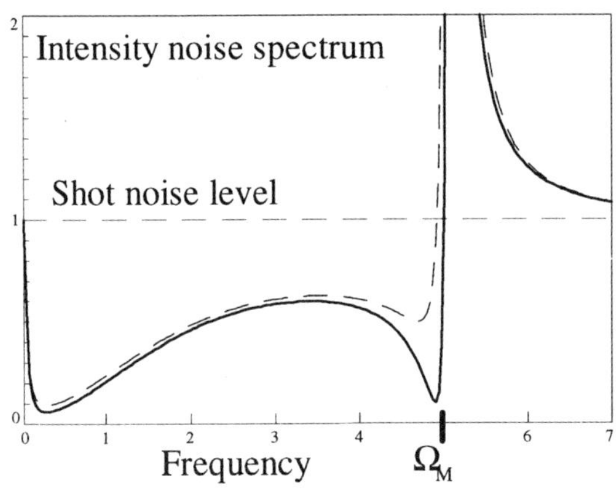

Figure 6.

We are presently studying the experimental implementation of such a device, using a crystal plate with reflecting coatings on both ends as a highly resonant mechanical and optical resonator : the system is no longer a harmonic oscillator and has several mechanical vibration modes, at frequencies Ω_M, $3\Omega_M$, $5\Omega_M$,.... The intensity noise spectrum of the light reflected by such a plate can be calculated by the same techniques. One can show that it is possible to get a significant quantum noise eating effect with this device in realistic conditions.

Acknowledgement

This work has been supported in part by the ESPRIT programme QUINTEC 6934.

References

[1] L. Lugiato, G. Strini, Optics Commun. 41, 67 (1982) ; S. Reynaud, C. Fabre, E. Giacobino, A. Heidmann, Phys. Rev. **A 40**, 1440 (1989).

[2] M. Reid, D. Walls, Phys. Rev. A 28, 332 (1983) ; L. Lugiato, Phys. Rev. **A 33**, 4079 (1986) ; M. Reid, Phys. Rev. **A 37**, 4792 (1988) ; L. Hilico, C. Fabre, S. Reynaud, E. Giacobino, Phys. Rev. **A 46**, 4397 (1992).

[3] A. Lambrecht, J.M. Courty, S. Reynaud, E. Giacobino "cold atoms : a new medium for quantum optics" submitted to.

[4] L. Lugiato, C. Oldano, C. Fabre, E. Giacobino, R. Horowicz, Nuovo Cimento **10**, 957 (1988) ; C. Fabre, E. Giacobino, A. Heidmann, L. Lugiato, S. Reynaud, M. Vadacchino, Wang Kaige, Quantum Optics **2**, 159 (1990).

[5] C. Richy, K. Petsas, E. Giacobino, C. Fabre "Observation of bistability and delayed bifurcation in a triply resonant optical parametric oscillator", submitted to J. Opt. Soc. of Am B (1994).

[6] A. Heidmann, S. Reynaud "Photon noise reduction by reflection from a movable mirror" to be published in Phys. Rev. A (1994).

[7] A. Dorsel, J. McCullen, P. Meystre, E. Vignes, H.Walther, Phys. Rev. Letters **51**, 1550 (1983) ; A. Gozzini, F. Maccarone, F. Mango, I. Longo, S. Barbarino, J. Opt. Soc. Am. **B 2**, 1841 (1985).

[8] C. Fabre, M. Pinard, S. Bourzeix, A. Heidmann, E. Giacobino, S. Reynaud, Phys. Rev. **A 49**, 1337 (1994).

STOCHASTIC SIMULATIONS OF DISSIPATION IN QUANTUM OPTICS: QUANTUM SUPERPOSITIONS

B.M. Garraway and P.L. Knight

Optics Section, The Blackett Laboratory,
Imperial College Prince Consort Road,
London, SW7 2BZ, UK.

Abstract

We show how nonclassical states can be constructed by interference amongst constituents in a superposition of basis states (chosen either from a Fock or coherent state basis). Such superpositions are fragile and highly susceptible to dissipation. We discuss the connection between ensemble averages and the evolution of individual realizations using quantum jump and quantum state diffusion models of conditioned measurements, each realization reflecting a particular measurement record. This allows us to distinguish between decoherence, localization and the deconstruction of superpositions into their components.

1 INTRODUCTION

We describe the formation of nonclassical states of light through interference in phase-space amongst constituent parts of quantum superpositions. This is illustrated with examples of superpositions of Fock states on the one hand, and coherent states on the other. Superpositions can generate interference fringes and regions of substantial negativity in the Wigner function quasi-probabilities. These features are dramatically sensitive to dissipation. We examine the evolution of single realizations of decaying superpositions, conditioned by the gain of information through the detection record and illustrate the use of jump methods to describe direct counting of decay products. We also illustrate the use of quantum state diffusion to describe heterodyne field measurements [1–3]. We will show how in some cases the local-

Quantum Communications and Measurement
Edited by V. P. Belavkin *et al.*, Plenum Press, New York, 1995

ization rate coincides with the decoherence rate, but give one example where these substantially differ.

2 CONSTRUCTION OF NONCLASSICAL FIELD STATES AS SUPERPOSITIONS

In the following, we will be interested in the construction of nonclassical states of the electromagnetic field, and so we first summarize the description of a quantized single mode of the field of frequency ω. The field mode operator $\hat{E}(z,t)$ for a field quantized in a box of volume V may be written as

$$\hat{E}(z,t) = \left(2\omega^2/\epsilon_0 V\right)^{\frac{1}{2}} \hat{q} \sin kz \tag{1}$$

where the operator $\hat{q}$ has the dimension of a length and can be expressed in terms of the annihilation operator $\hat{a}$ as

$$\hat{a} = (2\hbar\omega)^{-\frac{1}{2}} (\omega\hat{q} + i\hat{p}) \tag{2}$$

where $\hat{p}$ is the momentum operator conjugate to $\hat{q}$. This allows us to write $\hat{E}(z,t)$ as

$$\hat{E}(z,t) = \sqrt{2}\,\mathcal{E}_0 \sin kz \left\{\hat{a}\,\mathrm{e}^{-i\omega t} + \hat{a}^\dagger\,\mathrm{e}^{i\omega t}\right\} \tag{3}$$

where $\mathcal{E}_0$ is the "electric field per photon" given by

$$\mathcal{E}_0 \equiv (\hbar\omega/2\epsilon_0 V)^{\frac{1}{2}} . \tag{4}$$

It is sometimes more convenient to write the electric field operator in terms of the Hermitian quadrature operators $\hat{X}, \hat{Y}$ as

$$\hat{E}(z,t) = 2^{3/2}\mathcal{E}_0 \sin kz \left\{\hat{X}\cos\omega t + \hat{Y}\sin\omega t\right\} \tag{5}$$

where

$$\left[\hat{X},\hat{Y}\right] = i/2 \tag{6}$$

and

$$\hat{X} = \frac{1}{2}(\hat{a} + \hat{a}^\dagger), \qquad \hat{Y} = \frac{1}{2i}(\hat{a} - \hat{a}^\dagger). \tag{7}$$

The description of the field in terms of the X, Y (or p, q) phase-space is accomplished by introducing a phase-space quasiprobability [4]; we choose here to work with the Wigner function $W(q,p)$, which for a pure state described by a wavefunction $\Psi(q)$ is given by [5]

$$W(q,p) = \frac{1}{2\pi\hbar}\int \mathrm{d}q'\Psi^*(q+q'/2)\Psi(q-q'/2)\ \mathrm{e}^{ipq'/\hbar} . \tag{8}$$

The Wigner function may be negative in some region of phase-space; we take this as a useful measure of the nonclassical nature of the state Ψ. We note that [6, 7] for pure states only Gaussian wavefunctions (i.e. coherent or squeezed states) have Wigner functions which are positive everywhere. For example, a Fock state has a Laguerre polynomial Wigner function with striking fringe structure and negativity in substantial regions of phase space. We shall see in what follows that these fringes *and* the negativity are very sensitive to field mode damping.

We can construct nonclassical states (or other states) as superpositions of Fock states $|n\rangle$ as

$$|\Psi\rangle = \sum_n c_n |n\rangle \tag{9}$$

which are assembled as interfering Laguerre polynomial structures in phase space, or we may construct our state from superpositions of coherent states,

$$|\Psi\rangle = \sum_\alpha c(\alpha)|\alpha\rangle \tag{10}$$

which are assembled from Gaussian coherent states $|\alpha\rangle$. (A finite number of states $|\alpha\rangle$ may be used for the construction, or a continuous superposition may be employed in which case the sum (10) becomes an integral.) Which particular basis, which we can think of as a kind of "tiling" of phase space, is most useful is not entirely a matter of taste, as we shall see. The simplest example of a nonclassical superposition is [8] the superposition of the vacuum and the one-photon state so that

$$\Psi = c_0 |0\rangle + c_1 |1\rangle . \tag{11}$$

The Wigner function for one of these states is shown in Fig. 1a, where we note the existence of squeezing (the reduction of one quadrature uncertainty below the coherent state minimum for which $\Delta X = \Delta Y = \frac{1}{2}$), as well as a little negativity. In Fig. 1b, we show the corresponding Wigner function for the superposition of the vacuum and the two-photon state

$$\Psi = c_0 |0\rangle + c_2 |2\rangle \tag{12}$$

where we see more squeezing: this process is improved by superposing all the even photon numbers as in the classic squeezed vacuum [9].

More interesting nonclassical states can be generated by superposing coherent states, and the "Schrödinger cat state"

$$|\Psi\rangle = \frac{1}{N_+(\alpha)} \{|\alpha\rangle + |-\alpha\rangle\} , \tag{13}$$

where $N_+(\alpha)$ is a normalization factor, is the best studied example [10]. It is also interesting to examine the nature of the Wigner function for the state formed by a superposition of a number of coherent states as that number j increases, whilst maintaining each component on a ring of radius $|\alpha|$ [11]. For $j = 2$, we see a Gaussian hump for each constituent state, together with fringes whose period scales inversely with the separation between the components in phase space: the more macroscopic the superposition, the more rapid being the fringes. For $j = 4$ we see four constituent "humps" together with complicated interference structures between them. When the number of states is increased so that the individual components nearly, but not quite, fill the circumference of the ring we see a sequence of interference structures, large amounts of negativity in the Wigner functions and large fluctuations in the photon number. But when the number of coherent states is sufficiently increased to cover entirely the circumference of the ring (of radius $|\alpha|$), then a dramatic change occurs [11] with a collapse of the number variance as the superposition approaches ... a number state!

A continuous superposition of coherent states on a ring with an appropriate choice of phases is in fact a representation of a Fock state $|n\rangle$ [12,10]

$$|n\rangle = \frac{e^{|\alpha|^2/2}\sqrt{n!}}{2\pi|\alpha|^n} \int_0^{2\pi} d\phi \; e^{-in\phi} \left||\alpha|\, e^{i\phi}\right\rangle . \tag{14}$$

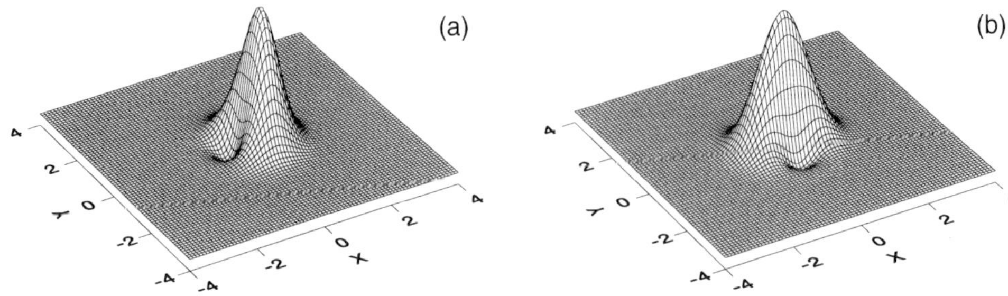

Figure 1: Wigner functions for superpositions of two Fock states. In (a) we depict the Wigner function for the state (11) with $c_1 = 1/2$. In the (b) the two-photon superposition (12) has $c_2 = \sqrt{1/2 - 1/\sqrt{6}}$ as given in [10].

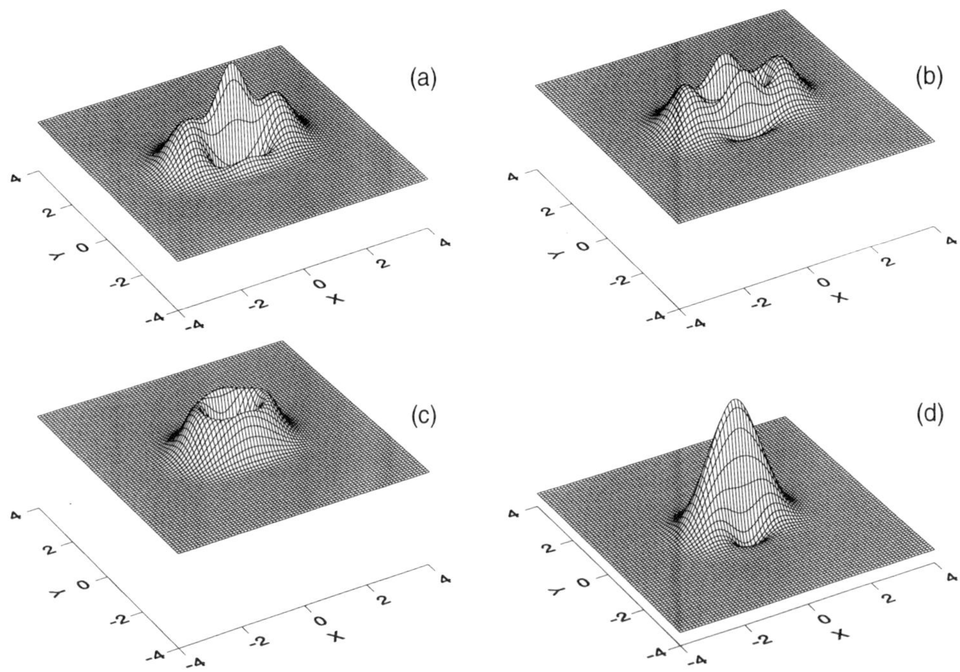

Figure 2: A "jumping cat" which was initially comprised of the coherent states $|\alpha\rangle$ and $|-\alpha\rangle$ with $\alpha = 2$. We show the Wigner functions before and after the first jump (at $\kappa t = 0.668$) in (a) and (b). In (c) and (d) we show the Wigner functions before and after the second jump (at $\kappa t = 1.884$).

It is amusing that coherent states, the most classical of all states, can be used to build Fock states using quantum interference. We have discussed elsewhere [13] how a sequence of conditional measurements in cavity QED may be used to realize the progression from coherent state to Fock state. If coherent states are continuously superposed on a line

$$|\Psi\rangle = C_F \int_{-\infty}^{\infty} F(\alpha, \Psi)|\alpha\rangle \, \mathrm{d}\alpha \; , \tag{15}$$

with $F(\alpha, \Psi)$ chosen to be a Gaussian function of α along the line, it is easy to demonstrate (by Fock decomposition and the interchange of summation and integration) that the resultant Ψ is a sum over even Fock states only and is in fact the squeezed vacuum [12].

3 DISSIPATION, DECOHERENCE AND DESTRUCTION

If the quantum system S (whose superposition states we are interested in) interacts with a reservoir R, then the energy and coherences decay for the system. Initially, the combined system-reservoir density operator $\rho_{S+R}(t=0)$ factorizes as

$$\rho_{S+R}(t=0) = \rho_S \otimes \rho_R \; . \tag{16}$$

At later times, the system reduced density operator can be described, in the Born-Markov approximation [14], by the master equation which for a field mode decaying into a zero-temperature bath is given by

$$\frac{\mathrm{d}\rho}{\mathrm{d}t} = \frac{\gamma}{2}\left(2\hat{a}\rho\hat{a}^\dagger - \hat{a}^\dagger\hat{a}\rho - \rho\hat{a}^\dagger\hat{a}\right) \tag{17}$$

where γ is the normal Golden Rule decay rate. The master equation emphasises the evolution of an entire ensemble [14]. Coherences are destroyed at a rate which is proportional to the degree of excitation of the system (much faster than energy decay) as the system relaxes to a statistical mixture of relevant "pointer bases" [15, 16]. For example, a field mode initially prepared in a Fock state relaxes very rapidly to a statistical mixture of Fock states, and the fringing structure in the Wigner function disappears at a rate $\sim n\gamma$. Such an ensemble-averaged approach does not easily address the question of how an individual member of an ensemble evolves in a dissipative environment, even though this issue is accessible to experiment [17]. A number of methods have been developed recently to simulate the evolution of single realizations. One centres on quantum jumps, reflecting the information gained from the detection of decay quanta, and involves a continuous evolution which is randomly interrupted by "instantaneous" jumps as the system state vector is conditioned by the information gained from the register of counts in detection of decay [18, 19].

Carmichael [19] has given a precise relationship between the conditioned density operator contingent on a precise sequence of detection events (a 'record') and the ensemble averaged density operator. He shows that if the master equation is written in Liouvillian form

$$\frac{\mathrm{d}\rho}{\mathrm{d}t} = \mathcal{L}\rho \tag{18}$$

and we split the Liouvillian action $\mathcal{L}$ as a sum of two terms, an anticommutator and a "sandwich" term,

$$\mathcal{L}\rho \;=\; -\frac{\gamma}{2}\left[\hat{a}^\dagger\hat{a} \, , \, \rho\right]_+ + \gamma\hat{a}\rho\hat{a}^\dagger$$

$$= \quad (\mathcal{L} - \mathcal{S}) + \mathcal{S} \tag{19}$$

then we may identify the "sandwich" term $\mathcal{S}$ as a jump operator. Equation (18) can be integrated formally as

$$\begin{aligned} \rho(t) &= \exp\left\{\left[(\mathcal{L}-\mathcal{S})+\mathcal{S}\right]t\right\}\rho(0) \\ &= \sum_{m=0}^{\infty}\int_0^t \mathrm{d}t_m \int_0^{t_m} \mathrm{d}t_{m-1} \ldots \int_0^{t_2} \mathrm{d}t_1 \\ &\quad \times \left\{ \mathrm{e}^{(\mathcal{L}-\mathcal{S})(t-t_m)}\mathcal{S}\,\mathrm{e}^{(\mathcal{L}-\mathcal{S})(t_m-t_{m-1})}\mathcal{S}\ldots\mathcal{S}\,\mathrm{e}^{(\mathcal{L}-\mathcal{S})t_1}\rho(0)\right\} \end{aligned} \tag{20}$$

where the quantity in curly brackets in Eq. (20) is labelled $\overline{\rho_c}(t)$ by Carmichael and is the conditioned density operator describing a specific "trajectory" or detection sequence. We can write $\overline{\rho_c}(t)$ in terms of the conditioned pure state projectors

$$\overline{\rho_c}(t) = |\overline{\Psi_c}(t)\rangle\langle\overline{\Psi_c}(t)| \,. \tag{21}$$

The component $\exp[(\mathcal{L}-\mathcal{S})\Delta t]$ propagates $\overline{\rho_c}(t)$ for a time Δt without a decay being recorded: for the conditioned state $|\overline{\Psi_c}(t)\rangle$

$$|\overline{\Psi_c}(t+\Delta t)\rangle = \exp\left[-iH_{\mathrm{eff}}\Delta t/\hbar\right]\,|\overline{\Psi_c}(t)\rangle \tag{22}$$

where the non-Hermitian effective Hamiltonian

$$H_{\mathrm{eff}} = H - i\frac{\gamma}{2}\hat{a}^{\dagger}\hat{a} \tag{23}$$

derives from the anticommutator in Eq. (19). Once a decay is registered, the gain in information is responsible for the jump

$$|\overline{\Psi_c}(t)\rangle \longrightarrow \hat{a}\,|\overline{\Psi_c}(t)\rangle \,. \tag{24}$$

In simulations, the jump time is chosen by utilizing a random number generator.

The other method to describe single trajectories involves both a quantum state *diffusion* where the individual quantum trajectory fluctuates through a Wiener process deriving from the environment, and *localization* to a coherent state, an eigenstate of the relevant Lindblad operator describing the coupling of the system to the environment [1].

Wiseman and Milburn [20] have related these two simulation methods to specific measurement schemes: the quantum jump method to continuous direct counting of decay quanta, and the quantum state diffusion method to a heterodyne measurement of oscillation amplitudes. Both methods, of course, lead to the *same* ensemble behaviour when averaged over many realizations [1, 18–21]. The new simulation schemes have been widely adopted for their practical utility in computing averages in a numerically efficient way on small computers. A quite separate issue is what, if any, meaning can be ascribed to individual trajectories, for these do differ in the different simulation schemes, reflecting the distinctly different conditionings applied, and reflecting the acquisition of different information in each simulation [19, 20].

Here, we examine single realizations of a superposition of coherent states of an oscillator (for example, a single electromagnetic field mode) interacting with a dissipative environment. We show that such realizations evolve very differently in the two simulation methods described above. The quantum jump, or state vector Monte-Carlo, method can lead to what we have called [22] a "jumping cat" where the conditioned state vector evolves through a series of jumps from one pure state "cat" (e.g. an even coherent state) to another (an odd coherent state) as individual quanta are removed at random by the dissipation. The quantum state diffusion model leads to a rapid localization (at essentially the decoherence rate $\gamma|\alpha|^2$) to one or other of the coherent state components of the initial superposition, reflecting the approximate projection onto one potential outcome of an amplitude measurement. The coherent states are an overcomplete set and form an interesting basis for an amplitude (heterodyne) measurement on any state: we exploit this in a study of individual trajectories of a chosen state (a Fock state) which localizes to one coherent state component (picked at random) of that state expressed in a coherent state basis.

In a typical dissipative process the master equation for ρ takes the form (with $\hbar = 1$)

$$\frac{\mathrm{d}}{\mathrm{d}t}\rho = -i\,[H,\rho] \; + \; \left(\hat{R}\rho\hat{R}^{\dagger} - \tfrac{1}{2}\hat{R}^{\dagger}\hat{R}\rho - \tfrac{1}{2}\rho\hat{R}^{\dagger}\hat{R}\right) \tag{25}$$

where the Lindblad reservoir operator $\hat{R}$ represents the influence of the environment. We have discussed elsewhere [21] the case of a two-photon absorber but $\hat{R}$ would typically be $\sqrt{\kappa}\hat{a}$ for ordinary single photon dissipation. In the case of a dissipative two-level system we would have $\hat{R} = \sqrt{\kappa}\sigma_-$ where σ_- is one of the Pauli spin matrices. In both of these cases the parameter κ signifies the loss rate of energy into the reservoir and the Hamiltonian H would represent the coherent evolution of the lossless system. The methods used to solve (25) depend specifically on the choice of operators $\hat{R}$ and the Hamiltonian H. Sometimes there is more than one reservoir operator, e.g. for a cavity in a thermal heatbath (characterised by a photon occupation $\bar{n}$) we have both $\sqrt{(\bar{n}+1)\kappa}\hat{a}$ and $\sqrt{\bar{n}\kappa}\hat{a}^{\dagger}$ as reservoir operators [23]. However, all the methods we describe below can be extended to cope with this kind of situation.

4 QUANTUM JUMPS

Each member, or sample, of an ensemble of state vectors undergoes a different evolution under the influence of a different sequence of random events. To obtain results that do not depend on a specific history of events and which can be compared with the density matrix results we average the observables over the ensemble (as would be done in an experiment). Thus given a set of state vectors at a time t and an observable $\hat{O}$, the averaged value may be denoted

$$\mathsf{M}\left\{ \, \langle\psi(t)|\hat{O}|\psi(t)\rangle \, \right\}, \tag{26}$$

where $|\psi(t)\rangle$ represents the conditioned realization. A sufficient number of samples are taken to ensure an acceptable average.

We now focus on single photon dissipation as simulated by quantum jumps, or the state vector Monte-Carlo, and we start by considering the two things that can happen to a single state vector over a very short time interval δt: there may be a quantum jump, or there may be no quantum jump. Our approach here most closely follows that of Ref. [18]. If we observe a quantum jump we project the operator $\hat{a}$ onto the state vector as in Eq. (24) because of the acquisition of information through the decay measurement. The probability of this quantum jump within a short time δt is

$$\Delta P = \langle\psi(t)|\hat{a}^{\dagger}\hat{a}|\psi(t)\rangle\delta t = \kappa\langle\hat{n}\rangle\delta t\,. \tag{27}$$

The state (given by 24) must be normalized to give the result

$$|\psi(t)\rangle \longrightarrow \frac{\hat{a}|\psi(t)\rangle}{\sqrt{\langle\psi(t)|\hat{a}^{\dagger}\hat{a}|\psi(t)\rangle}} = \frac{\hat{a}|\psi(t)\rangle}{\sqrt{\Delta P}}\sqrt{\delta t} \tag{28}$$

which is conditional on a quantum jump having taken place within δt.

Now we consider what happens to the wave function if we observe no quantum jump in the time interval δt. We propagate $|\psi(t)\rangle$ with the effective Hamiltonian Eq. (23). Because $H_{\rm eff}$ is not Hermitian the normalization of the state vector will shrink; this can be attributed to the possibility of quantum jumps and the state vector must be renormalized so that in a time step δt:

$$\begin{aligned} |\psi(t)\rangle \longrightarrow & \frac{{\rm e}^{-i\delta t H_{\rm eff}}|\psi(t)\rangle}{\sqrt{\langle\psi(t)|\,{\rm e}^{i\delta t H_{\rm eff}^{\dagger}}\,{\rm e}^{-i\delta t H_{\rm eff}}|\psi(t)\rangle}} \\ \sim & \frac{{\rm e}^{-i\delta t H_{\rm eff}}|\psi(t)\rangle}{\sqrt{\langle\psi(t)|\,{\rm e}^{-\delta t\hat{a}^{\dagger}\hat{a}}|\psi(t)\rangle}} \\ \sim & \frac{{\rm e}^{-i\delta t H_{\rm eff}}|\psi(t)\rangle}{\sqrt{1-\Delta P}}, \end{aligned} \tag{29}$$

where the approximations are correct to order δt.

The above forms a procedure for the discretised evolution of a state vector which describes a single member of the ensemble. Each time step δt proceeds in the following way. We calculate the small, but finite probability of a jump ΔP, Eq. (27). Then with a random number we decide with probability ΔP whether there will be a quantum jump or not. If there is a jump, the state vector is projected and renormalised according to Eq. (28). However, if there is to be no jump, the state vector evolves according to Eq. (29). Because the probability of jumping is small the usual decision at each time step is the non-linear evolution (29). When there is a jump a discontinuity in the state vector takes place and as a result the typical behaviour of observables for a single state vector is smooth evolution interrupted by jumps.

In Fig. 2 we illustrate the time evolution with some snapshots of the Wigner function of a single ensemble member as it undergoes two quantum jumps. The initial state was a Schrödinger "cat" state of the form $(|\alpha\rangle + |-\alpha\rangle)/N_{+}(\alpha)$, where $N_{+}(\alpha)$ is the normalization and $|-\alpha\rangle$ and $|\alpha\rangle$ are coherent states. During the time interval when there are no jumps the Schrödinger "cat" shrinks as $\alpha(t) = \alpha(0)\,{\rm e}^{-\kappa t/2}$ when the evolution (29) is considered. This process has taken place from the initial state up to Fig. 2(a). When there is a jump, then according to Eq. (28) we find that the Schrödinger "cat" jumps to a different type of "cat" with the form $(|\alpha(t)\rangle - |-\alpha(t)\rangle)/N_{-}(\alpha(t))$. The jump is seen as we go from Fig. 2(a) to Fig. 2(b); the jump from an "even" to an "odd" Schrödinger "cat" state is particularly visible in the Wigner functions as the interference fringes change sign. Subsequently the Schrödinger "cat" shrinks again as we evolve according to Eq. (29) in the absence of any jumps. This takes us to Fig. 2(c) where the two principal components of the "cat" are seen to have moved in towards the phase space origin. Then there is another quantum jump as we go from Fig. 2(c) to Fig. 2(d); the Wigner function fringes again change their sign.

5 QUANTUM STATE DIFFUSION

The second simulation method we consider is quantum state diffusion in which the evolution of $|\psi(t)\rangle$, representing one member of the ensemble, is determined by a non-linear stochastic differential equation [1, 24]. We write the equation (for the specific case of one-photon damping of a field state) in the Itô form involving drift and diffusion processes so that

$$\begin{aligned} |\psi(t+\delta t)\rangle &= (1 - i\delta t H)\,|\psi(t)\rangle + \kappa\left[\langle\hat{a}^\dagger\rangle\hat{a} - \tfrac{1}{2}\hat{a}^\dagger\hat{a} - \tfrac{1}{2}\langle\hat{a}^\dagger\rangle\langle\hat{a}\rangle\right]|\psi(t)\rangle\,\delta t \\ &+ \sqrt{\kappa}\,[\hat{a} - \langle\hat{a}\rangle]\,|\psi(t)\rangle\,\mathrm{d}\xi_t \end{aligned} \tag{30}$$

where $\mathrm{d}\xi_t$ is a random complex Wiener variable which perturbs the system. It varies randomly between each time step and each sample run so that when averaged:

$$\overline{\mathrm{d}\xi_t} = 0\,, \qquad \overline{\mathrm{d}\xi_t \mathrm{d}\xi_{t'}} = 0\,, \qquad \overline{\mathrm{d}\xi_t^* \mathrm{d}\xi_{t'}} = \delta_{t,t'}\,\delta t\,. \tag{31}$$

In consequence the time evolution is not smooth on a very small scale because of the discretised fluctuations of the state vector in its Hilbert space. This distinguishes the quantum state diffusion calculations from the state vector Monte-Carlo approach in a characteristic way; in the latter case the state vector (and hence the observables) jump only at certain places rather than at every time point. This manifestation of the constant jiggling of the state vector is clearly seen in the single trajectory examples of Fig. 3 (solid and dotted curves). However, as in the state vector Monte-Carlo the results for an ensemble average smooth out as we increase the number of samples.

It is interesting to note that if it happens that $|\psi(t)\rangle$ is an eigenstate of the Lindblad operator $\hat{a}$ the random fluctuations do not contribute to the system evolution in (30). This results in a vanishing of the random perturbations as can be seen in Fig. 3 as equilibrium is established. That $|\psi(t)\rangle$ is an eigenstate of $\hat{a}$ is not a sufficient condition for the steady state; an inspection of Eq. (30) shows that we also require $|\psi(t)\rangle$ to be an eigenstate of $H - \frac{1}{2}i\hat{a}^\dagger\hat{a}$ as is found in the state vector Monte-Carlo case.

The stochastic equation (30) amounts to a Langevin equation for the state vector (rather than the usual case of observables). As in the case of state vector Monte-Carlo simulations we can recover the master equation (25) when we average over the ensemble. Equation (30), constructed by Gisin and Percival, has recently been interpreted by Wiseman and Milburn as the state vector evolution in a balanced heterodyne detection scheme [20]. A related equation is found for the case of balanced homodyne detection by Carmichael [19]. In both cases a quantum jump theory is developed for the measurement process in the presence of a strong classical local oscillator. This quantum jump theory, unlike the single mode case presented above, contains two jump operators, i.e. one for each of the detectors. Using the fact that the local oscillator is strong and creates many jumps in the detector, the number of jumps in a short time interval can be determined by a stochastic function $\mathrm{d}\xi$ as found in Eq. (30) above. We can then treat the heterodyne case by smoothing out the fastest oscillations to obtain:

$$|\mathrm{d}\tilde{\psi}\rangle = \left[-\frac{\kappa}{2}\hat{a}^\dagger\hat{a}\mathrm{d}t \,+\, \kappa\hat{a}\langle\psi(t)|\hat{a}^\dagger|\psi(t)\rangle\mathrm{d}t \,+\, \sqrt{\kappa}\hat{a}\mathrm{d}\xi\right]\,|\tilde{\psi}(t)\rangle \tag{32}$$

where $|\tilde{\psi}(t)\rangle$ is an un-normalized state vector and $|\psi(t)\rangle$ is the normalized state vector. To obtain the equation of motion for the normalized state vector one has

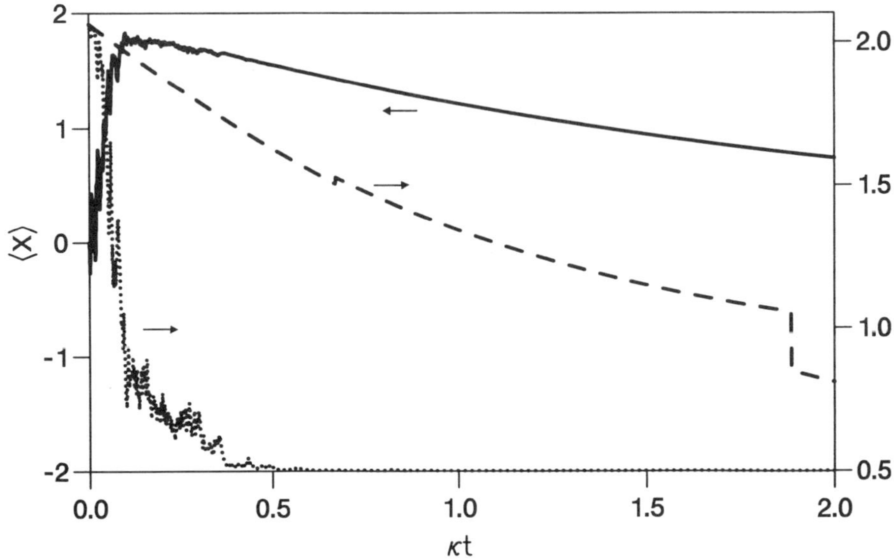

Figure 3: Localization of a Schrödinger "cat" state which is initially comprised of the coherent states $|\alpha\rangle$ and $|-\alpha\rangle$ with $\alpha = 2$. We show $\langle\hat{X}\rangle$ (solid curve—left-hand scale) and ΔX (dotted curve—right-hand scale). The localization is seen in the development of the non-zero value of $\langle\hat{X}\rangle$ and in the rapid reduction of the uncertainty. The dashed curve shows ΔX for the quantum jump trajectory of Fig. 2. The value of $\langle\hat{X}\rangle$ is zero in this case because there is no localization.

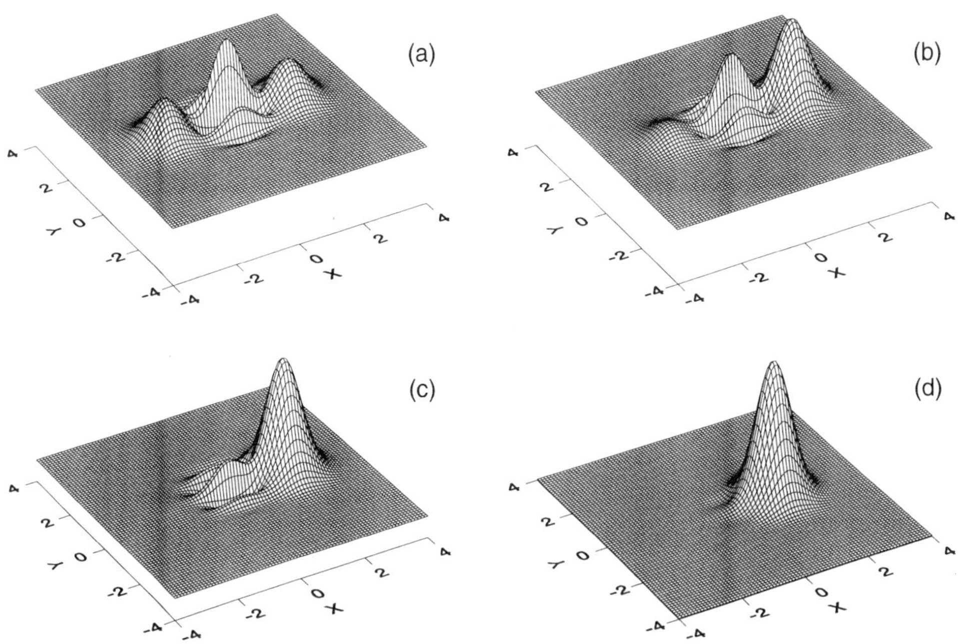

Figure 4: Wigner functions from the quantum state diffusion of a single sample We see $\kappa t = 0, 0.05, 0.2$ and 0.8 in (a-d). Localization is seen as the increasing dominance of one component of the "cat".

to be careful about differentiation because of the normalization of the noise term (Eq. (31)). The Itô calculus may be used under which, for example,

$$d(\langle\tilde{\psi}(t)|\tilde{\psi}(t)\rangle) = \langle\tilde{\psi}(t)|d\tilde{\psi}\rangle + \langle d\tilde{\psi}|\tilde{\psi}(t)\rangle + \langle d\tilde{\psi}|d\tilde{\psi}\rangle\,. \tag{33}$$

Furthermore, to obtain a connection with the Gisin-Percival equation (30) it is necessary to include a phase factor as well as renormalizing the state vector so that

$$|\psi(t)\rangle = \frac{e^{i\zeta(t)}|\tilde{\psi}(t)\rangle}{\sqrt{\langle\tilde{\psi}(t)|\tilde{\psi}(t)\rangle}}\,. \tag{34}$$

The phase factor is stochastic because

$$id\zeta = \frac{\langle\tilde{\psi}(t)|\hat{a}^\dagger|\tilde{\psi}(t)\rangle d\xi^* - \langle\tilde{\psi}(t)|\hat{a}|\tilde{\psi}(t)\rangle d\xi}{2\langle\tilde{\psi}(t)|\tilde{\psi}(t)\rangle} \tag{35}$$

resulting in the equation for the normalized $|\psi(t)\rangle$:

$$\begin{aligned}
|d\psi\rangle \;=\;& \kappa\Big[-\tfrac{1}{2}\hat{a}^\dagger\hat{a} + \hat{a}\langle\psi(t)|\hat{a}^\dagger|\psi(t)\rangle + \tfrac{1}{2}\langle\psi(t)|\hat{a}^\dagger\hat{a}|\psi(t)\rangle \\
&- \langle\psi(t)|\hat{a}^\dagger|\psi(t)\rangle\langle\psi(t)|\hat{a}|\psi(t)\rangle\Big]\,|\psi(t)\rangle dt \\
&+ \tfrac{1}{2}\kappa\Big[-\langle\psi(t)|\hat{a}^\dagger\hat{a}|\psi(t)\rangle + \langle\psi(t)|\hat{a}^\dagger|\psi(t)\rangle\langle\psi(t)|\hat{a}|\psi(t)\rangle\Big]\,|\psi(t)\rangle d\xi^* d\xi \\
&+ \sqrt{\kappa}\,[\hat{a} - \langle\psi(t)|\hat{a}|\psi(t)\rangle]\,|\psi(t)\rangle d\xi\,.
\end{aligned} \tag{36}$$

This equation yields the Gisin-Percival equation (30) if $d\xi^* d\xi \longrightarrow dt$. The only difference between these equations is that the state vector in Eq. (30) is only normalized for the ensemble average whilst in Eq. (36) it is always normalized. In practice this seems to result in no noticeable differences because whilst our computer program integrates Eq. (30) it renormalizes the state vector at every time step.

We should emphasize that the non-linear equations we consider here in no way violate the principles of quantum mechanics. Each kind of simulation is non-linear because it is conditional on a specific unravelling of the master equation (25) and when we ensemble average over a large number of state vectors we always recover the master equation (25) with the appropriate reservoir operators [18, 19, 1, 21].

In Fig. 4 we demonstrate the localization process which takes place in quantum state diffusion from an initial superposition by plotting the Wigner function. Interference fringes can be seen in the initial state, Fig. 4(a), which consists of a Schrödinger "cat" coherent superposition of two coherent states $|\alpha\rangle$ and $|-\alpha\rangle$ (with $\alpha = 2$ in this case), with the two components representing the live "cat" and the dead "cat". If we now examine the time sequence in Figs.4(a-d) we observe that one of the components of the "cat" diminishes as the other component grows. This is accompanied by a rapid loss of the fringes. The Wigner function clearly shows localization to one of the coherent state components of the original "cat", the choice of component being determined randomly according to the stochastic input. In Fig. 3 we show the values of the quadrature $\langle\hat{X}\rangle = \langle\psi(t)|(\hat{a}+\hat{a}^\dagger)|\psi(t)\rangle/2$ and uncertainty $\Delta X = (\langle\hat{X}^2\rangle - \langle\hat{X}\rangle^2)^{1/2}$ as a function of time and including the times at which the Wigner functions of Fig. 4 are appropriate. The localization manifests itself in $\langle\hat{X}\rangle$ through the rapid development of a non-zero value of the displacement which takes place as one of the component coherent states is chosen. Subsequently, the value of $\langle\hat{X}\rangle$ decays as the coherent state itself approaches the ground state. The localization is seen in ΔX as a rapid reduction (at essentially the decoherence rate

$\gamma|\alpha|^2$) to a value of one half, the uncertainty value for a coherent state. The value of one half is maintained as the coherent state dissipates to the origin. The picture would be quite different if we were to consider the density matrix approach. In that case the fringes in the Wigner function would also rapidly disappear, but we would be left with both of the components reflecting a statistical mixture of outcomes [25].

We have so far presented the heterodyne experiment with a simple choice of two components (outcomes) of a Schrödinger "cat". Any state of the oscillator can be represented as a superposition of coherent states, and the quantum state diffusion will describe localization to a single coherent state component. We next consider a complex superposition of coherent states: a Fock state. As mentioned earlier a of coherent states with any radius r can form a Fock state provided the correct phase factors $e^{-in\theta}$ are present. However, when localization takes place a coherent state is selected from that circle for which each component coherent state has nearly the same energy as the original Fock state. The process is illustrated in Fig. 5. The initial state in (a) contains nine quanta and its Wigner function, formed from a Laguerre polynomial, is made up of an outer rim with inner concentric fringes. As time progresses, (b,c), we observe that localization starts with the preferential orientation of the rim and fringes in one direction. This chosen direction is different for each realization. As the outer rim localizes, the inner fringes wash out with the vanishing of quantum interference between coherent state components. In (d) we see that the fringes have almost disappeared and the remaining localized state approaches a coherent state. In this case, the localization rate is much less than the decoherence rate: in order to localize, sufficient decay time has to elapse to generate a a range of photon numbers to support the full Poisson width of a coherent state component. During the subsequent time evolution the Wigner function becomes quite rounded and moves in towards the phase space origin. We have also investigated the localization from initially squeezed states, and again we see localization into components of Eq. (15) (and under some conditions, the formation of *Schrödinger "cats"*).

6 COMPARISON OF ENSEMBLE AVERAGED RESULTS

We have seen in the previous two sections how the state vector Monte-Carlo method and quantum state diffusion method of simulation lead to very different behaviour for individual realizations, whilst for an average over large samples they realise the results of the master equation (25). However, in all practical cases the number of samples must be finite. Then given that we may have a choice of the two methods for performing the simulation the question arises as to which to use: when ensemble averaged, both methods seem to give results with about the same accuracy. In order to check this we have calculated the mean fluctuations between simulated, and sample averaged, observables and the computed observables from direct integration of the master equation (17) in a similar fashion to the two photon processes examined in Ref. [21]. Here we consider one-photon decay and the fluctuations are averaged over the time evolution and are shown in Fig. 6 for the observable $\langle \hat{n} \rangle$ where the symbols $\times$ indicate the fluctuations with the state vector Monte-Carlo approach and the symbols $+$ show fluctuations in quantum state diffusion. The fluctuations decrease in size with sample number roughly as expected from the central limit theorem. However, these results also depend on the integration step size δt and can only be taken as representative. For example, if the

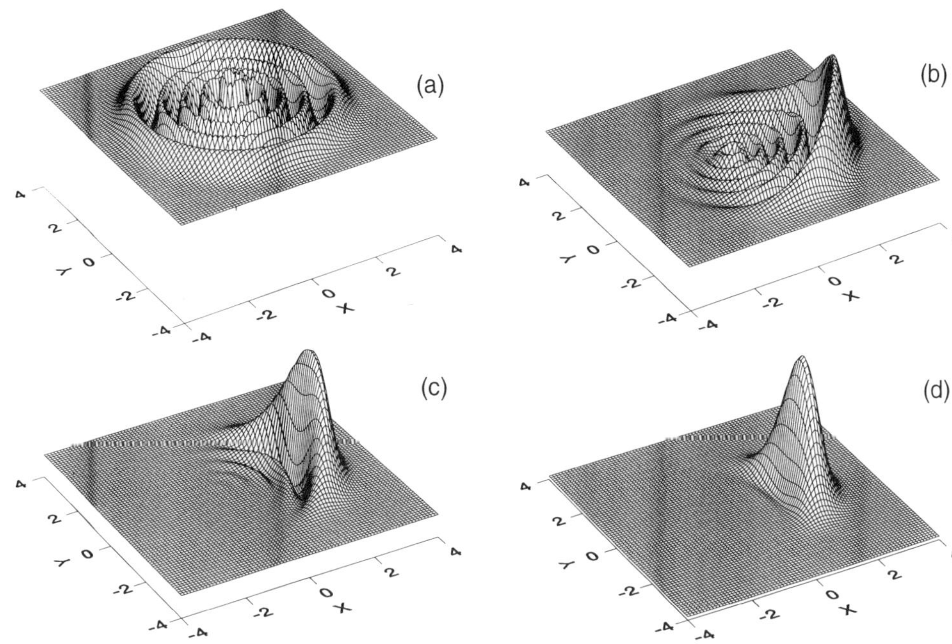

Figure 5: Localization of a Fock state $|n\rangle$ with $n = 9$. These results are for a single sample undergoing quantum state diffusion. The Wigner functions are shown for $\kappa t = 0, 0.1, 0.5$ and 1.0 in (a-d).

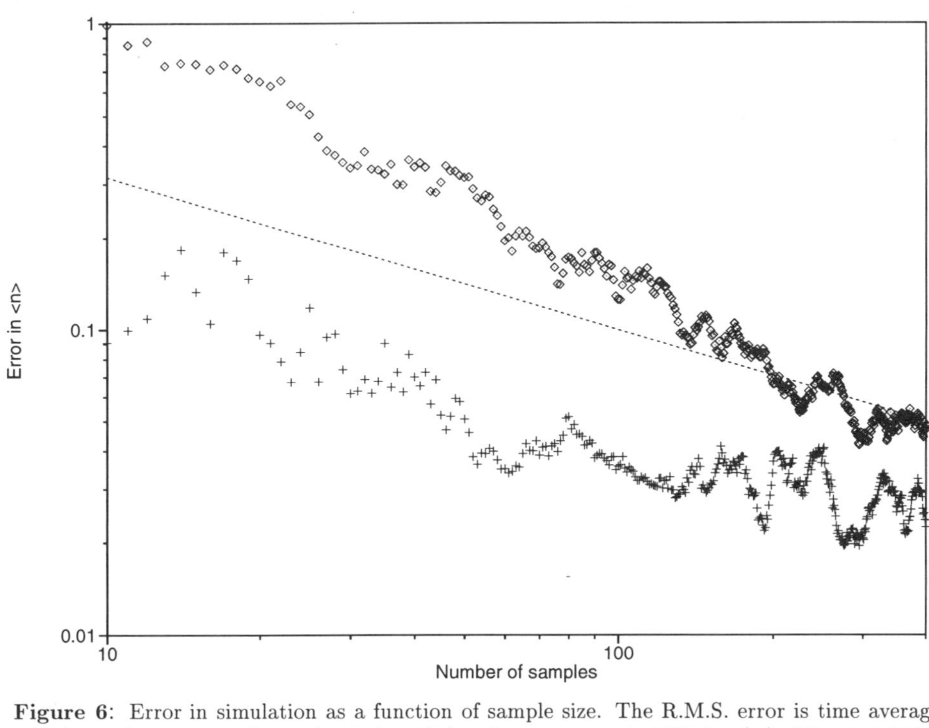

Figure 6: Error in simulation as a function of sample size. The R.M.S. error is time averaged and shown for quantum jump simulations ($\times$) and quantum state diffusion ($+$). The straight line indicates the trend of the central limit theorem, though the scaling factor is not known. In all the simulations the initial state was a 9-photon Fock state subjected to purely dissipative (1-photon) processes.

step size is reduced, then greater accuracy may be achieved with larger numbers of samples.

CONCLUSIONS

One can regard the evolution of single quantum trajectories as generating potential outcomes contingent on the acquisition of information in specific measurement schemes. The conditioning of the initial state vector reflects this time-dependent gain in information. The potential outcomes depend upon the measurement process: the pointer basis reflects this, and has to be chosen appropriately. If the level of excitation of the quantum system is monitored by quantum counting, then the appropriate basis reflects the successive loss of quanta observed in the decay process. But the quantum system may instead be studied by homodyne or heterodyne measurements where the local oscillator is very intense: then the effect of a single quantum jump is so small and the frequency of jumps is so high that the evolution appears to be a stochastic excursion.

Any initial state can be described in terms of its relevant phase space quasiprobability (for example, the Wigner function). The phase space can be built up in terms of a superposition of pointer state bases. Fock states tile the phase space by a set of dominant annulae centred on the origin, reflecting (at least in the semi-classical approximation) the lack of phase information in energy eigenstates. However, it may be preferable to employ a basis with amplitude *and* phase information, and this may be accomplished by tiling the phase space with a set of coherent states. The acquisition of information through observation conditions the wave function in a manner dependent on the specific measurement scheme, and the conditioning emphasises a particular tiling. Localization in the quantum state diffusion model reflects the coherent state tiling in a remarkable way.

Acknowledgments

This work was supported in part by the United Kingdom Science and Engineering Research Council, and by the European Community.

References

[1] N. Gisin and I.C. Percival, J. Phys. **A 25**, 5677 (1992); *ibid.* **26**, 2233 (1993); *ibid.* **26**, 2245 (1993).

[2] L. Diósi, J. Phys. **A 21**, 2885 (1988); Phys. Lett. **A 129**, 419 (1988).

[3] V.P. Belavkin, Usp. Math. Nauk. (Russian Math. Surveys) **47** (**1**), 53 (1992) and references therein.

[4] M. Hillery, R.F. O'Connell, M.O. Scully, and E.P. Wigner, Physics Reports **106**, 121 (1984).

[5] E.P. Wigner, Phys. Rev. **40**, 749 (1932).

[6] R.L. Hudson, Rep. Math. Phys. **6**, 249 (1974).

[7] N. Lütkenhaus, and S.M. Barnett, submitted to Phys. Rev. **A**.

[8] K. Wódkiewicz, P.L. Knight, S.J. Buckle, and S.M. Barnett, Phys. Rev. **A 35**, 2567, (1987).

[9] R. Loudon and P.L. Knight, J. Mod. Opt. **34**, 709 (1987).

[10] V. Bužek and P.L. Knight, Progress in Optics, (Ed. E. Wolf, North Holland, Amsterdam, 1994) to appear.

[11] J. Janszky, P. Domokos, and P. Adam, Phys. Rev. **A 48**, 2213 (1993).

[12] V. Bužek and P.L. Knight, Opt. Commun. **81**, 331 (1991).

[13] B.M. Garraway, B. Sherman, H. Moya-Cessa, P.L. Knight and G. Kurizki, Phys. Rev. **A 49**, 535 (1994).

[14] See, e g., W.H. Louisell, *Quantum Statistical Properties of Radiation*, John Wiley and Sons, Chichester, 1973.

[15] J.P. Paz, S. Habib and W.H. Zurek, Phys. Rev. **D 47**, 488 (1993); W.H. Zurek, Phys. Today **44**, 36 (1991); *ibid.* **46**, 84 (1993); W.H. Zurek, Prog. Theor. Phys. **89**, 281 (1993); W.H. Zurek, S. Habib and J.P. Paz, Phys. Rev. Lett. **70**, 1187 (1993) and references therein.

[16] E. Joos and H.D. Zeh, Z. Phys. **B 59**, 223 (1985); D.F. Walls and G.J. Milburn, Phys. Rev. **A 31**, 2403 (1985); S.J.D. Phoenix, Phys. Rev. **A 41**, 5132 (1990); G.J. Milburn, Phys. Rev. **A 36**, 5271 (1987).

[17] See, for example: N. Gisin, P.L. Knight, I.C. Percival, R.C. Thompson and D.C. Wilson, J. Mod. Opt. **40**, 1663 (1993) and references therein; M. Brune, S. Haroche, J.M. Raimond, L. Davidovich and N. Zagury, Phys. Rev. **A 45**, 5193 (1992).

[18] J. Dalibard, Y. Castin and K. Mølmer, Phys. Rev. Lett. **68**, 580 (1992); K. Mølmer, Y. Castin, and J. Dalibard, J. Opt. Soc. Am. **B 10**, 524 (1993).

[19] H.J. Carmichael, *An Open Systems Approach to Quantum Optics*, Lecture Notes in Physics (Springer-Verlag, Berlin, 1993).

[20] H.M. Wiseman and G.J. Milburn, Phys. Rev. **A 47**, 1652 (1993).

[21] B.M. Garraway and P.L. Knight, Phys. Rev. **A 49** 2785 (1994).

[22] B.M. Garraway and P.L. Knight, to appear in Phys. Rev. **A 50** (1994).

[23] T.P. Spiller, B.M. Garraway and I.C. Percival, Phys. Lett. **A 179**, 63 (1993).

[24] N. Gisin and I.C. Percival, Phys. Lett. **A 167**, 315 (1992).

[25] V. Bužek, A. Vidiella-Barranco and P.L. Knight, Phys. Rev. **A 45**, 6570 (1992).

CONTINUOUS-WAVE SQUEEZED-LIGHT GENERATION USING A TRIPLY RESONANT OPTICAL PARAMETRIC OSCILLATOR

Katsuyuki Kasai

Communications Research Laboratory,
Ministry of Posts and Telecommunications,
588-2 Iwaoka, Nishi-ku, Kobe-shi,
Hyogo 651-24, Japan

Abstract

Continuous-wave squeezed light is generated by using a triply resonant optical parametric oscillator (TRO) in a degenerate condition. Simultaneous resonance of the pump and signal (idler) is achieved by utilizing air dispersion, and the optical cavity is servo-locked to the pumping frequency by a polarization spectroscopy technique using the polarization anisotropy of the nonlinear crystal itself. The observed noise reduction is 15% relative to the vacuum noise level, which corresponds to an estimated noise reduction of 50% at the output of the TRO based on the observed squeezing and a simple model.

Keywords :
squeezed light, Nd:YAG ring laser, optical parametric oscillator, stabilization, balanced homodyne detector

1 INTRODUCTION

The quantum description of electromagnetic waves, including light, treats ideal laser light as a coherent state describable by two conjugate observables with equal quantum fluctuations, which obey the minimum uncertainty relation (Heisenberg uncertainty principle). Squeezed light is a non-classical type of light, where one of the conjugate observables has a smaller fluctuation than the other one. The minimum uncertainty relation still holds for squeezed light.

Squeezed light has been achieved by several methods. A typical experimental process for generating squeezed light is degenerate parametric down conversion. Ling-An Wu et al. demonstrated continuous-wave (cw) squeezed light generation by using degenerate parametric down conversion, but their experiment was very complicated since they used xenon gas and an rf sideband technique for effective cw parametric interaction [1].

This contribution describes cw squeezed-light generation from a triply resonant optical parametric oscillator (TRO) that used air as an intracavity dispersive element in order to obtain effective nonlinear coupling of the modes. The TRO was pumped by the second harmonic of a single-frequency Nd:YAG ring laser whose amplitude and frequency were stabilized simultaneously by using a reference cavity and an acousto-optic modulator [2]. The length of the optical cavity was servo-locked to the pumping frequency by a polarization spectroscopy technique using the polarization anisotropy of the crystal itself; this gave the TRO a simple feedback configuration. The design of the optical cavity of the TRO took into account refractive effects at the surface of the crystal, since the crystal length was not negligible compared to the cavity length. The squeezing was observed with a balanced homodyne detector, and the initial noise reduction at the output of the TRO was estimated by a simple model.

2 EXPERIMENTAL DETAILS

Figure 1 shows our experimental setup for generating and observing cw squeezed light. This setup consisted of a highly stabilized single-frequency cw Nd:YAG ring laser, a triply resonant optical parametric oscillator, and a balanced homodyne detector.

2.1 Stabilized Single-frequency CW Nd:YAG Ring Laser

A highly stabilized single-frequency cw Nd:YAG laser is necessary in order to generate squeezed light using an optical parametric oscillator (OPO). It is known that frequency fluctuations in the pumping laser cause unstable oscillations in the OPO, and amplitude fluctuations produce instabilities in crystal temperature through linear absorption.

To pump a cw OPO, both the amplitude and frequency of a single-frequency 1.064μm Nd:YAG ring laser pumped by a single krypton arc lamp must be simultaneously stabilized [2]. Our ring-laser cavity consisted of four mirrors, and the total optical length was 1.17 m. The optical polarization, which was selected by a Brewster-plate polarizer, was perpendicular to the cavity plane. The four mirror mounts and the other holders were carefully designed and constructed of duralumin on an Invar breadboard. We obtained the laser head from a commercial instrument (Spectron laser system Model SL903). A single transverse mode was achieved by means of an iris. A unidirectional traveling-wave oscillation was obtained by a Faraday rotator and a halfwave plate. The Faraday rotator, which rotated the polarization by 3°, consisted of an 11.5-mm FR5 glass (HOYA) and two magnets which provided a field of 2.2 kG. We judged that the laser output had a single frequency since spatial hole burning in the laser rod disappeared, but an etalon was necessary to eliminate etalon effects of the flat-ended laser rod. Without the etalon, this traveling-wave ring laser had two longitudinal modes.

An acousto-optic modulator (AOM: HOYA A-100), which removed the intensity fluctuation with the first-order diffracted beam, was placed inside the cavity of the Nd:YAG ring laser for amplitude stabilization. The intensity fluctuation was reduced to less than 0.1% by negative feedback, while the free-running intensity fluctuation was more than 2%.

The laser frequency was also actively stabilized by servo control to lock the laser frequency to the resonant reference cavity by using the dithering technique. The length of the ring-laser cavity was locked to the resonant frequency of the reference cavity with a feedback servo consisting of two PZT-mounted mirrors, integrators, and high-voltage amplifiers. One of the two PZT's, which had a small dynamic range and a high mechanical resonant frequency, was used for the fast feedback loop. The other, which had a large dynamic range, was used for the slow feedback loop. The measured frequency jitter under feedback control was 160 kHz rms, while the free-running frequency jitter was approximately 30 MHz.

Second harmonic generation (SHG) for pumping the TRO was achieved by placing a KTP crystal between the two lenses inside the ring cavity. The resulting output power of cw SHG was as much as 70 mW since self pulsing of more than 70 mW occurred.

2.2 Triply Resonant Optical Parametric Oscillator

In general, OPO's are known as tunable coherent light sources. It is difficult to obtain the low threshold of parametric oscillation with singly resonant or doubly resonant OPO's. Triply resonant OPO's can reach a low threshold, but a certain phase relationship between the modes is needed. In the degenerate condition, a relative phase shift of $\pi/2$ between the pumping wave and the signal (idler) wave is needed in a nonlinear crystal for efficient energy conversion. We created the triply resonant degenerate condition by adjusting the crystal temperature and taking advantage of the air dispersion between the nonlinear crystal and the cavity mirrors.

Our TRO cavity consisted of an $MgO:LiNbO_3$ crystal and two curved mirrors; it was also constructed on another Invar breadboard. The crystal was 25 mm $\times$ 4 mm $\times$ 3 mm and was heated by an oven for non-critical phase matching. The temperature was controlled by a PID controller within 0.01° at 128.05°C. The input mirror was highly reflecting at 1.064 μm but transmitted 4% at 0.532 μm, while the output mirror was highly reflecting at 0.532 μm but transmitted 4% at 1.064 μm. These two mirrors constituted a concave-concave resonant cavity.

We carefully designed the optical cavity of the TRO to achieve effective nonlinear coupling. For Gaussian beams, the optimum degree of focusing within the crystal was shown by Boyd and Kleinman, and the focusing parameter ξ and the double refraction parameter B were defined [3]. The ξ characterizes the degree of focusing within the crystal with the beam waist located in the center of the crystal, and $\xi = l/b$, where l is the crystal length and b is the confocal parameter. In our experiment we took B=0 since non-critical phase matching was used, so optimum interaction was obtained for $\xi = 2.84$. Although Boyd and Kleinman assumed the crystal to be embedded in an isotropic medium of the same refractive index as the crystal in order to eliminate the refractive effects at the surface, we additionally considered the refractive effects at the surface of the crystal. For our TRO we constructed two types of cavities: one with 50-mm-radius mirrors and the other with 100-mm-radius mirrors. The estimated optimum cavity lengths were 114 mm and 214 mm, respectively. Since these values were close to those of the concentric resonator condition, and therefore not so stable, we chose 106 mm and 210 mm for the respective cavity lengths. The thresholds of the TRO's were measured to be 40 mW and 20 mW, respectively. The buildup factor of the pump beam was measured to be 7 in the optical cavity.

Even though the optical cavity of the TRO was constructed carefully, it was affected by various perturbations such as acoustic vibrations, air flow, and thermal variation. It follows that the length of the optical cavity must be servo-locked to the pumping frequency by a feedback scheme. Hänsch and Couillaud demonstrated the polarization spectroscopy technique for laser frequency stabilization. In that technique, a linear polarizer or Brewster plate is placed inside the reference cavity, so the reflected light acquires a frequency-dependent elliptical polarization. A simple balanced polarization analyzer detects dispersion-shaped resonances, and the obtained error signal is used for the electronic feedback scheme [4]. In our experiment, we used the polarization anisotropy of the crystal instead of a linear polarizer, so the feedback configuration was very simple compared to the Pound-Drever rf sideband technique[5]. Figure 2 shows the error signal obtained by our polarization analyzer. This signal was used for the feedback servo loop.

Figures 3 (a) and (b) show oscilloscope traces of the cw output of the TRO with 50-mm-radius mirrors. Output of 0.2 mW was obtained for pump power of 50 mW. The output was unstable for low feedback gain, but stable for suitably high feedback gain. The signal and idler wavelengths observed by a monochromator (CVI Laser Corporation Digikrom 240) are shown in Fig. 4. Their wavelengths depend on the crystal temperature, and the degenerate condition was obtained at 128.05°C.

2.3 Balanced Homodyne Detector

The balanced homodyne detector was constructed on the same Invar breadboard as the TRO. It consisted of a 50:50 beam splitter and two InGaAs photodiodes (Fujitsu FID13S81WS, 80-μm diameter) with quantum efficiencies of nearly 90%. The local oscillator light illuminated the photodiodes as a uniform plane wave, while the squeezed light from the TRO was focused on the photodiodes through two lenses. The local and squeezed lights were mixed in the beam splitter, and the two mixed lights from the two output ports of the beam splitter were detected separately by the two photodiodes. The photodiode signals (shot noises) were amplified respectively by two homemade low-noise amplifiers, each composed of three transistors. The outputs of both amplifiers were connected to a differential amplifier, whose output noise was observed by a spectrum analyzer, as shown in Fig. 1.

The balancing of the balanced homodyne detector was very difficult, but we carefully tuned the gain of the amplifiers. For this tuning, the amplitude of the local laser light was modulated at the observation frequency, and both amplifier gains were adjusted so that the amplitude modulation could be cancelled by the differential amplifier. The resulting balance was better than 0.1 dB at the observation frequency.

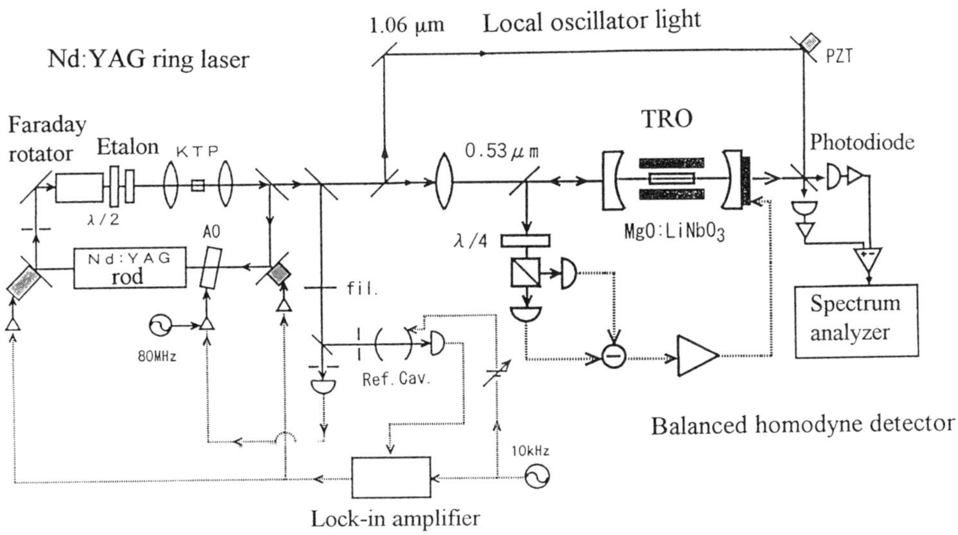

Figure 1: Experimental setup for generating cw squeezed light

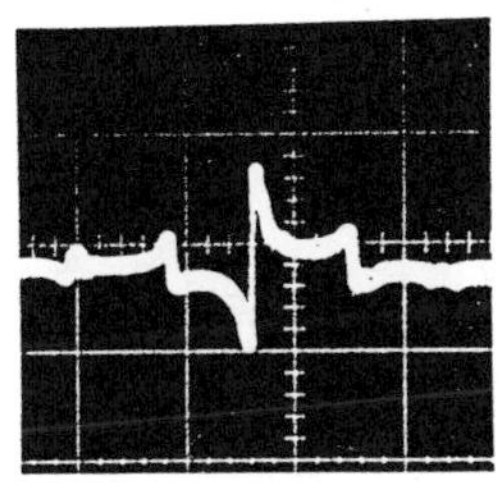

Figure 2: Output of the polarization analyzer

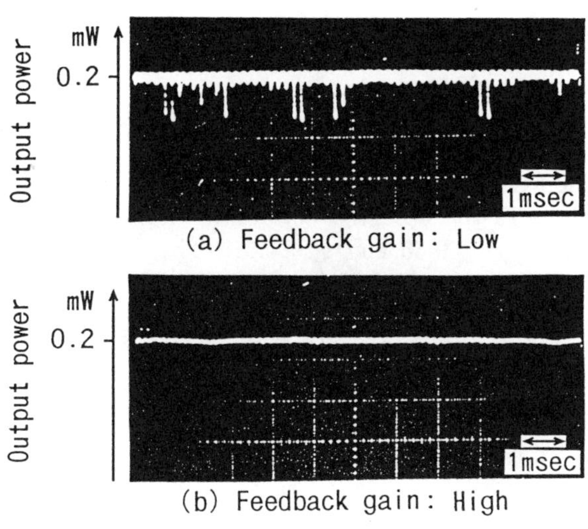

Figure 3: Output of the cw TRO

3 OBSERVATION OF SQUEEZED LIGHT

For the generation of squeezed light we used the 50-mm-radius mirrors cavity and set the pumping power just below the threshold, which was 40 mW. The cavity length and the crystal temperature were finely adjusted in order to achieve the perfect degenerate condition.

The squeezing was observed on the spectrum analyzer with the balanced homodyne detector. The experimental results are shown in Fig. 5. The frequency we observed was fixed at 1.6 MHz, i.e. zero span, while the local oscillator phase was swept at 2 Hz by a PZT. The resolution bandwidth of the spectrum analyzer was 100 kHz. The shot-noise level of vacuum fluctuations, which is shown by the dashed line in Fig. 5, was obtained by blocking the squeezed light. The maximum noise reduction was 15% (-0.71 dB) below the shot-noise level and the maximum noise increase was 30% (+1.14 dB) above it.

4 DISCUSSION

The initial noise reduction at the output of the TRO can be estimated from the observed squeezing. We considered the amplifier noise and the detection efficiency including optical losses, so that the squeezing asymmetry gave greater noise increase than noise reduction. The observed noise power of squeezing (S) is expressed by the following simple model [6],

$$S = \left[\eta \left(e^{2r} \cos^2 \phi + e^{-2r} \sin^2 \phi\right) + (1 - \eta)\right] P_{sh} + P_{amp} , \tag{1}$$

where $\eta = \eta_l \eta_h$, $\eta_h = \gamma \eta_d$. The following parameters are introduced:

η	=	total detection efficiency
η_l	=	optical losses
η_h	=	homodyne quantum efficiency
γ	=	homodyne detection parameter
η_d	=	photodiode quantum efficiency
r	=	squeezing parameter
ϕ	=	local oscillator phase
P_{sh}	=	shot noise
P_{amp}	=	amplifier noise.

By solving Eq. (1) with the observed noise reduction and increase and the ratio of P_{sh} to P_{amp}, we get $r = 0.35$ and $\eta = 0.61$. This result means that the noise reduction just at the output of the TRO is 50% (-3 dB) below the shot-noise level.

Moreover we examined whether the estimated total detection efficiency η of 61% was reasonable. The homodyne detection parameter γ depends on the mode matching between the local oscillator light and the squeezed light. In our experiment we must consider the mode matching between the uniform local beam and the Gaussian signal beam, and the maximum homodyne detection parameter $\gamma_{\max}$ should be 0.81 according to calculations. When we take $\eta_l = 0.97$, $\eta_d = 0.9$, and $\gamma_{\max} = 0.81$, the maximum η is 71%. The η estimated by the observed data was degraded by 10% from the calculated maximum, because the squeezed light was not focused to the optimum size on the photodiodes.

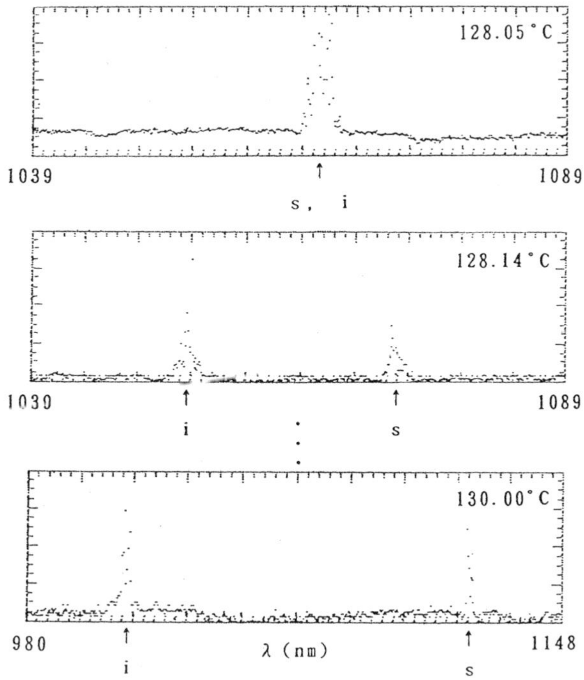

Figure 4: Dependence of signal (s) and idler (i) wavelengths on the crystal temperature

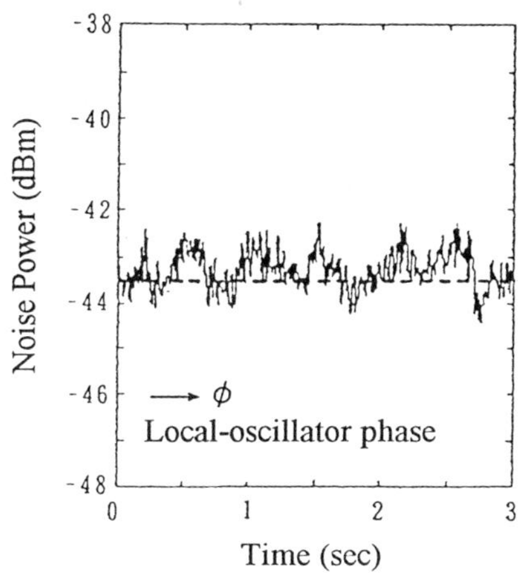

Figure 5: Noise power for the squeezed light as a function of local-oscillator phase

CONCLUSION

We have developed a highly stabilized single-frequency cw Nd:YAG ring laser, a cw triply resonant optical parametric oscillator, and a balanced homodyne detector. By integrating these components, we conducted experiments on cw squeezed-light generation. We observed 15% noise reduction below the shot-noise level, which corresponds to 50% noise reduction right at the output of the TRO considering the detection efficiency, optical losses, and amplifier noise.

To generate even more stable cw squeezed light, we plan to use a LD-pumped Nd:YAG laser, which is more stable, as the pumping laser. Then we will measure various characteristics of squeezed light.

Acknowledgments

I gratefully acknowledge helpful discussions with O. Hirota, C. Fabre, and H. Ito on several points in this work. I also thank M. Ishizu and J. Zhou for experimental support.

References

[1] Ling-An Wu, Min Xiao, and H. J. Kimble, J. Opt. Soc. Am. B4, 1465 (1987).

[2] K. Kasai and M. Ishizu, Jpn. J. Appl. Phys. 33, L230 (1994).

[3] G. D. Boyd and D. A. Kleinman, J. Appl. Phys. 39, 3597 (1968).

[4] T. W. Hänsch and B. Couillaud, Optics Commun. 35, 441 (1980).

[5] R. W. P. Drever, J. L. Hall, F. V. Kowalski, J. Hough, G. M. Ford, A. J. Munley, and H. Ward, Appl. Phys. B 31, 97 (1983).

[6] T. Hirano and M. Matsuoka, Appl. Phys. B 55, 233 (1992).

QUANTUM NOISE REDUCTION USING A NONLINEAR SAGNAC LOOP WITH POSITIVE DISPERSION

L. Boivin, C. R. Doerr, K. Bergman and H. A. Haus

Department of Electrical Engineering and
Computer Science and Research Laboratory
of Electronics
Massachusetts Institute of Technology
Cambridge, Massachusetts 02139 USA

Abstract

Recent experiments at MIT using light from a mode-locked Nd:YLF laser at $1.3\mu m$ in a fiber Sagnac loop have achieved noise levels 5.1dB below shot noise. In these experiments, squeezing is limited by the gaussian shape of the pump pulses. We show that by operating the interferometer away from $1.3\mu m$ that limitation can be overcome. For these wavelengths the group velocity dispersion of silica fibers creates negative correlations between the noise on different parts of the pump.

Motivations: Squeezed states of the radiation field were shown theoretically and experimentally to be relevant to high precision measurements. Sub-shot noise measurements of phase and amplitude modulations using squeezed radiation were demonstrated experimentally by a number of groups [1–3]. Squeezed states can also be used to overcome sensitivity limits in FM saturation spectroscopy [4]. On a more fundamental level, squeezed states offer a new set of initial conditions for the electromagnetic field with which the interaction of light and matter can be probed [5]. To study these phenomena, sources that can produce radiation with a large degree of squeezing over a broad bandwidth must be developed.

Sources of Squeezed Radiation: The most successful squeezer today use nonlinear optical media transforming coherent inputs into squeezed outputs. Sizman et al. [6] observed -2.2dB of squeezing in the second harmonic generated by a crystal enclosed in a cavity. Polzik et al. [4] on the other hand measured -5.2

Quantum Communications and Measurement
Edited by V. P. Belavkin *et al.*, Plenum Press, New York, 1995

±0.3 dB of squeezing in the degenerate down converted wave generated by a crystal pumped in a cavity. Taking into account the imperfection in the detection used in their experiment, one infers that the light incident on the balanced detector had fluctuations 6dB below the shot noise level. Quantum noise reduction of -5.1dB was recently obtained at MIT [7] using the $\chi^{(3)}$ nonlinearity of optical fibers in a Sagnac loop interferometer [8]. The quantum efficiency of the detectors in this experiment was estimated to be 0.85. This implies that the fluctuations of the maximally squeezed quadrature of the light incident on the balanced receiver were 7.3dB below the vacuum level. In this configuration one compensates for the relatively weaker $\chi^{(3)}$ nonlinearity by using short pulses with a large peak powers and a longer interaction region. The interferometer produces squeezed vacuum at one of its output ports and the pump pulse at the other port. This convenient separation of the pump and the squeezed radiation, in conjunction with the 5.1 dB of noise reduction reported, makes the Sagnac loop not only one of the most efficient squeezers, but also a natural source for enhancing the sensitivity of pulsed interferometers [3]. In this talk I will show that the Sagnac loop can generate light with an even larger degree of squeezing than reported when operated at a wavelength where the fiber exhibits group velocity dispersion.

Operation of the Nonlinear Sagnac Loop: The operating principle of the Sagnac loop squeezer is depicted on Fig. 1. In the linear Sagnac interferometer, a pulse incident on the input 50/50 beam splitter is reflected back into the same path it came from. This phenomenon is due to the destructive interference occurring at the other output port between the pulses traveling in opposite directions in the interferometer. Note that the vacuum fluctuations entering the empty input port are also reflected back. When a fiber is introduced in the Sagnac loop, photons and vacuum fluctuations forming either pulse 1 or 2 interact with each other due to the Kerr nonlinearity of the fiber. Transitions of photons between different frequencies occur independently within each counterpropagating pulse. When they recombine at the output coupler, their perfect phase correlation is partially destroyed, resulting in light exiting the previously empty output port. As fluctuations on each pulse are squeezed by the fiber, the output of this port is a squeezed vacuum. Squeezing at low frequencies is observed in homodyne detection with the local oscillator taken from the bright port of the Sagnac interferometer. The low frequency squeezing spectrum then arises from the fluctuations of the integrated photocurrent:

$$\hat{I} = I_0 \int d\tau \left\{ A^*(\tau) e^{i\theta} \hat{b}(\tau) + A(\tau) e^{-i\theta} \hat{b}^\dagger(\tau) \right\}. \tag{1}$$

In the experiment reported in [7], gaussian pulses from a mode-locked Nd:YLF laser operating at 1.3μm were used. Standard silica fibers have vanishing group velocity dispersion at that wavelength.

Pulse Shape Saturation The most important factor limiting the squeezing achieved in [7] is a saturation effect characteristic of the zero dispersion regime. When no dispersion is present, the various parts of the pulse evolve independently. The quantum noise traveling with any section of the pulse is then squeezed independently, to a degree controlled by the intensity of the pump in that region. The phase of the quadrature experiencing the most squeezing changes as well from one time interval to the next. The unique phase of the local oscillator is therefore a compromise between the optimum phases of each individual time slot operator. As a result, even though the squeezing at the center of the pulse can be as high as -20dB, the quantum noise reduction of the integrated photocurrent cannot exceed -8.7dB for gaussian pump pulses.

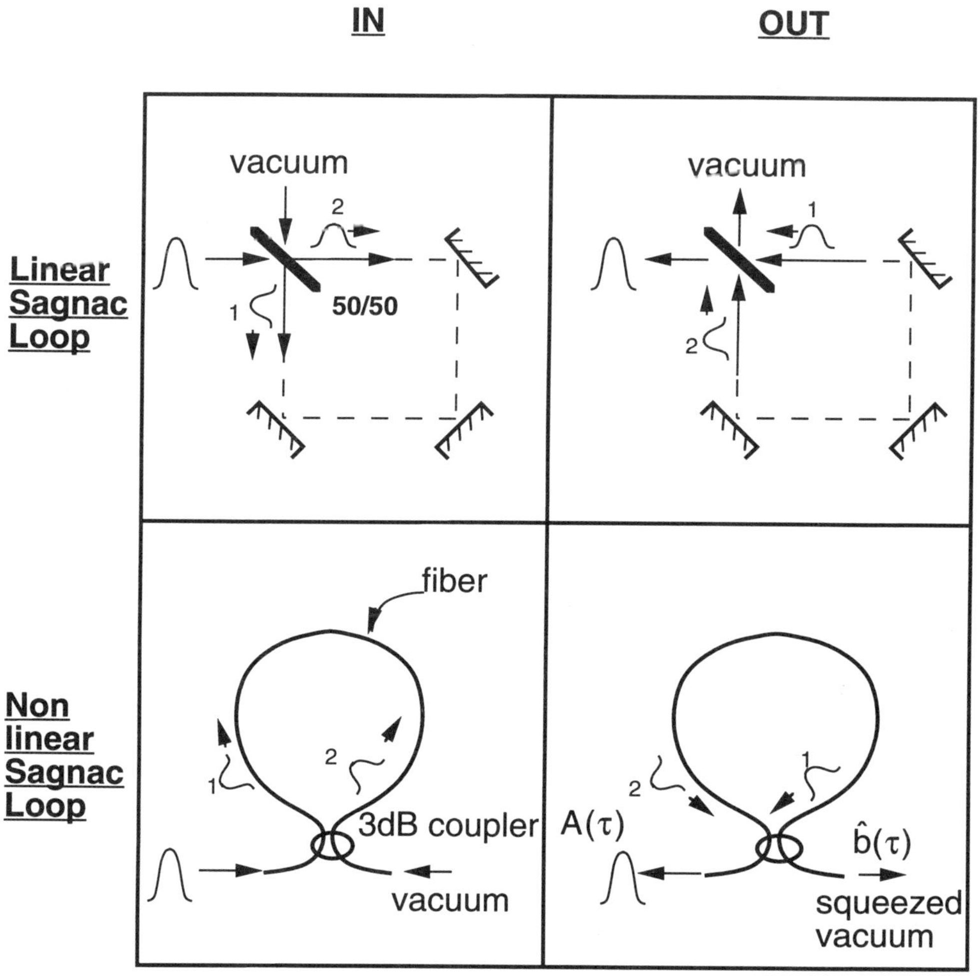

Figure 1: Principle of operation of the Sagnac loop squeezer

Dispersive Sagnac Loop In standard silica fibers, pump pulses with carrier frequencies different from 1.3μm are affected by group velocity dispersion. Although the equation describing the quantum evolution of the envelope in such media can be solved with the Bethe ansatz, this approach is not well suited for coherent inputs with large mean fields. In this case one prefers a simpler approach based on a linearized equation for the quantum noise [10]. When applied to optical solitons, this linearized analysis reveals that no saturation of the quantum noise reduction occurs in that case. This suggests that dispersion is beneficial to squeezing with pulses. Soliton lasers with high repetition rates and large peak powers are difficult to build. One would rather want to operate at Ti:sapphire wavelengths (700nm-900nm) where optical fibers exhibit a large amount of positive dispersion and large peak powers are routinely achieved. Below we use the linearized theory to analyze the quantum noise reduction performance of the nonlinear Sagnac loop in the presence of dispersion.

Linearized Theory The envelope operator is taken as the sum of a large classical excitation A and of an operator $\hat{b}$ measuring the small quantum fluctuations from pulse to pulse: $\hat{A} = A + \hat{b}$. The classical part, which is strongly affected by the nonlinearity, evolves according to:

$$\frac{\partial A}{\partial z} = -i\frac{\beta_2}{2}\frac{\partial^2 A}{\partial \tau^2} + i\kappa|A|^2 A, \tag{2}$$

where κ is the Kerr coefficient and β_2, which controls the group velocity dispersion, vanishes around 1.3μm in standard fibers. The noise operator is assumed too small to interact with itself. It rather evolves linearly on the background provided by the classical field:

$$\frac{\partial \hat{b}}{\partial z} = -i\frac{\beta_2}{2}\frac{\partial^2 \hat{b}}{\partial \tau^2} + 2i\kappa|A|^2\hat{b} + i\kappa A^2\hat{b}^\dagger. \tag{3}$$

From the linearity of this equation, the noise operator at the output of the fiber is linearly related to the noise operators in the input plane by Green's functions:

$$\hat{b}_{out}(\tau) = \int d\tau' \mu(\tau,\tau')\hat{b}_{in}(\tau') + \nu(\tau,\tau')\hat{b}^\dagger_{in}(\tau'). \tag{4}$$

Differential equations governing the evolution of μ and ν for a given classical background are obtained from (3) and solved numerically using the split step Fourier transform method [11].

Subpicosecond pulses with a duration comparable to the response time of the Raman oscillators are affected by localized independent sources of phase noise [12]. This effect is easily incorporated in the above formalism by introducing an additional Green's function for each Raman noise source. The degree of squeezing is not significantly affected by this additional noise even though significant frequency shifts are observed. Our analysis will be presented elsewhere.

Numerical Results Fig. 2 shows the degree of squeezing R for the quantum noise accompanying identical coherent sech input pulses of 100 fs duration. The three curves labelled (a), (b) and (c) give R at low frequencies as a function of the length of fiber in the Sagnac interferometer for the dispersion parameters $\beta_2 = 0$, $\beta_2 = 1\,ps^2/km$ and $\beta_2 = -20\,ps^2/km$. The energy of the pump was chosen for the counterpropagating pulses to be optical solitons in case (c). As Fig. 2 shows the pulse shape limit is overcome with both positive and negative dispersion. In the case where $\beta_2 = 1ps^2/km$, the pump develops into a flat disturbance with sharp leading and trailing edges.

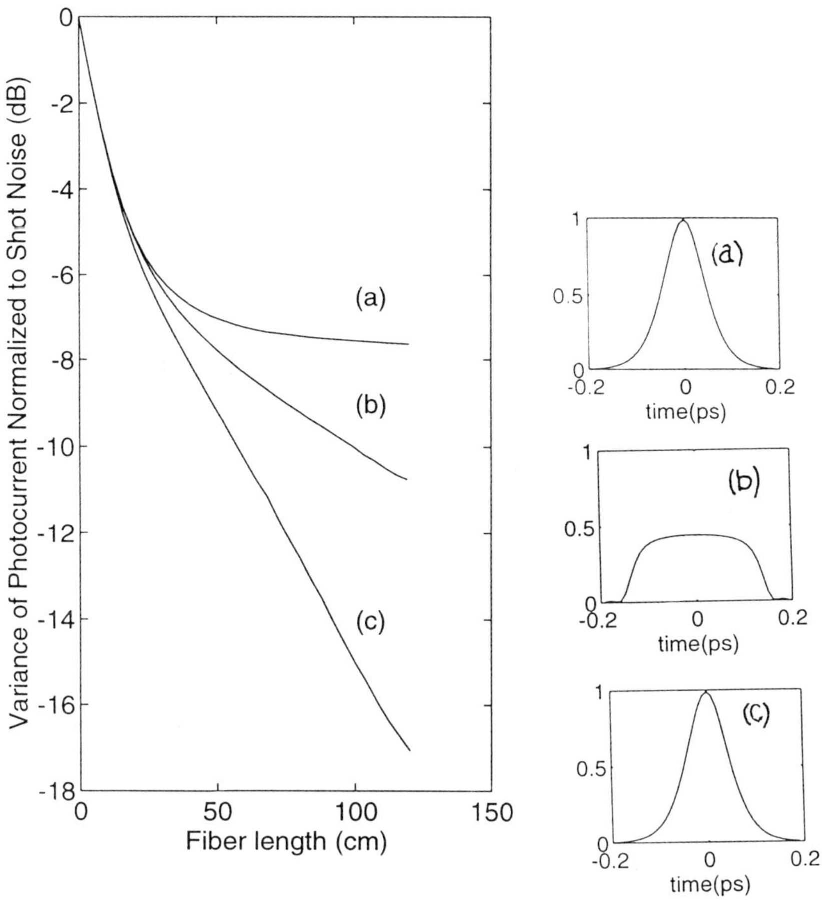

Figure 2: Numerical results from the linearized theory with (a) $\beta_2 = 0$, (b) $\beta_2 = 1\,ps^2/km$ and (c) $\beta_2 = -20\,ps^2/km$. In these three cases the input pulses were sech-like and coherent, with a duration of $100fs$. The energy of the pump was chosen so as to form two counterpropagating solitons in case (c). The insert give the output pump.

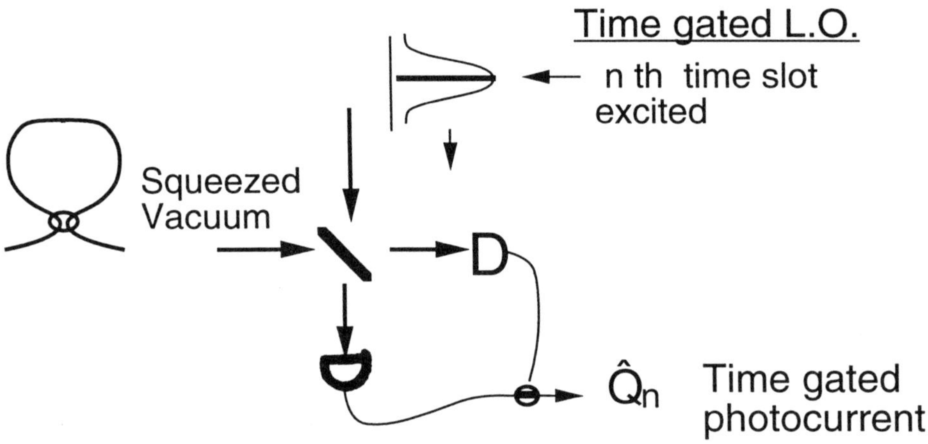

Figure 3: Time gated local oscillator. One can measure squeezing across the pulsed squeezed vacuum by varying the timing between the local oscillator and the signal.

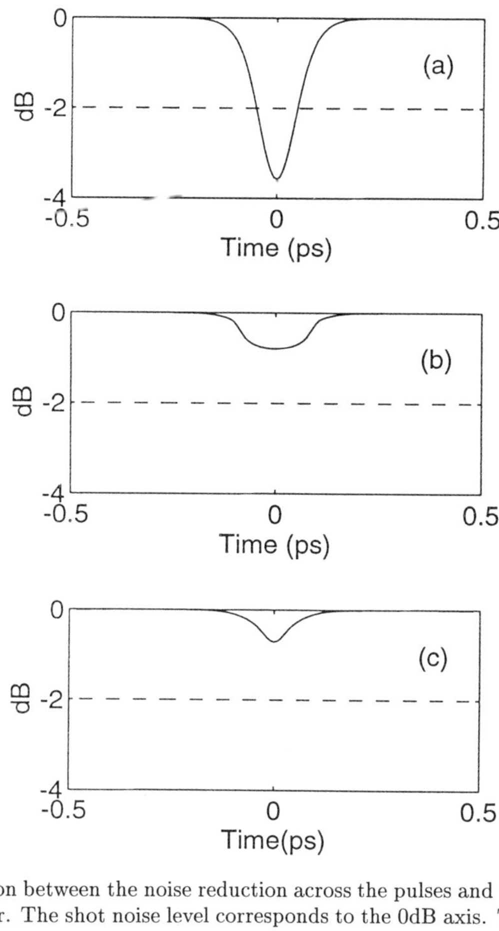

Figure 4: Comparison between the noise reduction across the pulses and the noise reduction with the full local oscillator. The shot noise level corresponds to the 0dB axis. The dashed lines are the squeezing levels with the full oscillator. (a) $\beta_2 = 0$, (b) $\beta_2 = 1ps^2/km$ and (c) $\beta_2 = -20ps^2/km$. The other parameters are as in figure 2.

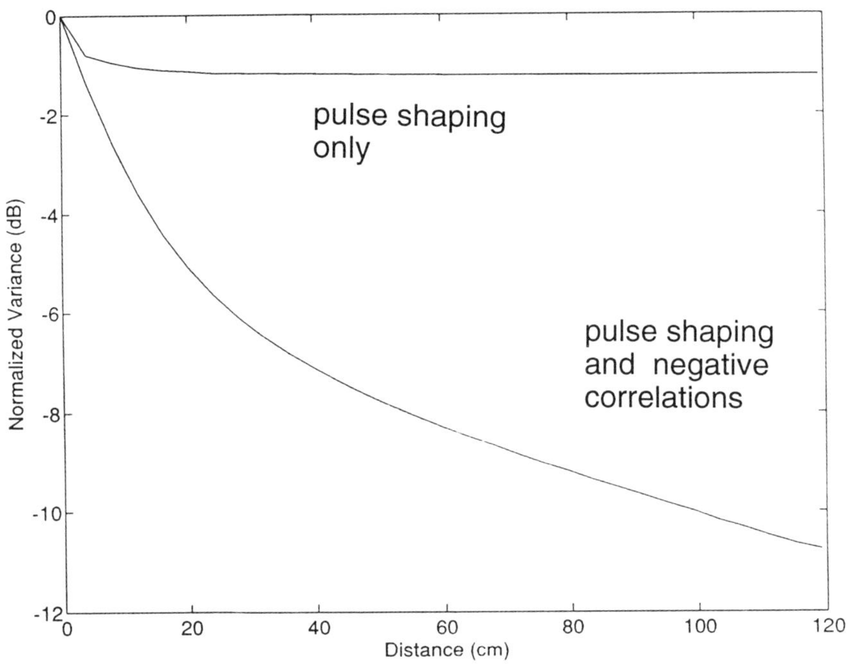

Figure 5: Relative importance of the two sums in (6).

Squeezing Across the Pulses We now investigate the reasons why dispersion increases QNR by using a series of time gated local oscillators as shown on Fig. 3. The balanced detector photocurrent is now:

$$\hat{Q}_n(\theta) = Q_0 \left\{ e^{i\theta} \hat{b}_n + e^{-i\theta} \hat{b}_n^\dagger \right\}, \tag{5}$$

where $\hat{b}_n$ is defined as $\hat{b}_n = T^{-1/2} \int_{\tau_n}^{\tau_n+T} d\tau \; \hat{b}(\tau)$. In this expression T is the local oscillator duration and τ_n is determined by the timing between the time gated local oscillator and the squeezed vacuum. Q_0 is an unimportant constant. The homodyne photocurrent is linear in the local oscillator amplitude. We then have:

$$\hat{I}(\phi) = \sum_n W_n \hat{Q}_n(\phi + \theta_n), \tag{6}$$

where $\hat{I}$ is the photocurrent (1) obtained when using the bright port of the Sagnac loop as the local oscillator. We also defined $W_n = |A(\tau_n)| I_0 / Q_0$ and θ_n is the phase of A. Using the above relationship between $\hat{I}$ and the $\hat{Q}_n$'s, one can express the fluctuations of the integrated photocurrent in terms of those of the field in each individual time slots. We find:

$$\frac{< \hat{I}^2(\phi) >}{< \hat{I}^2(l = 0) >} = \sum_n p_n < \hat{Q}_n^2 > + \sum_{n \neq m} \sqrt{p_n p_m} < \hat{Q}_n \hat{Q}_m + \hat{Q}_m \hat{Q}_n >, \tag{7}$$

where $p_n = |A_n|^2 / \sum_m |A_m|^2$ is the intensity of the pump in the nth time interval, normalized to $\sum_n p_n = 1$. The first term in (7) is the average of the variances of each time slot noise quadrature weighted by the relative pump intensity. From the positivity of p_n and $< \hat{Q}_n^2 >$, this term cannot be negative. The second term arises from correlations developing between the fluctuations of the field in different parts of the pulse. Unlike the first term, the correlators of this double sum can be negative.

Zero Dispersion When $\beta_2 = 0$, corresponding to the experiment reported in [7], the dynamics of the noise in the various time slots are independent from each other. The fluctuations of the integrated current is then simply the average given by the first sum and, consequently, must be larger than the fluctuations at the center of the pulse. This is apparent on Fig. 4(a) which shows , for a given fiber length, the minimized variance $< \hat{Q}_n^2 >$ as a function of time (i.e with τ_n varying continuously across the pulse). The dashed line corresponds to the noise level of the integrated current.

Negative Correlations The situation is different when the coupling between the time slots is restored. As quanta are exchanged between the time intervals, the level of fluctuations increases in the photocurrent $\hat{Q}_n$ from pulse to pulse. As a result squeezing across the pulse is considerably reduced as shown on Fig. 4(b) and 4(c). The second term in (7) however contributes to keep the fluctuations of the integrated current at a low level by taking negative values. These negative correlations indicate that for any given pulse, a local increase in the measured value of $\hat{Q}_n$ is matched to a certain degree by a decrease of the measured value of another quadrature somewhere else on the pulse. The local oscillator taken from the bright port of the Sagnac loop sums, to a certain degree, these negatively correlated quantities. Fig. 5 compares $< \hat{I}^2(\phi) > / < \hat{I}^2(l = 0) >$ as obtained from (7) with the average of the variances only (first sum of (7)). This plot confirms that negative correlations between the noise on different parts of the pulse is the main factor leading to an increase in squeezing for the integrated signal.

Conclusion We showed using a linearized theory for the quantum noise operator that the Nonlinear Sagnac interferometer can be a more efficient squeezer when operated away from the zero dispersion regime. We then presented an analysis based on a time gated local oscillator. We showed that the improvement in quantum noise reduction when dispersion was present was mainly due to negative correlations between the quadrature noise on different parts of the pulse.

References

[1] M. Xiao et al. Phys. Rev. Lett. **59**, 278 (1987); Opt. Lett. **13**, 476 (1988)

[2] P. Grangier et al. Phys. Rev. Lett. **59**, 2153 (1987)

[3] K. Bergman, C. R. Doerr, H. A. Haus and M. Shirasaki, Opt. Lett. **18**, 643 (1993)

[4] E. S. Polzik, J. Carri and H. J. Kimble, Phys. Rev. Lett., **68**, 3020 (1992)

[5] C. W. Gardiner, Phys. Rev. Lett. **56**, 1917 (1986); H. J. Carmichael et al., Phys. Rev. Lett., **58**, 2539 (1987); A. S. Parkins and C. W. Gardiner, Phys. Rev. A**40** 3796 (1989)

[6] A. Sizman et al., Opt. Commun, **80**, 138 (1990)

[7] K. Bergman, H. A. Haus, E. P. Ippen and M. Shirasaki, Opt. Lett. **19**, 290 (1994)

[8] M. Shirasaki and H. A. Haus, J. Opt. Soc. Am. B**7**, 30 (1990)

[9] F. X. Kärtner et al., Quant. Opt. **4**, 379, (1992); L. Joneckis and J. H. Shapiro, J. Opt. Soc. Am B **10**, 1102 (1993)

[10] H. A. Haus and Y. Lai, J. Opt. Soc. Am., B**7**, 386 (1990)

[11] C. R. Doerr, M. Shirasaki and F. I. Khatri, J. Opt. Soc. Am. B **11**, 145 (1994)

[12] F. X. Kärtner et al., J. Opt. Soc. Am.; L. Boivin et al., Phys. Rev. Lett., **73**, 240 (1994)

SUPPRESSION OF THERMAL SCATTERING IN NONLINEAR POLAR CRYSTALS

Ze Cheng

Department of Chemical Physics,
Weizmann Institute of Science,
Rehovot 76100, Israel

Abstract

The present paper describes the photonic superguiding state in polar crystals of a self-defocusing nonlinearity. The thermal scattering effects are suppressed in the superguiding state, and the traveling-wave superguiding state has the squeezing property. Possible applications of the photonic superguiding state to optical-communication systems are discussed.

1 INTRODUCTION

The author has established the photonic superguiding theory in polar crystals with a high nonlinearity [1]. In quantum theory it is shown that photons can sense an attractive effective interaction by exchange of virtual optical phonons. Such an interaction leads to the superguiding state. In the standing-wave superguiding state the photons with opposite wave vectors and spins are bound into pairs. In the traveling-wave superguiding state a propagating photon pair is the combination of the two photons with opposite transverse wave vectors and spins. We study the particle properties of the photonic superguiding state, and the most important property is that the system of photon pairs evolves without scattering attenuations. Quantum fluctuations of the standing-wave superguiding state exceed the vacuum fluctuations while the traveling-wave superguiding state has the squeezing property. We also investigate the wave properties of the photonic superguiding state. It is found that the polar crystals with a high nonlinearity are self-defocusing media. In the standing-wave superguiding state the system of photon pairs exists in the form of quantized vortices. In the traveling-wave superguiding state the system of photon pairs exists in the form of quantized temporal solitons.

2 PHOTONIC EFFECTIVE HAMILTONIAN

When the incident frequency ω_0 is assumed to be well below the electronic transition frequencies of the crystal, the photon-electron interaction and the related photon absorption by electrons can be neglected. Therefore we consider only the interaction between the electromagnetic (EM) field and the crystal. We also hypothesize that ω_0 is well above the transverse-optical phonon frequencies of the crystal, so that the interaction between photons and transverse-optical phonons and the related multiphonon absorption of a photon can be omitted. Under the influence of the macroscopic EM field, the charge center of the electron shell of an ion shifts relative to that of the nucleus and hence an electric-dipole moment is induced in the ion. For the isotropic medium, the interaction Hamiltonian involving the ionic deformations is quadratic in $\vec{E}$ and has the form [2]

$$H_I = -\frac{1}{2}\sum_{n,l}\int dt'\tilde{\alpha}_{nl}(t,t')\vec{E}(\vec{r}_n,t')\cdot\vec{E}(\vec{r}_n,t), \tag{1}$$

where $\tilde{\alpha}_{nl}(t,t')$ is the effective polarizability of the nlth ion.

In terms of the creation, annihilation operators $a^{\dagger}_{\vec{k}\sigma}$, $a_{\vec{k}\sigma}$ of circularly polarized photons with spin $\sigma = \pm 1$, the Hamiltonian of the EM field is quantized as

$$H_L = \sum_{\vec{k}\sigma}\hbar\omega_{\vec{k}}a^{\dagger}_{\vec{k}\sigma}a_{\vec{k}\sigma}\ . \tag{2}$$

For simplicity, the studied crystal is taken to be centrosymmetric. Polar modes in centrosymmetric crystals have odd parity and are Raman inactive. In the cubic system, the common polar crystals that are both centrosymmetric and Raman-active have the fluorite structure. At this point , the studied crystal is determined as a certain crystal of the fluorite structure, such as CaF_2 [3]. Since the following treatment has no relation to acoustic and polar modes, the ion vibration is limited to the Raman-active mode. When the phonons of the Raman-active mode are created or annihilated by the operators $b^{\dagger}_{\vec{q}}$ and $b_{\vec{q}}$, the Hamiltonian for lattice vibrations is quantized as

$$H_C = \sum_{\vec{q}}\hbar\omega_R b^{\dagger}_{\vec{q}}b_{\vec{q}}\ . \tag{3}$$

Furthermore, the incident intensity is assumed to be below the threshold of stimulated Brillouin scattering, so that the pressure fluctuations and Raman-active mode of the crystal are excited thermally. Near the equilibrium state of the ion system, the interaction Hamiltonian H_I is reduced to the two terms H_f and H_v:

$$H_f = \sum_{\vec{k}\sigma,\vec{q}} N_{\vec{k}\sigma}(\vec{q})\delta\alpha_{\vec{q}}a^{\dagger}_{\vec{k}+\vec{q},\sigma}a_{\vec{k}\sigma}\ , \tag{4}$$

$$H_v = \sum_{\vec{k}\sigma,\vec{q}} M_{\vec{k}\sigma}(\vec{q})a^{\dagger}_{\vec{k}+\vec{q},\sigma}a_{\vec{k}\sigma}(b_{\vec{q}} + b^{\dagger}_{-\vec{q}}). \tag{5}$$

The Hamiltonian of the coupled system becomes

$$H = H_L + H_C + H_f + H_v\ . \tag{6}$$

In the classical solid considered, temperature fluctuations at constant pressure travel diffusively and give rise to elastic Rayleigh scattering, but pressure fluctuations at constant entropy propagate ballistically as transverse and longitudinal sound waves and lead to spontaneous Brillouin scattering. Therefore the Hamiltonian H_f accounts for Rayleigh scattering and spontaneous Brillouin scattering. The effect of H_f on the photon system can be described by the Hamiltonian H'_f,

$$H'_f = \frac{<(\delta\alpha_n)^2>}{2Nk_BT} \sum_{\substack{\vec{k},\vec{k}',\vec{q} \\ \sigma,\sigma'}} N_{\vec{k}\sigma}(-\vec{q}) N_{\vec{k}'\sigma'}(\vec{q}) a^\dagger_{\vec{k}-\vec{q},\sigma} a^\dagger_{\vec{k}'+\vec{q},\sigma'} a_{\vec{k}'\sigma'} a_{\vec{k}\sigma} \ , \tag{7}$$

where H'_f represents a repulsive photon-photon interaction. The Hamiltonian H_v represents spontaneous Raman scattering by an optical phonon. The effect of H_v on the photon system can be described by the Hamiltonian H'_v,

$$H'_v = \sum_{\substack{\vec{k},\vec{k}',\vec{q} \\ \sigma,\sigma'}} \frac{M_{\vec{k}\sigma}(-\vec{q}) M_{\vec{k}'\sigma'}(\vec{q})\omega_R}{\hbar[(\omega_{\vec{k}} - \omega_{\vec{k}-\vec{q}})^2 - \omega_R^2]} a^\dagger_{\vec{k}-\vec{q},\sigma} a^\dagger_{\vec{k}'+\vec{q},\sigma'} a_{\vec{k}'\sigma'} a_{\vec{k}\sigma} \ . \tag{8}$$

The interaction matrix elements given by Eq.(8) can be either attractive or repulsive. If the states $\vec{k}$, $\vec{k} - \vec{q}$ are separated by an energy smaller than $\hbar\omega_R$, an attraction is present. The system will have to adjust itself to the presence of this attraction. The effective Hamiltonian of the photon system is $H_p = H_L + H'_f + H'_v$.

3 PHOTONIC SUPERGUIDING STATE

We first consider the standing-wave superguiding state. The studied crystal has rectangular faces and is enclosed by perfectly reflecting mirrors. In fact, we form a passive optical resonator. There is a small window in each mirror at $x = 0, y = 0, z = 0$. The external light field enters the resonator through the three windows within a time interval. The incident directions are along the x, y, z axes, and the three windows are closed at the end of the time interval. The macroscopic EM field induced by the incident light field is in a standing-wave configuration. By high nonlinearity, we mean that the repulsion in Eq.(7) can be overcome by the attraction in Eq.(8). The attractive effective interaction leads to bound photon pairs. In the standing-wave configuration a photon pair is stable only if the two photons have opposite wave vectors and spins. Thus one set $\vec{k}' = -\vec{k}$ and $\sigma' = -\sigma$ in Eqs.(7) and (8). Next we let $\vec{k}$ stand for $(\vec{k}, \sigma)$ and $-\vec{k}$ for $(-\vec{k}, -\sigma)$. Now the pair Hamiltonian of the photon system is

$$H_p = \frac{1}{2}\sum_{\vec{k}} \hbar\omega_{\vec{k}}(a^\dagger_{\vec{k}} a_{\vec{k}} + a^\dagger_{-\vec{k}} a_{-\vec{k}}) + \sum_{\vec{k},\vec{k}'} V_{\vec{k}\vec{k}'} a^\dagger_{\vec{k}'} a^\dagger_{-\vec{k}'} a_{-\vec{k}} a_{\vec{k}} \ , \tag{9}$$

$$V_{\vec{k}\vec{k}'} = \frac{|M_{\vec{k}}(\vec{k}' - \vec{k})|^2 \omega_R}{\hbar[(\omega_{\vec{k}} - \omega_{\vec{k}'})^2 - \omega_R^2]} + \frac{<(\delta\alpha_n)^2>}{2Nk_BT} |N_{\vec{k}}(\vec{k}' - \vec{k})|^2. \tag{10}$$

It will be supposed that $V_{\vec{k}\vec{k}'} = -V_0$ if $|\omega_{\vec{k}} - \omega_0|$ and $|\omega_{\vec{k}'} - \omega_0| < \omega_R$; otherwise $V_{\vec{k}\vec{k}'} = 0$.

The diagonalization of the pair Hamiltonian (9) can be performed by the Bogoliubov transformation

$$U = \exp\left[\frac{1}{2}\sum_{\vec{k}} \varphi_{\vec{k}} (a^\dagger_{\vec{k}} a^\dagger_{-\vec{k}} - a_{\vec{k}} a_{-\vec{k}})\right] , \tag{11}$$

$$a_{\vec{k}} = U^\dagger \alpha_{\vec{k}} U = \alpha_{\vec{k}} \cosh \varphi_{\vec{k}} + \alpha^\dagger_{-\vec{k}} \sinh \varphi_{\vec{k}} \ ,$$

$$a^\dagger_{\vec{k}} = U^\dagger \alpha^\dagger_{\vec{k}} U = \alpha^\dagger_{\vec{k}} \cosh \varphi_{\vec{k}} + \alpha_{-\vec{k}} \sinh \varphi_{\vec{k}} \ . \tag{12}$$

$\alpha^\dagger_{\vec{k}}$ and $\alpha_{\vec{k}}$ are the creation and annihilation operators of quasiparticles in the photon system. The state vector of photon pairs in the photon system may be constructed as $|G>=U|0>$, such that $\alpha_{\vec{k}}|G>=0$. Since quasiparticles are in a coherent state of many modes, we introduce the displacement operator

$$D(\eta) = \exp\left[\sum_{\vec{k}} (\eta_{\vec{k}} \alpha^\dagger_{\vec{k}} - \eta^*_{\vec{k}} \alpha_{\vec{k}})\right] . \tag{13}$$

The state vector of the photon system may be written as $|\eta>=D(\eta)|G>$, such that $\alpha_{\vec{k}}|\eta>=\eta_{\vec{k}}|\eta>$. $|\eta_{\vec{k}}|^2$ characterizes the mean quasiparticle number of a mode field and can be described by a distribution function $f_{\vec{k}}(u)=|\eta_{\vec{k}}|^2$, where the argument u is the average energy density of the EM field. In the superguiding state, photon pairs and single quasiparticles are present together. In the normal state, the photon pairs disappear and the quasiparticles become photons. We show that the photon system in the superguiding state has a lower energy than in the normal state and therefore the superguiding state is preferred. Furthermore, the superguiding ground state is most stable because of its maximum condensation energy.

In the normal state, the behavior of photons is governed by the Hamiltonian (6) of the coupled system. The normal photons suffer spontaneous Raman scattering by optical phonons as well as Rayleigh and spontaneous Brillouin scatterings. Since Rayleigh and spontaneous Brillouin scatterings are overcome by attractive photon-phonon-photon interaction, the resultant photon pairs incorporate the optical phonons and hence do not suffer any scatterings. However, individual quasiparticles experience scatterings like the normal photons. For inelastic scatterings of quasiparticles, the crystal both absorbs and emits energy but the absorbing ability is higher than the emitting one. The energy of the system of quasiparticles will decay out as time goes on. We find the time evolution of the EM energy density,

$$u(t) = u_s(u) + u_n(u) e^{-t/t_c}, \tag{14}$$

where t_c is the quasiparticle lifetime. The first and second terms represent the contributions from paired photons and quasiparticles, respectively. We let λ denote the nonlinear photon-phonon coupling strength. It is found that for $\lambda_0 = (2\omega_R/\omega_0)^{\frac{1}{2}}$, the superguiding state can occur only if $\lambda_0 < \lambda < 1$. The superguiding ground state corresponds to the ground-state energy density u_g. When the EM energy density u increases from u_g into the critical energy density u_c, the transition from the superguiding to the normal state takes place. As a result, another condition for superguiding is that $u_g \leq u < u_c$. Under the condition $\lambda_0 < \lambda < 1$, whether the standing-wave superguiding state will occur depends on the EM energy density u_0 at the initial instant.

Next we discuss the traveling-wave superguiding state. The traveling-wave configuration is determined by a cylindrical dielectric waveguide. The z axis coincides with the waveguide axis of symmetry. Now the incident light field is required to be a quasimonochromatic field of a central frequency ω_0. The incident light field excites a guided wave field in the waveguide. Each wave vector of the guided modes is separated into $\vec{k} = \vec{K} + \vec{Q}$, where $\vec{K}$ and $\vec{Q}$ are the components parallel to and transverse to the z axis. The axial wave vector $\vec{K}$ is real everywhere. $\vec{Q}$ is real in the core but imaginary in the cladding. For a fixed K, the values of allowed Q in the core are in the range $0 \leq Q < K \tan\theta_c$, where θ_c is the complement of the critical angle of total internal reflection. The axial wave number K is a function of the incident frequency ω. The function $K(\omega)$ must be specified by the Maxwell equations together with the boundary condition at the core-cladding interface. The waveguide supports only the fundamental mode and satisfies the weak-guidance approximation. The fundamental mode photons of wave vectors $\vec{K}+\vec{Q}$ and $\vec{K}-\vec{Q}$ always occur simultaneously and have an equal number. By exchange of virtual optical phonons, the fundamental mode photons can feel an attractive effective interaction. Such an interaction leads to the traveling-wave superguiding state, in which a propagating photon pair is the combination of $(\vec{K}+\vec{Q},\sigma)$ with $(\vec{K}-\vec{Q},-\sigma)$ and thus has the composite momentum $2\hbar\vec{K}$. Now, $\vec{K}+\vec{Q}$ is used for $(\vec{K}+\vec{Q},\sigma)$ and $\vec{K}-\vec{Q}$ for $(\vec{K}-\vec{Q},-\sigma)$. The state vector $|\eta>$ of the photon system may be constructed as

$$|\eta>= \exp\left[\sum_{\vec{K},\vec{Q}}\left(\eta_{\vec{K}+\vec{Q}}a^{\dagger}_{\vec{K}+\vec{Q}} - \eta^{*}_{\vec{K}+\vec{Q}}a_{\vec{K}+\vec{Q}}\right)\right]$$

$$\times \exp\left[\frac{1}{2}\sum_{\vec{K},\vec{Q}}\varphi_{\vec{K}+\vec{Q}}\left(a^{\dagger}_{\vec{K}+\vec{Q}}a^{\dagger}_{\vec{K}-\vec{Q}} - a_{\vec{K}+\vec{Q}}a_{\vec{K}-\vec{Q}}\right)\right]|0> . \tag{15}$$

Equation (15) demonstrates that the traveling-wave superguiding state has the squeezing property. Some other results are similar to those in the standing-wave superguiding state and will be neglected.

4 DISCUSSION

In the present optical communication systems, the transmitters all use coherent pulses of laser light, and the transmission media are highly flexible waveguides, i.e., optical fibers. There are three physical effects which limit the transmission of pulses in optical fibers: scattering, dispersion, and noise. Scattering reduces the intensity of pulses in optical fibers and so places an upper limit on the propagation distance of pulses. For a long haul system larger than the limiting distance, one needs repeaters to amplify the signal. Dispersion causes the pulses to spread out and eventually overlap to such an extent that all of the information is lost. This imposes an upper limit on the transmission rate of pulses. Noise degrades the signal and impairs the system performance. The quantum noise is a limitation on the signal-to-noise ratios. The present optical fibers are almost exclusively made from polar materials with a self-focusing nonlinearity. The scattering losses like Rayleigh scattering exist inevitably in these fibers. For example, a well-fabricated silica fiber has a loss of 0.2 dB/km at 1.53 μm, which almost corresponds to its intrinsic scattering-loss value. If we want to suppress the scattering losses of optical fibers it is necessary to develop the superwaveguides, which are made from polar materials with a self-defocusing nonlinearity. The propagation of light pulses in the superwaveguides gets rid of the dispersion effect. In fact there is no limitation on the transmitted power in the superwaveguides . If the incident light intensity exceeds the threshold of stimulated Brillouin scattering, initially the laser light field is in the coherent state. During propagation, the laser light field evolves into the traveling-wave superguiding state. The information is carried by the quadrature phase of the superguided light field with reduced quantum fluctuations. Utilization of the superwaveguides as the transmission media enables us to realize repeaterless optical communications with a high bit rate and a large signal-to-noise ratio. At the receivers of such optical communication systems, one can perform quantum nondemolition measurements without destroying the signal. The presentation of the photonic superguiding state is based on the experimental fact that infrared optical fibers have extremely low scattering losses compared with silica fiber. Halides are considered to be the most appropriate candidates for the superwaveguides because their band gap is large and their multiphonon absorption is located in a longer wavelength region than that for silica glass.

References

[1] Ze Cheng, Phys. Rev. Lett. **67**, 2788 (1991); Phys. Rev. A (to be published).

[2] M. Schubert and B. Wilhelmi, Nonlinear Optics and Quantum Electronics (Wiley, New York, 1986), p. 553.

[3] Crystals with the Fluorite Structure, edited by W. Hayes (Clarendon, Oxford, 1974).

INJECTION LOCKED LASERS AS AMPLIFIERS

C.M. Savage[1] and Yi Mu[2]

[1]Dept. of Physics & Theoretical Physics,
Australian National University,
Canberra, ACT 0200, Australia.
[2]Department of Computer Science,
University of Wollongong,
Wollongong, NSW 2522, Australia.

1 INTRODUCTION

This paper considers the gain and noise properties of injection locked, above-threshold, laser amplifiers. Injection locking makes the amplifiers phase-sensitive and hence able to amplify signals without degrading the signal to noise ratio. This paper is an elucidation of the work presented in reference [1]. The results are the same but here we emphasize their physical significance.

Our work follows on from theoretical and experimental work of Harris, Loudon, Mander, Vaughan, and Shepherd [2,3] on above-threshold laser amplifiers. We have added an injection locking field which frees the amplifier from adding the noise demanded by the quantum mechanics of phase-insensitive linear amplifiers [4].

As an amplifier our device is similar to the parametric amplifier, and may be a useful alternative for some applications. Reduced noise amplifiers are particularly attractive for systems of amplifier chains, such as occur in optical communication systems, since any added noise is further amplified down the chain [5]. Presently, however, the proposed applications of phase-sensitive amplifiers remain unrealized. This is due to their relative complexity and small bandwidth compared to conventional laser amplifiers.

The motivation for our work comes from Caves' amplifier theory [4]. This applies to amplifiers whose output electric fields are linearly related to their input electric fields. Above threshold laser amplifiers belong to this category. However some

Quantum Communications and Measurement
Edited by V. P. Belavkin *et al.*, Plenum Press, New York, 1995

important amplifiers such as electro-optical repeaters do not (see articles by A. Masalov and by H. Yuen, this volume). According to Caves, only phase-sensitive amplifiers can avoid adding noise to the signal. "Phase-sensitive" means that the amplifier gain depends on the phase of the signal relative to a reference field, which we take to be the injection locking field.

In the next section we discuss the self-heterodyne detection scheme that we use in our theory. Section 3 describes our fully quantum mechanical model of the above threshold laser amplifier. The final section presents and interprets our results.

2 SELF-HETERODYNE DETECTION

Figure 1 is a schematic diagram of the amplifier we consider. Besides the amplified signal and laser field outputs a so called "idler" field is also produced. This is detuned from the laser frequency by the same amount as the signal but with the opposite sign.

A simple and practical detection scheme is to detect the entire output of the device on a photodetector. The resulting radio-frequency photo-current can be filtered, for example by a spectrum analyzer, to obtain the sideband power. This is called "self-heterodyne detection" because the laser acts as a strong local oscillator for the weak detuned sidebands.

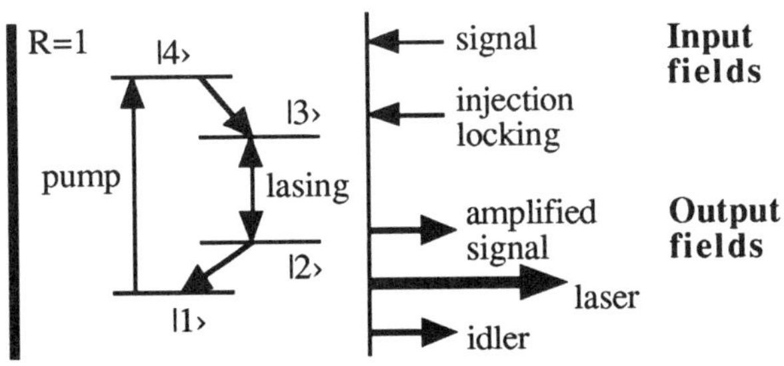

Figure 1: Schematic diagram of the injection locked, above-threshold, laser amplifier. The input fields are the injection locking field from a master oscillator, and the signal. The output fields are the amplified signal, the idler and the laser field. The lasing atoms have four levels and the lasing transition is $|2\rangle \leftrightarrow |3\rangle$.

The photo-current is proportional to

$$\left|E_1 e^{i\phi} + E_2 e^{i(\omega_2-\omega_1)t} + E_3 e^{-i(\omega_2-\omega_1)t}\right|^2, \tag{1}$$

where E_1, E_2, E_3 are the electric fields at the laser, signal, and idler frequencies respectively. These may be interpreted either as classical complex field amplitudes or as annihilation operators for the field modes. ϕ is the phase angle of the laser field with respect to the injected field, and $\omega_2 - \omega_1$ is the detuning between the laser and signal/idler fields. It is this frequency component of the photo-current that is of interest to us. The detected power *at this frequency* is therefore proportional to

$$\left|E_2 + E_3^* e^{i2\phi}\right|^2 = |X_\phi|^2, \tag{2}$$

where we have defined the heterodyne amplitude:

$$X_\phi \equiv E_2 + E_3^* e^{i2\phi}. \tag{3}$$

The heterodyne intensity is defined to be $|X_\phi|^2$, or for operator quantities $\hat{X}_\phi^\dagger \hat{X}_\phi$. The heterodyne intensity gain is defined to be:

$$G = \frac{|\hat{X}_\phi|^2_{\text{out}}}{|\hat{X}_\phi|^2_{\text{in}}} \quad . \tag{4}$$

The quantum mechanical noise associated with this signal is:

$$N = \langle \hat{X}_\phi \hat{X}_\phi^\dagger \rangle - \langle \hat{X}_\phi \rangle \langle \hat{X}_\phi^\dagger \rangle. \tag{5}$$

In the next section we describe the quantum mechanical model we use to calculate these quantities for the injection locked laser amplifier.

3 THE MODEL

Our amplifier model is based on the Haken-Risken [6] laser master equation for the reduced density operator $\hat{\rho}$. The amplifier consists of N_A four-level atoms interacting with three optical cavity modes: the laser, signal, and idler modes which we indicate by subscripts 1,2,and 3 respectively. Four-level atoms provide a more realistic model than two-level atoms, for which the theory is considerably simpler. This should allow easier comparison between our results and potential experiments.

The free Hamiltonian for the modes and atoms is:

$$\hat{H}_0 = \sum_{k=1}^{3} \hbar\omega_k \hat{a}_k^\dagger \hat{a} + \sum_{\nu=1}^{N_A} \hbar\omega_0 \hat{\sigma}_{23,\nu}^+ \hat{\sigma}_{23,\nu}^- \quad , \tag{6}$$

where $\hat{\sigma}_{23,\nu}^+$ and $\hat{\sigma}_{23,\nu}^-$ are the atomic raising and lowering operators for the lasing transition on the νth atom. The atoms interact with the field modes via the Jaynes-Cummings Hamiltonian:

$$\hat{H}_{JC} = i\hbar \frac{g}{\sqrt{N_A}} \sum_{\nu=1}^{N} \left\{ \hat{\sigma}_{23,\nu}^- \left(\hat{a}_1^\dagger e^{i\vec{k}_1 \cdot \vec{r}_\nu} + \hat{a}_2^\dagger e^{i\vec{k}_2 \cdot \vec{r}_\nu} + \hat{a}_3^\dagger e^{i\vec{k}_3 \cdot \vec{r}_\nu} \right) + \text{h.c.} \right\}, \tag{7}$$

where h.c. means the Hermitean conjugate of the preceding, g is a scaled interaction strength, $\vec{r}_\nu$ is the position of the νth atom, and the wave vectors $\vec{k}_i$ satisfy the phase matching condition $2\vec{k}_1 = \vec{k}_2 + \vec{k}_3$. The Hamiltonian for the input signal is

$$\hat{H}_s = i\hbar\sqrt{2\kappa} \left(\beta_2 \hat{a}_2^\dagger e^{-i\omega_2 t} - \beta_2^* \hat{a}_2 e^{i\omega_2 t} \right), \tag{8}$$

where β_2 is the field strength, such that an empty cavity with photon loss rate 2κ would be driven into a coherent state of amplitude $\beta_2\sqrt{2/\kappa}$. Similarly the Hamiltonian for the input injection locking field is:

$$\hat{H}_d = i\hbar\sqrt{2\kappa}\, \beta_1 \left(\hat{a}_1^\dagger e^{-i\omega_1 dt} - \hat{a}_1 e^{i\omega_1 t} \right). \tag{9}$$

We incorporate dissipation into our model by coupling to reservoirs and making Markovian assumptions [6]. We include cavity loss at rate 2κ and spontaneous emission from level $|4\rangle$ to level $|3\rangle$ at rate γ_{34} and from level $|2\rangle$ to level $|1\rangle$ at rate γ_{12}. Incoherent pumping of the laser is modelled as "inverse" spontaneous emission "upwards" from level $|1\rangle$ to level $|4\rangle$ at rate Γ. The resulting Schrödinger picture master equation is:

$$\frac{d\hat{\rho}}{dt} = \frac{1}{i\hbar}\left[\hat{H}_0 + \hat{H}_s + \hat{H}_d + \hat{H}_{JC}, \hat{\rho}\right] + L_c\hat{\rho} + L_p\hat{\rho} + L_{34}\hat{\rho} + L_{12}\hat{\rho}, \quad (10)$$

where we have defined the following dissipative terms:

$$L_c\hat{\rho} \equiv \kappa \sum_{k=1}^{3} \left(2\hat{a}_k\hat{\rho}\hat{a}_k^\dagger - \hat{a}_k^\dagger\hat{a}_k\hat{\rho} - \hat{\rho}\hat{a}_k^\dagger\hat{a}_k\right), \quad (11)$$

$$L_{ij}\hat{\rho} \equiv \frac{\gamma_{ij}}{2} \sum_{\nu=1}^{N_A} \left(2\hat{\sigma}_{ij,\nu}^-\hat{\rho}\hat{\sigma}_{ij,\nu}^+ - \hat{\sigma}_{ij,\nu}^+\hat{\sigma}_{ij,\nu}^-\hat{\rho} - \hat{\rho}\hat{\sigma}_{ij,\nu}^+\hat{\sigma}_{ij,\nu}^-\right), \quad (12)$$

$$L_p\hat{\rho} \equiv \frac{\Gamma}{2} \sum_{\nu=1}^{N_A} \left(2\hat{\sigma}_{14,\nu}^+\hat{\rho}\hat{\sigma}_{14,\nu}^- - \hat{\sigma}_{14,\nu}^-\hat{\sigma}_{ij,\nu}^+\hat{\rho} - \hat{\rho}\hat{\sigma}_{14,\nu}^-\hat{\sigma}_{14,\nu}^+\right). \quad (13)$$

We have solved this equation using the approach of Reid and Walls [7, 1]. The first step is to convert the master equation into a Fokker-Planck equation for the generalized P-function quasi-probability distribution. This requires a truncation at second order of a system size expansion in the inverse number of atoms $1/N_A$. The Fokker-Planck equation is statistically equivalent to a set of *stochastic differential equations* for c-number variables corresponding to system operators. The atomic operators are decomposed into *spatial fourier components*, and only those with the same wavevectors as one of the three field modes are retained. The next approximation is to keep only spatially phase matched interaction terms. Further, the sideband variables are assumed weak compared to the laser frequency variables so that any interaction terms higher than first order in them are neglected [7, 1].

Semiclassical steady states are obtained by setting the time derivatives and noise terms in the resulting stochastic differential equations equal to zero. These solutions determine the gain of the amplifier.

In order to solve for the noise on the amplified signal we *linearize* the stochastic differential equations. The fluctuations are assumed to be small enough that they can be linearized around the steady states.

Next the atomic degrees of freedom are assumed to relax rapidly compared to the field, which is the good cavity limit $\kappa \ll \gamma_{12}, \gamma_{34}$. Hence they are *adiabatically eliminated* to obtain stochastic differential equations for the c-number variables α_i corresponding to the field modes:

$$\frac{d\alpha_1}{dt} = -\kappa\alpha_1 + f_1(\tilde{n})\alpha_1 + \sqrt{2\kappa}\beta_1 + F_{\alpha_1}, \quad (14)$$

$$\frac{d\alpha_2}{dt} = -\kappa\alpha_2 + f_2(\tilde{n})\alpha_2 + f_3(\tilde{n})\alpha_3^\dagger + \sqrt{2\kappa}\beta_2 + F_{\alpha_2}, \quad (15)$$

$$\frac{d\alpha_3}{dt} = -\kappa\alpha_3 + f_4(\tilde{n})\alpha_3 + f_5(\tilde{n})\alpha_2^\dagger + F_{\alpha_3}. \quad (16)$$

There are three conjugate equations for the variables $\alpha_i^\dagger$ corresponding to the mode creation operators. The variables $\alpha_i, \alpha_i^\dagger$ correspond to the linearized fluctuations of the fields around the steady state values. The functions f_i of the steady state photon number in mode 1, $\tilde{n}$, are defined by reference to section III of reference [1]. We omit the definitions here because they are lengthy and do not illuminate the physics of the system. Similarly the correlations of the delta-correlated Gaussian noise terms F_{α_i} are fully defined in section III and appendix B of [1].

These equations can be written in vector form by defining the vector of field variables, $\vec{\alpha} \equiv \{\alpha_1, \alpha_1^\dagger, \alpha_2, \alpha_2^\dagger, \alpha_3, \alpha_3^\dagger\}$:

$$\frac{d\vec{\alpha}}{dt} = -\mathbf{A}\vec{\alpha} + \mathbf{D}^{1/2}\vec{\varepsilon}(t). \tag{17}$$

The matrices **A** and **D** are called the drift and diffusion matrices respectively. $\vec{\varepsilon}(t)$ is a vector of delta-correlated Gaussian noise terms. The spectral noise properties of the modes are calculated from the spectral covariance matrix $\mathbf{S}(\omega)$. This can be shown to be given by [8]

$$\mathbf{S}(\omega) = (\mathbf{A} - i\omega\mathbf{I})^{-1}\mathbf{D}\left(\mathbf{A}^T + i\omega\mathbf{I}\right)^{-1}, \tag{18}$$

where **I** is the 6×6 identity matrix.

We are interested in the variance of the heterodyne amplitude $\hat{X}_\phi$ which is expressed in terms of our modes by (c.f. Eq.(3)),

$$\hat{X}_\phi = \hat{a}_{2,\mathrm{out}} + \hat{a}_{3,\mathrm{out}}^\dagger e^{2i\phi} \quad . \tag{19}$$

The subscript "out" emphasizes that we are concerned with the mode output fields and hence must use the appropriate input/output theory [9]. We want the heterodyne intensity noise Eq.(5), that is the zero frequency covariance ($\omega = 0$),

$$N = 1 + 2\kappa\left[S_{\alpha_2,\alpha_2^\dagger} + S_{\alpha_3,\alpha_3^\dagger} + e^{-2i\phi}S_{\alpha_2,\alpha_3} + e^{2i\phi}S_{\alpha_2^\dagger,\alpha_3^\dagger}\right]. \tag{20}$$

This is the noise against which we wish to measure the amplified signal, $|\tilde{X}_\phi|^2$, where tildes denote semiclassical values. We define the *signal to noise ratio* $\mathcal{R}$:

$$\mathcal{R} \equiv \frac{|\tilde{X}_\phi|^2}{N} \quad . \tag{21}$$

An *input* signal consisting of mode 2 in a coherent state $|\beta_2\rangle$ and mode 3 in the vacuum state has $|\tilde{X}_\phi|^2 = |\beta_2|^2$ and $N = 1$. Hence the input signal to noise ratio is:

$$\mathcal{R}_{\mathrm{in}} = |\beta_2|^2. \tag{22}$$

The corresponding output signal to noise ratio is:

$$\mathcal{R}_{\mathrm{out}} = \frac{G\,|\beta_2|^2}{N} \quad , \tag{23}$$

where G is the intensity gain of the amplifier. Our fundamental figure of merit for the amplifier is the *signal to noise ratio degrading factor* L, defined to be the ratio of $\mathcal{R}_{\mathrm{out}}$ to $\mathcal{R}_{\mathrm{in}}$. From the preceding two equations:

$$L = \frac{\mathcal{R}_{\mathrm{out}}}{\mathcal{R}_{\mathrm{in}}} = \frac{G}{N}. \tag{24}$$

According to Caves [4] an ideal phase-insensitive amplifier, in the high gain limit, has L=0.5 while and ideal phase-sensitive amplifier can have L=1, hence not degrading the input signal to noise ratio at all.

4 RESULTS

We consider a single sided cavity, that is one with only one input/output port, Figure 1. This is because it has better noise properties than a two port cavity in which independent vacuum noise can enter each port. According to cavity input/output theory [9] the output heterodyne amplitude is given in terms of the cavity fields and input field by, $\sqrt{2\kappa}(\tilde{\alpha}_2 + e^{2i\phi}\tilde{\alpha}_3^\dagger) - \beta_2$. The heterodyne intensity gain Eq.(4) is therefore

$$G = \frac{\mid \sqrt{2\kappa}(\tilde{\alpha}_2 + e^{2i\phi}\tilde{\alpha}_3^\dagger) - \beta_2 \mid^2}{\mid \beta_2 \mid^2}, \tag{25}$$

where, as before, tildes denote semiclassical steady state values.

This gain is plotted in Figure 2(a) as a function of the signal detuning $\omega_2 - \omega_1$. The gain peak occurs at a detuning equal to the Rabi frequency for the lasing

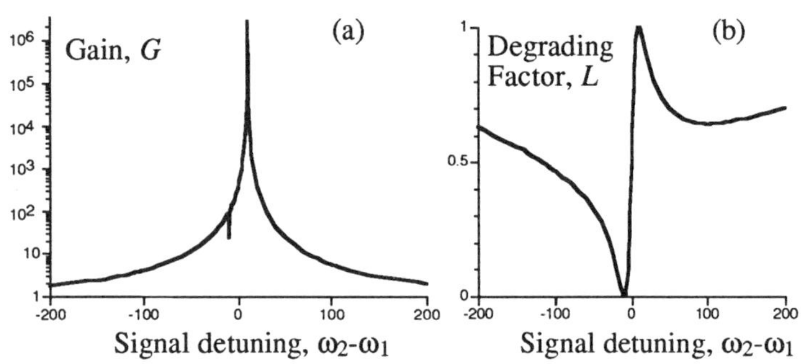

Figure 2: (a) Intensity heterodyne gain G versus signal detuning $\omega_2 - \omega_1$. (b) Signal to noise ratio degrading factor L versus signal detuning. Parameters: $\omega_0 - \omega_1 = 10, \Omega = 10.3, \kappa = 0.01, \beta_1 = 0.0071, \Gamma = 100, \gamma_{12} = 200, \gamma_{34} = 100$. All parameters are scaled by g, which hence has the value $g = 1$.

transition $\omega_2 - \omega_1 = \Omega \equiv \sqrt{(\omega_0 - \omega_1)^2 + I}$, where I, defined in reference [1], is proportional to the photon number in the laser mode $\tilde{n}$. When the detuning has the opposite sign there is absorption instead of gain (the absorption is not resolved in Figure 2(a)). This kind of gain/absorption spectrum was first predicted by Mollow [10,11]. Our gain and absorption peaks occur for the opposite signal detunings from Mollow's case because the lasing transition is inverted by the incoherent pumping.

The associated signal to noise ratio degrading factor L is plotted against signal detuning in Figure 2(b). The optimum performance is found when the signal detuning is near the atomic detuning. It almost reaches the maximum possible value of one at the gain maximum, at detuning $\omega_2 - \omega_1 = \Omega$. Consequently we have shown that the injection locked, above-threshold, laser amplifier can amplify a signal without degrading the signal to noise ratio.

C.S. acknowledges important discussions with Ping Koy Lam.

References

[1] Yi Mu and C.M. Savage, Phys. Rev. A **49**, 4093 (1994).

[2] M. Harris et al., Phys. Rev. Lett. **67**, 1743 (1991).

[3] R. Loudon et al., Phys. Rev. A **48**, 681 (1993).

[4] C.M. Caves, Phys. Rev. D **26**, 1817 (1982).

[5] Yi Mu and C.M. Savage, J. Opt. Soc. Am. B **9**, 65 (1992).

[6] H. Haken, in *Handbuch der Physik*, ed. by L. Genzel (Springer-Verlag, Berlin,1970), Vol.XXV/2c.

[7] M.D. Reid and D.F. Walls, Phys. Rev. A **34**, 4929 (1986).

[8] C.W. Gardiner, *Handbook of Stochastic Methods* (Springer-Verlag, Berlin,1985).

[9] M. Collett and C.W. Gardiner, Phys. Rev. A **30**, 1386 (1984).

[10] B.R. Mollow, Phys. Rev. A **5**, 2217 (1972).

[11] C. Cohen-Tannoudji et al., *Atom-Photon Interactions* (Wiley-Interscience, New York, 1992).

PHOTOCURRENT NOISE SUPPRESSION AND OPTICAL AMPLIFICATION IN NEGATIVE–FEEDBACK OPTO–ELECTRONIC LOOP

A.V.Masalov, A.A.Putilin, M.V.Vasilyev

Lebedev Physical Institute,
Leninsky pr.53, 117924 Moscow,
Russia

Abstract

Photocurrent shot-noise suppression is observed in negative-feedback opto-electronic loop below the photodetector limit: $(1 - \eta)$ of standard shot-noise level, which is imposed by conventional theory of photodetector with quantum efficiency η. Minimum noise spectral density is observed below the shot-noise level by a factor 260 (-24.2 dB), which is 52 times (-17.2 dB) below the photodetector limit. We suggest the concept of anticorrelation light state, where the light in the loop is described as a non-stationary state with Poissonian quantum fluctuations, and superimposed anticorrelated classical fluctuations.

The configuration of opto-electronic loop with a beam splitter has been used to amplify optical signals. Amplification factor of about 7 is realized, while the signal-to-noise ratio decreases only within 1.3 dB. Data on signal-to-noise ratio show improvement in comparison with -3 dB figire expected for conventional linear amplifier.

Keywords :
quantum optics, sub-Poissonian light, optical amplifier

INTRODUCTION

Negative-feedback opto-electronic loop is the device designed for generation of sub-Poissonian light. It consists of a light source, photodetector, negative-feedback electronic circuit and the element, controlling the light intensity via electric signal. Indeed, the authors [1–3] observed reduction of photocurrent noise in the loop. In [1,2] a loop consisting of semiconductor laser, photodiode and negative feedback controlling the light intensity via laser pump current the photocurrent noise spectral density was reduced by a factor of 5 (-7 dB) below the standard shot-noise level (SNL). A photodiode with quantum efficiency 0.8 was used in the feedback circuit. In [3] analogous device using He-Ne laser and photodiode with quantum efficiency about 0.8 was developed. Photocurrent noise reduction by a factor of 1.5 below the SNL was obtained. Describing the results of these experiment in terms of sub-Poissonian light one should mind the photodetector noise limit (PNL). This limit is the minimal photocurrent noise which might be obtained in the detection of light by photodiode with quantum efficiency η. In the theory of single-mode-field detection the PNL equals to $(1-\eta)$ of shot-noise level, and the photocurrent noise approaches this limit only when the light intensity has no quantum fluctuations. The authors [1,2] approached this maximum noise reduction where no intensity quantum fluctuations were expected, while the measurements [3] were far above the PNL. Theoretical description of the opto-electronic loop was given in [4]. It was found that photocurrent noise suppression in the loop depends on the feedback strength and could be done arbitrary large regardless of the photodetector quantum efficiency. This important result demonstrates that conventional relation between intensity fluctuations and photocurrent noise is not applicable to the opto-electronic loop. Up to now this result has no adequate interpretation. In our paper we study the opto-electronic loop in the regime of photocurrent noise suppression far below the photodetector limit. So large suppression can not be explained in terms of ordinary sub-Poissonian light. To describe the light in the loop the new concept of anticorrelation state is presented.

After experiments [1-3] the interest in the opto-electronic loop substantially decreased because it turned out that extraction of light with sub-Poissonian properties from the loop for further use was impossible. In this paper we present the results of measurements where the loop is used as an optical amplifier with the noise figure better than that of ordinary linear amplifier. Due to spontaneous emission in the linear amplifier the signal-to-noise ratio degrades at the output with a factor $1/(2-1/G) \approx 0.5 \rightarrow -3$ dB, where G is the amplifier gain. However, the -3 dB noise figure is not the limit; parametric amplifier may have 0 dB figure. We believe that optical amplifier based on opto-electronic loop is another example beating out the level of -3 db.

EXPERIMENTAL SETUP

Opto-electronic loop used in our experiments consisted of light beam from cw Ar^+ laser, electro-optical modulator operated in the amplitude modulation regime, *p-i-n* photodiode and electronic circuit including a low-noise amplifier (Fig.1). A single-mode single-frequency laser of 1 W output power at the wavelength 514 nm was used. Electro-optical modulator with half-wavelength voltage of about 170 V and frequency range 0÷150 MHz transmitted nearly a half of the incident light power in the regime of amplitude modulation. Quantum efficiency of the feedback photodiode was $\eta = 0.80 \pm 0.02$. DC output of photodiode was monitored, while AC signal was negatively fed back to the modulator through an amplifier. The amplifier of 80 dB gain had equivalent temperature of input noise less than 300 K.

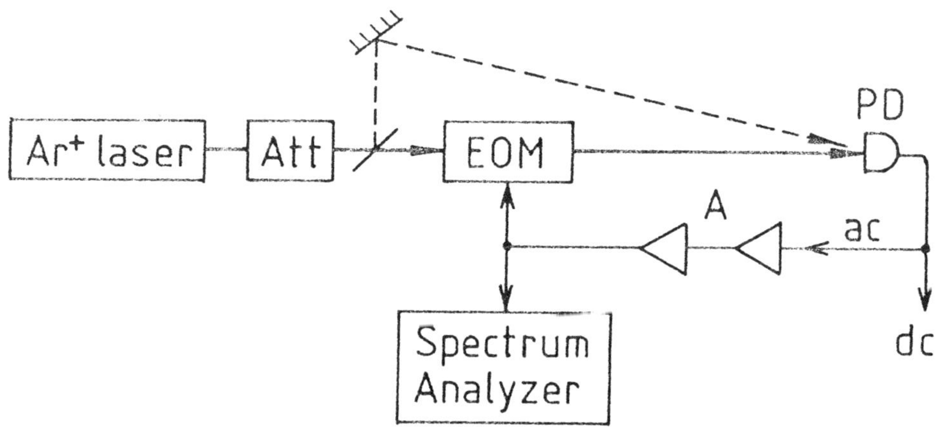

Figure 1: Experimental setup: *EOM* — electro-optical modulator, *PD* — photodiode, *A* — amplifiers, *Att* - attenuator.

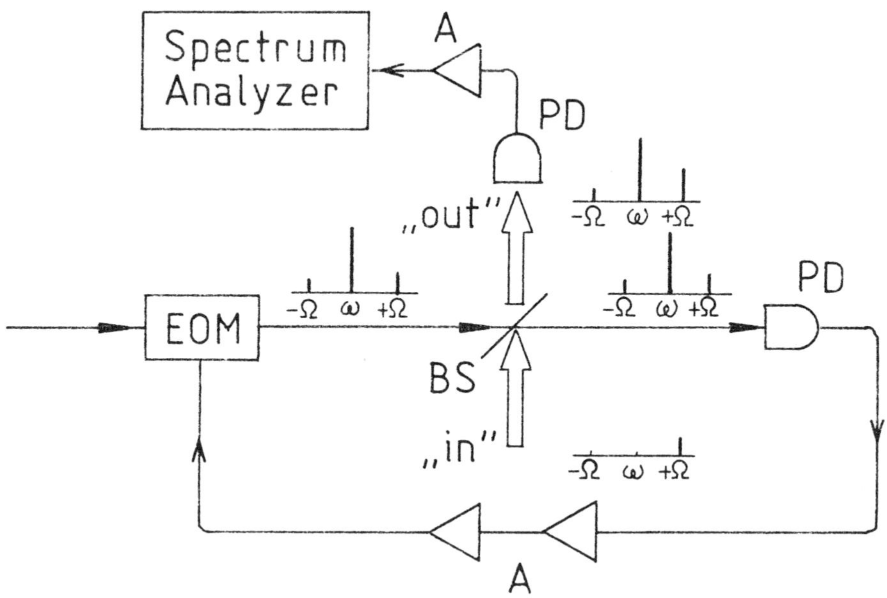

Figure 2: Scheme of amplification in opto-electronic loop: *EOM* — electro-optical modulator, *PD* — photodiodes, *A* — amplifiers, *BS* — beam splitter.

The small part of the signal was used for the noise measurements in the frequency range 1÷100 MHz by a spectrum analyzer.

The feedback strength A is the key parameter of feedback device, which is the amplification of photocurrent fluctuation after its round trip over the loop. In our measurements it is proportional to the light power in the loop or to the average photocurrent $\langle i \rangle$: $A/\langle i \rangle = 7.0\ \mathrm{mA}^{-1}$ in our case. This fact allowed us to study the loop at different values of the feedback strength by changing the light power.

The measured photocurrent noise power was compared with the SNL spectral density $2e\langle i \rangle$. The SNL density was obtained via illuminating the photodiode either by an incandescent lamp or by part of the laser beam (dashed line in the Fig.1). The noise levels in both cases did not differ within error of ± 0.1 dB and corresponded to the SNL value calculated from the spectrum analyzer sensitivity.

PHOTOCURRENT-NOISE SUPPRESSION

On the frequensy scale of photocurrent fluctuations the feedback sign periodically changes from negative to positive due to finite round-trip time of the loop. At $A = 1$ the loop approaches oscillation threshold in the frequensy ranges of positive feedback, and the current oscillations overload the feedback amplifier. To avoid oscillations at $A > 1$, a narrow-band filter with bandwidth of several megahertz, matched to the frequency of negative feedback 26.4 MHz, was inserted into the feedback electronic circuit. This improvement allowed us to observe the photocurrent noise suppression at $A \leq 16$.

An increase of the feedback strength A by increasing the light power results in reduction of the relative noise. The maximum noise suppression has been observed with a factor 260 ± 10 (-24.2 dB) below the SNL (Fig.3). This is about 52 times lower (-17.2 dB) than the photodetector limit. It is impossible to calculate the light intensity noise reduction from these data using a well-known relation:

$$\frac{\langle \delta I^2 \rangle}{\langle \delta I^2 \rangle_{SQL}} - 1 = \frac{1}{\eta}\left(\frac{\langle \delta i^2 \rangle}{\langle \delta i^2 \rangle_{SNL}} - 1\right), \tag{1}$$

where $\langle \delta I^2 \rangle_{SQL}$ is the standard quantum level (SQL) of intensity fluctuations, $\langle \delta i^2 \rangle_{SNL}$ is the shot-noise level of photocurrent. Such a calculation can be completed only for weak noise suppression: $\langle \delta i^2 \rangle / \langle \delta i^2 \rangle_{SNL} > 1-\eta$ (photocurrent noise is above the PNL). For strong suppression: $\langle \delta i^2 \rangle / \langle \delta i^2 \rangle_{SNL} < 1-\eta$ (photocurrent noise is below the PNL) the formula (1) gives the meaningless result: $\langle \delta I^2 \rangle / \langle \delta I^2 \rangle_{SQL} < 0$. In our measurements the latter happens at $\langle \delta i^2 \rangle / \langle \delta i^2 \rangle_{SNL} < 0.20$.

The minimum photocurrent noise level observed in our measurements is much lower than that for photodiode illuminated by ideal noiseless light. This fact brings up the question about the quantum state of light in the loop. The theory of the opto-electronic loop [4] does not elucidate this problem, the authors avoid discussion of the light state. Our concept of the light state in the loop is based on the analysis of the modulator influence upon the light passing through it. We consider the voltage on the modulator to be determined (classical) value, depending on time. In accordance with voltage fluctuations the modulator changes its transmission, i.e. the output light gains temporal variations of its average intensity (in the sense of quantum average). Except for these variations, the modulator does not affect the photon statistics of light provided that the statistics at the modulator input is Poissonian. (In general, an optical device with partial transmission distorts the photon

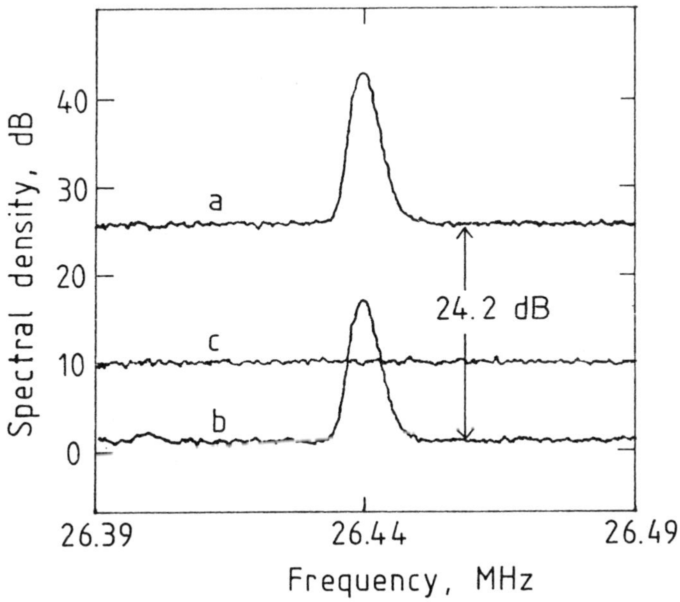

Figure 3: Spectral densities of input signal and noise registered by feedback photodiode: *a* — feedback is switched off, *b* — feedback is switched on ($A = 15.1$), *c* — amplifier noise at switch-off feedback.

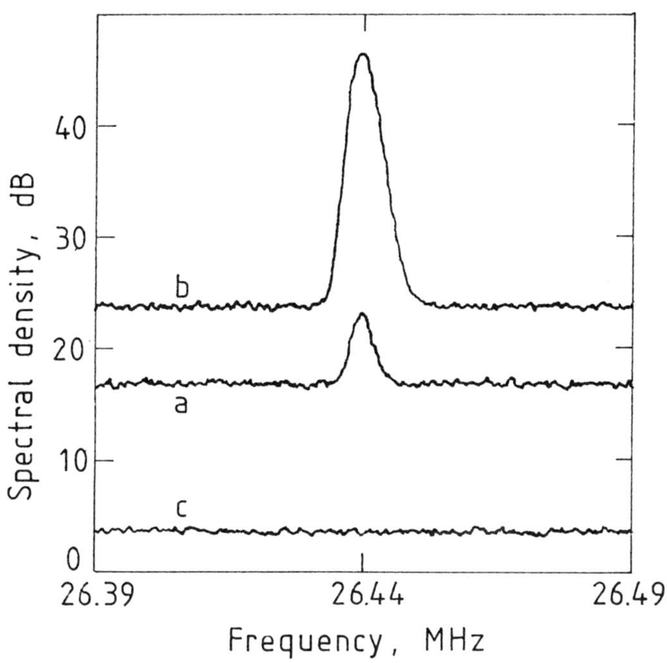

Figure 4: Spectral densities of output signal and noise: *a* — feedback is switched off, *b* — feedback is switched on ($A = 5.3$), *c* — amplifier noise.

statistics of any light state except coherent or thermal. However, in our case we may ignore this fact because of the coherent state of light at the modulator input.) The light intensity variations, caused by the modulator, can be treated as a signal. Thus, in addition to Poissonian intensity fluctuations (noise component of intensity) the light in the loop gains also the classical intensity variations induced by the modulator (modulation component). The negative feedback makes the classical variations opposite to Poissonian fluctuations. The feedback photodiode reproduces the destructive summation of the modulation and noise components as photocurrent shot-noise suppression. The difference between noise and modulation components in the photocurrent can be made arbitrary small by increasing the feedback strength, and there is no limit on photocurrent noise suppression. Thus, we treat the photocurrent noise suppression as the result of anticorrelation between classical and quantum fluctuations rather than the result of sub-Poissonian light generation. We call the state of light in the loop "anticorrelation state". This state is non-stationary one, and the relation (1) between photocurrent and intensity fluctuations is not applicable to description of the anticorrelation light detection, because this relation implies the average light intensity constant in time. One should note that the anticorrelation between noise and modulation components of light intensity is not complete. The accurate anticorrelation is imposed by feedback electronics between light modulation component and photocurrent noise, the latter consists of light noise contribution and noise due to non-unity detector quantum efficiency η. While the photocurrent noise suppression can be arbitrary large regardless of η, the degree of anticorrelation between the light noise and modulation components is limited by the quantum efficiency. The total intensity fluctuations can not be reduced below the limit allowed by η.

Idea on separation of photocurrent variations into the modulation and noise components was used in [5] for the analysis of the experiment on sub-Poissonian light generation in parametric down-conversion. In that experiment the photocurrent fluctuations from one of the parametric waves were fed back to the pump, while the other wave exhibited sub-Poissonian noise. The authors applied this idea only to photocurrents, and the light state was not discussed. However, we consider the description of light state to be necessary, because the photon statistics in the loop can be examined by quantum nondemolition measurements.

Detailed experimental data on photocurrent noise suppression in opto-electronic loop and on excess noise in the light extracted from the loop are presented in [6]. Calculations based on separation of photocurrent variations into modulation and noise components are in a good agreement with experimental data.

OPTICAL AMPLIFICATION

The scheme of optical amplifier based on opto-electronic loop is shown in Fig.2. The beam splitter (transmission $T = 6.3\%$ at 514 nm) was inserted into the loop. Optical signal to be amplified came to the system through the open port of this splitter. Output signal coincided with transmitted input signal in the beam extracted from the loop. Input signal was prepared from initial laser beam, and its frequency was shifted over $\Omega = 26.44$ MHz. Due to the beatings of in-loop light with the reflected part of input signal the feedback photodiode produces the photocurrent at frequency Ω, which modulates initial light intensity. One of two modulation sidebands at the modulator output coincides with input signal by frequency and phase, and comes out of loop as amplified signal. There is no spontaneous emission in this amplifier, and the noise characteristic is expected to be better than that of ordinary linear amplifier.

The input optical signal was registered by feedback photodiode in heterodyne regime. We did not separate the output signal from initial radiation, and it was registered by the second photodiode ($\eta_A = 0.69 \pm 0.02$), also in heterodyne regime. Heterodyning of signals gives explicit data on signal-to-noise ratio. Input signal and noise are shown in Fig.3 where the feedback has been switched "on" and "off". The feedback reduces both the noise and the signal, however the signal-to-noise ratio (SNR) remains unchanged. Data on Fig.3 are registered at maximum feedback strength where the noise suppression factor is about 260 ± 10 (-24.2 dB). One should note that the SNR presented in Fig.3 differs from that in the input optical signal by a factor $R\eta$, where $R = 93\%$ is reflection coefficient of beam splitter, and $\eta = 0.80$ is quantum efficiency of feedback photodiode.

Data on amplification of optical signal are shown in Fig.4. With feedback switched off (curve a) the registered SNR is significantly smaller than that of input light by factor $TT_F\eta_A$, where $T_F = 40\%$ is the transmission of filter placed in front of second photodiode (not shown in Fig.2). However, after switching the feedback on the signal grows up (curve b) and the SNR returnes close to initial value. The gain measured in heterodyne regime depends on feedback strength, and at large A it tends to $1/T$. Experimental data on gain are well fitted by formula

$$G_H = \frac{1}{T}\left(\frac{T+A}{1+A}\right)^2, \tag{2}$$

derived in the analysis of opto-electronic loop. It should be noted that in heterodyne regime both optical sidebands contribute to G_H (this is marked by subscript H). Genuine optical amplification of signal sideband G differs from G_H. The relation between G and G_H depends on modulator characteristics and has been found in special measurement procedure: $G - 1 = (G_H - 1)/2.2$. Maximum gain values achieved in our study are $G_H \approx 14$ and $G \approx 7$.

The noise levels measured by feedback photodiode and by that in output beam differ from each other. The feedback photodiode reproduces basically the Poissonian noise of initial radiation, while the second one reproduces the excess noise of light induced by modulator driven by shot-noise of feedback photocurrent. In this case quantum efficiency of second photodiode is less essential in SNR measurements than that of feedback.

The factor TSN describing the transfer of SNR from input to output: TSN $= (\mathrm{SNR})_{out}/(\mathrm{SNR})_{in}$, also depends on feedback strength. The increase of A results in the increase of TSN. Transfer of signal and noise in the amplifier is demonstrated in Fig.5 for two values of feeedback strength. Measured TSN factors are presented in the Table:

A	0.56	0.84	1.1	1.4	2.8	4.2
TSN ($\pm$10%)	0.56	0.71	0.75	0.94	0.86	1.0

At $A \geq 4.2$ the observed TSN factors are equal to unity within error of measurements. The data on TSN factors versus gain are summarized in Fig.6. The solid line in Fig.6 shows theoretical limit of conventional linear amplifier where the input optical signal is asumed to be in coherent state and the lower atomic level of active

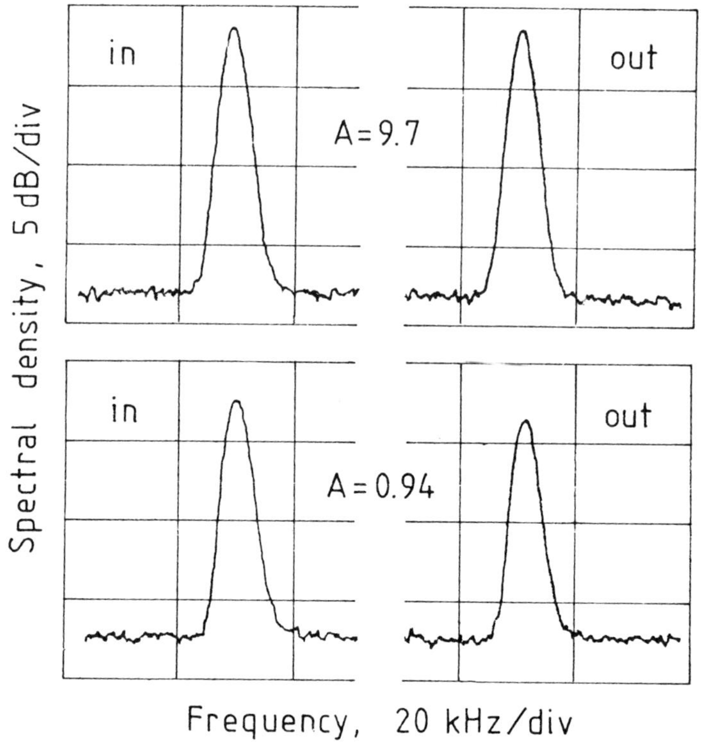

Figure 5: Spectral densities of input (left) and output (right) signal and noise in the frequency region near 26.44 MHz for two values of feedback strength: $A = 0.94$ (down) and $A = 9.7$ (up).

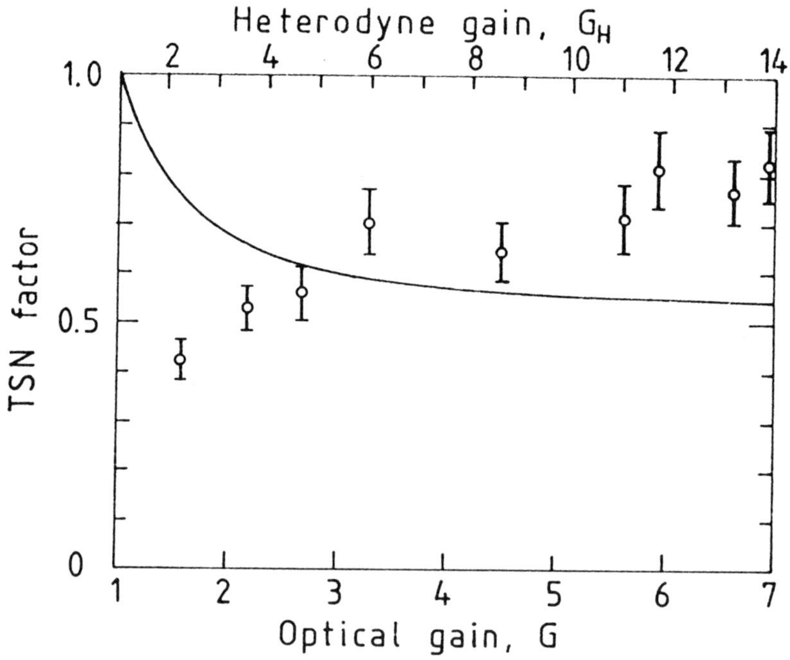

Figure 6: TSN factors of amplifier corrected for $T\eta$ versus gain (points). Conventional linear amplifier is represented by solid line (Eq.(3)).

medium is not populated:

$$\mathrm{TSN} = \left(2 - \frac{1}{G}\right)^{-1} . \tag{3}$$

TSN factor in Fig.6 are that from Table, multiplied by $T\eta$; in this case it characterizes the transfer of SNR from light to photocurrent. The basic contribution to TSN factors in Fig.6 is given by quantum efficiency of feedback photodiode $\eta = 0.8$. At $G \geq 3.4$ ($A \geq 1.4$) the TSN factors for the amplifier under study are higher than that for conventional one.

In distinction to conventional optical amplifier, the amplifier based on opto-electronic loop amplifies optical signals coherent in space to initial laser beam, and has the amplification bandwidth limited by round-trip time of loop.

CONCLUSIONS

The opto-electronic loop with negative feedback has been operated in the regime where the photocurrent noise is suppressed below the photodetector limit. In this case no stationary single-mode light state with sub-Poissonian photon statistics can explain observed noise suppression. The concept of anticorrelation state is presented to describe the light in the loop: temporal variations of the average light intensity, induced by the modulator, are anticorrelated to the quantum fluctuations of intensity. Destructive summation of these two parts of the intensity variations — quantum-noise and modulation components — is reproduced by photodetector as shot-noise suppression.

Optical amplifier based on negative-feedback opto-electronic loop is studied. At maximum optical gain $G \approx 7$ (heterodyne gain $G_H \approx 14$) the signal-to-noise transfer factor reaches the value -1.3 dB which is limited basically by quantum efficiency of feedback photodiode.

References

[1] S.Mashida, Y.Yamamoto. Optics Commun. **57**, 290 (1986).

[2] Y.Yamamoto, N.Imoto, S.Mashida. Phys. Rev. A **33**, 3243 (1986).

[3] Ya.A.Fofanov. Kvantovaya Elektronika (Russian) **16**, 2593 (1989).

[4] J.H.Shapiro, G.Saplakoglu, S.-T.Ho, P.Kumar, B.E.A.Saleh, M.C.Teich. JOSA B **4**, 1604 (1987).

[5] P.R.Tapster, J.G.Rarity, J.S.Satchell. Phys. Rev. A **37**, 2963 (1988).

[6] A.V.Masalov, A.A.Putilin, V.M.Vasilyev. J. of Mod. Opt. (1994) *(in press)*.

NONCLASSICAL FIELDS IN ONE-PHOTON MICROMASER ACTION

N. Nayak*, A. Kremid, B.V. Thompson and R.K. Bullough

Department of Mathematics,
UMIST, P.O.Box 88,
Manchester M60 1QD, U.K.
*On leave from S.N. Bose National Centre
for Basic Sciences, DB-17, Sector 1,
Salt Lake, Calcutta 700064, India.

1 INTRODUCTION

This paper is concerned with the open system formed by the one ^{85}Rb atom micromaser as it is currently in operation [1], and with the strictly quantum fields which can be created inside its maser cavity. We highlight the problems surrounding the actual *measurement* of these fields both practical and theoretical although some of such quantum measurements have already been carried out successfully [1]. It seems probable that the quantum fields created in this device could play a role in ultrasensitive, very few photon, microwave communication systems. However a dominant role in the current action of the micromaser [1] is played by the *very few* photons ($\overline{n}_{th} \leq \frac{1}{6}$, where $\overline{n}_{th}$ is the Planck function) of black-body radiation. The study of the effects of this black-body radiation is one of the concerns of this paper.

Well above threshold the field of a single mode *laser* approximates to a coherent state field: the probability P_n of finding n photons in the mode is thus Poissonian; and the variance v defined in terms of the expectation values $\langle n \rangle = \sum_{n=0}^{\infty} nP_n$, and $< n^2 >$ by

$$v^2 = \left\{< n^2 > - (< n >)^2\right\} / < n > \tag{1}$$

has the value one for the coherent state field. Sub-Poissonian fields are characterised

Quantum Communications and Measurement
Edited by V. P. Belavkin *et al.*, Plenum Press, New York, 1995

by $v^2 < 1$ and are evidently substantially 'non-classical'.[1‡] An extreme example is the pure Fock state, the n-photon state $|n\rangle$ of n photons, in the single mode field: $P_m = \delta_{nm}$ for such fields and $v^2 = 0$. Sub-Poissonian fields with $v^2 \geq 0.3$ were observed in the one ^{85}Rb atom micromaser[1]. This micromaser has been in operation at the Max Planck Institute at Garching near Munich, Germany for some ten years [1–3], and attempts are currently in hand to generate pure Fock states this way[4].

We need some practical details:- The micromaser is truly microscopic: at any one time it can involve, with probability O(1), one single, ^{85}Rb atom in interaction with one, single, cavity mode of the field. The individual ^{85}Rb atoms in the form of an atomic beam successively enter a super-cooled maser cavity of Niobium metal at a temperature $T \leq 0.5°K$ [1,4]. They enter in the high Rydberg $63P_{\frac{3}{2}}$ state, in spectroscopic language, and the cavity is tuned to microwave frequency transitions to the energetically lower $61D_{\frac{5}{2}}$ state at 21.456 GHz. The atoms crossing the cavity are strongly velocity selected so that the interaction time t_{int} for the atom and the cavity field is well defined: $t_{int} \approx 35\mu\,\text{sec}$ in the recently reported experiments [1]. The atoms nevertheless enter stochastically with a mean rate $R \equiv T_p^{-1}$ per sec, but R is small enough (T_p is large enough) so that the probability $p = (1 - e^{-Rt_{int}})$ of finding one atom in the cavity during t_{int} is 0(1) compared with the probability of finding two or more atoms in that time. In the experiments $T_p \approx 7,000\mu\,\text{sec}$ so that for 99.5% of the time the cavity is *empty* of atoms and the cavity field evolves under its own dynamics. This is not unitary evolution because, despite the very low temperatures, a small amount of black-body radiation must also be present in the cavity. At $T = 0.5°K$, $\overline{n}_{th} \approx 0.15$ i.e. about 1/6. We [5]and others [6,7] have shown how this 1/6 photon on average, of energy $\sim 1/6 \times 10^{-23}J$, *qualitatively* (not just quantitatively) changes the dynamics of the maser action. In this paper we report the use of an analytical perturbation theory, to reinforce this view. We also investigate the assumption in Refs. 6,7 of unitary evolution during the *short* time t_{int}.

The change in question is from the corresponding action in an *ideal* cavity which has a cavity quality factor $Q = \infty$ and there is no black-body radiation. The cavity Q in the experiments [1] was $Q = 3 \times 10^{10}$, i.e. quite exceptionally large: nevertheless this high Q is not sufficient for the device to follow the dynamics predicted for $Q = \infty$. For $Q = \infty$ the one 2-level atom coupled to a single cavity mode is modelled precisely by the Jaynes-Cummings model [8]. The dynamics of maser action based on this model is investigated by Cummings and Rajagopal [9]and by ourselves [5,10] (for example). This dynamics, without additional black-body radiation, is 'trapping state' dynamics (§4) and if for $Q = \infty$ black body radiation is initially present the dynamics is not steady (apparently [5,10]). This dynamics is very different from the steady [2‡] steady-state dynamics described first by Meystre and colleagues [6], and then by Lugiato, Scully and Walther [7], and subsequently by ourselves [5]. This steady [2‡] steady-state theory *at the chosen values of Q and T* seems to be borne out by the experimental measurements [1].

At larger Q and lower T one expects trapping state dynamics to dominate [4]. Trapping states $|n_0\rangle$ (say) are pure Fock states towards which the cavity field evolves at $Q = \infty$. These states include the vacuum state $|0\rangle$ of the field (of little interest)

[1‡] N. Lütkenhaus and S.M. Barnett offer a quantitative classification of 'non-classical' at this meeting.

[2‡] 'Steady' means independent of the number of inverted atoms which have passed through the cavity.

but also the very non-classical Fock states $|n_0\rangle$ with $n_0 > 0$ — of considerable interest if they can be realised. Access to them is controlled by the level of black-body radiation, so it is of some importance to extend $Q = \infty$ theory to include the effects of coupling the cavity mode to a finite temperature 'ambient' heat-bath at $Q < \infty$. The model system is then an open system pumped by the incident inverted atoms and coupled to the heat-bath.

2 THE MICROMASER AS AN OPEN SYSTEM: EQUATIONS OF MOTION

The *interaction* between a single 2-level atom with upper and lower states $|a\rangle$, $|b\rangle$ (say) and with transition frequency ω_0 between them, and a single mode of the radiation field at frequency ω in an ideal cavity can be taken as the Jaynes-Cummings interaction [8]

$$\hat{H} = \Delta\omega S^z + g(S^+a + S^-a^\dagger) \; ; \qquad (2)$$

$\Delta\omega \equiv \omega_0 - \omega$ and g is the coupling constant. The single atom has $\hat{H}_a = \omega_0 S^z$ and the field has $\hat{H}_f = \omega a^\dagger a$ (and $\hbar = 1$) so the total Hamiltonian is $\omega(a^\dagger a + S^z) + \hat{H}$: $\left[a, a^\dagger\right] = 1$, so a is the annihilation operator for photons in the field, and $S^\pm, S^z$ are the usual pseudo-spin operators satisfying the $su(2)$ Lie algebra $[S^\pm, S^z] = \mp S^\pm, [S^+, S^-] = 2S^z$. For one atom total spin $S = \frac{1}{2}$.

In the frame rotating at frequency ω the equation of motion for the density operator ρ for the combined matter-field system can be taken in Schrödinger representation to be [11,12]

$$\begin{aligned}
\frac{d\rho}{dt} = -i[\hat{H}, \rho] \quad & - \quad \kappa\{(1 + \overline{n}_{th})(a^\dagger a\rho - 2a\rho a^\dagger + \rho a^\dagger a) \\
& \quad + \overline{n}_{th}(aa^\dagger\rho - 2a^\dagger\rho a + \rho aa^\dagger)\} \\
& - \quad \gamma\{(1 + \overline{n}_{th})(S^+S^-\rho - 2S^-\rho S^+ + \rho S^+S^-) \\
& \quad + \overline{n}_{th}(S^-S^+\rho - 2S^+\rho S^- + \rho S^-S^+)\}. \qquad (3)
\end{aligned}$$

The effect of heat-baths at the temperature $T > 0$ is introduced through the terms in the Planck function $\overline{n}_{th}$. These are invariant under rotation at ω and it is plain that e.g.

$$\left(a^\dagger a\rho - 2a\rho a^\dagger + \rho a^\dagger a\right) = -\left[a, \rho a^\dagger\right] - \left[a\rho, a^\dagger\right] \; , \qquad (4)$$

so that the master equation eq.(3) is in the Linblad-Kossakowski [13–15] strictly Markovian form as discussed by several authors at this meeting [16,17].

This form eq.(3) can be derived in 'weak-coupling' [11–13,18–20] to each of the two baths - the heat-bath at the temperature T for the single cavity mode and the continuous spread of e.m. field modes whose wave-numbers embrace the atomic resonance wave-number $\omega_0 c^{-1}$ for the atom: the e.m. field modes are supposed in equilibrium with a large bath of black-body radiation at the temperature T. 'Weak coupling' is then equivalent to a second order perturbation-theoretical coupling in which each of the baths is large and so does not couple back to the smaller system of interest, the cavity field mode or the atom. If we write eq.(3) as $d\rho/dt = \mathcal{L}\rho$, $\rho(t) = \exp \mathcal{L}t$, and ρ evolves under the semi-group with generator $\mathcal{L}$; but for this to *be* a semi-group it is necessary but not sufficient that the baths are infinitely large [13]. However, by coupling the *atom* to the black-body radiation we find from operator reaction field theory [20] that the system is strongly coupled to the e.m. field modes and the master equation (3) is non-Markovian and more complicated. In these circumstances it seems *inappropriate* to take the form eq.(3) as an axiom for an open system quantum mechanics [13,17]. Still for $\kappa = 0$ and $\gamma \neq 0$ eq.(3) describes exponential decay into black-body radiation with γ a measure of the A-coefficient the actual value of which depends on the geometry of the e.m. field modes. For low-Q cavities $\gamma = \eta\gamma_0$ where γ_0 is the A-coefficient and η is Purcell's cavity factor [21]. The A-coefficient for Rydberg transitions is small ($\lesssim 10^3$) but enhancement by η can mean that γ exceeds decay rates from the two Rydberg states to the much lower atomic states - features discussed by Wang [22] and others. On this ground the term in γ should be retained in eq.(3).

However for the Garching micromaser we are concerned with very *high* Q cavities coupled to a heat-bath and this explains why a and $a^\dagger$ appear in the terms in κ in eq.(3): the parameter κ, 'the cavity band width', relates to Q as $\kappa = \omega/2Q$ and the cavity lifetime $T_c = (2\kappa)^{-1}$: $T_c \approx 0.2\,\text{sec}$ and $Q = 3 \times 10^{10}$ in the experiments [1] as noted. At these high Q all of the accessible e.m. field modes must be inside the cavity and for cavities of microwave dimensions the dense set of modes is not available. Still we shall maintain $\gamma \neq 0$ in eq.(3). In §3 we sketch an analysis which allows us to see under what conditions (i.e. for what values of γ) spontaneous and stimulated processes across the two atomic levels as described by the second curly bracket in eq.(3) can be safely neglected. In actuality we consider only the spontaneous processes in the §3.

The mathematical model we have adopted attempts of course to mimic the real physical situation in the micromaser in which the 2-level atoms enter the maser cavity in their upper states $|a\rangle$. Thus during t_{int} the initial condition is $\rho = |a\rangle\langle a| \otimes |\psi\rangle\langle\psi|$ and $|\psi\rangle$ is the state of the cavity field at that time (as atoms successively pass through the cavity the state $|\psi\rangle$ is successively built up: $|\psi\rangle\langle\psi| \equiv \rho_f$ where $\rho_f = Tr_a\rho$: ρ is the density operator of matter and field and Tr_a means trace over atomic states : $Tr_a\rho = \sum_{\alpha=a,b} \langle\alpha|\rho|\alpha\rangle$). With this initial condition, ρ evolves in the frame rotating at ω as eq.(3) during t_{int} and $\hat{H}$ is eq.(2). Then, for $T_p - t_{int}$, $\hat{H} \equiv 0$ in eq.(3), the initial condition is ρ *at that stage of evolution* and then ρ evolves with $\hat{H} \equiv 0$, $\gamma \equiv 0$ but $\kappa \neq 0$. As noted in §1 we shall find that the action of the black-body radiation during the long time $T_p - t_{int}$ qualitatively changes the theory as compared with the theory for the ideal cavity $Q = \infty$.

Finally we note in this section that the photon statistics, the $P_n's$, are given by $P_n = \langle n\,|Tr_a\rho|\,n\rangle$. We study these P_n in §3. We point out that in order to *measure* these P_n for the field only the statistical states of the outcoming *atoms* is available [1-3,5,10]. This means we must use $\langle \alpha\,|Tr_f\rho|\,\alpha\rangle = \langle \alpha|\sum_n \langle n|\rho|n\rangle\,|\alpha\rangle$ to calculate the probabilities of the atoms leaving the micromaser in the states $|\alpha\rangle = |a\rangle$ or $|b\rangle$ and infer back. Results which have been obtained on the same lines as the calculations for the P_n sketched in §§3,4 will be reported elsewhere. Successful experiments reporting measurements of v^2, eq.(1), for the field obtained via the atomic statistics are already reported in Ref.1.

3 PHOTON STATISTICS

For insight into the action of the micromaser coupled to the heat-bath throughout the motion and not simply during the empty-of-atoms time $T_p - t_{int}$ [1–3,5,10] we can try to follow a 'dressed state' analysis with secular approximation [3]‡. In practice we must introduce an assumption of 'course-graining' [4]‡ as this was already adopted in Ref.5. This blurs the distinction between the times $0 < t < t_{int}$ and $t_{int} < t < T_p$. In contrast an 'exact' analysis for $\gamma = 0$ but $\kappa \neq 0$ is summarised briefly in §4. We work at exact resonance as in the experiments [1] so $\Delta\omega = 0$.

The Hamiltonian interaction eq.(2) commutes with the photon number operator $\hat{N} = S^z + a^\dagger a + \frac{1}{2}$, and at exact resonance the interaction eq.(2) splits the two atom-field states $|a, n\rangle = |a\rangle |n\rangle$, $|b, n+1\rangle = |b\rangle |n+1\rangle$ at fixed photon number $n+1$ of $\hat{N}$ to the doublets [8]

$$|\pm, n\rangle = \frac{1}{\sqrt{2}} \{\pm |a, n\rangle + |b, n+1\rangle\} \; ; \tag{5}$$

the energy splitting is $2g\sqrt{n+1}$. If with $n = 0, 1, 2, ...$ we use the 'dressed states $|\pm, n\rangle$ as basis we find the diagonal elements $\langle\pm, n|\rho|\pm, n\rangle$ rotate at zero frequency in the rotating frame but the off-diagonal elements $\langle\pm, n|\rho|\mp, n\rangle$ rotate at the splitting frequency $2g\sqrt{n+1}$. The *coupling* between diagonal and off-diagonal elements is proportional to κ. In the secular approximation, for which $\kappa \ll g$ must be satisfied, the diagonal and off-diagonal elements can be decoupled to $0(\kappa g^{-1})$. (Compare Ref.24).

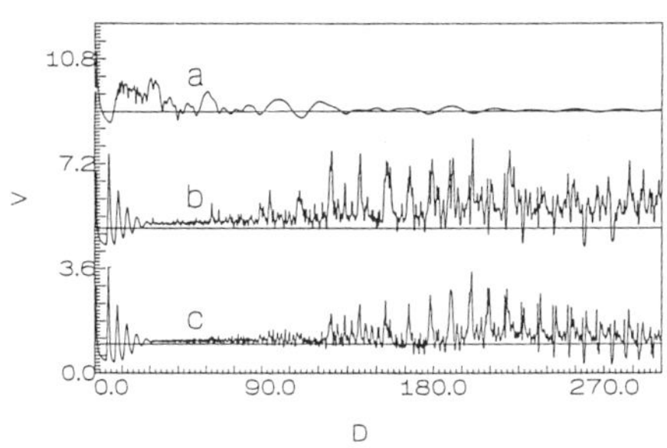

Figure 1: Variation of v with D: $N = 100.0$, $\overline{n}_{th} = 0.1$ and $\kappa/g = 0.00001$. The three Figures support an interpretation in terms of a laser-micromaser formalism with a random pump: $\gamma/g = 0.1(a)$, $0.001(b)$, and $0.0(c)$. For clarity curves (b) and (c) are shifted upwards by 4.0 and 8.0 respectively. The horizontal lines are each the lines $v = 1.0$. The quantity $N = T_c T_p^{-1}$.

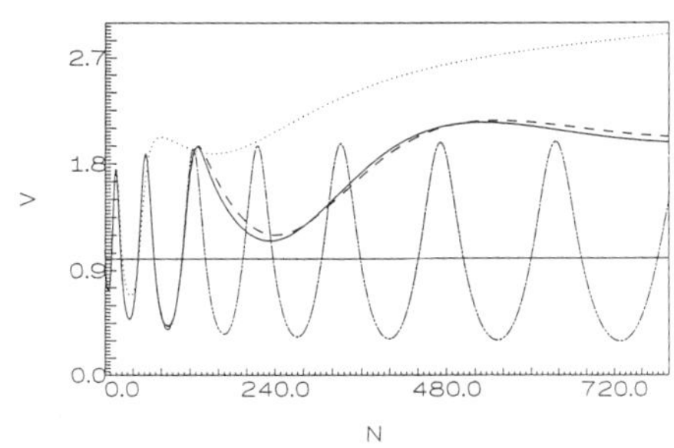

Figure 2: Variance versus N for the experimental situation in Ref.1: $t_{int} = 35\mu\,\text{sec}$, $g = 44kHz$, $Q = 3 \times 10^{10}$ and $T = 0.5^o K$. The reported experimental result $v = 0.5477$ *near* $N = 30$ is recovered for $\gamma/g \leq 0.0001$. The Figure displays $\gamma/g = 0.1$ (dotted), 0.0001 (dashed) and 0.0(full). Also contrasted, are the results (dash-dot-dot) from Refs. 6 and 7 in which $\kappa = \gamma = 0.0$ during t_{int}.

[3]‡ When $\gamma \equiv 0$ the calculations of this §3 are already reported in Ref.23

[4]‡ This course-grained theory is the 'steady' theory referred to in §1 which is independent of the number of atoms which have passed through the cavity.

It is now straightforward to solve the master equation eq.(3) during t_{int} for the elements of the field density matrix found by tracing over the two atomic states $|a\rangle$, $|b\rangle$. In particular, we find equations for $P_n \equiv Tr_a \langle n|\rho|n\rangle$ which take the form[23]

$$\dot{P}_n \mid_{gain} = X_n P_{n-1} + (Y_n - R)P_n + Z_n P_{n+1} \tag{6}$$

We solve numerically for the P_n, in the steady state in the 'coarse graining' assumption [5] which simply adds gain from Eq.(6) during t_{int} to the loss during $T_p - t_{int}$. We then use these P_n to evaluate the normalised variance v, eq(1). In Fig.1 we plot v against a parameter we call the 'pump parameter'

$$D = \sqrt{N} g t_{int} \tag{7}$$

for different values of γ. If $v < 1$, the cavity field is sub-Poissonian. In Fig.1(a) v remains close to 1.0 for $D \gtrsim 100$: the parameters γ, κ here satisfy $t_s < t_{int} \ll T_p$ $(t_s = \gamma^{-1})$. Recall that the single mode laser theory [26], which is essentially a single atom theory, is based on this parameter range and the photon statistics is close to the coherent state statistics ($v = 1$). We also note that the smaller the value of γ the better is the *sub*-Poissonian character of the cavity field - as might be expected.

This is shown more clearly in Fig.2. Here for $\gamma g^{-1} \lesssim 10^{-4}$ atomic decay has negligible influence on the photon statistics (compare the dashed and full curves). Fig.2 also shows the influence of cavity damping: the dash-dot-dot curve is for $\kappa = \gamma = 0$ during t_{int} as in Refs.6,7. Furthermore one can see that the cavity damping through κ can indeed be neglected during t_{int} for $N < 50$ while the results of the experiments [1] which have $N \approx 30$ very much support Meystre's formula [1,6] which is a special case of our results. However, this is only true for the chosen $\kappa g^{-1} = 0.81 \times 10^{-5} (Q = 3 \times 10^{10})$. For smaller Q, N will be still lower than 50 before damping through κ can be neglected during t_{int} in agreement with our numerical work [5,10]. An analytical expression with this feature seems to be difficult to get and only numerical results like those in the Fig. 2 can show the range of N for the chosen Q at which damping via κ during t_{int} can be neglected.

Evidently the Figs.1 and 2 tend to support our conclusion in Refs. 5,10 that reducing T_p (increasing N) tends to induce ideal cavity dynamics at values of Q substantially less than $Q = \infty$, these values of Q decreasing as T_p decreases. In principle we analyse this proposition more completely in the next section, §4.

4 TRAPPING STATE DYNAMICS: SEARCH FOR THE NUMBER STATE

In a $Q = \infty$ cavity without initial black-body radiation one finds [5,9] that

$$P_n(k+1) = a(n)P_n(k) + b(n-1)P_{n-1}(k) \tag{8}$$

where $P_n(k)$ is the probability of finding n photons in the cavity field *after k inverted atoms have entered and passed through the cavity.* The $b(n), a(n)$ are *a priori* probabilities of a photon entering ($b(n)$) or not entering ($a(n)$) the cavity field : $a(n) + b(n) = 1$ and $b(n) = \sin^2 g\sqrt{n+1}t_{int}$ and the $a(n)$, $b(n)$ depend only on n and not on k. The zeros of $b(n-1)$ are of evident importance. If $b(n-1) = 0$, $P_n(k+1) = a(n)P_n(k)$, and if initially, i.e. for $k = 0$, $P_n(k) = 0$ for all $n > n_0$ where $b(n_0) = 0$, $P_n(k+1) = 0$ for all $n > n_0$ and probability never builds up. The Ref.9 proves then that probability builds up at $n = n_0$ and, since total probability is bounded and equal to one, $P_n(k) \to \delta_{nn_0}$ as $k \to \infty$ i.e. the Fock state $|n_0\rangle$ builds up in the cavity as $k \to \infty$. The zeros n_0 of $b(n)$ are thus 'upward going trapping states' a phraseology due to Meystre[27]. Earlier studies of trapping states based on 'coarse-grained theory' (see below) are in Ref.28.

In general if the atoms enter the cavity in the coherent superposition $|\psi\rangle = \alpha\,|a\rangle + \beta\,|b\rangle$, $|\alpha|^2 + |\beta|^2 = 1$, then with $P_n(k) \equiv \rho_{nn}(k)$

$$\begin{aligned} P_n(k+1) \;=\; & P_n(k)\,|\alpha|^2\,a(n) + P_{n+1}(k)\,|\beta|^2\,b(n) \\ & + P_n(k)\,|\beta|^2\,a(n-1) + P_{n-1}(k)\,|\alpha|^2\,b(n-1) \\ & - i\alpha\beta^*\rho_{n\,n+1}\,(k)a^{\frac{1}{2}}(n)b^{\frac{1}{2}}(n) \\ & - i\alpha^*\beta\;\rho_{n\,n-1}\,(k)a^{\frac{1}{2}}(n-1)b^{\frac{1}{2}}(n-1) \\ & + i\alpha\beta^*\rho_{n-1n}\,(k)a^{\frac{1}{2}}(n-1)b^{\frac{1}{2}}(n-1) \\ & + i\alpha^*\beta\;\rho_{n+1n}\,(k)a^{\frac{1}{2}}(n)b^{\frac{1}{2}}(n). \end{aligned} \tag{9}$$

The case $|\alpha| = 1$, $|\beta| = 0$ regains eq.(8) and $b(n_0) = 0$ is an upward going trapping state. The case $|\alpha| = 0$, $|\beta| = 1$ yields

$$P_n(k+1) = a(n-1)P_n(k) + b(n)P_{n+1}(k) \tag{10}$$

and $b(n) = 0$ defines a downward going trapping state[5]‡.

These results are all $Q = \infty$ results. However, we have used 'exact' numerical matrix continued fraction methods as described in Ref.5 and references to study the following situation at $Q = 2 \times 10^{10} < \infty$ and $T = 0.5^oK$: $\kappa \neq 0$ but $\gamma = 0$. We

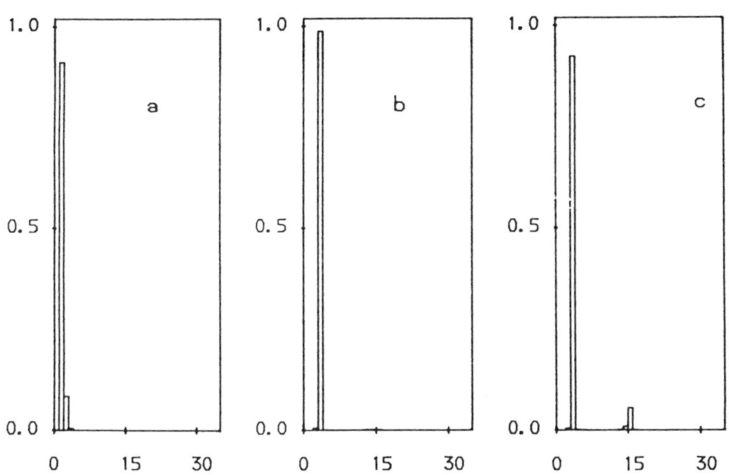

Figure 3: Calculated histogram of P_n after 1(a), 40(b) and 500(c) atoms have passed through the cavity; $\Delta\omega = 0$, $t_{int} = \pi/2g$, $\overline{n}_{th} = 0.15$, and $\kappa/g = 0.0000747$

suppose the maximum possible value of $N = T_p^{-1}T_c$ (~ 6000) for given t_{int} - namely

[5]‡ The trapping states $(r, n+1)$ satisfying $gt_{int}\sqrt{n+1} = r\pi$ are dense! They have the number of the rational fractions r/q, and $q^2 = n+1$.

$T_p = t_{int}$ (we remarked in §1 that now the probability of more than one atom in the cavity during t_{int} is large but we do not consider that complication). We choose $gt_{int} = \frac{1}{2}\pi$ ($t_{int} \approx \frac{3}{2}/44 \times 10^3 \sim 3 \times 10^{-5}$). So there is the $Q = \infty$ trapping state at $n_0 = 3$ and there are subsequent zeros at $n_1 = 15$, $n_2 = 35$, etc. Note that the empty-of-atoms time for the cavity is here zero. By the 'exact' numerical methods described in Ref.5 we find, for these conditions, the Fig.3. The results in Fig.3 depend very much on the number of atoms k which have passed through the maser cavity. Indeed starting from black-body radiation, $P_n \approx \delta_{nn_0}$ at $n_0 = 3$ after some $k = 15$ atoms have passed (precisely, $P_3 = 0.988$), so there is almost the Fock state at $n_0 = 3$ then. However, probability has already begun to transfer to $n_1 = 15$ so that after some 500 atoms have passed $P_3 = 0.924$, $P_{15} = 0.056$ and P_{35} is building up. Evidently this is $Q = \infty$ 'trapping state dynamics' with initial black-body radiation[5,9], but at the experimentally realised $Q = 2 \times 10^{10}$ ($Q = 3 \times 10^{10}$ is realised in Ref.1).[6]‡

A strong conclusion is that the coarse-grained theories describe a *qualitatively* different situation from the situation at large enough Q and low enough T where trapping state dynamics dominates. To study this dynamics the exact studies of the type we have so far achieved only numerically become necessary. The coarse-grained theories describe the 'steady' situation actually realised in the experiments[1]. However the trapping state dynamics occuring at low enough T is not adequately described by these coarse grained theories as $\overline{n}_{th} \to 0$. The forthcoming experiments [4] must take this into account.

Finally we note that in so far as open systems as such represent quantum measurements [13] the micromaser is already a remarkable and experimentally realised example of such quantum measurement.

References

[1] G. Rempe, F. Schmidt-Kaler and H. Walther, *Phys.Rev.Lett* .**64**, 2783(1990); Gerhard Rempe and Herbert Walther, *Phys.Rev.A* **42**, 1650(1990).

[2] D. Meschede, H. Walther and G. Müller, *Phys.Rev.Lett.* **54**, 551 (1985)

[3] .G. Rempe, H. Walther and N. Klein, *Phys.Rev.Lett.* **58**, 353(1987).

[4] To this end cavity $Q's \sim 5 \times 10^{10}$ and temperatures $T \sim 0.07^o K$ are reached at Garching. Further experiments depend on improved velocity selection (private communication from Herbert Walther, 1993).

[5] R.K. Bullough, N.Nayak and B.V. Thompson in: Recent Developments in Quantum Optics, R. Inguva ed., Plenum Press, New York (1993) pp273-288.

[6] P. Filopowicz, J. Javainen and P. Meystre, *Phys.Rev.A* **34**, 3077(1986).

[7] L.A. Lugiato, M.O. Scully and H. Walther, *Phys.Rev.A* **36**, 740(1987).

[8] E.T. Jaynes and F.W. Cummings, *Proc.IEEE* **51**, 89(1963).

---

[6]‡ Results in Fig.3 still depend on small values of $k \lesssim 500$. Only now are we getting numerical results for much larger k.

[9] F.W. Cummings and A.K. Rajagopal, *Phys.Rev.A* **39**, 3414(1989).

[10] R.K. Bullough, N.M. Bogoliubov, N.Nayak and B.V. Thompson in: "Quantum Measurements in Optics", P. Tombesi and D.F. Walls eds. *NATO ASI Series B: Physics Vol.* **282**, Plenum Press, New York (1992) pp.129-149.

[11] F. Haake, "Statistical Treatment of Open Systems by Generalized Master Equations", Springer Tracts in Modern Physics **66**, Springer-Verlag, Berlin (1973).

[12] G.S. Agarwal, "Quantum Optics", Springer Tracts in Modern Physics **70**, Springer-Verlag, Berlin (1974).

[13] G. Linblad, *Comm.math.Phys.***48**, 119 (1976).

[14] A. Kossakowski, *Rep.Math.Phys.***3**, 247 (1972).

[15] V. Gorini, A. Frigerio, M. Verri, S. Kossakowski and E.C.G. Sudarshan, *Rep.Math.Phys.* **13,** 149 (1978).

[16] For example K. Haubold, Invited Lecture, this meeting.

[17] I.C. Percival, Invited Lecture, this meeting.

[18] H. Haken, *Hanb.Phys.* **25**, 2c Springer-Verlag, Berlin (1970).

[19] E.B. Davies, *Comm.math.Phys.* **39**, 91 (1974).

[20] S.S. Hassan, M.R. Wahiddin, R. Saunders and R.K. Bullough, "Master Equation for the Dicke Model in a Broad Band Squeezed Vacuum" submitted to J.Mod.Optics (1992).

[21] G.P. Hildred, R.R. Puri, S.S. Hassan and R.K. Bullough, *J.Phys B: At.Mol.Phys.* **17**, L535 (1984), and references.

[22] L.Z. Wang, *Phys.Rev.* A **45**, R27 (1992).

[23] N.Nayak and D. Das, *Phys.Rev.A* **48**, 2475 (1993); N. Nayak, unpublished.

[24] S. Haroche and J.M. Raimond, in *Advances in Atomic and Molecular Physics,* Vol. 20, edited by G.R. Bates and B. Bederson (Academic, New York, 1985), p.347.

[25] L. Davidovich, S.-Y. Zhu, A.S. Khoury and C. Su, *Phys. Rev. A* **46**, 1630 (1992).

[26] M.O. Scully and Willis E. Lamb, *Jr. Phys. Rev.* **159**, 208 (1967).

[27] J.J. Slosser, P. Meystre and S.L. Braunstein, *Phys.Rev.Lett.* **63**, 934 (1989).

[28] P. Meystre, G. Rempe and H. Walther, *Opt. Letts.* **13**, 1078 (1987).

INDEX